NMR AND THE PERIODIC TABLE

NMR and the Periodic Table

Edited by

ROBIN K. HARRIS

University of East Anglia

and

BRIAN E. MANN

University of Sheffield

ACADEMIC PRESS · 1978

LONDON · NEW YORK · SAN FRANCISCO

A Subsidiary of Harcourt Brace Jovanovich, Publishers

ACADEMIC PRESS INC. (LONDON) LTD
24/28 Oval Road
London NW1 7DX

United States Edition published by
ACADEMIC PRESS INC.
100 Fifth Avenue
New York, New York 10003

Library of Congress Catalog Number: 78-52091
ISBN 0-12-327650-0

FILMSET BY COMPOSITION HOUSE LIMITED, SALISBURY, WILTSHIRE
PRINTED IN GREAT BRITAIN BY WHITSTABLE LITHO LTD., KENT

CONTRIBUTORS

Dr. C. Brevard Bruker Spectrospin S.A., 34 rue de l'Industrie, B.P. 56, 67160 Wissembourg, France

Dr. R. W. Briggs Department of Chemistry, University of Arkansas, Fayetteville, Arkansas 72701, U.S.A.

Prof. S. Forsén Physical Chemistry 2, Chemical Centre, University of Lund, P.O.B. 740, S-220 07 Lund 7, Sweden

Dr. R. J. Goodfellow Department of Inorganic Chemistry, The University, Bristol BS8 1TS, England

Dr. R. K. Harris School of Chemical Sciences, University of East Anglia, Norwich NR4 7TJ, England

Prof. J. F. Hinton Department of Chemistry, University of Arkansas, Fayetteville, Arkansas 72701, U.S.A.

Dr. J. D. Kennedy Department of Structural and Inorganic Chemistry, The University, Leeds LS2 9JT, England

Prof. R. G. Kidd Faculty of Graduate Studies, Stevenson–Lawson Building, The University of Western Ontario, London, Ontario N6A 5B8, Canada

Dr. J. P. Kintzinger Institut de Chimie, Université Louis Pasteur, 1 rue Blaise Pascal, B.P. 296/R8, 67008 Strasbourg-Cedex, France

Dr. B. Lindman Physical Chemistry 1, Chemical Centre, University of Lund, P.O.B. 740, S-220 07 Lund 7, Sweden

Dr. B. E. Mann Department of Chemistry, University of Sheffield, Sheffield S3 7HF, England

Dr. H. C. E. McFarlane Department of Chemistry, Sir John Cass School of Science and Technology, City of London Polytechnic, 21 Jewry Street, London EC3N 2EY, England

Dr. W. McFarlane Department of Chemistry, Sir John Cass School of Science and Technology, City of London Polytechnic, 21 Jewry Street, London EC3N 2EY, England

Dr. C. A. Rodger School of Chemical Sciences, University of East Anglia, Norwich NR4 7TJ, England

Dr. G. J. Schrobilgen Department of Chemistry, University of Guelph, Guelph, Ontario N1G 2W1, Canada

Dr. D. Shaw Oxford Instruments Ltd., Osney Mead, Oxford OX2 0DX, England

Prof. N. Sheppard School of Chemical Sciences, University of East Anglia, Norwich NR4 7TJ, England

Dr. G. A. Webb Department of Chemical Physics, University of Surrey, Guildford, Surrey GU2 5XH, England

PREFACE

Nuclear magnetic resonance (NMR) spectroscopy is firmly established as one of the major methods available to the chemist for structure determination and for obtaining physico-chemical information about molecular dynamics. Few industrial or academic chemistry laboratories of any size have no NMR spectrometers. In popularity it vies with the longer established infrared and ultraviolet spectroscopies. This position has been reached as the result of studies of the proton and, more recently, of the carbon-13 nucleus. However, this course of events is largely a result of chemical priorities and of historical accident. Most elements of the Periodic Table have at least one nuclide that is reasonably suitable for NMR studies. This fact is now becoming better appreciated, and a boom in multinuclear studies may be anticipated; indeed, it is already under way. Such a development partly follows instrumental improvements and partly stimulates them.

At the time of writing, the NMR spectra of most elements have at least received some chemical investigation. Therefore we consider it to be an appropriate moment at which to summarise the situation. Since NMR studies are firmly linked to particular nuclides, it is rewarding to examine the results in terms of the Periodic Table; hence the basic organisation of the present book. We doubt whether in future any single volume will be able to give such a comprehensive account as has been attempted here. Indeed, the decision had to be made to give only brief summaries of the situation for the most popular nuclei, viz. ^{1}H, $^{10,11}B$, ^{13}C, $^{14,15}N$, ^{19}F, ^{31}P.

It is hoped that this book will be used as a primary source of NMR information for many years to come. Later reviews for individual nuclei or groups of nuclei can act to update the information herein. It is also hoped that it will stimulate interest in multinuclear NMR—for chemists of all sections of the subject (organic, inorganic, organometallic, physical, theoretical and biochemical)—since the potential of the NMR technique is certainly not yet fully exploited. In line with the way in which NMR has been applied so far, the book is mainly concerned with the liquid or solution state, However, multinuclear NMR high-resolution studies of solids may be expected to grow rapidly in importance in the future, though this is probably still a few years away at the moment.

Before describing in this book what has been achieved to date, we wish to draw the reader's attention to those elements which, as far as we can ascertain, have not been the subject of chemical NMR studies so far, though we believe them to be of potential importance. The most surprising case where there seems to have been no direct study is that of ^{103}Rh, which has $I = \frac{1}{2}$ and $N = 100\%$, placing it in a very favourable category; however, the magnetogyric ratio (and hence resonance frequency) is very low (see Table 1.1). Moreover ^{103}Rh has been the subject of a number of indirect studies by heteronuclear double resonance. Several transition elements possessing only quadrupolar nuclei have yet to be studied by NMR for chemical purposes, though the relevant spin properties are known. The nuclei in question are:

Nuclide	Spin	Natural abundance N/%	Magnetic moment μ/μ_N	Quadrupole moment $Q/10^{-28}$ m^2	NMR frequency Ξ/MHz	Receptivity D^C
^{61}Ni	$\frac{3}{2}$	1.19	−0.9665	[a]	8.94	0.241
^{91}Zr	$\frac{5}{2}$	11.23	−1.5415	[a]	9.33	6.04
^{99}Ru	$\frac{3}{2}$	12.72	−0.367	[a]	3.39	0.141
^{101}Ru	$\frac{5}{2}$	17.07	−0.82	[a]	4.9	1.4
^{105}Pd	$\frac{5}{2}$	22.23	−0.756	[a]	4.58	1.41
^{177}Hf	$\frac{7}{2}$	18.50	0.69	3	3.1	0.67
^{179}Hf	$\frac{9}{2}$	13.75	−0.52	3	1.9	0.17
^{191}Ir	$\frac{3}{2}$	37.3	0.1859	1.5	1.72	0.054
^{193}Ir	$\frac{3}{2}$	62.7	0.2024	1.5	1.87	0.117
^{197}Au	$\frac{3}{2}$	100	0.1852	0.59	1.71	0.143

[a] Not available

Among the lanthanides the spin-$\frac{1}{2}$ nuclei ^{169}Tm (N = 100%) and ^{171}Yb (N = 14.31%) also await chemical NMR studies. Perhaps some of these nuclei are currently under study by NMR somewhere. If so, we are not aware of it. If not, we predict it will not be long before such studies are made. At any rate, all the main-group elements with stable magnetic nuclei have been the subject of chemical NMR investigations already, though in a few cases little has been achieved.

The present book is the result of the efforts of many people. As editors we have attempted to harmonise these contributions to produce a coherent volume. Some variations are, however, inherent in the NMR uses to which different nuclei are suited. The first chapter attempts to set the scene in a general way. Chapters 2 and 3 give a basic discussion of experimental and theoretical aspects respectively. Chapter 4 summarises the situation for the "popular nuclei." The rest of the book considers other elements, in order of the groups in the Periodic Table. This format should provide the reader with an easy way of finding material of interest in a given area of chemistry. The commonly used symbols are listed on pp. xix to xxv. The SI system is employed throughout, though, except for some of the equations in chapter 3, this has few effects.

In planning this book we have deliberately allowed for much tabular material in the belief that, at any rate for the less popular nuclei for which no recent reviews exist, this will be much appreciated by readers and will cause the book to be used, inter alia, as a reference. In considering the three types of parameter (chemical shifts, coupling constants and relaxation times) we have given most emphasis to chemical shifts because to date relaxation effects have been of lesser chemical significance while coupling constants (at least to the "popular" nuclei) have in many cases been available for a considerable time, since their measurement does not require direct observation of the resonances of the less popular nuclei.

Finally, we would like to express our hope that readers find as much interest in this book as we found in preparing it.

September 1977

ROBIN K. HARRIS
BRIAN E. MANN

ACKNOWLEDGEMENTS

As with any book, many people have had a hand in the preparation and production of "NMR and the Periodic Table." At this point we wish to record our indebtedness to many NMR spectroscopists for provision of unpublished results, or for results communicated to us prior to publication, or for valuable discussions. Among those who have helped in this way we would mention the following: E. de Boer, J. D. Cargioli, A. M. Chippendale, G. Engelhardt, P. Granger, H. G. Hertz, J. H. Holloway, H.-J. Kroth, P. Laszlo, O. Lutz, H. C. Marsmann, A. I. Popov, D. S. Rycroft, J. Schraml, A. Schwenk, M. D. Sefcik, F. W. Wehrli, H. Weingärtner, E. A. Williams and K. M. Worvill. We also gratefully acknowledge permission to reproduce some published figures, as indicated in the relevant captions.

CONTENTS

SYMBOLS

a_N	hyperfine (electron-nucleus) interaction constant
B	magnetic induction field (magnetic flux density)
B_0	static magnetic field of an NMR spectrometer
B_1, B_2	r.f. magnetic fields associated with ν_1, ν_2
B_L	local fluctuating magnetic field
C_X	spin-rotation coupling constant of nucleus X
D	self-diffusion coefficient
D_X^C	receptivity of nucleus X with respect to ^{13}C
D_X^p	receptivity of nucleus X with respect to ^1H
D_u	*d*-electron "imbalance" at a given atom
e	magnitude of the charge on the electron
E	(*i*) electric field
	(*ii*) energy or eigenvalue of $\hat{\mathscr{H}}$
f	microviscosity factor
$\overline{F^2}$	mean square electric field due to dispersion forces
$\hat{\mathscr{H}}$	Hamiltonian operator—subscripts indicate its nature
$\hat{\mathbf{I}}_i$	nuclear spin operator for nucleus i
$\hat{I}_{ix}, \hat{I}_{iy}, \hat{I}_{iz}$	components of $\hat{\mathbf{I}}_i$
I_i	magnetic quantum number associated with $\hat{\mathbf{I}}_i$
I	moment of inertia
^nJ	nuclear spin-spin coupling constant through n bonds (in Hz).
J_r	reduced splitting observed in a double resonance experiment
$J(\omega)$	motional spectral power density
$\mathscr{J}$	Coulomb integral
^nK	reduced nuclear spin-spin coupling constant
$\mathscr{K}$	exchange integral
L	linewidth factor for quadrupolar nuclei
m_i	eigenvalue of $\hat{I}_{iz}$ (magnetic component quantum number)
M_0	equilibrium macroscopic magnetization of a spin system in the presence of B_0
M_x, M_y, M_z	components of macroscopic magnetization
$P_{\mu\nu}$	element of bond-order, charge-density matrix (see Eq. 3.33)
P_u	*p*-electron "imbalance" at a given atom (see Eq. 3.36b)
q	electric field gradient (subscripts refer to tensor component direction)
Q	(*i*) nuclear quadrupole moment
	(*ii*) quality factor for an r.f. coil
Q_{AB}	Bond-order, charge-density term (see Eq. 3.32)
s_A	valence *s*-orbital of atom A
$S_A^2(0)$	electron density in S_A at nucleus A
S	(*i*) electron (or, occasionally, nuclear) spin
	(*ii*) ordering parameter for oriented systems
t_p	pulse duration

T	temperature
T_c	coalescence temperature for an NMR spectrum
T_1^X	spin-lattice relaxation time of the X nuclei (further subscripts refer to the relaxation mechanism)
T_2^X	spin-spin (transverse) relaxation time of the X nuclei (further subscripts refer to the relaxation mechanism)
$T_{1\rho}^X$, $T_{2\rho}^X$	spin-lattice and spin-spin relaxation times of the X nuclei in the frame of reference rotating with B_1
T_2'	inhomogeneity contribution to dephasing time for M_x or M_y
T_2^*	total dephasing time for M_x or M_y; $(T_2^*)^{-1} = T_2^{-1} + (T_2')^{-1}$
T_{10}^X	spin-lattice relaxation time of nucleus X due to all mechanisms except dipole-dipole interactions between X and another specified nuclear type
T_d	pulse delay time
T_{ac}	acquisition time
T_p	period for repetitive pulses (= interpulse time = $T_{ac} + T_d$ if t_p is negligible)
T_q	relaxation time for a quadrupolar nucleus ($T_{1q} = T_{2q}$)
$W_{1/2}$	full width (in Hz) at half-height of a resonance line in the absorption mode
W'	maximum-to-minimum peak-to-peak separation for a resonance line in the derivative mode
Z_A	atomic number of atom A
α	(*i*) nuclear spin wavefunction (eigenfunction of $\hat{I}_z$) for a spin-$\frac{1}{2}$ nucleus (*ii*) angle of rotation of magnetization caused by a r.f. pulse (*iii*) fine structure constant (= $e^2/\hbar c$)
α_A^2	*s*-character of hybrid orbital at atom A
β	nuclear spin wavefunction (eigenfunction of $\hat{I}_z$) for a spin-$\frac{1}{2}$ nucleus
γ_X	magnetogyric ratio of nucleus X
$1 + \gamma_\infty$	Sternheimer antishielding factor
δ_X	chemical shift for a resonance of a nucleus of element X (positive when the sample resonates to high frequency of the reference). Given in ppm, with respect to the resonance of a specified compound
δ_{ij}	Kronecker delta (= 1 if $i = j$, and = 0 otherwise)
$\delta(r_{KA})$	Dirac delta operator
Δ	spectral width
ΔE	electronic excitation energy (often an ill-defined average value is intended)
$\Delta\delta$	change or difference in δ
$\Delta\sigma$	(*i*) anisotropy of $\boldsymbol{\sigma}$ (for axial symmetry $\Delta\sigma = \sigma_\parallel - \sigma_\perp$) (*ii*) differences in σ for two different situations
$\Delta\chi$	(*i*) susceptibility anisotropy (for axial symmetry $\Delta\chi = \chi_\parallel - \chi_\perp$) (*ii*) difference in electronegativities
η	(*i*) nuclear Overhauser effect or enhancement (*ii*) asymmetry factor (e.g. in NQCC) (*iii*) viscosity
μ	magnetic dipole moment
μ_0	permeability of a vacuum
μ_B	Bohr magneton

μ_N	nuclear magneton
ν_i	Larmor precession frequency of nucleus i (in Hz)
ν_0	(*i*) spectrometer operating frequency (*ii*) Larmor precession frequency (general, or of bare nucleus)
ν_1	frequency of "observing" r.f. magnetic field
ν_z	frequency of "irradiating" r.f. magnetic field
Ξ_X	resonance frequency for the nucleus of element X in a magnetic field such that the protons in TMS resonate at exactly 100 MHz
ρ_N	unpaired electron spin density at nucleus
ρ_{s_A}	spin density in s_A
σ_i	shielding constant of nucleus i (used sometimes in tensor form, $\boldsymbol{\sigma}$). Given in ppm. Subscripts may alternatively indicate contributions to σ or components of $\boldsymbol{\sigma}$
$\sigma_{\parallel}, \sigma_{\perp}$	components of $\boldsymbol{\sigma}$ parallel and perpendicular to a molecular symmetry axis
σ^d	diamagnetic contribution to σ
σ^p	paramagnetic contribution to σ
τ	(*i*) pre-exchange lifetime of molecular species (*ii*) time between r.f. pulses (general symbol)
τ_0	correlation time (general symbol)
τ_c	correlation time for molecular tumbling. Other subscripts are used in special cases
τ_{sr}	correlation time for spin-rotation interaction
τ_J	angular momentum correlation time
χ	(*i*) magnetic susceptibility (*ii*) electronegativity (*iii*) nuclear quadrupole coupling constant ($= e^2qQ/h$). Subscripts refer to component directions
ω_c	carrier frequency (in rad s^{-1})
$\omega_i, \omega_0, \omega_1, \omega_2$	as for $\nu_i, \nu_0, \nu_1, \nu_2$ but in rad s^{-1}

ABBREVIATIONS

Physical Properties

i.d.	inside diameter
mol. wt.	molecular weight
o.d.	outside diameter
ppm	parts per million
r.f.	radio frequency
r.m.s.	root mean square
ADC	analogue-to-digital converter
AEE	average energy approximation
ARP	adiabatic rapid passage
CI	configuration interaction
CNDO	complete neglect of differential overlap
CPMG	Carr–Purcell pulse sequence, Meiboom–Gill modification
CW	continuous wave
DD	dipole-dipole (interaction or relaxation mechanism)
DEFT	driven equilibrium Fourier transform
EFG	electric field gradient
ENDOR	electron-nucleus double resonance
ESR	electron spin resonance
FID	free induction decay
FPT	finite perturbation theory
FRD	fully random distribution
FT	Fourier transform
HFSC	hyperfine splitting constant
INDO	intermediate neglect of differential overlap
INDOR	internuclear double resonance
IR	infrared
IUPAC	International Union of Pure and Applied Chemistry
LCAO	linear combination of atomic orbitals
MB	molecular beam
MINDO	modified intermediate neglect of differential overlap
MO	molecular orbital
MW	microwave
NMR	nuclear magnetic resonance
NOE	nuclear Overhauser effect
NQCC	nuclear quadrupole coupling constant
NQR	nuclear quadrupole resonance
PRFT	partially relaxed Fourier transform
QCPE	Quantum Chemistry Program Exchange
QF	quadrupole moment/field gradient (interaction or relaxation mechanism)
QLRF	quasilattice random flight

SA	shielding (chemical shift) anisotropy
SC	scalar (interaction or relaxation mechanism)
SCF	self-consistent field
S/N	signal-to-noise ratio
SR	spin-rotation (interaction or relaxation mechanism)
TC	time constant
TIP	temperature-independent paramagnetism
UV	ultraviolet
VB	valence bond

Chemical Species

The eight common group abbreviations: Me (methyl), Et (ethyl), Pr (propyl), Bu (butyl), Ac (acetyl), Cp (cyclopentadienyl), Ph (phenyl), and Vi (vinyl)—are used together with superscripts n (normal), i (iso), s (secondary) and t (tertiary) for Pr and Bu where appropriate.

The following abbreviations are also used. In general, use of lower case letters implies reference to a ligand.

acac	acetylacetonato
acacen	*N*,*N*′-ethylenebis (acetylacetonato)
aq	aquated
ATP	adenosine triphosphate
ATPase	adenosine triphosphatase
ba	benzylamine
bmip	1,3-bis (biacetylmono-oximeimino) propane
BSA	bovine serum albumin
bzac	benzoylacetonato
cod	1,5-cyclooctadiene
DBC	dibenzo-18-crown-6
dbzm	dibenzoylmethanato
DCC	dicyclohexyl-18-crown-6
DEA	*N*,*N*-diethylacetamide
DG	diglyme
dipy	2,2′-dipyridyl
DMA	*N*,*N*-dimethylacetamide
DME	1,2-dimethoxyethane
DMF	*N*,*N*-dimethylformamide
dmg	dimethylglyoximato
DMSO	dimethylsulphoxide
DPL	dipalmitoyl lecithin
DPPC	dipalmitoylphosphatidylcholine
edta	ethylenediaminetetraacetato
EDTA	ethylenediaminetetraacetic acid
en	ethylene diamine
FEP	fluoroethylene-propylene copolymer
HMPT	hexamethylphosphorotriamide
inaa	isonitrosoacetylacetonato
NADH	nicotinamide adenine dinucleotide
NEA	*N*-ethylacetamide

NEF	*N*-ethylformamide
NMF	*N*-methylformamide
NTA	$N(CH_2CO_2H)_3$
ox	oxinate ≡ 8-hydroxyquinolinate
phen	1,10-phenanthroline
PMS	poly(dimethylsiloxane)
pn	propylenediamine
ppd	1-phenyl-1,3-propanedionato
PVP	poly (vinyl pyrrolidone)
py	pyridine
salen	*N,N'*-ethylenebis (salicylaldiminato)
TANOL	4-hydroxy-2,2,6,6-tetramethylpiperidine-1-oxyl
tfa	trifluoroacetato
TFA	trifluoracetic acid
tfac	1,1,1-trifluoroacetylacetonato
tfthbd	1,1,1-trifluoro-4-(2-thienyl)-buta-2,4-dionato
TG	triglyme
THF	tetrahydrofuran
TMS	tetramethylsilane
TMU	tetramethylurea
TNB	trinitrobenzene
TTG	tetraglyme

1

INTRODUCTION

ROBIN K. HARRIS, University of East Anglia, England

1A.1 General

Periodicity in the chemical properties of the elements has been of interest to chemists for well over a century, and still forms an important part of the attraction of chemistry for students at all levels. The principles underpinning the Periodic Table are fundamental to the unity and integrity of chemistry as a science. Another, more recent, vital strand in chemistry is the availability of a wide variety of spectroscopic techniques for studying chemical structure and dynamics. Consequently one might anticipate that many textbooks would be written to evaluate the uses of spectroscopy as viewed from the standpoint of the Periodic Table. However, this is not really the case. One reason for this is that two of the most widespread forms of spectroscopy, infrared and ultraviolet absorption, are, strictly speaking, based on molecular, rather than atomic properties. Nuclear magnetic resonance, on the other hand, is very firmly based on magnetic properties of individual isotopes, though it gives much molecular information. Moreover, nearly all elements have at least one potentially valuable isotope for NMR. Various review articles[1–7] have, indeed, been published giving NMR data on a variety of elements, but these have invariably been fragmented, with very many elements omitted. No coherent, comprehensive description of NMR and the Periodic Table has hitherto existed. The reason is simple, namely that the overwhelming majority of NMR studies published up to 1970 concerned a single nucleus, the proton, and most of the remaining ones involved fluorine, phosphorus or boron. However, the improvements in NMR technique in the last decade, particularly due to the advent of Fourier transform operation (see ch. 2), have led to a great increase in the use of other nuclei for NMR. This is especially the case for the ^{13}C nucleus, which now rivals the proton as the most important NMR isotope. A substantial amount of NMR work has also been carried out with the two isotopes of nitrogen, ^{14}N and ^{15}N. The eight isotopes ^{1}H, ^{10}B, ^{11}B, ^{13}C, ^{14}N, ^{15}N, ^{19}F and ^{31}P will be referred to as the common NMR nuclides, and will be treated somewhat differently from the others in this book (see ch. 4) because of the large amount of published material involving their use and because, in general, adequate reviews about these nuclei exist.

The remainder of this chapter will be devoted to a general description of NMR and of the relevant properties of the elements of the Periodic Table. Chapters 2 and 3 will describe Experimental Methods and Background Theory of NMR Parameters respectively, while the remaining chapters, forming the bulk of the book, will discuss NMR data for the various groups of the Periodic Table.

1A.2 Nuclear Magnetic Resonance

There are many general texts[8–14] on NMR, to which the reader can be referred, and also several review series[15–19], which enable one to be fully informed about the status of most aspects of the subject. Thus, only a very brief description is warranted here.

The NMR phenomenon depends on the existence of a property which is conveniently described as the spin of a nucleus. Each isotope of an element can be assigned a ground-state nuclear spin quantum number I (which may be integral or half-integral, and is positive, including the possibility of zero), such that the associated angular momentum is $\hbar[I(I+1)]^{1/2}$. Excited states with different quantum numbers exist, and are utilised in Mössbauer spectroscopy, but are of much higher energies and are irrelevant for NMR. Some ground-state values of I are zero (whenever both the mass number and the charge number of the nucleus are even), e.g. for ^{12}C, ^{16}O and ^{32}S. These give no NMR spectra themselves, and have little influence on NMR properties. However, when I is non-zero the nucleus has a magnetic moment, μ_I, given by*:

$$\mu_I = \gamma_I \hbar[I(I+1)]^{1/2} \tag{1.1}$$

where γ_I is a constant characterising the nucleus in question and is known as the magnetogyric ratio. The value of γ_I may be either positive or negative, depending on whether the magnetic moment vector is parallel or anti-parallel to the angular momentum vector. Values of I for nuclei of current NMR usage range up to $\frac{9}{2}$. Oddly enough very few useful NMR nuclei have integral spins (only ^{2}H, ^{6}Li and ^{14}N with $I = 1$, and ^{10}B with $I = 3$).

When a nucleus with a non-zero I is placed in a strong magnetic field B_0 (usually between 1 and 10 T) the orientation of its spin axis becomes quantised, each possible orientation having a different energy. Suitable radiofrequency radiation will cause transitions, which may be detected. These form the basis of NMR spectroscopy.

1A.3 NMR Parameters

For mobile fluids under certain simplifying conditions (the first-order approximation), stationary state energies, E, in NMR obey the equation:

$$h^{-1}E = -\sum_{A} \nu_A m_A + \sum_{A<B} J_{AB} m_A m_B \tag{1.2}$$

The first term (the Zeeman term) is due to interactions between the spins and B_0, while the second term is due to coupling between spins in pairs. The quantities m_A and m_B are orientational magnetic quantum numbers. The Larmor frequencies ν_A are given by:

$$\nu_A = \frac{\gamma_A}{2\pi} B_0(1 - \sigma_A) \tag{1.3}$$

The parameter σ_A is the shielding (screening) constant of nucleus A (a dimensionless number usually quoted in ppm), which depends upon the electronic environment of A. The parameter J_{AB} is the (scalar) spin-spin coupling constant between nuclei A and B. It is not normally feasible to measure shielding constants directly, but changes in shielding may be found. These lead to chemical shifts in the resonance frequency, which are generally reported in ppm relative to the resonance of some suitable reference compound.

* Many lists (e.g. that in ref. 20a) of nuclear spin properties actually quote the maximum observable value of μ_I, which is $\gamma_I \hbar I$.

Chemical shifts and coupling constants are two of the types of parameter which may be obtained by NMR and which may be related to chemical structure.

The third type of measurable NMR parameter of value in chemistry is the relaxation time for a nucleus, which governs the rate at which it recovers its equilibrium properties after being disturbed. In fact, there are two distinct relaxation times for a nucleus even in simple cases (for multi-spin systems there are further complications), viz. T_1, the spin-lattice (longitudinal) relaxation time, which relates to magnetisation parallel to B_0, and T_2, the spin-spin (transverse) relaxation time, which refers to magnetization perpendicular to B_0. For non-viscous fluids $T_1 = T_2$ in general, but they may be separately measured. Relaxation times give information about, *inter alia*, molecular dynamics.

This book is principally about chemical shifts (shielding), coupling constants and relaxation times for the various elements of the Periodic Table and about their interpretation in chemical terms. However, the emphasis is on direct observation of the resonance of the various nuclides, whereas much information about the coupling constants J_{AB} involving nucleus A can be obtained from the resonance of nucleus B. Coupling constants involving at least one of the common nuclides have been well-known for a number of years and therefore relatively little emphasis is placed on them here. For many nuclei (particularly those with $I = \frac{1}{2}$) relaxation times have only come to be of chemical significance relatively recently. Therefore the greatest importance in this book is given to chemical shift information, and numerous tables of shifts are given. However, the situation does vary extensively from nucleus to nucleus. Thus, relaxation information is of particular significance for deuterium (since chemical shifts and coupling constants are usually better obtained from proton studies), for the alkali metals, and for Cl, Br and I. More values of T_1 appear to have been obtained for ^{29}Si than for any nuclide other than the "common" ones, and so greater space is devoted to relaxation work for ^{29}Si than for the other cases.

Some other nuclear spin parameters, such as the quadrupole coupling constant (see ch. 3 and 5) and the spin-rotation interaction constant (see ch. 3) are involved in certain nuclear spin relaxation mechanisms but they are not easy to obtain accurately by NMR because molecular motional characteristics also contribute to the relevant physical measurables.

NMR is also used to obtain rate constants and equilibrium constants for chemical reactions, but a discussion of such parameters is clearly beyond the scope of this book.

1A.4 Nuclear Spin Properties of the Elements

Several nucleus-dependent factors affect the ease with which NMR signals may be observed; they also influence the values of NMR parameters. As far as the spin quantum number, I, is concerned, the most important division is into nuclei with zero spin (which will be ignored here), those with $I = \frac{1}{2}$ and those with $I > \frac{1}{2}$. Nuclei in the last category possess nuclear electric quadrupole moments, Q, which shorten relaxation times and broaden lines, rendering the nuclei less suitable, in general, for some NMR purposes than nuclei with $I = \frac{1}{2}$.

The magnetogyric ratio, γ, for a nucleus is of very great importance, since it affects both the frequency and the intensity of the resonance. Its units (in SI) are rad T^{-1} s^{-1}, and values range in magnitude from 0.4582×10^7 (for ^{197}Au) to 28.5335×10^7 (for ^3H), giving a consequent variation of nearly two orders of magnitude in NMR frequencies at constant applied field. The values of γ and I determine the magnetic moment of a nucleus, usually quoted in units of the nuclear magneton, $\mu_N = 5.05095 \times 10^{-27}$ J T^{-1}.

It has been shown that at constant B_0, signal strengths (peak heights) obtained in continuous wave experiments using optimum radiofrequency powers (i.e. to get maximum signal height, which involves partially saturating the resonance) are proportional[20a] to $\gamma^3 NI(I+1)$, where N is the natural abundance of the isotopic species concerned, if it is assumed that $T_1 = T_2$. This quantity, which will be designated here the *receptivity* of the nucleus, can be taken as a very crude guide to the ease of obtaining an observable signal for a given concentration of the relevant atoms in the solution at a constant magnetic field B_0, neglecting any frequency-dependence of electronic noise in the spectrometer. It is helpful to quote such receptivities relative to that of the proton or of ^{13}C. These relative receptivities are given the symbols D^p and D^C respectively, and for nucleus X are defined by:

$$D_X^p = \left|\frac{\gamma_X^3}{\gamma_p^3}\right| \frac{N_X I_X(I_X+1)}{N_p I_p(I_p+1)} = 6.9664 \times 10^{-28} \gamma_X^3 N_X I_X(I_X+1)/\mathrm{rad}^3\,\mathrm{T}^{-3}\,\mathrm{s}^{-3}\% \tag{1.4}$$

$$D_X^C = \left|\frac{\gamma_X^3}{\gamma_C^3}\right| \frac{N_X I_X(I_X+1)}{N_C I_C(I_C+1)} = 3.9472 \times 10^{-28} \gamma_X^3 N_X I_X(I_X+1)/\mathrm{rad}^3\,\mathrm{T}^{-3}\,\mathrm{s}^{-3}\% \tag{1.5}$$

where N_X is to be used as a %.

Values of N, μ, γ, D^p, D^C, and Q for the nuclides discussed in this book are given* in Tables 1.1 and 1.2. In many cases the data are uncorrected for shielding, e.g. $\gamma(1-\sigma)$ is often listed rather than γ itself.

1A.5 Chemical Shifts

Two general features of chemical shifts need discussion at this point, viz. their signs and the origins (zero points) of their scales. The question of sign is a vexed one because the term "chemical shift" is ambiguous[22] in that it may refer either to changes in shielding or to changes in resonance (Larmor) frequency. According to Eq. 1.3, $(\nu_A - \nu_B) \propto -(\sigma_A - \sigma_B)$ at constant B_0, so increase in the resonance frequency implies *de*crease in shielding. In the older literature both possible sign conventions have been used, often without proper definition. Now, however, IUPAC have recommended[23] the universal adoption of the convention that chemical shift signs should relate to resonance frequencies (but expressed in ppm). This high-frequency-positive convention has been used throughout for the present book whenever experimental data are reported, and the symbol δ has been adopted for such shifts. Whenever it is essential to emphasise shielding properties (e.g. in ch. 3) the symbol σ is used, and ambiguity is avoided. A conscious effort is made to drop the outmoded terms "high field" and "low field", since most modern instruments do not use the field-sweep mode of operation. The terms low frequency/high frequency or shielding/deshielding are used instead.

The problem of a suitable reference compound for each nucleus is a more difficult one, since solvent effects can appreciably change the shielding of the reference if used internally to the solution under study, particularly if there is any incipient complex formation. One can avoid this problem by using an external reference in a concentric tube arrangement, but then bulk susceptibility corrections[8,11] are needed. In any case chemical shifts should ideally be reported for infinite dilution in an inert solvent (if one can be found!). Unfortunately, because of solvent effects, which are particularly large for the shielding of the heavier nuclei, such ideal circumstances have seldom occurred for NMR

* Mostly based on the data of ref. 20a. Reference 20b contains a more up-to-date list of μ and Q (together with discussion of the corrections for shielding), but it is relatively inaccessible.

Isotope	Natural abundance N/%	Magnetic moment[b] μ/μ_N	Magnetogyric ratio[c] $\gamma/10^7$ rad T^{-1} s^{-1}	NMR frequency[d] Ξ/MHz	Standard	Relative receptivity[e]	
						D^p	D^C
^{1}H	99.985	4.8371	26.7510	100.000 000	Me_4Si	1.000	5.68×10^3
^{3}H	—	5.1594	28.5335	(106.663)	Me_4Si-t	—	—
^{3}He	1.3×10^{-4}	−3.6848	−20.378	(76.178)	—	5.75×10^{-7}	3.26×10^{-3}
^{13}C	1.108	1.2162	6.7263	25.145 004	Me_4Si	1.76×10^{-4}	1.00
^{15}N	0.37	−0.4901	−2.7107	10.136 783	$MeNO_2$ or NO_3^-	3.85×10^{-6}	2.19×10^{-2}
^{19}F	100	4.5506	25.1665	94.093 795	CCl_3F	0.8328	4.73×10^3
^{29}Si	4.70	−0.9609	−5.3141	19.867 184	Me_4Si	3.69×10^{-4}	2.09
^{31}P	100	1.9581	10.829	40.480 720	85% H_3PO_4	0.0663	3.77×10^2
^{57}Fe	2.19	0.1563	0.8644	(3.231)	$Fe(CO)_5$[f]	7.39×10^{-7}	4.19×10^{-3}
^{77}Se	7.58	0.9223	5.101	19.091 523	Me_2Se	5.26×10^{-4}	2.98
^{89}Y	100	−0.2370	−1.3106	(4.899)	$Y(NO_3)_3$aq.[f]	1.18×10^{-4}	0.668
^{103}Rh	100	−0.1522	−0.8420	3.172 310	mer-$[RhCl_3(SMe_2)_3]$	3.12×10^{-5}	0.177
^{109}Ag[g]	48.18	−0.2251	−1.2449	4.653 623	Ag^+aq.	4.86×10^{-5}	0.276
^{113}Cd[h]	12.26	−1.0728	−5.9330	22.193 173	$CdMe_2$	1.34×10^{-3}	7.59
^{119}Sn[i]	8.58	−1.8029	−9.9707	37.290 662	Me_4Sn	4.44×10^{-3}	25.2
^{125}Te[j]	6.99	−1.528	−8.453	31.549 802	Me_2Te	2.21×10^{-3}	12.5
^{129}Xe	26.44	−1.3380	−7.3995	(27.661)	$XeOF_4$	5.60×10^{-3}	31.8
^{169}Tm	100	−0.400	−2.21	(8.272)	—	5.66×10^{-4}	3.21
^{171}Yb	14.31	0.8520	4.7117	(17.613)	—	7.81×10^{-4}	4.44
^{183}W	14.40	0.2013	1.1131	4.161 733	WF_6	1.04×10^{-5}	5.89×10^{-2}
^{187}Os	1.64	0.1114	0.6161	2.282 343	OsO_4[f]	2.00×10^{-7}	1.14×10^{-3}
^{195}Pt[k]	33.8	1.0398	5.7505	21.414 376	$[Pt(CN)_6]^{2-}$	3.36×10^{-3}	19.1
^{199}Hg	16.84	0.8623	4.7690	17.910 841	Me_2Hg neat	9.54×10^{-4}	5.42
^{205}Tl[l]	70.50	2.7914	15.438	57.633 833	$TlNO_3$aq.	0.1355	7.69×10^2
^{207}Pb	22.6	1.0120	5.5968	20.920 597	Me_4Pb	2.07×10^{-3}	11.8

[a] A complete list, excluding radioactive nuclei (except tritium). [b] The data are modified from those collated by K. Lee and W. A. Anderson[20a], except where otherwise stated. (The magnetic moments listed therein are the maximum observable values and the reader is referred to the original literature to see what substance was studied and whether or not a diamagnetic correction has been applied—see also ref. 20b). [c] Given by Eq. 1.1. [d] Scaled such that the ^{1}H resonance of Me_4Si is at exactly 100 MHz. The values quoted are for the resonances of the standards listed in the next column; the literature references are quoted, together with any alternative values of Ξ, in the relevant chapter. Values in brackets are calculated from the magnetogyric ratios given in the preceding column. [e] Receptivity is defined in the text; D^p is the receptivity relative to that of ^{1}H, whereas D^C is relative to ^{13}C. [f] The only compound studied. [g] Also ^{107}Ag, N = 51.82%, $\gamma = -1.0828 \times 10^7$ rad T^{-1} s^{-1}, $D^p = 3.44 \times 10^{-5}$. The values of μ and γ for ^{107}Ag and ^{107}Ag are from ref. 21(j). [h] Also ^{111}Cd, N = 12.75%, $\gamma = -5.6720 \times 10^7$ rad T^{-1} s^{-1}, $D^p = 1.22 \times 10^{-3}$. The values of μ and γ for ^{111}Cd and ^{113}Cd are from ref. 21(k). [i] Also ^{115}Sn, N = 0.35%, $\gamma = -8.7475 \times 10^7$ rad T^{-1} s^{-1}, $D^p = 1.22 \times 10^{-4}$, and ^{117}Sn, N = 7.61%, $\gamma = -9.5301 \times 10^7$ rad T^{-1} s^{-1}, $D^p = 3.44 \times 10^{-3}$. [j] Also ^{123}Te, N = 0.87%, $\gamma = -7.011 \times 10^7$ rad T^{-1} s^{-1}, $D^p = 1.57 \times 10^{-4}$. [k] The values of μ and γ for ^{195}Pt (in H_2PtCl_6) are from ref. 21(l). [l] Also ^{203}Tl, N = 29.50%, $\gamma = 15.288 \times 10^7$ rad T^{-1} s^{-1}, $D^p = 5.51 \times 10^{-2}$.

TABLE 1.2 A partial list of the spin properties of quadrupolar nuclei

Isotope[a]	Spin	Natural abundance[b] N/%	Magnetic moment[b] μ/μ_N	Magnetogyric ratio $\gamma/10^7$ rad T^{-1} s^{-1}	Quadrupole moment[c] $Q/10^{-28}$ m^2	NMR frequency Ξ/MHz	Usual reference	Relative receptivity[b] D^p	D^C
^{2}H[d]	1	0.015	1.2125	4.1064	2.73×10^{-3}	15.351	—	1.45×10^{-6}	8.21×10^{-3}
^{6}Li	1	7.42	1.1624	3.9366	-8×10^{-4}[e]	14.716	Li^{+}aq.[i]	6.31×10^{-4}	3.58
^{7}Li	$\frac{3}{2}$	92.58	4.2035	10.396	-4.5×10^{-2}[e]	38.864	Li^{+}aq.[i]	0.272	1.54×10^{3}
^{9}Be	$\frac{3}{2}$	100	−1.5200	−3.7594	5.2×10^{-2}	14.053	Be^{2+}aq.[i]	1.39×10^{-2}	78.8
(^{10}B)	3	19.58	2.0793	2.8748	7.4×10^{-2}	10.747	—	3.89×10^{-3}	22.1
^{11}B	$\frac{3}{2}$	80.42	3.4702	8.5827	3.55×10^{-2}	32.084	—	0.133	7.54×10^{2}
^{14}N[d]	1	99.63	0.5706	1.9324	1.6×10^{-2}	7.224	—	1.00×10^{-3}	5.69
^{17}O	$\frac{5}{2}$	0.037	−2.2398	−3.6266	-2.6×10^{-2}	13.557	—	1.08×10^{-5}	6.11×10^{-2}
^{21}Ne	$\frac{3}{2}$	0.257	−0.8539	−2.1118	—[f]	7.894	—	6.32×10^{-6}	3.59×10^{-2}
^{23}Na	$\frac{3}{2}$	100	2.8610	7.0760	0.12[e]	26.451	Na^{+}aq.[i]	9.25×10^{-2}	5.25×10^{2}
^{25}Mg	$\frac{5}{2}$	10.13	−1.0110	−1.6370	0.22[e]	6.120	Mg^{2+}aq.[i]	2.71×10^{-4}	1.54
^{27}Al	$\frac{5}{2}$	100	4.3051	6.9706	0.149	26.057	$Al(H_2O)_6^{3+}$	0.206	1.17×10^{3}
^{33}S	$\frac{3}{2}$	0.76	0.8296	2.0517	-6.4×10^{-2}	7.670	—	1.71×10^{-5}	9.73×10^{-2}
^{35}Cl	$\frac{3}{2}$	75.53	1.0598	2.6212	-7.89×10^{-2}	9.798	Cl^{-}aq.[i]	3.55×10^{-3}	20.2
^{37}Cl	$\frac{3}{2}$	24.47	0.8821	2.182	-6.21×10^{-2}	8.156	Cl^{-}aq.[i]	6.64×10^{-4}	3.77
^{39}K	$\frac{3}{2}$	93.1	0.5047	1.2484	5.5×10^{-2}[e]	4.666	K^{+}aq.[i]	4.73×10^{-4}	2.69
(^{41}K)	$\frac{3}{2}$	6.88	0.2770	0.6852	6.7×10^{-2}[e]	2.561	K^{+}aq.[i]	5.78×10^{-6}	3.28×10^{-2}
^{43}Ca	$\frac{7}{2}$	0.145	−1.4914	−1.7999	0.2 ± 0.1[e]	6.728	Ca^{2+}aq.[i]	9.27×10^{-6}	5.27×10^{-2}
^{45}Sc	$\frac{7}{2}$	100	5.3851	6.4989	−0.22	24.294	$Sc(ClO_4)_3$aq.	0.301	1.71×10^{3}
^{47}Ti	$\frac{5}{2}$	7.28	−0.9313	−1.5079	—[f]	5.637	TiF_6^{2-}/48%HF aq.	1.52×10^{-4}	0.864
^{49}Ti	$\frac{7}{2}$	5.51	−1.2498	−1.5083	—[f]	5.638	TiF_6^{2-}/48%HFaq.	2.07×10^{-4}	1.18
^{51}V	$\frac{7}{2}$	99.76	5.827	7.032	0.3[e]	26.29	$VOCl_3$ neat	0.381	2.16×10^{3}
^{53}Cr	$\frac{3}{2}$	9.55	−0.6113	−1.5120	—[f]	5.652	CrO_4^{2-}	8.62×10^{-5}	0.490
^{55}Mn	$\frac{5}{2}$	100	4.075	6.598	0.55	24.67	$KMnO_4$aq.	0.175	9.94×10^{2}
^{59}Co	$\frac{7}{2}$	100	5.2344	6.3171	0.40	23.614	$Co(CN)_6^{3-}$	0.277	1.57×10^{3}
^{63}Cu	$\frac{3}{2}$	69.09	2.8668	7.0904	−0.16	26.505	$Cu(CN)_4^{3-}$[g]	6.43×10^{-2}	3.65×10^{2}
^{65}Cu	$\frac{3}{2}$	30.91	3.0711	7.5958	−0.15	28.394	$Cu(CN)_4^{3-}$[g]	3.54×10^{-2}	2.01×10^{2}
^{67}Zn	$\frac{5}{2}$	4.11	1.0330[e]	1.6726[e]	0.15	6.252	—	1.17×10^{-4}	0.665
(^{69}Ga)	$\frac{3}{2}$	60.4	2.596	6.421	0.178	24.00	$Ga(H_2O)_6^{3+}$	4.18×10^{-2}	2.37×10^{2}

^{71}Ga	$\frac{3}{2}$	39.6	3.2984	8.1578	0.112	30.495	$Ga(H_2O)_6^{3+}$	5.62×10^{-2}	3.19×10^2
^{73}Ge	$\frac{9}{2}$	7.76	−0.9693	−0.9332	−0.2	3.488	—	1.09×10^{-4}	0.617
^{75}As	$\frac{3}{2}$	100	1.8524	4.5816	0.3	17.127	$K\,AsF_6$	2.51×10^{-2}	1.43×10^2
(^{79}Br)	$\frac{3}{2}$	50.54	2.7098	6.7021	0.33	25.054	Br^-aq.[i]	3.97×10^{-2}	2.26×10^2
^{81}Br	$\frac{3}{2}$	49.46	2.9210	7.2245	0.28	27.006	Br^-aq.[i]	4.87×10^{-2}	2.77×10^2
^{83}Kr	$\frac{9}{2}$	11.55	−1.069	−1.029	0.15	3.848	—	2.17×10^{-4}	1.23
(^{85}Rb)	$\frac{5}{2}$	72.15	1.5952	2.5829	0.247[e]	9.655	Rb^+aq.[i]	7.58×10^{-3}	43.0
^{87}Rb	$\frac{3}{2}$	27.85	3.5391	8.7532	0.12[e]	32.721	Rb^+aq.[i]	4.88×10^{-2}	2.77×10^2
^{87}Sr	$\frac{9}{2}$	7.02	−1.2043	−1.1594	0.36[e]	4.334	Sr^{2+}aq.[i]	1.89×10^{-4}	1.07
^{93}Nb	$\frac{9}{2}$	100	6.7919	6.5387	−0.2	24.443	NbF_6^-/48%HFaq.	0.482	2.74×10^3
^{95}Mo	$\frac{5}{2}$	15.72	1.076	1.743	0.12	6.515	MoO_4^{2-}[g]	5.07×10^{-4}	2.88
(^{97}Mo)	$\frac{5}{2}$	9.46	−1.099	−1.780	1.1	6.652	MoO_4^{2-}[g]	3.25×10^{-4}	1.84
(^{113}In)	$\frac{9}{2}$	4.28	6.0761	5.8496	1.14	21.867	$In(H_2O)_6^{3+}$	1.48×10^{-2}	83.8
^{115}In[h]	$\frac{9}{2}$	95.72	6.0892	5.8622	1.16	21.914	—	0.332	1.89×10^3
^{121}Sb	$\frac{5}{2}$	57.25	3.9537	6.4016	−0.5	23.930	$SbCl_6^-$	9.15×10^{-2}	5.20×10^2
(^{123}Sb)	$\frac{7}{2}$	42.75	2.8726	3.4668	−0.7	12.959	$SbCl_6^-$	1.95×10^{-2}	1.11×10^2
^{127}I	$\frac{5}{2}$	100	3.3056	5.3521	−0.69	20.007	I^-aq.[i]	9.34×10^{-2}	5.30×10^2
^{131}Xe[d]	$\frac{3}{2}$	21.18	0.8869	2.1935	−0.12	8.200	—	5.84×10^{-4}	3.31
^{133}Cs	$\frac{7}{2}$	100	2.9076	3.5089	-3×10^{-3}	13.117	Cs^+aq.[i]	4.74×10^{-2}	2.69×10^2
(^{135}Ba)	$\frac{3}{2}$	6.59	1.0745	2.6575	0.18[e]	9.934	Ba^{2+}aq.[i]	3.23×10^{-4}	1.83
^{137}Ba	$\frac{3}{2}$	11.32	1.2020	2.9729	0.28[e]	11.113	Ba^{2+}aq.[i]	7.77×10^{-4}	4.41
^{139}La	$\frac{7}{2}$	99.911	3.1312	3.7789	0.21	14.126	—	5.92×10^{-2}	3.36×10^2
^{181}Ta	$\frac{7}{2}$	99.988	2.653	3.202	3	11.97	TaF_6^-[g]	3.60×10^{-2}	2.04×10^2
(^{185}Re)	$\frac{5}{2}$	37.07	3.7197	6.0227	2.8	22.514	ReO_4^-[g]	4.94×10^{-2}	2.80×10^2
^{187}Re[h]	$\frac{5}{2}$	62.93	3.7578	6.0844	2.6	22.744	ReO_4^-[g]	8.64×10^{-2}	4.90×10^2
^{189}Os[d]	$\frac{3}{2}$	16.1	0.8392	2.0756	0.8	7.759	OsO_4[g]	3.76×10^{-4}	2.13
^{201}Hg[d]	$\frac{3}{2}$	13.22	−0.7138	−1.7655	0.50	6.600	—	1.90×10^{-4}	1.08
^{209}Bi[h]	$\frac{9}{2}$	100	4.4652	4.2988	−0.4	16.070	—	0.137	7.77×10^2

[a] Nuclei in brackets are considered to be not the most favourable for the element concerned. The ^{99}Tc nucleus ($I = \frac{9}{2}$, $\gamma = 6.0211 \times 10^7$ rad T^{-1} s^{-1}, $Q = 0.3 \times 10^{-28}$ m^2, $\Xi \simeq 22.508$ MHz) has also been used, but it is radioactive (half-life 2.12×10^5 yr) and of negligible natural abundance. Other radioactive nuclei (e.g. ^{40}K, ^{41}Ca) have been examined by NMR but are unimportant for chemical studies. [b] Footnotes *b*, *c* and *e* of Table 1.1 apply. [c] The sign of Q relates to the distribution of charge within the nucleus. The data are from ref. 20a, except where otherwise stated. It should be noted that reported values of Q may be in error by as much as 20–30%. [d] A useful isotope of $I = \frac{1}{2}$ exists. [e] Ref. 21. [f] Not reported. [g] The only compound studied. [h] Radioactive, with a long half-life ($>10^{10}$ years). [i] By extrapolation to infinite dilution.

work with less common nuclides because of the lack of adequate signal intensity in most cases. Consequently comparisons of shift data from different sources need to be made only very cautiously, and apparent variations of several parts per million may sometimes be of no real significance. Choice of a suitable reference compound for each nuclide is itself often a difficulty; in many instances several such compounds have been used for the same nucleus (e.g. for chemical convenience, such as is necessitated by the insolubility of tetramethylsilane in water). But it is clearly desirable for purposes of comparison to have a single primary standard for each nucleus, and to convert shifts based on other, secondary, standards as necessary. The data presented in this book have all had this principle in mind, and the reference compound used for each nuclide is given in Tables 1.1 and 1.2. It may be noted that the single compound tetramethylsilane (TMS) is the accepted reference standard for three nuclei: ^{1}H, ^{13}C and ^{29}Si.

Actually it would theoretically be preferable to reference measurements to the appropriate bare nucleus. Although this is not directly feasible experimentally, it is possible[3,24] to deduce such data after appropriate calculations. A more accessible alternative to the system of having one reference compound for each nuclide is to use an absolute frequency method to indirectly link the scales for the different nuclides. This may be done using heteronuclear double resonance[25], and is particularly simple[26] for modern FT spectrometers which operate using a heteronuclear lock (usually ^{2}H or ^{19}F). In such a system all Larmor frequencies are scaled to be appropriate for a standard applied field, B_0. The situation generally chosen is that for which the protons in TMS (in principle under closely defined solvent conditions) resonate exactly at 100 MHz. Absolute frequencies on this scale are denoted Ξ and are defined by Eq. 1.6.

$$\Xi_X = \frac{100\gamma_X(1 - \sigma_X)}{\gamma(^1H)[1 - \sigma^{TMS}(^1H)]} \tag{1.6}$$

Values of Ξ for various reference standards of spin-$\frac{1}{2}$ nuclei are listed in Table 1.1. Where available, accurate directly-measured values of Ξ are quoted; in other cases approximate data are given, calculated from the values of μ in the compilation of Lee and Anderson[20a]. The values of Ξ for quadrupolar nuclei in Table 1.2 are approximate, as calculated from data on μ. It seems clear that more direct measurements of Ξ are needed, and the values quoted probably need to be checked by several laboratories. Chemical shifts for a given compound, B, from a chosen standard, A, for a given nucleus can be found by measuring Ξ(B) and subtracting Ξ(A), then converting to ppm. Indeed, one can go further, and do away with separate reference substances for each nucleus, even as secondary standards, since it is feasible to define an arbitrary frequency as the reference zero in each case, as has been done for Pt in chapter 8.

Finally, it is worth commenting that care needs to be taken[27] before comparing frequency shifts (or field shifts at constant frequency) with shielding changes in cases where the variations are large.

1A.6 Coupling Constants

Only two general features of coupling constants need to be mentioned at this point. The first is that they may be either positive or negative. By definition a positive sign implies that the coupling interaction stabilises anti-parallel rather than parallel spins, i.e. the term $J_{AB}m_Am_B$ in Eq. 1.2 is negative when m_A and m_B have opposite signs. *Relative* signs of coupling constants can be measured by double resonance methods[28] or even by spectral analysis[8,10,11] in favourable cases, but absolute signs are more difficult to obtain experimentally. When quoted, absolute signs are often based on the well-founded assumption that all values of $^1J(^{13}C, ^1H)$ are positive. Signs are naturally very important for theoretical interpretation of coupling constants, and conversely theory can give signs with confidence in favourable cases. In many cases signs of coupling constants are actually not measured, though this is often not stated in the literature so that the absence of a sign may either mean it is positive or that it is unknown. When numerical values of coupling constants are given in this book plus and minus signs will be given explicitly when known. Absence of a sign implies that $|J|$ is being listed, though for ease of reading the modulus symbols will normally be omitted. Throughout this book, when coupling constants are mentioned, the convention[29] has been adopted that the heavier nuclide involved is written first, e.g. $J(^{29}Si, ^{13}C)$ is preferred to $J(^{13}C, ^{29}Si)$.

Secondly, as will be discussed in chapter 3, coupling constants are proportional to the product of the relevant magnetogyric ratios. For comparison purposes when different nuclei are involved (i.e. for studies in connection with the Periodic Table) it is therefore convenient to define[9,29] a *reduced* coupling constant, given the symbol K, by:

$$J_{AB} = h\left(\frac{\gamma_A}{2\pi}\right)\left(\frac{\gamma_B}{2\pi}\right)K_{AB} \tag{1.7}$$

In the SI system K appears in N A^{-2} m^{-3} whereas in c.g.s. the units are cm^{-3}. Sometimes it is mistakenly supposed that in SI the units must be m^{-3}, but if the definition of eqn. (1.7) is adhered to, without the explicit use of the permeability constant, the conversion of the units is not so simple. In fact, 1 N A^{-2} m^{-3} (SI) $\equiv$ 10 cm^{-3} (c.g.s.). Reduced coupling constants are of the order of 10^{20} to 10^{23} N A^{-2} m^{-3}. Some conversion factors from J to K are given in Table 1.3.

Since magnetogyric ratios themselves may be positive or negative, signs of reduced coupling constants are not necessarily the same as for J values themselves (this is an additional reason for using K's rather than J's for comparison purposes). It should be noted that double resonance experiments yield directly the relative signs of K values, not those of the J's.

1A.7 Relaxation Effects

Values of relaxation times, T_1 and T_2, vary from ca. 0.1 ms to ca. 1000s, even when only mobile liquids are considered. The major factor differentiating T_1 for different nuclei lies in the value of the spin quantum number, I. Nuclei with $I = \frac{1}{2}$ tend to have relatively long relaxation times (in the absence of paramagnetic additives), whereas values of T_1 and T_2 for nuclei of $I > \frac{1}{2}$ tend to be short. This is because, as mentioned earlier, if $I > \frac{1}{2}$ the nucleus possesses an electric quadrupole moment, and electric field gradients at the nucleus lead to relaxation. This quadrupolar relaxation mechanism is, of course, in addition to contributions from other mechanisms, which operate both for nuclei of $I = \frac{1}{2}$ and those of $I > \frac{1}{2}$. Spin-lattice relaxation *rates* due to different mechanisms, j, are additive in simple cases, i.e.:

$$T_1^{-1} = \sum_j T_{1j}^{-1} \tag{1.8}$$

Short relaxation times, such as are often found for quadrupolar nuclei, allow rapid pulse repetition in FT experiments, which helps to improve the S/N. However, when T_2 is short the resonance lines are broad, which keeps resolution low. For single lines in

TABLE 1.3 Factors for reducing coupling constants involving only the proton and the first two rows of the Periodic Table[a,b]

X	^{1}H	^{9}Be	^{11}B	^{13}C	^{14}N	^{15}N	^{17}O	^{19}F	^{29}Si	^{31}P	Y
	0.08325	−0.59242	0.25949	0.33113	1.15224	−0.82146	−0.61412	0.08850	−0.41910	0.20567	^{1}H
		4.21553	−1.84649	−2.35624	−8.19910	5.84531	4.36993	−0.62972	2.98224	−1.46347	^{9}Be
			0.80880	1.03208	3.59138	−2.56037	−1.91412	0.27583	−1.30628	0.64103	^{11}B
				1.31700	4.58282	−3.26719	−2.44254	0.35198	−1.66690	0.81800	^{13}C
					15.94704	−11.36897	−8.49940	1.22478	−5.80038	2.84642	^{14}N
						8.10517	6.05940	−0.87317	4.13521	−2.02927	^{15}N
							4.52998	−0.65278	3.09147	−1.51707	^{17}O
								0.09407	−0.44549	0.21861	^{19}F
									2.10975	−1.03532	^{29}Si
										0.50806	^{31}P

[a] Conversion factor Q given so that in SI $K_{XY}/NA^{-2}\,m^{-3} = Q \times 10^{19} \times J_{XY}/Hz$ (in c.g.s. $K_{XY}/cm^{-3} = Q \times 10^{20}\, J_{XY}/Hz$). [b] Data involving $^{2/3}H$, He, Li, ^{10}B, Na, Mg, Al, S and Cl are omitted. A more complete list can be obtained from Dr. R. Grinter, School of Chemical Sciences, University of East Anglia, Norwich NR4 7TJ, England.

simple cases, when inhomogeneity in B_0 is negligible, the lineshape is Lorentzian and has full width, $W_{1/2}$, at half height:

$$W_{1/2} = (\pi T_2)^{-1} \tag{1.9}$$

where $W_{1/2}$ is in Hz if T_2 is in s. However, for spectra recorded in the obsolete CW derivative mode, linewidths are generally quoted as maximum-to-minimum peak-to-peak separations, W', which are (when in Hz) related to T_2 by Eq. 1.10. In fact, in such cases, values are often given in magnetic field units, which may be converted to Hz by use of Eq. 1.3, usually ignoring σ.

$$W' = (\sqrt{3}\pi T_2)^{-1} \tag{1.10}$$

A factor of some importance in modern FT NMR spectroscopy is the nuclear Overhauser effect (NOE) which arises[30] in double-resonance experiments when, say, nucleus I is observed while species S is irradiated. In general the integrated intensity of the I spectrum will differ from its value for a simple single-resonance I spectrum. The NOE is defined as the ratio of the double-resonance intensity to the single-resonance intensity. It depends[30] on the balance between relaxation mechanisms (see ch. 3). Under normal circumstances the NOE lies between 1 (no effect) and a maximum value given by:

$$\mathrm{NOE}_{\mathrm{max}} = 1 + \frac{\gamma_S}{2\gamma_I} \tag{1.11}$$

For the very common ^{13}C-{^{1}H} experiment (the irradiated nucleus is placed in curly brackets after the observed nucleus) this maximum NOE is therefore ca. 3, i.e. there is an enhancement of the intensity. Such an enhancement is given the symbol η defined by:

$$\mathrm{NOE} = 1 + \eta \tag{1.12}$$

Thus:

$$\eta_{\mathrm{max}} = \frac{\gamma_S}{2\gamma_I} \tag{1.13}$$

Since the abbreviation for nuclear Overhauser enhancement may also be NOE, some care is necessary in discussing numerical results. If either γ_S or γ_I is negative, there will normally be a diminution of integrated intensity under double-resonance conditions compared to single resonance. In fact, as is clear from the form of the expression, if in addition $|\gamma_S| > 2|\gamma_I|$ the signal itself may go negative or it may, for a given balance of relaxation mechanisms, be zero (the "null signal" problem). Since most FT experiments are carried out under ^{1}H- or ^{19}F-irradiation conditions (in order to observe spectra decoupled from ^{1}H or ^{19}F respectively) the negative NOE situation can be a problem for the observation of ^{15}N and the other preferred spin-$\frac{1}{2}$ nuclei (^{29}Si, ^{89}Y, ^{103}Rh, ^{109}Ag, ^{113}Cd, ^{119}Sn, ^{125}Te, ^{129}Xe and ^{169}Tm) with negative magnetogyric ratios. However, nuclear Overhauser effects are rarely encountered for nuclei with $I > \frac{1}{2}$ because quadrupolar interactions provide the dominant relaxation mechanisms.

1A.8 NMR and the Periodic Table

Figure 1.1 shows the Periodic Table up to radon (excluding the lanthanides), with elements indicated for which chemical NMR studies (solution-state) have been reported. Use of spin-$\frac{1}{2}$ and spin $> \frac{1}{2}$ nuclei is clearly differentiated. It may be seen that the majority of the elements have received some attention by now, and others are also suitable, though

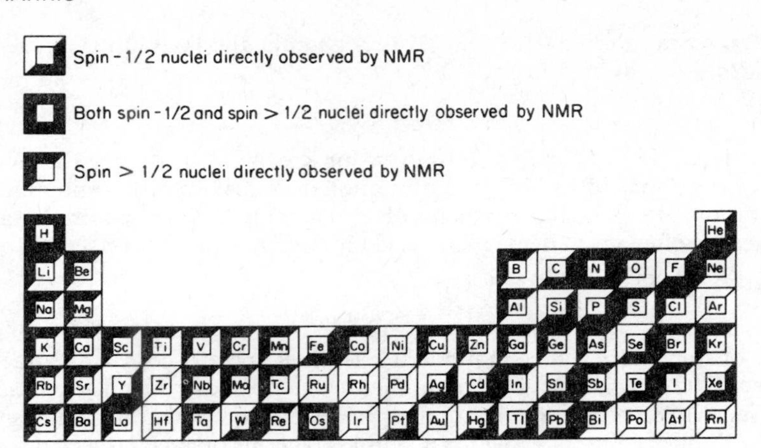

Fig. 1.1 The Periodic Table up to radon (excluding the lanthanides) showing the chemical NMR usage of the elements.

in some cases (e.g. Os, Fe) extensive use is perhaps not to be expected. In other cases (e.g. Sc, Y, In, Ne and the lanthanides) the lack of chemical interest has inhibited NMR work. In fact all elements up to and including lead have suitable stable isotopes for NMR except Tc and Pm (neither of which have any stable isotopes at all), Ar (which is unimportant chemically) and Ce. Many nuclides were studied in the early days of NMR, but then for some years the vast majority of applications came to be in proton NMR, until FT methods were developed. However, for many elements most NMR studies have occurred in the last few years, since FT spectrometers came into widespread use. Undoubtedly this trend to multinuclear magnetic resonance will continue and will be of particular importance for inorganic chemistry. Inspection of Fig. 1.1 and Tables 1.1 and 1.2 shows that only six elements have potentially useful nuclei of both the $I = \frac{1}{2}$ and $I > \frac{1}{2}$ types, namely hydrogen (^{1}H and ^{3}H have $I = \frac{1}{2}$; ^{2}H has $I = 1$), nitrogen (^{15}N has $I = \frac{1}{2}$; ^{14}N has $I = 1$), xenon (^{129}Xe has $I = \frac{1}{2}$, ^{131}Xe has $I = \frac{3}{2}$), ytterbium (^{171}Yb has $I = \frac{1}{2}$, ^{173}Yb has $I = \frac{5}{2}$), osmium (^{187}Os has $I = \frac{1}{2}$; ^{189}Os has $I = \frac{3}{2}$) and mercury (^{199}Hg has $I = \frac{1}{2}$; ^{201}Hg has $I = \frac{3}{2}$). For hydrogen and nitrogen choice of the best nucleus for study depends on the problem in hand (see ch. 4 and 5). Osmium and ytterbium are scarcely used for NMR so the question of choice has not really arisen, and there is little reason to study ^{201}Hg or ^{131} Xe rather than the corresponding spin-$\frac{1}{2}$ isotopes.

The principle factors which limit the applicability of NMR, apart from the extent of chemical interest in the element, are two, viz. (a) the availability of diamagnetic compounds (since paramagnetic species often give very broad, sometimes unobservable, resonances—this is one reason why NMR studies of iron are unlikely to be useful), and (b) considerations of signal intensity. As already mentioned, signal intensity depends on isotopic abundance, magnetogyric ratio and (to a limited extent) spin quantum number. Nuclei may therefore be usefully grouped in several ways, viz. (a) spin-$\frac{1}{2}$ vs. spin $> \frac{1}{2}$, (b) according to N—rare ($N < 10\%$), of medium abundance ($10\% \leq N \leq 90\%$) and abundant ($N > 90\%$)—and (c) according to magnetic strength—strong ($|\gamma|/10^7 > 10$ rad T^{-1} s^{-1}), medium (10 rad T^{-1} s^{-1} $\geq |\gamma|/10^7 \geq 2.5$ rad T^{-1} s^{-1}) and weak ($|\gamma|/10^7 < 2.5$ rad T^{-1} s^{-1}) The divisions within (b) and (c) are, of course, arbitrary but convenient. Under these definitions strong nuclei generally resonate with $\Xi > 38$ MHz, whereas for weak nuclei $\Xi < 9.6$ MHz. Factors (b) and (c) will be considered in more detail in the next two sections, which deal with $I = \frac{1}{2}$ and $I > \frac{1}{2}$ nuclei in turn. Of course it is possible to change N by artificial enrichment (or depletion), and this has frequently proved useful for ^{2}H, ^{13}C, ^{15}N and ^{17}O. The radioactive tritium nucleus ($I = \frac{1}{2}$) has also been rendered suitable for NMR by "enrichment" (see ch. 5).

1A.9 Spin-$\frac{1}{2}$ Nuclei

Only 24 elements of the Periodic Table have non-radioactive nuclei of spin-$\frac{1}{2}$; their NMR properties are given in Table 1.1. NMR chemical studies of 21 (i.e. excluding Rh, Tm and Yb) have now been carried out. Helium-3 is not very suitable for *chemical* studies. Chemically the remainder may be grouped as follows:

Non-metals	H, C, N, F, Si, P, Se, Te, Xe
Group B metals	Sn, Tl, Pb
Transition metals	Fe, Y, Rh, Ag, Cd, W, Os, Pt, Hg
Lanthanides	Tm, Yb

The first category has provided the overwhelming majority of all NMR studies to date, and will undoubtedly continue to do so because of the popularity of ^{1}H and ^{13}C NMR (for chemical reasons). The receptivities of ^{57}Fe and ^{187}Os are so low that it is unlikely their resonances will be greatly used for chemical studies. It is of interest to note that there are no spin-$\frac{1}{2}$ nuclei among the alkali and alkaline earth metals, and only one halogen (fluorine) has $I = \frac{1}{2}$. This means the simplest ionic solutions are not readily amenable to NMR as far as the observation of small chemical shifts are concerned. Fortunately, for these nuclei shifts (even solvent effects) are often large. In contrast, four out of the five group IVB elements (C, Si, Sn and Pb) have suitable spin-$\frac{1}{2}$ nuclei. Several of the elements on the above list have more than one spin-$\frac{1}{2}$ nuclide (e.g. ^{107}Ag/^{109}Ag, ^{111}Cd/^{113}Cd, ^{115}Sn/^{117}Sn/^{119}Sn, ^{203}Tl/^{205}Tl) though in each case there is little doubt which nuclide is the better for NMR (that with the higher receptivity). Since primary isotopic effects on chemical shifts are generally negligible (shift differences in ^{117}Sn NMR are the same as those in ^{119}Sn NMR), there is no reason for studying the less favourable nuclei of these pairs or trios of spin-$\frac{1}{2}$ isotopes. The more favourable isotopes will be referred to as the preferred nuclides, and the relationship between Ξ and D^{C} for these is illustrated in Fig. 1.2. For hydrogen, however, for certain specialised applications the radioactive tritium nucleus may be preferred to the abundant proton (see ch. 5).

The spin-$\frac{1}{2}$ nuclei are classified according to natural abundance and magnetic strength in Table 1.4. It is interesting to note that there are only six spin-$\frac{1}{2}$ nuclei of essentially 100% natural abundance, and three of these (^{1}H, ^{19}F, ^{31}P) are strongly magnetic and of great chemical interest, whereas the other three (^{89}Y, ^{103}Rh and ^{169}Tm) are very weakly magnetic and of relatively little chemical interest. Nature has favoured the chemist in this regard! The fact that ^{1}H, ^{19}F and ^{31}P are highly favourable for NMR explains their early popularity. As is shown in Table 1.1 the receptivity of ^{31}P is already only 6.6% that of ^{1}H and most other spin-$\frac{1}{2}$ nuclei are at least an order of magnitude less favourable than this, the only exceptions being ^{19}F and ^{205}Tl. The only nucleus which has a higher value of γ than ^{1}H is tritium (with $\Xi = 106.526$ MHz), but this is radioactive and the natural abundance is negligible.

Carbon-13 is typical of the rare spins, and it is necessary to point out that although there is a penalty in terms of low signal intensity to studying such a nuclide, its very rarety does bring one considerable advantage, namely that in most circumstances the chances of getting two such spins in the same molecule can be ignored. This means that no (^{13}C,

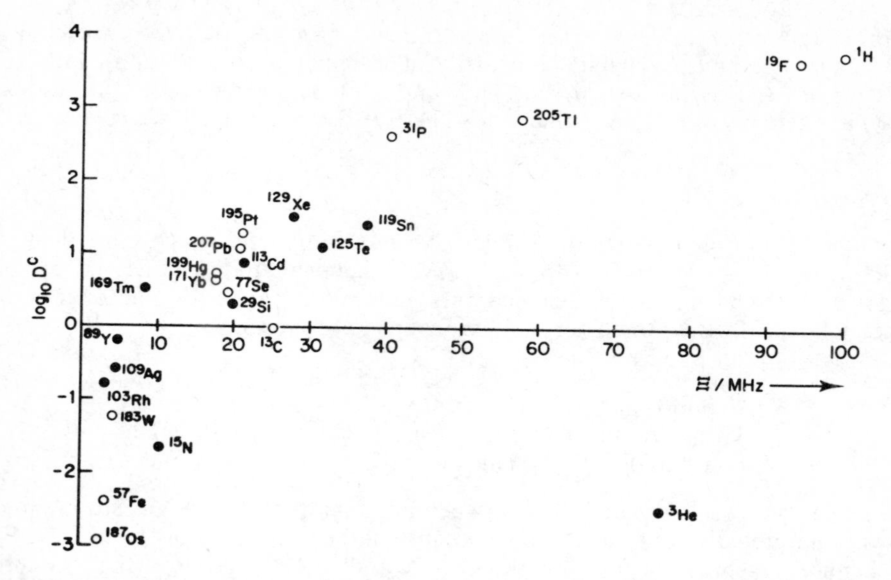

Fig. 1.2 Relative NMR receptivities, D^C, for spin-$\frac{1}{2}$ nuclei as a function of the standardised NMR frequencies Ξ. For each element only the most favourable nuclide has been included. The open circles represent nuclei with positive magnetogyric ratios; filled circles indicate nuclei with negative γ.

TABLE 1.4 Classification of spin-$\frac{1}{2}$ nuclei according to natural abundance[a] and magnetic strength[a]

Magnetic strength	Natural abundance: High (>90%)	Medium	Low (<10%)
Strong	^{1}H, ^{19}F, ^{31}P	(^{203}Tl), ^{205}Tl	^{3}He
Medium		(^{111}Cd), ^{113}Cd, ^{129}Xe, ^{171}Yb, ^{195}Pt, ^{199}Hg, ^{207}Pb	^{13}C, ^{15}N, ^{29}Si, ^{77}Se, (^{115}Sn), (^{117}Sn), ^{119}Sn, (^{123}Te), ^{125}Te
Weak	^{89}Y, ^{103}Rh, ^{169}Tm	(^{107}Ag), ^{109}Ag, ^{183}W	^{57}Fe, ^{187}Os

[a] See the definitions in the text. The less favourable nuclei for a given element are listed in brackets. Tritium is a strong nuclide but is radioactive (negligible natural abundance).

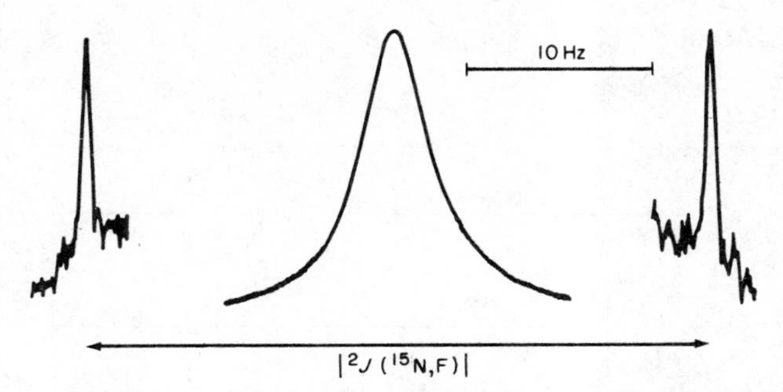

Fig. 1.3 The 94.1 MHz ^{19}F spectrum of 2,6-difluoro-3,4,5-trichloropyridine, showing the weak, sharp ^{15}N satellites (at increased amplitude) on either side of the intense broad resonance due to molecules containing ^{14}N. The width of the central line is due to the incompleteness of ^{14}N-relaxation-induced collapse of splittings due to (^{19}F, ^{14}N) coupling. (Reproduced with permission from ref. 31.)

^{13}C) coupling effects can usually be seen in a spectrum, and, under conditions of proton noise decoupling (and in the absence of other spin-$\frac{1}{2}$ nuclei of high or medium natural abundance) spectra show a series of single lines, one for each non-equivalent carbon site in the molecule being studied. This simplifies spectral interpretation considerably, though it means values of J_{CC} are difficult to obtain. The above situation is particularly important for the non-metals C, N, Si and Se, whose chemistry is replete with compounds containing many such atoms. Such is not normally the case for the metals, and the advantage of rarity is seldom of significance for quadrupolar nuclei.

The abundant nuclei ^{1}H, ^{19}F and ^{31}P, on the other hand, give spectra which yield very useful spin-coupling data, such as J_{HH}, J_{FF} and J_{PP}, and the spectra are uncomplicated by the existence of other isotopes. Nuclei of medium abundance, such as ^{195}Pt can give complicated spectra. For example a compound containing two non-equivalent platinum sites gives three principal ^{195}Pt NMR lines for each site. The central one of these (66.2% of the intensity for the site) is due to molecules in which the second site is occupied by ^{194}Pt, ^{196}Pt or ^{198}Pt, whereas the outer two lines (each 16.9% of the intensity) arises from molecules in which ^{195}Pt occupies the second site. The outer lines would be referred to as "satellites" and have separation $|J(^{195}\text{Pt}, ^{195}\text{Pt})|$. Such satellites also occur in spectra of other nuclides, e.g. one can observe ^{195}Pt satellite lines in ^{13}C spectra. Figure 1.3 shows ^{15}N satellite lines in a ^{19}F spectrum; in this case the central line is broad because it is due to molecules containing the quadrupolar ^{14}N nucleus. For elements with more than one spin-$\frac{1}{2}$ nuclide there will be several types of satellite e.g. tin satellites would have component lines due to the presence of ^{117}Sn (and, in principle, ^{115}Sn) as well as the more abundant ^{119}Sn, and the separations of the appropriate pairs of lines would be slightly different since coupling constants are proportional to magnetogyric ratios.

1A.10 Quadrupolar Nuclei

The spin properties of quadrupolar nuclei discussed in this book are given in Table 1.2, and they are classified in Table 1.5. With the definition adopted here, only ^{7}Li is a "strong" magnetic nucleus (it is the only quadrupolar nucleus with $\Xi > 33$ MHz). However, the definition of "strong" is rather arbitrary, particularly as it relates to γ (and hence Ξ), rather than μ. In fact most of the elements listed in Table 1.2 have at least one nuclide

TABLE 1.5 Classification of quadrupolar nuclei according to natural abundance[a] and magnetic strength[a]

Magnetic strength	Natural abundance: High (>90%)	Medium	Low (<10%)
Strong	^{7}Li		
Medium	^{9}Be, ^{23}Na, ^{27}Al, ^{45}Sc, ^{51}V, ^{55}Mn, ^{59}Co, ^{75}As, ^{93}Nb, ^{115}In, ^{127}I, ^{133}Cs, ^{181}Ta, ^{209}Bi	(^{10}B), ^{11}B, ^{35}Cl, ^{63}Cu, ^{65}Cu, (^{69}Ga), ^{71}Ga, (^{79}Br), ^{81}Br, (^{85}Rb), ^{87}Rb, ^{121}Sb, (^{123}Sb), ^{137}Ba, ^{139}La, (^{185}Re), ^{187}Re	^{2}H, ^{6}Li, ^{17}O, ^{21}Ne, (^{113}In), (^{135}Ba)
Weak	^{14}N, ^{39}K	^{25}Mg, ^{37}Cl, ^{83}Kr, ^{95}Mo, ^{131}Xe, ^{189}Os, ^{201}Hg	^{33}S, (^{41}K), ^{43}Ca, ^{47}Ti, ^{49}Ti, ^{53}Cr, ^{67}Zn, ^{73}Ge, ^{87}Sr, (^{97}Mo)

[a] See the definitions in the text. The less favourable nuclei (where the choice is clear-cut) for a given element are listed in brackets. Technetium-99 is a nucleus of moderate magnetic strength, but it is radioactive (negligible natural abundance).

with a receptivity greater than that of ^{13}C (exceptions being O, S, Ca, Zn and Ge, with ^{2}H also having a very low receptivity). Over a dozen quadrupolar nuclides (including ^{7}Li, ^{11}B, ^{23}Na, ^{25}Mg, ^{51}V, ^{55}Mn, and ^{59}Co) have receptivities greater than that of ^{31}P. However, such comparisons between quadrupolar and spin-$\frac{1}{2}$ nuclei are not very significant because of the linewidth problems for the former.

The ^{2}H and ^{14}N nuclei are special cases because of the existence of alternative spin-$\frac{1}{2}$ isotopes for these elements. The halogens and the alkali metals are well-represented by moderately strong magnetic nuclei, and it is clear that studies of simple ionic solutions (including biochemical situations) will depend heavily on such nuclei (see ch. 6 and 13). Boron-11 has been for many years a relatively popular nucleus (see ch. 4)—^{10}B is less favourable. Oxygen-17 is in principle an important nucleus (see ch. 12) but the lines can be quite broad and the natural abundance is very low, giving considerable problems (much of the work reported in the past has used material enriched in ^{17}O). The ^{33}S is even less favourable for NMR. However, an appreciable number of transition metals have useful quadrupolar NMR nuclei (especially ^{55}Mn and ^{59}Co), which can be of importance when the chemistry involves a substantial number of diamagnetic compounds.

The principal difficulty in working with quadrupolar nuclei is that linewidths can be very large. This prevents resolution of lines due to chemically similar sites, even when total chemical shift ranges are large, and it precludes very accurate measurement of chemical shifts. It also reduces S/N availability, though in the FT mode the fact that T_1 is short as well as T_2 allows rapid pulsing and therefore a compensatory improvement in signal intensity (providing T_2 is not so short that "dead time" in the electronics following the pulse becomes a problem).

For a single quadrupolar nucleus of spin I the relaxation times, T_{1q} and T_{2q}, are given,[11,12,32] when molecular motion can be characterised by an isotropic tumbling correlation time τ_c (usually in the range 1–10 ps for non-viscous liquids), by:

$$T_{1q}^{-1} = T_{2q}^{-1} = \frac{3\pi^2}{10} \frac{(2I + 3)}{I^2(2I - 1)} \chi^2(1 + \tfrac{1}{3}\eta^2)\tau_c \quad (1.14)$$

where χ is the nuclear quadrupole coupling constant (NQCC, measured in Hz if T_1 or T_2 and τ_c are in s). The NQCC is defined by:

$$\chi = \frac{e^2 q_{zz} Q}{h} \quad (1.15)$$

where q is the electric field gradient at the nucleus, with q_{zz} as its biggest component, e is the charge on the electron and Q is the nuclear electric quadrupole moment. In Eq. 1.14 η is the asymmetry parameter for q, given by Eq. 1.16; it lies in the range $0 < \eta < 1$.

$$\eta = \frac{q_{yy} - q_{xx}}{q_{zz}} = \frac{\chi_{yy} - \chi_{xx}}{\chi_{zz}} \quad (1.16)$$

Linewidths are related to T_{2q} by Eq. 1.9 or 1.10 if the quadrupolar term is dominant (as it usually is). Thus linewidths are given by Eq. 1.17:

$$W_{1/2} = \frac{3\pi}{10} \frac{(2I + 3)}{I^2(2I - 1)} \chi^2(1 + \tfrac{1}{3}\eta^2)\tau_c \quad (1.17)$$

The dependence on q_{zz} leads to a wide variation of $W_{1/2}$ for different compounds containing the same nuclide (variations in τ_q also have an effect of course). For a sufficiently symmetrical molecule q_{zz} will be small. Thus the ^{14}N NMR linewidth of Me_4N^+ is <4.5 Hz, whereas that for aniline is 1300 Hz.[33] The effect is even more marked for ^{35}Cl. The linewidth for aqueous NaCl (with a spherical Cl^- ion if the environment is ignored) is only ca. 10 Hz, whereas for CCl_4 it is ca. 10 kHz (see ch. 8). Quadrupolar linewidths for different nuclides depend on the relative Q values. Thus $|Q|$ for ^2H, ^6Li and ^{133}Cs are rather small ($<10^{-30}$ m^2), and these species tend to give relatively sharp lines. The values of $|Q|$, and hence the linewidths, increase markedly in the series ^{35}Cl, ^{79}Br, ^{127}I so that the iodine nuclide is very difficult to study by NMR. When there are two or more quadrupolar nuclides for a given element, the one preferred for NMR studies may sometimes be the one with the smaller Q even if it has the lower receptivity (e.g. when the parameter being studied is the chemical shift rather than the relaxation time, ^6Li may be preferred[34] to ^7Li). However, for most elements where there is a choice, the preferred nucleus is obvious, viz. the one with the higher receptivity and lower Q. In cases where the choice is clear, it is indicated in Table 1.2.

Equation 1.17 shows that the spin I also influences linewidths, and the factor of importance (for constant electric field gradients) when comparing different nuclides is:[35]

$$L = \frac{(2I + 3)Q^2}{I^2(2I - 1)} \quad (1.18)$$

The spin term $f(I) = (2I + 3)/I^2(2I - 1)$ varies as follows:

I	1	$\frac{3}{2}$	$\frac{5}{2}$	3	$\frac{7}{2}$	$\frac{9}{2}$
$f(I)$	5	$\frac{4}{3}$	$\frac{8}{25}$	$\frac{1}{5}$	$\frac{20}{147}$	$\frac{2}{27}$

Thus from this point of view the higher spins tend to give the sharper spectra; the effect of I can be substantial.

REFERENCES

1. P. C. Lauterbur, Ch. 7 of "Determination of Organic Structures by Physical Methods", Vol. 2 (F. C. Nachod and W. D. Phillips, eds), Academic Press, New York (1962)
2. P. R. Wells, Ch. 5 of "Determination of Organic Structures by Physical Methods", Vol. 4 (F. C. Nachod and J. J. Zuckerman, eds), Academic Press, New York (1971)
3. R. R. Sharp, *Prog. Anal. Chem.* **6**, 123 (1973)
4. A. L. van Geet, *Prog. Anal. Chem.* **6**, 155 (1973)
5. R. K. Harris, *Chem. Soc. Rev.* **5**, 1 (1976)
6. R. K. Harris, Ch. 4 of "Molecular Spectroscopy" (Proc. 6th. Conf., Inst. of Petroleum, Durham, 1976) (A. R. West, ed), Heyden and Son Ltd, London (1977)
7. R. H. Cox, *Magn. Resonance Rev.* **1**, 271 (1972); **3**, 207 (1974)
8. J. A. Pople, W. G. Schneider and H. J. Bernstein, "High-resolution Nuclear Magnetic Resonance", McGraw-Hill, New York (1959)
9. R. M. Lynden-Bell and R. K. Harris, "Nuclear Magnetic Resonance Spectroscopy", Nelson, London (1969)
10. R. J. Abraham, "Analysis of High Resolution NMR Spectra", Elsevier (1971)
11. "High-resolution NMR Spectroscopy", J. Emsley, J. Feeney and L. H. Sutcliffe, Pergamon Press, Oxford (1966)
12. "The Principles of Nuclear Magnetism", A. Abragam, Clarendon Press, Oxford (1961)
13. "High Resolution NMR: Theory and Chemical Applications", E. D. Becker, Academic Press, New York (1969)
14. "Fourier Transform NMR Spectroscopy", D. Shaw, Elsevier, Amsterdam (1976)
15. "Progress in NMR Spectroscopy" (J. Emsley, J. Feeney and L. H. Sutcliffe, eds), Pergamon Press, Oxford, Vols 1–12
16. "Advances in Magnetic Resonance" (J. S. Waugh, ed), Academic Press, New York, Vols 1–9
17. "Annual Reports on NMR Spectroscopy" (E. F. Mooney and G. A. Webb, eds), Academic Press, London, Vols 1–7
18. "NMR: Basic Principles and Progress" (P. Diehl, E. Fluck and R. Kosfeld, eds), Springer-Verlag, Berlin, Vols 1–15
19. "Specialist Periodical Reports on NMR" (R. J. Abraham and R. K. Harris, eds), The Chemical Society, London, Vols 1–7
20. (a) K. Lee and W. A. Anderson (1967), "Handbook of Chemistry and Physics", 55th Edn., p. E-69 (R. C. West, ed). Chemical Rubber Co., Cleveland (1974); (b) G. H. Fuller, *J. Phys. Chem. Ref. Data* (National Bureau of Standards, Washington) **5**, 835 (1976)
21. (a) O. Lutz, A. Schwenk and A. Uhl, *Z. Naturforsch.* **30a**, 1122 (1975); (b) R. Neumann, F. Träger, J. Kowalski and G. zu Putlitz, *Z. Physik A* **279**, 249 (1976); (c) K. Murakawa, *J. Phys. Soc. Japan* **11**, 422 (1956); (d) G. zu Putlitz, *Z. Physik* **175**, 543 (1963); (e) G. zu Putlitz, *Ann. Physik* **11**, 248 (1963), (f) I-J. Ma and G. zu Putlitz, *Z. Physik A* **277**, 107 (1176); (g) G. W. Canters and E. DeBoer, *Mol. Phys.* **27**, 665 (1974); (h) S. Garpman, I. Lindgren, J. Lindgren and J. Morrison, *Phys. Rev. A* **11**, 758 (1975); (i) Nuclear Data Tables, Vol. **5**, no. 5–6 (1969), U.S. Atomic Energy Commission, Washington; (j) C.-W. Burges, R. Koschmieder, W. Sahm and A. Schwenk, *Z. Naturforsch.* **28a**, 1753 (1973); (k) H. Krüger, O. Lutz, A. Nolle, A. Schwenk and G. Stricker, *Z. Naturforsch.* **28a**, 484 (1973); (l) R. J. Goodfellow and W. Hackbusch, private communication; (m) P. W. Spence and M. N. McDermott, *Phys. Lett.* **24A**, 430 (1967).
22. R. K. Harris, *Educ. Chem.* **14**, 44 (1977)
23. "Recommendations for the Presentation of NMR Data for Publication in Chemical Journals", *Pure Appl. Chem.* **29**, 627 (1972); **45**, 217 (1976)
24. J. Mason, *Adv. Inorg. Chem. Radiochem.* **18**, 197 (1976)
25. W. McFarlane, *Proc. Roy. Soc.* (*London*) **A306**, 185 (1968)

26. R. K. Harris and B. J. Kimber, *J. Magn. Reson.* **17**, 174 (1975)
27. F. H. A. Rummens, *Org. Magn. Reson.* **2**, 209 (1970)
28. R. A. Hoffman and S. Forsén, *Prog. NMR Spectry* **1**, 15 (1966)
29. R. Grinter, Ch. 2 of Specialist Periodical Reports on NMR, Vol. 4 (R. K. Harris, ed), The Chemical Society, London (1975)
30. J. R. Lyerla and D. M. Grant, Ch. 5 of MTP internat. Rev. Sci. (*Phys. Chem.* Series 1), Vol. 4 (C. A. McDowell, ed), Butterworths (1972)
31. A. V. Cunliffe and R. K. Harris, *Mol. Phys.* **15**, 413 (1968)
32. B. Lindman and S. Forsén, *NMR Basic Principles and Progress* **12**, 1 (1976)
33. J. M. Lehn and J. P. Kintzinger, in "Nitrogen NMR" (M. Witanowski and G. A. Webb, eds), Table 3.5, Plenum Press, London (1973)
34. F. W. Wehrli, *Org. Magn. Reson.* **11**, 106 (1978)
35. J. W. Akitt, *Ann. Rep. NMR Spectry* **5A**, 465 (1972)

2

EXPERIMENTAL METHODS

DEREK SHAW, Oxford Instruments Ltd., England

2A.1 Introduction

Nuclear magnetic resonance spectroscopy developed from the successful attempts in 1946 of two groups (those of Professors Bloch[1] at Stanford University and Purcell[2] at Harvard University), to measure the magnetogyric ratio of the proton. Improved instrumentation led, four years later, to the discovery of chemical shifts, and coupling constants were observed the year after.

In the early sixties commercial proton spectrometers became available and NMR developed, almost explosively, into a major spectroscopic technique, finding its main applications in structural organic chemistry. The major limitation of NMR is its low (by spectroscopic standards) sensitivity. Study was therefore concentrated on the more sensitive nuclei such as ^{1}H, ^{19}F, and ^{31}P, despite the large number of suitable nuclei within the Periodic Table. In the early 1970s the introduction of Fourier transform methods and wide-band decoupling techniques dramatically increased NMR sensitivity. This in turn led to an upsurge in interest in less sensitive, but chemically very significant, nuclei, such as ^{13}C and ^{15}N. The study of each individual nucleus however required a considerable amount of hardware specific to that nucleus, e.g. r.f. source, receiver, etc. The high costs involved in studying an "obscure" nucleus therefore was a deterrent to wandering over the Periodic Table. In the last few years commercial spectrometers that are truly multinuclear, or, as they are frequently termed, wideband, have become available. With these instruments the difficulties and costs of observing a very wide range of nuclei have dramatically decreased.

Now is therefore an appropriate time to consider NMR as a tool, not limited to the study of a small group of nuclei, but as one which can span the whole Periodic Table. This chapter will discuss the experimental aspects of NMR with particular reference to those areas, mainly instrumental, where the study of "other nuclei" requires a different approach to be taken than that appropriate to the study of the more common and sensitive nuclei like the proton.

2A.2 Direct Observation Methods

An NMR spectrum is a plot of nuclear magnetism as a function of frequency $M(\nu)$. The behaviour of $M(\nu)$, is best described by the set of differential equations developed by Bloch.[3,4] These show that $M(\nu)$ is complex, consisting of two components, one out of phase with the exciting frequency, which gives a Lorentzian absorption at the resonance condition (when the Larmor equation $2\pi\nu = \gamma B_0$ is met). This is called the v mode signal; the other component is the in-phase one, giving the associated dispersion (u mode)

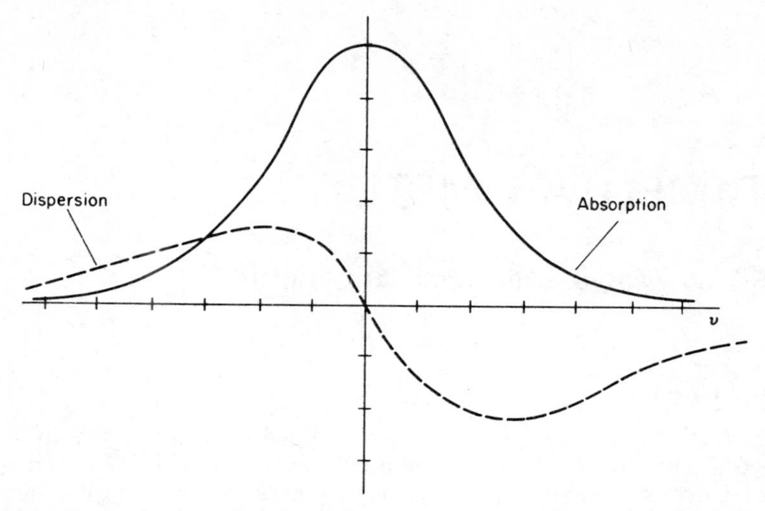

Fig. 2.1 The absorption and dispersive components of an NMR line.

signal. The two modes are illustrated in Fig. 2.1. It is the absorption signal which is normally plotted as the NMR spectrum since this is integratable, the area being proportional to the number of nuclei resonating. The v mode signal is described by:

$$v = \frac{\gamma M_0 B_1 T_2}{1 + 4\pi^2 T_2^2 \cdot (\nu_0 - \nu_2)^2 + \gamma^2 B_1^2 T_1 T_2} \tag{2.1}$$

It follows therefore that in order to record a spectrum directly, it is necessary to have a basic magnetic field and a range of frequencies with which to excite the sample, plus the electronics, etc. required to detect any changes induced in nuclear magnetisation of the sample. For any individual nucleus the range of the frequencies required will be small, 10–10 000 ppm. On the other hand the range of frequencies required to study all the nuclei of interest is large, as can be seen in Tables 1.1 and 1.2.

a. Sweep Methods

The original and most obvious way of providing the spread of frequencies required to observe the full spectrum of a nucleus is to sweep the exciting frequency through the appropriate range. The frequency must be swept linearly and the value of the nuclear magnetisation is plotted. In order to obtain the true spectrum the sweep rate, dv/dt, must be slow enough to maintain equilibrium, i.e. $dv/dt < (T_1 \cdot T_2)^{-1}$. If the power is low enough to avoid saturation, i.e. $\gamma^2 B_1^2 T_1 T_2 < 1$ and the out of phase component of $M(\nu)$ is detected, a Lorentzian line is produced (see Eq. 2.1). The peak of this line occurs at the resonance frequency of the corresponding nuclei; its linewidth is characterised by the life time of the excited state (see section 2A.7*a*) and its area by the number of resonant nuclei. If an exciting field whose power is higher than required by the above inequality is applied then saturation of the spin system occurs and the signal intensity is no longer proportional to the applied field but decreases with increasing field. Useful spectra can be obtained using high r.f. powers providing the sweep rate is fast.[5] This is termed the adiabatic rapid passage (ARP) method. The ARP method, due to its use of high power, has a sensitivity higher than the slow sweep method. The lines in an ARP experiment are subject to considerable line broadening coupled with frequency shifts and hence the method is of little use for recording high resolution spectra.

The advantage of the sweep approach is its experimental simplicity and consequent low cost. It is, however, an inefficient method since all the resonances are detected sequentially at a slow rate. The method can in theory be improved in two ways: firstly, by only sweeping the areas of interest, skipping the rest, and secondly by sweeping below saturation power at a rate faster than $(T_1 \cdot T_2)^{-1}$. The distortions introduced are subsequently corrected for. The first alternative is obviously not very practical since in an unknown sample the areas of interest are unknown.

The second approach is generally referred to as correlation spectroscopy.[6,7] The transient effects introduced by using a fast scan (10–50 Hz s^{-1}) can be easily expressed analytically and corrected for, thus producing a normal undistorted spectrum. The correction procedure uses cross-correlation techniques and hence leads to the name correlation spectroscopy. The output of the spectrometer during the rapid sweep is digitised and stored in a computer. The cross-correlation is achieved by using some properties of Fourier transforms.[8] Specifically that cross-correlation of two functions in one domain is equivalent to the complex multiplication of their Fourier transforms in the co-domain. The data is subjected to a Fourier transform then multiplied by the Fourier transform of the excitation. The result of this multiplication is a signal analogous to the free induction decay of the sample (see next section). After any required weightings etc., the resulting signal is transformed for a second time to give the normal spectrum.

Correlation spectroscopy is considerably more efficient than adiabatic slow passage methods (by a factor of at least 5) but requires more instrumentation, e.g. digitisers, and computers. The method is little used except in proton NMR where its ability to sweep only specific regions (or avoid certain undesirable regions, e.g. solvent resonances) is an advantage. The method also has potential for nuclei with a very large chemical shift range, where producing a pulse with sufficient power to generate a uniform excitation field can present a problem.

b. Pulse Excitation

The most efficient method of recording an NMR spectrum is to record the responses of all the nuclei simultaneously. This objective can be achieved by means of a short pulse of radio frequency energy.[10] The shorter the pulse, by the uncertainty principle, the wider the frequency produced. The power from a pulse, $t_p s$ long, is distributed as $(\sin \pi t_p\{\nu_0 - \upsilon_0\})/\pi t_p(\nu_0 - \upsilon_0)$ i.e. it has a maximum value at ν_0 and has a zero value at t_p^{-1} Hz from ν_0. If the pulse is placed in the middle of the spectrum, and pulse width is made small compared with the spectral width, then effectively equal excitation is achieved over the total spectral width. In practical terms this means that a pulse width of t_p secs. will equally excite (to within 2%) all nuclei over a range of $(2t_p)^{-1}$ Hz., i.e. 10 μs pulse will cover nuclei within ± 25 kHz. from the pulse frequency.[9]

The effect of the excitation pulse on the nuclear magnetism is best described by a classical vector model. In a frame of reference rotating at the Larmor frequency, the nuclear magnetism of the sample can be represented by a static vector along the axis, i.e. parallel to the magnetic field. During the presence of the excitation along an axis orthogonal to the static field (conventionally the x axis) the magnetisation vector rotates through an angle α given by:

$$\alpha = \gamma B_1 t_p \tag{2.2}$$

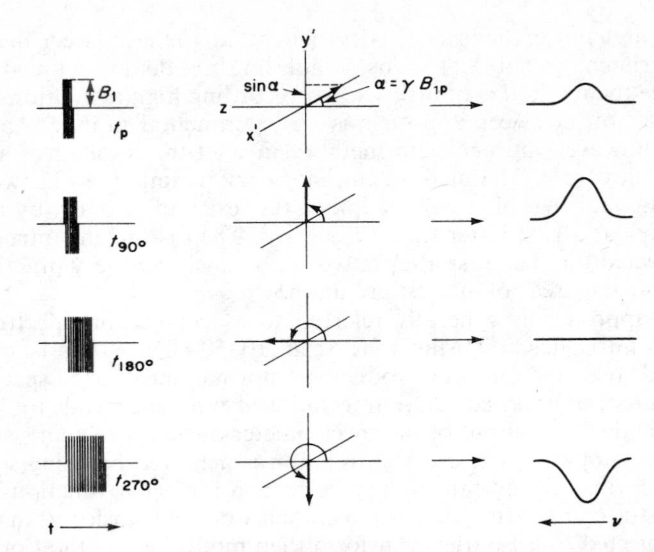

Fig. 2.2 The effect increasing pulse width on the nuclear magnetic vector. The right hand column is the Fourier transform of the signal induced in the receiver coil by the nuclei following the pulse, from ref. 13, with permission.

The process is illustrated in Fig. 2.2. After the termination of the pulse the nuclei relax towards equilibrium by exponential processes which are characterised by the spin-spin relaxation time (T_2). During this period they induce voltages in the receiver coil of the spectrometer yielding the time domain spectrum of the sample, which is normally referred to as the free induction decay (FID). Fourier transform of the FID yields the normal, or frequency domain, spectrum,[11] as is illustrated in Fig. 2.3.

The above equation leads to the description of pulses being given in terms of an angle α. Pulses in excess of 180° are infrequently used, thus the power of the excitation pulse should be sufficient to give a 180° rotation in less than $(2 \times \text{spectral width})^{-1}$ s. As can be seen in Fig. 2.2 the magnitude of the signal produced by a single pulse is proportional to $\sin \alpha$; the maximum signal follows a 90° pulse, a zero signal follows a 180° pulse, etc. In the more common case where a regular series of pulses is applied, the situation is not so simple. If a second pulse is applied before the magnetisation has returned via spin lattice relaxation to its full equilibrium value, then the signal from the second pulse will be smaller than that from the first pulse. An interval between the pulses (T_d) in the order of $4T_1$ (for 98 % recovery) is therefore indicated. The detected signal is, however, decaying as T_2^* (i.e. by natural T_2 processes plus any effect of magnet homogeneity, see section 2A.7*e*) which is in general less than T_1. The signal will therefore have decayed into the noise long before the time of $4T_1$ required above.

In order to achieve maximum efficiency, the pulses are usually applied more frequently than $4T_1$ when a complex steady state is set up.[10,12] The pulse interval used is normally that required to acquire the data from the FID (see section 2.5). Under these conditions the optimum pulse angle can be expressed in terms of the pulse interval T_p, by[10,12,13]

$$\cos \alpha = \exp - \frac{T_p}{T_1} \tag{2.3}$$

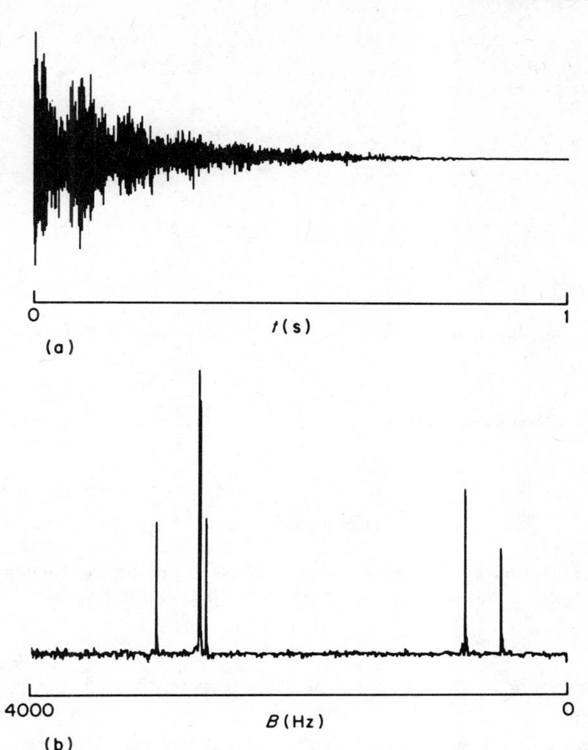

Fig. 2.3 The free induction decay from the ^{13}C nuclei in ethyl benzene following pulse excitation (a) and the Fourier transform into a conventional frequency spectrum (b).

which as expected gives α tending to 90° when $T_p \gg T_1$ and decreasing to 68° when $T_p = T_1$. For a more detailed consideration of these problems see refs 12 and 13.

The advantage of pulse excitation over swept excitation is its efficiency,[10] the former method exciting all the nuclei simultaneously as opposed to sequentially, thus gaining the Fellgett or multi-channel advantage. Using a simultaneous or multichannel approach of the type shown in Fig. 2.4 is more efficient than a single channel method by a factor of $n^{1/2}$ where n is the number of channels used. This gain results from the property that all signals are coherent, whereas noise is, by its very nature, incoherent. There must obviously be a maximum value for n or an infinite signal to noise ratio is possible. The maximum value n can take is the total number of spectral elements within the spectrum, i.e. the total spectral width divided by the typical line width. The gain achievable by multi-channel detection is typically 10–20 for protons where normal spectra extend over 1 kHz (at 2.3T) and greater for nuclei like ^{31}P and ^{13}C where larger spectral widths are encountered. The gain increases with increasing field. The relative gains obtained by using pulse excitation are thus most significant when studying nuclei with large chemical shift ranges, as is quite often the case for the less common nuclei. The gain in efficiency achievable by using pulse excitation is, however, obviously lower when broad lines are encountered, for example when studying quadrupolar nuclei, but it is none the less always finite. For this reason pulse excitation is now preferred to swept excitation for all except

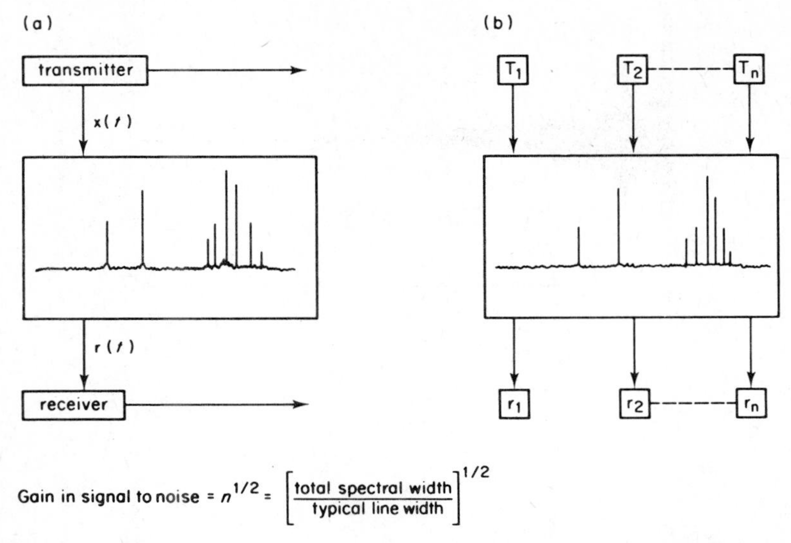

$$\text{Gain in signal to noise} = n^{1/2} = \left[\frac{\text{total spectral width}}{\text{typical line width}}\right]^{1/2}$$

Fig. 2.4 Single channel excitation (a) can theoretically be improved in efficiency by increasing the number of channels (b) up to maximum of one channel per line width, from ref. 13, with permission.

routine proton work, where sensitivity is of little importance and spectrometer cost is a prime consideration.

The spectra produced by sequential excitation of the spectrum and those produced by simultaneous excitation of all the spins are only totally equivalent under conditions where the spin system behaves in a linear manner. The term linear basically means that the output is proportional to the input and that the proportionality constant is time independent. The first condition is only met for small angles where α is proportional to $\sin \alpha$; fortunately however it can be shown that for normal spectra there is complete equivalence between spectra produced by both methods below saturation.[14]

When higher excitation fields are used the linearity condition is violated and spectra showing different intensity patterns are produced by the two methods. Within the limitations above spectra produced during chemical exchange are equivalent. However those recorded during chemically induced dynamic nuclear polarisation experiments differ if pulse angles above 20° are used. Above these angles the multiplet effect is averaged to zero.[14]

In situations where the spin-lattice relaxation time far exceeds the instrumental line width, e.g. when T_1 codes are well over 100 seconds, as can be seen from the above, Fourier Transform methods became very inefficient. There are basically two techniques which can be employed to overcome this dilemma. One technique is to use driven equilibrium techniques (DEFT).[15] In this method the relaxed transverse components of the magnetization are forcibly returned on to the axis for use during the second phase.

The second approach is to establish a steady state, i.e. not wait for equilibrium to be restored and to record spectra within the steady state. The steady state condition has been analysed in detail by Freeman and Hill.[16] A steady state technique has been successfully used along with a four pulse sequence which eliminates anomalies associated with the steady state in the study of nuclei such as silver where very long T_1's are often found. This technique has been given the name Quadriga Fourier transform.[17]

c. Stochastic and Tailored Excitation

The use of a pulse of r.f. energy is not the only possible method of obtaining a spread of frequencies, and hence exciting all the nuclei in the sample simultaneously. A second method, called stochastic excitation, is to modulate the basic frequency with either white,[18] or pseudo-random noise.[19] The most common method is the latter one; in this method the output of a suitable coded shift register is used to phase -modulate the carrier. The width of the noise modulation depends on the clock frequency used to drive the shift register. The response of the sample of this type of excitation is more complex than that obtained using simple pulse excitation, but since the input noise is known it can be unscrambled using cross-correlation techniques of the type outlined above (see ch. 2A.2*a*).[20] The main advantage of stochastic excitation lies in its more efficient use of available r.f. power (see section 2A.5*c*) and hence its superior ability to deal with large spectral widths.[19]

It has been assumed so far that a uniform excitation is required over the total spectral region. This may not always be the case, e.g. in homo decoupling experiments where one multiplet should be subjected to stronger irradiation than the rest of the spectrum, or when, for example in order to avoid dynamic range effects, a solvent line should not be excited at all. Such a desired or tailored excitation spectrum can be obtained by digitally transforming the desired frequency spectrum into the time domain and interpreting each of the resulting time domain values as a pulse.[21] Tailored excitation is therefore a sequence of pulses at least 1000 long (the longer the sequence the better defined and more selective the pulse sequence is). As a rule stochastic modulation is simultaneously applied to improve the power usage. Tailored excitation permits many novel experiments for nuclei such as ^{1}H and ^{19}F, which have complex spectra with many homonuclear interactions.[22] The various types of NMR excitation and their uses are summarised in Table 2.1.

TABLE 2.1 Types of NMR experiments

Type	Description	Use and advantages
Continuous wave	Slow low power sweep gives spectrum directly.	Low cost routine spectrometer.
Correlation spectroscopy	Fast frequency sweep using cross correlation to produce spectrum.	Used in proton work to avoid exciting solvent signal. Gives increase in sensitivity compared with slow sweep techniques.
Simple pulsed (FT)	Repetitive pulse sequence used for excitation. Resulting time domain is Fourier transformed to give normal NMR spectrum.	High sensitivity due to exciting all nuclei simultaneously. Gives easy access to relaxation data. Normally used in multinuclear spectrometers.
Stochastic	Uses pseudo-random noise instead of pulses and cross correlation prior to Fourier transformation.	Similar to above but using pseudo-random pulses. Requires less peak power.
Tailored excitation	Uses pulse sequence calculated to give desired power spectrum followed by cross-correlation and Fourier transformation.	Has the sensitivity of simple pulsed excitation but has the advantage of placing r.f. power only where it is required. Multiple resonance can be done with one transmitter.

TABLE 2.2 Classification of double resonance experiments

Amplitude of B_2 (Hz)	Effects observed	Terminology
$\gamma B_2 \gg \Sigma n \cdot J$ (n depends on the multiplicity of the line)	Spectral simplification	Spin decoupling
$\gamma B_2 \sim J$	Selective removal of spin coupling	Selective decoupling
$\gamma B_2 \sim (T_2^*)^{-1}$	Additional splitting	Spin tickling
$\gamma^2 B_2^2 (T_1 T_2) \approx 1$	Changes in relative intensities	Generalised Overhauser effect

2A.3 Double Resonance

The previous section was concerned with the excitation applied to the nuclei being observed. In many experiments it is necessary to excite two or more groups of nuclei differentially; such experiments are referred to as double resonance experiments. The basic spectrum is excited by one of the methods described in the preceding section. The effect of the second radio frequency depends on its power (see Table 2.2). Double resonance experiments can be either homo- or heteronuclear; for most nuclei (except ^1H and perhaps ^{19}F) it is heteronuclear decoupling which is of most interest.

a. Coherent Decoupling

The simplest double resonance experiment is the coherent irradiation of one multiplet. In the rotating frame the equilibrium magnetisation is along the z axis; the addition of resonant radio frequency causes rapid precession of the vectors of that part of the spin multiplet about the x axis. This precession has the effect, as viewed from the other part of the spin multiplet, of averaging away the splittings caused by spin coupling. In order to achieve complete spin decoupling, the decoupling field (B_2) must be powerful enough to affect all the components of the spin multiplet equally, i.e. $\gamma B_2 \gg \Sigma nJ$. If lower powers are used, more complex spectra are produced. The above is a very much simplified description of the phenomenon of double resonance. For fuller treatment the reader is referred to reference 24. Spin decoupling can be used for spectrum simplification and assignment.[25]

b. Off Resonance Decoupling

The so called "off resonance" decoupling experiment is essentially the same as the previous one, except that the second radio frequency is deliberately offset from the true resonance frequency of the multiplet which is to be perturbed. The result is a spectrum not showing multiplet collapse, but one which has the same multiplicity as the spectrum, with, however, residual splittings (J_r). The residual splitting can be related (to a first approximation) to the true splitting (J), the decoupler power (γB_2), and the offset of the decoupler from resonance $|v_0 - v_2|$ thus:[26]

$$J_r = \frac{2\pi |v_0 - v_2|}{\gamma B_2} \tag{2.4}$$

Experiments of this type are of great value in spectral assignment, especially in ^{13}C NMR.[27] The carbon spectrum is recorded and the multiplicity, whilst applying an off-resonance proton frequency, gives the number of protons directly bonded to each carbon. At the same time it is possible from the value of the residual splittings to deduce the shift of the directly bonded protons from the decoupler frequency. In this way we can map out the proton spectrum of the sample from the carbon spectrum and valuable assignment information becomes available.

c. Wide Band Decoupling

The spectra of nuclei other than the proton are frequently complicated, often impossibly so, by spin coupling to the proton. These complications can be removed by spin decoupling with a consequential gain in spectrum simplicity. Furthermore, since the total area under an NMR spectrum is constant no matter how many lines it contains, reducing the multiplicity of a spectrum will increase the intensity of those lines which remain and hence increase the sensitivity of the experiment. To achieve this goal it is necessary to irradiate all the protons simultaneously with an adequate power.[26] The required power distribution is normally obtained by modulating the basic decoupler frequency, which is set in the middle of the spectrum to be decoupled, with either pseudo random noise[26] (see section 2A.2*c*) or a simple square wave.[28] Analysis of the "noise decoupling" experiment is not simple:[26] the randomly fluctuating fields provide, via the scalar coupling, additional relaxation processes for the observed spins. Consequently some line broadening may be induced in the observed spectrum, especially if too little power is used, and measurements of T_2 become very difficult. T_1 measurements are not affected by these processes; in fact wide-band decoupling is essential when studying nuclei coupled to protons, in order to minimise the complications arising from cross-relaxation.[29]

d. The Nuclear Overhauser Effect

The application of a second radio frequency to a sample does not only cause changes in multiplicity as discussed so far, but it can also induce intensity changes due to the nuclear Overhauser effect,[30] (see ch. 1). The nuclear Overhauser enhancement, η, only has its maximum value when dipole/dipole relaxation dominates the relaxation of the observed nucleus; this is often the case for nuclei with a spin of a half with directly bonded protons, e.g. ^{13}C, ^{15}N, and ^{29}Si, but rarely for quadrupole nuclei, e.g. ^{14}N, and ^{10}B, where quadrupole relaxation dominates. The result is that η varies from its maximum value to zero, even within one spectrum, depending on the relative contributions of the various possible relaxation mechanisms (see ch. 3D). Consequently integrals measured during double resonance experiments must be treated with caution. For ^{13}C spectra the nuclear Overhauser effect is frequently near its maximum value of 2 and hence provides a welcome increase in the achievable sensitivity.

A second complication in spectra recorded with wide-band proton decoupling is that the sign of η depends on the sign of the magnetogyric ratio. For nuclei like ^{13}C it is positive, but for nuclei such as ^{15}N and ^{29}Si it is negative, as is shown in Fig. 2.5. Thus for ^{29}Si for example, signals can have intensities ranging from approximately -2 to $+1$ depending on η, which in turn depends on the relevant relaxation processes. One embarrassing value is zero!

Fortunately it is possible to suppress the Overhauser effect if required. Suppression can be achieved in two basic ways, either by using instrumental (see next section) or chemical methods. In the second method a so-called relaxation reagent is added to the solution in order to "short circuit" dipole/dipole relaxation and hence eliminate the

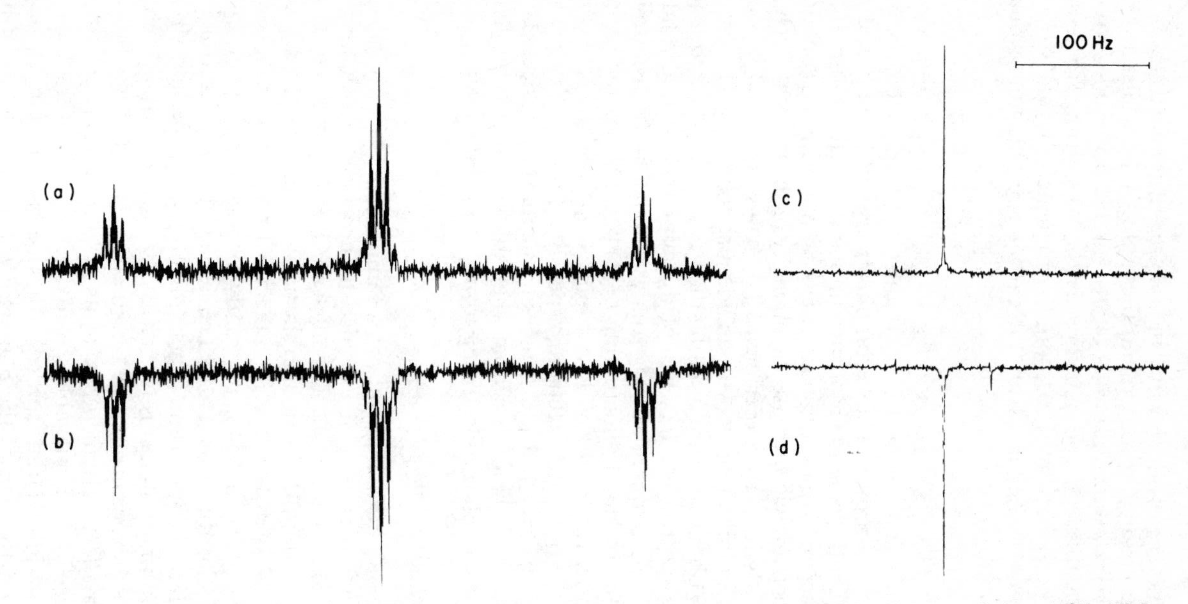

Fig. 2.5 The ^{29}Si Spectrum of Ph_2SiH_2. (a) Coupled, no NOE; (b) Coupled, with NOE; (c) Decoupled, no NOE; (d) Decoupled, with NOE.
Note: (a) and (b) were recorded with 300 transients whereas only 14 were necessary for (c) and (d). From B. J. Kimber, Ph.D. thesis, University of East Anglia, 1974.

NOE; relaxation agents are paramagnetic transition metal ions with an isotropic g tensor, usually chromium or ferric trisacetonylacetonates.[31] The fluctuating fields generated by these ions provide an exceedingly efficient and hence dominant relaxation mechanism. Since only dipole/dipole relaxation between the irradiated and observed spin can give rise to an Overhauser enhancement no NOE is observed if another relaxation mechanism dominates.

The obvious disadvantage of this method of NOE suppression is chemical contamination. A less obvious disadvantage is that any specific complexation can cause non-uniform effects which, if unsuspected, can produce misleading results.

e. Gated Decoupling

The time scales associated with decoupling and the nuclear Overhauser effects are different. It is this difference which provides the basis of the instrumental method of separating them.[32]

On the application of a second radio frequency field decoupling occurs immediately; the NOE however builds up slowly. The growth is exponential with a time constant in the order of the spin lattice time T_1. If the second rf is therefore switched on or off immediately before data collection, then either decoupled or undecoupled spectra will be recorded. The status of the NOE, however, depends on the state of the decoupler in the time prior to the exciting pulse. If it had been off (or on) for a period in the order of at least $3T_1$ prior to the pulse then there would be zero (or maximum) NOE in the resultant spectrum. A timing diagram for such experiments is given in Fig. 2.6. The

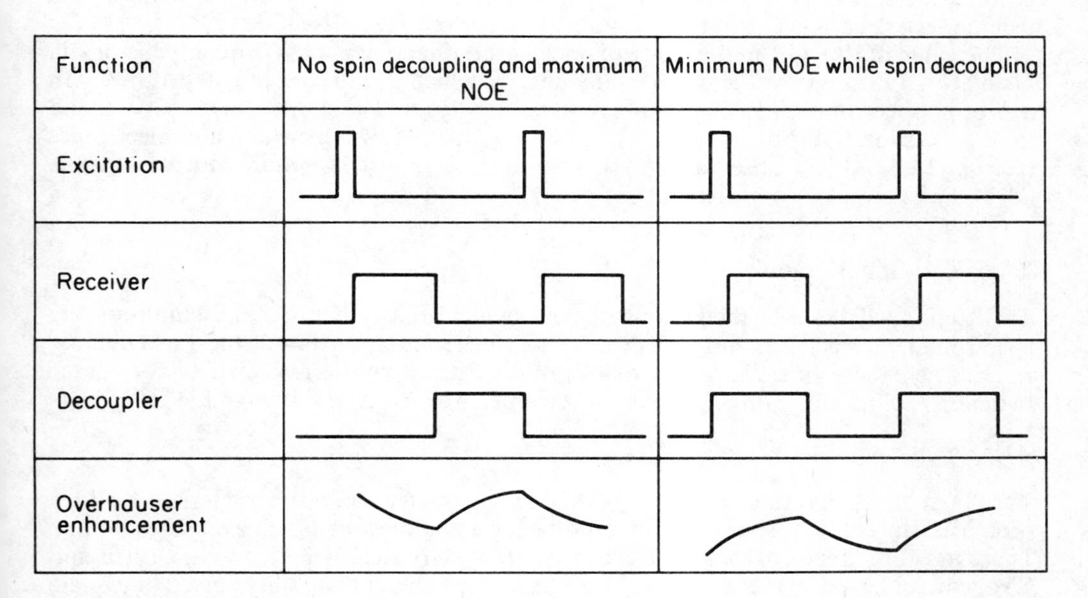

Fig. 2.6 Timing diagram showing the switching sequences used during homonuclear decoupling with pulse excitation.

behaviour of the NOE during data acquisition has no effect on the intensities of the spectrum (as can easily be shown by Fourier theory).[33] Gated decoupling experiments can therefore be used either to enhance sensitivity (for pairs of nuclei with positive NOEs) whilst retaining full coupling information and the NOE, or to eliminate the NOE but keeping the simplicity of a decoupled spectrum. Fig. 2.5 shows the four possible situations for ^{29}Si NMR, i.e. the two possible gated decoupling experiments and the spectra run with and without wide band decoupling.

2A.4 Indirect Methods of Observation

It is sometimes easier to study nuclei by indirect means, i.e. by investigating their effect on the spectrum of another nucleus to which they are spin coupled. The normal example of this is the inter nuclear double resonance (INDOR) experiment.[34] This experiment consists of monitoring the amplitude of one transition (usually in the proton spectrum of the molecule being studied) while slowly sweeping a second low power excitation through the frequency range appropriate to the heteronucleus. When the second frequency stimulates a transition with an energy level in common with the monitor line intensity changes are induced in the monitor line. It is thus possible to map out the energy level diagram for the spin system and deduce the magnitude and signs of the coupling constant involved and also, more significantly, the chemical shift of the heteronucleus. Homonuclear INDOR is also possible and is used frequently in assignment problems.[25]

There is no direct equivalent to INDOR while using pulse excitation; equivalent results can however be obtained using spectral subtraction techniques. The spectrum obtained without any secondary excitation is subtracted from the spectrum obtained either while continuously irradiating the transition equivalent to the monitor line in the INDOR experiment with a low power ($\gamma B_2 \sim T_2^*$)[35] or after having inverted its population with a selective 180° pulse.[36,37] The spectra obtained by either of these two techniques are equivalent to those obtained by conventional INDOR.

The value of INDOR in the study of nuclei other than the proton is mainly historical. Using INDOR it is possible to study any nucleus which shows coupling to protons with a proton spectrometer (with its inherent sensitivity) and a simple (low cost) second frequency source. With the advent of pulse spectrometers and wide-band electronics (see ch. 2A.5*c*) direct observation is becoming much more common with its obvious advantages.

2A.5 The Spectrometer

This section will describe the various elements which go together to make a multinuclear spectrometer. Special attention will be paid to the requirements of the less common nuclei where these differ from those of the proton. Pulse excitation only will be considered in detail. A block diagram of a typical spectrometer of this type shown in Fig. 2.7.

a. The Magnet

The magnet, which is the heart of any NMR spectrometer, can be characterised by three properties, its field, the size of the largest tube it can handle and its achievable resolution. There are three types of magnet used in NMR spectrometers, permanent, electro and cryogenic. The permanent magnet with its high inherent stability, given adequate temperature control, is ideal for routine proton work but is in general limited to fields up to 2.1T (90 MHz for ^{1}H) and 5 mm tubes, more powerful magnets becoming excessively

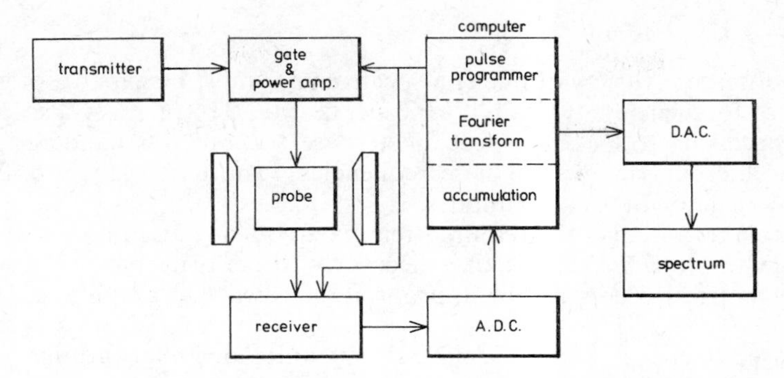

Fig. 2.7 Block diagram of a typical high resolution pulse spectrometer.

bulky. Electromagnets provide the field for the majority of the present generation of multinuclear spectrometers working in the 1.8 to 2.3T (80–100 MHz) region and taking tubes up to 20 mm or so in diameter. The cryogenic magnets were originally (and still are) developed for the high fields they can achieve, up to 13.1T (600 MHz). Latterly a new generation of superconducting spectrometers has emerged using cryo magnets, not to produce ultra high fields but to produce economically a large volume of field in the 4.2 ~ 4.6T (180–200 MHz) region

The criteria of magnet performance for proton work are somewhat different from those applicable to spectrometers mainly used for other nuclei. The proton spectrometer requires the highest field achievable (for the forseeable future, at least!) both for increased line separation and sensitivity. Line separation is not as critical for other nuclei, since the chemical shift ranges found are normally much larger than those of the proton whereas, with the exception of quadrupolar nuclei, the linewidths are more comparable. Furthermore, sensitivity does not increase with field for many nuclei above about 4T due to increased contributions to line broadening from chemical shift anisotropy. Linewidths for the heavier nuclei are generally larger than those of the proton; magnet resolution is therefore not such a demanding and critical factor in producing good spectra.

The field of a spectrometer can also affect the spectrum of many nuclei in quite a complex way (apart from modifications resulting from changing the second order nature of the spectrum), once the extreme narrowing limit is exceeded. In simple terms the extreme narrowing limit is exceeded once the Larmor frequency of the nucleus being detected, or any nucleus to which it is spin coupled, e.g. ^{1}H, approaches or exceeds the reciprocal of the molecular correlation time. Under these conditions, which are increasingly encountered as magnet fields increase, relaxation characteristics change, the NOE decreases, and can even invert in sign,[30] T_1 will exceed T_2 and consequently the efficiency of NMR decreases.

None of the types of magnets mentioned above are capable, unaided, of producing a magnetic field of sufficient homogeneity that does not contribute to the line width of the recorded spectrum. Typical raw homogeneities are in the order of 1 part in 10^6, whereas for many nuclei the natural line width is in the order of 1 in 10^8. The required improvements are brought about firstly by spinning the sample in order to average inhomogeneities in the *xy* plane, and secondly by the use of shim coils. The latter are small coils, usually wound onto formers attached to the pole pieces or the probe, which produce specific field gradients with up to 12 being used for some large magnets. By adjusting the currents in these coils, most of the basic inhomogeneities in the magnet can be corrected.

b. Frequency Generation

In order to observe a variety of nuclei a spectrometer must be capable of generating a wide range of frequencies, typically 7–100 MHz (at 2.3T); it must also produce the frequencies necessary to lock the spectrometer (see section 2A.5*e*) and perform any required decoupling experiments. All these frequencies are normally generated from a single source by frequency synthesis techniques.

By using this technology, the required relative frequency and phase stability can be achieved over the long periods of time necessary to perform many experiments. The spectrometer must also contain the appropriate hardware to amplify, gate, etc., these frequencies.

The increased interest in other nuclei NMR has stimulated many groups of workers to modify their existing spectrometers to permit easier multinuclear operation. The basic principle used in nearly all these modifications is to use a frequency synthesizer, in conjunction with a mixing scheme, to make a range of nuclei "appear like" (i.e. have the same frequency following the mixer) a nucleus already available in the spectrometer, e.g. ^{13}C. The details of these modifications depend on the original spectrometer. The reader is referred to ref. 38 and references therein for further details.

c. The Probe

The probe is the key area of an NMR spectrometer both in terms of the achievable sensitivity and flexibility. There are two basic probe designs, the single and the crossed coil. The former, which is illustrated in Fig. 2.8, is the one most commonly used in multinuclear spectrometers since it makes the most efficient use of the power available.

The requirements placed on a probe in a pulse spectrometer are very demanding. In order to achieve a high sensitivity the coil must be tightly coupled to the sample and have a high Q. However, in order to respond correctly to short pulses and recover quickly, so that fast relaxing nuclei such as those with the quadrupole moments can be studied, a low Q is required. Some very ingenious methods have been used to facilitate these conflicting requirements (see section 3-21 of ref. 37). The probe also has to carry out simultaneously the second, field/frequency or lock experiment. In the case of an internal lock

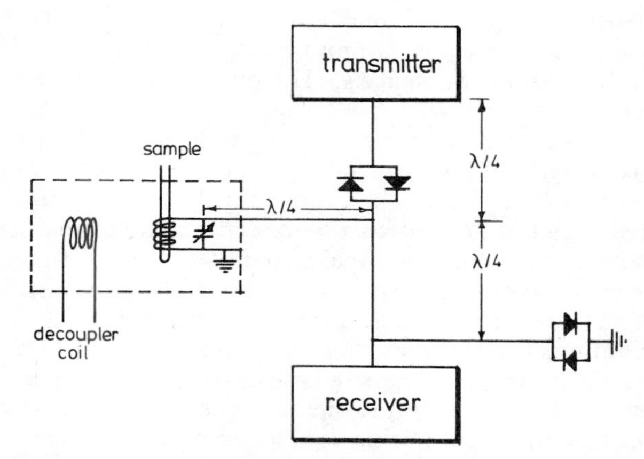

Fig. 2.8 Schematic diagram of a single coil probe, from ref. 13, with permission.

(see ch. 2A.5*e*) this is achieved by double tuning the coil used for normal observation. The decoupling frequency is normally applied by a separate orthogonal coil.

Until very recently it was only possible to tune probes over a limited range. It was therefore necessary to use separate probes (or inserts in the case of "Universal" probes) for each nucleus, or for each narrow frequency range, of interest. New designs enable one probe to be used over a very wide range, e.g. between ^{14}N and ^{31}P. The proton and ^{19}F, because of their very different frequency, still usually require a separate probe or insert. These developments in wide-band probe design are of very great significance for multinuclear work, not only as they reduce the costs involved, but probably more significantly they dramatically reduce the speed of changeover between nuclei. Furthermore, many nuclei within the sample can be studied without taking the sample from the probe or even losing the lock. In this way it is possible to ensure that the spectra are run under identical conditions, not an insignificant advantage when performing kinetic or complex relaxation experiments.

Also housed within, or very close by, the probe is the pre-amplifier whose function is to amplify the signal by a small, but significant, amount before it is transmitted to the main r.f. transmitter. The main amplifier detects the signal and boosts it to the level required to drive the recorder or ADC. These radio frequency elements of the spectrometer must either be capable of handling the appropriate discrete frequency or be of a wide-band design. In the detection stages the signals are referred to the frequency of the pulse. The output of the receiver following an exciting pulse is thus a mixture of decaying audio frequencies. The frequencies are the chemical shifts, i.e. offsets of the spectral lines from the pulse frequency.

d. The Field/Frequency Lock

The magnetic field and the exciting frequency cannot be made independently stable enough to permit the long term time averaging that is necessary for the study of weak nuclei. It is therefore standard practice to perform a second NMR experiment in order to stabilise, or lock, the two variables together. The locking channel normally uses the dispersion resonance of a nucleus not under investigation, most frequently deuterium. In this mode the signal changes sign on either side of resonance and for small changes is proportional to the displacement from resonance. A simple feedback circuit is used to ensure a constant field/frequency ratio.

The origin of the locking signal is used to classify the type of lock used. The first distinction is between the homo and heteronuclear type. The former type is infrequently used in FT work, mainly due to dynamic range problems (see later) which can arise when the lock nucleus is of the same species as the one being detected. The second type utilises the resonance of another (hetero) nucleus. Heteronuclear locks, for chemical simplicity, use the deuterium resonance of the solvent. A second classification is that of internal or external lock. The external lock uses a separate sample (usually of either H_2O or D_2O) which is built into the probe independently of the analytical sample. This technique has great experimental convenience, since it places no chemical constraints on the samples and the lock is not lost on removal of the sample. It does however have a limited stability (0.5 Hz h^{-1}) as there is a physical separation between the analytical and control samples. External locks are frequently used for nuclei other than ^1H where their instabilities are frequently lost in the line width of the observed nucleus and the chemical convenience is especially valuable.

The internal lock has much greater stability than the external lock but requires the presence within the analytical samples of nuclei which will produce a suitable signal.

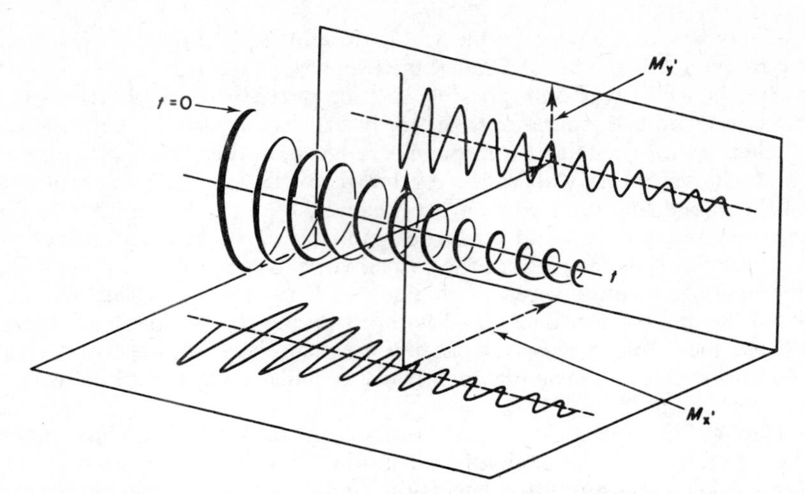

Fig. 2.9 The decay and precession of the nuclear magnetism, from ref. 13, with permission.

As mentioned above ^2H is frequently used, and sometimes proton resonances, but obviously the latter cannot be used when proton noise decoupling is required. A hybrid approach is sometimes used whereby the locking material is contained in a small concentric tube within the analytical sample, thus gaining most of the stability of an internal lock without its chemical problems. The limitations here are resolution, sensitivity (which is lower because of the extra space taken up by glass and locking material) and inconvenience in sample handling.

e. Digitisation

Once the signal from the sample has been detected, amplified, etc., it must be digitised in order to be averaged in the spectrometer's computer. Sampling theory tells us that to digitise a sinusoidal signal unambiguously a sampling rate of twice the highest frequency present is necessary. FID's consist of decaying sinusoidal frequencies ranging from zero Hz to half the spectral width being studied. An analogue to digital converter (ADC) working at a speed of twice the spectral width being studied is therefore necessary to record the spectrum accurately. For some nuclei this can require ADCs with rates between 50 and 100 kHz. It is common practice in present day spectrometers to detect the two components necessary to describe the free induction decay (see Fig. 2.9). To do this the signal from the spectrometer is detected by two phase sensitive detectors running in quadrature (90° out of phase). The values from these detectors are sampled simultaneously and stored separately in the computer. Working in this mode each ADC samples at a rate equal to the spectral width with the pulse frequency placed in the middle. If for any reason, e.g. limited computer storage, only a partial spectrum is recorded, complications due to fold-over occur. If a frequency which exceeds half the sampling rate is present at the digitisation stage, it appears in the spectrum at an apparent frequency which is equal to the spectral width minus the true frequency. The presence of alaised or folded lines can be confusing. It is normal, therefore, to avoid such frequencies reaching the ADC either by recording the total spectrum or by filtering out the unwanted frequencies with analogue (usually Butterworth) filters.

A second property of the digitisation stage is important, that of accuracy, i.e. the word length of the ADC, since the wordlength determines the accuracy of the digitisation (e.g. a 10 bit ADC is accurate to 1 in 2^{10}, i.e. about 0.1 %), and, perhaps more importantly, the dynamic range it can handle. As is to be expected the requirements of conversion speed and accuracy are divergent. The problem of dynamic range is mainly encountered in proton NMR where intense solvent signals are encountered along with the much weaker solute resonances. Consequently in proton NMR the longest wordlength ADC possible, usually 13 bits, is used. However, in some cases special experimental techniques have to be adopted in order to decrease the dynamic range of signals. These techniques either are based on saturating the solvent resonance by means of the spin decoupling or involve the different relaxation times frequently found between solvent and solute. The relaxation times of solvents are usually longer than those of the solute for systems where dynamic range is a problem, e.g. aqueous solutions of biomolecules. If the interval in a $180°$-τ-$90°$ sequence (see section 2A.7*b*) is chosen as $\ln 2T_1$ (solvent), no solvent signal will be detected, leaving only the solute signals to be digitised. Where spectral widths are large and dynamic range is no problem, shorter wordlength ADCs are frequently used.

Having digitised the sample it is necessary to store the data, usually in computer core memory, and perform the necessary arithmetic for the averaging of the data from successive FIDs. Wordlength is also important at this stage. Computers with 16 or 24 bit words are normally used with double precision arithmetic being employed if further dynamic range is required. The size of the computer core is determined by the resolution required, as can be deduced either by considering the number of data points required to describe a line, or by using the principle that the digital resolution is given by (sampling time)$^{-1}$.

These considerations give the following pair of equations.

$$\text{spectral width} = \frac{\text{no. of data points}}{2 \times \text{acquisition time}} \tag{2.5}$$

$$\text{resolution} = (\text{acquisition time})^{-1} \tag{2.6}$$

The consequences of these rules are summarised in Table 2.3.

f. Problems Associated with Studying Broad Lines

When studying nuclei with quadrupole moments, broad lines are frequently encountered. Broad lines, with their fast relaxation rates place differing demands on a pulse spectrometer than do nuclei producing narrow lines. These extra requirements fall into

TABLE 2.3 Effect of data table size on resolution

	4096 data points		8192 data points		32 768 data points	
Spectra width (Hz)	Acquisition time (s)	Resolution (Hz)	Acquisition time (s)	Resolution (Hz)	Acquisition time (s)	Resolution (Hz)
50	40.96	0.02	81.92	0.01	327.68	0.003
100	20.48	0.05	40.96	0.02	163.84	0.006
500	4.10	0.24	8.19	0.12	32.76	0.03
1000	2.05	0.49	4.10	0.24	16.38	0.06
5000	0.41	2.44	0.82	1.22	3.28	0.30
10 000	0.21	4.88	0.41	2.44	1.64	0.61

three general areas; firstly the amplifier must be powerful enough to produce a 90° pulse short enough to give a uniform power distribution over the spectrum, secondly the spectrometer must have a fast enough ADC to record the decay, i.e. a digitisation rate of at least 4–5 times the line width. These two conditions can frequently be met by "high resolution" pulse spectrometers for lines up to 5–10 kHz wide. The third area, that of receiver recovery, is the one which gives rise to the greatest strain for high resolution spectrometers. Before digitisation can commence the receiver must be allowed to recover from the effect of the exciting pulse, if the FID decays significantly during this period then distorted spectra will result. Because of their high linearity and large dynamic range high resolution spectrometer receivers tend to recover slowly (typically 20–50 μs) limiting the linewidths which can be handled to about 5 kHz.

g. Spin Decoupling

The concepts of double resonance are discussed in section 2A.3. From the practical viewpoint double resonance experiments require the ability to apply a second radio frequency field to the sample. This frequency is generated in the normal manner from the same master clock as is used to generate the observing and locking frequencies, in order that it will have the required long term stability, etc.

For homo decoupling experiments only low powers are required (~10–20 Hz) and the frequency is normally applied to the sample via the transmitter coil. Care has to be taken to ensure that the decoupler frequency is not allowed to pass directly into the receiver and hence into ADC where its presence will cause overload problems. Direct leakage is minimised by alternating the switching of the decoupler and the ADC in such a way that the two are never "on" at the same time.[40] This technique, known as time sharing, is illustrated in Fig. 2.10. In the heteronuclear case problems of direct leakage into the receiver do not occur as the two frequencies are too far apart. Heteronuclear decoupling often requires higher powers than those used for homo decoupling, typically 5–10 kHz which are normally applied via a special decoupler coil, orthogonal to the receiver coil. The use of a separate coil has the advantage that it can be easily

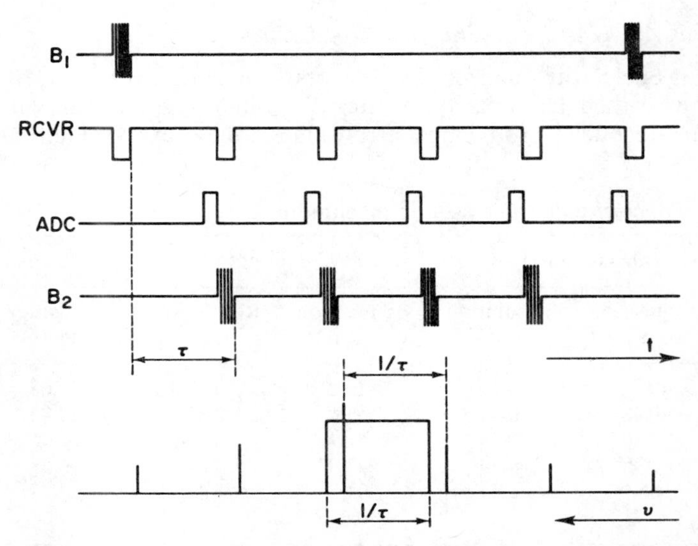

Fig. 2.10 Timing diagram for the two gated decoupling experiments, from ref. 13, with permission.

tuned to the decoupling frequency, independent of the nucleus under observation. If wide band heteronuclear decoupling is to be used then ν_2 must be suitably modulated, e.g. by a square wave signal, or more commonly, by the output from a pseudo random shift register in order to "spread" it over the desired area.

h. Data Manipulations

The digitised time domain spectrum (free induction decay) is subjected to many manipulations, as well as the obvious Fourier transformation, before the final spectrum is plotted. The first of these is the application of a so-called weighting or filter function.[41]

The signal component decays exponentially with time; the noise component on the other hand is constant. It follows therefore that if the FID is weighted in such a way that more significance is attached to the early part of the signal without distorting the spectrum the signal to noise ratio of the resultant spectrum will be enhanced. It can easily be shown that this goal can be best achieved by multiplying the FID with a negative exponential function of the form $\exp(-t/\mathrm{TC})$. Ideally from the viewpoint of optimising the signal to noise ratio the function should have a time constant (TC) equal to T_2^*. The use of a negative weighting function does induce broadening of the resonance line without distortion of the lineshape, e.g. the use of a time constant equal to T_2^* doubles the linewidth in the final spectrum. Inverse weighting i.e. $\exp(t/\mathrm{TC})$ is also possible. In this case the signal to noise ratio is degraded but the resolution of the spectrum is enhanced. In order to differentiate, i.e. resolve, two similar frequencies it is necessary to observe them for a time in the order of the reciprocal of their frequency difference. It follows therefore that to achieve the best spectral resolution possible it is necessary to pay special attention to the latter part of the FID, hence the use of a positive exponential weighting function for the purpose.

Theoretically the free induction decay should be studied to infinity; obviously this is impractical and digitisation is terminated at the end of the acquisition time as previously discussed. If the signal has decayed into the noise at the end of the data acquisition then no effects of the discontinuity, or truncation, are observable in the final spectrum. If, however, the truncation is significant either, for example, due to the use of too short an acquisition time, or as the result of using a positive weighting function, then the peaks in the final spectrum are distorted, showing oscillation both before and after the peak. Distortions due to truncation effects are attentuated by using apodizing functions which forcibly decrease the last part of the data towards zero.

Following the Fourier transformation a mathematically complex frequency domain spectrum will be produced. Ideally the real or cosine part of the transform gives the v mode, the imaginary (sine) part giving the u mode. Unfortunately, the phase of the spectrum will almost certainly be distorted as a result of the spectrometer's electronics, i.e. the real part will not be the pure v mode that is desired, it will be a mixture of v and u modes. These phase shifts can be corrected for and the required v mode spectrum plotted by outputting linear combinations of the real and imaginary data, as is illustrated in Fig. 2.11. The additional parameters are normally adjusted via two knobs on the spectrometer and are optimised with respect to a visual display of the spectrum on an oscilloscope.

2A.6 The Sample

Sample preparation is a critical part of recording an NMR spectrum, and extra care at this stage is very rewarding. High resolution spectra are usually recorded in the liquid phase though it is possible to record spectra in both the gaseous and solid phases.

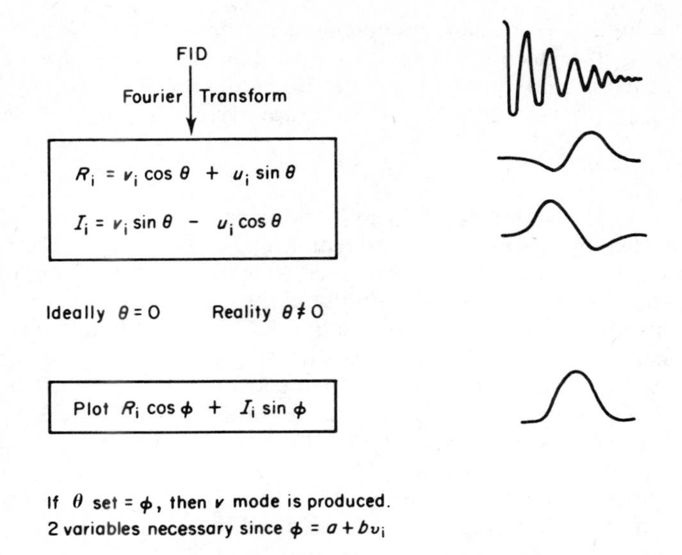

Fig. 2.11 The principle of phase correction for Fourier NMR spectra.

Gaseous spectra are only recorded for the more sensitive nuclei, usually protons, because of the problems of getting a high enough sample concentration into the spectrometer. The lines in gaseous spectra are usually broader than those of the liquid state due to the efficiency of spin rotational relaxation processes. Solid state samples require complex multi-pulse sequences and though interest in their study is growing, work of this nature lies outside the scope of this chapter.

a. Choice of Solvent and Tube Size

The choice of a suitable solvent is usually straightforward. The only constraints, apart from the obvious one that it should dissolve the sample, are that it should not contain resonances which overlap with the solute and, if an internal deuterium lock is to be used, it should be deuterated. Sufficient solution to produce a column of liquid 2 cm high in the sample tube being used is necessary to avoid "end effects" degrading the resolution; excess length, however, is in general wasted as it lies outside the receiver coil. The choice of sample tube size, assuming the spectrometer has facilities to permit a choice, should be approached from one of two viewpoints, depending on whether one is limited by solubility or sample. If solubility limits the amount of solute that can be accommodated within the active volume of the receiver coil then the obvious rule is to use the largest available sample tube. If the quantity of solute is the limiting factor, then it should be dissolved in the minimum quantity of solvent and the resulting solution placed in the smallest tube which gives a sample height greater than 2 cm and for which the spectrometer has an insert/probe. If the solution volume is too small to give the required height in any available tube it must be diluted until it does, otherwise "end effects" will seriously affect the spectrometer's resolution. A micro-cell could alternatively be used. Working at the highest possible concentration minimises artifacts due to the solvent, e.g. impurities, and side-band, and small inserts give greater efficiency than do large tubes.

b. *Referencing*

The magnetic field experienced by a sample depends to a small, but often significant, extent on many things such as the characteristics of the sample tube, and the bulk susceptibility of the solvent. Since there is no convenient method, except for NMR of measuring or controlling magnetic fields to anything like the required degree of accuracy, a circular argument ensues. The loop is opened by defining a point of reference (e.g. the resonance frequency for the protons in tetramethylsilane (TMS)) and measuring chemical shifts from the condition. i.e.

$$\delta/\text{ppm} = \left(\frac{\nu_x - \nu_{\text{ref}}}{\nu_{\text{ref}}}\right) \times 10^6 \quad (2.7)$$

When working with "other nuclei" problems arise in defining a suitable reference compound. The reference compound should ideally have only one resonance for the nucleus for which it is being used, be chemically stable, inert, and since the reference must be in the same physical state as the sample it must be soluble in most common solvents. Apart from the use of TMS as a reference for hydrogen, carbon, and silicon there is only limited agreement, to date, as to the appropriate references for other nuclei. The reader should note the references used in the following chapters which are collected in Tables 1.1 and 1.2. The chemical problems of referencing can be eased by using an external reference. Alternatively, especially for nuclei where there is no suitable reference compound available, the locking material used to stabilise (hence define) the magnetic field, or TMS can be used as what may be termed a "second order reference", (see ch. 1 A.5).

2A.7 The Measurement of Relaxation Times

There are five parameters which may be determined from an NMR experiment and subsequently used to provide information of chemical interest; these are chemical shifts, coupling constants, intensities, and the two relaxation times T_1 and T_2. The first three, which mainly give information of a structural nature, can be measured directly from a spectrum recorded as described above. In order to measure relaxation times, which give information on molecular motion, etc. additional experiments are usually required; these extra experimental techniques are discussed in this section.

a. *Definition of* T_1 *and* T_2

Basically T_1 and T_2 describe the return to equilibrium of the nuclear magnetism following a perturbation, e.g. a pulse; the relaxation is assumed to be a simple first order exponential process and the relaxation time is simply the time constant for the decay. The spin-lattice relaxation time characterises the return of the M_z component of the magnetisation to its equilibrium value, i.e. the leakage of the excitation energy absorbed by the spin system back into the surroundings, or lattice. T_2 on the other hand measures the rate of decay of the transverse magnetisation which is an entropy effect; frequently these are equal but under certain circumstances T_2 can be much shorter than T_1. These relaxation processes take place via interactions between the nuclei and locally fluctuating magnetic (and, in the case of quadrupolar nuclei, also electric) fields, hence their ability to give information about molecular motion.

Study of relaxation times of the individual lines of a high relaxation spectrum originally proposed in 1968[42] was delayed due to practical and theoretical difficulties. The practical difficulties have been solved with the advent of Fourier techniques which can record a

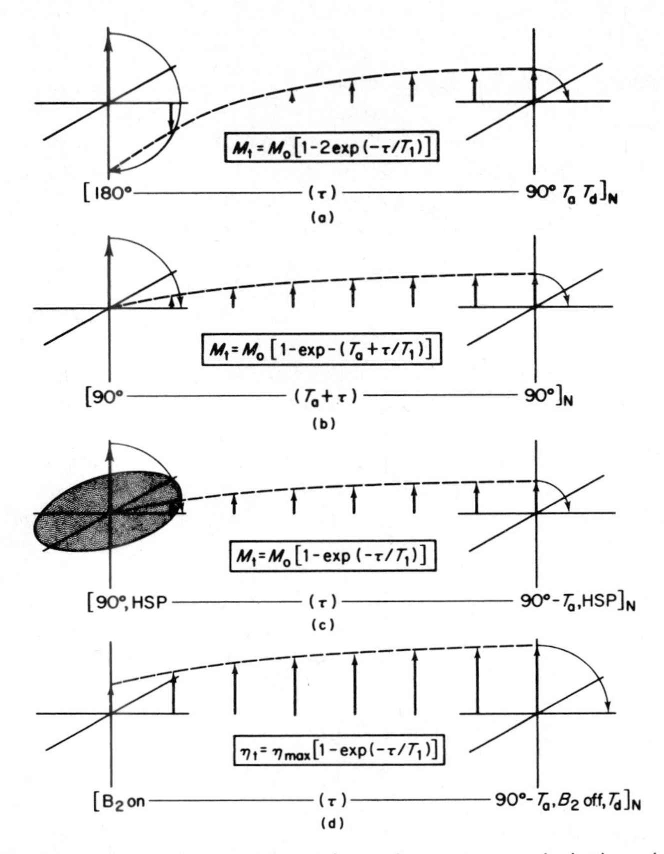

Fig. 2.12 The four pulse sequences commonly used to measure spin-lattice relaxation times.

full high resolution spectrum in a time less than the relaxation time being measured. The theoretical problems, mainly associated with cross-correlation within spin systems showing strong homonuclear coupling, have been eased by the access to data from magnetically dilute spins, e.g. ^{13}C, ^{15}N, ^{29}Si, and the use of wide-band decoupling techniques to remove the effects of the protons.

b. *Methods of Measuring T_1*

The basic method of measuring T_1 is to perturb the system and then sometime (τ) later examine the status of the spin system. By varying τ, T_1 can be calculated.

(*i*) *The Inversion Recovery Method*. The conceptually simplest and probably the generally most applicable and accurate method of measuring T_1 is the inversion recovery method. This sequence is shown diagrammatically in Fig. 2.12(*a*). The magnetisation vector is first inverted with a 180° pulse from $+M_0$ to $-M_0$ and allowed to recover for τ s. After this period the sample is subjected to a 90° pulse and the resulting FID digitised etc. A delay time of at least $4T_1$ must then be left to allow for a complete return to equilibrium before the sequence can be repeated. The 2 pulse sequence 180°-τ-90° is repeated

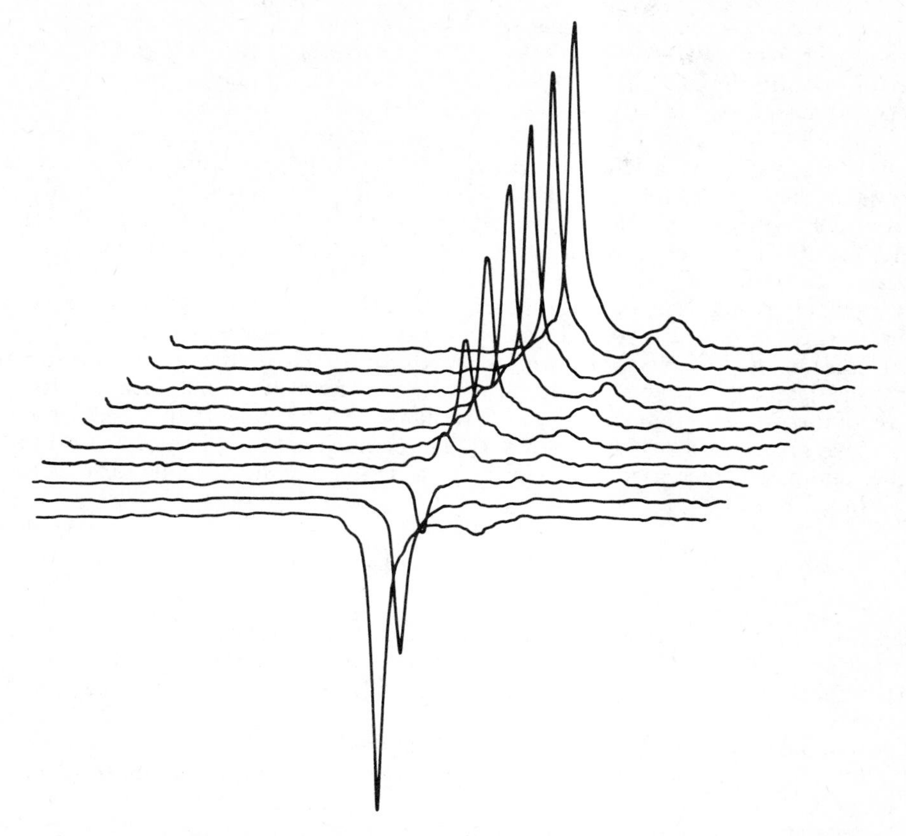

Fig. 2.13 A series of inversion recovery spectra for the ^{27}Al resonances of a concentrated solution $KAl(SO_4)_2$. Note the slower relaxation of the dominant symmetrical $Al^{3+} + 6H_2O$; F. W. Wehrli, unpublished results.

for as many times as is necessary to achieve the desired signal to noise ratio and the FID is then weighted and transformed in the normal way. A series of spectra of the type shown in Fig. 2.13 result. These spectra are in effect "snap shots" of the spin system as it returns to equilibrium. For small values of τ the spectrum contains inverted lines, a negative M value gives a negative M_y component after the 90° pulse and hence an inverted signal. When τ is equal to ln $2T_1$ no signal will appear since M_z is zero. Choosing this condition is the basis of one of the solvent elimination techniques mentioned previously. For longer values of τ, positive signals are obtained and when τ is equal to 4 or $5T_1$ a signal of normal intensity is observed.

(*ii*) *The Progressive Saturation Method*. A second method of measuring T_1 is based on the principle of saturation. If a sample is subjected to a series of 90° pulses more frequently spaced than $3T_1$, the line intensities are detectably reduced by saturation. By studying line intensities as a function of the interval between the pulses T_1 can be measured.[43] The main limitation of this method is that the smallest interval between the pulses is limited by the data acquisition time. When using the progressive saturation method no data should be recorded until a steady state has been established, typically after 4 or 5 pulses, as the first transients are atypical. The sequence is shown in Fig. 2.12(*b*).

(*iii*) *The Saturation Recovery Method.* T_1 may also be measured by studying the recovery of the spin system from saturation. The saturation is achieved by applying a 90° pulse followed by a homogeneity spoiling pulse i.e. applying a large field gradient to the sample for a few ms during which the transverse magnetisation is completely destroyed.[44] The system then returns towards equilibrium; at a time τ seconds after the first 90° pulse a second 90° pulse is applied and the resulting FID is recorded, transformed, etc. Any residual transverse magnetisation is destroyed by a second homo spoiling pulse. The sequence is shown in Fig. 2.12(*c*).

The advantage of this method is that it does not require a delay of $4T_1$ in the sequence before the next sequence is started, as does the inversion recovery method.

(*iv*) *The Method of NOE Growth.* A fourth approach to the measurement of T_1, shown in Fig. 2.12(*d*), only really applicable to nuclei which couple to the proton, is in the use of the property that the nuclear Overhauser effect (under wide-band decoupling conditions) grows with a time constant T_1. The sequence is to switch on the proton decoupler and as τ is increased, η increases towards the maximum value for that sample (η_{max}) thus

$$\eta = \eta_{max}\left\{1 - \exp\left(\frac{-\tau}{T_1}\right)\right\} \tag{2.8}$$

A delay of $4T_1$ is again necessary before the sequence can be repeated. This sequence is especially valuable for ^{29}Si and ^{15}N signals which have a negative NOE, whereas for the other sequences, which have the decoupler on all the time, a value of $\eta_{max} \sim -1$ is possible!

A further advantage of the method is that the nuclear Overhauser can be measured at the same time. Thus the contribution of nuclear dipole/dipole relaxation to the total relaxation, i.e. the measured T_1, can be determined simultaneously.

The relative merits of these sequences is summarised in Table 2.4.

c. *Errors in Measurement of T_1*

The measurement of T_1 values is very prone to systematic error. These errors, which are summarised in Table 2.5 can be divided into two main classes. The first source of error arises from chemical sources. Unqualified contributions to relaxation can be induced by any paramagnetic impurities. Nuclear relaxation takes place via interaction with fluctuating magnetic fields and since the magnetic moment of the electron is much larger (650 times) than that of the nucleus, even small traces of paramagnetic species in the sample can dramatically reduce the measured relaxation time. The most common sources of these impurities are transition metal ions, e.g. from a nickel spatula, ions leached from glass cleaned in chromic acid, and dissolved oxygen. Before measuring relaxation times care must be taken to remove these impurities, e.g. by precipitation or degassing, especially when measuring T_1 values larger than 10 seconds or so, or when specific complexation is expected between the paramagnetic species and the solute. The latter situation is important, since relaxation has inverse sixth power dependence on the distance between interacting dipoles, and hence any complex formation can have a dramatic effect. Even in the absence of paramagnetic ions, it cannot be assumed *a priori* that the measured T_1 values correspond to those for the molecule or ion at infinite dilution. T_1 values, especially for quadrupolar nuclei in ions, can be sensitive to concentration and

TABLE 2.4 Advantages and disadvantages of various T_1 measurement sequences

Sequence	Advantages	Disadvantages
Inversion recovery	Applicable to all ranges of T_1 independent of T_a	Prior knowledge of T_1 (to set delay)
		Requires 180° pulses (i.e. puts more demands on B_1 power)
	Insensitive to accuracy of pulses (if slope used to determine T_1)	Sensitive to off-resonance effects
	Gives quick, easy qualitative information on T_1	
Saturation recovery	Greater sensitivity than IR on long T_1 (especially where high S/N possible).	Requires extra hardware. Can interfere with field/frequency lock, autoshim operation
	Requires only 90° pulses	
	Independent of T_a	
Progressive saturation	Most efficient sequence (on long runs)	Inapplicable to short $T_1 (T_1 > T_2)$
	Requires only 90° pulse and will work with any basic spectrometer	Sensitive to accuracy of 90° pulse. Lowest value of limited by T_a
NOE Growth	Applicable to nuclei with negative magnetogyric ratios. Determines the dipole/dipole contribution directly	Only applicable to nuclei which exhibit an NOE. Prior knowledge of T_1 (to set delay)

TABLE 2.5 Factors which influence the accuracy of T_1 measurements

(*a*) *Chemical*

Concentration, mainly via viscosity which effects τ_c, paramagnetics, principally dissolved O_2 (only for $T_1 > 60$ s).

Molecular diffusion, important if B_1 is inhomogeneous and/or short samples are used.

(*b*) *Instrumental*

Coil geometry, B_1 inhomogeneity effects.

Errors in pulse angles, varies with pulse sequence.

Temperature stability, mainly via viscosity changes.

Resolution/sensitivity stability, particularly over long periods.

Digitisation, intensity errors due to the finite number of data points per line.

Choice of pulse sequence.

Method of calculation.

other diamagnetic species present. Since chemical effects always shorten the measured T_1 value, if there are two conflicting values the longest one is the most likely to be correct.

The second group systematic errors found in T_1 experiments can generally be termed "instrumental and digital". Instrumental errors arise from the generation of inaccurate 90° and 180° pulses either due to timing errors in the spectrometer or the inherent change in power level across a spectrum when any pulse longer than a delta function is used. The effects of these errors depend on the pulse sequence and will not be discussed further here. (see ref. 45). A second source of error lies in data manipulation; the normal method of extracting T_1 values from experimental data is to perform a linear least squares fit to a semi log plot. This method is sensitive to errors in the value of M_0 used. The value of M_0 should always be measured at least at the beginning and end of the experiment to correct for any spectrometer drift (preferably more frequently). A more accurate, though more complex, method which is insensitive to errors in M_0 is to perform a least squared fit on the exponential delay itself.[46]

When reporting T_1 data care must be taken to minimise errors arising from effects discussed above and given in Table 2.5. If absolute values are to be given then parameters such as exact simple concentration and temperature, which are frequently not reported with NMR data, must be included.

d. The Measurement of Spin-Spin Relaxation Times

The measurement of spin-spin relaxation times is more difficult than the spin-lattice relaxation times discussed previously. The major cause of the increased difficulty, and, ironically, the major interest in T_2, is that spin-spin relaxation is sensitive to low and zero frequency magnetic fields. Any inhomogeneity in the basic magnetic field (and no magnet is perfectly homogeneous) results in an unwanted contribution to T_2. This contribution is referred to as T_2' and the normally observed value as T_2^* thus:

$$(T_2^*)^{-1} = (T_2)^{-1} + (T_2')^{-1} \tag{2.9}$$

where T_2 is the "natural" relaxation time resulting from the mechanisms discussed in ch. 3. The value of T_2^* can be measured directly from any spectrum as the line width at half height of an NMR line and is given by (πT_2^*). If T_2' is insignificant compared with T_2^*, as in the broad line of a quadrupolar nucleus, then T_2 can be directly calculated. The classical technique to overcome the unwanted effects of T_2' is the spin echo sequence originally proposed by Carr and Purcell.[47] An NMR sample can be thought of as consisting of an assembly of extremely small regions called isochromats. Within an isochromat magnetic field inhomogeneities are negligible. Following a 90° pulse each isochromat will precess at its own specific Larmor frequency depending on its own local magnetic field. As time goes on the isochromats, which initially were all together, fan out; this is a T_2 process and can be corrected for since the process is reversible by the use of a 180° pulse as is shown in Fig. 2.14. A 180° pulse has the effect of rotating all the isochromats to a mirror image position within the $x'y'$ plane. Those isochromats which had precessed faster than the mean are now behind the mean, and vice-versa. If the time between the 90° and 180° pulse is τ, then a further τ later refocussing will occur and a spin echo results whose amplitude has decayed as T_2 alone.

After the first echo the isochromats lose phase coherence again, and can be refocussed by a further 180° pulse τs later. Thus if a chain of pulses at times $(2n + 1)\tau$ is applied a train of echoes (called a Carr–Purcell train) at time $(2n + 2)\tau$ are formed. Amplitude of this echo train decays as T_2 provided that (a) τ is kept short enough that significant

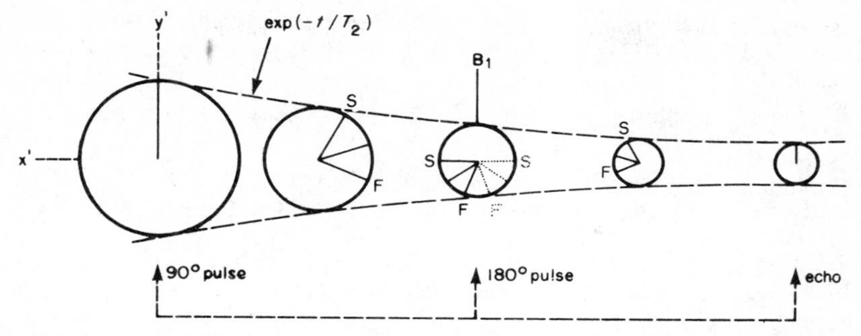

Fig. 2.14 The formation of a spin echo. All the spin isochromats are in phase following the 90° pulse. They then precess, those in a higher local magnetic field precess faster (F) than the mean, those in a lower field slower (S). A 180° pulse reverses the order of the isochromats and then refocus and produce a spin echo, from ref. 13, with permission.

diffusion does not occur between pulses, as this would inhibit echo formation, and (b) the r.f. pulses are homogeneous.

If the duration between the pulses is decreased to zero, the experiment becomes the so called "spin-locking" technique. Here the spins are rotated with a $90°x'$ pulse onto the y' axis; then a field which is strong with respect to local magnetic inhomogeneities is applied along the y' axis, i.e. there is a 90° phase shift with respect to the pulse. In the presence of the locking field, $T_{1\rho}$ type relaxation occurs, termed spin-lattice relaxation in the rotating frame or $T_{1\rho}$.[48] On removal of the locking field normal relaxation (T_2^*) takes place. The value of $T_{1\rho}$ under high resolution conditions is equal to T_2.

Spin-spin relaxation times of individual lines can be obtained by maintaining a locking field (or a Carr–Purcell–Meiboom–Gill pulse train) for a variable time (τ), after which the refocussing pulses are stopped and a normal FID is obtained, which is digitised, transformed, etc. Peak heights in a series of spectra with an increasing duration of refocussing pulses decay as T_2.

REFERENCES

1. F. Bloch, W. W. Hansen and M. Packard, *Phys. Rev.* **69**, 127 (1946)
2. E. M. Purcell, H. C. Torrey and R. V. Pound, *Phys. Rev.* **69**, 37 (1946)
3. F. Bloch, *Phys. Rev.* **70**, 1946 (1964)
4. R. K. Wangsness and F. Bloch, *Phys. Rev.* **89**, 728 (1953)
5. P. C. Lauterbur, *Ann. N.Y. Acad. Sci.* **70**, 841 (1958)
6. J. Dadok and R. F. Sprecher, *J. Magn. Res.* **13**, 243 (1974)
7. R. K. Gupta, J. A. Ferretti and E. D. Becker, *J. Magn. Res.* **13**, 275 (1974)
8. "Fourier Transforms and their Physical Applications", D. C. Champeney, Academic Press (1973)
9. P. Meakin and J. P. Jesson, *J. Magn. Res.* **10**, 296 (1973)
10. R. R. Ernst and W. A. Anderson, *Rev. Sci. Instruments* **37**, 93 (1966)
11. I. J. Lowe and R. E. Norberg, *Phys. Rev.* **107**, 46 (1957)
12. R. Freeman and H. D. W. Hill, *J. Magn. Res.* **4**, 366 (1971)
13. D. Shaw, "Fourier Transform N.M.R. Spectroscopy", Section 5-2, Elsevier, Amsterdam (1976)
14. S. Schaublin, A. Hopener and R. R. Ernst, *J. Magn. Res.* **13**, 196 (1974)

15. E. D. Becker, J. A. Ferretti and T. C. Farrar, *J. Amer. Chem. Soc.* **91**, 7784 (1969)
16. R. Freeman and H. D. W. Hill, *J. Magn. Res.* **4**, 366 (1971)
17. A. Schwenk, *J. Magn. Res.* **5**, 376 (1971)
18. R. Kaiser, *J. Magn. Res.* **3**, 28 (1970)
19. R. R. Ernst, *J. Magn. Res.* **3**, 10 (1970)
20. Ref. 13, chapter 4
21. B. L. Tomlinson and H. D. W. Hill, *J. Chem. Phys.* **59**, 1775 (1973)
22. R. Freeman, H. D. W. Hill, B. L. Tomlinson and L. D. Hall, *J. Chem. Phys.* **61**, 4466, (1974)
23. F. Bloch, *Phys. Rev.* **94**, 496 (1954)
24. R. A. Hoffman and S. Forsén, *Prog. in NMR Spectry* **1**, 15 (1966)
25. W. Von Philipsborn, *Angew. Chem. Internat. Ed.* **10**, 472 (1971)
26. R. R. Ernst, *J. Chem. Phys.* **45**, 3845 (1966)
27. F. W. Wehrli and T. Wirthlin, "Interpretation of Carbon-13 NMR Spectra", Heyden (1976)
28. J. B. Grutzner and R. E. Santini, *J. Magn. Res.* **19**, 173 (1975)
29. I. D. Campbell and R. Freeman, *J. Magn. Res.* **11**, 143 (1973)
30. J. H. Noggle and R. E. Schirmer, "The Nuclear Overhauser Effect: Chemical Applications", Academic Press, New York (1971)
31. G. C. Levy and R. A. Komoroski, *J. Am. Chem. Soc.* **96**, 678 (1974)
32. R. Freeman and H. D. W. Hill, *J. Magn. Res.* **5**, 278 (1971)
33. Ref. 13, chapter 5
34. E. D. Baker, *J. Chem. Phys.* **37**, 911 (1962)
35. J. Feeney and P. Partington, *J.C.S. Chem. Comm.* 611 (1973)
36. A. A. Chalmers, K. G. R. Packer and P. L. Wessels, *J. Magn. Res.* **15**, 419 (1974)
37. T. C. Farrar and E. D. Becker, "Pulse and Fourier Transform", Academic Press, New York (1973)
38. D. F. Hoult, "Nuclear Magnetic Resonance" (Specialist periodical report) Vol. 5, 148 (1976)
39. F. W. Wehrli, "Solvent Suppression Techniques in Pulsed NMR", Varian Applications Note NMR 75-2.
40. J. P. Jesson, P. Meakin and G. Kneissel, *J. Amer. Chem. Soc.* **95**, 618 (1973)
41. R. R. Ernst, "Advances in Magnetic Resonance", **2**. 1 (1957)
42. R. L. Vold, J. S. Waugh, M. P. Klein and D. E. Phelps, *J. Chem. Phys.* **48**, 3831 (1968)
43. R. Freeman and H. D. W. Hill, *J. Chem. Phys.* **54**, 3367 (1971)
44. G. G. MacDonald and J. S. Leigh Jr., *J. Magn. Res.* **9**, 358 (1973)
45. R. Freeman, H. D. W. Hill and R. Kaptien, *J. Magn. Res.* **1**, 327 (1972)
46. L. A. Malachan, *J. Magn. Res.* **26**, 223 (1977)
47. H. Y. Carr and E. M. Purcell, *Phys. Rev.* **94**, 630 (1954)
48. I. Solomon, *Compt Rend.* **248**, 92 (1959)

3
BACKGROUND THEORY OF NMR PARAMETERS

G. A. WEBB, University of Surrey, England

3A INTRODUCTION

3A.1 General

Both the positions and the profiles of the resonances in a high resolution NMR spectrum contain chemically interesting information. Chemical shifts and spin-spin coupling constants can be obtained, as discussed below, from an analysis of the resonance positions and their relative intensities, while relaxation contributes to bandshapes and affects the evolution of intensities under non-equilibrium situations.

Chemical shifts, coupling constants and the interactions leading to nuclear relaxation are intimately related to the electronic environments of the nuclei concerned. Thus a study of the NMR spectra of several different nuclei in a given molecule can contribute to an understanding of the electronic distribution in, and structure of, that molecule. Some of the nuclei occupying skeletal positions, such as ^{13}C, ^{14}N, ^{15}N, ^{17}O, ^{29}Si and various metals, can provide more useful information than those in peripheral sites such as ^{1}H and ^{19}F.

Molecular orbital (MO) and valence bond (VB) calculations can provide estimates of NMR parameters (including some of the factors, such as nuclear quadrupole coupling constants, which affect relaxation times) and thus assist in reaching an understanding of molecular structure. A satisfactory theory of NMR parameters will have a number of chemical applications; in particular the identification of structure or conformation of the species present in a given sample by comparison of the observed and calculated NMR parameters. As yet the theoretical aspects are not sufficiently well advanced to achieve this in the general case. However, significant advances in both *ab initio* and semi-empirical calculations of NMR parameters have been made in recent years. Relaxation times are strongly dependent on the nature of molecular motion, and the theory of such motion is therefore also required to explain data on T_1 and T_2. Conversely, when the theory is adequate, relaxation times yield information on molecular dynamics.

A paramagnetic species can induce contact and dipolar shifts in NMR spectra as well as provide contributions to the nuclear relaxation rates. The shifts are related to the bonding and molecular geometry involving the paramagnetic centre, whereas line broadening can be gainfully employed in establishing binding sites and determining the kinetics of processes involving the paramagnetic species.

Thus, taken together, the NMR parameters can yield various types of molecular information. This Chapter is devoted to a discussion of the theoretical relationship between the NMR and molecular data. It will be illustrated mainly with examples taken from NMR studies on ^{13}C, ^{14}N, ^{15}N, ^{19}F and ^{31}P in order to provide a balance with contributions in later chapters, but the theory is applicable to all nuclei.

3A.2 Spectral Analysis

For a system of nuclei in an external magnetic field, B_0, applied in the z direction, the corresponding spin Hamilton operator $\hat{\mathscr{H}}$ may be written:

$$\hat{\mathscr{H}} = -\sum_{A} \gamma_A B_0 \hbar (1 - \sigma_A) \hat{I}_{Az} + \sum_{A<B} h J_{AB} \hat{\mathbf{I}}_A \cdot \hat{\mathbf{I}}_B \quad (3.1)$$

where the first and second terms on the right-hand side are due to Zeeman (chemical shift) and spin-coupling interactions respectively. The symbols have their usual meanings (see the Preface and ch. 1). This expression reduces to Eq. 1.2 when all chemical shift differences are much greater than the coupling constants. The problem of obtaining chemical shifts and coupling constants from complicated (second order) spectra is usually approached by first writing Eq. 3.1 in matrix form using estimated spin parameters for the system in question (with chemical shifts referenced to some arbitrary zero), then solving it. The procedures for evaluating the matrix elements of $\hat{\mathscr{H}}$ in terms of the chemical shifts and coupling constants are adequately described in standard NMR texts (see ch. 1 for relevant references). Diagonalisation techniques then yield the eigenvalues and eigenvectors of Eq. 3.1. The differences in appropriate eigenvalues yield the transition frequencies, and the relative intensities are obtained from the coefficients of the eigenvectors. Various factorisation procedures and selection rules may be used to simplify the problem.

Usually the mathematical processes are tackled by computation for multi-spin systems. The simulated spectra are compared with the experimental spectra, and the spin parameters varied in the computation until an acceptable fit is obtained. The spin parameters used in the computation are then taken to be those appropriate for the system. Iterative programs such as LAOCN and UEA NMR are available from the Quantum Chemistry Program Exchange organisation* and from the U.K. Science Research Council NMR Computer Program Library.† Different programs incorporate factorisation for magnetic equivalence, for various symmetries of the spin system, and for a variety of experimental conditions.

3B NUCLEAR SCREENING

3B.1 General

The magnetic field, B, experienced by a nucleus in a molecule differs from the applied field, B_0. In the absence of medium effects this difference is attributable to the nuclear screening produced by the electrons in the molecule. In an NMR experiment these two fields may be related by:

$$B = B_0(1 - \sigma) \quad (3.2)$$

* Information from QCPE, Chemistry Building, Indiana University, Bloomington, Indiana 47401, U.S.A.

† Information from Dr. R. K. Harris, School of Chemical Sciences, University of East Anglia, Norwich NR4 7TJ, England.

Due to rapid rotation and random motion in non-viscous media, NMR measurements provide an estimate of the scalar screening constant σ in Eq. 3.2, which is the trace of the second rank tensor $\boldsymbol{\sigma}$. Measurements on liquid crystals and solids can, in principle, yield values for the individual components of the screening tensor and the screening anisotropy, $\Delta\sigma$. For linear and symmetric top molecules;

$$\Delta\sigma = \sigma_{\parallel} - \sigma_{\perp} \tag{3.3}$$

where $\sigma_{\parallel}$ refers to the screening component along the major molecular axis and $\sigma_{\perp}$ is that in the direction perpendicular to it. The average value is then:

$$\sigma = \tfrac{1}{3}(\sigma_{\parallel} + 2\sigma_{\perp}) \tag{3.4}$$

For less symmetrical molecules $\Delta\sigma$ is usually defined by:

$$\Delta\sigma = \sigma_{\alpha\alpha} - \tfrac{1}{2}(\sigma_{\beta\beta} + \sigma_{\gamma\gamma}) \tag{3.5}$$

where the σ_{ii}'s are the three principal tensor components taken in accordance with the convention $\sigma_{\alpha\alpha} \geq \sigma_{\beta\beta} \geq \sigma_{\gamma\gamma}$. The average is obtained as:

$$\sigma = \tfrac{1}{3}(\sigma_{\alpha\alpha} + \sigma_{\beta\beta} + \sigma_{\gamma\gamma}) \tag{3.6}$$

In order to correlate with experimental results, theoretical values of chemical shift trends for a given nucleus in a number of chemically different environments are usually produced as differences in calculated values of σ. Thus many of the inaccuracies introduced into the theoretical treatment of nuclear screening may cancel when considering chemical shifts.[1] A more rigorous examination of a theoretical investigation is obtained by comparing the calculated absolute values of σ, and their anisotropies, with experimental values.

It would be unreasonable to anticipate the exact reproduction of experimental chemical shifts by any theoretical treatment. The model employed in most theoretical studies is that of an isolated molecule, whereas experimental values are usually reported for liquid samples in which various media interactions may occur. For broad resonances the experimental chemical shifts may show appreciable errors. Another factor for consideration is the dependence of the chemical shift upon the relative populations of the rotational and vibrational levels of the molecule. A screening difference of several ppm from that calculated for the equilibrium geometry may occur.[2]

3B.2 Basic Theoretical Considerations

Nuclear screening is usually discussed in the terminology developed by Ramsey.[3] This leads to the definition of "diamagnetic", σ^d and "paramagnetic", σ^p, components of the screening constant:

$$\sigma = \sigma^d + \sigma^p \tag{3.7}$$

where σ^d involves the free rotation of electrons about the nucleus in question and σ^p describes the hindrance to this rotation caused by other electrons and nuclei in the molecule.

By placing the nucleus of interest and the zero point of the vector potential due to the external magnetic field used in the NMR experiment at the origin, the contributions to σ^d and σ^p are given by Eq. 3.8 and 3.9 respectively:

$$\sigma^d_{\alpha\beta} = \frac{\mu_0 e^2}{8\pi m} \langle 0 | \sum_k r_k^{-3}(r_k^2 \delta_{\alpha\beta} - r_{k\alpha} r_{k\beta}) | 0 \rangle \tag{3.8}$$

$$\sigma^p_{\alpha\beta} = -\frac{\mu_0 e^2}{8\pi m^2} \sum_n \Big[\langle 0 | \sum_k r_k^{-3} \hat{L}_{k\alpha} | n \rangle \langle n | \sum \hat{L}_{k\beta} | 0 \rangle + \langle 0 | \sum_k \hat{L}_{k\beta} | n \rangle \langle n | \sum_k r_k^{-3} \hat{L}_{k\alpha} | 0 \rangle \Big] (E_n - E_0)^{-1} \tag{3.9}$$

where α and β are subscripts labelling the cartesian components (x, y or z), μ_0, e and m have their usual significance, r_k is the position of the kth electron compared with the nucleus of interest, $\hat{L}_k$ is the corresponding orbital angular momentum operator, $\delta_{\alpha\beta}$ is the Kronecker delta function ($=1$ if $\alpha = \beta$, otherwise $=0$). $|0\rangle$ and $|n\rangle$ refer to the electronic ground and excited states of the molecule with energies E_0 and E_n respectively. The summation over n in Eq. 3.9 is taken to include all the excited states together with the continuum.

In general little is known about either the discrete high energy or the continuum states of a molecule. The few detailed calculations at present available are on small molecules. These suggest that screening contributions from the continuum may be at least as important as those from the discrete molecular states.[4,5] Consequently Eq. 3.9 is impractical for the majority of molecules and it may be simplified by means of the closure relation:

$$\sum_{n=0}^{\infty} |n\rangle\langle n| = 1 \tag{3.10}$$

This relation holds for a complete orthogonal set of functions $|n\rangle$. A further simplification involves choosing an average excitation energy (AEE), ΔE, for all excited states such that for two operators $\hat{P}$ and $\hat{Q}$:

$$\sum_{n=1}^{\infty} \langle 0 | \hat{P} | n \rangle \langle n | \hat{Q} | 0 \rangle (E_n - E_0)^{-1}$$

$$\approx \Delta E^{-1} \sum_{n=1}^{\infty} \langle 0 | \hat{P} | n \rangle \langle n | \hat{Q} | 0 \rangle \tag{3.11}$$

$$= \Delta E^{-1} \left[\sum_{n=0}^{\infty} \langle 0 | \hat{P} | n \rangle \langle n | \hat{Q} | 0 \rangle - \langle 0 | \hat{P} | 0 \rangle \langle 0 | \hat{Q} | 0 \rangle \right] \tag{3.12}$$

$$= \Delta E^{-1} \left[\langle 0 | \hat{P} \left(\sum_{n=0}^{\infty} |n\rangle\langle n| \right) \hat{Q} | 0 \rangle - \langle 0 | \hat{P} | 0 \rangle \langle 0 | \hat{Q} | 0 \rangle \right] \tag{3.13}$$

The last term in Eq. 3.13 vanishes if either $\hat{P}$ or $\hat{Q}$ is not totally symmetric. In the case under consideration $\hat{P} = \Sigma_k r_k^{-3} \hat{L}_{k\alpha}$ and $\hat{Q} = \sum_k \hat{L}_{k\beta}$ and the final term of Eq. 3.13 becomes zero by symmetry.

By combining Eq. 3.10 and 3.13 we find:

$$\sum_{n=1}^{\infty} \langle 0 | \hat{P} | n \rangle \langle n | \hat{Q} | 0 \rangle (E_n - E_0)^{-1} \approx \Delta E^{-1} \langle 0 | \hat{P}\hat{Q} | 0 \rangle \tag{3.14}$$

Application of Eq. 3.14 to Eq. 3.9 gives:

$$\sigma^{p}_{\alpha\beta} = -\frac{\mu_0 e^2}{4\pi m^2} \Delta E^{-1} \left[\langle 0 | \sum_k r_k^{-3} \hat{L}_{k\alpha} \sum_k \hat{L}_{k\beta} | 0 \rangle \right] \tag{3.15}$$

Equation (3.15) is more accessible than Eq. (3.9) since only the ground state function is involved. However it is usually not possible to obtain a suitable value of ΔE by independent means.

A further drawback to Ramsey's formulation is that for molecules of a reasonable size, σ^d and σ^p become large and of opposite sign. Thus, σ becomes the relatively small difference between two large and variable terms and is consequently subject to considerable error.

It should be remembered that the relative values of σ^d and σ^p depend upon the origin chosen for the calculation.[1] Although this should not affect the value of σ it does present hazards when comparing different authors' results. Moreover, due to the practical aspects of choosing atomic orbital functions, σ is often found to depend upon the choice of origin also. This has been illustrated in the case of the nitrogen screening in ammonia where theoretical values of σ varying between 176.2 and 272.35 ppm[6–8] are compared with an experimental result of 270.4 ppm.[9,10] Most of the variation in the calculated results arises from the choice of atomic basis set, the dependence of σ upon the origin becoming less significant as the basis set is enlarged. Since it is a fundamental physical requirement that σ be invariant to a gauge transformation[11] calculations employing Ramsey's approach are at present restricted to small molecules containing only light atoms.

Calculations giving a gauge-invariant value for σ still produce gauge-variant σ^d and σ^p contributions. Thus it would appear that a suitable choice of gauge could result in the removal of the more difficult "paramagnetic" term.[12] In this case σ could be estimated from the relatively simple "diamagnetic" term only. However, this conclusion is not generally valid since under given symmetry conditions both σ^d and σ^p can become accidentally gauge-invariant.[13]

3B.3 Independent Electron Model

The problems associated with gauge-variant calculations of σ can be obviated by using an approach in which each MO is composed of a linear combination of gauge-dependent atomic orbitals as demonstrated by Pople.[14–17] Unfortunately this approach has been referred to in the literature as a gauge-*in*dependent atomic orbital (GIAO) method. The dependence of the atomic orbitals used, χ_μ, on the gauge is given by:

$$\chi_\mu = \phi_u \exp\left[-\frac{ie}{\hbar} \mathbf{A}_\mu(\mathbf{r}) \cdot \mathbf{r} \right] \tag{3.16}$$

where $\mathbf{A}_\mu(\mathbf{r})$ is the vector potential for the electron associated with atomic orbital ϕ_μ at a distance r from the nucleus. The screening data obtained by using χ_μ are gauge-independent.

Pople developed his MO theory within the framework of an independent electron model, consequently all of the explicit two-electron terms vanish. Using this approach the nuclear screening is expressed as the sum of local, non-local and interatomic contributions:[14,17,18]

$$\sigma = \sigma^d(\text{loc}) + \sigma^d(\text{non-loc}) + \sigma^d(\text{inter}) + \sigma^p(\text{loc}) + \sigma^p(\text{non-loc}) + \sigma^p(\text{inter}) \tag{3.17}$$

The various "diamagnetic" and "paramagnetic" terms in Eq. 3.17 are not directly comparable with the expressions bearing these names in Ramsey's equations.

The local terms, σ^d(loc) and σ^p(loc), arise from electronic currents localised on the atom containing the nucleus of interest. Similarly σ^d(non-loc) and σ^p(non-loc) are derived from currents on neighbouring atoms. Finally, σ^d(inter) and σ^p(inter) are due to screening currents not localised on any atom in the molecule, e.g. ring currents. These latter two terms usually produce a screening contribution of at most a few ppm, which is negligible compared with the chemical shift ranges of several hundred ppm found for nuclei other than those of hydrogen.

The non-local contributions are usually considered by assuming that the induced magnetic moments in the electrons on a neighbouring atom can be replaced by a point dipole.[15,19] The effect of this dipole on the screening of the nucleus under investigation is usually negligible for singly-bonded first row nuclei. Due to the smaller chemical shift range of hydrogen nuclei the non-local contributions can be significant[20,21] as well as for multiply bonded first row nuclei, as shown in Table 3.1. The data are calculated from Pople's model using the INDO/S-MO parameterisation.[22]

Thus within the confines of the independent electron model the major screening contributions arise from σ^d(loc) and σ^p(loc). Various theoretical levels of approach to the evaluation of these terms are available. They mainly differ in the choice of basis functions and method of evaluating the appropriate integrals.[1] Whilst *ab initio* calculations can give reasonable screening results for small molecules containing light atoms, considerably more computer time is required than by semi-empirical methods.[8,23] Since the number of integrals to be evaluated expands rapidly with increases in atomic and molecular size, only semi-empirical calculations are practical at present for heavier nuclei and larger molecules.

3B.4 Semi-empirical Calculations

a. General

By assuming that only *s* and *p* atomic orbitals are important, the rotationally averaged values of the local terms for nucleus A, σ_A^d and σ_A^p, are given by Eq. 3.18 and 3.19 respectively:

$$\sigma_A^d = \frac{\mu_0 e^2}{12\pi m} \sum_{\mu}^{A} P_{\mu\mu} \langle \mu | r_{A\mu}^{-1} | \mu \rangle \tag{3.18}$$

$$\begin{aligned} \sigma_A^p = -\frac{\mu_0 e^2 \hbar^2}{6\pi m^2} \langle r^{-3} \rangle_{np} \sum_{j}^{\text{occ}} \sum_{k}^{\text{unoc}} (E_k - E_j)^{-1} \\ \times \Big[(C_{y,A_j} C_{z,A_k} - C_{z,A_j} C_{y,A_k}) \sum_{B} (C_{y,B_j} C_{z,B_k} - C_{z,B_j} C_{y,B_k}) \\ + (C_{z,A_j} C_{x,A_k} - C_{x,A_j} C_{z,A_k}) \sum_{B} (C_{z,B_j} C_{x,B_k} - C_{x,B_j} C_{z,B_k}) \\ + (C_{x,A_j} C_{y,A_k} - C_{y,A_j} C_{x,A_k}) \sum_{B} (C_{x,B_j} C_{y,B_k} - C_{y,B_j} C_{x,B_k}) \Big] \end{aligned} \tag{3.19}$$

where $P_{\mu\mu}$ is the charge density in the atomic orbital μ which is at an average distance of $r_{A\mu}$ from nucleus A, C_{x,A_j} is the LCAO coefficient of the P_x orbital on atom A in the MO j etc. The summation over B includes A. It is apparent that σ_A^p will be zero unless both

TABLE 3.1[22] Some calculated values of σ^d(loc), σ^d(non-loc), σ^p(loc) and σ^p(non-loc) in ppm using the INDO/S parameterisation scheme

Molecule	Nucleus	σ^d(loc)	σ^d(non-loc)	σ^p(loc)	σ^p(non-loc)
CO	C	259.36	0.03	−206.24	−4.92
	O	395.39	0.64	−367.39	−7.25
$(CH_3)_2CO$	C	257.60	−0.12	−212.14	4.99
	O	398.47	0.10	−580.07	2.42
H_2CO	C	257.77	−0.10	−208.24	5.61
	O	397.80	0.10	−651.47	1.85
CO_2	C	255.16	−0.25	−126.79	−6.94
	O	397.32	0.12	−269.51	−3.55
CH_3CN	C	258.94	0.07	−159.96	−5.91
	N	326.75	0.07	−300.57	−5.27
HCN	C	259.08	0.11	−157.53	−6.10
	N	326.70	0.06	−301.30	−5.04
N_2	N	324.39	0.24	−382.84	−8.62

A and B possess valence p electrons. From the ordering of the LCAO coefficients in Eq. 3.19 it follows that $\pi \rightarrow \pi^*$ transitions do not contribute to σ_A^p.

E_k and E_j refer to the energies of the MO's k and j which are unoccupied and occupied respectively. The excitation energies, $E_k - E_j$, are those for excited singlet states which are mixed with the ground state by an external magnetic field. Thus:

$$E_k - E_j = \varepsilon_k - \varepsilon_j - \mathscr{J}_{jk} + 2\mathscr{K}_{jk} \tag{3.20}$$

where ε_k and ε_j are eigenvalues of the unperturbed molecule, $\mathscr{J}_{jk}$ and $\mathscr{K}_{jk}$ are respectively the Coulomb and exchange integrals:

$$\mathscr{K}_{jk} = \iint \psi_j^*(1)\psi_k^*(2)\frac{1}{r_{12}}\psi_k(1)\psi_j(2)d\tau_1\, d\tau_2 \tag{3.21}$$

and:

$$\mathscr{J}_{jk} = \iint \psi_j^*(1)\psi_k^*(2)\frac{1}{r_{12}}\psi_j(1)\psi_k(2)d\tau_1\, d\tau_2 \tag{3.22}$$

The value of $\langle r^{-3}\rangle_{np}$ is usually obtained from the expression:

$$\langle r^{-3}\rangle_{np} = \frac{1}{3}\left(\frac{Z_{np}}{na_0}\right)^3 \tag{3.23}$$

The variation of this term with atomic number is presented in Fig. 3.1.[24] The matrix elements in Eq. 3.18 are obtained from:

$$\langle\mu|r_A^{-1}|\mu\rangle = \frac{Z_\mu}{n^2 a_0} \tag{3.24}$$

where Z_μ is the effective nuclear charge for the atomic orbital μ and a_0 is the Bohr radius. Equations 3.23 and 3.24 represent a way of including the effects of the different molecular environments on these atomic integrals.

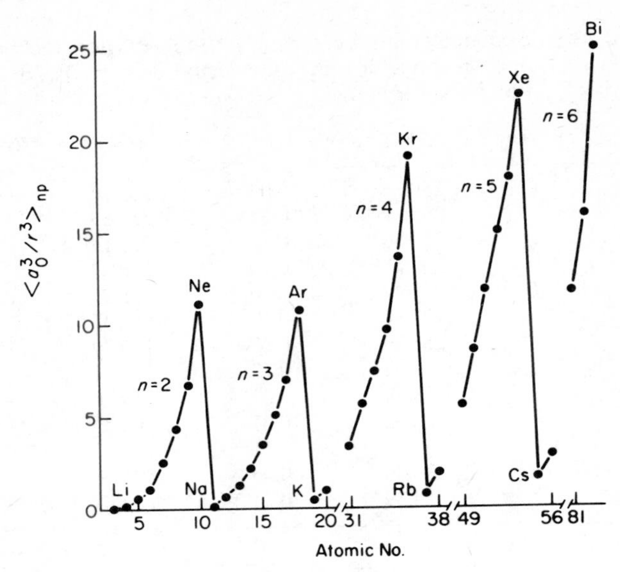

Fig. 3.1 Variation of $\langle r^{-3} \rangle_{np}$ with atomic number.[24]

The value of Z_μ may be obtained from Slater's rules[25] according to which the 2*s* and 2*p* orbitals of first row atoms experience the same effective nuclear charge, given by:

$$Z_{2s} = Z_{2p} = Z^0 + 0.35q_A^{net} \tag{3.25}$$

where Z^0 is the effective nuclear charge for the free atom and q_A^{net} is the net charge on atom A. Equation 3.25 may also be written in terms of atomic charge densities as:

$$Z_{2p} = A - 0.35P_{2s2s} - 0.35(P_{2p_x2p_x} + P_{2p_y2p_y} + P_{2p_z2p_z}) \tag{3.26}$$

Alternatively Z_μ can be found from Burns' rules[26] as:

$$Z_{2s} = B - 0.4P_{2s2s} - 0.35(P_{2p_x2p_x} + P_{2p_y2p_y} + P_{2p_z2p_z}) \tag{3.27}$$

and:

$$Z_{2p} = C - 0.5P_{2s2s} - 0.35(P_{2p_x2p_x} + P_{2p_y2p_y} + P_{2p_z2p_z}) \tag{3.28}$$

where the values of the constants A, B and C for some first row atoms are given in Table 3.2.

TABLE 3.2 Values of the constants A, B and C as used in Eq. 3.26 to 3.28 for some first row nuclei

Nucleus	B	C	N	O	F
A	3.65	4.65	5.65	6.65	7.65
B	3.4	4.4	5.4	6.4	7.4
C	3.35	4.35	5.35	6.35	7.35

Equations 3.18 and 3.19 together with Eq. 3.20 to 3.28 have been used to obtain values of σ_A^d and σ_A^p for several first row nuclei.[1] The standard semi-empirical MO parameterisation schemes; CNDO/2, CNDO/S, INDO, INDO/S and MINDO/3, all show that for a given first row nucleus in a range of molecular environments σ_A^d remains approximately constant[22,27] and that screening differences arise from changes in σ_A^p.[1,22]

As anticipated from Eq. 3.18 the value obtained for σ_A^d is very similar to that of the free atom, differences arising only from changes in atomic charge densities. Equation 3.24 shows that the free atom value of σ_A^d runs parallel with changes in the effective nuclear charge as demonstrated in Table 3.3.[28]

Calculations based upon Eq. 3.18 and 3.19 and standard CNDO/S parameters provide a satisfactory account of the chemical shifts of ^{11}B, ^{13}C, ^{14}N, ^{17}O and ^{19}F in a variety of molecular environments.[1,29–31] The results of some calculated ^{13}C chemical shifts with respect to benzene are compared with experimental data in Fig. 3.2.[30] The standard deviation is 9.95 ppm and the correlation coefficient 0.9736. Reasonable agreement between the measured and calculated values of σ and its anisotropy, $\Delta\sigma$, is also demonstrated in Table 3.4.[30] In some cases the agreement is closer than that obtained between the experimental and *ab initio* results.

TABLE 3.3[28] Some free atom values of σ_A^d in ppm

Nucleus	σ_A^d	Nucleus	σ_A^d	Nucleus	σ_A^d
He	59.90	Ga	2638.57	Nd	6511.53
Li	101.45	Ge	2757.10	Pm	6663.32
Be	149.26	As	2877.18	Sm	6816.48
B	201.99	Se	2998.37	Eu	6971.01
C	260.74	Br	3121.19	Gd	7126.94
N	325.47	Kr	3245.61	Tb	7284.20
O	395.11	Rb	3366.80	Dy	7442.80
F	470.71	Sr	3489.16	Ho	7604.09
Ne	552.27	Y	3614.45	Er	7765.77
Na	628.90	Zr	3741.61	Tm	7926.75
Mg	705.60	Nb	3870.35	Tb	8090.72
Al	789.88	Mo	4000.64	Lu	8250.49
Si	874.09	Tc	4132.52	Hf	8412.72
P	961.14	Ru	4265.53	Ta	8574.32
S	1050.47	Rh	4400.14	W	8736.83
Cl	1142.64	Pd	4536.25	Re	8900.35
Ar	1237.64	Ag	4673.85	Os	9064.44
K	1329.36	Cd	4813.03	Ir	9229.54
Ca	1422.85	In	4948.72	Pt	9395.58
Sc	1521.35	Sn	5085.55	Au	9561.41
Ti	1622.72	Sb	5223.43	Hg	9729.07
V	1726.62	Te	5362.00	Tl	9894.16
Cr	1833.02	I	5501.64	Pb	10060.92
Mn	1942.10	Xe	5642.32	Bi	10227.13
Fe	2052.91	Cs	5780.22	Po	10393.86
Co	2166.38	Ba	5918.99	At	10559.93
Ni	2282.32	La	6064.83	Rn	10728.20
Cu	2400.71	Ce	6212.43		
Zn	2521.67	Pr	6361.59		

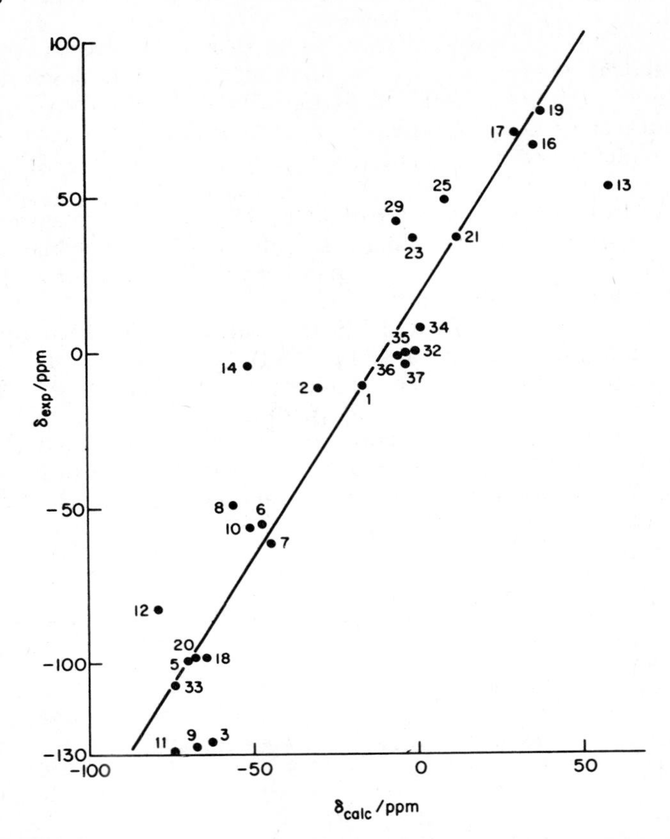

Fig. 3.2 Comparison of some ^{13}C chemical shifts, with respect to benzene, and those calculated from Pople's method using the CNDO/S parameterisation scheme.[30]

1 = HCN; 2 = CH_3C^*N; 3 = $^*CH_3CN$

5 = $^*CH_3NC$; 6 = HCCH; 7 = $CH_3—C{\equiv}C^*—H$

8 = $CH_3—C^*{\equiv}C—H$; 9 = $^*CH_3—C{\equiv}C—H$; 10 = $CH_3—C^*{\equiv}CCH_3$

11 = $^*CH_3—C{\equiv}C—CH_3$; 12 = CH_3OH; 13 = CO

14 = CO_2; 16 = H_2CO; 17 = $CH_3{}^*CHO$

18 = $^*CH_3CHO$; 19 = $(CH_3)_2{}^*CO$; 20 = $(^*CH_3)_2CO$

21 = $HCONH_2$; 23 = HCOOH; 25 = $CH_3{}^*COOH$

29 = $CH_3{}^*COOCH_3$; 32 =; 33 = $^*CH_3$—

34 = CH_3—*; 35 = CH_3— *; 36 = CH_3— *

37 = CH_3— *

TABLE 3.4 Comparison of some experimental values of the average and the anisotropy of the ^{13}C screening tensor with results obtained from CNDO/S calculations[30]

Molecule	Calculated/ppm		Experimental/ppm	
	σ	$\Delta\sigma$	σ	$\Delta\sigma$
HCN	85.98	260.42	76 ± 10	282 ± 20
$CH_3{*}CN$	89.91	265.2	76 ± 10	311 ± 30
$*CH_3CN$	132.64	10.8	193 ± 14	5 ± 10
$H-C\equiv CH$	116.82	215.7	120 ± 10	245 ± 20
$CH_3-{*}C\equiv C-CH_3$	119.9	202.63	117 ± 10	201 ± 10
$*CH_3-C\equiv C-CH$	143.43	9.13	190 ± 10	14 ± 10
CH_3OH	148.83	13.9	146 ± 10	63 ± 10
CO	10.67	372.66	12 ± 10	384 ± 10
$CH_3{*}CHO$	38.71	183.3	-6 ± 10	168 ± 10
$*CH_3CHO$	134.77	13.1	162 ± 10	46 ± 10
$(CH_3{*}CO)_2O$	92.85	115.0	28 ± 10	82.5 ± 6
$(*CH_3CO)_2O$	145.43	9.7	173 ± 10	19 ± 10

Since Eq. 3.19 only relates to atomic p orbitals it is inappropriate for second row nuclei where the inclusion of d orbitals becomes important. An analogous expression including d orbitals is available[24] which has been used together with the AEE approximation to estimate ^{31}P chemical shifts.[32] Some ^{29}Si chemical shifts have also been considered.[33]

In general the methods described in this section appear to be capable of providing a reasonable account of chemical shift trends for first row nuclei. For second row nuclei the inclusion of d orbitals has yet to be considered comprehensively. Another problem is that, with the exception of CNDO/2, the semi-empirical MO schemes are not generally parameterised beyond the first row as yet. This is mainly due to the absence of appropriate electronic spectral data.

b. *Atom-dipole Model*

Flygare has proposed an atom-dipole model for estimating the "diamagnetic" contribution to the nuclear screening.[34] Electronic eigenfunctions are not explicitly involved in this model whose expression for σ_A^d is given by:

$$\sigma_A^d = \sigma_A^d(\text{free atom}) + \frac{\mu_0 e^2}{12\pi m} \sum_{B \neq A} \frac{Z_A}{r_{AB}} \tag{3.29}$$

where r_{AB} is the separation between nucleus A and its neighbours B. Calculations show that even when the summation is restricted to nuclei directly bonded to A, the value of σ_A^d is considerably enhanced as both the number of bonded atoms and of electrons in the molecules increases.[27] These increments arise from the final, non-local, term in Eq. 3.29. Thus values of σ_A^d obtained from this equation are not those of a local term as defined in Pople's theory but are comparable to those found by Ramsey's approach. Hence in discussing chemical shift trends in terms of changes in Pople's σ_A^p parameter, the corresponding value of σ_A^d must be obtained from Eq. 3.18 rather than from Eq. 3.29.

c. Finite Perturbation Method

Finite perturbation theory within the framework of the INDO parameterisation has been employed to evaluate σ_A^d and σ_A^p for ^{13}C nuclei in various chemical environments.[20,35,36] This approach avoids the necessity of requiring detailed information about all of the excited MO's and their energies. However, the use of standard INDO parameters does not provide a good account of the ^{13}C chemical shifts of hydrocarbons, fluorocarbons and nitrogen-containing molecules. By reparameterisation better agreement between the calculated and observed chemical shifts is obtained.[20,35,36] The modified INDO parameters are chosen to exaggerate the level of electronic polarization associated with the presence of electronegative atoms. Thus, although reparameterisation can reproduce gross ^{13}C chemical shift trends some other molecular properties such as equilibrium distances are less well accounted for by the "new" INDO parameters.[37] Consequently, the finite perturbation method seems at present to be less satisfactory than Pople's model, employing Eq. 3.18 and 3.19, which reproduces chemical shift trends fairly satisfactorily using standard CNDO/S and INDO/S parameters.

d. Average Excitation Energy (AEE) Method

The AEE approximation is often invoked to simplify Eq. 3.19. Since overlap is neglected in the semi-empirical MO methods employed, it follows that:

$$\sum_j^{\text{occ}} C_{\mu j} C_{\lambda j} + \sum_k^{\text{unocc}} C_{\mu k} C_{\lambda k} = \delta_{\mu\lambda} \tag{3.30}$$

By incorporating Eq. 3.30 and the AEE approximation (Eq. 3.11), Eq. 3.19 becomes:

$$\sigma_A^p = -\frac{\mu_0 e^2 \hbar^2}{8\pi m^2} \frac{\langle r^{-3} \rangle_{np}}{\Delta E} \sum_B Q_{AB} \tag{3.31}$$

where the summation over B includes A and all the other atoms in the molecule, not merely those bonded to A. The bond-order, charge-density terms, Q_{AB}, are given by:

$$Q_{AB} = \tfrac{4}{3}\delta_{AB}(P_{x_A x_B} + P_{y_A y_B} + P_{z_A z_B}) - \tfrac{2}{3}(P_{x_A x_B} P_{y_A y_B} + P_{x_A x_B} P_{z_A z_B} + P_{y_A y_B} P_{z_A z_B}) + \tfrac{2}{3}(P_{x_A y_B} P_{x_B y_A} + P_{x_A z_B} P_{x_B z_A} + P_{y_A z_B} P_{y_B z_A}) \tag{3.32}$$

where the P's are defined by:

$$P_{\mu\nu} = 2 \sum_j^{\text{occ}} C_{\mu j} C_{\nu j} \tag{3.33}$$

For first row nuclei a combination of Eq. 3.18 and 3.24 gives:

$$\sigma_A^d/\text{ppm} = 17.7501 \sum_\mu^{A} P_{\mu\mu} Z_\mu n^{-2} \tag{3.34}$$

where $n = 1$ for the $1s$ orbitals and $n = 2$ for the $2s$, $2p$ orbitals. Similarly, Eq. 3.24 and 3.31 produce:

$$\sigma_A^p \Delta E/\text{ppm eV} = 30.1885 Z_{2p}^3 \sum_B Q_{AB} \tag{3.35}$$

Equations 3.34 and 3.35 have been extensively used to relate semi-empirical MO data to chemical shift trends. A simplified form of Eq. 3.32, including the d orbital term, is often employed. This involves restricting the summation of Q_{AB} to valence orbitals centred on the atom in question. The resulting expression for σ^p is given[24] by:

$$\sigma^p_A = -\frac{\mu_0 e^2 \hbar^2}{6\pi m^2 \Delta E}(\langle r^{-3}\rangle_{np} P_u + \langle r^{-3}\rangle_{md} D_u) \quad (3.36a)$$

Where *np* and *md* refer to the appropriate valence *p* and *d* electrons. The expression for P_u is:

$$P_u = (P_{xx} + P_{yy} + P_{zz}) - \tfrac{1}{2}(P_{xx}P_{zz} + P_{yy}P_{xx} + P_{yy}P_{zz}) + \tfrac{1}{2}(P_{xy}P_{yx} + P_{xz}P_{zx} + P_{zy}P_{yz}) \quad (3.36b)$$

D_u is the corresponding term for the *d* electrons. Equation 3.36b expresses the concept of *p*-electron "imbalance" about the nucleus in question. Equations 3.36 have been widely used although they are not very satisfactory from a theoretical point of view. In practice it seems likely that some of the error introduced by ignoring the valence orbitals centred on other nuclei in the molecules becomes absorbed into ΔE.

AEE calculations of various degrees of sophistication have been reported including π electron calculations taken together with σ bond polarization effects, as well as all valence electron methods such as extended Hückel, CNDO/2, INDO and MINDO. In general these methods are reasonably successful in accounting for gross chemical shift trends in series of closely related molecules. However their value is limited by the necessity of choosing ΔE, for which no *a priori* method is available. Atoms with lone-pair electrons tend to have ΔE dominated by relatively low energy $n \rightarrow \pi^*$ transitions. For other atoms the significant contributions to ΔE are usually confined to those from $\sigma \rightarrow \pi^*$, $\pi \rightarrow \sigma^*$ and $\sigma \rightarrow \sigma^*$ transitions since $\pi \rightarrow \pi^*$ transitions are non-contributors.

An analysis of the AEE description of nitrogen chemical shifts reveals that in many cases significant changes occur in both $\langle r^{-3}\rangle_{2p}$ and $\sum_B Q_{AB}$ in passing along a series of related molecules, changes in the latter term being, in general, more pronounced than in the former. Equations 3.23 and 3.32 show that both of these terms depend upon atomic charge density, the dependence being greatest when the atom bears a lone-pair of electrons and its sigma bonds are significantly polarised.

The difficulties associated with a choice of ΔE in the *AEE* approach are demonstrated in the case of the ^{14}N chemical shift difference between pyridine and the pyridinium ion. *AEE* descriptions of the relative deshielding of the pyridine nitrogen attribute it to a contribution from a low-lying $n \rightarrow \pi^*$ transition together with a substantial contribution from a $\sigma \rightarrow \pi^*$ transition. In contrast to this, CNDO/S parameterised calculations, based upon Eq. 3.19, reveal that the pyridine deshielding is due to $n \rightarrow \pi^*$ and $n \rightarrow \sigma^*$ transitions only.[1] The $\sigma \rightarrow \sigma^*$, $\sigma \rightarrow \pi^*$ and $\pi \rightarrow \sigma^*$ transitions all produce shielding contributions for the pyridine nitrogen relative to that in the pyridinium ion.[1]

For series of structurally related molecules, in which ΔE and the Q_{AB} terms are reasonably constant, a simple proportionality exists between nuclear shielding differences and changes in atomic charge densities. This is found to apply to the ^{13}C chemical shifts of several series of molecules, the proportionality constant being a characteristic of a given series.[1] Such relationships are not very revealing of electronic structure, and interpretations based upon a more complete description, such as Eq. 3.19, are to be preferred.

3B.5 Chemical Shift Ranges

A summary of some of the chemical shifts of ^{13}C, ^{14}N, ^{17}O and ^{19}F, relative to the bare nuclei, is presented in Fig. 3.3.[38] The ranges of observed shifts for these nuclei are about

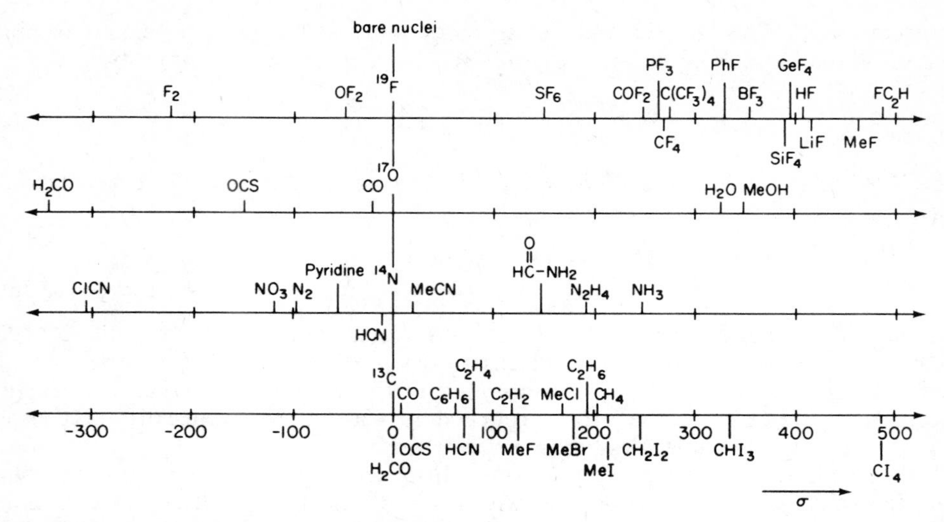

Fig. 3.3 Comparison of the screening of ^{13}C, ^{14}N, ^{17}O and ^{19}F in various molecular environments with that of the bare nuclei.

650, 930, 700 and 800 ppm respectively, whilst that for boron is about 200 ppm. A somewhat similar situation arises amongst the second row nuclei, where the approximate ranges for ^{27}Al, ^{29}Si, ^{31}P, ^{33}S and ^{35}Cl are 270, 400, 700, 600 and 820 ppm. Thus although the nuclei with the lower values of $\langle r^{-3}\rangle_{np}$ tend to have the smaller chemical shift ranges in each row, the ranges do not follow the trends of $\langle r^{-3}\rangle_{np}$ shown in Fig. 3.1. The greater ranges shown by ^{14}N, ^{17}O and ^{19}F, compared with ^{11}B and ^{13}C, reflect the importance of lone-pair electrons. In general these electrons give rise to low energy $n \rightarrow \pi^*$ transitions and thus to large values of σ^p. If the lone-pair becomes involved in bonding their contribution to σ^p is removed resulting in increased nuclear screening.

This effect is well demonstrated in nitrogen NMR where the ^{14}N signal of pyrrole occurs about 170 ppm to low frequency of that in pyridine. Similarly the high frequency shift of 250 ppm observed for NO_2^-, compared with NO_3^-, is largely due to a low energy excitation involving a non-bonding orbital in NO_2^-. The large range of bonding situations available to nitrogen is reflected in its chemical shift range, which is the largest amongst the first row nuclei.

The much smaller ranges observed for ^{7}Li and ^{23}Na, about 10 and 15 ppm respectively, imply that only σ^d is of importance for these nuclei, since σ^p requires the presence of p valence electrons.

Some of the heavier nuclei have very large chemical shift ranges; for example ^{59}Co and ^{205}Tl have ranges of about 12 000 and 34 000 ppm respectively. These undoubtedly reflect the greater polarisability of the higher p and d valence orbitals and their varied commitments to chemical bonding.

3C SPIN-SPIN COUPLING

3C.1 General

Nuclear spin-spin coupling is produced by an indirect interaction between the spins of neighbouring nuclei via the valence electrons of the molecule. The first nucleus perturbs the electrons and these in turn produce a magnetic field at the second nucleus. The energy of the interaction between nuclei A and B, E_{AB}, is given by:

$$E_{AB} = hJ_{AB}\mathbf{I}_A \cdot \mathbf{I}_B \tag{3.37}$$

In Eq. 3.37 J_{AB}, the (A, B) spin-spin coupling constant, is a scalar quantity which like σ is one-third of the trace of a second-rank tensor. Unlike σ, J is independent of the applied magnetic field; thus the gauge problem does not enter into discussions of nuclear spin-spin interactions.

Measurements on oriented samples can provide an estimate of the anisotropy of J, though there are few reported values. However, the situation may be complicated by the presence of direct, dipole-dipole, interactions which only average to zero in relatively non-viscous isotropic media. From a knowledge of the expected structure of the oriented molecule it is possible to calculate the dipole-dipole coupling anisotropy. By subtracting this from the observed coupling anisotropy an estimate of the anisotropy of the indirect coupling is obtained.

Unlike σ, values of J are obtained directly from NMR spectra. However the experimental values of J may incorporate uncertainties due to line broadening, media effects and variable populations of particular molecular rotational, vibrational and electronic levels. These may have to be taken into account when comparing theoretical and experimental values of J.

3C.2 Basic Theoretical Considerations

The calculation and interpretation of indirect nuclear spin-spin interactions is usually based on Ramsey's formulation.[39] Ramsey described the interaction by means of the Hamiltonian, $\mathscr{H}$, where:

$$\hat{\mathscr{H}} = \hat{\mathscr{H}}_1^{(a)} + \hat{\mathscr{H}}_1^{(b)} + \hat{\mathscr{H}}_2 + \hat{\mathscr{H}}_3 \tag{3.38}$$

The first two terms in Eq. 3.38 account for the interaction between the nuclear magnetic moments, depicted by their magnetogyric ratios γ, and the field produced by the orbital motion of the electrons. Thus these are called the *orbital* terms; $\hat{\mathscr{H}}_1^{(a)}$ and $\hat{\mathscr{H}}_1^{(b)}$ are mono- and binuclear respectively:

$$\hat{\mathscr{H}}_1^{(a)} = \frac{\mu_0 \mu_B \hbar}{2\pi i} \sum_A \sum_K \gamma_A r_{KA}^{-3} \hat{\mathbf{I}}_A \cdot (\mathbf{r}_{KA} \times \hat{\nabla}_K) \tag{3.39}$$

and:

$$\hat{\mathscr{H}}_1^{(b)} = \frac{\mu_0 e\hbar \mu_B}{4\pi} \sum_A \sum_B \sum_K \gamma_A \gamma_B r_{KA}^{-3} r_{KB}^{-3} [(\hat{\mathbf{I}}_A \cdot \hat{\mathbf{I}}_B)(\mathbf{r}_{KA} \cdot \mathbf{r}_{KB}) - (\hat{\mathbf{I}}_A \cdot \mathbf{r}_{KB})(\hat{\mathbf{I}}_B \cdot \mathbf{r}_{KA})] \tag{3.40}$$

where μ_B is the Bohr magneton, $\hat{\mathbf{I}}$ is the nuclear spin operator, r_{KA} refers to the separation between electron K and nucleus A, and $\hat{\nabla}_K$ is del, the vector operator for the electrons.

The contribution $\hat{\mathscr{H}}_2$ accounts for the dipole-dipole interaction between the nuclear and electron spins. This term is only applicable to extra-nuclear electrons and is referred to as the *dipolar* contribution:

$$\hat{\mathscr{H}}_2 = \frac{\mu_0 \mu_B \hbar}{2\pi} \sum_A \sum_K \gamma_A \left[\frac{3(\hat{\mathbf{S}}_K \cdot \mathbf{r}_{KA})(\hat{\mathbf{I}}_A \cdot \mathbf{r}_{KA})}{r_{KA}^5} - \frac{\hat{\mathbf{S}}_K \cdot \hat{\mathbf{I}}_A}{r_{KA}^3} \right] \tag{3.41}$$

where $\hat{\mathbf{S}}$ is the electron spin operator.

The final contribution, $\hat{\mathscr{H}}_3$, arises from electrons which are in orbitals having a finite density at the nucleus (although also basically due to dipolar-type interactions, $\hat{\mathscr{H}}_3$

	contact	dipolar	orbital	
	$\hat{\mathcal{H}}_3$	$\hat{\mathcal{H}}_2$	$\hat{\mathcal{H}}_1^{(a)}$	$\hat{\mathcal{H}}_1^{(b)}$
contact $\hat{\mathcal{H}}_3$	isotropic	anisotropic	0	0
dipolar $\hat{\mathcal{H}}_2$	anisotropic	isotropic + anisotropic	0	0
orbital $\hat{\mathcal{H}}_1^{(a)}$	0	0	isotropic + anisotropic	0
$\hat{\mathcal{H}}_1^{(b)}$	0	0	0	isotropic + anisotropic

Fig. 3.4 Symmetries of the various components contributing to the nuclear spin-spin coupling interaction.

has no simple macroscopic counterpart). This contribution is expressed in terms of the Dirac delta operator, $\delta(\mathbf{r}_{KA})$, which picks out the value of a function at the nucleus in an integration over the coordinates of electron K. Since a contribution from $\hat{\mathcal{H}}_3$ implies that the nucleus and electrons concerned are in direct contact with each other, $\hat{\mathcal{H}}_3$ is referred to as the *contact* term:

$$\hat{\mathcal{H}}_3 = \tfrac{4}{3}\mu_0 \mu_B \hbar \sum_A \sum_K \gamma_A \delta(\mathbf{r}_{KA})\hat{\mathbf{S}}_K \cdot \hat{\mathbf{I}}_A \tag{3.42}$$

Ramsey[39] expressed the total nuclear spin-spin coupling energy by means of second order perturbation theory:

$$E_{AB} = -\sum_n |\langle 0|\hat{\mathcal{H}}|n\rangle\langle n|\hat{\mathcal{H}}|0\rangle|(E_n - E_0)^{-1} \tag{3.43}$$

By combining Eq. 3.38 and 3.43 the various contributions to the spin-spin coupling energy are obtained. The excited states which give non-zero contributions from $\hat{\mathcal{H}}_1^{(a)}$ and $\hat{\mathcal{H}}_1^{(b)}$ give zero contributions for $\hat{\mathcal{H}}_2$ and $\hat{\mathcal{H}}_3$, and vice versa. Thus as shown in Fig. 3.4 the only cross-term contribution is that between $\hat{\mathcal{H}}_2$ and $\hat{\mathcal{H}}_3$;[40] since this is anisotropic it is averaged to zero by frequent intermolecular collisions. However, it can be an important contributor to the spin-spin coupling anisotropy of oriented molecules.

For isotropic, non-viscous, fluids the total nuclear spin-spin interaction energy is:

$$E_{AB} = E_{AB}^{(a)} + E_{AB}^{(b)} + E_{AB}^{2} + E_{AB}^{3} \tag{3.44}$$

where Eq. 3.37 relates E_{AB} to the measured value of J_{AB} in Hz. Thus, within Ramsey's model, the calculation of J_{AB} involves the evaluation of the various terms in Eq. 3.44.

For protons the only significant contribution is due to the contact energy, E_{AB}^3. This is also usually the dominant term for other nuclei although the remaining terms in Eq. 3.44 can provide significant contributions for heavier nuclei as well as in cases of multiple bonding. Consequently most attention has been focused on the calculation of E_{AB}^3.

Equation 3.43 shows that it is necessary to consider infinite summations over excited electronic (molecular and continuum) states. The difficulties associated with these summations are often removed in the same manner as described by Eq. 3.10 and 3.14 in dealing with the screening tensor, namely by invoking the closure relationship and the AEE approximation. This naturally gives rise to the problem of choosing a suitable value for ΔE.

It is noteworthy that the other difficulties associated with Ramsey's model of nuclear screening, namely the gauge dependence problem and the necessity of considering small differences between large and variable terms are absent from his formulation of spin-spin coupling constants. Thus the requisite calculations for indirect nuclear coupling interactions are almost entirely based on Ramsey's approach.

To obtain an understanding of the relationship between nuclear spin-spin coupling and molecular electronic structure it is convenient to use reduced coupling constants, K, as defined in ch. 1. The sign of K depends only upon the electronic environment of the nucleus. If either γ_A or γ_B is negative then K_{AB} and J_{AB} are of opposite sign; this occurs with several nuclei, e.g. ^{15}N, ^{17}O and ^{29}Si.

By analogy with Eq. 3.37 and 3.44 the reduced coupling constant may be expressed in terms of contributions from the various indirect nuclear spin-spin interactions:

$$K_{AB} = K_{AB}^{1(a)} + K_{AB}^{1(b)} + K_{AB}^{2} + K_{AB}^{3} \tag{3.45}$$

The contributions to K_{AB} are usually obtained from semi-empirical MO or VB calculations.[41,42]

Some *ab initio* calculations of spin-spin coupling constants for small molecules have been reported. However, in general, the results are disappointing.[43] The majority of calculations to date are semi-empirical, involving the sum-over-states and finite perturbation methods.

3C.3 Sum-over-states Calculations

a. General

Using an independent electron model and a minimum basis set of valence shell atomic orbitals Pople and Santry[44] have developed the necessary expressions for sum-over-states calculations. All of the excited states obtained by simple electron excitations from occupied to unoccupied MO's are included in the treatment and the summation in Eq. 3.43 is carried out explicitly. In keeping with the nature of the independent electron model only one-centre integrals are included in the calculations.

Within this framework $K_{AB}^{1(a)}$ depends only on the electronic ground state of the molecule and it becomes negligible due to the r_B^{-3} factor in Eq. 3.46:

$$K_{AB}^{1(a)} = \frac{\mu_0 e^2}{3\pi m} \sum_{j}^{\text{occ}} \sum_{\lambda\mu} C_{j\lambda} C_{j\mu} \langle \lambda | \mathbf{r}_A \cdot \mathbf{r}_B (r_A^{-3} r_B^{-3}) | \nu \rangle \tag{3.46}$$

where λ and μ are p orbitals on atom A at a mean distance of r_A whose LCAO coefficients are $C_{j\lambda}$ and $C_{j\mu}$ respectively in the j occupied MO's.

The corresponding expression for $K_{AB}^{1(b)}$ is:

$$K_{AB}^{1(b)} = -\frac{4\mu_0 \mu_B^2}{3\pi} \sum_{j}^{\text{occ}} \sum_{k}^{\text{unocc}} ({}^1\Delta E_{j\to k})^{-1} \sum_{\lambda\mu\nu\sigma} C_{j\lambda} C_{k\mu} C_{j\nu} C_{j\sigma} \times \langle \lambda | r_A^{-3} \hat{M}_A | \mu \rangle \langle \nu | r_B^{-3} \hat{M}_B | \sigma \rangle \tag{3.47}$$

where ν and σ refer to p orbitals on atom B and $\hat{M}_A\hbar$ is the orbital angular momentum operator describing the motions of the p electrons about A, ${}^1\Delta E_{j\to k}$ refers to the singlet excitation energy between the empty and filled MO's, j and k respectively.

The *dipolar contribution is:*

$$K_{AB}^2 = -\frac{\mu_0\mu_B^2}{3\pi}\sum_j^{occ}\sum_k^{unocc}({}^3\Delta E_{j\to k})^{-1}\sum_{\lambda\mu\nu\sigma}C_{j\lambda}C_{k\mu}C_{k\nu}C_{j\sigma}$$
$$\times\langle\lambda|(3r_{A\alpha}r_{A\beta}-r_A^2\delta_{\alpha\beta})r_A^{-5}|\mu\rangle\langle\nu|(3r_{B\alpha}r_{B\beta}-r_B^2\delta_{\alpha\beta})r_B^5|\sigma\rangle \quad (3.48)$$

where α and β refer to the components along the x, y or z directions.

The expression for the *contact* term is given by:

$$K_{AB}^3 = -\tfrac{64}{9}\pi\mu_0\mu_B^2\pi\langle s_A|\delta(\mathbf{r}_A)|s_A\rangle\langle s_B|\delta(\mathbf{r}_B)|s_B\rangle\sum_j^{occ}\sum_k^{unocc}({}^3\Delta E_{j\to k})^{-1}C_{js_A}C_{ks_A}C_{ks_B}C_{js_B} \quad (3.49)$$

where s_A and s_B refer to s orbitals on atoms A and B respectively. Since inner-shells are relatively unimportant in MO formation the most important contributions to Eq. 3.49 are those involving valence shell s functions.

Equations 3.46, 3.47 and 3.48 show that the orbital and dipolar contributions to spin-spin coupling depend upon the atoms concerned having valence p electrons. Consequently these contributions vanish when a hydrogen nucleus is involved in the spin-spin coupling.

Since both the orbital and dipolar terms depend upon the expectation value of the inverse cube of the radius of the orbitals on atoms A and B it is to be anticipated that they will be more important when multiple bonding occurs. They may also be expected to vary periodically in the manner shown in Fig. 3.1, as demonstrated in Table 3.5.

A contraction in the s orbital radius with increase in effective nuclear charge is likely to produce a larger s electron density at the nucleus as its mass increases as shown in Table 3.5. Thus K_{AB}^3 is expected to become very large for heavier nuclei together with a periodically dependent increase in K_{AB}^1 and K_{AB}^2. This is reflected in the very large values

TABLE 3.5 Some values of the atomic parameters $S^2(0)$ and $\langle r^{-3}\rangle_p$ in atomic units

Nucleus	$S^2(0)$	$\langle r^{-3}\rangle_p$
^{11}B	1.408	0.775
^{13}C	2.707	1.692
^{15}N	4.770	3.101
^{17}O	7.638	4.974
^{19}F	11.966	7.546
^{29}Si	3.807	2.041
^{31}P	5.625	3.319
^{33}S	7.919	4.814
^{35}Cl	10.643	6.710
^{75}As	12.460	7.111
^{129}Xe	26.710	17.825

of J involving thallium, for example, $^{1}J(^{205}Tl, ^{13}C)$ is found[45] to be 10 718 Hz in phenylthallium (III) trifluoroacetate.

Equation 3.49 may be written in terms of the mutual polarisability of the s orbitals on A and B, Π_{AB}:

$$K_{AB}^{3} = -\tfrac{16}{9}\pi\mu_0\mu_B^2 S_A^2(0)S_B^2(0)\Pi_{AB} \tag{3.50}$$

where $S_A^2(0)$ and $S_B^2(0)$ represent the s electron densities at the nuclei A and B. If $^{3}\Delta E_{j\rightarrow k}$ is represented by the difference $(\varepsilon_k - \varepsilon_j)$ of the one-electron energies, then the mutual polarisability is:

$$\Pi_{AB} = 4\sum_{j}^{\text{occ}}\sum_{k}^{\text{unocc}}(\varepsilon_k - \varepsilon_j)^{-1}C_{js_A}C_{ks_A}C_{js_B}C_{ks_B} \tag{3.51}$$

Provided that the contact interaction makes the dominant contribution to the spin-spin coupling, Eq. 3.50 demonstrates the existence of a relationship between the coupling constant and the amount of s electron character in hybridised orbitals on the atoms concerned.

Equations 3.47, 3.48 and 3.49 have been used with various semi-empirical MO parameter sets. In general the calculated values of the coupling constants are smaller than the experimental ones. This is most probably due to the fact that the $^{1}\Delta E$ and $^{3}\Delta E$ excitation energies are overestimated by the parameterisation schemes used. However, the calculations reproduce the experimental trends reasonably well in most cases; in particular experimental ^{1}J data are well accounted for.

b. Inclusion of Electron-electron Interactions

An extension of the independent electron model permits electron-electron interactions to occur in calculating both the ground and excited state eigenfunctions. The inclusion of one-centre exchange integrals stabilizes a pair of parallel spins in different orbitals, relative to a pair of antiparallel spins. This may be accomplished by means of the INDO parameterisation scheme. Configuration interaction between singly excited states is likely to provide a more realistic estimate of the excitation energies, $^{1}\Delta E$ and $^{3}\Delta E$.

Consequently calculations based upon INDO-MO's are preferable to those using the CNDO/2 method, whilst the inclusion of configuration interaction and a variation in atomic charge with calculated electron densities is found to provide better agreement between the theoretical and experimental data[46–48] as shown in Table 3.6.[46]

The importance of the orbital and dipolar contributions to the total spin-spin interaction is demonstrated in Table 3.7. The data given are derived from INDO calculations employing configuration interaction and variation in atomic charge. In all cases considered the orbital and the dipolar terms are not negligible; for couplings involving ^{19}F they frequently dominate the contact contribution.[46] A similar conclusion has been reached from a study of *geminal* and *trans* $(^{19}F, ^{19}F)$ couplings in some fluoro-olefins.[49] If the fluorine nuclei are separated by more than four bonds the contact term becomes the most important one.[49]

c. AEE Method

The AEE approximation can be applied to Eq. 3.47, 3.48 and 3.49 in an attempt to remove the difficulties associated with the calculation of excitation energies. By means of this

TABLE 3.6[46] Some values of one bond coupling constants in Hz for first row nuclei calculated by the sum-over-states method, compared with experimental data

Coupling constant	Molecule	INDO Without CI Fixed charge	INDO Without CI Varied charge	INDO With CI Fixed charge	INDO With CI Varied charge	CNDO/2 Without CI Fixed charge	CNDO/2 Without CI Varied charge	CNDO/2 With CI Fixed charge	CNDO/2 With CI Varied charge	Exp
$^1J(^{13}C, ^{13}C)$	C_2H_2	73.53	70.74	79.42	76.40	77.58	74.09	79.23	75.66	171.5
	C_2H_4	22.34	22.44	26.79	26.92	23.71	23.24	24.30	23.83	67.6
	C_2H_6	7.78	8.11	11.26	11.74	8.34	8.31	8.75	8.72	34.6
$^1J(^{19}F, ^{13}C)$	CH_3F	−111.4	−118.1	−116.5	−123.5	−61.01	−62.34	−79.39	−81.12	−157.5
	CH_2F_2	−116.6	−135.3	−121.2	−140.7	−61.12	−67.58	−80.53	−89.09	−234.8
	CHF_3	−101.7	−126.7	−112.1	−139.7	−45.05	−53.22	−69.40	−82.08	−274.3
	CF_4	−60.37	−80.04	−71.71	−103.1	−7.27	−8.87			−259.2
	HFCO	−287.7	−331.9	−291.1	−335.8	−198.9	−220.8	−220.1	−244.5	−369.0
$^1J(^{19}F, ^{10}B)$	BF_3	30.38	40.74	32.88	44.09	42.86	55.17	39.50	50.89	6.7
$^1J(^{19}F, ^{14}N)$	NF_3	−63.70	−70.36	−71.04	−78.49	−48.28	−52.32	−52.53	−56.96	−160

TABLE 3.7[46] Values of the various components, in Hz, of some coupling constants between first row nuclei calculated by the sum-over-states method using the INDO parameterisation together with configuration interaction and variation in atomic charges.

Coupling constant	Molecule	Contact	Orbital	Dipolar	Total
$^1J(^{13}C, ^{13}C)$	C_2H_2	69.43	4.13	2.84	76.40
	C_2H_4	29.75	−3.76	0.93	26.92
	C_2H_6	11.99	−0.55	0.30	11.74
$^1J(^{19}F, ^{13}C)$	CH_3F	−126.30	−4.15	6.92	−123.5
	CH_2F_2	−132.64	−12.54	4.50	−140.7
	CHF_3	−125.53	−17.51	3.30	−139.7
	CF_4	−85.33	−20.17	2.40	−103.1
	FHCO	−32.13	−15.83	0.15	−335.8
$^1J(^{15}N, ^{13}C)$	H_2CN_2	8.21	−1.36	0.24	7.09
$^1J(^{15}N, ^{15}N)$	H_2CN_2	−6.12	−0.42	0.30	−6.24
$^1J(^{19}F, ^{10}B)$	BF_3	48.88	−5.12	0.33	44.09
$^1J(^{19}F, ^{15}N)$	NF_3	−80.68	−3.77	5.96	−78.49
$^2J(^{19}F, ^{19}F)$	CH_2F_2	2.90	43.72	23.72	70.35
	CHF_3	−1.27	22.65	19.76	41.14
	CF_4	−1.09	6.20	16.73	21.84
	F_2CO	19.18	−26.88	12.54	4.84
	BF_3	−1.42	64.18	3.72	−61.88
	NF_3	−0.02	58.72	34.25	92.95
$^2J(^{15}N, ^{13}C)$	H_2CN_2	−0.51	−1.30	0.38	1.43
$^1J(^{19}F, ^{19}F)$	F_2	−470.9	1631.8	903.5	2064.4

approximation, employed within the LCAO framework, the following expressions are obtained for contributions to K_{AB}:

$$K_{AB}^{1(b)} = \frac{2\mu_0\mu_B^2}{3\pi}\langle r_A^{-3}\rangle_p\langle r^{-3}\rangle_p(^1\Delta E)^{-1}$$
$$\times [P_{x_Ax_B}P_{y_Ay_B} + P_{y_Ay_B}P_{z_Az_B} + P_{z_Az_B}P_{x_Ax_B}$$
$$- P_{x_Ay_B}P_{y_Ax_B} - P_{y_Az_B}P_{z_Ay_B} - P_{z_Ax_B}P_{x_Az_B}] \tag{3.52}$$

$$K_{AB}^2 = \frac{\mu_0\mu_B^2}{25\pi}\langle r_A^{-3}\rangle_p\langle r_B^{-3}\rangle_p(^3\Delta E)^{-1}$$
$$\times [2(P_\sigma^2 + P_\pi^2 + P_{\pi'}^2) + 3(P_\sigma P_\pi + P_\sigma P_{\pi'} + P_\pi P_{\pi'})] \tag{3.53}$$

where P_σ, P_π and $P_{\pi'}$ are the σ and π bond orders for the bonding between A and B, and:

$$K_{AB}^3 = \tfrac{16}{9}\pi\mu_0\mu_B^2(^3\Delta E)^{-1}S_A^2(0)S_B^2(0)P_{s_As_B}^2 \tag{3.54}$$

Equation 3.54 is analogous to Eq. 3.50 and was first presented by McConnell.[50] Application of 3.54 can only produce positive contributions to the coupling constant. This is a serious drawback since negative values are often obtained experimentally.

Equation 3.52 demonstrates that the orbital contribution is non-zero for directly bonded nuclei only if they are multiply bonded. This arises since if, for example, the x axis is chosen to lie along the A—B bond direction, the cross term in Eq. 3.52 will either be zero in molecules with high symmetry or small in other cases. Thus $K_{AB}^{(b)}$ will vanish unless one of the π bond orders $P_{y_A y_B}$ or $P_{z_A z_B}$ is non-zero. The data in Table 3.7 demonstrate that this conclusion is erroneous. For example the orbital term contributes more than 20 Hz to the single bond $^1J(^{19}F, ^{13}C)$ couplings of CF_4 and F_2CO.

Therefore it may be concluded that coupling constants calculated by means of the AEE approximation may be unreliable both in terms of sign and with respect to the significance of non-contact contributions.

3C.4 Finite Perturbation Calculations

The sum-over-states method is not always easy to apply satisfactorily on account of problems associated with the construction of reliable excited state wavefunctions. In addition there are some serious cancellation difficulties to contend with. This latter problem also arises in an analogous VB treatment of spin-spin couplings.[51,52] A different semi-empirical approach involves the use of finite perturbation theory.[53,54]

The finite perturbation approach is equivalent, in principle, to a complete self-consistent perturbation calculation involving the sum over all appropriate excited states. A perturbation corresponding to one of the three coupling mechanisms is inserted during the calculation of the SCF wavefunction for the molecule. The calculation of the excited state wavefunctions is not required, which is an advantage in employing the finite perturbation approach.

Within this framework, Blizzard and Santry[55] have used the approximations inherent to the INDO-MO method to obtain expressions for the various contributions to K_{AB}. The orbital term appears as:

$$K_{AB}^1 = \frac{2\mu_0\mu_B^2}{3\pi}\langle r_A^{-3}\rangle_p\langle r_B^{-3}\rangle_p[P_{x_B y_B} + P_{z_B y_B} + P_{x_B z_B}] \tag{3.55}$$

where the P's refer to the first order elements of the charge-density, bond-order matrix.

The dipolar term is given by:

$$K_{AB}^2 = \frac{\mu_0 \hbar^2 \mu_B^2}{5\pi}\langle r_A^{-3}\rangle_p\langle r_B^{-3}\rangle_p[2P_{z_B z_B}^{\alpha\alpha} - P_{x_B x_B}^{\alpha\alpha} - P_{y_B y_B}^{\alpha\alpha} + 3P_{x_B z_B}^{\alpha\beta} - 3Q_{y_B z_B}^{\alpha\beta}] \tag{3.56}$$

where the superscripts α and β refer to the spin-orbitals with α and β spin respectively and $Q^{\alpha\beta}$ is the imaginary part of the first order matrix.

Finally, the contact term appears as:

$$K_{AB}^3 = \tfrac{32}{9}\pi\mu_0\mu_B^2 S_A^2(0)S_B^2(0)P_{S_A S_B}^{\alpha} \tag{3.57}$$

Equations 3.55, 3.56 and 3.57 have been used to demonstrate the importance of all three contributions to (F, F) and (F, C) couplings.[55] The integrals $S^2(0)$ and $\langle r^{-3}\rangle_p$ are treated as parameters which are evaluated from the best least-squares agreement between the calculated and experimental coupling constants. The values obtained for $S_C^2(0)$, $S_F^2(0)$,

TABLE 3.8[56] Values of the various components, in Hz of some (^{19}F, ^{19}F) coupling constants for a series of fluorofurans from a perturbation treatment using the INDO parameterisation scheme

Molecule	Positions	Contact	Orbital	Dipolar	Total	Exp.
1,2,3,4-tetrafluorofuran	1,2	−5.5	−17.6	4.6	−18.5	15.5
(F_1, F_2, F_3, F_4 on furan ring, O)	1,3	−3.3	−4.9	0.0	−8.2	13.5
	1,4	−4.5	−37.9	−2.1	−44.5	20.5
	2,3	−2.7	8.5	−0.4	5.4	5.5
1,2,3-trifluorofuran	1,2	−5.5	−14.9	4.1	−16.3	15.1
	1,3	−3.1	−2.4	0.9	−4.6	9.9
	2,3	−2.1	8.7	0.8	7.4	7.4
1,2-difluorofuran	1,2	−5.6	−13.7	4.2	−15.1	13.5
1,4-difluorofuran	1,4	−4.4	−30.9	3.3	−32.0	25.0
2,3-difluorofuran	2,3	−2.1	8.2	0.7	6.8	8.8

$\langle r_C^{-3}\rangle_p$ and $\langle r_F^{-3}\rangle_p$ are larger than the atomic values given in Table 3.5. The changes are attributed to differences between the atomic and molecular environments of the C and F nuclei.

As in the case of sum-over-states calculations orbital and dipolar contributions can predominate over those from the contact term for (F, F) couplings. This is illustrated in Table 3.8 for some fluorofurans.[56] The (F, F) couplings operate predominantly through the σ bonds but this does not preclude the possibility of the contact term being influenced by induced σ spin-densities transmitted through the π system due to $\sigma \rightarrow \pi$ polarisation procedures. A definite π involvement is noted for the orbital term of the (F_2, F_3) coupling in the fluorofurans as well as in the dipolar contributions for all of the (F, F) couplings considered.[56]

For (C, C) couplings in saturated molecules, and all long-range (C, C) interactions, the contact term is sufficient to account for the experimental data. However the orbital and dipolar contributions are significant for multiply bonded carbon atoms.[57] For the C_1—C_3 bond of bicyclobutane these terms are more important than the contact term and produce a negative value for $^1J(^{13}C_1, ^{13}C_3)$, which has recently been confirmed experimentally.[58]

Comparable conclusions are reached on the relative importance of the three contributions to (N, C) coupling constants.[59] In this case the lone-pair electrons may play a very important role in determining the magnitude of the contact term. This may be illustrated by comparison of the $^1J(^{15}N, ^{13}C)$ data for pyridine and the pyridinium ion. For the former the contact contribution is calculated to be −0.7 Hz whereas −13.7 Hz is obtained for the latter. This difference is largely due to a low energy transition from the highest filled non-bonding orbital in pyridine, which provides a large positive contribution to the contact term and thus largely cancels other, negative, contributions. The absence of the lone-pair in the pyridinium ion removes this effect and results in a large and negative contact contribution.

Calculations of the contact contribution alone appear to adequately describe (^{31}P, ^{13}C), (^{31}P, ^{17}O), (^{31}P, ^{15}N), (^{31}P, ^{31}P) and (^{35}Cl, ^{31}P) spin-spin couplings.[60–63] In this respect phosphorus differs from the other nuclei mentioned in this section,

TABLE 3.9[64] Values of the various components, in Hz, of the isotropic and anisotropic parts of the coupling constants between the directly bonded nuclei in MeF calculated from the finite perturbation method using the INDO parameterisation scheme, compared with experimental data

Coupling constant	Isotropic part				
	Contact	Orbital	Dipolar	Total	Exp
$^1J(^{13}C, ^1H)$	147.0	0.0	0.0	147.0	148.8
$^1J(^{19}F, ^{13}C)$	−97.0	−15.0	15.0	−97.0	−161.9

Coupling constant	Anisotropic part				
	Contact − Dipolar	Orbital	Dipolar	Total	Exp
$^1J(^{13}C, ^1H)$	−19.0	0.0	0.0	−19.0	1890 ± 130
$^1J(^{19}F, ^{13}C)$	208.0	27.0	26.0	261.0	700 ± 130

especially ^{19}F and to a lesser extent ^{15}N and ^{13}C. Table 3.5 shows that $\langle r^{-3} \rangle_p$ for atomic phosphorus is larger than in the case of ^{15}N and ^{13}C; thus the magnitude of this parameter is unlikely to account for the apparent insignificance of the orbital and dipolar contributions to phosphorus coupling constants. However, the phosphorus atoms are singly bonded to the spin-coupled nucleus in most of the cases considered. Consequently ^{31}P may be similar to ^{15}N and ^{13}C in that the non-contact terms assume a greater degree of importance when multiple bonding occurs.

NMR measurements employing nematic solvents permit the estimation of spin-spin coupling anisotropies. The results of some finite perturbation calculations using INDO data are compared with experimental results for the directly coupled nuclei in MeF in Table 3.9.[64] Although the calculated values of the isotropic coupling constants are in reasonable agreement with experiment the anisotropies agree less well. However, solvent effects may be responsible for the apparent discrepancies.[65]

It is noteworthy that the major contribution to the (F, C) coupling anisotropy arises from the contact-dipolar cross term. Table 3.9 also shows that the (F, C) coupling anisotropy is much greater than that of the (C, H) coupling. Although the dipolar term is zero for this latter coupling, the cross term with the contact interaction provides the anisotropy. For similar reasons (F, F) coupling constants usually have much larger anisotropies than (F, H) couplings.[66]

The sensitivity of the calculated value of the contact term to choice of molecular geometry has been emphasised.[59] Similar comments are applicable to the results obtained from sum-over-states calculations.[46] This point should be held firmly in mind when attempting to relate the results of MO calculations to experimental values of spin-spin coupling constants. Geometries obtained from standard bond angles and bond lengths can sometimes provide a more realistic approach to molecular structure in solution than can X-ray diffraction data obtained on the solid state. Thus the choice of a suitable geometry can be a critical one for calculations of spin-spin couplings. However, in general, coupling constant trends in series of related molecules are not as susceptible to geometry as are the absolute values of coupling constants.

3C.5 Correlation of Coupling Constants with other Molecular Properties

Spin-spin coupling constants are usually dependent on various molecular parameters such as hybridisation, electronegativity of substituents and dihedral bond angles.

a. Hybridisation effects

In those cases where the orbital and dipolar terms are unimportant Eq. 3.49, 3.54 and 3.57 show that the spin-spin coupling depends upon the amount of *s*-character in the bond joining the nuclei. Empirical relationships, such as Eq. 3.58:

$$80|{}^1J({}^{15}\mathrm{N}, {}^{13}\mathrm{C})| = \%s_\mathrm{N}\%s_\mathrm{C} \tag{3.58}$$

between the coupling constant and percentage *s*-character of the atomic orbitals forming the σ bond between the nuclei have been established. Finite perturbation calculations of the contact contribution for ${}^1J({}^{15}\mathrm{N}, {}^{13}\mathrm{C})$ show[67] that the proportionality constant in Eq. 3.58 should be 94 and that ${}^1J({}^{15}\mathrm{N}, {}^{13}\mathrm{C})$ is negative.

Some exceptions to the linearity of this relationship are found even when the orbital and dipolar contributions are not significant. These are attributed to the effect of lone-pair electrons in *s* orbitals on the carbon or nitrogen atoms. The presence of lone-pairs in *p* orbitals does not appear to cause significant deviations from the relationship predicted by Eq. 3.58. It seems probably that this situation will arise in the consideration of other coupling constants such as ${}^1J({}^{15}\mathrm{N}, {}^1\mathrm{H})$ and ${}^1J({}^{31}\mathrm{P}, {}^{13}\mathrm{C})$.

The dependence of the contact contribution to ${}^1J({}^{31}\mathrm{P}, {}^{13}\mathrm{C})$, calculated from CNDO/2 data including *d* orbitals in the finite perturbation approach, on the phosphorus 3*s*-carbon 2*s* bond order is demonstrated for a variety of molecules in Fig. 3.5.[68] With the

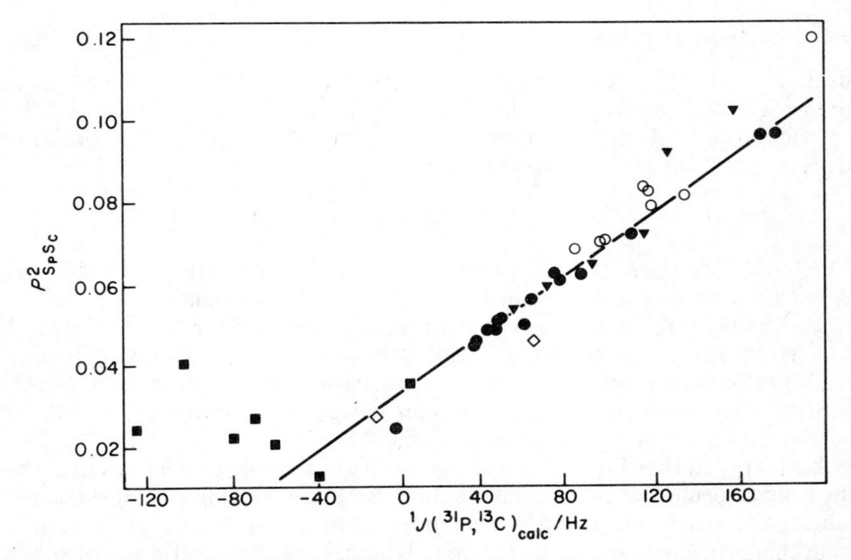

Fig. 3.5 Dependence of the ${}^{31}\mathrm{P}(3s)$-${}^{13}\mathrm{C}(2s)$ bond orders ($P_{S_PS_C}$) on the calculated value of ${}^1J({}^{31}\mathrm{P}, {}^{13}\mathrm{C})$ including d orbitals. The classes of compounds studied are: ■, phosphines; ◇, phosphoranes; ●, phosphonium cations; ○, phosphine oxides and sulphides and ▼, alkylidenephosphoranes.[68]

exception of the phosphines a correlation coefficient of 0.995 is obtained for the least-squares fit. The presence of lone-pair *s* electrons in the phosphines could be responsible for their deviation from the linear plot. However, the absence of one-centre exchange integrals in the CNDO/2 parameterisation scheme may also play a role in this observation.

In general (X, H) coupling constants show a reasonable correlation with the amount of *s*-character in the X—H bond. Substitution may cause large changes in effective nuclear charge in which case exact correlations are not anticipated. For example $^1J(^{13}C, ^1H)$ is 125 Hz for CH_4 and 209 Hz for $CHCl_3$ whereas the hybridisation of the carbon atom is nominally the same in both molecules. The relationship between $^1J_{CH}$ and the C—H bond order is relatively unsuccessful in accounting for substituent effects, whereas comparison of the coupling constant with the full contact term is more satisfactory.[69,70]

Finite perturbation calculations of the contact contribution to $^1J(^{13}C, ^1H)$ have been reported for a wide variety of compounds.[71] Good agreement is found for molecules with $-I^+$ substituents such as F, OR, NR_2 and =O, but the situation is less satisfactory for molecules with $-I^-$ substituents such as CF_3, —C(O)X, NO_2 and CN. It is possible that "through space" effects may be important for some of these substituents.

The relative importance of σ and π electron contributions to the contact term for coupling in conjugated systems can be estimated by comparing the results of INDO and CNDO calculations. Since the CNDO parameterisation omits the one-centre exchange integrals necessary to describe $\sigma \rightarrow \pi$ interactions on intervening carbon atoms, only σ contributions are estimated in a CNDO calculation, whereas the INDO scheme produces both σ and π electron effects. In some cases INDO calculations in which the appropriate one-centre exchange integrals are set to zero provide a more reasonable estimate of the σ electron contribution than does the CNDO parameterisation.[72,73]

b. Dihedral Angle Dependence

In saturated X—C—C—Y fragments the value of $^3J_{XY}$ depends critically upon the X—C—C—Y dihedral angle, ϕ. This dependence is primarily produced by the interaction of vicinal orbitals directed towards X and Y[74] and is adequately described by a Karlpus-type equation of the general form:

$$^3J_{XY} = A \cos 2\phi + B \cos \phi + C \qquad (3.59)$$

where A, B and C are empirical constants; generally B is negative and $|A| > |B|$. Thus $^3J_{XY}$ values are usually at a maximum at $\phi = 180°$ and a minimum when ϕ is close to 90°.

A typical example of a Karplus-type relationship is shown for $^3J(^{31}P, ^1H)$ in Fig. 3.6 for the ethyl phosphonium cation. The calculated value of the contact contribution to $^3J(^{31}P, ^1H)$ is obtained within the finite perturbation approach using CNDO/2 parameters with and without *d* orbitals.[68] Similar behaviour is found for $^3J(^{31}P, ^{13}C)$ in a number of tetravalent phosphorus compounds.[68]

It has been argued that Eq. 3.59 may be used to predict dihedral angles in compounds of unknown conformation by comparing their coupling constants with those predicted from values of A, B and C determined from model compounds. However, there are serious dangers in this procedure since Eq. 3.59 is only applicable in the absence of substituents of different electronegativities and of changes in bond angles and bond lengths at the substitution centre. Empirical modifications may be employed to account for substituent effects but these provide no justification for the use of Eq. 3.59 in determining angles to within a few degrees.

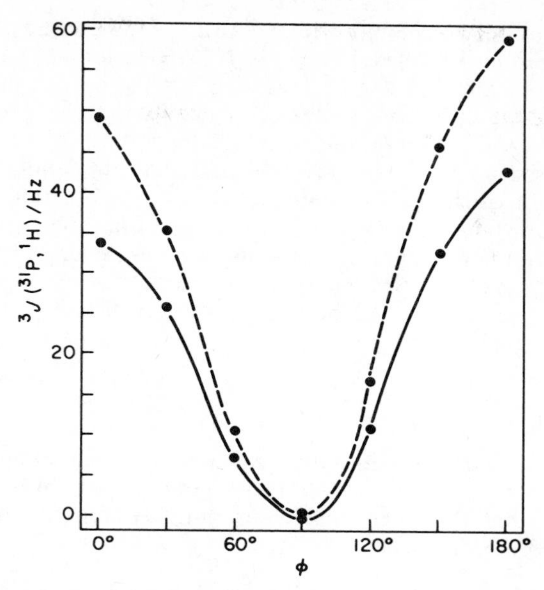

Fig. 3.6 Calculated values of $^3J(^{31}P, ^1H)$, compared with the P-C-C-H dihedral angle, ϕ, for the ethylphosphonium cation. The solid and dashed lines correspond to calculations with and without d orbitals respectively.[68]

c. Through-space Coupling

This is most often invoked to account for (F, F) interactions where nuclei are formally separated by several chemical bonds. A semi-empirical approach to this interaction is provided by Eq. 3.60:

$$K_{AB} = \text{constant} \times S_A^2(0)S_B^2(0)(S_{s_A s_B})^2 \qquad (3.60)$$

where $S_{s_A s_B}$ is the overlap integral between the valence s orbitals of atoms A and B and the constant is determined experimentally.[75]

Although Eq. 3.60 may be suitable for (^{1}H, ^{1}H) interactions it does not appear to satisfactorily account for (F, F) through-space contributions which depend upon the overlap of lone-pair p orbitals.[76] In the case of (^{19}F, ^{1}H) through-space couplings, both positive and negative contributions are found.[77] These cannot be accounted for by Eq. 3.60 where the sign of the constant is determined by the product of the magnetogyric ratios of the interacting nuclei. The (^{19}F, ^{1}H) couplings in question may be rationalised by means of calculations of the contact term which show the presence of several large contributions of either sign.[77] Since the total contact interaction is obtained by summing these contributions its sign will depend critically on the occurrence of conformational averaging.

3D NUCLEAR SPIN RELAXATION

3D.1 General

The relaxation times T_1 and T_2 are controlled by interactions both within the relaxing spin system and between this system and its surroundings. The rate at which a relaxing spin system comes into thermal equilibrium with its surroundings (the "lattice") is characterised by the spin-lattice relaxation time, T_1. Relaxation of magnetization in the plane perpendicular to B_0 affects linewidths and is described by the spin-spin relaxation time, T_2. For solids and highly viscous liquids T_2 is smaller than T_1, whereas in relatively non-viscous samples T_1 and T_2 are usually numerically equal.

At any given time some molecules in a liquid are moving slowly and others very rapidly. The Fourier component of these motions at the resonance angular frequency, ω_0, is responsible for nuclear spin-lattice relaxation (but see Eq. 3.66 to 3.68). Hence, the magnitude of this component, together with the coupling energy for the interaction between the spin system and the molecular motions, determines the value of T_1. In the case of T_2 low-frequency motions also contribute to the relaxation. The general situation for T_1 is expressed[78] by Eq. 3.61:

$$T_1^{-1} = \frac{\mu_0 \gamma^2 \langle B_L^2 \rangle \tau_0}{6\pi(1 + \omega_0^2 \tau_0^2)} \tag{3.61}$$

where $\langle B_L^2 \rangle$ is the mean-square average of the local fluctuating magnetic field produced by molecular motions, and τ_0 is the *correlation time* characterising these motions.

Equation 3.61 shows that T_1 goes through a minimum as a function of τ_0, at $\omega_0 \tau_0 = 1$. When rapid molecular motions occur:

$$\omega_0^2 \tau_0^2 \ll 1 \tag{3.62}$$

which is referred to as the motional narrowing limit. Under these conditions T_1 becomes frequency-independent and T_1 and T_2 are equal.

The frequency distribution of B_L is expressed by the spectral density function, $J(\omega)$, Fig. 3.7 shows the variation of $J(\omega)$ with frequency for two different values of the correlation time τ_0. Since $J(\omega)$ represents the power of the local fluctuating magnetic field it

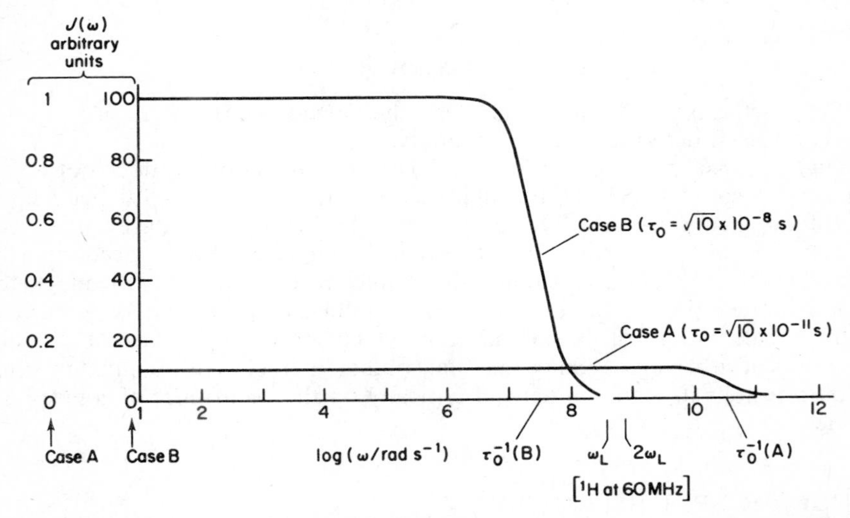

Fig. 3.7 Variation of the spectral density function, $J\omega$), with frequency for two different values of the correlation time τ_0. In the case of rotational (molecular tumbling) correlation case A ($\tau_0 = \sqrt{10} \times 10^{-11}$ s) would imply a small molecule, high temperature, low viscosity situation, whereas case B ($\tau_0 = \sqrt{10} \times 10^{-8}$ s) would refer to a macromolecule, low temperature and high viscosity.

follows that the most efficient relaxation occurs when the molecular frequencies are comparable to ω_0, as noted previously.

Usually the translational and rotational motions of molecules have components at the correct frequencies to induce spin-lattice relaxation whereas vibrational motions are normally too rapid to make an effective contribution.

The correlation time is often difficult to calculate and experimental measurements are either difficult or impossible in most cases. The temperature dependence of T_1 is normally determined by that of τ_0.

In contrast to the correlation time, the other parameters in Eq. 3.61 are usually well understood. Various relaxation mechanisms may contribute to B_L. Relaxation rates given by different mechanisms are additive in most circumstances of interest, as indicated by Eq. 1.8. The different mechanisms are discussed in the following sections, where it is shown that they are related to various molecular parameters which are usually relatively temperature independent. Consequently the analysis of experimental T_1 values is able to provide molecular information.

3D.2 Correlation Times

Molecular rotation is perhaps the best understood of the processes occurring in liquids. It is most easily interpreted by assuming that the rotational motions proceed by a small-step Brownian process with many steps per radian. In the general case it is necessary to consider a rotational diffusion tensor whose components are related to particular molecular axes.[79] To obtain solutions to this problem it is usually necessary to consider the rotating molecule as a rigid top of known symmetry.

By assuming a spherical solute molecule in a medium of viscosity η the rotational correlation time, τ_c, is given by:

$$\tau_c = \frac{\eta V f}{kT} \tag{3.63}$$

where f is a microviscosity factor which is about 0.16 for pure liquids and V is the volume of the solute molecule. A value for V is often found from the density of the solute by assuming it to be hexagonally close packed.

Equation 3.63 demonstrates that τ_c increases with molecular size and viscosity as well as with decreasing temperature and an increasing amount of molecular association.

Translational correlation times are more difficult to estimate. Molecular translation is usually discussed in terms of molecular translational diffusion constants. For a spherical molecule of radius r in a spherical solvent of viscosity η, the translational diffusion constant, D, is obtained from:

$$D = \frac{kT}{6\pi\eta r f} \tag{3.64}$$

Since the translational correlation time depends on the reciprocal of D it follows that its dependence on viscosity, molecular size and temperature will be similar to that of τ_c.

The interaction of the nuclear magnetic moment and the rotational magnetic moment of the molecule containing the nucleus provides a mechanism for the transfer of nuclear spin energy to the molecular rotation. This is known as the *spin-rotation interaction* and is characterised by a correlation time τ_{sr}.

For spherical molecules in a diffusion controlled process, τ_{sr} is related to the rotational correlation time and the moment of inertia, I, by Hubbard's relation:[80]

$$\tau_c \tau_{sr} = \frac{I}{6kT} \tag{3.65}$$

It follows from Eq. 3.63 and 3.65 that the temperature dependence of τ_{sr} is inversely proportional to that of η. For liquids η invariably decreases markedly as the temperature is raised, and so τ_{sr} lengthens.

The three correlation times discussed in this section are those required for a consideration of the various processes leading to nuclear spin relaxation. Additional times are required if motion is anisotropic, and if relaxation by scalar coupling is important.

3D.3 Relaxation by Magnetic Dipole Interactions

Although the coupling energy between two magnetic dipoles at a fixed distance is averaged to zero by molecular rotations it can still cause nuclear relaxation. For spin one-half nuclei this is often the dominant contributer to spin relaxation. It may involve interactions with other nuclei or with unpaired electrons.

a. *Intramolecular Interactions*

For a nucleus I being relaxed by a nucleus S at a distance r the transition probabilities, W, for dipole-dipole interactions are:

$$W_0 = \frac{\mu_0^2 \gamma_I^2 \gamma_S^2 \hbar^2}{120\pi^2 r^6} \frac{\tau_c S(S+1)}{[1 + (\omega_I - \omega_S)^2 \tau_c^2]} \tag{3.66}$$

$$W_1 = \frac{\mu_0^2 \gamma_I^2 \gamma_S^2 \hbar^2}{80\pi^2 r^6} \frac{\tau_c S(S+1)}{(1 + \omega_I^2 \tau_c^2)} \tag{3.67}$$

$$W_2 = \frac{\mu_0^2 \gamma_I^2 \gamma_S^2 \hbar^2}{20\pi^2 r^6} \frac{\tau_c S(S+1)}{[1 + (\omega_I + \omega_S)^2 \tau_c^2]} \tag{3.68}$$

where the subscript on W refers to the change in the total spin quantum number when the transition described by W occurs. For example W_0 indicates a mutual spin flip for I and S, from $\alpha\beta$ to $\beta\alpha$ or vice versa. If the spin system S is saturated then the intramolecular dipole-dipole relaxation time for I, T_{1dd}(intra), is given as,

$$T_{1dd}^{-1}(\text{intra}) = W_0 + 2W_1 + W_2 \tag{3.69}$$

If I and S belong to the same species then Eq. 3.69 is not relevant, but it can be shown that:

$$T_{1dd}^{-1}(\text{intra}) = \frac{\mu_0^2 \gamma_I^4 \hbar^2 I(I+1)}{40\pi^2 r^6} \left[\frac{\tau_c}{1 + \omega_I^2 \tau_c^2} + \frac{4\tau_c}{1 + 4\omega_I^2 \tau_c^2} \right] \tag{3.70}$$

A similar expression is obtained for $1/T_{2dd}$(intra):

$$T_{2dd}^{-1}(\text{intra}) = \frac{\mu_0^2 \gamma_I^4 \hbar^2 I(I+1)}{80\pi^2 r^6} \left[3\tau_c + \frac{5\tau_c}{1 + \omega_I^2 \tau_c} + \frac{2\tau_c}{1 + 4\omega_I^2 \tau_c^2} \right] \tag{3.71}$$

Hence in the general case T_{1dd}(intra) and T_{2dd}(intra) are different and both are frequency dependent, with a minimum near the condition $\omega_I \tau_c = 1$. Under the extreme narrowing condition the frequency dependence vanishes and Eq. 3.70 and 3.71 are identical:

$$T_{1dd}^{-1}(\text{intra}) = T_{2dd}^{-1}(\text{intra}) = \frac{\mu_0^2 \gamma_I^4 \hbar^2 I(I+1)\tau_c}{8\pi^2 r^6} \tag{3.72}$$

When I and S belong to different nuclear species then by substitution from Eq. 3.66 to 3.68, Eq. 3.72 becomes[98]:

$$T_{1dd}^{-1}(\text{intra}) = \frac{\mu_0^2 \gamma_I^2 \gamma_S^2 \hbar^2 S(S+1)\tau_c}{12\pi^2 r^6} \tag{3.73}$$

In the more general case of dipolar interactions with several S spins, Eq. 3.73 becomes:

$$T_{1dd}^{-1}(\text{intra}) = \frac{\mu_0^2 \gamma_I^2 \gamma_S^2 \hbar^2 S(S+1)\tau_c}{12\pi^2 \sum_i r_{IS_i}^6} \tag{3.74}$$

Due to the inverse sixth power dependence upon internuclear separation, nuclei which are directly bonded to other dipolar nuclei are much more efficiently relaxed by this process than are those bonded to non-magnetic nuclei. For example, in specifically ^{13}C enriched neopentane, $^{13}C(^{12}CH_3)_4$, the ^{13}C nucleus has a T_1 value of about 1 minute whereas in adamantane the ^{13}CH and $^{13}CH_2$ ^{13}C relaxation times are 20.5 and 11.4 s respectively.

Since T_{1dd}(intra) is controlled by the rotational correlation time it is expected to increase at higher temperatures and decrease in the presence of intermolecular association as indicated by Eq. 3.63. Carbon-13 data show the expected temperature dependence for some carboxylic acids and methyl esters.[81] For the ethylenic carbons in fumaric acid the values of T_1 are 2.0, 1.8 and 0.81 s at concentrations of 0.01 M, 0.05 M and 3.0 M in d_6-DMSO respectively. The dramatic decrease in T_1 for the 3 M solution probably denotes an increase in intermolecular association although an increase in viscosity also occurs for the 3 M solution.[81]

Equations 3.70 to 3.74 are only valid for isotropic molecular motion. If preferred rotation about a given axis occurs (a situation which includes the case where internal rotation is important) then the rotational diffusion is not described by a single value of τ_c, and Eq. 3.70 to 3.74 require modification.[82] Examples of anisotropic rotational diffusion and its effect on ^{13}C NMR spectra are provided by some monosubstituted benzenes.[83] Rotation around the C_2 axis of these molecules is usually appreciably faster than around axes perpendicular to it. Since the former motion does not influence relaxation of the *para* ^{13}C nucleus its value of T_1 is shorter than those of the *ortho* and *meta* ^{13}C nuclei.

b. Intermolecular Interactions

When the two interacting spins belong to different molecules then their intermolecular dipole-dipole interactions cause relaxation via the translational motions of the molecules. The resulting relaxation time, T_{1dd}(inter) is given by:[84]

$$T_{1dd}^{-1}(\text{inter}) = \frac{\mu_0^2 N_S \gamma_I^2 \gamma_S^2 \hbar^2 S(S+1)}{90\pi D a} \tag{3.75}$$

where N_S is the concentration of spins per unit volume, a is the distance of closest approach and D is the mutual translational self-diffusion constant of the molecules containing I and S.

The dependence of T_{1dd}(inter) on temperature and viscosity is shown by a comparison of Eq. 3.64 and 3.75 to be the same as T_{1dd}(intra) whereas the dependence upon intermolecular separation appears to be much less critical than it is for T_{1dd}(intra).

c. *Nuclear Overhauser Effect*

Experimentally the NOE is determined by a change in the integrated NMR intensity of the signal from a nuclear spin system when the absorption of a second spin system is saturated. The nuclei concerned may be of different species or chemically shifted spins of the same species.

The NOE operates by cross-relaxation between the two spin systems by intramolecular dipole-dipole coupling as a rule. However, in some cases intermolecular dipole-dipole coupling may be important. Scalar coupling modulated by chemical exchange (see section 3D.7) can also be a contributor.[84]

If spin I relaxes only by dipole-dipole coupling with S then the nuclear Overhauser enhancement, η_{max}, is given by:

$$\eta_{max} = \frac{\gamma_S(W_2 - W_0)}{\gamma_I(W_0 + 2W_1 + W_2)} \tag{3.76}$$

As shown by Eq. 3.66 to 3.68 η_{max} is, in general, frequency-dependent However, in the motional narrowing limit it becomes frequency-independent and Eq. 1.13 is obtained.

In practice other processes may also contribute to the relaxation of I. Thus η_{max} represents the maximum NOE enhancement obtainable in any given case. The experimental value of the NOE, η, may be related to η_{max} by:

$$\eta = \frac{\eta_{max} T_1}{T_{1dd}} \tag{3.77}$$

where T_1 is the measured spin-lattice relaxation time. Hence if $\eta < \eta_{max}$ this is an indication of the presence of non-dipolar relaxation processes. The contribution to the relaxation rate from the other processes may be given (see Eq. 1.8) as:

$$T_1^{-1} = T_{1dd}^{-1} + T_{1o}^{-1} \tag{3.78}$$

where T_{1o} arises from the non-dipolar mechanisms such as spin-rotation and nuclear screening anisotropy.

For nuclei with negative magnetogyric ratios, dipole-dipole relaxation produces a negative NOE. When $\eta = -1$ signal nulling occurs. This can be a particular hazard in ^{15}N NMR spectroscopy,[85,86] for example in the case of 2-methyl-2-nitro-1,3-propanediol.[86]

d. *Paramagnetic Interactions*

The presence of paramagnetic centres induces shifts and nuclear spin relaxation.[87] At the motional narrowing limit the spin-lattice relaxation time becomes:

$$T_1^{-1} = \frac{\mu_0^2 \gamma_I^2 \gamma_S^2 \hbar^2 S(S+1)\tau_c}{12\pi^2 r^6} + \frac{\mu_0^2 \gamma_S^2 a_N^2 S(S+1)\tau_e}{24\pi^2} \tag{3.79}$$

The first term in on the RHS of Eq. 3.79 represents the dipolar contribution and the second term corresponds to the contact contribution. I and S refer to the nuclear and electronic spins respectively, a_N is the nuclear-electron hyperfine interaction constant and τ_c and τ_e are correlation times related to those characterising the rotational motion of the separation vector between the nucleus and electron, the electron spin-lattice relaxation and any chemical exchange process occurring.[87]

Comparison with Eq. 3.70 and 3.72 shows that paramagnetic centres not only produce enhanced nuclear relaxation by means of larger magnetogyric ratios but also through the contact interaction. Consequently paramagnetic centres are frequently used to reduce very long spin-lattice relaxation times as well as to remove the NOE in cases where it is not required. This is achieved by over-riding the nuclear dipole-dipole relaxation process. It is particularly useful for nulled signals such as may occur with nuclei having negative magnetogyric ratios, for example ^{15}N and ^{29}Si.[85,86]

3D.4 Relaxation by Electric Quadrupole Interactions

Any nucleus with $I > \frac{1}{2}$ has an electric quadrupole moment in addition to its magnetic dipole moment. The interaction of a nuclear quadrupole moment, eQ, with an electric field gradient, eq, provides a very efficient process for nuclear relaxation via the molecular rotation.

Although this interaction is electrical, rather than magnetic, in nature its general features may be described by Eq. 3.61. In the motional narrowing limit the appropriate longitudinal and transverse relaxation rates, T_{1q}^{-1} and T_{2q}^{-1} are identical, and are given by Eq. 1.14, which involves a nuclear quadrupole coupling constant and a field gradient asymmetry, defined by Eq. 1.15 and 1.16 respectively.

As a rule the relaxation rates of nuclei with $I > \frac{1}{2}$ are dominated by the quadrupolar mechanism, the resulting relaxation times being very short. As shown by Eq. 1.14 η and χ play major roles in determining T_{1q}. Since both of these parameters depend upon the electronic environment of the nucleus it follows that changes in this environment will be reflected in the quadrupolar relaxation time. For example T_1 of ^{14}N ($I = 1$) is 22 ms for MeCN whereas it is in excess of 50 s for the tetrahedral ammonium ion in aqueous solution. For MeCN the large electric field gradient at the nitrogen nucleus produces rapid quadrupolar relaxation whereas in the ammonium ion the field gradient is effectively zero and the nitrogen relaxation is controlled by the (N,H) dipole-dipole interaction. The effect of variation in the electric field gradient on the ^{14}N linewidth is demonstrated in Fig. 3.8.[88]

Since T_{1q} depends upon the rotational correlation time its relation to temperature and viscosity is the same as for T_{1dd}.

3D.5 Relaxation by Spin-rotation Interactions

Whereas the transfer of nuclear spin energy to the molecular rotation is indirect by the dipole-dipole and quadrupole mechanisms, in the case of spin-rotation interactions it is direct. As a consequence of this the nuclear relaxation rate $(T_{1sr})^{-1}$ is directly proportional to temperature;[80] in this sense it is unique amongst the relaxation mechanisms under consideration. Assuming that the motion is isotropic:

$$T_{1sr}^{-1} = \frac{2IkTC^2\tau_{sr}}{3\hbar^2} \qquad (3.80)$$

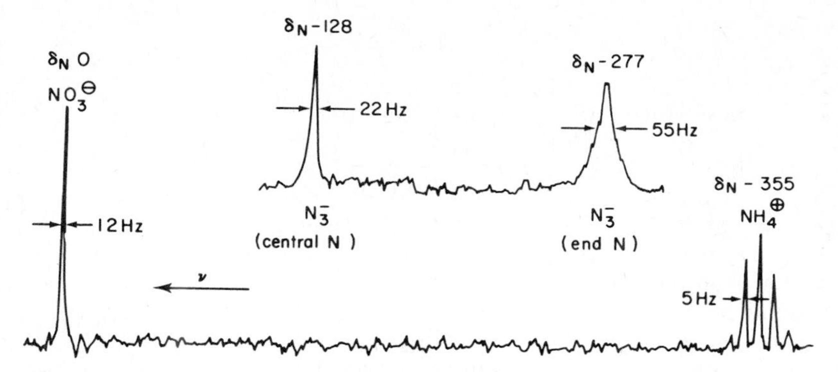

Fig. 3.8 Nitrogen-14 NMR spectra (at 7.226 MHz) of aqueous azide, with aqueous acidified ammonium nitrate as external reference.[88]

where I represents the molecular moment of inertia, τ_{sr} is given by equation 3.65 and C^2 is the average of the square of the spin-rotation tensor. The value of C^2 has been measured in only a few cases and mostly for diatomic molecules. However, in some instances C^2 may be estimated from chemical shift measurements.[89]

It may be anticipated from Eq. 3.80 that spin-rotation relaxation will be most efficient for small, rapidly tumbling molecules at elevated temperatures. Thus this mechanism is particularly important in low viscosity liquids and the vapour phase.

Both the spin-rotation and the dipole-dipole interactions may be operative in relaxing spin one-half nuclei. When they occur simultaneously the dipole-dipole mechanism will tend to dominate at lower temperatures while the spin-rotation process is more important at higher temperatures. Thus T_1 will have a maximum at an intermediate temperature. This has been observed, for example, in the ^{13}C relaxation of methyl iodide.[90]

Spin-rotation relaxation is important for the ^{19}F and ^{31}P nuclei in fairly symmetrical molecules. For example the ^{19}F relaxation times of SF_6 ($T_1 = 0.5$ s)[91] and MoF_6 ($T_1 = 0.85$ s)[92] are most probably dominated by this process. A similar situation exists for the ^{31}P relaxation of PCl_3 ($T_1 \sim 1.5$ s)[93] and $POCl_3$ ($T_1 \sim 5$ s).[93]

The ^{31}P relaxation times of several organo-phosphorous compounds have major contributions from spin-rotation interactions.[94,95] For example at 29°C the NOE, η, of ^{31}P in triethylphosphine is 0.2 and it increases along with T_1 at lower temperatures. This shows that at 29°C the spin-rotation mechanism dominates the ^{31}P relaxation process, becoming less important as the temperature is decreased.[94]

For ^{13}C nuclei spin-internal rotation is often an important relaxation mechanism for freely rotating groups such as Me and CF_3. The large molecule, to which these groups belong, need not be rotating rapidly for spin-rotation to be a significant relaxation process.

3D.6 Relaxation by Nuclear Screening Anisotropy

Modulation of the nuclear screening tensor by Brownian motion provides a fluctuating magnetic field at the nucleus. Thus, an anisotropic screening tensor may act as a source of nuclear spin relaxation dependent upon the rotational correlation time. The contributions from the screening anisotropy to T_{1sa} and T_{2sa} are given by:

$$T_{1sa}^{-1} = \frac{\mu_0}{60\pi} \gamma_I^2 B_0^2 \Delta\sigma^2 \left[\frac{2\tau_c}{1 + \omega_I^2 \tau_c^2}\right] \tag{3.81}$$

and:

$$T_{2sa}^{-1} = \frac{\mu_0}{360\pi} \gamma_I^2 B_0^2 \Delta\sigma^2 \left[\frac{6\tau_c}{1 + \omega^2 \tau_c^2} + 8\tau_c \right] \tag{3.82}$$

Thus in the general case these contributions to T_1 and T_2 are frequency dependent. In the motional narrowing limit this vanishes but $T_1 \neq T_2$; since:

$$T_{1sa}^{-1} = \frac{\mu_0 \gamma_I^2 B_0^2 \Delta\sigma^2 \tau_c}{30\pi} \tag{3.83}$$

and:

$$T_{2sa}^{-1} = \frac{7\mu_0 \gamma_I^2 B_0^2 \Delta\sigma^2 \tau_c}{180\pi} \tag{3.84}$$

comparison of Eq. 3.83 and 3.84 show that at this limit T_{1sa}: $T_{2sa} = 7:6$.

The dependence of T_{1sa} and T_{2sa} upon the applied magnetic field is noteworthy and provides a means of detecting the presence of this relaxation process when other, more efficient, ones are also operating at the motional narrowing limit.

It is expected that nuclear screening anisotropy will be most effective as a relaxation mechanism at low temperatures and high magnetic fields. Few unambiguous cases of the importance of this mechanism for nuclear spin relaxation have been reported. However, the ^{19}F relaxation of $CHFCl_2$ at $-142°C$ and 56.4 MHz does appear to have a significant contribution from this process.[96] Carbon-13 relaxation time data of hen egg-white lysozyme taken at 6.34 T and 1.42 T demonstrate the dominance of nuclear screening anisotropy as a relaxation mechanism for the nonprotonated carbons at field strengths above 4 T, whereas the protonated carbons appear to be dominated by (^{13}C, ^{1}H) dipolar relaxation.[97]

3D.7 Relaxation by Scalar Coupling

If the spin-spin coupling constant for the interaction between the two nuclei becomes time dependent as a result of chemical exchange or internal rotation nuclear relaxation can be induced. This is referred to as *scalar relaxation of the first kind* and its presence is usually predictable from a knowledge of the chemistry of the sample under investigation.

Similarly, if the relaxation rate of nucleus S is fast compared to $2\pi J$ then *scalar relaxation of the second kind* may occur for nucleus I. This often arises if nucleus S is quadrupolar. The resulting scalar coupling contribution to T_1 is given as:[98]

$$T_{1sc}^{-1} = \tfrac{8}{3}\pi^2 J^2 S(S+1) T_1^S [1 + (\omega_I - \omega_S)^2 (T_1^S)^2]^{-1} \tag{3.85}$$

where T_1^S refers to the relaxation time of nucleus S. A similar expression is available for the scalar coupling contribution to T_2:

$$T_{2sc}^{-1} = \tfrac{1}{2} T_{1sc}^{-1} + \tfrac{4}{3}\pi^2 J^2 S(S+1) T_1^S \tag{3.86}$$

Equations 3.85 and 3.86 become frequency independent if the motional narrowing limit is reached.

The scalar coupling contribution to T_2 can be important when $(T_1^S)^{-1}$ is comparable to $2\pi J$. An example of this is afforded by the broad resonances observed for protons which are spin-spin coupled to quadrupolar nuclei such as ^{14}N.

To date there appear to be very few examples in which scalar coupling makes a significant contribution to T_1. In PBr_3 the ^{79}Br and ^{81}Br nuclei are reported to induce scalar coupling relaxation of the ^{31}P nucleus, although spin-rotation relaxation is also important.[99] The value of T_1 for the quaternary carbon in *p*-bromobenzonitrile is found to be 47 s whereas the carbon bonded to bromine exhibits a value of 8.8 s. Scalar coupling has been ascribed as the most important relaxation mechanism for the latter carbon.[100]

The possibility of relaxation via the coupling of the anisotropy of J and molecular motion should also be considered. However, in practice, this anisotropy tends to be 1000 times or more smaller than the dipole-dipole interaction; thus it may be discounted as a likely source of nuclear spin relaxation.

3D.8 Separation of Contributions from various Relaxation Mechanisms

From a knowledge of the chemistry of the system under investigation it is usually apparent whether quadrupolar or paramagnetic interactions are present. If either of them are then it customarily dominates the nuclear spin relaxation.

Since relaxation by scalar coupling makes a significant contribution in only very few cases, it is normally considered to be absent.

As demonstrated by Eq. 3.77 and 3.78, measurement of the NOE and T_1 serves to separate the dipole-dipole contribution from those due to spin-rotation and screening anisotropy interactions. In which case:

$$T_{1o}^{-1} = T_{1sr}^{-1} + \left(\frac{\mu_0}{30\pi}\right)\gamma_I^2 B_0^2 \Delta\sigma^2 \tau_c \qquad (3.87)$$

Thus by measuring T_1 at two different magnetic field strengths, and subtracting the two values of T_{1o}^{-1} obtained, a value for $\Delta\sigma$ is found. This can be used together with Eq. 3.83 in order to determine the T_{1sa}^{-1} contribution and thus the contribution from T_{1sr}^{-1} by means of Eq. 3.87. As noted previously, the spin-rotation mechanism is also identifiable from its temperature dependence (section 3D.5).

Experiments of this type have been performed on the ^{15}N nuclei in nitrobenzene and pyridine.[101,102] At low frequencies spin-rotation interactions dominate the ^{15}N relaxation of nitrobenzene and at lower temperatures intermolecular dipole-dipole interactions become important.[101] At 32 MHz the predominant relaxation process is due to nuclear screening anisotropy. For pyridine screening anisotropy and intermolecular dipole-dipole interactions dominate at low temperatures whereas spin-rotation interactions become more important at higher temperatures.[102]

REFERENCES

1. K. A. K. Ebraheem and G. A. Webb, *Prog. NMR Spectry.* **9** (1977); **11**, 149 (1977)
2. W. T. Raynes, in "Specialist Periodical Reports on NMR", Vol. 3, p. 1. Chemical Society, London (1974)
3. N. F. Ramsey, *Phys. Rev.* **78**, 689 (1950)
4. H. F. Hameka, *J. Chem. Phys.* **40**, 3127 (1964)
5. L. C. Snyder and R. G. Parr, *J. Chem. Phys.* **34**, 827 (1961)
6. R. Ditchfield, D. P. Miller and J. A. Pople, *J. Chem. Phys.* **54**, 4186 (1971)
7. G. P. Arrighini, M. Maestro and R. Moccia, *Chem. Phys. Letters* **7**, 351 (1970)
8. B. R. Appleman, T. Tokuhiro, G. Fraenkel and C. W. Kern, *J. Chem. Phys.* **60**, 2574 (1974)
9. T. D. Gierke and W. H. Flygare, *J. Amer. Chem. Soc.* **94**, 7277 (1972)
10. S. G. Kukolich and S. C. Wofsky, *J. Chem. Phys.* **52**, 5477 (1970)

11. D. F. O'Reilly, *Prog. NMR Spectry.* **2**, 1 (1967)
12. S. I. Chan and T. P. Das, *J. Chem. Phys.* **37**, 1527 (1962)
13. A. Okninski and A. J. Sadlej, *Mol. Phys.* **26**, 1545 (1973)
14. J. A. Pople, *Discuss. Faraday Soc.* **34**, 7 (1962)
15. J. A. Pople, *J. Chem. Phys.* **37**, 53 (1962)
16. J. A. Pople, *J. Chem. Phys.* **37**, 60 (1962)
17. J. A. Pople, *Mol. Phys.* **7**, 301 (1964)
18. A. Saika and C. P. Slichter, *J. Chem. Phys.* **22**, 26 (1954)
19. H. M. McConnell, *J. Chem. Phys.* **27**, 226 (1957)
20. P. D. Ellis, G. E. Maciel and J. W. McIver, *J. Amer. Chem. Soc.* **94**, 4069 (1972)
21. R. Ditchfield and P. D. Ellis, in "Topics in ^{13}C NMR Spectroscopy", Vol. 1, p. 1, (G. C. Levey, ed.). Wiley, New York (1974)
22. K. A. K. Ebraheem, M. Jellali-Hervai and G. A. Webb, Unpublished results
23. R. Ditchfield, *Mol. Phys.* **27**, 789 (1974)
24. C. J. Jameson and H. S. Gutowsky, *J. Chem. Phys.* **40**, 1714 (1964)
25. J. C. Slater, *Phys. Rev.* **36**, 57 (1930)
26. G. Burns, *J. Chem. Phys.* **41**, 1521 (1964)
27. K. A. K. Ebraheem, G. A. Webb and M. Witanowski, *Org. Magn. Reson.* **8**, 317 (1976)
28. G. Malli and C. Froese, *Intl. J. Quant. Chem.* **1**, 95 (1967)
29. K. A. K. Ebraheem and G. A. Webb, *J. Magn. Reson.* **25**, 399 (1977)
30. K. A. K. Ebraheem and G. A. Webb, *Org. Magn. Reson.* **9**, 241 (1977)
31. K. A. K. Ebraheem and G. A. Webb, *Org. Magn. Reson.* **9**, 248 (1977)
32. J. M. Letcher and J. R. Van Wazer, in "Topics in Phosphorus Chemistry", Vol. 5, p. 75 (M. Grayson and E. J. Griffiths, eds.) Interscience, New York (1967)
33. R. Wolff and R. Radeglia, *Z. Phys. Chem. Leipzig* **257**, 181 (1976); **258**, 145 (1977)
34. W. H. Flygare and J. Goodisman, *J. Chem. Phys.* **49**, 3122 (1968)
35. H. Fukui, *Bull. Chem. Soc. Japan* **47**, 751 (1974)
36. G. E. Maciel, J. L. Dallas, R. L. Elliott and H. C. Dorn, *J. Amer, Chem. Soc.* **95**, 5857 (1973)
37. M. Kondo, I. Ando, R. Chujo and A. Nishioka, *J. Magn. Reson.* **24**, 315 (1976)
38. W. H. Flygare, *Chem. Rev.* **74**, 653 (1974)
39. N. F. Ramsey, *Phys. Rev.* **91**, 303 (1953)
40. H. Nakatsuji, H. Kato, I. Morishima and T. Yonezawa, *Chem. Phys. Letters* **4**, 607 (1970)
41. J. N. Murrell, *Prog. NMR Spectry.* **6**, 1 (1970)
42. M. Barfield and B. Chakraborti, *Chem. Rev.* **69**, 757 (1969)
43. R. Grinter, in "Specialist Periodical Reports on NMR", Vols 1-5. Chemical Society, London (1972)–(1976)
44. J. A. Pople and D. P. Santry, *Mol. Phys.* **8**, 1 (1964)
45. J. T. Lallemand and M. Duteil, *Org. Magn. Reson.* **8**, 328 (1976)
46. A. D. C. Towl and K. Schaumburg, *Mol. Phys.* **22**, 49 (1971)
47. R. Ditchfield and J. N. Murrell, *Mol. Phys.* **14**, 481 (1968)
48. R. Ditchfield, *Mol. Phys.* **17**, 33 (1969)
49. K. Hirao, H. Nakatsuji, H. Kato and T. Yonezawa, *J. Amer. Chem. Soc.* **94**, 4078 (1972)
50. H. M. McConnell, *J. Chem. Phys.* **24**, 460 (1956)
51. M. Barfield, *J. Chem. Phys.* **48**, 4458 (1968)
52. M. Barfield, *J. Chem. Phys.* **48**, 4463 (1968)
53. J. A. Pople, J. W. McIver and N. S. Ostlund, *J. Chem. Phys.* **49**, 2960 (1968)
54. J. A. Pople, J. W. McIver and N. S. Ostlund, *J. Chem. Phys.* **49**, 2965 (1968)
55. A. C. Blizzard and D. P. Santry, *J. Chem. Phys.* **55**, 950 (1971)
56. I. Brown and D. W. Davies, *Chem. Phys. Letters* **37**, 132 (1976)
57. P. Lazzeretti, F. Taddei and R. Zanasi, *J. Amer. Chem. Soc.* **98**, 7989 (1976)
58. J. M. Schulman and M. D. Newton, *J. Amer. Chem. Soc.* **96**, 6295 (1974)
59. J. M. Schulman and T. Venanzi, *J. Amer. Chem. Soc.* **98**, 4701 (1976)
60. G. A. Gray and T. A. Albright, *J. Amer. Chem. Soc.* **98**, 3857 (1976)
61. R. K. Safiullin, Y. Y. Samitov and R. M. Aminova, *Teor. Eksp. Khim.* **10**, 829 (1974)
62. R. K. Safiullin, R. M. Aminova and Y. Y. Samitov, *Zhur. Strukt. Khim.* **16**, 42 (1975)

63. J. P. Allbrand, H. Faucher, D. Gagnaire and J. B. Robert, *Chem. Phys. Letters* **38**, 521 (1976)
64. H. Nakatsuji, K. Hirao, H. Kato and T. Yonezawa, *Chem. Phys. Letters* **6**, 541 (1970)
65. J. Bulthius and C. M. Maclean, *J. Magn. Reson.* **4**, 148 (1971)
66. C. W. Haigh and S. Sykes, *Chem. Phys. Letters* **19**, 571 (1973)
67. J. M. Schulman and T. Venanzi, *J. Amer. Chem. Soc.* **98**, 6739 (1976)
68. T. A. Albright, *Org. Magn. Reson.* **8**, 489 (1976)
69. P. D. Ellis and G. E. Maciel, *J. Amer. Chem. Soc.* **92**, 5829 (1970)
70. P. D. Ellis and G. E. Maciel, *Mol. Phys.* **20**, 433 (1971)
71. G. E. Maciel, J. W. McIver, N. S. Ostlund and J. A. Pople, *J. Amer. Chem. Soc.* **92**, 1 (1970)
72. M. Bacon and G. E. Maciel, *Mol. Phys.* **21**, 257 (1971)
73. W. F. Reynolds, I. R. Peat and G. K. Hamer, *Canad. J. Chem.* **52**, 3415 (1974)
74. M. Barfield and M. Karplus, *J. Amer. Chem. Soc.* **91**, 1 (1969)
75. A. D. Buckingham and J. E. Cordle, *J. C. S. Faraday II* **70**, 994 (1974)
76. J. Hilton and L. H. Sutcliffe, *J. C. S. Faraday II* **71**, 1395 (1975)
77. T. Schaefer, K. Chum, K. Marat and R. E. Wasylishen, *Canad. J. Chem.* **54**, 800 (1976)
78. C. P. Slichter, "Principles of Magnetic Resonance", p. 153. Harper, New York (1963)
79. W. T. Huntress, *Adv. Magn. Reson.* **4**, 2 (1970)
80. P. S. Hubbard, *Phys. Rev.* **131**, 1155 (1963)
81. G. C. Levy and D. Terpstra, *Org. Magn. Reson.* **8**, 658 (1976)
82. D. E. Woessner, *J. Chem. Phys.* **36**, 1 (1962); D. E. Woessner, B. S. Snowden and G. H. Meyer, *J. Chem. Phys.* **50**, 719 (1969)
83. G. C. Levy, J. D. Cargioli and F. A. L. Anet, *J. Amer. Chem. Soc.* **95**, 1527 (1973)
84. J. H. Noggle and R. E. Schirmer, "The Nuclear Overhauser Effect", Academic, New York, (1971)
85. M. Witanowski, L. Stefaniak and G. A. Webb, Ann. Repts. NMR Spectry. **7**, 118 (1977)
86. G. C. Levy, C. E. Holloway, R. C. Rosanske, J. M. Hewitt and C. H. Bradley, *Org. Magn. Reson.* **8**, 643 (1976)
87. G. A. Webb, *Ann. Repts. NMR Spectry.* **6A**, 1 (1975)
88. N. Logan, in "Nitrogen NMR", p. 319 (M. Witanowski and G. A. Webb, eds). Plenum, London (1973)
89. C. Deverell, *Mol. Phys.* **18**, 319 (1970)
90. K. T. Gillen, M. Schwartz and J. H. Noggle, *Mol. Phys.* **20**, 899 (1970)
91. W. S. Hackleman and P. S. Hubbard, *J. Chem. Phys.* **39**, 2688 (1963)
92. P. Rigney and J. Virlet, *J. Chem. Phys.* **47**, 4645 (1967)
93. D. W. Aksnes, M. Rhodes and J. G. Powles, *Mol. Phys.* **14**, 333 (1968)
94. N. J. Koole, A. J. De Koning and M. J. A. De Bie, *J. Magn. Reson.* **25**, 375 (1977)
95. R. K. Harris and E. M. McVicker, *J. C. S. Faraday II* **72**, 2291 (1976)
96. E. L. Mackor and C. Maclean, *Prog. NMR Spectry.* **3**, 152 (1967)
97. R. S. Norton, A. O. Clause, R. Addleman and A. Allerhand, *J. Amer. Chem. Soc.* **99**, 79 (1977)
98. A. Abragam, "The Principles of Nuclear Magnetism", Ch. 9. Oxford University Press (1961)
99. M. Rhodes, D. W. Aksnes and J. H. Strange, *Mol. Phys.* **15**, 541 (1968)
100. R. Freeman and H. D. W. Hill, "Molecular Spectroscopy", Institute of Petroleum, London (1971)
101. D. Schweitzer and H. W. Spiess, *J. Magn. Reson.* **16**, 243 (1974)
102. D. Schweitzer and H. W. Spiess, *J. Magn. Reson.* **15**, 529 (1974)

4

THE COMMON NUCLEI

BRIAN E. MANN, University of Sheffield, England

4A INTRODUCTION

A detailed discussion of the NMR properties of the common NMR nuclei, i.e. ^{1}H, ^{10}B, ^{11}B, ^{13}C, ^{14}N, ^{15}N, ^{19}F, and ^{31}P, has been omitted from this book as there are many other sources of information available. The purpose of this chapter is three-fold: firstly to provide typical data on the common nuclei as a back-cloth for the rest of the book, secondly to direct the reader towards these other sources of information, and thirdly to discuss referencing for each nucleus.

For most elements, compounds with p_π-p_π double bonding and catenation are unusual or unknown. The majority of NMR data quoted in this book depends on the nature of the hetero atoms attached to the nucleus being observed, and consequently emphasis is placed in this chapter on the dependence on heteroatoms while little attention is placed on compounds, e.g. benzene, which do not have analogues among the heavier elements. The discussion of coupling constants is restricted to the common nuclei, as coupling constants involving the less common nuclei will be discussed with these nuclei in the later chapters of this book. Within this chapter the convention is adopted that coupling between two different nuclei will be included in the section on the heavier nucleus. Thus $^{1}J(^{13}C, ^{11}B)$ is included in the carbon section, (see ch. 4D.2).

No attempt is made to catalogue all the books and reviews that have been published on the common nuclei as this is done on an annual basis in Specialist Periodical Reports on NMR,[1] but a guide is given to the major texts in each area. Unfortunately the wealth of information published on these nuclei has deterred the compilation of recent comprehensive or even representative tabulations of data, except in the case of nitrogen and fluorine. In other cases where comprehensive tabulations exist, they are now out of date, e.g. ^{31}P chemical shifts up to 1969.[2]

Even for the common nuclei, referencing and sign convention can provide serious problems. There is now reasonable agreement as to what reference to use for the common nuclei, but problems still arise. With the exception of Me_4Si for ^{1}H where changes are small, the shifts of the references are dependent on solvent and temperature, but it is rare to find conditions of use strictly defined. When an external reference is used, it is rare to find a paper where it is stated whether or not a susceptibility correction has been made. At present sign convention also provides a problem for ^{11}B, ^{14}N, ^{19}F, and ^{31}P NMR data, and much of the data in subsequent chapters. Originally low frequency was taken as being positive. Subsequently IUPAC recommended[3] the reverse convention, high frequency positive, as used in this book. The IUPAC convention has been adopted by many authors, but at present there are many papers appearing where the sign convention is not stated, and it is left to the reader to decide whether a given shift is to high or low frequency of the reference. It is to be hoped that the correct guesses have been made in accumulating the data for the tables.

4B HYDROGEN, ^{1}H

Although hydrogen possesses three nuclei of interest to the NMR spectroscopist, ^{1}H, ^{2}H, and ^{3}H, by far the greatest attention has been paid to ^{1}H NMR spectroscopy, virtually to the exclusion of the ^{2}H and ^{3}H nuclei. Consequently ^{2}H and ^{3}H are not considered to be common nuclei and are discussed separately, see ch.5.

Of the spectroscopic tools available to the preparative chemist, ^{1}H NMR spectroscopy is arguably the most useful. Hydrogen is present in the vast majority of compounds, often at strategic positions. Consequently ^{1}H NMR spectroscopy will often yield information about the structure of molecules. It is therefore now standard to measure the ^{1}H NMR spectrum of any new diamagnetic hydrogen containing compound. Each year there is a mass of information published, with the result that now there is no comprehensive listing of data. The last reasonable comprehensive listing of organic compounds[4] appeared over 10 years ago and even that is very difficult to use with a plethora of reference compounds. For the organic chemist, the situation is, however, quite good with a number of compilations of data with the hydrogen atom in most conceivable situations.[5] For the inorganic chemist there is no general compilation of data, although there are a number of reviews on specific aspects which include specialised compilations of data.

4B.1 ^{1}H Chemical Shifts

The proton chemical shift scale is frequently quoted as extending from $\delta 0$, defined as the shift of internal Me_4Si, to $\delta 10$. For most compounds this is indeed true, however, diamagnetic compounds are known with chemical shifts to as low a frequency as $\delta -50$ and as high as $\delta 20$. If discussion is extended to paramagnetic compounds then $\delta -500$ to $\delta 500$ would be nearer the truth. For the purposes of this book, discussion will be restricted to diamagnetic compounds and readers interested in paramagnetic compounds are referred to a recent book[6] on the subject.

In general the nuclear screening may be split into two components

$$\sigma = \sigma_{\text{intra}} + \sigma_{\text{inter}} \tag{4.1}$$

where σ_{intra} is the intraatomic screening as discussed in ch. 3B, while σ_{inter} is the interatomic screening. The σ_{inter} arises from near neighbour anisotropy effects. σ_{inter} is generally less than 2 ppm but in exceptional cases e.g. the transition metal hydrides it may be as large as 50 ppm. For all other elements, it is customary to neglect σ_{inter} as σ_{intra} is large in comparison although this may not be true for atoms attached to a transition metal. In the case of ^{1}H NMR spectroscopy as σ_{inter} is comparable to σ_{intra} it cannot be neglected and causes an extra source of information and complication in ^{1}H chemical shifts.

TMS is now accepted as the reference for ^{1}H NMR spectroscopy. It is solvent[7] and temperature dependent,[8] but the effects are generally small and for most purposes can be neglected. As it is insoluble in water, a derivative $[Me_3SiCH_2CH_2CH_2SO_3]^- Na^+$ is used for aqueous solutions.

Some ^{1}H chemical shifts are given in Table 4.1 for hydrogen attached to various hetero-atoms. Considerable caution is necessary in using these data as in many cases

TABLE 4.1 A comparison of ^{1}H, ^{11}B, ^{13}C, $^{14,15}N$, ^{19}F and ^{31}P chemical shifts as a function of substituent. Reference compounds used are TMS (^{1}H), $Et_2O \cdot BF_3$ (^{11}B), TMS (^{13}C), $[NO_3]^-$/$MeNO_2$ ($^{14,15}N$), $CFCl_3$ (^{19}F) and 85% H_3PO_4 (^{31}P).

In each case high frequency is positive. Note that many of the quoted shifts are medium dependent. This is particularly true for the ^{1}H chemical shifts where gas phase shifts are given when possible. The observed nucleus is given at the head of each column.

^{1}H	^{11}B	^{13}C	$^{14,15}N$	^{19}F	^{31}P
H_2 4.06	$[BH_4]^-$ -40	CH_4 -2.3	NH_3 -383	HF -221	PH_3 -240
$[BH_4]^-$ -2.0		BMe_3 14.8		$[BF_4]^-$ -150	
CH_4 -0.24	BMe_3 86.0	CMe_4 31.5	NMe_3 -365	CF_4 -70	PMe_3 -62
NH_3 -0.31	$B(NMe_2)_3$ 27.3	NMe_3 47.5	N_2H_4 -312	NF_3 145	$P(NMe_2)_3$ 121.5
H_2O 0.31	$B(OMe)_3$ 18.3	Me_2O 59.4	$[NO_2]^-$ 237	F_2O 248	P_4O_6 113
HF 1.85	BF_3 9.4	MeF 75.4	NF_3 -7	F_2 422	PF_3 97
$[AlH_4]^-$ 0.0				$[AlF_4]^-$ -153	
SiH_4 2.75	$Me_3SiB(NMe_2)_2$ 36.1	$SiMe_4$ 0.0		SiF_4 -174	$P(SiMe_3)_3$ -251.2
PH_3 1.21	$Et_2PB(NMe_2)_2$ 35.9	PMe_3 20.5		PF_3 -34	P_4 -461
H_2S -0.17	$B(SBu^n)_3$ 66.2	SMe_2 19.5	$[S_4N_3]^+$ -12; 0	SF_6 49	$P(SMe)_3$ 125
HCl -0.73	BCl_3 47	MeCl 25.1	ClNO 224	ClF -448	PCl_3 220
GeH_4 2.60		$GeMe_4$ -0.8		GeF_4 -178	$P(GeMe_3)_3$ -228
AsH_3 1.47				AsF_3 -41	
SeH_2 -2.31				SeF_4 64	
HBr -4.58	BBr_3 39	MeBr 10.2	BrNO 352	BrF_3 -38; -23	PBr_3 227
SnH_4 3.87	$Me_3SnB(NMe_2)_2$ 39.0	$SnMe_4$ -9.3		$[SnF_6]^{2-}$ -156	$P(SnMe_3)_3$ 330
SbH_3 1.38				SbF_3 -53	
H_2Te -7.08				TeF_4 -126	
HI -13.49	BI_3 6	MeI -20.5		IF_5 14; 59	PI_3 178
PbH_4 6.8		$PbMe_3$ -3.4			
PhH 7.27	BPh_3 60	PhMe 21.3	$NHPh_2$ -289	PhF -113.1	PPh_3 -6

the chemical shift is very dependent on conditions. It will be observed that there is a crude relationship between electronegativity of the heterogroup and the ^1H chemical shift although in many cases this is masked by near neighbour anisotropy effects. A far more marked effect is the movement to high frequency down a group e.g. CH_4 $\delta -0.24$; SiH_4 $\delta 2.75$; GeH_4 $\delta 2.60$, SnH_4 $\delta 3.87$, and PbH_4 $\delta 6.8$. The factors leading to the dependence of chemical shift on heteroatom have been discussed.[9]

4B.2 ^1H Coupling Constants

As ^1H NMR spectroscopy is so common, coupling to a wide variety of nuclei has been observed, but these coupling constants will be discussed with the heteronucleus. In this section, only $J(^1H, ^1H)$ will be discussed. Thus apart from $^1J(^1H, ^1H)$ for H_2 of 278 Hz only couplings across a heteroatom are relevant.

By far the greatest information on $^2J(^1H, ^1H)$ comes from $-XH_2$ groups where the protons are prochiral, but some information comes from using Eq.1.7 to calculate $^2J(^1H, ^1H)$ from $^2J(^2H, ^1H)$. In general $^2J(^1H, ^1H)$ is negative with -14 Hz being typical for a sp^3 hybridised CH_2, PH_2 groups and H_2S. Substituents can have a marked effect with -20.3 Hz for $CH_2(CN)_2$ and $+6.3$ for $\overline{CH_2CH(CO_2H)O}$. When the central atom is sp^2 hybridised there is a marked increase. Thus for $H_2C{=}CHX$, $^2J(^1H, ^1H)$ ranges from -3.2 Hz (X = F) to $+7.1$ Hz (X = Li) and for NH_2COR ca.2.5 Hz.

In general couplings over more than two bonds are very sensitive to molecular geometry leading to $^3J(^1H, ^1H)$ often being larger than $^2J(^1H, ^1H)$; a phenomenon also found in coupling between other nuclei. For hydrogen, and, it would appear, for a number of other nuclei, there is the very useful Karplus relationship[10] between $^3J(^1H, ^1H)$ and the dihedral angle, see Eq. 3.59. The values for the constants depends on the electronegativity of the substituents and the nature of the intermediate atoms, but once the values for the constants are established, Eq. 3.59 provides a very useful tool in the determination of the conformations of compounds in solution.

4B.3 ^1H Relaxation Studies

Extensive use has been made of ^1H relaxation measurements in the study of molecular motion and interactions. It is easy to obtain reliable relaxation times when there is only one type of proton in a molecule, although intermolecular effects can be significant. The problems arise when there are more than one type of proton in the molecule. There is then cross-relaxation between the different types of protons leading to oversimplified conclusions.[11] This problem has been solved in a number of ways:

(*i*) a detailed study of the over-all relaxation process,[11]
(*ii*) double resonance effects,[12]
(*iii*) specific deuteriation.[13]

However, it has been shown that in the absence of strong coupling, the errors[14] introduced by ignoring cross-relaxation is only of the order of 10%. The multitudinous applications of relaxation measurements have been reviewed.[15] Much work has been done on ^1H relaxation yielding information on a wide variety of molecular behaviour, e.g. correlation times, activation energies of rotation and exchange, and molecular dynamics. Even unresolved coupling may be determined. Thus $^2J(Cl, H)$ has been measured[16] as 5.25 Hz in $CHCl_3$.

On account of the difficulties of cross-relaxation coupled with the relatively recent availability of Fourier transform methods, little use has been made of ^{1}H relaxation in complex spectra. Recently some applications have been reviewed.[17] It has been found that T_1 for protons can be as long as 100 s although values between 1 and 10 s are more common. For a series of sugars, it was possible to differentiate between isomers on the basis of spin-lattice relaxation times, with values differing by over a factor of two being found. Individual spin-spin relaxation measurements have also been made.[18] A related technique is the measurement of *J*-spectra.[19] Thus for 3-bromothiophene-2-aldehyde it was possible to measure coupling constants of 0.051 Hz and 1.356 Hz.

4C BORON, $^{10,11}B$

There are two boron nuclei available for investigation, ^{10}B and ^{11}B. There is however very little interest in ^{10}B NMR spectroscopy. This arises from its lower natural abundance, 19.58 %, than 80.42 % for ^{11}B and its inherent low sensitivity. As a consequence, it has a receptivity of 22.1 relative to ^{13}C while ^{11}B has a receptivity of 754. There is therefore very little incentive in observing ^{10}B NMR spectra. From Eq. 1.18 it can be shown that the linewidth of a ^{10}B line is 0.65 the linewidth of a ^{11}B line assuming quadrupole relaxation domination. However this gain is not reflected as better resolution with the separation of resonances and coupling constants being reduced by $\gamma(^{10}B): \gamma(^{11}B) = 0.335$. It is possible that in cases among the boron hydrides and carboranes where broadening due to coupling is important that $^{10}B\{^{1}H, ^{11}B\}$ spectroscopy will produce better resolution than $^{11}B\{^{1}H\}$ spectroscopy. So far the major practical use of ^{10}B NMR spectroscopy has involved Eq.1.17 and the linewidths to separate quadrupolar and coupling broadening.

The reviews of boron NMR spectroscopy are out of date,[20] but since 1968, a complete listing of all papers published on boron NMR spectroscopy has been available.[21]

4C.1 $^{10,11}B$ Chemical Shifts

Although in principle ^{10}B and ^{11}B can have different chemical shifts, this is not likely to cause much practical concern. Thus for the pair of ions $[BH_4]^-$ and $[BF_4]^-$ there is a primary isotope effect[22] of 0.11 ± 0.03 ppm. For the vast majority of boron compounds 0.11 ppm is negligible compared to the line width and ^{10}B and ^{11}B chemical shifts may for most purposes be considered to be identical. $Et_2O \cdot BF_3$ is generally accepted as the external reference for boron NMR spectroscopy although a number of other references, e.g. $B(OMe)_3$, and $[BH_4]^-$ have been used. There appear to have been no checks on the reliability of this reference and consequently discrepancies in measured chemical shifts are not unusual. The shift of $Et_2O \cdot BF_3$ relative to TMS does not appear to have been measured but Ξ for $[^{10}BH_4]^-$ is 10 743 230 Hz and for $[^{11}BH_4]^-$ is 32 082 695 Hz. There is a considerable mass of information on the chemical shifts of boron derivatives. For simple boron compounds, there is a close relationship between their chemical shifts and the ^{13}C chemical shifts of the isoelectronic species.[23] It has been shown that ^{11}B chemical shifts are sensitive to the electronegativity and π-bonding character of the ligand.[24] Thus for a series of compounds R_nBF_{3-n}, a good linear relationship has been found between $\delta(^{11}B)$ and the boron 2p gross orbital population.[25] Subsequently, it has been shown that for a considerable number of boron compounds a linear relationship exists between the calculated total or *p* charge densities and $\delta(^{11}B)$.[26] Some representative ^{11}B chemical shifts are given in Tables 4.1 and 4.2.

TABLE 4.2 Some representative ^{11}B chemical shifts referenced to $BF_3 \cdot OEt_2$ (R, R^1, R^2 = alkyl)

	δ/ppm		δ/ppm
RBF_2	28 to 30	R_2BF	59 to 60
$RBCl_2$	62 to 64	R_2BCl	76 to 78
$RBBr_2$	62 to 66	R_2BBr	79 to 82
$PhBF_2$	24.8	Ph_2BF	47.4
$PhBCl_2$	54.8	Ph_2BCl	61.0
$RB(NMe_2)_2$	33.5 to 34.2	R_2BNMe_2	44 to 46
$RB(OMe)_2$	29 to 32	$R_2B(OMe)$	53 to 54
$[B(OMe)_4]^-$	3	$[B(OH)_4]^-$	1.1
$[BR_4]^-$	−16 to −20	$[BPh_4]^-$	−6.3
$[BF_4]^-$	−2.2	$[BCl_4]^-$	6.6
$[BBr_4]^-$	−24.1	$[BI_4]^-$	−128
$[B(NO_3)_4]^-$	−86.6	$[B(OPh)_4]^-$	2.
$(R^1BNR^2)_3$	23 to 37	$B(NR_2)_2X$	20 to 30

4C.2 $^{10,11}B$ Coupling Constants

In general only direct coupling constants need to be considered. Longer range coupling constants, when detected, are small and are usually obscured by broadening due to boron relaxation. As a consequence of this broadening and $^1J(^{10}B, X)$ being 0.335 $^1J(^{11}B, X)$, coupling to the 19.58% abundant ^{10}B is normally lost between the lines due to ^{11}B coupling.

For terminal boron-hydrogen bonds, $^1J(^{11}B, ^1H)$ is between 100 and 190 Hz, but when a hydrogen bridge is involved $^1J(^{11}B, ^1H)$ is less than 80 Hz. Thus it is possible to use these coupling constants to determine the coordination at each boron in a boron hydride or carborane when resolution permits.

It is unusual to resolve $^1J(^{11}B, ^{11}B)$ and, in cases where it has been resolved, it has been necessary to use resolution enhancement techniques, see ch. 2A.5*h*. In the cases where it has been resolved, values between 14 and 28 Hz are found.[27–29] For the compounds 1-XB_5H_8, there is a very marked dependence on X.[30] Recently it has proved possible to calculate $^1J(^{11}B, ^{11}B)$ for a number of boron hydrides and carboranes.[31]

4C.3 $^{10,11}B$ Relaxation Studies

It is in relaxation studies that ^{10}B NMR spectroscopy provides useful information when used in conjunction with ^{11}B relaxation data. If boron relaxation is dominated by quadrupolar relaxation then from Eq. 1.17 it may be shown that

$$\frac{T_1(^{10}B)}{T_1(^{11}B)} = \frac{T_2(^{10}B)}{T_2(^{11}B)} = 1.534 \tag{4.2}$$

for boron nuclei in identical chemical environments. This relationship has indeed been found for B_2H_6, B_5H_9 and solid *ortho*-carborane,[32,33] and was subsequently applied to deduce the magnitude of $J(^{11}B, ^{11}B)$ in some boron hydrides. The T_2 contribution to the linewidth was measured, additional broadening was attributed to coupling, and by line-fitting, $J(^{11}B, ^{11}B)$ was deduced.[29] It would appear that T_1 and T_2 for ^{11}B are in the range 10 to 200 ms with values of 1 s being found in favourable cases. In spite of ^{11}B NMR spectroscopy being frequently used with over 100 papers being published in 1976 where it was used, very little attention has been paid to ^{11}B relaxation. Yet markedly different values are found for T_1 within one molecule. Thus for B_4H_{10}, the BH groups, B(1,3), have $T_1(^{11}B) = 62.6$ ms and the BH_2 groups, B(2,4) have T_1 $(^{11}B) = 13.1$ ms while for B_5H_{11}, the values are T_1 $(^{11}B_1) = 54.9$ ms, T_1 $(^{11}B_{2,5}) = 7.5$ ms and T_1 $(^{11}B_{3,4}) = 20.7$ ms. It is clear that further work in this area is required.

4D CARBON, ^{13}C

Although carbon has only one NMR nucleus, ^{13}C, which is very insensitive compared with ^{1}H, it is now second in importance to ^{1}H. Consequently considerable attention has been devoted to most aspects of ^{13}C NMR spectroscopy in various books,[34] a collection of spectra,[35] and a review series.[36] In addition a large number of reviews on various aspects of ^{13}C NMR spectroscopy, too numerous to catalogue here, have appeared. The reader is again referred to the annual compilation of NMR books and reviews.[1]

The great popularity of ^{13}C NMR spectroscopy arises from the key role played by carbon in the skeleton of most molecules, the relatively large chemical shift range, 650 ppm, and the simplicity of the proton decoupled spectra. Consequently it is common to find separate signals for each type of carbon environment in molecules of moderate molecular weight. The low natural abundance, 1.1 %, and ^{1}H decoupling produce very simple spectra. As effects due to near neighbour anisotropy are negligible compared to the spectral width, the chemical shifts can be readily correlated with structure. Other bonuses, such as the facile measurement of T_1 also come from low natural abundance of ^{13}C nuclei.

4D.1 ^{13}C Chemical Shifts

Carbon is one of the few nuclei where there is complete agreement on a common reference substance, Me_4Si. Unfortunately, the ^{13}C shifts of Me_4Si are both solvent[37] and temperature dependent[38] although in general these effects are small. These shifts may account for the discrepancy in the two reported[39,40] values of Ξ for Me_4Si of 25 145 004 Hz and 25 144 995 Hz.

A huge mass of information has been accumulated over this decade and no attempt can be made to compile a representative list of chemical shifts. A diagrammatic representation of organic ^{13}C chemical shifts is given in Fig.4.1. It will be observed that most ^{13}C chemical shifts lie in the region $\delta 0$ to $\delta 240$, but some lie outside this range giving a much more extended range of $\delta -292.5$ (CI_4) to $\delta 362$ $\{(OC)_5CrC(OMe)Me\}$. For saturated carbon atoms the usual range of shifts is $\delta 0$ to $\delta 100$. The chemical shift depends mainly on the electronegativity of the substituent, E. For simple methyl derivatives it has been shown[41] that

$$\delta(^{13}CH_3) = -105.5 + 55E - 13m \qquad (4.3)$$

where m is the number of lone pairs on the central atoms. Hence electronegative substituents generally cause high frequency shifts. However, substituent effects are not additive and may even change sign. Thus for $CH_{4-n}Br_n$, the effect of increasing n from 0 to 4 produces shifts of $+12.3$, $+11.4$, -9.3, and -16.6 ppm for each replacement of a hydrogen atom by a bromine atom. For small selections of compounds additive rules

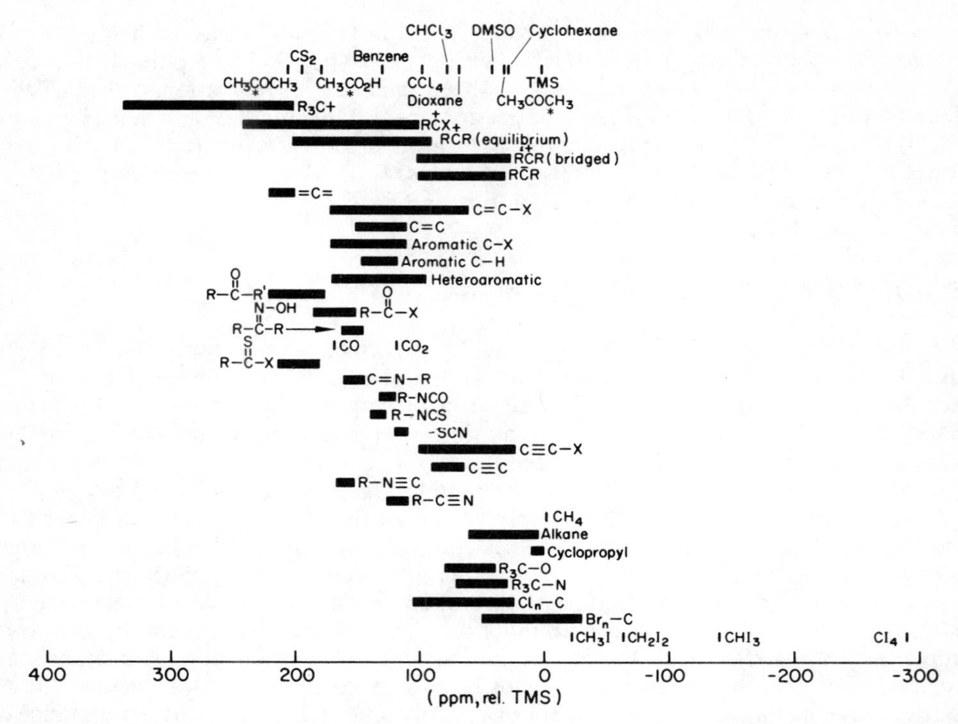

Fig. 4.1 Ranges of ^{13}C chemical shifts, from G. C. Nelson and G. L. Levy, "Carbon-13 Nuclear Magnetic Resonance for Organic Chemists", Wiley-Interscience, New York (1972).

may be fairly effectively laid down. For example, the chemical shifts of alkanes may be calculated to quite a high degree of accuracy by use of an additive relationship[42] based on 22 parameters. Similar relationships have been established in a number of other cases.

Tricoordinate carbon atoms generally fall in the range $\delta 80$ to $\delta 240$ and can usually be readily identified. Once again carbon atoms bearing electronegative groups are moved to high frequency. The major difficulty encountered in this area is resonance effects. Thus although δ(ethene) = 122.8 for $MeOC^1H{=}C^2H_2$, $\delta(^{13}C^1) = 153.2$ and $\delta(^{13}C^2) = 84.1$. Unfortunately the movement to high frequency does not continue on going to dicoordinate carbon atoms. Alkynes are generally found from $\delta 20$ to $\delta 110$ depending on the substituent while the central carbon atom of allene is at $\delta 213.5$.

In general the ^{13}C NMR signals of carbon atoms attached to metal atoms fit into the above trends after making allowance for the electropositive nature of the metals. After the large ^1H NMR shifts to low frequency found for transition metal hydrides, see ch. 4B.1, it is at first sight surprising that no large low frequency shifts are found. This probably reflects our inability to predict *a priori* a ^{13}C chemical shift of a carbon atom attached to a transition metal to better than 20 ppm rather than the absence of such shifts. The ^{13}C chemical shifts of organometallic compounds are in general badly understood with metal carbonyls falling in the ranges $\delta 290$ to $\delta 170$ and metal carbenes $\delta 370$ to $\delta 250$. When an olefin or arene is π-bonded to a metal there is generally a low frequency shift from the position of the free ligand,[43] but the shift may be very small, ca. 0 ppm or relatively large, 100 ppm depending on the bonding situation.

4D.2 ^{13}C Coupling Constants

$^{1}J(^{13}C, ^{1}H)$ is known to cover the range 96 Hz in $[MeLi]_4$ to 320 Hz in $[HC{\equiv}NH]^+$. It depends on both the hybridisation of and charge on the carbon atom. For simple hydrocarbons, $^{1}J(^{13}C, ^{1}H)$ is related to hybridisation by the simple relationship

$$J_{CH} = 500\alpha_C^2 \qquad (4.4)$$

where α_C^2 is the *s*-electron character of the carbon orbital used in the bond. Thus for ethane, $^{1}J(^{13}C, ^{1}H) = 125$ Hz, ethylene, 156.2 Hz, and, acetylene, 248.7 Hz. Electronegativity of substituent can have a similar effect, e.g., for MeF, $^{1}J(^{13}C, ^{1}H) = 149.1$ Hz, for CH_2F_2, $^{1}J(^{13}C, ^{1}H) = 184.5$ Hz, and for CHF_3, $^{1}J(^{13}C, ^{1}H) = 239.1$ Hz.

In favourable cases,[44] $^{1}J(^{13}C, ^{7}Li)$ can be observed for $[LiMe]_4$, +15 Hz, for $LiBu^t$, 11 Hz, for $LiBu^n$, 14 Hz, and for $(C_6H_4CH_2NH_2)_4Li_2Ag_2$, 7.2 Hz, see ch. 6F.3. $^{1}J(^{13}C, ^{9}Be)$ has been observed[45] for $(C_5H_5)BeBr$ as 1.1 Hz. There have been a number of measurements of $^{1}J(^{13}C, ^{11}B)$ with values[46] varying between 76 Hz for $MeB(OMe)_2$ to 18 Hz for $C_2B_3H_5$ and it has been claimed that the equation[47]

$$(\%\,s\ ^{11}B)(\%\,s\ ^{13}C) = 21\ J(^{13}C, ^{11}B) \qquad (4.5)$$

applies. This equation certainly seems to be a useful rule of thumb, but like $^{1}J(^{13}C, ^{1}H)$ electronegativity is important with 76 Hz for $^{1}J(^{13}C, ^{11}B)$ in $MeB(OMe)_2$ and 45 Hz for Me_3B.

$^{1}J(^{13}C, ^{13}C)$ has been subjected to intensive study in spite of the inherent experimental difficulties and like most direct coupling constants it is sensitive to both hybridisation and electronegativity. Thus for C_2H_6, $^{1}J(^{13}C, ^{13}C)$ is 34.6 Hz, C_2H_4 is 67.2 Hz, C_2H_2 is 170.6 Hz which are approximately consistent with hybridisation while electronegativity effects can be illustrated by $^{1}J(^{13}C, ^{13}C)$ for **C**H_3**C**OEt, 38.4 Hz, and **C**H_3**C**O_2Et, 58.8 Hz. The observation of $^{1}J(^{13}C, ^{13}C)$ clearly offers a valuable structural tool, especially in the assignment of signals. Unfortunately the low abundance of ^{13}C makes such work very difficult without enrichment.

4D.3 ^{13}C Relaxation Measurements

Since the arrival of Fourier transform ^{13}C NMR spectroscopy, ^{13}C spin-lattice relaxation measurements have become routine although ^{13}C spin-spin relaxation measurements[18] are far from routine. The spin-lattice relaxation time for ^{13}C is generally between 1 and 100 s. For most compounds, dipole–dipole relaxation (see ch. 3D.3) is the dominant mechanism, leading to a marked dependence of T_1 on the number of ^{1}H nuclei attached to a molecule, e.g. for iso-octane, 4.I

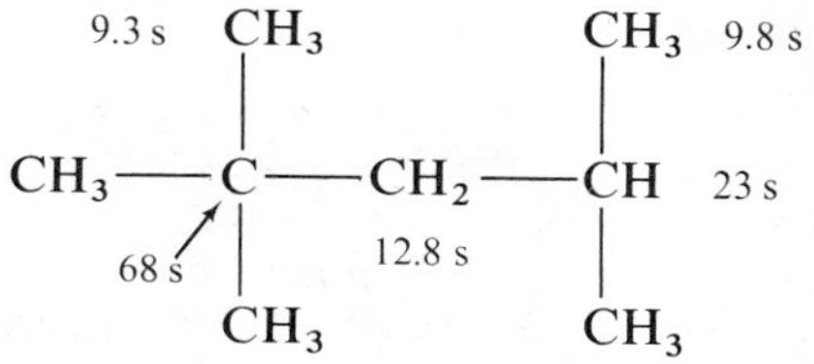

```
       9.3 s  CH3               CH3  9.8 s
              |                 |
CH3 ——— C ——— CH2 ——— CH   23 s
        68 s ↗ |    12.8 s       |
              CH3               CH3
```

4.I

The ^{13}C nucleus with no protons attached has by far the longest T_1 while for the other ^{13}C nuclei T_1 is approximately proportional to n^{-1} where n is the number of directly attached hydrogen atoms.[48] There is also a dependence on the speed of motion, the faster the motion the longer T_1. For phenyl acetylene, 4.II

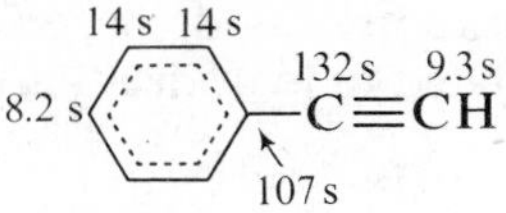

$$\text{14 s, 14 s (ring); 8.2 s (para); 107 s (ipso); 132 s } \text{C}\equiv\text{; 9.3 s } \equiv\text{CH}$$

4.II

there is preferential rotation of the molecule about the Ph—C≡CH axis causing a greater motion for the *ortho*- and *meta*-carbon atoms than for the *para*- and acetylenic CH carbon atoms on the axis of rotation leading to different T_1 values. An investigation of these values can give activation energies for rotations.[49] Even when used qualitatively, T_1 values can be a valuable aid to assignment. The major snag is that a T_1 measurement takes at least ten times as long as the measurement of a routine ^{13}C NMR spectrum.

4E NITROGEN, ^{14,15}N

Nitrogen possesses two isotopes of interest to NMR spectroscopists. Nevertheless both isotopes present considerable problems. Nitrogen-14 is 99.63% abundant with a receptivity of 5.69, but as $I = 1$ the nucleus possesses a quadrupole moment. Consequently ^{14}N signals are generally easily detected but the lines are usually broad with linewidths often between 100 and 1000 Hz, i.e., 14 to 138 ppm at 2.35 T. Such linewidths cause severe resolution problems when there are more than one type of nitrogen environment in the molecule. Although the total chemical shift range is large, ca. 1000 ppm, the chemical shift range for a given class of compounds is often only 50 ppm. Nitrogen-15 NMR spectroscopy would appear to be the better method for structural investigations as $I = \frac{1}{2}$ but unfortunately it is only 0.36% abundant with the consequently low receptivity and γ is negative leading to a negative NOE see ch. 1A.7 and 2A.3*d*. Until very recently ^{15}N NMR spectroscopy was restricted to neat low molecular weight or enriched samples. Recently the advent of wide-bore superconducting magnet systems has made ^{15}N NMR spectroscopy more accessible. It is to be anticipated that within the near future ^{15}N NMR spectroscopy will become an important structural tool, while the relaxation behaviour of ^{14}N nuclei will be exploited to obtain further information, especially on molecular dynamics.

There are a number of books and reviews on nitrogen NMR spectroscopy, resulting in up to date coverage.[50–52]

4E.1 ^{14,15}N Chemical Shifts

The use of two isotopes raises the possibility of a primary isotope shift giving rise to fundamentally different chemical shifts, but it has been shown that for $[Me_4N]^+$, $PhCH_2NC$, $MeONO_2$ and Me_2NNO_2 the chemical shifts are no greater than the experimental error, 0.2 ppm, and no attention need be paid to whether a chemical shift is determined using ^{14}N or ^{15}N NMR spectroscopy.[53] Reference compounds have not been fully decided, but it is generally agreed that $MeNO_2$ (organic solvents) and $[NO_3]^-$ (polar solvents) should be used. Sharp resonances are found and the relative shift between these two references is negligible for all practical purposes.[54] Unfortunately it has been shown that the chemical shift of $Me^{15}NO_2$ is markedly solvent and concentration

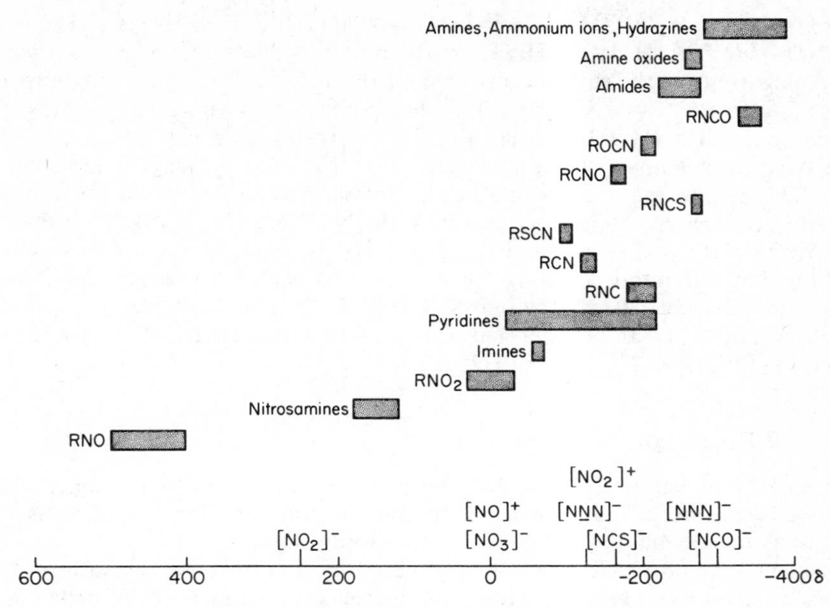

Fig. 4.2 Ranges of ^{14,15}N chemical shifts.

dependent, varying by up to 7 ppm,[51] and it would therefore be preferable to use TMS as the reference either directly or as a shift with respect to $MeNO_2$ as 0.1 M in $CDCl_3$ where $\Xi(^{15}N)$ is 10 136 783 Hz and $\Xi(^{14}N)$ is 7 226 329 Hz. The uncertainties in referencing nitrogen chemical shifts may not appear to be severe as the chemical shift range is of the order of 1000 ppm but for a given class of compounds the range is normally only 50 ppm. By far the greatest attention has been paid to organic compounds but there is also a reasonable number of investigations of inorganic compounds. Some typical data are given in Table 4.1 and Fig. 4.2. It will be observed that amines and ammonium ions are at lowest frequency while double bonding and electronegative substituents cause high frequency movements with nitroso compounds being at highest frequency.

4E.2 ^{14,15}N Coupling Constants

On account of the rapid relaxation of ^{14}N, it is unusual to observe resolved coupling to ^{14}N although broadening due to unresolved coupling is common. ^{14}N Coupling is normally only resolved for ammonium salts and RNC compounds. For convenience discussion will be restricted to ^{15}N coupling, but the corresponding ^{14}N coupling constant is given by

$$^{1}J(^{14}N, X) = -0.713\ ^{1}J(^{15}N, X) \tag{4.6}$$

The negative sign arises from the negative gyromagnetic ratio of ^{15}N.

$^{1}J(^{15}N, ^{1}H)$ is dependent on hybridisation and electronegativity. The dependence on hybridisation is not fully agreed with two equations[55,56] being given

$$\%\,s = 0.43\ ^{1}J(^{15}N, ^{1}H) - 6 \tag{4.7}$$

$$\%\,s = 0.34\ ^{1}J(^{15}N, ^{1}H) \tag{4.8}$$

Thus for $[NH_4]^+$, $^1J(^{15}N, ^1H)$ is 73.3 Hz, $O{=}C(NH_2)_2$ $^1J(^{15}N, ^1H)$ is 89 Hz, while for $HC{\equiv}^{15}NH^+$ $^1J(^{15}N, ^1H)$ is 134 Hz. Unfortunately this relationship breaks down when the electronegativity of the substituent is changed. This may be achieved by introducing a lone pair, e.g., $Ph_2C{=}^{15}NH$, 51.2 Hz and $^{15}NH_3$, 61.2 Hz, or an electronegative substituent, e.g., $Ph^{15}NHOH$, 79.3 Hz and $Ph^{15}NHNH_2$, 89.6 Hz.

There have been many determinations of $^1J(^{15}N, ^{13}C)$, principally assisted by enrichment. There appears to be reasonable agreement between hybridisation and $^1J(^{15}N, ^{13}C)$, see Eq. 3.57. Some typical values are $[Me_4N]^+$, 5.8 Hz; $[CN]^-$, 5.9 Hz; EtNC, 7.3 Hz; $MeCH_2NC$, 10.2 Hz; and EtCN, 16.4 Hz. It is dangerous to make the usual assumption that one bond coupling constants are considerably larger than two bond coupling constants, e.g. $^2J(^{15}N, ^{13}C)$ for $^{13}CH_3CO^{15}NHPh$ is 9.5 Hz.

The observation of $^1J(^{15}N, ^{15}N)$ is rare but two values are $Ph^{15}N{=}^{15}N(O)Ph$, 13.7 Hz, and $(PhCH_2)_2^{15}N^{15}NO$, 19 Hz.

4E.3 $^{14,15}N$ Relaxation Measurements

As ^{14}N and ^{15}N nuclei present very different problems in relaxation measurements they are discussed separately. The quadrupole moment is dominant for ^{14}N while as ^{15}N has $I = \frac{1}{2}$ its relaxation behaviour is similar to that found for ^{13}C.

Nitrogen-14 relaxation, $T_1 = T_2$, is dominated by quadrupolar relaxation except in some special cases like $[NR_4]^+$. This leads to relaxation times of the order of 0.1 to 10 ms with values as large as 2 s for symmetric ions such as $[NMe_4]^+$. For high resolution work such relaxation times are a nuisance with the consequential overlap of resonances. However it does offer a valuable tool to investigate chemical exchange, equilibria, and molecular rotation, and such applications have been reviewed.[57]

On account of its low sensitivity ^{15}N relaxation has received little investigation. The available data suggest that there will be a great similarity between ^{13}C and ^{15}N relaxation. Thus T_1 values of the order 1 to 100 s are usually found. In the case of ^{15}N relaxation conditions play a vital role. Thus the $T(^{15}N)$ of glycine[58] is pH dependent varying between 12 s and 1 s as the pH is increased from -0.5 to 8.1. Similarly $T_1(^{15}N)$ for acetamide is concentration dependent, decreasing from 21.6 s for a 0.6 M solution in water to 11.9 s for a 6 M solution. This behaviour was attributed to changes in hydrogen bonding.[59]

4F FLUORINE, ^{19}F

Fluorine possesses only one stable isotope, ^{19}F, which with $I = \frac{1}{2}$ and receptivity 4730 relative to natural abundance ^{13}C is ideally suited for high resolution NMR investigations. Consequently ^{19}F NMR spectroscopy has attracted a great deal of attention. Fortunately a number of recent books and reviews on the subject provide an excellent collection of data.[60–62] Although originally many references were used, $CFCl_3$ is now generally adopted as the internal reference. It is unreactive and very volatile making it ideal when compound recovery is important. Unfortunately it does suffer from the disadvantage of having a somewhat broad signal[63] and the shifts are both temperature and solvent dependent.[61] Consequently it is better to reference to ^1H in TMS. For conversion[64] $\Xi(CFCl_3)$ is 94 093 795 Hz.

With Fourier spectrometers, the observation of ^{19}F NMR spectra can provide problems. The full chemical shift range is 800 ppm which at 2.35 T is 75 kHz. On account of the speed of the ADC, most spectrometers are restricted to spectral widths of 25 or

50 kHz. Fortunately most compounds have a ^{19}F spectral width of less than 25 kHz but a second problem arises. A resolution of say 0.2 Hz is normally computer limited to a spectral width of ca. 1 kHz. There are then problems of fold-over, see ch. 2A.5*e*. For most heteronuclei it is standard to broad-band proton decouple. In the case of ^{19}F, the proximity of the ^{19}F and ^{1}H signals in addition to relatively large values for $^{2}J(^{19}F, {}^{1}H)$ of 50 Hz makes it difficult or impossible to use broad-band decoupling. Consequently CW is often preferable to FT for the observation of ^{19}F NMR spectra.

4F.1 ^{19}F Chemical Shifts

Fluorine-19 chemical shifts cover a very wide range, ca. $\delta -250$ to $\delta 550$. This very wide range causes considerable experimental problems in measurement of accurate chemical shifts. Consequently considerable discrepancies are found between literature chemical shifts. Further problems are created by the use of both sign conventions for chemical shifts, often without definition.

Fluorine-19 chemical shifts are very sensitive not only to directly attached substituents but also to long range effects. In Table 4.1, ^{19}F NMR shifts are given for a wide range of binary fluorides. It will be observed that the signal moves to high frequency with increasing electronegativity and oxidation number, thus following the usual trends. There is a marked effect of sterochemistry e.g. for the square-pyramidal molecule BrF_5, the axial fluorine is at $\delta 142$ while the equatorial fluorines are at $\delta 272$. Charge has a much smaller effect e.g. SiF_4 at δ-165 and $[SiF_6]^{2-}$ at $\delta -128$. Even within the much more restricted range of organic fluorine compounds, ^{19}F chemical shifts cover a very wide range, and some representative shifts are given in Table 4.3. The great sensitivity of ^{19}F

TABLE 4.3 Some representative ^{19}F chemical shifts referenced to $CFCl_3$

	δ/ppm		δ/ppm		δ/ppm
MeF	−271.9	$CFBr_3$	7.4	FCH=CH_2	−114
EtF	−213	CF_2Br_2	7	F_2C=CH_2	−81.3
CF_2H_2	−1436	CFH_2Ph	−207	F_2C=CF_2	−135
CF_3R	−60 to −70	CF_2Cl_2	−8	C_6F_6	−163
AsF_5	−66	$[AsF_6]^-$	−69.5	$[BeF_4]^-$	−163
BF_3	−131	ClF_3	116; −4	ClF_5	247; 412
IF_7	170	MoF_6	−278	ReF_7	345
SeF_6	55	$[SbF_6]^-$	−109	SbF_5	−108
$[SiF_6]^{2-}$	−127	TeF_6	−57	WF_6	166
XeF_2	258	XeF_4	438	XeF_6	550

chemical shifts has made them very useful probes to changes in molecular structure. Thus if WCl_6 and WF_6 are mixed, the ^{19}F NMR spectrum shows well resolved signals for the 12 different fluorine environments in the 9 fluorine containing isomers $WF_{6-n}Cl_n$, $n = 0$ to 5. Fluorine-19 chemical shifts are also sensitive to isotope effects,[63,65] 0.05 ppm for $[^{10}BF_4]^-$ compared to $[^{11}BF_4]^-$ and 0.009 ppm for $C^{35}Cl_3F$ compared to $C^{35}Cl_2^{37}ClF$.

4F.2 ^{19}F Coupling Constants

There is an extensive compilation of fluorine coupling constants available for data published up to 1972.[61]

There have been a multitude of measurements of $J(^{19}F, ^{1}H)$. For HF, $^{1}J(^{19}F, ^{1}H)$ is 530 Hz. $^{2}J(^{19}F, ^{1}H)$ depends on the HF distance, HCF angle and the heteroatom, but values of 50 Hz are common.[66] As for other nuclei, $^{3}J(^{19}F, ^{1}H)$ obeys the Karplus equation with values between 0 and 45 Hz being found with only a small dependence on substituent.[67]

$^{1}J(^{19}F, ^{9}Be) = 33$ Hz in $[BeF_4]^{2-}$ has been observed[68] while $^{1}J(^{19}F, ^{11}B)$ is frequently observed. The value may be very small, e.g. for $Na^+[BF_4]^-$ 1.4 Hz or as large as 189 Hz (BBrFI). The values are very substituent dependent e.g. 84 Hz for BF_2H but 15 Hz for BF_3. $^{1}J(^{19}F, ^{13}C)$ also is very sensitive to substituent and structural effects e.g. 372 Hz for $CFBr_3$ and 158 Hz for CFH_3. It is probable that it is negative in all cases. There is only limited information available for $^{1}J(^{19}F, ^{14}N)$, but some representative values are 328 Hz for $[FN{\equiv}N]^+$, 234 Hz for $[NF_4]^+$, 160 Hz for NF_3 and 113 Hz for FNO_2.

Like other fluorine coupling constants, $^{2}J(^{19}F, ^{19}F)$ is sensitive to structure and substituents. Some representative values are 154 Hz ($CBrF_2CBrCl_2$), and 270 Hz (CF_3CF_2CFICl).

4F.3 ^{19}F Relaxation Measurements

The investigation of ^{19}F relaxation, like ^{1}H relaxation, is often made difficult by cross-relaxation and consequently work on coupled systems has generally been deterred. With the recent successes on ^{1}H relaxation it is to be hoped that this area will be explored further in the near future. Attention has been turned to problems where there is only a single resonance, and a few examples will be given to illustrate the power of the method.

The ^{19}F relaxation rate[69] has been measured in $D_2^{17}O/D_2O$ and in $HD^{16}O/D_2O$. It was concluded that the ^{19}F, ^{1}H interaction is much stronger than ^{19}F, ^{17}O interaction. It was therefore concluded that the hydration is $F^- \cdots H{-}OH$ rather than $F^- \cdots OH_2$.

Both ClF_3 and $ClOF_3$ have two different fluorine environments and there is rapid exchange between the two environments. From an investigation[70] of T_1, T_2, and $T_{1\rho}$ as a function of field strength it was possible to deduce the chemical shift difference and rate of exchange between the two fluorine environments for both compounds and $\langle {}^{1}J_{ClF}{}^{2}\rangle^{1/2}$ as 260 Hz for ClF_3 and 195 Hz for $ClOF_3$.

4G PHOSPHORUS, ^{31}P

The observation of ^{31}P NMR spectra is relatively easy and consequently there is a huge mass of data in the literature. There are two major difficulties. Firstly phosphorus compounds are generally reactive and no suitable internal reference compound has been found for general use. Normally 85% H_3PO_4 is used as an external reference, and ^{31}P chemical shifts quoted in this book are referenced in this way. This reference is far from ideal with the reference signal being broad. Subsequently the use of external neat P_4O_6 as an external reference has been suggested.[71] P_4O_6 gives a sharp signal but the high reactivity and poor commercial availability has deterred many workers from its use. Frequently no correction is made for susceptibility and little is known about the temperature dependence of these external references. As a consequence it is common to find reported ^{31}P chemical shifts differing by up to 5 ppm. With the modern generation of NMR spectrometers, it is preferable to use the ^{2}H internal lock signal to obtain accurate ^{31}P chemical shifts, referenced to Me_4Si or to an idealised 85% H_3PO_4 shift. There is little agreement on Ξ for 85% H_3PO_4 with values of 40 480 720 Hz, 40 480 740 Hz, and 40 480 754 Hz being given in the literature.[72] These variations are beyond those to be expected, and require reinvestigation. The discrepancies cannot be attributed to temperature effects in the case of 85% H_3PO_4 as it is generally agreed that the shift is temperature invariant.[73] This temperature invariance is not general with shifts of up to 0.065 ppm/K having been reported.

Two books have been published on ^{31}P NMR spectroscopy.[74] The earlier one contains both an authorative discussion of the theory and a compilation of data while the latter is essentially an up-dating to 1969 of the compilation of data. Both texts are very weak on metal containing derivatives but this has been reviewed separately.[75] Unfortunately the mass of data that has subsequently appeared seems to have deterred further compilations of data. However, an extensive compilation of coupling constants has appeared.[76]

4G.1 ^{31}P Chemical Shifts

^{31}P Chemical shifts are known over the range $\delta 250$ to $\delta -461$ but the range found depends on the oxidation state. Thus for P^{III} compounds the complete range is found with P_4 at $\delta -461$ and PF_2Me at $\delta 245$ showing that the chemical shifts are very substituent dependent. Electronegative substituents, e.g., NMe_2, OMe, and halogens, reduce this range to ca. $\delta 100$ to $\delta 250$. The chemical shifts found for P^V compounds cover a much smaller range $\delta -50$ to $\delta 100$ in general. For example while on going from PMe_3 to PBu^t_3 the chemical shift changes from $\delta -62$ to $\delta 63$, i.e. 125 ppm but on going from $[PMe_4]^+$ to $[PHBu^t{}_3]^+$ the chemical shift only changes from $\delta 25$ to $\delta 58.3$.

Phosphorus-31 chemical shifts may be reasonably well interpreted following the treatment of VanWazer *et al.* In effect three variables are used, the bond angles, the electronegativity of the substituents and the π-bonding character of the substituent. When only one parameter is being changed at a time then useful correlations exist. Thus for alkyl substituents the order $Me < Et < Pr^i < Bu^t$ is found; Bu^t substituted phosphorus compounds being invariably found at higher frequency than methyl substituted compounds. Consequently there are additive relationships for alkyl substitution, e.g. for tertiary phosphines.[77] Additive relationships fail when π-bonding or electronegative substituents are involved, e.g., PCl_3, $\delta 220$; PCl_2F $\delta 224$; $PClF_2$ $\delta 176$ and PF_3 $\delta 97$.

In Table 4.1, only data on phosphorus(III) compounds have been given while in Table 4.4 there are data on phosphorus(V) compounds for comparison.

4G.2 ^{31}P Coupling Constants

There have been numerous reports of coupling constants involving phosphorus and only typical values can be given. More extensive compilations can be found elsewhere. Only direct coupling constants will be considered here. In general $^{1}J(^{31}P, X)$ depends on the oxidation number. Thus for P^{III} compounds $^{1}J(^{31}P, ^{1}H)$ is ca. 200 Hz while for P^V, $^{1}J(^{31}P, ^{1}H)$ is 400 to 1100. For the P^V compounds $^{1}J(^{31}P, ^{1}H)$ increases with increasing electronegativity of substituents on the phosphorus.

TABLE 4.4 Some representative ^{31}P chemical shifts referenced to 85% H_3PO_4

(*a*) Phosphorus(III) compounds

	δ/ppm		δ/ppm
PMe_3	−62	$PMeF_2$	245
PEt_3	−20	$PMeH_2$	−163.5
$PPr^n{}_3$	−33	$PMeCl_2$	192
$PPr^i{}_3$	19.4	$PMeBr_2$	184
$PBu^n{}_3$	−32.5	PMe_2F	186
$PBu^i{}_3$	−45.3	PMe_2H	−99
$PBu^s{}_3$	7.9	PMe_2Cl	96.5
$PBu^t{}_3$	63	PMe_2Br	90.5

(*b*) Phosphorus(V) compounds

	δ/ppm		δ/ppm
Me_3PO	36.2	Me_3PS	59.1
Et_3PO	48.3	Et_3PS	54.5
$[Me_4P]^+$	24.4	$[Et_4P]^+$	40.1
$[PO_4]^{3-}$	6.0	$[PS_4]^{3-}$	87
PF_5	−80.3	$[PF_6]^-$	−145
PCl_5	−80	$[PCl_4]^+$	86
$MePF_4$	−29.9	$[PCl_6]^-$	−295
Me_3PF_2	−158	Me_2PF_3	8.0

A number of reports of $^1J(^{31}P, ^{11}B)$ have produced values in the range 30 to 100 Hz. Far more extensive data is available for $^1J(^{31}P, ^{13}C)$. For P^{III} compounds, $^1J(^{31}P, ^{13}C)$ is generally negative, −25 Hz to 0 Hz while for P^V compounds $^1J(^{31}P, ^{13}C)$ is positive, +45 Hz to 300 Hz, but the usual maximum is 100 Hz. In contrast, $^1J(^{31}P, ^{19}F)$ is insensitive to the oxidation number of the phosphorus with a common range 800 Hz to 1500 Hz. Generally for P^{III} compounds $^1J(^{31}P, ^{19}F)$ increases with increasing substituent electronegativity, e.g., ca. 1140 Hz for PF_2H and ca. 1420 Hz for PF_3 while for P^V compounds the situation is confused depending on the series of compounds chosen.

The measurement of $^1J(^{31}P, ^{31}P)$ has attracted considerable attention, either directly by synthesising unsymmetrical compounds or by analysis of the spectra of the type $[AX_n]_2$ for symmetric compounds.[78] As with all phosphorus couplings, the oxidation state of the phosphorus is important but some representative values are −443 Hz for $Me_3P{=}PCF_3$, −226.5 to −230.3 Hz for $F_2P{-}PF_2$, −179.7 Hz for $Me_2P{-}PMe_2$, 18.7 Hz for $Me_2P(S)P(S)Me_2$, and +465 Hz for $[O_2P(O)P(O)HO]^{3-}$. Data are also available[78] for $^2J(^{31}P, ^{31}P)$ and $^3J(^{31}P, ^{31}P)$.

4G.3 ^{31}P Relaxation Measurements

The use of ^{31}P relaxation data has been principally restricted to the investigation of molecular motion in solution. Some examples are:

(*i*) A study of the temperature dependence[79] of ^1H, ^2H, and ^{31}P relaxation of PH_2Ph, PD_2Ph, $PH_2(C_6D_5)$, and $PD_2(C_6D_5)$ has enabled the determination of the energy of rotation of the P—C bond.

(*ii*) From an investigation[80] of T_1 and T_2 for PCl_3 and PBr_3 it was possible to deduce $^1J(^{35}Cl, ^{31}P)$ as 127 Hz and $^1J(^{79}Br, ^{31}P)$ as 296 Hz. T_1 was of the order of 5 s but T_2 was of the order of 0.01 s for PCl_3 and 0.1 s for PBr_3 being very temperature dependent. The small values of T_2 are attributable to coupling with the halogen.

(*iii*) One of the more chemically relevant investigations is that of the T_1 dependence on size of the ring and chain phosphates.[81] For example in D_2O, T_1 for cyclic metaphosphates depends on ring size reducing from 55 s for a 3 membered ring to 13 s for an 8 membered ring. Similar smaller effects are found in H_2O. The values were found to be independent of pH.

REFERENCES

1. "NMR Spectroscopy", Specialist Periodical Reports, (R. K. Harris, ed., Vols 1–5, and R. J. Abraham, ed. Vol. 6), The Chemical Society, London (1972–1977)
2. G. Mavel, *Ann. Rep. NMR Specty. 5B* (1973)
3. *Pure Applied Chem.* **29**, 627 (1972); **45**, 219 (1976)
4. J. W. Emsley, J. Feeney and L. H. Sutcliffe, "High Resolution Nuclear Magnetic Resonance Spectroscopy", two Volumes. Pergamon Press, Oxford (1966)
5. "NMR Spectra Catalogue", compiled by N. S. Bhacca, D. P. Hollis, L. F. Johnson, E. A. Pier, and J. N. Shoolery, Varian Associates, USA, 1962 and 1963; L. M. Jackman and S. Sternhall, "Applications of Nuclear Magnetic Resonance Spectroscopy in Organic Chemistry", Vol. 5 of "International Series of Monographs in Organic Chemistry", (D. H. R. Barton and W. Doering, eds). Pergamon Press, Oxford (1969); N. F. Chamberlain, "The Practice of NMR Spectroscopy with Spectra-Structure Correlations for Hydrogen-1", Plenum Press, New York and London (1974)
6. "NMR of Paramagnetic Molecules", (G. L. LaMar, W. DeW. Horrocks, jun., and R. H. Holm, eds). Academic Press, New York and London (1973)
7. P. Laszlo, *Prog. NMR Spectry.* **3**, 231 (1967)
8. A. K. Jameson and C. J. Jameson, *J. Amer. Chem. Soc.* **95**, 8559 (1973)
9. J. Mason, *J. C. S. Dalton* 1422 (1975)
10. M. Karplus, *J. Chem. Phys.* **30**, 11 (1959)
11. A. A. Brooks, J. D. Cutnell, E. O. Stejskal and V. W. Weiss, *J. Chem. Phys.* **49**, 1571 (1968)
12. J. H. Noggle and R. E. Schirmer, "The Nuclear Overhauser Effect", Academic Press, New York (1971)
13. J. D. Cutnell, S. B. W. Roeder, S. L. Tignor and R. S. Smith, *J. Chem. Phys.* **62**, 879 (1975)
14. I. D. Campbell and R. Freeman, *J. Magn. Reson. 11*, 143 (1973)
15. N. Boden, in "Determination of Organic Structures by Physical Methods", Vol. 4, p. 51, (F. C. Nachod and J. J. Zuckerman, eds). Academic Press, New York and London (1971)
16. Dinesh and M. T. Rogers, *Chem. Phys. Letters* **12**, 352 (1971)
17. L. D. Hall, *Chem. Soc. Rev.* **4**, 401 (1975)
18. R. Freeman and H. D. W. Hill, in "Dynamic NMR Spectroscopy", p. 131, (L. M. Jackman and F. A. Cotton, eds). Academic Press, New York and London (1975)
19. R. Freeman and H. D. W. Hill, *J. Chem. Phys.* **54**, 301 (1971)
20. W. G. Henderson and E. F. Mooney, *Ann. Rev. NMR. Spectry.* **2**, 219 (1969); G. R. Eaton and W. N. Lipscomb, "NMR Studies of Boron Hydrides and Related Compounds", Benjamin, New York (1969)
21. "Spectroscopic Properties of Inorganic and Organometallic Compounds", Ch. 1, Vols 1 to 10. The Chemical Society, London (1968–1977)
22. W. McFarlane, *J. Magn. Reson.* **10**, 98 (1973)

23. B. F. Spielvogel, W. R. Nutt and R. A. Izydore, *J. Amer. Chem. Soc.* **97**, 1609 (1975); H. Nöth and B. Wrackmeyer, *Chem. Ber.* **107**, 3070: 3089 (1974) and references therein.
24. R. J. Thompson and J. C. Davis, *Inorg. Chem.* **4**; 1464 (1965)
25. N. J. Fitzpatrick and N. J. Mathews, *J. Organometallic Chem.* **94**, 1 (1975)
26. J. Kroner and B. Wrackmeyer, *J. C. S. Faraday II*, **72**, 2283 (1976)
27. J. W. Akitt and C. G. Savory, *J. Magn. Reson.* **17**, 122 (1975)
28. J. D. Odom, P. D. Ellis, D. W. Lowman and M. H. Gross, *Inorg. Chem.* **12**, 95 (1973)
29. E. J. Stampf, A. R. Garber, J. D. Odom and P. D. Ellis, *J. Amer. Chem. Soc.* **98**, 6550 (1976)
30. D. W. Lowman, P. D. Ellis and J. D. Odom, *Inorg. Chem.* **12**, 681 (1973)
31. J. Croner and B. Wrackmeyer, *J. C. S. Faraday II*, **72**, 2283 (1976)
32. A. Allerhand, J. D. Odom, and R. E. Molls, *J. Chem. Phys.* **50**, 5037 (1969)
33. A. J. Leffler, M. N. Alexander, P. L. Sagalyn and N. Walker, *J. Chem. Phys.* **63**, 3971 (1975)
34. G. C. Nelson and G. L. Levy, "Carbon-13 Nuclear Magnetic Resonance for Organic Chemists". Wiley-Interscience, New York (1972); J. B. Stothers, "^{13}C NMR Spectroscopy". Academic Press, London and New York (1972); E. Breitmaier and W. Voelter, "^{13}C NMR Spectroscopy: Methods and Applications", (Monographs in Modern Chemistry, Vol. 5), Verlag Chemie, Weinheim, Germany (1974); J. T. Clerc, E. Pretsch, and S. Sternhell, "Methods of Analysis in Chemistry", Vol. 16, "^{13}C NMR Spectroscopy", Akad. Verlagsges, Frankfurt (1973); F. Wehrli and T. Wirthlin, "Interpretation of Carbon-13 Nuclear Magnetic Resonance Spectra", Heyden, London (1976)
35. J. F. Johnson and W. C. Jankowski, "^{13}C NMR Spectra: a Collection of Assigned, Coded and Indexed Spectra", Wiley-Interscience, New York (1972)
36. "Topics in ^{13}C NMR Spectroscopy", (G. C. Levy, ed). Wiley-Interscience, New York, Vol. 1 (1974); Vol. 2 (1976)
37. M. Bacon, G. E. Maciel, W. K. Musker and R. Scholl, *J. Amer. Chem. Soc.* **93**, 2537 (1971)
38. H. J. Schneider, W. Freitag and M. Schommer, *J. Magn. Reson.* **18**, 393 (1975)
39. R. K. Harris and B. J. Kimber, *J. Magn. Reson.* **17**, 174 (1975)
40. W. McFarlane, *Proc. Roy. Soc.*, **306A**, 185 (1968)
41. P. Bucci, *J. Amer. Chem. Soc.* **90**, 252 (1968)
42. L. P. Lindeman and J. Q. Adams, *Anal. Chem.* **43**, 1245 (1971)
43. B. E. Mann, *Adv. Organometallic Chem.* **12**, 135 (1974); M. H. Chisholm and S. Godleski, *Prog. Inorg. Chem.* **20**, 299 (1975)
44. P. Fischer, J. Stadelhafer and J. Weidlein, *J. Organometallic Chem.* **116**, 65 (1976)
45. L. D. McKeever and R. Waack, *Chem. Comm.* 750 (1969); L. D. McKeever, R. Waack, M. A. Doran and E. B. Baker, *J. Amer. Chem. Soc.* **90**, 3244 (1968); W. McFarlane and D. S. Rycroft, *J. Organometallic Chem.* **64**, 303 (1974); A. J. Leusink, G. Van Koten, J. W. Marsman and J. G. Noltes, *J. Organometallic Chem.* **55**, 419 (1973)
46. D. E. Axelson, A. J. Oliver and C. E. Holloway, *Org. Magn. Reson.* **6**, 64 (1974); **5**, 255 (1973); J. D. Jerome, L. W. Hall and P. D. Ellis, *Org. Magn. Reson.* **6**, 360 (1974); L. W. Hall, D. H. Lowman, P. D. Ellis and J. D. Odom, *Inorg. Chem.* **14**, 580 (1975); W. McFarlane, B. Wrackmeyer and H. Nöth, *Chem. Ber.* **108**, 3831 (1975); V. V. Negrebetski, V. S. Bogdanov, A. V. Kessenikh, P. V. Petrovskii, Yu. N. Bubnov and B. M. Mikhailov, *Zhur. obshchei Khim.* **44**, 1882 (1974)
47. T. Onak and E. Wan, *J. C. S. Dalton* 665 (1974)
48. See for example G. C. Nelson and G. L. Levy, "Carbon-13 Nuclear Magnetic Resonance for Organic Chemists", p. 187. Wiley-Interscience, New York (1972)
49. G. Hertz, *Prog. NMR Spectry.* **3**, 159 (1967); B. E. Mann, *Prog. NMR Spectry.* **11**, 99 (1977) and references therein
50. E. F. Mooney and P. H. Winson, *Ann. Rep. NMR Spectry.* **2**, 125 (1969); E. W. Randall and D. G. Gillies, *Prog. NMR Spectry.* **6**, 119 (1971); R. L. Lichter, in "Determination of Organic Structures by Physical Methods", Vol. 4, p. 195 (J. J. Zuckerman and F. C. Nachod, eds). Academic Press, New York and London (1972); M. Witanowski and G. A. Webb, *Ann. Rep. NMR Spectry.* **5A**, 395 (1972)
51. "Nitrogen NMR", (M. Witanowski and G. A. Webb, eds). Plenum Press, London and New York (1973)

52. M. Witanowski, G. A. Webb, and L. Stefaniak, *Ann. Rep. NMR Spectry.* **7**, 117 (1977)
53. E. D. Becker, R. B. Bradley and T. Axenrod, *J. Magn. Reson.* **4**, 136 (1971)
54. M. Witanowski and H. Januszewski, *J. Chem. Soc. B*, 1063 (1967); M. Witanowski, L. Stefaniak, H. Januszewski and G. A. Webb, *Tetrahedron* **27**, 3129 (1971)
55. G. Binsch, J. B. Lambert, B. W. Roberts and J. D. Roberts, *J. Amer. Chem. Soc.* **86**, 5564 (1964)
56. A. J. R. Bourn and E. W. Randall, *Mol. Phys.* **8**, 567 (1964)
57. J. M. Lehn and J. P. Kintzinger, ref. 51, p. 80
58. T. K. Leipert and J. H. Noggle, *J. Amer. Chem. Soc.* **97**, 269 (1975)
59. D. D. Giannini, I. M. Armitage, H. Pearson, D. M. Grant, and J. D. Roberts, *J. Amer. Chem. Soc.* **97**, 3416 (1975)
60. J. W. Emsley and L. Phillips, *Prog. NMR Spectry.* **7**, (1971)
61. J. W. Emsley, L. Phillips and V. Wray, *Prog. NMR Spectry.* **10**, 83 (1976)
62. E. F. Mooney and P. H. Winson, *Ann. Review NMR Spectry.* **1**, 244 (1968); K. Jones and E. F. Mooney, *Ann. Reports NMR Spectry.* **3**, 261 (1970); **4**, 391 (1971); R. Fields, *Ann. Reports NMR Spectry.* **5A**, 99 (1972); L. Cavalli, *Ann. Reports NMR Spectry.* **6B**, 43 (1976); C. H. Dungan and J. R. Van Wazer, "Compilation of Reported ^{19}F Chemical Shifts". Wiley-Interscience, New York and London (1970); E. F. Mooney, "An Introduction to ^{19}F NMR Spectroscopy", Heyden, London (1970)
63. P. R. Carey, H. W. Kroto and M. A. Turpin, *Chem. Comm.* 188 (1969)
64. W. McFarlane, A. M. Noble and J. M. Winfield, *J. Chem. Soc. A*, 948 (1971)
65. K. Kuhlmann and D. M. Grant, *J. Phys. Chem.* **68**, 3208 (1964)
66. M. S. Gopinathan and P. T. Narasimhan, *Mol. Phys.* **21**, 1141 (1971)
67. G. Govil, *Mol. Phys.* **21**, 953 (1971)
68. J. C. Kotz, R. Schaeffer and A. Clouse, *Inorg. Chem.* **6**, 620 (1967)
69. H. G. Hertz and C. Rädle, *Ber Bunsengesellschaft phys Chem.* **77**, 521 (1973).
70. M. Alexandre and P. Rigny, *Canad. J. Chem.* **52**, 3676 (1974)
71. A. C. Chapman, J. Homer, D. J. Mowthorpe and R. T. Jones, *Chem. Comm.* 121 (1965)
72. W. McFarlane, *Ann. Rev. NMR Spectry.* **1**, 148 (1968); R. K. Harris, M. I. M. Wazeer, O. Schlak and R. Schmutzler, *J. C. S. Dalton* 1912 (1974); P. L. Goggin, R. J. Goodfellow, J. R. Knight, M. G. Norton and B. F. Taylor, *J. C. S. Dalton* 2220 (1973)
73. M. M. Crutchfield, C. H. Dungan and J. R. Van Wazer, "Topics in Phosphorus Chemistry" Vol. 5, p. 46, (M. Grayson and E. J. Griffith, eds). Interscience (1967); M. D. Gordan and L. D. Quin, *J. Magn. Reson.* **22**, 149 (1976)
74. M. M. Crutchfield, C. H. Dungan, J. H. Letcher, V. Mark and J. R. Van Wazer, "Topics in Phosphorus Chemistry", Vol. 5, (M. Grayson and E. J. Griffith, eds). Interscience Publishing (1967); G. Mavel, *Ann. Rep. NMR Spectry.* **5B**, 1 (1973)
75. J. Nixon and A. Pidcock, *Ann. Rep. NMR Spectry.* **2**, 345 (1969)
76. J. R. Llinas, E.-J. Vincent and G. Peiffer, *Bull. Soc. Chim. France* 3209 (1973)
77. S. O. Grim and W. McFarlane, *Nature* **208**, 995 (1965); S. O. Grim, W. McFarlane and E. F. Davidoff, *J. Org. Chem.* **32**, 781 (1967)
78. E. G. Finer and R. K. Harris, *Prog. NMR Spectry.* **6**, 61 (1971)
79. S. J. Seymour and J. Jones, *J. Magn. Reson.* **8**, 376 (1972)
80. A. D. Jordan, R. G. Cavell and R. B. Jordan, *J. Chem. Phys.* **56**, 1573 (1972)
81. T. Glonek, P. J. Wang and J. R. Van Wazer, *J. Amer. Chem. Soc.* **98**, 7968 (1976)

5
DEUTERIUM AND TRITIUM

C. BREVARD, Bruker Spectrospin S.A., France
J. P. KINTZINGER, Université Louis Pasteur, France

Naturally-occurring hydrogen is predominantly ^{1}H, which is very favourable for NMR because it has $I = \frac{1}{2}$ and a high receptivity. For many years proton NMR was the predominant form of the technique (see ch. 4), but two other isotopes merit attention by NMR spectroscopists, viz. deuterium and tritium. The former has $I = 1$, and chemical species enriched in this isotope are rather readily available. The latter is radioactive, but can be handled without great difficulty; it has $I = \frac{1}{2}$ and an even higher receptivity than the proton. Clearly ^{2}H and ^{3}H NMR are only to be used for special chemical problems for which ^{1}H NMR is unsuitable, as will be shown below.

5A DEUTERIUM

5A.1 General

The deuterium nucleus is very often used in NMR experiments as a "spinless" nucleus. For example, in proton NMR its use avoids solvent peak detection (deuterated solvents), allows a distinction to be made between inter and intra molecular dipole–dipole relaxation mechanisms (spin dilution) or greatly simplifies proton coupled spectra (spin labelling). Due to the magnetic constants of the deuteron (Table 1.2), deuterium NMR was thought to be hopeless from the chemist's point of view, and in a recent review[1] the author emphasised the scarcity of chemically significant results obtained via deuterium NMR.

It appears that the low intrinsic NMR sensitivity and low resonance frequency of ^{2}H is no longer a real problem for its detection, the deuterium lock channel having even been used as an observation channel to record ^{2}H spectra.[2] Fourier transform NMR has widened the field of applications by lowering the usable ^{2}H concentration in the sample down to the natural abundance and by permitting T_1 measurements using appropriate pulse sequences. The aim of the discussion here is to show how deuterium NMR can give, or already has given, valuable results which cannot be obtained by proton NMR.

5A.2 Chemical Shifts

a. Introduction

Deuterium homonuclear coupling constants being very small, ^{2}H NMR spectra will appear, under broad-band ^{1}H decoupling, as "stick" spectra, each anisochronous deuteron giving rise to one resonance peak. The corresponding chemical shifts are then easy to measure, paralleling the proton ones, if no primary or secondary isotope effect is detectable. Due to the low ^{2}H resonance frequency, these isotope shifts are of the same order of magnitude as the experimental errors[3] (0.02 ppm = 0.26 Hz at 2.21 T) and can be obscured by magnetic susceptibility or solvent effect differences between protio and deutero compounds.[4] However, these secondary isotope shifts are easily seen[4,5] in the following homologues: $—CH_2{}^2H$, $—CH^2H_2$, $—C^2H_3$. Some even more important isotopic shifts have been measured for hydrogen-bonded deuterons[1] and the phytyl chain terminal methyl group in d_{72}-chlorophyll.[6]

b. Spectral Simplification

Deuterium NMR was first used by spectroscopists for simplifying such problems as the determination of deuterium content in ^{17}O enriched water[7] or precise chemical shifts measurements for strongly coupled-^{1}H spectra. One typical example is shown in Fig. 5.1. With ^{1}H NMR, ethyl benzene, benzaldehyde and butyl iodide spectra are complex. Natural abundance ^{2}H spectra of these compounds, under proton broad-band decoupling conditions, allow direct chemical shift measurement for each magnetically non-equivalent deuteron.[8] Similarly, Grimaud *et al.*[9] were able to determine precise chemical shifts for axial and equatorial positions of 4-*t*-butyl cyclohexanols and 4-*t*-butylcyclohexylamines.

These examples show how useful ^{2}H NMR can be for accurate chemical shift measurement, thence giving the possibility of calculating strongly-coupled ^{1}H spectra to

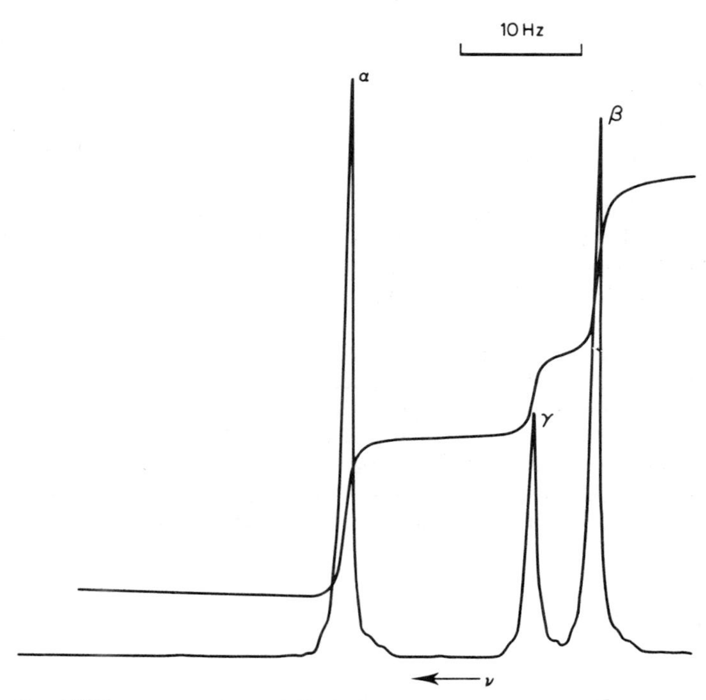

Fig. 5.1 Deuterium NMR spectrum of neat 99% perdeuterated pyridine (single pulse; deuterium external lock)

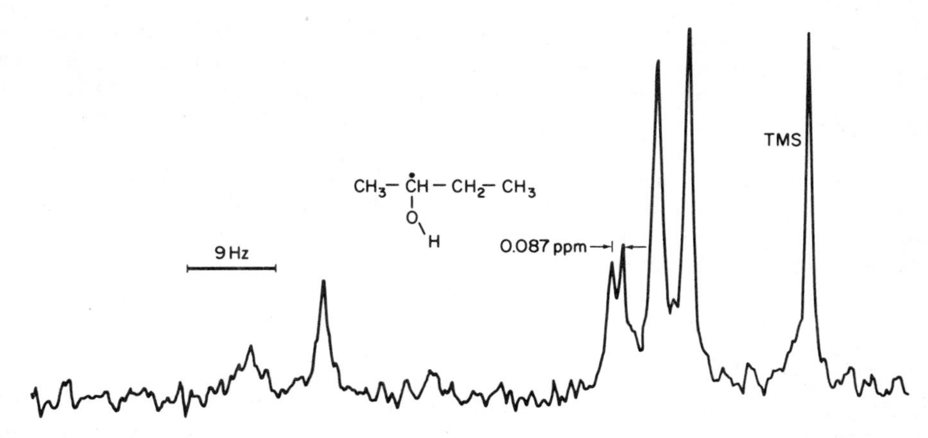

Fig. 5.2 Natural abundance broad-band ^{1}H-decoupled ^{2}H spectrum of 2-butanol (60% by volume in $CFCl_3$), showing clearly the two non-equivalent deuterons due to the C-2 asymmetric centre. (10 000 pulses; the total time taken is 5 hours; fluorine internal lock)

determine (^{1}H, ^{1}H) coupling constants via iterative methods. This situation remains favourable even with ^{2}H spectra being less dispersed than the corresponding ^{1}H spectra by a factor $\gamma(^1H)/\gamma(^2H) = 6.51$, everything else being kept constant. Fig. 5.2 exemplifies this fact.

c. *Reaction Mechanisms*

Each dynamic process which involves an inter or intra molecular hydrogen exchange can be followed by ^{2}H NMR much more easily than by studying the corresponding ^{1}H or ^{13}C spectra. For example, the stereochemistry of the cyclopentyl paratoluene sulphonate solvolysis could not be determined by ^{1}H NMR even by adding paramagnetic reagents to the solution. The answer was given by ^{2}H NMR at 33.8 MHz under ^{1}H broad-band decoupling conditions.[10] The methyl group redistribution in the $TiMe_4$-$[AlMe_3]_2$-Et_2O system in hexane solution is difficult to study with ^{1}H or ^{13}C NMR because of solvent peaks. However, the system may be investigated by ^{2}H NMR.[11] A ^{2}H NMR study of the thermal decomposition of $(\eta^5\text{-}C_5H_5)Fe(CO)(PPh_3)(1, 1\text{-}d_2\text{-}n\text{-butyl})$ established clearly that both butane isomerization and deuteron statistical scrambling occurred during the reaction.[12]

A great many kinetic processes have already been studied by ^{2}H NMR. For example, adamantanone homoenolisation;[13] deuterium exchange between methylamine amino and methyl groups;[5] deuteration kinetics of *trans*-1-decalone;[14] 1,2-cyclohexadiene dimerisation;[15] the coupling reaction between 1-chlorocycloheptene or 1-chlorocyclohexene with phenyllithium;[16] benzvalene formation from cyclopentadienyl anion and dichloromethane;[17] benzvalene isomerisation with metals or metallic cations[18,19] coal hydrogenation;[20] etc...

d. *Paramagnetic Substances*

Paramagnetic-induced chemical shifts should be the same (in ppm) in ^{2}H and ^{1}H NMR. This chemical shift equality is generally observed along the $M(acac)_3$ series of compounds, in which M represents a paramagnetic metal.[21] The only deviations larger than experimental error possibly arise either from isotopic effects on induced shifts[22] or from internal referencing problems. Line-broadening induced by paramagnetic species in ^{2}H spectra is $\gamma^2(^1H)/\gamma^2(^2H)$ smaller (in Hz) than for proton resonance. Thus ^{2}H NMR spectra gain a theoretical resolution enhancement factor $\gamma(^1H)/\gamma(^2H)$, which is 6.51, over proton spectra.[23,24]

Generally, the enhancement factor does not reach the theoretical value, though it is greater than 1. Such a situation has been observed with the perdeuterated radical ions of diphenyl,[25] 1,3,5-triphenylbenzene,[26] fluorenone and phenanthrene.[27] This fact will be dealt with in detail in section 4*c*.

5A.3 Spin-Spin Coupling Constants

Consideration of reduced coupling constants, as expressed in Eq. 1.7, make it clear that $J(^2H, ^2H)$ and $J(X, ^2H)$ are reduced by factors of 42.38 and 6.51 from the corresponding $J(^1H, ^1H)$ and $J(X, ^1H)$ respectively (see Fig. 5.3). The homonuclear deuterium coupling constants are normally too small to be detected; on the other hand, ^{2}H or X NMR easily give $J(X, ^2H)$. The existence of (^{2}H, ^{1}H) couplings allows deuterium spectra to be recorded via ^{1}H-{^{2}H} INDOR techniques,[28] and enables (^{1}H, ^{1}H) coupling between magnetically equivalent protons to be measured (supposing $K(^1H, ^1H)$ and $K(^2H, ^1H)$ to be equal).

Generally, no isotopic effect has been detected between $J(X, ^2H)$ and $J(X, ^1H)$ ($X = ^{13}C, ^{31}P, ^{15}N, \ldots$).[4,29] In contrast, an accurate study of $^2J(^1H, ^1H)$ and the corresponding $^2J(^2H, ^1H)$ shows that the $J(^1H, ^1H) = 6.51\, J(^2H, ^1H)$ relationship is only an approximation.[30] No other useful information can normally be gained from (^{2}H, ^{1}H) couplings, and ^{2}H spectra are often recorded under broadband proton-decoupling conditions. However, the ^{2}H spectrum of monodeuteroethene[31] shows the sign of $^2J_{gem}$ to be positive.

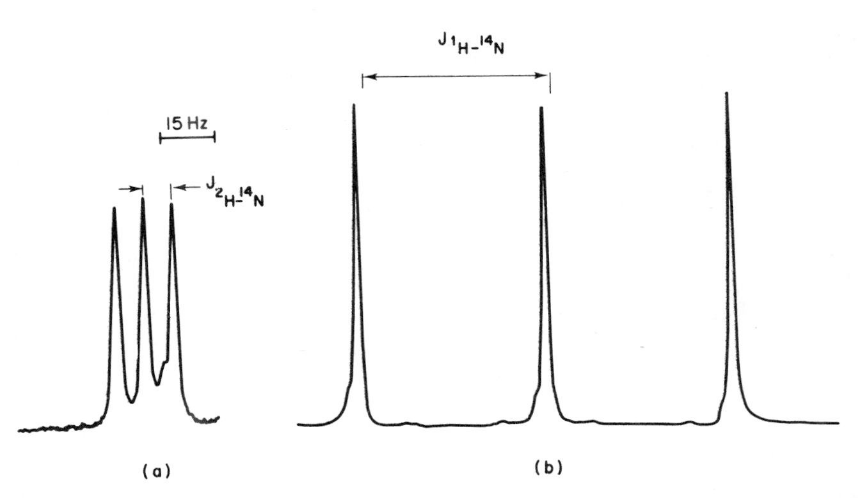

Fig. 5.3 (a), ^{2}H and (b), ^{1}H spectra of an aqueous acidified solution of ammonium chloride, exemplifying the coupling constant dependency upon ^{1}H and ^{2}H magnetogyric ratios. For spectrum (a) the solvent contained 99.5% D_2O.

5A.4 Relaxation

Due to the quadrupolar properties of the deuteron, the quadrupole contribution to its relaxation is normally dominant. Therefore the beginning of this section will be devoted to a critical survey of deuterium nuclear quadrupole coupling constants (NQCC). The relevant data are given in Tables 5.1 and 5.2.

a. Quadrupole Coupling Constants

The quadrupole coupling constant and the associated asymmetry parameter are defined by Eq. 1.15 and 1.16. They involve the electric field gradient at the relevant nucleus. The deuterium field gradient does not arise from local asymmetric electronic charge distributions but is due to imperfect mutual cancelling of nuclear and electronic contributions from all the atoms of the molecule.[32] This situation gives rise to weak field gradients around the deuteron nucleus and consequently, because of the smallness of the deuteron quadrupole moment, deuteron NQCC values do not exceed 300 kHz. The NQCC can be obtained from gas, liquid or solid state measurements via a number of techniques but we shall focus on the ^{2}H NMR one. Gas-phase methods are described by Kukolich.[33] Direct determination by solid state NQR has been pioneered by the Clarendon Laboratory Group[34] and double resonance NQR by Ragle *et al.*[35] Except for the O—D type of bond (see Table 5.1), gas, liquid and solid state NQCC are identical within experimental errors (cf. Table 5.2 for C^2H_3CN and $C_6{}^2H_6$ values). For hydrogen-bonded deuterons, there is a relationship between the O$\cdots{}^2$H bond length and the corresponding NQCC.[35]

Within the X^2H_n series of hydrides (cf. Table 5.1) one can find a linear correlation between the NQCC and the X-^{2}H bond strength constant[36,37] or the X electronegativity.[36] Such a correlation between NQCC and electronegativity rationalizes the ^{2}H NQCC dependence upon carbon hybridization:[38]

$$\chi(sp) > \chi(sp^2) > (sp^3).$$

In the XC^2H_3 series, known NQCC values are centred around 180 kHz (cf. Table 5.2), with small deviations arising from X electronegativity changes.[39,40] In the same way, the $C^2H_2X_2$ and C^2HX_3 χ values are smaller than the C^2H_2X one. A similar effect is also noticeable for deuteron-bearing sp^2 carbons,[41] or for deuterium nuclei engaged in charge-transfer or hydrogen-bonded complexes.[32]

The data concerning asymmetry parameters are very scarce, mainly because of the smallness of this term.[42] In fact η will be equal to zero only if the X-^{2}H bond acts as a three fold symmetry axis for the molecule itself. Thus assuming $\eta = 0$ for a C_{2v} symmetry leads to overestimating the corresponding NQCC value.[43]

b. Relaxation Time Measurements

One can get the ^{2}H T_2 values from deuteron linewidths, and T_1 values from adiabatic fast sweep or pulsed methods with or without Fourier transform treatment. Actually, FT measurements are the most popular, due to the ease with which they are performed on modern FT spectrometers. However, when sensitivity problems arise, one can remember that if the relation $0.1 \leq 10\pi|J|T_1 \leq 10$ holds, then a bandshape analysis of a spin-$\frac{1}{2}$ nucleus J-coupled with a deuteron allows an elegant determination of T_1 to be made. For liquids of low viscosity, like water or benzene, direct measurement of T_2 via a Carr-Purcell sequence demonstrates the equality between T_1 and T_2,[44] but this equality no longer holds for supercooled glycerine.[45] Typical values of $T_1(^2H)$ are given in Table 5.3.

c. *Relaxation Mechanisms*

For a spin-1 nucleus, under extreme narrowing conditions, the quadrupolar spin-lattice relaxation time T_1 and spin–spin relaxation time T_2 are identical[46] ($T_1 = T_2 = T_q$), and Eq. 1.14 becomes:

$$T_q^{-1} = \tfrac{3}{2}\pi^2\chi^2\left(1 + \frac{\eta^2}{3}\right)\tau_c \tag{5.1}$$

For a typical NQCC value of 180 kHz, assuming $\eta = 0$ and correlation times spanning from 10^{-10} to 10^{-13} s, Eq. 5.1 gives T_q values ranging from 17 ms to 17 s (Table 5.3). The quadrupolar relaxation mechanism is thus very efficient, the only feasible competing one being the electronic relaxation pathway. Deuterium NMR of paramagnetic species[27,47] supports this assumption, but in that case it is difficult to ascertain the relative contributions because the ^2H NQCC value can be modified by the paramagnetic centre. One can encounter the same situation for ^2H NMR of radicals, for which the broad deuterium lines have been explained by a drastic NQCC increase.[26] Another striking example of paramagnetic–quadrupolar relaxation competition can be found in the ^{13}C spectrum of C^2HCl_3 with $Cr(acac)_3$ added.[48] The broadening of the ^{13}C triplet comes from an electronic contribution to the ^2H relaxation mechanism.

On the other hand, acetone-d_6 solutions doped with $Eu(fod)_3$ show a decrease of the acetone ^2H relaxation time; in that case, the drop in T_q comes from a slower reorientational behaviour of the [acetone-d_6-$Eu(fod)_3$] complex compared to free acetone (increase in τ_c), not from a paramagnetic contribution.[49] The validity of this assumption is proved via the studies of the diamagnetic $La(fod)_3$ complex, for which the same decrease in T_q is observed.

d. *Applications of T_q measurements in liquids*

Deuterium relaxation times are very easy to interpret when compared to proton T_1 data. Generally speaking, a deuterium T_1 measurement gives a straightforward estimamation of the corresponding reorientational correlation time, cf. Eq. 5.1, provided η and χ are known. Even ^{13}C T_1 values are more difficult to handle since other mechanisms (dissolved oxygen, intermolecular dipole-dipole, spin-rotation, chemical shift anisotropy, scalar relaxation) can compete with pure intramolecular dipolar (^{13}C, ^1H) relaxation. The small range of the ^2H NQCC values for most deuterated organic compounds makes up for the limited availability of precise liquid-phase χ values.

(*i*) *Rotational Diffusion Model—Validity.* Relation 5.1 is based upon a rotational diffusion model, the correlation time τ_c being a function of viscosity and temperature as in Eq. 3.68. The parameter $f(0 \le f \le 1)$ was first introduced to take into account[50,51] the discrepancies found between values of τ_c obtained from Eq. 5.1 and those calculated from Eq. 3.68 for the same compound with $f = 1$ (Debye-B.P.P. formula). This parameter can be deduced from the Wirtz microviscosity theory[52] or the Hill mutual viscosity,[53] which introduce molecular inertial or volume considerations. These two treatments succeeded in matching experimental τ_c values with calculated ones within broad experimental conditions, especially for dodecane,[54] phenylacetylene,[55] acetonitrile, and methylacetylene[56] solutions, chloroform, liquid ammonia at various temperatures,[57] and for most deuterated organic solvents.[58]

However, a generalised treatment of micro or mutual viscosity has been given by Kivelson,[59] f being then equal to the ratio between the intermolecular torques experienced respectively by solute and solvent molecules; in fact f is now defined as proportional to the intermolecular interaction potential. Needless to say, the f values as defined above are very difficult to extract from an anisotropic potential; the value is nearly zero for isotropic molecules (e.g. CH_4) and reaches 1 for the H_2O molecule.

One can also interpret the f value as measuring the coupling between rotational and translational motions of the molecule. Studying *para*-monodeuterated substituted benzenes, Jonas *et al.* found[60] a good correlation between f and the volume of the substituent. A pressure-dependent (up to 4 kbar) study of these compounds and the dioxane-water system[61] has shown an important decrease of f for small f values and relatively smaller changes for high f. These results are corroborated by a pressure- and temperature-dependent experiment on C^2HCl_3.[62]

Another test for checking the reorientational diffusional behaviour of the molecule is the comparison of the reorientational correlation time τ_c to the correlation time τ_J which defines the interval between two molecular collisions (often taken as equal to τ_{sr}). For minute reorientational steps the Hubbard relationship,[63] Eq. 3.70, applies. This equation represents the limiting case of the J-diffusion model of Gordon.[64]

Variable temperature and pressure studies on d_5-fluorobenzene,[65] d_3-trifluorobenzene and d_3-acetonitrile[66,67] show that the J-diffusion model is valid for these compounds, equation (3.70) being verified even at low temperature. In contrast, for P^2H_3,[68] the product $\tau_c . \tau_J [I/(6\,kT)]^{-1}$ deviates greatly from the theoretical value from 130 to 300 K.

In order to get more information from deuterium relaxation times, several reorientational models have been proposed: a two-site model for ^2H_2O,[69] a conditional inertial orientational model for N^2H_3,[70] and a quasi-lattice random-flight model for CH_3O^2H,[71] $C^2H_2Cl_2$ and d_5-pyridine.[72] T_q variations for various water-organic solvent binary systems have been discussed in terms of water structure[73] and microheterogeneity.[74,75]

All the above proposed models show how difficult it is to set up a unified theory dealing with the dynamical structure of a liquid. However, taking into account the f parameter, the rotational diffusional model gives the chemist a powerful tool for predicting the deuterium relaxation times.

(*ii*) *Molecular Anisotropic Reorientation.* Equation 5.1 is valid only for isotropic motion, which is an unlikely situation for ellipsoidal molecules such as acetonitrile or benzene. Shimizu[76] and Woessner[77] first introduced the necessary corrections. Huntress[78] has derived the complete treatment for quadrupolar nuclei. In that case, the correlation time τ_c in Eq. 5.1 is replaced by a $f(\Omega, D)$ function which depends on the field-gradient orientation with respect to the principal axis of the rotational diffusion tensor. Any molecular reorientation is then described by three rotational diffusional constants $D_i (i = x, y, z)$, the corresponding τ_i being defined as $6D_i^{-1}$.

The deuterium nucleus is ideally suited for such studies, as one can, in theory, substitute any proton in a molecule by deuterium; however, this nucleus alone cannot always give a definite answer and it is then necessary to switch to (an)other molecular probe(s) to reach a clear conclusion.

Significant results are compiled[79–85] in Table 5.4 for ellipsoidal species; it is clear from these data that the fastest reorientational motion (lowest activation energy) is always associated with the highest symmetry axis, which does not always coincide with the lowest inertial axis. No change arising from a significant dipolar moment axis can be detected on $D_{\parallel}$ or $D_{\perp}$ when the dipolar vector is collinear with the symmetry axis. On the other hand, the orthogonality of these two vectors leads to a noticeable decrease in the reorientational anisotropy.

The $D_{\parallel}$ and $D_{\perp}$ variations associated with the NQCC values used are critically discussed in refs 79, 80, 82 and 85. Varying χ leads to a mutual shift of the $D_{\parallel}$, $D_{\perp}$ vs. T^{-1} graphs without affecting the activation energy values.

The microdynamical reorientational behaviour of DMF,[86] quinoline and isoquinoline[87] has been defined as anisotropic; in contrast, $CF_3{}^2H$ motions are isotropic.[88] Acetonitrile reorientational anisotropy is greatly favoured by a pressure increase, the $\Delta V_{\perp}^{\neq}$ activation volume being much more important than the $\Delta V_{\parallel}^{\neq}$ one.[89]

The influence of complexing agents has also been followed, but this introduces one more parameter to deal with, viz. complexation-dependence of χ.[32] In the C^2H_3CN-metallic cations[80] and C^2HCl_3-benzene[82] systems, $D_{\perp}$ is reduced, not $D_{\parallel}$. Water-pyridine mixtures behave quite differently compared to neat pyridine: the $D_{\parallel}$ parameter being reduced while the motion around the electrical dipole axis speeds up.[90]

(*iii*) *Internal Rotations.* The effect of internal motions on relaxation has been thoroughly dealt with.[91–97] An effective correlation time τ_c is defined, which arises from a combination between τ_M (overall correlation time) and τ_i (local correlation time).

For example, the deuteron τ_c for a deuterium-labelled methyl group in a large isotropically reorientating molecule is given by:

$$\tau_c = 0.11\tau_M + 0.89\left(\frac{1}{\tau_M} + \frac{1}{\tau_i}\right)^{-1} \tag{5.2}$$

In this case τ_i is the time interval between two 120° methyl jumps. The time τ_M can be extracted from a deuterium-labelled site in the molecular frame whereas τ_c is measured directly via a ^2H deuteromethyl relaxation time. One can then get τ_i from Eq. 5.2. This treatment has been applied to substituted aromatic molecules,[98–100] toluene,[101–102] phenylphosphine,[103] isobutylene,[104] methylamines,[105] tertiary butanol,[106] tertiary butylchloride[107] and DMF.[86] For chloroform solutions, the methyl effective correlation time of toluene and 2-methylfluorene[99] equals the 0.11 τ_M value predicted from Eq. 5.2. For $—CH^2HX$ substituents (X = OAc, OH, Cl), the local correlation time τ_i increases and equals the overall correlation time τ_M.

Variable temperature studies on toluene give a methyl activation barrier of 3.3 kJ mol^{-1} compared to 8.4 kJ mol^{-1} for the molecule itself,[101] these two activation parameters being pressure dependent.[102] For isobutylene,[104] *t*-butylchloride,[107] and *t*-butanol,[106] the barrier to internal rotation was found to be 5.9 kJ mol^{-1}, 2.9 kJ $mole^{-1}$ and 14.2 kJ mol^{-1}, respectively. Deuteron relaxation times of deuteromethylamines[105] have been interpreted with quasilattice random flight (QLRF) theory;[71] from tri- to mono-methylamine, the activation energies for methyl group rotation and overall molecular reorientation are 8.4 and 5.0; 11.7 and 13.4; 7.1 and 11.7 kJ mol^{-1} respectively.

As pointed out by the authors,[105] an anisotropic diffusion-reorientation model leads to unrealistic results for trimethylamine because internal rotations are not taken into account. However, in this case, the QLRF and the isotropic diffusional model give identical results provided internal rotations are introduced. The only difference, then, comes from the interpretation one gives to the reorientational correlation times.

When the deuteron is labelling a chain terminal group, each link of the chain participates in the internal motions, with its own τ_i fulfilling Eq. 5.2.[95] This situation is well exemplified by perdeuterated ethanol.[108]

The same situation applies also for large deuterated molecules such as oxytocin[109] or chlorophyll,[6] for which the deuterium signal is sharper than expected. Similarly, deuteromethyl groups have been used for monitoring cyanoferri-, carbonmonoxy- and oxy-myoglobin conformational changes.[110]

(*iv*) *Molecular Associations.* When the deuterium nucleus is engaged in a dynamic process such as:

$$\underset{k_f}{A(^2H) + B} \underset{k_d}{\overset{k_a}{\rightleftharpoons}} \underset{k_g}{[A(^2H)B]} \tag{5.3}$$

one can follow the complexation molecular dynamics via ^{2}H NMR; k_a and k_d are the association and dissociation constants, while $k_f = \tau_f^{-1}$; $k_g = \tau_g^{-1}$ where τ_f and τ_g represent the correlation times of the free and complexed deuterated species.

When the exchange is fast with respect to the A and AB chemical shift difference, and when k_a and k_d are long compared to k_g and k_f,[111] then the measured relaxation times (T_1 or T_2) depend upon the complexed fraction, x:

$$(T_q)^{-1}_{\text{obs}} = (1 - x)(T_q^{-1})_{\text{free}} + x(T_q)^{-1}_{\text{complexed}} \tag{5.4}$$

Equation 5.4 has been used for studying deuterium relaxation times of heavy water-electrolyte systems,[112] organic solvents,[73,74] biopolymer-polyelectrolyte systems,[113] and biopolymer-denaturating agents.[114] Hydration spheres of paramagnetic metallic cations such as Ni^{II},[115] Cr^{III},[116] and V^{IV} [117] can be studied by ^{2}H NMR, and exchange rates as well as correlation times of the different species are obtained by combining ^{1}H and ^{2}H chemical shifts with T_2 measurements.

A further step forward is reached with rather complicated biological systems on which ^{2}H NMR gives unique information even if extreme narrowing conditions are no longer valid. For example, biomolecule–substrate interactions have been studied with α chymotrypsin–N-acetyltryptophan,[118] horse liver alcohol dehydrogenase–2-propanol[119] or neurophysine–oxytocin.[120]

The idea of assuming that the correlation time τ_g obtained from deuterium T_1 measurement of the A-labelled part represents the overall [A(^{2}H)B] complex correlation time, implies no mutual flexibility at all for the two components. Such a hypothesis is ruled out in an elegant study of cyclodextrine-inclusion complexes with *p*-methylsuccinate, *m*-methylsuccinate and *p*-*t*-butylcinnamate.[121]

The complete mathematical treatment of equilibrium 5.3 as a function of k_a, k_d, k_g, k_f[122] has been developed and applied to different kinds of equilibria. Thus, for trinitro benzene (TNB)–fluorene, TNB–methyl fluorene and TNB–methylenephenanthrene[123] charge-transfer complexes, the k_g value was found to be close to k_a and k_d but much greater than k_f. This situation strongly favours a "collision complex" model for these systems.

One can refine the mathematical frame by including in the equations the reorientational anisotropy of one of the two components of the equilibrium 5.3[124] or by taking into account the difference between the ^{2}H NQCC of free and complexed A molecules.[124,125] Polyvinyl–pyrrolidone (PVP) or polystyrene–chloroform,[126] pyrrole–camphor, pyrrole–camphene and pyrrole–PVP[127] associations have been investigated in such a way.

(*v*) *Determination of relaxation pathways of other nuclei; liquid-phase NQCC measurements.* By measuring the deuteron τ_c of a —C^2H_n molecular fragment, one can calculate the proton dipolar contributions to ^{13}C or ^{1}H relaxations without NOE determinations. This method permitted the determination of the spin-rotation contribution to proton relaxation in acetonitrile,[81] phenylphosphine,[103] toluene, 2-butyne and 2,4-hexadiyne;[128] a similar approach was used for ^{13}C relaxation in toluene[129] and several organic liquids.[130]

Conversely, deuteron NQCC can be determined from ^{1}H or ^{13}C relaxation time measurements. However, these liquid-phase NQCC values can be well in error if no care has been taken in analysing the different contributions to the measured ^{1}H or ^{13}C relaxation times or if the molecular reorientational anisotropy has been ignored.[131,132]

When these potential errors were eliminated, determination of ^{13}C (benzene, phenylacetylene,[133] cyanoacetylene,[134] nitrobenzene[135]), ^{1}H (1,4-diethynyl benzene, chlorobenzene[136]) or ^{35}Cl (methylene chloride[137]) relaxation times led to very accurate NQCC values, which compare very well with the ones obtained from direct methods. Deuterochloroform and deuterobromoform ^{2}H liquid-phase NQCC's have been obtained from the corresponding T_q and τ_c, the correlation time being extracted from Raman lineshape analysis.[138] The deuterium NQCC of absorbed water in zeolites was determined from the minimum of the $T_1 = f(1/T)$ graph.[139]

Lastly, one can cite the ^{14}N NQCC determination from ^{2}H relaxation measurements of glycine[140] as an example of the great range of deuterium relaxation applications.

5A.5 Physical States other than Isotropic Fluids

Deuterium NMR is widely used for studying heterogeneous, nematic, solid or mesomorphic phases, areas which are somewhat beyond the general scope of this book. A brief survey of the actual results will be attempted. These subjects are being reviewed periodically.[141]

a. Heterogeneous and Nematic Phases.

Deuterium relaxation is very useful for studying these systems, as the ^{2}H relaxation time is not affected by translational motions. By comparing ^{1}H and ^{2}H T_1 data Resing was able to differentiate rotational and translational motions of zeolite-absorbed water;[139] rather similar work was done on silica-absorbed methanol.[142] In these systems, deuterium T_1 and T_2 values are different. A study of a glass beads-water mixture showed that this difference between T_1 and T_2 arises from the magnetic field micro-inhomogeneities around the beads, not from an exchange process between free and adsorbed water.[143]

In nematic phases, T_1 should vary according to the relation:[144]

$$T_1 = f(\nu_0^{1/2}) \tag{5.5}$$

Equation 5.5 was indeed verified for the hydroxyl group deuterium of *p*-hexylbenzoic acid-d_1[145] but not for the ring deuterium of di-alkoxyazoxybenzenes.[146]

Selective T_1 measurements on the alkyl chain of 4 cyano-4-d_{17}-*n*-octylbiphenyl[147] have been undertaken in order to verify the relation $T_1 \cdot q_{\parallel}^2 = \text{constant}$, in which $q_{\parallel}$ represents the electrical field gradient component parallel to the nematic phase privileged axis

b. The Solid State

In a solid, the quadrupolar Hamiltonian is not, in general, averaged to zero by molecular motions. The deuterium NMR spectrum then appears as a doublet whose line-separation $\Delta\nu_Q$ is given by:[148]

$$\Delta\nu_Q = \tfrac{3}{4}\chi[(3\cos^2\theta - 1) + \eta\sin^2\theta\cos^2\varphi] \tag{5.6}$$

in which χ and η represent the deuterium NQCC and asymmetry parameter; θ and φ define the deuterium field gradient orientation with respect to the external magnetic field. When working on monocrystals, variation of θ and φ allows an unambiguous determination of the quadrupolar interaction.[149]

A statistical averaging of θ and φ is obtained for a powdered sample. One can then show[43] that the deuterium spectrum is composed of two maxima whose separation is given by

$$\Delta\nu_Q = \tfrac{3}{4}\chi \tag{5.6}$$

A detailed bandshape analysis once more allows the determination of χ and η.

When the deuterium nucleus is labelling a rapidly reorientating group, $\Delta\nu_Q$ is obtained from

$$\Delta\nu_Q = \tfrac{3}{4}\chi(3\cos^2\theta - 1)/2 \tag{5.7}$$

where θ measures the angle between the group rotation axis and the field-gradient direction. By lowering the sample temperature in order to slow down the reorientation, the ^2H NQCC can be easily measured.[150] Vice-versa, molecular motions in the solid state can be reached via $\Delta\nu_Q$ variations with temperature. Inorganic salt hydrates, clathrates,[151a] insertion compounds such as $TiS_2 \cdot N^2H_3$, $TaS_2 \cdot N^2H_3$,[151b] $LaNi_5{}^2H_x$,[152] or deuterium in niobium foils[153] have been explored with this method. Even when the molecular motions average the NQCC to zero, the minimum of the $T_1 = f(T)$ graph still allows NQCC calculation.[154]

c. Mesomorphic Phases and Membranes

Deuterium NMR spectra of mesomorphic phases or membranes show a doublet-like pattern[155] with

$$\Delta\nu_Q = \tfrac{3}{4}\chi S \tag{5.8}$$

The S term is an "order parameter" varying with phase ordering, deuterium position in the molecule under study, and molecular dynamics of the medium. When working on molecules dissolved in a mesomorphic phase, the knowledge of S (extracted from the proton spectra) and $\Delta\nu_Q$ allows χ to be determined. However, the precision of the method suffers serious limitations due to

- variations in temperature and phase composition[156]
- guessed molecular geometry
- solute molecular geometry variations caused by the solvent itself[157]
- non-equivalence between proton and deuterium S values[158]
- bond length variations[159]
- random molecular orientations[160]

More details will be found in books by Diehl and Khetrapal[161] and by Emsley and Lindon.[162]

TABLE 5.1 Deuterium nuclear quadrupole coupling constants in X—^2H bonds

Compound	State[a]	Method[b]	NQCC/kHz	η^c/%	Ref.
$^2H_2$	G	MB	225 ± 0.1	0	168
Li ^2H	G	MB	33 ± 1	0	169
Na B^2H_4	SP	NMRQ	94 ± 4	0	170
$Me_3NB^2H_3$	SP	NMRQ	105 ± 1	(0)	171
$B_{10}H_4{}^2H_{10}$	SP	NMRQ	180 ± 2	16 ± 5	172
$B_{10}{}^2H_4H_{10}$	SP	NMRQ	211 ± 2	30 ± 5	172
$CH_3{}^2H$	G	MB	191.48 ± 0.77	0	173
$NH_2{}^2H$	G	MW	290.6 ± 0.7	2	174
N^2H_3	SP	NMRQ	217 ± 3	(0)	150, 151
$C_4H_4N^2H$	SP	NQR	216.5	(0)	175
$N^2H_4^+NO_3$	SP	NMRQ	159 ± 8	(0)	170
$N^2H_4^+Cl^-$	SC	NMRQ	180 ± 1	0	176
Me CON^2H_2	SP	NQR	194.7 ± 0.5	18 ± 1	34
			195.9 ± 0.5	16 ± 1	
$^-OOCCHR\ N^2H_3^+$	SP	NQR	140 – 190[d]	2 – 7[d]	177
^2HNCO	G	MB	345 ± 2	0	178
H^2HO	G	MB	307.9 ± 0.14	13.5 ± 0.7	179
^2H_2O	SC	NMRQ	213.2 ± 0.8	10	180
			216.4 ± 1	~10	
	SP	NQR	213.4 ± 0.3[e]	11.2 ± 0.5	181
$^2H_3O^+ClO_4^-$	SP	NMRQ	212 ± 5	(0)	182
Me O^2H	G	MB	303 ± 12	(0)	183
	SP	NMRQ	192 ± 15	(0)	71
$HCOO^2H$	G	MB	272 ± 3	7.5 ± 1	184
	SC	NMRQ	165.1 ± 2.7	12.5 ± 3	149
F^2H	G	MB	340 ± 40	0	185
Li Al^2H_4	SP	NMRQ	72	0	186
Si^2H_4	SP	NMRR	95	0	154
Me Si^2H_3	LN	NMRQ	91 ± 2	(0)	187
Ph Si^2H_3	LN	NMRQ	91 ± 2	(0)	36
$P^2H_4^+I^-$	SP	NMRR	91.6	0	188
Ph P^2H_2	LN	NMRQ	115 ± 2	(0)	36
^2H_2S	SP	NMRQ	142 ± 9	~10	189
H^2HS	G	MB	154.7 ± 1.6	12 ± 13	190
Ph S^2H	LN	NMRQ	146 ± 3	(0)	36
^2HCl	SP	NMRQ	190	(0)	191
	G	MB	187.4 ± 0.4	0	192
Ge^2H_4	SP	NMRR	82 ± 5	0	193
Me Ge^2H_3	LN	NMRQ	82 ± 2	(0)	187
^2HBr	SP	NMRQ	153	(0)	191
^2HI	SP	NMRQ	113	(0)	191
$(\pi\text{-}C_5H_5)_2Mo^2H_2$	SP	NMRQ	52 ± 3	(0)	194
$(\pi\text{-}C_5H_5)_2W^2H_2$	SP	NMRQ	54 ± 4	(0)	194
$^2H\ Mn(CO)_5$	SP	NQR	68.3	0	37

[a] G = gas phase; SP = solid phase, polycrystalline sample; SC = solid phase, single crystal; LN = liquid phase, nematic solution. [b] MB = molecular beam; MW = microwave; NMRQ = deuterium quadrupolar splitting in NMR; NMRR = deuterium relaxation time in NMR; NQR = nuclear quadrupole resonance. [c] ( ) indicates an assumed value for the asymmetry parameter. [d] depending upon the substituent *R*. [e] for hexagonal ice.

TABLE 5.2 Deuterium nuclear quadrupole coupling constants in C—^{2}H bonds

Compound	State[a]	Method[b]	NQCC/kHz	η[c]/%	Ref.
sp^3 *hybridised carbon*					
methane-d_1	G	MB	191.48 ± 0.77	0	173
ethane-d_6	SP	NMRQ	168 ± 3	(0)	195
cyclohexane-d_{12}	SP	NMRQ	173.7 ± 1.7	<1	43
cyclopropane-d_6	LN	NMRQ	184 ± 20	(0)	38
bromoethane-2-d_1	LN	NMRQ	182.4 ± 3	(0)	159
ethanol-2-d_3	LN	NMRQ	174.7 ± 15	(0)	196
trichloroethane-d_3	SP	NQDR	174.3 ± 1.6	(0)	39
methylfluoride-d_3	LN	NMRQ	133 ± 7	3 ± 3	197
methylchloride-d_3	SP	NQDR	177.45	(0)	32
bromoethane-1-d_2	LN	NMRQ	175.7 ± 12		159
methylbromide-d_3	LN	NMRQ	171 ± 4		38, 40, 158
methyliodide-d_3	LN	NMRQ	180 ± 5		38, 40, 158
methanol-d_3	SP	NMRQ	141 ± 15	(0)	71
methylsilane-d_3	LN	NMRQ	176 ± 4	(0)	187
methylgermane-d_3	LN	NMRQ	175 ± 4	(0)	187
nitromethane-d_3	SP	NMRQ	164 ± 15	(0)	71
acetonitrile-d_3	G	MB	167.5 ± 4	(0)	198
	LN	NMRQ	171 ± 17	(0)	38
	LN	NMRQ	165 ± 5	(0)	40
	LN	NMRQ	172.5 ± 1.5	(0)	158
methylacetylene-d_3	G	MW	176 ± 15	(0)	199
toluene-d_3	SP	NMRQ	165	(0)	155
acetone-d_6	SP	NMRQ	150 ± 15	(0)	71
acetone-d_3	LN	NMRQ	174.5 ± 3	(0)	156
acetic acid-d_3	SC	NMRQ	167; 174	(3.3); (0)	149
dimethylformamide-d_6	LN	NMRQ	192; 138	(0)	200
methylene chloride-d_2	SP	NQDR	176 ± 3	5.8 ± 0.2	201
	LN	NMRQ	160	(0)	202
	L	NMRR	150	(0)	137
methylene fluoride-d_2	G	MB	186 ± 10	15 ± 5	203
glycine-d_2	SP	NMRQ	166.4 ± 1.3	<4	204
malonic acid-d_2	SC	NMRQ	165 ± 2 to 181.9 ± 0.2	1 to 13	205
fluoroform-d_1	G	MB	170.8 ± 2	0	206
chloroform-d_1	SP	NQDR	166.9 ± 0.1	0	32
chloroform-d_1-ether	SP	NQDR	149.1 ± 0.5	0	207
chloroform-d_1-acetone	SP	NQDR	151.4 ± 0.4	0	207
chloroform-d_1-mesitylene	SP	NQDR	162 ± 0.5	0	207
bromoform-d_1	L	NMRR	177 ± 5		138
sp^2 *hybridised carbon*					
ethene-d_1	LN	NMRQ	175.3 ± 1.3	3.9 ± 0.1	31
benzene-d_1	LN	NMRQ	196.5 ± 1.3	(0)	208
benzene-d_6	SC	NMRQ	186.6 ± 1.6	(0)	209
	SP	NMRQ	180.7 ± 1.5	4.1 ± 0.7	43
	SP	NMRQ	193 ± 2.6	(0)	155
	LN	NMRQ	194 ± 4	(0)	40
	LN	NMRQ	183 ± 10	(0)	38
	L	NMRR	200 ± 10	(0)	210
	L	NMRR	186 ± 9	(0)	133

TABLE 5.2 (*Continued*)

Compound	State[a]	Method[b]	NQCC/kHz	η^{c}/%	Ref.
anthracene-d_{10}	SC	NMRQ	181 ± 3	6.4 ± 1.3	211
toluene-d_5	SP	NMRQ	179.9 ± 1.7	5.6 ± 0.2	43
antimony triphenyl-d_{15}	SP	NQDR	180.4 ± 0.7	4.4	32
ferrocene-d_{10}	SP	NMRQ	198 ± 2	(0)	212
chlorobenzene-d_5	SP	NMRQ	180 ± 15	(0)	71
bromobenzene-d_5	SP	NMRQ	179	(0)	213
nitrobenzene-d_5	SP	NMRQ	175 ± 15	(0)	71
	L	NMRQ	180 ± 10	(0)	165
	L	NMRR	190 ± 3	(0)	135
			(2, 6-d)		
nitrobenzene-2,4,6-d_3	LN	NMRQ	203 ± 10	(0)	41
			(2, 6-d)		
			192 ± 4		
			(4-d)		
p-dibromobenzene-d_4	SP	NMRQ	179	(0)	213
1,3,5-trifluorobenzene-d_3	LN	NMRQ	180 ± 3	5 ± 1	197
1,3,5-trichlorobenzene-d_3	SP	NQDR	181.9 ± 0.3	≈6 to 8	214
1,3,5-trinitrobenzene-d_3	LN	NMRQ	181 ± 4	(0)	41
pyridine-d_5	SP	NMRQ	178 ± 1.2	3.9 ± 0.6	43
pyridine-4-d_1	LN	NMRQ	186.9 ± 1	(0)	215
pyridine-d_5-chloroform	SP	NQDR	178.9 ± 0.2	1 to 2	32
pyrrole-d_4	SP	NQR	187	(0)	175
triazine-d_3	LN	NMRQ	200 ± 10	(0)	84
3,6-dichloropyridazine-d_2	SP	NQDR	180 ± 0.5	2	216
			175.5 ± 0.3		
formaldehyde-d_1	G	MW	171 ± 3	2	217
acetaldehyde-d_1	LN	NMRQ	144.8 ± 3	(0); (3.3)	218
			170 ± 3		
formic acid-d_1	SC	NMRQ	161 ± 2	4 ± 2.5	149
formate-d_1	SC	NMRQ	167.6 ± 1	4.3 ± 0.6	219
			170.3 ± 1	2.5 ± 0.6	
formylfluoride-d_1	G	MB	205 ± 4	(0)	220
ketene-d_2	G	MW	120 ± 12	(0)	221
sp hybridised carbon					
acetylene-d_1	G	MB	200 ± 10	0	222
	LN	NMRQ	198 ± 7	0	38
methylacetylene-d_1	G	MW	208 ± 10	(0)	223
	LN	NMRQ	199.4 ± 2	(0)	38
phenylacetylene-d_1	SP	NMRQ	215 ± 5	(0)	212
	L	NMRR	227 ± 4	(0)	133
chloroacetylene-d_1	G	MW	225 ± 18	0	223
fluoroacetylene-d_1	G	MW	212 ± 10	0	223
cyanoacetylene-d_1	L	NMRR	200 ± 2	0	134
^{2}HCN	LN	NMRQ	199 ± 3	0	38
	G	MW	290	0	224
^{2}HCP	G	MW	233 ± 40	0	225

[a] G = gas phase; SP = solid phase, polycrystalline sample; SC = solid phase, single crystal; L = liquid phase; LN = liquid phase, nematic solution. [b] MB = molecular beam; MW = microwave; NMRQ = deuterium quadrupolar splitting in NMR; NMRR = deuterium relaxation time in NMR; NQR = nuclear quadrupole resonance; NQDR = nuclear quadrupole double resonance. [c] ( ) indicates an assumed value for the asymmetry parameter.

TABLE 5.3 Typical ^2H spin-lattice relaxation times for some neat liquids and solutions

Compound	Solution	Temperature/K	Method[a]	T_1/s	Ref.
methyliodide-d_3	neat	304	ARP	5.84	58
methylenechloride-d_2	neat	304	ARP	3.18	58
chloroform-d_1	neat	304	ARP	1.60	58
methanol-d_3	neat	304	ARP	6.10	58
ethanol-1,1-d_2	neat	298	pulse	1.14	74
ethanol-2,2,2-d_3	neat	298	pulse	0.94	74
acetonitrile-d_3	neat	304	ARP	6.90	58
dimethylsulphoxide-d_6	neat	304	ARP	0.72	58
dimethylformamide-d_7	neat	303	FT		130
C^2H_3-cis				3.0	
C^2H_3-trans				1.6	
C^2H				0.95	
cyclohexane-d_{12}	neat	304	ARP	1.47	58
benzene-d_6	neat	304	ARP	1.45	58
toluene-methyl-d_3	neat	304	ARP	5.11	58
ring-d_5	neat	304	ARP	1.08	58
benzylfluoride-α-d_1	acetone (molar)	213	BSA	0.38	100
naphthalene-1,4,5,8-d_4	5% CCl_4	303	FT	0.52	130
fluorene-9d_1	$CDCl_3$ (molar)	306	BSA	0.22	98, 99
phenylphosphine-P^2H_2	neat	298	ARP	4.21	103
-ring-d_5	neat	298	ARP	0.65	103
pyridine-d_5	neat	304	ARP	1.11	58
2,6-d_2	neat	298	pulse	1.14	85
3-d_1	neat	298	pulse	1.17	85
4-d_1	neat	298	pulse	1.20	85
pyrrole-N-d_1	neat	295	pulse	0.27	127
-d_4	neat	295	pulse	0.43	127
water-d_2	neat	303	pulse	0.53	69
ammonia-d_3	neat	299	pulse	6.3	70
ethylamine-N-d_2	neat	304	ARP	2.0	58
aniline-N-d_2	neat	304	ARP	0.17	58
trifluoroacetic acid-d_1	neat	304	ARP	0.11	58

[a] ARP = adiabatic rapid passage; BSA = bandshape analysis for a spin-$\frac{1}{2}$ nucleus coupled to ^2H; FT = Fourier transform.

TABLE 5.4 Rotational Diffusion Constants and activation energies for some symmetric tops

Molecule	$D_{\parallel}{}^{a}$ $/10^{10}\,s^{-1}$	$D_{\perp}{}^{a}$ $/10^{10}\,s^{-1}$	T /°K	$Ea_{\parallel}$ /kJ mol^{-1}	$Ea_{\perp}$ /kJ mol^{-1}	Ref.
methylacetylene	183	13	243	2.1	7.1	79
acetonitrile	120	13.5	298	3.3	7.1	80
	156	15	298	3.0	8.4	81
chloroform	18	9.6	293	2.9	6.7	82
benzene	38	7	300	1.25	11.3	83
s-triazine	31.4	7.4	364	1.25	14.2	84
pyridine[b]	13	8.5	303	4.2	13.0	85

[a] $D_{\parallel}$ and $D_{\perp}$ are related to diffusional reorientational steps along the highest molecular symmetry axis and perpendicular to it, respectively. [b] Although it is a planar asymmetric top, pyridine behaves like a symmetric top molecule with a symmetry axis orthogonal to the molecular plane.

The structure of the mesomorphic phase itself can be tackled by ^{2}H NMR. In fact, if one knows the χ value, it is then easy to determine the S factor for each deuterium-labelled molecular site, thus studying

—mutual orientation between molecular and phase axes
—S variations along an alkyl chain
—influence of the phase composition on the nature of the phase itself.

Focussing on ^{2}H NMR, potential experimenters have to be aware that double quantum transitions which appear in the CW mode, if too large an observation field B_1 is used,[163] can lead to erroneous interpretations; these forbidden transitions appear as a singlet, which would alternatively imply some isotropic motional averaging.

One can study the variations in the widths of the doublet components with magnetic field gradients in order to gain some information on the phase spatial homogeneity.[164]

The electric-field-induced alignment of nitrobenzene[165] or benzene molecules dissolved in nitrobenzene[166] has also been measured, and the double quantum transitions discussed.[167]

5B TRITIUM

Tritium (^{3}H) is a radioactive isotope of hydrogen whose half-life is 12.3 years, generating soft β radiation (0.018 MeV). Tritium-enriched samples can be handled safely if stored in sealed NMR tubes, as the β emission is stopped by the glass walls, but the laboratory must be equipped to face rapidly any tube breaking because of the then excessive local radioactivity[226] (maximum dose: 5×10^{-6} μCi cm^{-3} of air for 40 hours exposure per week). From an NMR point of view, ^{3}H is a very interesting isotope due to its nuclear spin value of $\frac{1}{2}$. The ^{3}H sensitivity and resonance frequency are higher than those of the proton (cf. table 1.1).

The first tritium NMR spectrum was recorded by Tiers *et al.*[227] on a tritiated ethyl benzene sample of very high radioactivity (10 Curies!). For tritiated water, an acceptable one-scan spectrum is obtained with a 1 Curie/ml sample[228] (0.038%). The advent of FT NMR has lowered considerably this limit, down to the milliCurie level.

At least three laboratories have tritium FT facilities, viz. those of Elvidge and of Altman, and the Service des Molécules Marquées, Commissariat à l'Energie Atomique, Saclay.

Isotopic effects can be easily studied with tritium. A survey of 26 monotritiated substrates showed the $\delta(^3H) = \delta(^1H)$ relation to be valid.[229] As for deuterium, a 0.021 ppm secondary isotopic shift exists when going from $CH_2{}^3H$ to CH^3H_2.[229] Significant differences have been noticed between experimental and calculated $^1J(^{13}C, ^3H)$ when the latter is taken from:

$$^1J(^{13}C, ^3H) = 1.0664\ ^1J(^{13}C, ^1H)$$

Deviations of ± 0.2 Hz have been found, except for chloroform for which a -1.52 Hz discrepancy exists.[230]

$(^1H, ^3H)$ couplings are normally eliminated by proton broad-band decoupling, which gives "stick" spectra for low isotopically-enriched samples. Consequently, a ^3H-$\{^1H\}$ NOE is predictable. After some controversy,[231,232] the existence and value of this NOE are now clearly demonstrated.[233] This effect appears to be important if ^3H NMR is used for determining overall ^3H molecular labelling, especially if different tritium-labelled sites have unequal relaxation times and NOE factors.

For medium sized molecules such as purine bases[234] or steroids,[235] ^3H NMR could be very useful for precise line assignments. In fact ^3H NMR is faster and safer[236,237,238] than classical chemical analysis.

Actually, the development of dedicated FT spectrometers using superconducting magnets will certainly lead to an increase of ^3H NMR of large biomolecules, for which ^2H or ^1H lines are too broad or seriously overlapping.

REFERENCES

1. P. Diehl, In "Nuclear Magnetic Resonance Spectroscopy of Nuclei other than Protons". (T. Axenrod and G. H. Webb, eds). J. Wiley and Sons, New York and London (1974)
2. R. E. Santini, *Anal. Chem.* **43**, 801 (1971)
3. H. Jensen and K. Schaumburg, *Acta Chim. Scand.* **25**, 663 (1971)
4. P. Diehl and Th. Leipert, *Helv. Chim. Acta* **47**, 545 (1964)
5. J. D. Halliday and P. E. Bindner, *Can. J. Chem.* **54**, 3775 (1976)
6. R. C. Dougherty, G. D. Norman and J. J. Katz, *J. Amer. Chem. Soc.* **87**, 5801 (1966)
7. S. Meiboom, *J. Chem. Phys.* **34**, 375 (1961)
8. J. M. Briggs, L. F. Farnell and E. D. Randall, *J. C. S. Chem. Comm.* 70 (1973)
9. R. Wylde and J. Grimaud, *C. R. Acad. Sci., Ser. C.* **271**, 597 (1970)
10. K. Humski, V. Sendijarerié and V. J. Shiner Jr, *J. Amer. Chem. Soc.* **98**, 2865 (1976)
11. L. S. Bresler, A. S. Khachaturov and I. Ya. Podburnyi, *J. Organometal. Chem.* **64**, 335 (1974)
12. D. L. Reger and E. C. Culbertson, *J. Amer. Chem. Soc.* **98**, 2789 (1976)
13. J. B. Stothers and C. T. Tan, *J. C. S. Chem. Comm.* 738 (1974)
14. E. Casadevall and P. Metzger, *Tetrahedron Lett.* 4199 (1970)
15. W. R. Moore, P. D. Mogolesko and D. D. Traficante, *J. Amer. Chem. Soc.* **94**, 4753 (1972)
16. L. K. Montgomery, A. O. Clouse, A. M. Crelier and L. E. Applegate, *J. Amer. Chem. Soc.* **89**, 3453 (1967)
17. U. Burger and F. Mazenod, *Tetrahedron Lett.* 2881 (1976)
18. U. Burger and F. Mazenod, *Tetrahedron Lett.* 2885 (1976)
19. U. Burger and F. Mazenod, *Tetrahedron Lett.* 1757 (1977)

20. F. D. Schweighardt, B. C. Bockrath, R. A. Friedel and Retcofsky, *Anal. Chem.* **48**, 1254 (1976)
21. A. Johnson and G. W. Everett Jr, *J. Amer. Chem. Soc.* **92**, 6705 (1970); **94**, 1419 (1972); **94**, 6397 (1972)
22. R. R. Horn and G. W. Everett Jr, *J. Amer. Chem. Soc.* **93**, 7173 (1971)
23. G. Laukien and F. Noack, *Z. Physik*, **159**, 311 (1960)
24. J. Reuben and D. Fiat, *J. Amer. Chem. Soc.* **91**, 1242 (1969)
25. G. W. Canters, B. M. P. Hendriks and E. de Boer, *J. Chem. Phys.* **53**, 445 (1970)
26. J. A. M. Van Broekhoven, B. M. P. Hendriks and E. de Boer, *J. Chem. Phys.* **54**, 1988 (1971)
27. G. W. Canters, B. M. P. Hendricks, J. W. M. de Boer and E. de Boer, *Mol. Phys.* **25**, 1135 (1973)
28. J. R. Campbell, L. D. Hall and P. R. Steiner, *Can. J. Chem.* **50**, 504 (1972)
29. E. Breitmaier, G. Jung, W. Voelter and L. Pohl, *Tetrahedron* **27**, 361 (1971)
30. R. R. Fraser, M. A. Petit and M. Miskow, *J. Amer. Chem. Soc.* **94**, 3253 (1972)
31. J. Kowalewski, T. Lindblom, R. Vestin and T. Drakenberg, *Mol. Phys.* **31**, 1669 (1976)
32. J. L. Ragle, M. Mokarram, D. Presz and G. Minott, *J. Magn. Reson.* **20**, 195 (1975)
33. S. G. Kukolich, *Mol. Phys.* **29**, 249 (1975)
34. D. T. Edmonds, M. J. Hunt and A. L. Mackay, *J. Magn. Reson.* **11**, 77 (1973); **20**, 505 (1975)
35. G. Soda and T. Chiba, *J. Chem. Phys.* **50**, 439 (1968)
36. B. M. Fung and I. Y. Wei, *J. Amer. Chem. Soc.* **92**, 1497 (1970)
37. P. S. Ireland and T. L. Brown, *J. Magn. Reson.* **20**, 300 (1975)
38. F. S. Millett and B. P. Dailey, *J. Chem. Phys.* **56**, 3249 (1972)
39. J. L. Ragle and K. L. Sherk, *J. Chem. Phys.* **50**, 3553 (1969)
40. W. J. Caspary, F. Millett, S. Reichbach and B. P. Dailey, *J. Chem. Phys.* **51**, 623 (1969)
41. I. Y. Wei and B. M. Fung, *J. Chem. Phys.* **52**, 4917 (1970)
42. R. Bersohn, *Mol. Phys.* **27**, 605 (1974)
43. R. G. Barnes and J. W. Bloom, *J. Chem. Phys.* **57**, 3082 (1972)
44. U. Haeberlen, H. W. Spiess and D. Schweitzer, *J. Magn. Reson.* **6**, 39 (1972)
45. P. W. Drake and R. Meister, *J. Chem. Phys.* **54**, 3046 (1971)
46. A. Abragam, "Les Principes du Magnétisme Nucléaire", Presses Universitaires de France, Paris (1961)
47. B. M. Fung, *J. Chem. Phys.* **58**, 192 (1973)
48. G. Fronza, R. Mondelli and E. W. Randall, *J. C. S. Chem. Comm.*. 195 (1974)
49. J. Wooten, G. B. Savitsky and J. Jacobus, *J. Amer. Chem. Soc.* **97**, 5027 (1975)
50. P. Debye, "Polare Molekeln", Verlag von S. Hirzel, Leipzig (1929)
51. N. Bloembergen, E. M. Purcell and R. V. Pound, *Phys. Rev.* **73**, 679 (1948)
52. A. Spernol and K. Wirtz, *Z. Naturforsch.* **8a**, 522 (1953); T. Gierer and K. Wirtz, *Z. Naturforsch.* **8a**, 532 (1953)
53. N. E. Hill, *Proc. Phys. Soc.* **67**, 149 (1954)
54. D. E. Woesnner, B. S. Snowden, R. A. McKay and E. T. Strom, *J. Magn. Reson.* **1**, 105 (1969)
55. J. Jonas and T. M. Digennaro, *J. Chem. Phys.* **50**, 52 (1969)
56. R. A. Assink and J. Jonas, *J. Phys. Chem.* **73**, 2445 (1969)
57. K. T. Gillen and J. H. Noggle, *J. Chem. Phys.* **53**, 801 (1970)
58. J. A. Glasel, *J. Amer. Chem. Soc.* **91**, 4569 (1969)
59. a. R. E. D. McClung and D. Kivelson, *J. Chem. Phys.* **49**, 3380 (1969); b. D. Kivelson, M. G. Kivelson and I. Oppenheim, *J. Chem. Phys.* **52**, 1810 (1972)
60. R. A. Assink, J. Dezwaan and J. Jonas, *J. Chem. Phys.* **56**, 4975 (1972)
61. Y. Lee and J. Jonas, *J. Chem. Phys.* **59**, 4845 (1973)
62. D. L. Vanderhart, *J. Chem. Phys.* **60**, 1858 (1974)
63. P. S. Hubbard, *Phys. Rev.* **131**, 1155 (1967)
64. R. G. Gordon, *J. Chem. Phys.* **44**, 1830 (1966)
65. R. A. Assink and J. Jonas, *J. Chem. Phys.* **57**, 3329 (1972) J. Dezwaan, R. J. Finney and J. Jonas, *J. Chem. Phys.* **60**, 3223 (1974)
66. T. E. Bull, *J. Chem. Phys.* **59**, 6173 (1973)
67. T. E. Bull, *J. Chem. Phys.* **62**, 222 (1974)
68. D. W. Sawyer and J. G. Powles, *Mol. Phys.* **21**, 83 (1971)

69. J. C. Hindman, A. J. Zielen, A. Svirmickas and M. Wood, *J. Chem. Phys.* **54**, 621 (1971)
70. P. W. Atkins, A. Loewenstein and Y. Margalit, *Mol. Phys.* **17**, 329 (1969)
71. D. E. O'Reilly and E. M. Peterson, *J. Chem. Phys.* **55**, 2155 (1971)
72. D. E. O'Reilly, E. M. Peterson and E. L. Yasaitis, *J. Chem. Phys.* **57**, 890 (1972)
73. J. C. Hindman, A. Svirmickas and M. Wood, *J. Phys. Chem.* **72**, 4188 (1968)
74. E. V. Goldammer and H. G. Hertz, *J. Phys. Chem.* **74**, 3734 (1970)
75. E. Tomchuck, J. J. Czubryt, E. Bock and N. Chatterjee, *J. Magn. Reson.* **12**, 20 (1973)
76. H. Shimizu, *J. Chem. Phys.* **37**, 765 (1962); **40**, 754 (1964)
77. D. E. Woessner, *J. Chem. Phys.* **37**, 647 (1962)
78. W. T. Huntress Jr, *J. Chem. Phys.* **48**, 3524 (1968); W. T. Huntress Jr, *Adv. Magn. Reson.* **4**, 1 (1970)
79. J. Jonas and T. M. Digennaro, *J. Chem. Phys.* **50**, 2392 (1969)
80. T. T. Bopp, *J. Chem. Phys.* **47**, 3621 (1967)
81. D. E. Woessner, B. S. Snowden Jr and E. T. Strom, *Mol. Phys.* **14**, 265 (1968)
82. W. T. Huntress Jr, *J. Phys. Chem.* **73**, 103 (1969)
83. K. T. Gillen and J. E. Griffiths, *Chem. Phys. Lett.* **17**, 359 (1972)
84. J. P. Kintzinger and J. M. Lehn, *Mol. Phys.* **27**, 491 (1974)
85. J. P. Kintzinger and J. M. Lehn, *Mol. Phys.* **22**, 273 (1971)
86. D. Wallach and W. T. Huntress Jr, *J. Chem. Phys.* **50**, 1219 (1969)
87. R. Zinsius, Thèse 3° cycle, Strasbourg (1970)
88. J. W. Harrell Jr, *J. Magn. Reson.* **23**, 235 (1976)
89. T. E. Bull and J. Jonas, *J. Chem. Phys.* **53**, 3315 (1970)
90. J. P. Kintzinger, *Mol. Phys.* **30**, 673 (1975)
91. E. O. Stejskal and H. S. Gutowsky, *J. Chem. Phys.* **28**, 388 (1958)
92. D. E. Woessner, *J. Chem. Phys.* **36**, 1 (1962); **42**, 1855 (1965)
93. H. G. Hertz, *Prog. NMR Spectry.* **3**, 159 (1967)
94. M. D. Zeidler, *Ber. Bunsenges. Phys. Chem.* **69**, 659 (1965)
95. D. Wallach, *J. Chem. Phys.* **47**, 5258 (1967)
96. J. H. Noggle, *J. Phys. Chem.* **72**, 1324 (1968)
97. D. E. Woessner, B. S. Snowden, Jr and G. H. Meyer, *J. Chem. Phys.* **50**, 729 (1969)
98. C. Brevard, J. P. Kintzinger and J. M. Lehn, *Chem. Comm.* 1193 (1969); *Tetrahedron* **28**, 2429 (1972)
99. C. Brevard, J. P. Kintzinger and J. M. Lehn, *Tetrahedron* **28**, 2447 (1972).
100. G. Beguin and R. Dupeyre, *Mol. Phys.* **32**, 699 (1976)
101. D. E. Woessner and B. S. Snowden, Jr, *Adv. Mol. Relax. Processes* **3**, 181 (1972)
102. D. J. Wibur and J. Jonas, *J. Chem. Phys.* **62**, 2800 (1975)
103. S. J. Seymour and J. Jonas, *J. Magn. Reson.* **8**, 376 (1972)
104. R. A. Assink and J. Jonas, *J. Chem. Phys.* **53**, 1710 (1970)
105. A. Loewenstein and R. Waiman, *Mol. Phys.* **25**, 49 (1973)
106. Y. Margalit, *J. Chem. Phys.* **55**, 3072 (1971)
107. D. E. O'Reilly, E. M. Peterson, C. E. Scheie and E. Seyfarth, *J. Chem. Phys.* **59**, 3576 (1973)
108. Teng Ko Chen, A. L. Beyerlein, G. B. Savitsky, *J. Chem. Phys.* **63**, 3176 (1975)
109. J. A. Glasel, J. F. Kinley, V. J. Hruby and A. F. Spatola, *Ann. N.Y. Acad. Sci.* **222**, 778 (1974)
110. O. Oster, G. W. Neireiter, A. O. Clouse and F. R. N. Curd, *J. Biol. Chem.* **250**, 7990 (1975)
111. J. R. Zimmerman and W. E. Brittin, *J. Phys. Chem.* **61**, 1328 (1957)
H. G. Hertz, *Ber. Bunsenges. Phys. Chem.* **71**, 979, 999 (1967)
112. H. G. Hertz and M. D. Zeidler, *Ber. Bunsenges. Phys. Chem.* **68**, 821 (1964)
G. Engel and H. G. Hertz, *Ber. Bunsenges. Phys. Chem.* **72**, 808 (1968)
113. J. A. Glasel, *J. Amer. Chem. Soc.* **92**, 375 (1970); *Nature* **220**, 1524 (1968)
114. J. A. Glasel, *J. Amer. Chem. Soc.* **92**, 372 (1970)
115. J. Granot, A. M. Achlama (Chmelnick) and D. Fiat, *J. Chem. Phys.* **61**, 3043 (1974)
116. A. M. Achlama (Chmelnick) and D. Fiat, *J. Chem. Phys.* **59**, 5197 (1973)
117. J. Reuben and D. Fiat, *J. Amer. Chem. Soc.* **91**, 4652 (1969)
118. J. T. Gerig and R. A. Rimerman, *J. Amer. Chem. Soc.* **94**, 7565 (1972)
119. T. Drakenberg, *J. Magn. Reson.* **15**, 354 (1974)

120. J. A. Glasel, V. J. Hruby, J. F. McKelvy and A. F. Spatola, *J. Mol. Biol.* **79**, 555 (1973)
121. J. P. Behr and J. M. Lehn, *J. Amer. Chem. Soc.* **98**, 1743 (1976)
122. D. Beckert and H. Pfeiffer, *Amer. Phys.* **16**, 262 (1965); J. E. Anderson and P. A. Fryer, *J. Chem. Phys.* **50**, 3784 (1969); J. E. Anderson, *J. Chem. Phys.* **51**, 3578 (1969); H. Sillescu, *Adv. Mol. Relax. Processes* **3**, 91 (1972)
123. C. Brevard and J. M. Lehn, *J. Amer. Chem. Soc.* **92**, 4987 (1970)
124. R. G. Brüssau and H. Sillescu, *Ber. Bunsenges. Phys. Chem.* **76**, 31 (1972)
125. A. G. Marshall, *J. Chem. Phys.* **52**, 2527 (1970)
126. H. Rockelmann and H. Sillescu, *Zeit. Phys. Chem. Neue Folge* **92**, 263 (1974)
127. R. Voelkel and H. Sillescu, *Zeit. Phys. Chem. Neue Folge* **95**, 73 (1975)
128. R. G. Parker and J. Jonas, *J. Magn. Reson.* **6**, 106 (1972)
129. H. W. Spiess, D. Schweitzer and U. Haeberlen, *J. Magn. Reson.* **9**, 444 (1973)
130. H. Saitô, H. H. Mantsch and I. C. P. Smith, *J. Amer. Chem. Soc.* **95**, 8453 (1973); H. H. Mantsch, H. Saitô, L. C. Leitch and I. C. P. Smith, *J. Amer. Chem. Soc.* **96**, 256 (1974)
131. G. Bonera and A. Rigamonti, *J. Chem. Phys.* **42**, 175 (1965)
132. M. D. Zeidler, *Ber. Bunsenges. Phys. Chem.* **69**, 659 (1965)
133. L. M. Jackman, E. S. Greenberg, N. M. Szeverenyi and G. K. Schnorr, *J. C. S. Chem. Comm.* 141 (1974)
134. N. Szeverenyi, R. R. Vold and R. L. Vold, *Chem. Phys.* **18**, 23 (1976)
135. R. E. Stark, R. L. Vold and R. R. Vold, *Chem. Phys.* **20**, 337 (1977)
136. R. A. Assink and J. Jonas, *J. Magn. Reson.* **4**, 347 (1971)
137. M. P. Klein, D. Gill and H. Kotowicz, *Chem. Phys. Lett.* **2**, 677 (1968)
138. D. A. Wright and M. T. Rogers, *J. Chem. Phys.* **63**, 909 (1975)
139. H. A. Resing, *J. Phys. Chem.* **80**, 186 (1976)
140. A. Tzalmona and E. Loewenthal, *J. Chem. Phys.* **61**, 2637 (1974)
141. Specialist Periodical Reports: Nuclear Magnetic Resonance, The Chemical Society, Burlington House, London
142. S. J. Seymour, M. I. Cruz and T. T. Fripiat, *J. Phys. Chem.* **77**, 2847 (1973)
143. J. A. Glasel, *J. Amer. Chem. Soc.* **96**, 970 (1974)
144. P. G. De Gennes, "The Physics of Liquid Crystals," Clarendon Press (1974)
145. B. Deloche and B. Cabane, *Mol. Cryst. Liq. Cryst.* **19**, 25 (1972)
146. R. D. Orwoll and C. G. Wade, *J. Chem. Phys.* **63**, 986 (1975)
147. J. W. Emsley, J. C. Lindon and G. R. Luckhurst, *Mol. Phys.* **32**, 1187 (1976)
148. M. H. Cohen and F. Reif, *Solid State Phys.* **5**, 321 (1957)
149. G. J. Adriaensens and J. L. Bjorkstam, *J. Chem. Phys.* **56**, 1223 (1972)
150. S. N. Rabideau and P. Waldstein, *J. Chem. Phys.* **45**, 4600 (1960)
151. a. J. A. Ripmeester, *Can. J. Chem.* **54**, 3677 (1976); b. B. G. Silbernagel and F. R. Gamble, *J. Chem. Phys.* **65**, 1914 (1976)
152. R. G. Barnes, W. C. Harper, S. O. Nelson, O. K. Thome and D. R. Torgeson, *J. Less Common Metals* **49**, 483 (1976)
153. H. Lutgemeier, H. G. Bohn and R. R. Arons, *J. Magn. Reson.* **8**, 80 (1972)
154. U. Zäuteenmäki, L. Niemelä and P. Pyykkö, *Phys. Lett.* A**25**, 4 (1967)
155. J. C. Rowell, W. D. Phillips, L. R. Melby and M. Panar, *J. Chem. Phys.* **43**, 3442 (1965)
156. J. W. Emsley, J. C. Lindon and J. Tabony, *Mol. Phys.* **26**, 1499 (1973); **31**, 1617 (1976)
157. R. Ader and A. Loewenstein, *Mol. Phys.* **30**, 199 (1975)
158. P. Diehl and C. L. Khetrapal, *J. Magn. Reson.* **1**, 524 (1969); P. Diehl, M. Rheinbold and A. S. Tracey, *J. Magn. Reson.* **19**, 405 (1975)
159. J. W. Emsley and J. Tabony, *J. Magn. Reson.* **17**, 233 (1975)
160. P. Diehl, M. Rheinhold, A. S. Tracey and E. Wullschleger, *Mol. Phys.* **30**, 1781 (1975)
161. P. Diehl and C. L. Khetrapal, *N.M.R.: Basic Principles and Progress*, **1**, 1 (1969)
162. J. W. Emsley and J. C. Lindon, "N.M.R. Spectroscopy using Liquid Crystal Solvents," Pergamon Press (1975)
163. H. Wennerström, N. O. Perssonn and B. Lindman, *J. Magn. Reson.* **13**, 348 (1974)
164. L. A. McLachlan, Natusch and R. H. Newman, *J. Magn. Reson.* **10**, 34 (1973)
165. C. W. Hilbers and C. MacLean, *Mol. Phys.* **17**, 433 (1969)

166. J. Biemond and C. MacLean, *Mol. Phys.* **28**, 571 (1974)
167. J. Biemond, J. A. B. Lohman and C. MacLean, *J. Magn. Reson.* **16**, 402 (1974)
168. N. J. Harrick, R. G. Barnes, P. J. Bray and N. F. Ramsey, *Phys. Rev.* **90**, 260 (1953)
169. L. Wharton, L. P. Gold and W. Klemperer, *J. Chem. Phys.* **37**, 2149 (1962)
170. P. Pyykkö, *Ann. Universitatis Turkuensis, Ser. A.* **103**, 1 (1967)
171. S. Z. Merchant and B. M. Fung, *J. Chem. Phys.* **50**, 2265 (1969)
172. J. Wirschel, Jr and B. M. Fung, *J. Chem. Phys.* **56**, 5417 (1972)
173. S. C. Wofsy, J. S. Muenter and W. Klemperer, *J. Chem. Phys.* **53**, 4005 (1970)
174. S. G. Kukolich, *J. Chem. Phys.* **49**, 5523 (1968); **50**, 4601 (1969)
175. D. T. Edmonds, in ref. 127
176. M. Linzer and R. A. Forman, *J. Chem. Phys.* **47**, 4690 (1967)
177. M. J. Hunt and A. L. Mackay, *J. Magn. Reson.* **15**, 402 (1974); **22**, 295 (1976)
178. S. G. Kukolich, A. C. Nelson and B. S. Yamanashi, *J. Amer. Chem. Soc.* **93**, 6769 (1971)
179. J. Verhoeven, A. Dynamus and H. Bluyssen, *J. Chem. Phys.* **50**, 3330 (1969)
180. P. Waldstein, S. W. Rabideau and J. A. Jackson, *J. Chem. Phys.* **41**, 3407 (1964)
181. D. T. Edmonds and A. L. Mackay, *J. Magn. Reson.* **20**, 515 (1975)
D. T. Edmonds, S. D. Goren, A. L. Mackay, A. A. L. White and W. F. Sherman, *J. Magn. Reson.* **23**, 505 (1976)
182. D. E. O'Reilly, E. M. Peterson and J. M. Williams, *J. Chem. Phys.* **54**, 96 (1971)
183. K. H. Casleton and S. G. Kukolich, *Chem. Phys. Lett.* **22**, 331 (1973)
184. D. J. Ruben and S. G. Kukolich, *J. Chem. Phys.* **60**, 100 (1974)
185. H. M. Nelson, J. A. Leavitt, M. R. Barker and N. F. Ramsey, *Phys. Rev.* **122**, 856 (1961)
186. P. Pyykkö and B. Pedersen, *Chem. Phys. Lett.* **2**, 297 (1968)
187. R. Ader and A. Loewenstein, *J. Amer. Chem. Soc.* **96**, 5336 (1974)
188. P. Pyykkö, *Chem. Phys. Lett.* **2**, 559 (1968)
189. D. E. O'Reilly and J. H. Eraker, *J. Chem. Phys.* **57**, 2407 (1970)
190. P. Thaddeus, L. C. Krishner and J. H. N. Loubser, *J. Chem. Phys.*, **40**, 257 (1964)
191. D. J. Genin, D. E. O'Reilly, E. M. Peterson and T. Tsang, *J. Chem. Phys.* **48**, 4525 (1968)
192. E. W. Kaiser, *J. Chem. Phys.* **53**, 1686 (1970)
193. V. Hovi, U. Lähteenmäki, and R. Tuulensuu, *Phys. Lett. A.* **29**, 520 (1969)
194. I. Y. Wei and B. M. Fung, *J. Chem. Phys.* **55**, 1486 (1971)
195. L. J. Burnett and B. H. Muller, *J. Chem. Phys.* **55**, 5829 (1971)
196. J. W. Emsley, J. C. Lindon and J. Tabony, *Mol. Phys.* **26**, 1485 (1973)
197. P. K. Bhattacharyya and B. P. Dailey, *J. Chem. Phys.* **63**, 1336 (1975)
198. S. G. Kukolich, D. J. Ruben, J. H. S. Wang and J. R. Williams, *J. Chem. Phys.* **58**, 3155 (1973)
199. R. L. Shoemacker and W. H. Flygare, *J. Amer. Chem. Soc.* **91**, 5417 (1969)
200. E. T. Samulski and H. J. C. Berendsen, *J. Chem. Phys.* **56**, 3920 (1972)
201. G. L. Minott and J. L. Ragle, *J. Magn. Reson.* **21**, 247 (1976)
202. G. Gill, M. P. Klein and G. Kotowycz, *J. Amer. Chem. Soc.* **90**, 6870 (1968)
203. A. C. Nelson, S. G. Kukolich and D. J. Ruben, *J. Mol. Spectry.* **51**, 107 (1974)
204. R. G. Barnes and J. W. Bloom, *Mol. Phys.* **25**, 493 (1973)
205. W. Derbyshire, T. C. Gordin and D. Warner, *Mol. Phys.* **17**, 401 (1969)
206. S. G. Kukolich, A. C. Nelson and D. J. Ruben, *J. Mol. Spectry.* **40**, 33 (1971)
207. J. L. Ragle, G. Minott and M. Mokarram, *J. Chem. Phys.* **60**, 3184 (1974)
208. P. Diehl and C. L. Khetrapal, *Can. J. Chem.* **47**, 1411 (1969)
209. P. Pyykkö, *Ann. Universitatis Turkuensis, Ser. A*, **88**, 93 (1966)
210. J. G. Powles, M. Rhodes and J. H. Strange, *Mol. Phys.* **11**, 515 (1966)
211. D. M. Ellis and J. L. Bjorkstam, *J. Chem. Phys.* **46**, 4460 (1967)
212. P. L. Olympia, Jr, I. Y. Wei and B. M. Fung, *J. Chem. Phys.* **51**, 1610 (1969)
213. M. Rinne, J. Depireux and J. Duchesne, *J. Mol. Struct.* **1**, 178 (1967)
214. J. L. Ragle and M. Mokarram, *J. Chem. Phys.* **62**, 3361 (1975)
215. J. W. Emsley, J. C. Lindon and J. Tabony, *J. C. S. Faraday II*, 579 (1975)
216. J. Barocas, J. L. Ragle and H. D. Stidham, *J. Chem. Phys.* **58**, 4034 (1973)
217. W. H. Flygare, *J. Chem. Phys.* **41**, 206 (1969)
218. J. W. Emsley, J. C. Lindon and J. Tabony, *J. C. S. Faraday II*, 586 (1975)

219. M. G. C. Dillon and J. A. S. Smith, *J. C. S. Faraday II*, 2183 (1972)
220. S. G. Kukolich, *J. Chem. Phys.* **55**, 610 (1971)
221. W. H. Flygare and V. W. Weiss, *J. Amer. Chem. Soc.* **87**, 5317 (1965)
222. J. N. Pinkerton, Thesis Harvard, 1961, in E. A. C. Lucken "Nuclear Quadrupole Coupling Constants," Academic Press, London and New York (1969)
223. V. W. Weiss and W. H. Flygare, *J. Chem. Phys.* **45**, 8 (1966)
224. R. L. White, *J. Chem. Phys.* **23**, 253 (1955)
225. S. L. Hartford, W. C. Allen, C. L. Norris, E. F. Pearson and W. H. Flygare, *Chem. Phys. Lett.* **18**, 153 (1973)
226. "Hazards in the Chemical Laboratory." (G. D. Mair ed.). The Royal Institute of Chemistry, London (1972)
227. G. V. D. Tiers, C. A. Brown, R. A. Jackson and T. N. Lahr, *J. Amer. Chem. Soc.* **86**, 2526 (1964)
228. J. Bloxsidge, J. A. Elvidge, J. R. Jones and E. A. Evans, *Org. Magn. Reson.* **3**, 127 (1971)
229. J. M. A. Al-Rawi, J. P. Bloxsidge, C. O'Brien, D. E. Caddy, J. A. Elvidge, J. R. Jones and E. A. Evans, *J. C. S. Perkin II*, 1635 (1974)
230. J. M. A. Al-Rawi, J. A. Elvidge, J. R. Jones and E. A. Evans, *J. C. S. Perkin II*, 449 (1975)
231. J. M. A. Al-Rawi, J. P. Bloxsidge, J. A. Elvidge, J. R. Jones, V. E. M. Chambers and E. A. Evans, *Steroids* **28**, 359 (1976)
232. L. J. Altman and N. Silberman, *Anal. Biochem.* **79**, 302 (1977); *Steroids* **29**, 557 (1977)
233. J. P. Bloxsidge, J. A. Elvidge, J. R. Jones and R. B. Mane, *J. Chem. Research* 258 (1977)
234. J. A. Elvidge, J. R. Jones, C. O'Brien, E. A. Evans and H. C. Sheppard, *J. C. S. Perkin II*, 174 (1974)
235. J. M. A. Al-Rawi, J. A. Elvidge, R. Thomas and B. J. Wright, *J. C. S. Chem. Comm.* 1031 (1974)
236. J. M. A. Al-Rawi, J. A. Elvidge, D. K. Jaiswal, J. R. Jones and R. Thomas, *J. C. S. Chem. Comm.* 220 (1974)
237. J. M. A. Al-Rawi, J. A. Elvidge, J. R. Jones, V. M. A. Chambersand and E. A. Evans, *J. Label. Comp. Radiopharm.* **12**, 265 (1976)
238. J. M. A. Al-Rawi, J. P. Bloxsidge, J. A. Elvidge, R. J. Jones, W. M. A. Chambers and E. A. Evans, *J. Label. Comp. Radiopharm.* **12**, 295 (1976)

Note added in proof: A review of the applications of ^{2}H NMR in chemistry, physics and biology has recently appeared (H. H. Mantsch, H. Saito and I. C. P. Smith, *Prog. NMR Spectry.* **11**, 211 (1977)).

6
THE ALKALI METALS

BJÖRN LINDMAN, University of Lund, Sweden
STURE FORSÉN, University of Lund, Sweden

6A CHEMICAL AND BIOLOGICAL SIGNIFICANCE

The Group Ia elements, i.e. lithium, sodium, potassium, rubidium, caesium and francium, commonly termed the alkali metals, are found in widely varying proportions in the earth's surface. Francium, the least abundant element in the series, has no known stable isotope—the most long-lived one, ^{223}Fr, has a half-life of about 22 min. The presence of francium on earth is therefore entirely due to its formation in the radioactive decay of ^{227}Ac. Francium will not be considered further in this chapter.

At room temperature the pure elements are soft low melting solids, beautifully silvery when freshly cut. Caesium metal will melt on a hot day ($>28.5°C$). The thermal and electrical conductivities of the metals are very high. Sodium and potassium have found use as heat exchange liquids in breeder type power reactors, rubidium and caesium are considered for use in magnetohydrodynamic generators of electricity since the atoms are easily ionised at higher temperatures.

Much of the chemistry of the Group Ia elements can be rationalized in terms of their electronic structure with one valence shell s-electron and filled inner shells. The first ionisation energy is small as is evident from Table 6.1. Thus they all easily form ionic compounds although more nearly covalent compounds, hydrides, oxides, carbides and nitrides are also well known.

The small radius of Li^+ makes its interaction with the oxygen atom in water very strong. The enthalpy of hydration of Li^+ is nearly twice that of Cs^+ (cf. Table 6.1). Most of the solution chemical differences between the alkali metal ions can be interpreted as the result of their different ionic radii.

Organolithium, and to a minor extent organosodium and organopotassium compounds have found large uses in synthetic organic chemistry, both on a laboratory and an industrial scale. Lithium alkyls for example may be used as catalysts for the production of stereospecific polymers. Sodium alkyls are also used in polymerisation reactions ("living polymers").

At least two of the Group Ia elements, sodium and potassium, are essential for the normal function of plants and animals. Lithium may also be essential for the growth of some special plants like tobacco. The human need for common salt has been recognized since the early days of man and this has left its trace in the form of proverbs and phrases in most languages. The word "salary" is a reminder that salt was once a means of payment.

The extracellular and intracellular concentrations of sodium and potassium are different in animals. The concentration differences are upheld by energy consuming, "active" transport processes across the cell membranes. The concentration differences determine a number of physiological functions like osmotic pressure regulation and the transmission of nerve impulses.

A large number of enzymes are selectively activated by potassium and sodium. The interior of cells contains polyelectrolytes like DNA and RNA and the connective tissues consist largely of anionic mucopolysaccharides. To what extent these polyanions interact selectively with potassium or sodium is largely unknown.

A number of cyclic and open chain compounds with the ability of forming highly stable complexes with alkali ions have been discovered during the last decade and are currently the subject of intense studies.

Potassium has long been recognised as an essential nutrient for plant growth and is selectively absorbed through the root system.

Rubidium and caesium cannot substitute well for potassium in animals—the ingestion of larger amounts will cause severe physiological disturbances and ultimately death.

The physiological effect of lithium salts has found therapeutical applications in the treatment of manio-depressive states.

The chemistry and biochemistry of the Group Ia elements involve many challenging problems and NMR spectroscopy will in all likelyhood become an increasingly important research tool in this area.

It is not possible in the present treatment to cover all aspects of alkali metal NMR. In particular we will not discuss systems of amphiphilic compounds, solutions of macromolecules and complex biological systems as these will be covered in a forthcoming article.[1] Furthermore solid state studies will be excluded.

6B NUCLEAR PROPERTIES AND EXPERIMENTAL ASPECTS

In Table 1.1 and 6.1 we have summarised a number of nuclear and atomic properties of relevance in NMR studies of the Group Ia elements. All magnetic nuclei have $I > \frac{1}{2}$. The magnetic moments of ^{6}Li, ^{7}Li and ^{23}Na have been obtained for the gas phase, the other magnetic moments refer to the cation in dilute H_2O or D_2O solutions.

The determination of nuclear electric quadrupole moments from fine structure in electron spectra of atoms involves quantum chemical calculations similar to those leading to the Sternheimer antishielding factors (cf. Table 6.1). For some nuclei the quadrupole moments reported by different groups deviate considerably. To emphasize that some caution must be exercised in the use of quadrupole moment values we have listed values from different sources in Table 1.1.

The Sternheimer antishielding factor, $1 + \gamma_\infty$, is a measure of the effective field gradient produced at an atomic *nucleus* as a result of polarization effects in the electron cloud induced when the *atom as a whole* is exposed to an electric field gradient. It therefore plays an important role in the theoretical calculation of quadrupole relaxation of ions in solution. Values of $1 + \gamma_\infty$ calculated by different workers in the field differ quite considerably as is evident from Table 6.1. The values listed refer to the free cations. In so far as the electronic wave function of the ion becomes significantly altered as a result of strong ion-ion or ion-solvent interactions the value of $1 + \gamma_\infty$ will change.

Some of the alkali metal nuclei, particularly ^{7}Li and ^{23}Na, were extensively studied by CW NMR methods already in the mid 1950's, but for the more general applicability

TABLE 6.1 Some properties of the Group IA elements

Property	Li	Na	K	Rb	Cs	Ref.
Ionic radius: Pauling (nm)	0.060	0.095	0.133	0.148	0.169	[a]
Ionic radius: Gourary & Adrian (nm)	0.094	0.117	0.149	0.163	0.186	[b]
Absolute standard thermodynamic functions of hydration of M^+ at 298 K						
$-H_h^0$(kJ/mol)[c]	552	443	359	334	301	[d]
$-S_h^0$(kJ/mol, K)	0.141	0.110	0.074	0.062	0.059	[d]
$-G_h^0$(kJ/mol)	510	410	337	316	282	[d]
Sternheimer antishielding factor $(1 + \gamma_\infty)$ of free cation	0.74 to 0.75	4.66 to 6.18	13.1 to 20.2	48.2 to 56.0	103.7 to 111.4	[e,f]

[a] L. Pauling, "The Nature of the Chemical Bond," 3rd Edn. Itacha, Cornell University Press, (1960). [b] G. S. Gourary and F. J. Adrian, *Solid State Phys.* **1**, 127 (1960). [c] On this scale the enthalpy of hydration of H^+ is -1128 kJ/mol. [d] J. E. Desnoyers and C. Jolicoeur, in "Modern Aspects of Electrochemistry," Vol. 5. (J. O. Bockris and B. Conway, eds). Butterworths, London (1969). [e] E. A. C. Lucken, "Nuclear Quadrupole Coupling Constants." Academic Press, London (1969). [f] K. D. Sen, and P. T. Narasimhan, "Advances in Nuclear Quadrupole Resonance," **1**, 277 (1974).

the FT technique is a prerequisite. (For a general discussion of the relative merits of CW and FT methods in the study of broad NMR signals cf. ch. 2A.2)

For several alkali nuclei, in particular ^{23}Na, ^{39}K, ^{85}Rb and ^{87}Rb, the quadrupole moments are big enough to make quadrupole effects dominate the NMR spectra for most systems, while for other nuclei like ^{6}Li and ^{7}Li quadrupole effects can often be neglected. Thus even in partly covalent alkyl lithium compounds the quadrupole relaxation of ^{7}Li is relatively inefficient and natural linewidths are often less than 1 Hz.[1a] In solutions ^{6}Li behaves more like a spin $\frac{1}{2}$ nucleus and the relaxation of $^{7}Li^+$ is not always dominated by quadrupolar interactions.* The natural linewidths of the $^{23}Na^+$ and $^{39}K^+$ NMR signals of aqueous solutions are of the order of a few Hertz and interpulse spacing in FT studies can be fairly short. Reasonable signal-to-noise ratios can usually be obtained rather quickly even on dilute solutions. It may be of some interest to compare data obtained on aqueous solutions in our laboratory and other laboratories at B_0-fields of the order of 2 Tesla and sample diameters between 9 and 12 mm. It is reasonably easy (acquisition times of the order of 30 min) to obtain good signal-to-noise on 0.5 mM Na^+ solutions. From this we can estimate that it would be feasible to study $^{39}K^+$ signals on ca. 0.1 M potassium solutions. Studies of $^{7}Li^+$ may require long interpulse spacings but to obtain signals from solutions less than 1 mM should presumably present no greater problems. FT studies of $^{6}Li^+$ will however present some problems since the longitudinal relaxation time may be of the order of a few hundred seconds.[2]

Linewidths of the $^{85}Rb^+$ and $^{87}Rb^+$ signals in aqueous solutions are of the order of 150 Hz. $^{85}Rb^+$ signals on 50 mM solutions have in our laboratory been obtained with an old wideline NMR spectrometer ($\phi \simeq 15$ mm, $B_0 \simeq 1.4$ T) and FT techniques could presumably improve on this.

An extrapolation to the potassium isotopes ^{40}K and ^{41}K is somewhat more uncertain. A signal from $^{40}K^+$ in 31 molal KNO_2 in D_2O has been obtained after $3.2 \cdot 10^6$ pulses

* It may be pointed out that in considerations of the efficiency of the quadrupolar relaxation of ions in solution the value of the product $Q(1 + \gamma_\infty)$ is more relevant than is Q alone.

and 10 hours using the rapid pulsing "Quadriga" FT technique.[3] This technique is suitable only for the location of a NMR signal since the linewidth is independent of the natural linewidth but inversely dependent on the pulse spacing used. This experiment is so far the only observation of ^{40}K NMR in bulk.

Several representative NMR spectra of the alkali nuclei, as well as of many other nuclei, may be found in Ref. 34.

6C AQUEOUS SOLUTIONS OF INORGANIC SALTS

6C.1 Introduction

It is, of course, natural both from a chemical and an NMR point of view that aqueous solutions of simple inorganic salts during a long period of time stood in the foreground in investigations of the magnetic resonances of the alkali nuclei. Since these solutions have been well characterised by a manifold of physicochemical methods, they constitute an ideal starting point for the chemist. For the NMR spectroscopist they have several qualities: In contrast to other systems, such as solids or solutions in other solvents, broadening of the signals is relatively insignificant and only a single peak is obtained and, furthermore, it is possible to attain high concentrations. For the interpretation and theoretical development it is most significant that, with a suitable choice of counterion, the effect of ion-ion interactions on NMR parameters can be minimised.

Mainly to avoid line-broadening even at reasonably high concentrations it was natural that both the early detections of the alkali resonances as well as recent high-precision determinations of the nuclear magnetic moments have been concerned with aqueous solutions of simple salts.[4–29] It is of historical interest to mention that the first observations of the alkali resonances were the following: ^{6}Li, refs. 16 and 30; ^{7}Li, refs 4 and 31–33; ^{23}Na, refs 5–7; ^{39}K, ref 15; ^{40}K, ref 3; ^{41}K, ref. 23; ^{85}Rb, refs 8 and 10–12; ^{87}Rb, refs 8–10 and 12; ^{133}Cs, refs 8 and 10–12. Very precise determinations of the resonance frequencies of the alkali nuclei have been performed by in the first place the Tübingen group.[27–29, 35–37] As will be considered below these studies have been of great importance in establishing absolute shielding scales.

In our discussion of alkali NMR studies of aqueous solutions we will follow a path of increasing complexity, treating the various NMR parameters in parallel. Thus we will set out by treating the free hydrated aqueous ion, the NMR parameters of which are obtained by extrapolating to infinite dilution. Next NMR studies concerned with the interaction of alkali ions with simple anions like halide ions, $[NO_3]^-$, $[SO_4]^{2-}$, etc. are presented. The special shielding and relaxation effects due to paramagnetic ions are then discussed as well as the hitherto sparse diffusion studies which terminate this chapter. The more complex systems, i.e. alkali ions in mixed solvents and the complexes of alkali ions with small molecules or ions are outlined in the sections to follow.

6C.2 Relaxation of the Free Aqueous Ions

In principle the rate of nuclear magnetic relaxation of an ion in solution can be assumed to contain direct and valuable information, not easily obtainable by other methods, on the ion-solvent interactions and on the structure of the solvation sheath. However, the extraction of this information is by no means a simple process but is associated with several problems. First, the experimental determination of the relaxation rate at sufficiently high dilution may be difficult for nuclei with low NMR receptivities. For aqueous solutions one is, however, greatly helped by the fact that for several salts there is only a weak concentration dependence of relaxation thus making the extrapolation error small. For the case where there is an appreciable variation of relaxation with concentration it is advisable to base the extrapolation on the theoretical considerations of Richards and co-workers.[38,39] The second main problem concerns the relaxation mechanism. Disregarding presently the paramagnetic systems, it is straightforward to establish that the rapid relaxation of $^{23}Na^+$, $^{85}Rb^+$ and $^{87}Rb^+$ can only be due to quadrupolar effects, while rather detailed considerations are required for example for the Li isotopes. The next step will be to construct a theory of relaxation based on a model of the microdynamics and structure of the solutions. In so doing for the case of quadrupole relaxation, a major problem lies in the fact that several potential sources of the time-modulated electric field gradients have to be considered. In the final step a comparison between experimental and theoretical results is made to judge the quality of the different models.

a. Experimental Results of the Relaxation in Infinitely Dilute Aqueous Solutions

The solvation of small ions has been a field of intense research, both experimental and theoretical, during the last decades but it has been a general difficulty to reconcile results obtained by different experimental approaches and to develop theoretical models rationalizing more than a limited set of experimental observations. In the last few years, however, three approaches to the problem have been quite successful, i.e. X-ray and neutron diffraction,[40] quantum mechanical calculations[41] and gas phase equilibria studies.[42] All these methods have been found to give reliable information on solvation energetics and structure but are limited in their application to low molar ratios of solvent to ion and the relevance of the results to more dilute, and for many problems more interesting, solutions can be questioned. NMR quadrupole relaxation studies are found to be a good complement: This method is applicable to quite low concentrations and provides, given a satisfactory relaxation theory, quite significant information on both structural and dynamic aspects of ion solvation.

To use alkali ion quadrupole relaxation to study hydration phenomena has been considered during a long period of time and the feasibility of determining the infinite dilution relaxation rate in water by extrapolation of experimental results was also established in early work. Until quite recently the majority of experimental investigations was line width studies and it is a general experience that these in many cases overestimate the relaxation rates significantly.[43,44] In the collection of "best" experimental results given in Table 6.2 we have therefore preferred to give only direct relaxation time determinations by pulse techniques. The values listed in Table 6.2 should be correct within ca. 10% or better.

b. Contributions from Different Relaxation Mechanisms for the Aqueous Alkali Ions

Before an interpretation of the data given in Table 6.2 can be attempted it is important to achieve a separation of observed relaxation rates into contributions from different relaxation mechanisms. For the case of the aqueous $^{23}Na^+$ ion, for which an analysis was made by Eisenstadt and Friedman,[45] $^{39}K^+$, $^{41}K^+$, $^{85}Rb^+$ and $^{87}Rb^+$ even rough estimates tell that quadrupole relaxation is greatly dominating over other mechanisms. For example, the magnetic dipole-dipole mechanism can be excluded by considering the quite sizeable relaxation rates in relation to rather small nuclear magnetic moments. Furthermore, in most of these cases direct demonstration of the absence of appreciable dipole-dipole and spin rotation relaxation has been obtained from investigations[3,45–48]

TABLE 6.2 Quadrupole relaxation of alkali ions at infinite dilution in water at ca. 25°C

	$(1/T_1)_{obs}$, s^{-1}	$1 + \gamma_\infty$	$[r_0]10^{-10}$ m	$(1/T_1)_{FRD}$, s^{-1}	$(1/T_1)_{SS}$, s^{-1}
$^7Li^+$	0.027[a]	0.74	2.08	0.037	0.20
$^{23}Na^+$	15.8–17.55[b,c,d]	5.1	2.40	4.4	24.2
$^{39}K^+$	17, 18[e]	18.3	2.73	13.3	
$^{85}Rb^+$	420[c]	48.2	2.88	200	
$^{133}Cs^+$	0.071–0.080[c,d,f]	111	3.07	0.055	

Comment: The table is adapted after Hertz.[62] For details on correlation times used and for the calculation of $(1/T_1)_{SS}$, the relaxation rate in the presence of a distinct first solvation sphere, the reader is referred to this article.

[a] Ref. 53. In D_2O. In H_2O 0.55 s^{-1} due to contribution from dipole-dipole relaxation. [b] Refs 45, 61, 73, 74. [c] Ref. 48. [d] Ref. 47. [e] Refs 3, 74. [f] Refs 58, 59; 30°C.

of the H_2O/D_2O isotope effect as well as of the temperature dependence of relaxation.[45,50–52] The remaining three nuclei, ^6Li, ^7Li and ^{133}Cs, are characterised by quite small relaxation rates of the aqueous ions and the contribution from non-quadrupolar relaxation mechanisms has to be carefully examined. For $^7Li^+$ several studies have demonstrated that the relaxation is considerably more effective in H_2O than in D_2O. For pure quadrupolar relaxation one expects the relaxation rate to be slightly (ca. 15–20%) larger in D_2O than in H_2O due to a slightly longer water correlation time. On the basis of this correlation time effect Hertz et al.[53] have achieved a separation of the total relaxation rate into two contributions, one amounting to 0.031 s^{-1} (at 25°C) due to dipole-dipole interactions and another amounting to 0.024 s^{-1} due to quadrupolar interactions. Similar considerations have been presented also by others[33,54,55] while in a number of cases the interpretation has been made only in terms of quadrupolar effects.[56,57]

The $^6Li^+$ ion has been very little studied and it is only in a recent investigation[2,58] of Wehrli that a good understanding of its relaxation behaviour is obtained. The experimental relaxation rate of $^6Li^+$ in a 3.9 M LiCl solution in H_2O is only 0.0059 s^{-1} at 30°C while it is as low as 0.0012 s^{-1} in D_2O solution. (The values obtained in D_2O solution at higher temperature are probably the longest relaxation times reported for liquid systems.) From the $^7Li^+$ quadrupolar relaxation rate and the ^6Li and ^7Li quadrupole moments, the quadrupole relaxation is found to be totally negligible in the case of $^6Li^+$. In two independent ways, i.e. from studies of the nuclear Overhauser enhancement and from a consideration of the ^7Li dipole-dipole relaxation rate[53] in combination with the nuclear magnetic moments of the two isotopes, it could be established that dipole-dipole relaxation accounts for ca. 86% of the relaxation rate at 25°C while it falls off to 48% at 100°C. From the temperature dependence of the nondipolar contribution it was suggested to be mainly spin-rotation relaxation.

The relaxation of the remaining ion, $^{133}Cs^+$, has been investigated in both H_2O and D_2O solution[47,48] and from these studies it could be shown that in spite of the small relaxation rate dipole-dipole relaxation does not make an appreciable contribution. This conclusion is confirmed by Wehrli's recent study[58,59] of the nuclear Overhauser enhancement and we will consider the quadrupole relaxation as the sole significant mechanism for $^{133}Cs^+$.

c. *Electrostatic Model of Ion Quadrupole Relaxation*

In the "electrostatic" approach to the quadrupole relaxation of monoatomic ions, the field gradients are considered to be created by the distribution of charges set up by surrounding species. The field gradients fluctuate with time as a result of molecular motion and cause relaxation. To simplify the analysis, the surrounding ions and molecules are treated as point charges and dipoles and the effect of distortion of the closed electron shell from spherical symmetry is taken into account by using the Sternheimer antishielding factor. The electrostatic theory has been worked out during a long period of time by Hertz[48,60–63] and Valiev and co-workers;[64–71] the different presentations differ in details and in completeness while the general ideas are the same. We will consider here the recent work by Hertz[48,62,63] which is the most elaborate treatment. It will not be possible to penetrate in details different possible situations treated and simplifying assumptions made. Instead the most relevant theoretical expressions will be given with a few brief comments.

For an ion at infinite dilution in a solvent the field gradients arise from the distribution of the solvent dipoles around the relaxing ion. Several motions may modulate these field gradients, but Hertz considers in particular the reorientation of the solvent dipoles and of the ion-solvent molecule vector. From Hertz' analysis it appears that the lateral distribution and orientation of the solvent molecules in the first solvation sphere play an important role; for a rigid well-ordered solvation complex of cubic symmetry, no contribution from the first solvation sphere results. It is found to be useful to distinguish between three states of solvation[72] and we will here consider two of these. The "fully random distribution" (FRD) model assumes both a random distribution and a random dipole orientation of the solvent molecules, i.e. there is no distinct first solvation sphere in this model. The resulting relaxation rate is obtained to be (in the limit of extreme narrowing)

$$\frac{1}{T_1} = \frac{1}{T_2} = \frac{24\pi^3(2I+3)}{5I^2(2I-1)}\left(\frac{eQ}{h}\right)^2 \{\mu(1+\gamma_\infty)P\}^2 \frac{C_s\tau_s}{r_0^5}. \tag{6.1}$$

Here μ is the solvent dipole moment, P a polarisation factor (assumed to be 0.5), C_s the solvent concentration, τ_s the solvent correlation time, generally taken to be the reorientational correlation time of the pure solvent. r_0 is the distance between the relaxing ion and an adjacent solvent molecule and is taken to be the sum of the ionic radius and the radius of the solvent molecule. Eq. 6.1, which should apply in the limit of weak solvation, is well adapted for testing although the low accuracy of especially the Sternheimer antishielding factor is a limiting factor.

Quadrupole relaxation rates of the aqueous alkali ions calculated by means of Eq. 6.1 are given in Table 6.2 and in view of the simplifying assumptions made and the inaccuracy of some parameters used, the agreement must in general be considered to be most satisfactory.

Turning now to Hertz' "fully oriented solvation" (FOS) model, this is characterized by a distinct first solvation sphere in which the electric dipole moments have radial orientation. For a solvation sphere of tetrahedral or octahedral symmetry, the relaxation rate is for the extreme narrowing situation given by

$$\frac{1}{T_1} = \frac{1}{T_2} = \frac{2\pi^2(2I+3)}{I^2(2I-1)}\left(\frac{eQ}{h}\right)^2 \{\mu(1+\gamma_\infty)P\}^2 \frac{n_s}{15}\left(\frac{9}{r_0^4}\right)^2 \cdot\left\{1 + \frac{1}{2}\sum_{j=2}^{n_s}(3\cos^2 v_{0j} - 1)e^{-6\lambda}\right\}\tau_c^* + \frac{12\pi}{5}\frac{C_s}{r^{*5}}\tau_s \tag{6.2}$$

The symbols introduced have the following meaning: n_s is the number of solvent molecules in the first solvation layer, τ_c^* the reorientational correlation time of the ion-solvent molecule vector and r^* the distance between the relaxing ion and a solvent molecule in the second solvation layer. ν_{0j} is an angle describing the relative orientation of the solvent molecules in the first solvation layer and λ a parameter describing the width of the angular distribution in the first solvation sphere: $\lambda \to 0$ corresponds to solvent molecules on tetrahedral or octahedral sites and $\lambda \to \infty$ corresponds to random lateral distribution. For the case of cubic symmetry, Eq. 6.2 can be seen to become identical to Eq. 6.1 with r_0 replaced by r^*. The main contributions to quadrupole relaxation then come from the second solvation sphere.

As shown by Hertz[62] the electrostatic theory of ion quadrupole relaxation can be said to provide a good rationalization of the results for the alkali ions at infinite dilution in water. A major step forward would, of course, result if the field gradient term and the correlation time (and not only their product) could be separately determined experimentally. Recently a successful investigation of this type was presented by Geiger and Hertz,[51] working with glycerol and low-temperature aqueous systems. These results provide rather convincing support for the electrostatic model (see ch. 6D.1*a*).

d. Electronic Distortion Model of Ion Quadrupole Relaxation

It is mainly Deverell[75,76] who has been an advocate of the electronic distortion model and has developed appropriate expressions although the original idea seems to go back to Itoh and Yamagata[77] and Hertz[62,48] recently improved the theoretical treatment. Deverell in his approach considers the origin of the field gradients causing relaxation to be distortions from spherical symmetry of the electron cloud which are produced in collisions by short-range overlap repulsive forces. If so, there should be a relation between the field gradients and overlap integrals involving the outer orbitals of the relaxing ion. To estimate the field gradients, Deverell made use of the fact that these overlap integrals are related to the paramagnetic shielding term, σ_p, according to the Kondo-Yamashita theory (see ch. 6C.5*b*). Deverell obtained the following result for extreme narrowing

$$\frac{1}{T_1} = \frac{1}{T_2} = \frac{3\pi^2}{1000} \frac{2I + 3}{I^2(2I - 1)} \left(\frac{e^2 Q}{h}\right)^2 \left(\frac{\sigma_p \Delta E}{\alpha^2}\right) \tau_c \tag{6.3}$$

Here $\alpha = e^2/\hbar c$, the fine structure constant. The correlation time should be a collision time though Deverell identifies it as the water reorientational correlation time. Partly due to the difficulty in estimating τ_c but also for other reasons Deverell's theory is difficult to test.

e. Dipole-Dipole Relaxation

For $^6Li^+$, dipole-dipole effects dominate relaxation while for $^7Li^+$ it accounts for around half of the relaxation rate. Hertz *et al.*[53] made a thorough attempt to rationalise the dipole relaxation of $^7Li^+$ at infinite dilution in water. Thereby they divided the dipole-dipole relaxation into two contributions, one due to the first hydration sphere and the other due to the remaining water molecules. The following expression was used in the analysis

$$\frac{1}{T_1} = \gamma_{Li}^2 \gamma_H^2 \hbar^2 \frac{\tau_c}{b^6} n^* + \frac{8\pi}{15} \gamma_H^2 \gamma_{Li}^2 \hbar^2 \frac{C_{H_2O}}{aD_{LiH}} \tag{6.4}$$

Here b is the distance between the relaxing Li^+ ion and the water protons in the first hydration sphere and n^* the number of protons at distance b. The correlation time of the contribution from the first hydration layer depends both on the reorientational correlation time of the Li^+-water proton vector and on the residence time of the water molecule in the first hydration sphere. C_{H_2O} is the water concentration, a the distance of closest approach between the Li^+ ion and a water molecule outside the first hydration sphere and D_{LiH} the mean of the Li^+ and H_2O self-diffusion coefficients. The second term of Eq. 6.4 is calculated in a straightforward manner by Hertz *et al.* and then they are able to determine τ_c. From a comparison of this value with the correlation time of the H-H vector of a water molecule in the first hydration layer it could be suggested that there is an appreciable rotation of the water molecules around the Li-O axis.

6C.3 Absolute Shielding Values of Alkali Metal Ions in Aqueous Solutions

To gain any deeper understanding of the physical nature of the interactions that affect the chemical shifts of ions in solutions it would clearly be of greatest importance to have access to *absolute* values of the shielding constants. In the case of alkali metal ions in aqueous solutions the situation is more fortunate than for most other elements. The free atoms may be readily produced and their nuclear magnetic moments have been determined to a high accuracy through atomic beam magnetic resonance or optical pumping experiments. Comparison of these values with values obtained through NMR studies on the ions in aqueous solutions have yielded the difference in shielding constants between the free atoms and the ions in solutions, $\sigma_{M^+,aq} - \sigma_{M,free}$. To obtain the absolute shielding constant $\sigma_{M^+,aq}$ i.e. the shielding relative to the bare nucleus, the shielding constant $\sigma_{M,free}$ must be known. The calculation of the latter is however in principle very simple and involves only the ground state wave function of the atom. The accuracy of the calculated value of $\sigma_{M,free}$ is primarily dependent on the accuracy of the atomic wave function employed.

In Table 6.3 we have summarised recent experimental data on the differences $\sigma_{M^+,aq} - \sigma_{M,free}$.* In a special type of optical pumping experiments, charge exchange collisions $M^+ + {}^*M \rightarrow {}^*M^+ + M$ couple the nuclear spin polarization of atoms and ions and make possible the determination of the magnetic moment also of the free M^+ ion. From these experiments it is then possible to calculate $\sigma_{M,free} - \sigma_{M^+,free}$. These data are also included in Table 6.3.

The differences between $\sigma_{M,free}$ and $\sigma_{M^+,free}$ can on theoretical grounds be estimated to be small, especially for the heavier atoms since the major contribution to the shielding arises from the inner shell electrons. This is born out in Table 6.4 where we have listed theoretical values of nuclear shielding constants for the free alkali metal atoms and ions. The data for the heavier atoms in this table should be regarded with some caution since they are actually obtained by extrapolation from values calculated for the lighter atoms and ions for which Hartree-Fock type wave functions are available. The experimental data for the differences $\sigma_{Rb,free} - \sigma_{Rb^+,free}$ and $\sigma_{Cs,free} - \sigma_{Cs^+,free}$ (cf. Table 6.3) are not very accurate but differences appear to be small. No significant difference in shielding between the pairs of lithium and rubidium isotopes is observable. Since the diamagnetic shift changes are small the difference $|\sigma_{M^+,aq} - \sigma_{M^+,free}| \simeq |\sigma_{M^+,aq} - \sigma_{M,free}|$ may be regarded as the paramagnetic contribution to the shielding of the alkali ions caused by the

* The shielding differences refer to the ions at infinite dilution in D_2O or H_2O solutions. A small solvent isotope effect on the shielding has been observed for the heavier alkali metal nuclei, cf. ch. 6C.6.

TABLE 6.3 Values of $\sigma_{M^+, aq} - \sigma_{M, free}$ and $\sigma_{M, free} - \sigma_{M^+, free}$ obtained through atomic beam resonance or optical pumping experiments

M	$\sigma_{M^+,aq} - \sigma_{M,free}$	$\sigma_{M,free} - \sigma_{M^+,free}$	Comments	Ref.
^6Li	$(-11.4 \pm 0.8) \cdot 10^{-6}$	—	D_2O; 24 ± 1°C	78
^7Li	$(-11.0 \pm 0.7) \cdot 10^{-6}$	—	D_2O; 24 ± 1°C	78
^{23}Na	$(-60.5 \pm 1.0) \cdot 10^{-6}$	—	D_2O; 24 ± 1°C	78
^{39}K	$(-105.1 \pm 1.0) \cdot 10^{-6}$	—	H_2O; 25°C	3
^{40}K	$(-100 \pm 300) \cdot 10^{-6}$	—	H_2O; 25°C	3
^{41}K	$(-105.4 \pm 1.2) \cdot 10^{-6}$	—	H_2O; 25°C	3
^{85}Rb	$(-211 \pm 2) \cdot 10^{-6}$	—	D_2O; 28 ± 2°C	36
^{87}Rb	$(-211.6 \pm 1.2) \cdot 10^{-6}$	—	D_2O; 24 ± 1°C	78
		$(3.8 \pm 2.6) \cdot 10^{-6}$		79
^{133}Cs	$(-344.0 \pm 1.5) \cdot 10^{-6}$	—	D_2O; 24 ± 1°C	80
	$(-344.3 \pm 5.8) \cdot 10^{-6}$	—	D_2O; 24 ± 1°C	78
		$(14 \pm 13) \cdot 10^{-6}$		81
		$(14 \pm 12) \cdot 10^{-6}$		82

TABLE 6.4 Theoretical values of nuclear shielding constants for free alkali atoms, $\sigma_{M,free}$, and ions, $\sigma_{M^+,free}$

M	$\sigma_{M,free} \cdot 10^6$	$\sigma_{M^+,free} \cdot 10^6$	Ref.
Li	101.45	95.40	[b]
	(110.71)[a]	95.45	[c]
Na	628.87	623.82	[b]
	(639.65)[a]	634.34	[c]
K	1329.34	1325.40	[b]
	(1351.66)[a]	1341.81	[c]
Rb	(3338.19)[a]	(3346.36)[a]	[b]
	(3400.45)[a]	(3390.83)[a]	[c]
Cs	(5776.5)[a]	(5687.98)[a]	[b]
	(5822.41)[a]	(5815.06)[a]	[c]

[a] These values have been obtained through extrapolation from values calculated for the lighter elements using Hartree-Fock or Hartree-Fock-Slater type wavefunctions. [b] G. Malli and S. Fraga, *Theoret. Chim. Acta.* **5**, 275 (1966). [c] K. M. S. Saxena and P. T. Narasimhan, *Inst. J. Quant. Chem.* **1**, 731 (1967).

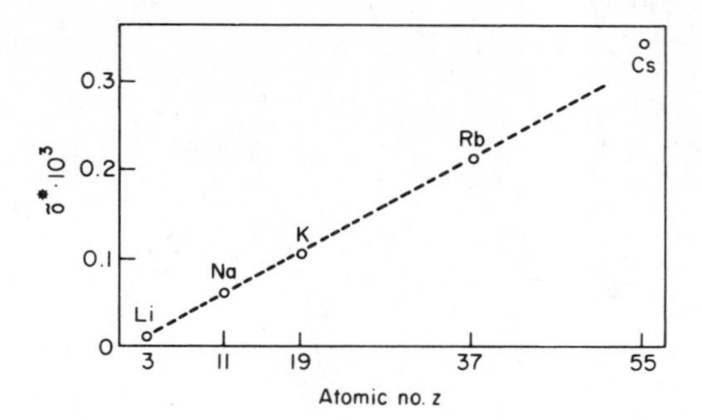

Fig. 6.1 Shielding constants, $|\sigma_{M^+,aq} - \sigma_{M,free}|$, of the aqueous alkali ions at infinite dilution relative to the free atoms as a function of the atomic number (reproduced with permission from ref. 3).

solvation sphere of water molecules. These differences are as first noted by Sahm and Schwenk[3] close to a linear function of the atomic number Z (cf. Fig. 6.1.)

6C.4 Effects of Ion-Ion Interactions on Relaxation

a. *Experimental Studies of Aqueous Alkali Halide Solutions*

The interest in the ion distribution and the dynamics of ion encounters has been the motivation behind a large number of investigations of alkali ion relaxation in simple aqueous electrolyte solutions. Originally, a simple relation between the magnitude of the relaxation rate change with concentration and the degree of ion-ion interaction was assumed but, as will be shown below, symmetry effects may have a great influence and have to be analysed in detail. For several reasons, suitability for theoretical analysis, frequent occurrence in chemical and biological studies, high solubility, etc., the alkali halides are by far the most studied salts.

A large number of accurate investigations are available in the literature as regards $^7Li^+$ and $^{23}Na^+$, and to some extent also $^{85}Rb^+$ and $^{133}Cs^+$, while for the other cases information available is rather sparse. The only relevant study of $^6Li^+$ is that of Wehrli[2] but here the object was not to obtain the ion-ion contribution to relaxation. Although experimentally difficult, studies of the concentration dependence of $^6Li^+$ relaxation can be expected to be a good complement to $^7Li^+$ relaxation because of the different weights of different relaxation mechanisms in the two cases. The concentration dependence of $^7Li^+$ relaxation in alkali halide solutions has been studied by several workers.[33,47,48,51,53–57,61,83–89] According to the investigations of Hertz and co-workers[48,53,61] exemplified in Fig. 6.2, the $^7Li^+$ quadrupole relaxation rate for the lithium halides increases in the series $I^- < Br^- < Cl^-$ and the effect of F^- should be markedly greater.[86] Hertz *et al.*[53] made a separation of their results for H_2O solutions into quadrupole and dipole-dipole contributions. While the total $^7Li^+$ dipole-dipole relaxation rate is only little concentration dependent, it appears from their analysis that especially for LiCl the contribution from the first hydration sphere decreases with increasing concentration.

The concentration dependence of $^{23}Na^+$ relaxation in aqueous sodium halide solutions has not surprisingly been extensively studied;[26,45,47,48,52,61,83–85,87,90–95] however, in most of the linewidth studies there is a considerable experimental line-broadening. The most reliable data are given in a number of pulsed NMR studies,[26,45,

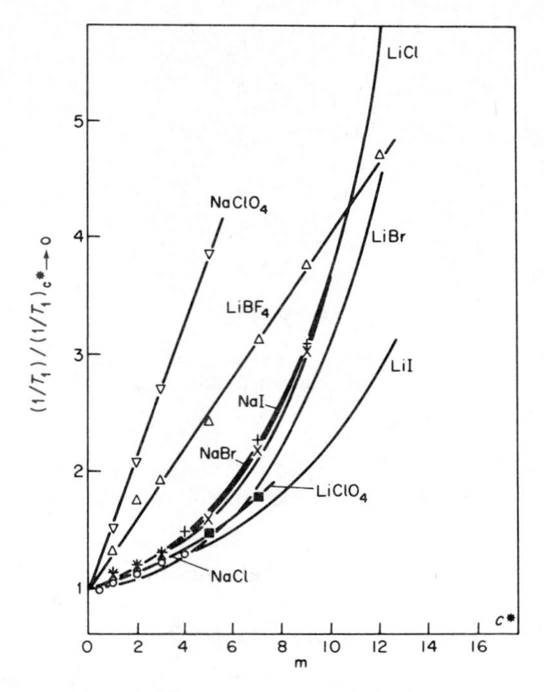

Fig. 6.2 Relaxation rates of $^{7}Li^{+}$ and $^{23}Na^{+}$ in aqueous solutions of different salts at 25°C. Data given relative to the infinite dilution values (reproduced with permission from ref. 48).

[48,61,73,87] and are illustrated in Fig. 6.2. From the work of Eisenstadt and Friedman,[26] the $^{23}Na^{+}$ relaxation rate appears to follow the sequence $Cl^{-} < Br^{-} \simeq I^{-} < F^{-}$ and the results of refs. 48, 73, 87, are consistent with this. (The position in the sequence of Br^{-} and I^{-} varies between different experimenters and may also change with concentration.)

The only study concerned with the concentration dependence of K^{+} relaxation is that of Sahm and Schwenk[3] who found using ^{39}K NMR the relaxation rate sequence $Cl^{-} < I^{-} < Br^{-} < F^{-}$.

We illustrate Rb^{+} relaxation with the ^{87}Rb T_1 and ^{85}Rb linewidth data of Hertz *et al.*[48] given in Fig. 6.3 and as can be seen the sequence of increasing relaxation is $Cl^{-} < Br^{-} < I^{-} < F^{-}$. The pronounced minimal relaxation rate at intermediate concentrations of RbCl may be noted. Other reports of the relaxation of $^{85}Rb^{+}$ and $^{87}Rb^{+}$ in alkali halide solutions are given in refs. 47, 50, 61 and 96.

For $^{133}Cs^{+}$ relaxation, the effect of the halide ions follows the sequence $Br^{-} < I^{-} \simeq Cl^{-} < F^{-}$ according to the detailed study of Hertz *et al.*,[48] and other investigations are reported in refs. 47, 51, 58 and 84.

Summarising the concentration dependence of alkali ion quadrupole relaxation in aqueous alkali halide solutions, it can only be stated that except for F^{-}, which gives a moderate increase in relaxation rate, the effects are quite small. The same relaxation rate changes are obtained closely for Cl^{-}, Br^{-} and I^{-} and the sequence of these ions bears no simple relationship with ion size.

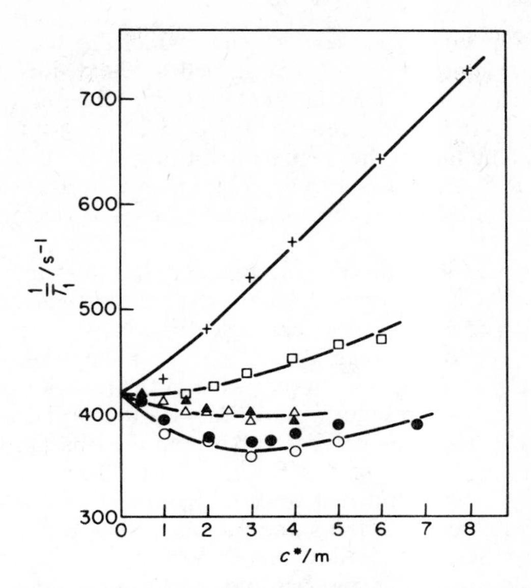

Fig. 6.3 Rb^+ relaxation rates in aqueous solutions of rubidium halides at 25°C. Open symbols give ^{85}Rb linewidth data and filled ones ^{87}Rb T_1 data converted to ^{85}Rb relaxation. The following salts are given: RbF (+), RbCl (○, ●), RbBr (△, ▲) and RbI (□) (reproduced with permission from ref. 48).

b. Electrostatic Treatment of Ion-Ion Effects on Quadrupole Relaxation

The electrostatic approach to the ion-ion contributions to the quadrupole relaxation of an ion involves taking into account the time-dependent field gradients produced by other ions diffusing relative to the relaxing ion.[48,60,61,63,69,70] The ion-ion effects can be expected to depend on the concentration and valency of the ions causing relaxation, the distance of closest approach between ions, and the ions' translational mobilities. In early theoretical work on this problem, it was found that even after taking into account a modification of the ion-water term due to the added ions, the theoretically calculated relaxation rates were far too large. A most successful theoretical analysis of the very difficult problem of ion-ion contribution to quadrupole relaxation has recently been put forward by Hertz.[43,48,63] Here only a few remarks on the problems arising can be given.

In a schematic way, the ion relaxation rate may be written

$$\frac{1}{T_1} = \frac{1}{T_2} = \pi^2 \frac{2I + 3}{I^2(2I - 1)} \left[\frac{eQ(1 + \gamma_\infty)}{h}\right]^2 (A + B + C + D) \qquad (6.5)$$

Here A is the same term that appeared in the FRD model (ch. 6C.2*c*) but we have now to take into account a change with concentration of in particular C_{H_2O} and τ_{H_2O}. This in general causes no great difficulties as τ_{H_2O} may be estimated from separate NMR relaxation studies on the water molecules. *B* corresponds to another indirect ionic effect, i.e. the effect of other ions on the relative orientation of water molecules around the relaxing ion and hence on the water-water correlation. This effect can be expected to be particularly important for ions having an approximately symmetrical first solvation layer via a partial unquenching of the field gradients which are due to the water molecules. C, which is a pure ionic contribution, is determined by the distance of closest approach between two ions and the relative translational motion of the ions relative to each other. An important new aspect of Hertz' recent treatment[63] is the finding that ion-ion correlations cause a strong quenching of the ionic contribution to the field gradient through an ion cloud type effect. D finally takes into account ion-water cross-correlation effects. In ref. 48 the theory is extended to the examination of higher ion-ion and ion-water correlations.

As the complete expression for the ion-ion contributions to an ion's quadrupole relaxation becomes intractable, the path of analysis of experimental results must pass through simplifying considerations. For example, for weakly hydrated ions it may be a good first approach to only consider terms A and C, i.e. to neglect local structural effects. The relevance of this can be judged from the value of the ion-cloud parameter obtained in the analysis. To test for the presence of short range structural effects, it is found useful to consider a quantity Δ being the difference between the observed relaxation rate and the solvent contribution according to A in Eq. 6.5. Thus if the ion contribution to relaxation corresponds to a random ion distribution around a cation-anion pair then this contribution should be essentially the same for both ions in a solution. From widely different values of Δ for alkali and halide ions, from results on solutions containing two alkali halides as well as some other observations, Hertz *et al.*[48] demonstrate the importance of higher ion-ion correlations for the alkali ions. The observation of maxima in the ion contribution as a function of composition for mixed electrolyte solutions may be explained in terms of a perturbation of the ion quenching effect due to the presence of two different anions. For Li^+, Δ is very much larger than that of the halide ions and this is discussed in terms of distortions of the tetrahedral water symmetry around the Li^+ ion on approach of a halide ion; this effect is also important for Na^+.

c. Electronic Distortion Theory of Ion-Ion Contributions

In the approach due to Deverell,[75,76] the field gradients are due to distortions of the electron cloud of the ion due to ion-ion collisions. In a similar way as for the ion-solvent contribution, Deverell relates the ion-ion term to the paramagnetic shielding and obtains

$$T_1^0/T_1 = (1 + 2\Delta\sigma_p^{\text{ops}}/\sigma_p^0)\Sigma\tau_c^k/\tau_c^0. \tag{6.6}$$

Here superscript 0 refers to infinite dilution, $\Delta\sigma_p^{\text{ops}}$ is the observed shielding relative to infinite dilution and τ_c^k the correlation time which should characterize the ion-ion collisions. Deverell's electronic distortion approach has been difficult to test mainly because of difficulties in estimating τ_c^k reliably. The counter-ion sequences in relaxation and shielding show strong deviations and a further argument against this treatment can be found in observations of maxima in the relaxation rate for mixed electrolyte systems. Thus quenching effects should not occur according to Deverell's electronic distortion theory. This and some other aspects of the electronic contributions to relaxation have been treated by Hertz *et al.*[48]

d. Studies of Other Aqueous Inorganic Electrolytes

In addition to the alkali halides, the alkali ion relaxation has been investigated for aqueous solutions of *inter alia* the following simple salts:

$^{7}Li^{+}$: $[OH]^{-}$,[88,97] $[NO_3]^{-}$,[32,33,85,88,89] $[SO_4]^{2-}$,[88] $[ClO_4]^{-}$,[48,88] $[BF_4]^{-}$;[48,98]
$^{23}Na^{+}$: $[OH]^{-}$,[26,92,95,99] $[NO_3]^{-}$,[26,73,85,100] $[ClO]^{-}$,[92] $[ClO_4]^{-}$,[26,45,73,85,95] $[B(C_6H_5)_4]^{-}$,[26,95] $[H_2PO_4]^{-}$,[92] aluminate,[101,102] $[ReO_4]^{-}$,[26,103] $[ClO_3]^{-}$,[26] $[BrO_3]^{-}$,[26] $[IO_3]^{-}$,[26] $[SCN]^{-}$,[26] $[IO_4]^{-}$,[26] $[BF_4]^{-}$,[26] $[PF_6]^{-}$,[26] $[HSO_4]^{-}$,[26] $[SO_4]^{2-}$,[26] $[S_2O_3]^{2-}$,[26] $[PO_4]^{3-}$;[26]
$^{39}K^{+}$: (all ref. 3) $[Fe(CN)_6]^{3-}$, $[Fe(CN)_6]^{4-}$, $[C_2O_4]^{2-}$, $[HSO_4]^{-}$, $[NO_3]^{-}$, $[OH]^{-}$, $[NO_2]^{-}$.

It may also be added that the effect of certain cations on the $^{23}Na^{+}$ and $^{85}Rb^{+}$ relaxation has been investigated[61,104] and that early ^{23}Na linewidth studies[90,91] included a large number of systems.

In considering these data, which are generally much less detailed and systematic than those of alkali halide solutions, the large effect of some polyatomic anions on the relaxation of Li^{+} (e.g. $[BF_4]^{-}$) and Na^{+} (e.g. $[ClO_4]^{-}$, $[BF_4]^{-}$ and $[PF_6]^{-}$) is particularly striking but no satisfactory explanation to these findings has yet been presented.

e. Temperature Dependence of Relaxation

Studies of the temperature dependence of an ion's relaxation can, in view of the information they provide on the potential barriers associated with the relaxation process, be expected to be helpful in discussions of relaxation mechanisms and also to give additional information on the dynamics of the system. An interesting approach is that of Hertz and co-workers[50] who compared the Arrhenius' activation energies of ion quadrupole relaxation with those for the reorientation of the water molecules in the first hydration layer. For $^{85}Rb^{+}$ the activation energy of ion relaxation is 10.5 kJ/mol which is slightly lower than that of hydration water molecule rotation, while for $^{23}Na^{+}$ the activation energy of ion relaxation (10.5 kJ/mol[45]) is much smaller than the corresponding result for the hydration water. A possible explanation to this result lies in terms of a partial unquenching of the water contribution to the field gradients leading to a slower decrease with temperature of the relaxation rate than of the water molecule reorientation.

A recent variable temperature study of Geiger and Hertz[51] contains most important information on the relaxation mechanism. $^{7}Li^{+}$ and $^{133}Cs^{+}$ relaxation were studied down to very low temperatures (ca. $-100°C$) achievable by the use of high salt concentrations, and the temperature dependences compared with those of ^{1}H relaxation. As the non-extreme narrowing condition was reached, the correlation times could be determined and it was established that the correlation time of water molecule reorientation corresponds closely to that of ion quadrupole relaxation. In view of the earlier described quantitative correspondence of observed quadrupole relaxation rate with that predicted by the electrostatic theory, it can be concluded on the basis of these findings that both correlation times and field gradients are well rationalised by Hertz' theory (at least for Cs^{+} and Li^{+}), which is a major progress.

6C.5 Effects of Ion-Ion Interactions on Chemical Shifts

a. Survey of Experimental Findings

Since both the magnitude of the chemical shift changes and the line-broadening due to relaxation, varies strongly among the alkali ions, the feasibility of doing accurate studies of the shielding is very different for different cases. Our general knowledge of the dependence of chemical shift on atomic number in a series tells us that the chemical shift range should widen very much from Li^+ to Cs^+. This taken in combination with the above quoted relaxation rates explains why early attempts to observe chemical shifts for aqueous Li^+ and Na^+ ions were unsuccessful and why Cs^+ has received an unproportionally great attention in this respect.

Experimental studies of the chemical shifts of ions in aqueous solutions of inorganic salts have mainly been concerned with establishing the relative effects of different ions and another point of research has concerned the exact shape of the chemical shift vs. concentration curve. In all cases investigated, monotonous changes with concentration were observed but the degree of curvature varies to a large extent.

Somewhat surprisingly, $^7Li^+$ chemical shifts for aqueous solutions have been reported only in two early articles[86,105] where the experimental conditions were somewhat imperfect. The shielding was found to change insignificantly for the lithium halides after applying a susceptibility correction except that a slight low frequency shift is noted in the case of Br^- and I^- for very high salt concentrations (δ around 0.55 ppm for 0.5 mole fraction of LiI).

For $^{23}Na^+$, the chemical shifts are an order of magnitude bigger and may be studied rather easily.[102,106,107] From the data of refs 106 and 107 the sequence of increased shielding is obtained to be $I^- < Br^- < Cl^- < H_2O$ (infinite dilution) $< [NO_3]^- < [ClO_4]^-$. In addition to this, $^{23}Na^+$ chemical shifts were used in an attempt to study the structure of aqueous sodium aluminate solution.[102]

Three accurate and systematic studies of the chemical shift of the aqueous $^{39}K^+$ ion have been reported[3,106,108] and hereby the following sequence of increasing shielding emerges:

$$\begin{aligned}[Fe(CN)_6]^{4-} &< I^- < Br^- < [CN]^- < [PO_4]^{3-} < [OH]^- < Cl^- < [CNO]^- \\ &< [CO_3]^{2-} < [CNS]^- < [CH_3COO]^- < F^- < [N_3]^- < [NO_2]^- \\ &\simeq H_2O < [HSO_4]^- < [SO_4]^{2-} < [CrO_4]^{2-} < [NO_3]^- \\ &< [Cr_2O_7]^{2-}.\end{aligned}$$

Observations of Rb^+ chemical shifts are given in Refs. 109, 106 (^{87}Rb) and 36 (^{85}Rb), the two latter reports giving accurate and detailed concentration dependences leading to the shielding sequence $I^- < Br^- < Cl^- < [CO_3]^{2-} < [SO_4]^{2-} < H_2O < [NO_3]^-$.

$^{133}Cs^+$ chemical shifts are quite large and this in combination with a small relaxation rate explains the large number of $^{133}Cs^+$ chemical shift studies reported.[28,38,39,75,106,109–112] From the detailed studies of Lutz[28,112] increasing $^{133}Cs^+$ shielding is found to follow the sequence $I^- < Br^- < Cl^- < [CO_3]^{2-} < [SO_4]^{2-} < F^- < [HSO_4]^- < H_2O < [NO_3]^-$. In Fig. 6.4 $^{133}Cs^+$ results are illustrated.[106] The effect of cation-cation interactions on $^{133}Cs^+$ shielding has been studied using mixed electrolyte solutions for the following diamagnetic cations: Li^+,[75,110] Na^+,[39,75,110] K^+,[75,110] Mg^{2+},[39] Ca^{2+},[28,39,112] and La^{3+}.[28] Small shifts in the $^{133}Cs^+$ signal were documented by Deverell[75] on partial substitution of another alkali ion for Cs^+ in aqueous solutions of $CsNO_3$. The sequence of increased shielding in $Cs^+ < K^+ < Na^+ < Li^+$.

In conclusion, the qualitative effects of various anions on the shielding are, except for Li^+, well established and it seems that anion sequences are very closely the same for Na^+, K^+, Rb^+ and Cs^+. Another problem of great interest in connection with ion chemical shifts concerns the functional relation between chemical shift and concentration or another appropriate variable. As can be seen in Fig. 6.4, and as is generally found, the concentration dependence is considerably nonlinear and by extending their accurate $^{133}Cs^+$ chemical shift investigations to low concentrations (<10 mM) Richards and coworkers[38,39] established a good basis for theoretical work on this problem.

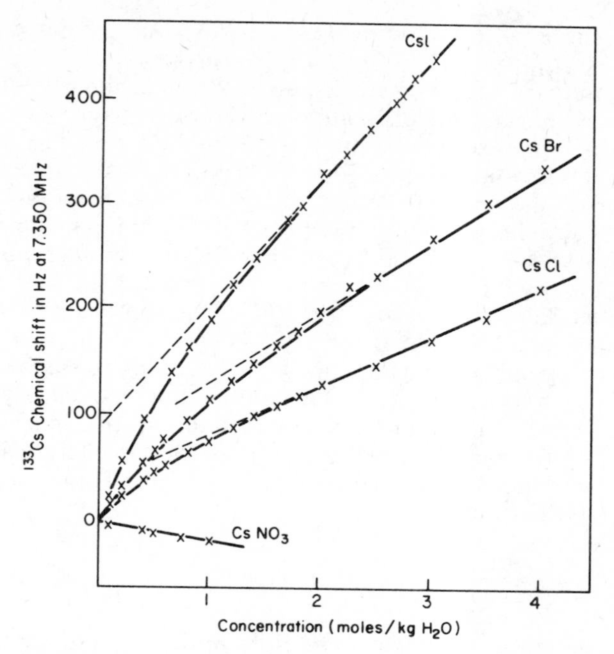

Fig. 6.4 $^{133}Cs^{+}$ chemical shifts in aqueous solutions (reproduced with permission from ref. 106).

b. Aspects on the Interpretation

As a function of time there is, due to the molecular motion, a fluctuation of the shielding due to ion-ion and ion-water interactions and the observed chemical shift is an average over the different possible configurations around the ion. To rationalise the chemical shift it is therefore essential to consider both the probability of various configurations, i.e. ion-water and ion-ion distances, around the studied ion and the magnitude of the shielding contributions exerted by other species when situated at various distances. These two problems may be considered separately as the shape of the concentration dependence of the shielding can be interpreted from the radial distribution functions of water molecules and other ions without assigning a particular physical mechanism behind the shielding. As was noted above, the contribution to the shielding from ion-water interactions is large and it is essential to realize that even small relative changes in this term may appreciably affect observed chemical shifts.

The concentration dependence problem has been examined at considerable depth by Richards and co-workers.[38,39] These authors assumed that the concentration dependence of the chemical shift is essentially determined by cation-anion interactions and, therefore, the concentration dependence of the chemical shift should follow the probability of cation-anion contact. By taking into account the influence of the ionic atomosphere on the probability of ion-ion contact through a Debye-Hückel type of treatment, the $^{133}Cs^{+}$ chemical shift data are accurately represented from the lowest up to very high concentrations. The later work[39] treats, *inter alia,* the difficulties encountered in view of the nonlinear concentration dependences to obtain reliable infinite dilution shifts, as well as the choice of dielectric constant. It is no doubt that this work deserves further attention as it presents a most interesting possibility for testing radial distribution functions for electrolyte solutions.

Several physical interactions have been discussed in connection with the concentration dependence of the chemical shifts of small ions. The shift changes observed can most certainly be considered to arise mainly from changes in the paramagnetic shielding term. Effects on this from electrostatic polarisation due to ion-induced dipole or van der Waals interactions and from the repulsive overlap of the ion wavefunctions with the wavefunctions of other species have been discussed.[38,75,76,106,113] It is not an easy problem to unambiguously choose the correct model, but quantitative estimates on the basis of the different models,[38] as well as other arguments have lead to the overlap model being preferred by most workers in the field.[38,75,106,108] This approach has been elaborated by Kondo and Yamashita[114] and by Ikenberry and Das[113,115] for alkali halide crystals and one important argument for its applicability to solutions has been its successful rationalisation of the shifts for the crystals in combination with closely the same interionic sequences for the crystals as for the aqueous solutions. In a schematic way we can express the paramagnetic shielding in terms of sums of overlap intergrals (A) between the outer p-orbitals of the studied alkali ion (M) and the orbitals of other cations, anions (X) or water molecules (W) as follows

$$\sigma_p - \sigma_p^0 = -\frac{4\mu_B^2\mu_0}{\pi\Delta E}\langle r^{-3}\rangle_p(A_{M-M} + A_{M-X} + A_{M-W} - A_{M-W}^0) \tag{6.7}$$

Here superscript 0 denotes quantities at infinite dilution and $\langle r^{-3}\rangle_p$ the expectation value of r^{-3} for an outer p-electron of the alkali ion.

It is probably a good approximation to assume that ΔE and $\langle r^{-3}\rangle_p$ are essentially independent of concentration and counter-ion for a given alkali ion and that the chemical shift changes may be attributed mainly to the overlap integrals.[75] A comparison between the chemical shift ranges encountered for the aqueous ions reveals that the magnitude of the chemical shift (for comparable conditions) multiplied by $\Delta E/\langle r^{-3}\rangle_p$ increases strongly with increasing atomic number of the alkali ion.[75] This observation as well as the finding of a markedly increased σ_p value with increasing atomic number of the halide ion provides good qualitative support for the overlap model provided the probability of cation-anion contact is not highly variable. (From the studies cited above it appears that the influence of cation-cation overlap is relatively insignificant for the alkali ions and it appears also that the alkali ion-water term changes only little.[38]) The particularly small chemical shifts for Li^+ (and to some extent also Na^+) may be due to the strong hydration reducing considerably the probability of ion-ion contact.

This discussion may be concluded by noting that ion-ion contributions to the shielding are generally easily obtained experimentally and certainly contain highly relevant information on the structure of electrolyte solutions. However, from a theoretical point of view the situation is less satisfactory and the applicability of the Kondo-Yamashita model is not satisfactorily established. It seems that theoretical work on these problems should be most rewarding not the least by providing a firm basis of the studies of more complex systems.

6C.6 Effects of Water Solvent Isotope on Ion Shielding

The chemical shifts of many cation and anion resonances have been found to be different in H_2O and D_2O solutions and this type of solvent isotope effect has also been observed for the heavier alkali metal ions. The experimental results are summarised in Table 6.5.

TABLE 6.5 Water solvent isotope effects on the shielding of alkali metal ions

	$\sigma_{D_2O} - \sigma_{H_2O}$ ppm	Sample	Ref.
^7Li	0.01 ± 0.02	0.5–1.0 M LiCl	121
^{23}Na	0.00 ± 0.1	0.1–0.4 M NaBr	121
^{39}K	0.15 ± 0.10	KNO_2 soln.[a]	3
^{87}Rb	0.29 ± 0.06	0.25–1.00 M RbBr	121
^{133}Cs[b]	0 to 1.5	0.1 to 4.0 M CsCl[c]	44, 118, 121

[a] Nearly independent of the concentration of the salt. [b] $\sigma(H_2\,^{18}O) - \sigma(H_2\,^{16}O) = 0.8 \pm 0.4$ ppm[121] or 0 ppm.[118] [c] Data have also been reported for $C_7H_{15}CO_2Cs$.[120]

In the case of ^{133}Cs, the data of different groups are in apparent conflict and there seems to be hidden systematic errors in some of the measurements.

The D_2O/H_2O isotope effect on the shielding of an ion has been interpreted in slightly different terms by different workers in the field. Deverell *et al.*[116] take the differences as due to changes in the average excitation energy—small frequency shifts (ca. 1%) in UV absorption bands between D_2O and H_2O solutions have been observed in many systems.[117] Halliday *et al.*[118] take the absence of a $H_2^{16}O/H_2^{18}O$ isotope effect on the ^{133}Cs resonance as an indication that the D_2O/H_2O shifts are mainly caused by different distances of closest approach—and thus different overlap interactions—between ion and solvent, as a result of the different strength of hydrogen bonds formed by D_2O compared with H_2O.

In one study[118] of the change with concentration in the water solvent isotope effect on the shielding of an ion it was observed that with increasing concentration of CsCl in water the isotope effect is virtually constant up to ca. 0.5 M while at higher concentrations it decreases to become unobservable above ca. 4 M. The reduction in isotope effect was referred to the breaking of hydrogen bonds. This type of study appears to provide a valuable complement to other methods for studies of the effect of ions on water structure. Gustavsson and Lindman recently suggested the use of the D_2O/H_2O solvent isotope effect on ^{133}Cs shifts as a probe for changes in ion hydration for example in mixed solvent systems or in surfactant solutions.[119,120]

6C.7 Effects of Water Solvent Isotope on Ion Shielding

In especially ^1H NMR the spin-echo method to study self-diffusion has become quite popular for studies of chemical problems while in the case of other nuclei progress has been rather slow. As the applicability of the method is favoured by a large magnetic moment while the relaxation may not be too rapid, the $^7Li^+$ ion should be the best adapted to diffusion studies among the alkali ions. Indeed, the three NMR studies[53,122,123] of alkali ion diffusion in aqueous solutions of simple salts all concern the $^7Li^+$ ion. Hertz *et al.*[53] observed the self-diffusion coefficient (D_{Li^+}) of Li^+ to decrease with increasing concentration of LiCl, LiBr and LiI and discussed the data in terms of Li^+ hydration. Weiss and Nothnagel[122] studied $^7Li^+$ self-diffusion in concentrated solutions of LiCl and Li_2SiF_6 as a function of temperature (in the range 0–100°C) and evaluated the activation energies of Li^+ diffusion. Hertz *et al.*[123] studied $^7Li^+$ self-diffusion in aqueous solutions containing 1.0 M LiBr and varying amounts of KBr or NaCl or 1.0 M LiI solution with varying concentration of KI. Interestingly enough, and in correlation with the water structure-breaking effect, D_{Li^+} displays a marked maximum in the case of KI and KBr addition but a steady decrease on addition of NaCl. A model was developed which explains the effect of the added anion in terms of altered hydrogen-bonding between water molecules.

6C.8 Interactions between Alkali Ions and Paramagnetic Ions

Chemical shifts, generally attributed to Fermi contact interactions, of magnetic nuclei like ^1H, ^{13}C, ^{14}N, ^{15}N, ^{17}O and ^{19}F in atoms or molecules directly liganded to paramagnetic ions have been observed in a number of liquid and solid systems (for a review cf. ref. 124). Somewhat surprisingly, changes in chemical shifts, generally towards higher frequency, have also been observed for $^7Li^+$, $^{23}Na^+$, ^{85}Rb and ^{133}Cs in aqueous solutions containing paramagnetic cations like Fe^{3+}, Mn^{2+}, Ni^{2+}, Co^{2+}, Cu^{2+} and $[VO]^{2+}$.[28,112,125–131] For ^{133}Cs the shifts can become quite substantial and in a solution containing 3M CsCl and 1M $FeCl_3$ amount to 370 ppm. For a particular transition metal cation the magnitude of the alkali ion shift is dependent on the nature of the anions in the solution.

In contrast to the pronounced shift changes produced by the paramagnetic transition metal ions, the trivalent paramagnetic lanthanides cause only small shifts not necessarily paramagnetic in origin since the diamagnetic lanthanide La^{3+} produces similar shift changes.[28]

Through relaxation studies of the alkali metal resonances in the presence of transition metal ions it has been inferred that scalar contact interactions are responsible for the shift changes.[128–131] In the case of Cs^+ in presence of Fe^{3+} with Cl^- or F^- as the common anion the results may be rationalized if it is assumed that Cs^+ and $[FeX]^{2+}$ or $[FeX_2]^+$ ($X{=}Cl^-$ or F^-) form a transient "ion-pair" with the two cations sharing the halogen ligand(s).[130]

Shporer, Poupko and Luz[130] have estimated the fraction Cs^+ nuclei bonded in such an ion pair to be 0.1–0.2 in solutions containing 1M F^- and 0.3M Fe^{3+}.

It may be mentioned that ligand shared ion pairs analogous to the type discussed above have been proposed to be important in electron transfer reactions.

6D NON-AQUEOUS AND MIXED SOLVENT SOLUTIONS OF SIMPLE SALTS

6D.1 Relaxation Due to Ion-Solvent Interactions in Non-aqueous Systems

One important conclusion of the discussion of aqueous solutions was that the ion-water contribution to relaxation is rather well understood. The theories used are of such a nature that they are directly applicable also to non-aqueous systems and one important objective of studies of alkali ion relaxation in non-aqueous solutions is consequently to investigate ion solvation phenomena. Conversely, by allowing extensive variation in certain parameters entering the theories, examination of the non-aqueous systems may also be fruitfully performed to test the theories.

In our subsequent discussion it is convenient to separately consider quadrupole and dipole relaxation effects. As for aqueous solutions, quadrupole relaxation dominates greatly for the diamagnetic non-aqueous solutions except for ^{7}Li and, in particular, ^{6}Li.

a. Quadrupole Relaxation

Studies of an alkali ion's NMR signal in non-aqueous solvents were first reported by Jardetzky and Wertz,[91] who observed a relatively narrow ^{23}Na signal in acetone but were unable to observe the ^{23}Na resonance in some other solvents. This was attributed to strong line-broadening. An early quantitative relaxation study was performed by Craig and Richards[56] who examined the $^{7}Li^{+}$ ion in MeOH, HCO_2H, and DMF. Most studies, however, date from the last few years and are mainly concerned with Na^{+} and Li^{+}. Thus a variety of solvents have been used in the study of $^{7}Li^{+}$,[51,72,88,98,141,142] $^{23}Na^{+}$,[72,132–134,136–140] $^{39}K^{+}$,[3] $^{89}Rb^{+}$,[72] and $^{133}Cs^{+}$ relaxation.[51,72,132]

The rather extensive experimental material awaits a complete analysis and intercorrelation but the fundamental work of Hertz and co-workers[51,72] clearly illustrates the outcome of a detailed interpretation of the quadrupole relaxation results. Hertz and Melendres[72] (see also ref. 190) performed the type of analysis, which was schematized above for aqueous solutions, of the infinite dilution quadrupole relaxation rates in terms of the electrostatic model. It could then be established that the electrostatic model well accounts for alkali ion quadrupole relaxation also in nonaqueous solutions. Among the interesting results of the analysis may be mentioned the demonstration of a rather tightly packed first solvation sphere of all the alkali ions in MeOH and of Li^{+} also in a number of other solvents.[143]

A most important investigation is the variable temperature (and frequency) study of $^{7}Li^{+}$ and $^{133}Cs^{+}$ spin-lattice relaxation in glycerol.[51] (Li^{+} relaxation was studied also in perdeuterated glycerol to enable an estimate of dipolar contributions to relaxation.) Additionally, the proton relaxation of the solvent was investigated. As can be inferred from Fig. 6.5, illustrating the results, both solvent and ion relaxation shows a distinct maximum and this is due to the correlation time being comparable to the inverse angular nuclear precession frequency. A detailed analysis, complicated by the theoretically predicted non-exponentiality of relaxation for the spin $\frac{3}{2}$ ($^{7}Li^{+}$) and $\frac{7}{2}$ ($^{133}Cs^{+}$) nuclei, enabled the authors to estimate the correlation time of ion quadrupole relaxation at different temperatures. As different models of ion quadrupole relaxation are characterised by different time scales of the relevant motion such information is most valuable. From this analysis good support for the electrostatic model was obtained, while the electronic distortion model and a model involving shortlived distortion of the first solvation shell are characterised by much shorter correlation times than those observed. Significant support for the electrostatic theory was also provided from a comparison of the correlation times with those of the solvent relaxation studies. Thus the two correlation times were closely the same over the entire temperature range investigated, the small differences explainable in terms of a structure-making influence of Li^{+} and a structure-breaking one of Cs^{+}. As regards the deduced values of the field gradients they are well rationalised in terms of the fully oriented solvation model (see ch. 6C.2*c*) for both ions with some quenching due to symmetry effects for Li^{+}.

b. Dipole-Dipole Relaxation

The dipole-dipole contribution to the relaxation of $^{7}Li^{+}$ in non-aqueous media has been carefully examined in a few cases.[51,141,142] The work of Hertz *et al.*[142] on HCOOH solutions involves an elaborate and very successful attempt to elucidate in detail the

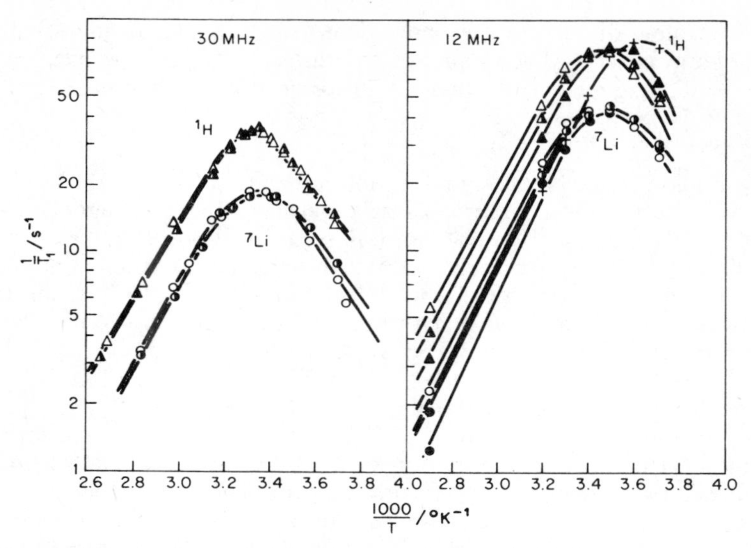

Fig. 6.5 $^7Li^+$ and ^1H longitudinal relaxation rates in solutions of LiCl in glycerol as a function of the inverse temperature. LiCl concentrations, in moles per 55.5 moles of solvent, were: 11 (○, △), 9.3 (◑, ▲), 6.9 (▲), 2.9 (●) and 1 (+) (reproduced with permission from ref. 51).

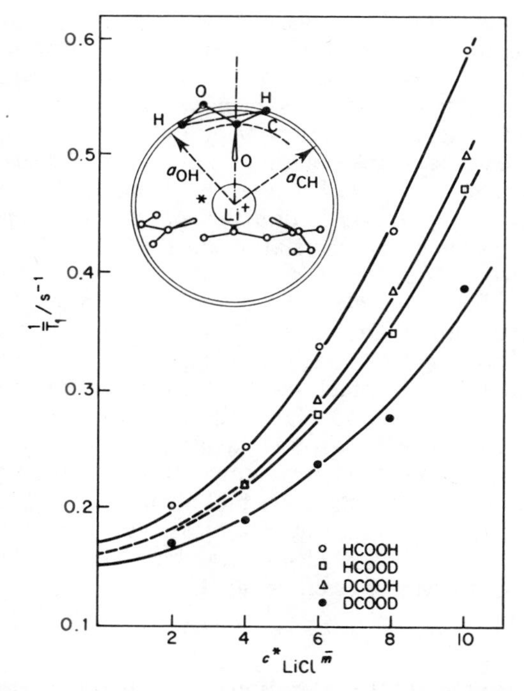

Fig. 6.6 Concentration dependence of the $^7Li^+$ relaxation of LiCl in various isotopic species of formic acid. Concentrations given in moles salt per 55.5 moles solvent; temperature 25°C. The insert indicates the deduced solvation structure of Li^+ in formic acid (reproduced with permission from ref. 142).

geometry of the solvation complex. As can be seen in Fig. 6.6, Li^+ relaxation is different for all the different H/D formic acid isotopic species. By considering the difference between DCOOH and HCOOD at infinite dilution, in relation to the expression for dipole-dipole relaxation (cf. Eq. 6.4), information is deduced on distances between the Li^+ ion and $-OH$ and $-CH$ protons. The solvation structure of Li^+ dissolved in HCOOH as obtained by Hertz *et al.*[142] is shown in the insert of Fig. 6.6. In a similar vein, Mishustin and Kessler[141] could deduce from a comparison between Li^+ relaxation in CH_3OH and CD_3OH that the hydroxylic proton is much nearer to Li^+ than the methyl protons. Similarly, these authors found for DMSO and DMF no contribution to Li^+ relaxation from the methyl protons. Hertz and Geiger[51] in their study of $^7Li^+$ relaxation in glycerol deduced a distance between Li^+ and glycerol protons from the separated dipolar contribution which is in good agreement with expectation.

6D.2 Ion-Ion Contributions to Quadrupole Relaxation in Non-Aqueous Media

Because of generally lower solvating ability, the ion-ion interactions for non-aqueous solvents can be expected to be relatively much more important than for water and it can also be anticipated that the quadrupole relaxation rate of an ion should provide a very sensitive probe of these interactions. The few investigations reported on the relaxation of ^7Li,[51,72,88,98,141] ^{23}Na,[72,133,138] ^{39}K,[3] ^{87}Rb,[72] and ^{133}Cs,[51,72] have indeed amply verified that the concentration dependence is in general much stronger in an organic solvent than in water.

As regards the ionic contribution to quadrupole relaxation, only Melendres and Hertz[72] have approached the difficult problem of a quantitative rationalisation. In addition to the terms encountered for the aqueous solutions {see ch. 6C.2*c* and 6C.4*b*}, one has now also to consider the contribution to relaxation from ion-pair formation, i.e. ion-ion structures of a life-time which is long compared to the correlation time. The corresponding term was derived to be[72]

$$(T_1^{-1})_{\text{ion-pair}} = \frac{3\pi^2}{10}\,\frac{2I+3}{I^2(2I-1)}\left[\frac{PeQ(1+\gamma_\infty)}{\mathrm{h}}\right]^2 P_a\left(\frac{2e}{R^3}\right)^2 \tau_c^* \tag{6.8}$$

P_a is the probability for the relaxing ion to occur in an ion-pair, R the ion-ion separation in the ion-pair and τ_c^* the rotational correlation time of the ion-pair; the other quantities have been defined above. As can be seen, Eq. 6.8 assumes the field gradient to be of electrostatic origin and the model, therefore, neglects covalent binding. From the analysis of Melendres and Hertz, it appears that the often observed one order of magnitude greater ion-ion contribution to relaxation compared to aqueous systems may be attriuted either to incomplete dissociation of ion-pairs or to a quenching effect at low concentrations being eliminated.

Partially similar ideas, although in a less detailed way, have been put forward by Mishustin and Kessler[141] who set forth the hypothesis that the significant solvation symmetry effect should make it possible to perform reliable discriminations between contact and solvent-separated ion pairs. Mishustin and Kessler also discussed pronounced differences in $^7Li^+$ relaxation between the two counter-ions Cl^- and $[ClO_4]^-$ and were able to correlate these results with predicted differences in the abilities of the solvents to dissolve the two anions.

It is, of course, of no surprise that the ion-ion contributions to relaxation are much more cumbersome to treat theoretically than the ion-solvent effects. Thus, although in principle the quadrupole relaxation method should be a most appropriate approach for studying ion-pair formation, the reliable evaluation of equilibrium constants has to await further development of our understanding of other contributions to ion-ion effects. More near-lying presently may be to use quadrupole relaxation rates in combination with extraneously obtained equilibrium constants to deduce information on ion-pair structure. Some further aspects on ion-pair structure are noted in ch. 6G.

6D.3 Chemical Shifts in Non-Aqueous Solutions

The chemical shifts of the alkali ions are very sensitive both to the solvent and to salt concentration and anion present in non-aqueous solvents and in principle alkali ion NMR should be a very useful method for studying ion-solvent and ion-ion interactions in these systems. Indeed the method has become very popular in recent years even if in certain respects the insight provided is somewhat disappointing as a result of the lack of a theory permitting quantitative predictions of chemical shifts. As has already been discussed above the physical mechanisms behind the chemical shifts are still under debate. Thus the infinite dilution chemical shifts, certainly containing significant information on ion solvation, have hitherto mainly been used in correlations with other physicochemical parameters believed to be related to quantities entering in the theory of the chemical shifts. On the basis of such a discussion some qualitative information on ion solvation has been deduced. As regards the elucidation of ion-ion interactions one is somewhat less hampered by the deficient theoretical understanding: Firstly, the functional variation of the chemical shift with electrolyte concentration contains important information on the structure of the solution[38,39] mainly radial distribution functions obtained in Debye-Hückel type theories and the association into ion-pairs and other complexes. From the magnitude of the concentration dependence of the chemical shifts, or from its dependence on the anion, significant qualitative information on ion-pair structure may be deduced, for example a distinction between solvent-separated and contact ion-pairs. It should be stressed that in many cases ion-pair formation is so strong that ion-pairs dominate already at the lowest concentrations accessible to study, thus precluding the possibility of obtaining the chemical shift of the free solvated ion by extrapolation to zero concentration. However, by comparing different salts, possible ambiguities in this respect may be resolved.

The large number of non-aqueous solution systems studied precludes us from presenting an exhaustive compilation but instead a schematic overview of the various applications will be given.

The first systematic investigation on the applicability of alkali ion NMR in non-aqueous environments is due to Maciel *et al.*[144] who determined $^{7}Li^{+}$ chemical shifts of LiBr and $LiClO_4$ in eleven simple organic solvents and also discussed the origin of the chemical shift differences. Of later studies[86,145–149] of $^{7}Li^{+}$ chemical shifts those of Popov and co-workers[146,148,149] deserve special consideration. By a systematic variable concentration study of $LiClO_4$, LiCl, LiBr, LiI, LiI_3 and $LiBPh_4$ in a number of solvents, the infinite dilution chemical shifts, exemplified in Table 6.6, were obtained. In agreement with the study of Maciel *et al.*,[144] the range of chemical shifts amounts to 5–6 ppm.

The prospects revealed by the work of Bloor and Kidd,[152] dealing with NaI in several organic solvents, have stimulated strongly the interest in using $^{23}Na^{+}$ NMR on non-aqueous systems.[135–138,140,146,149,153–160] To illustrate the variation with solvent we reproduce in Table 6.6 some of the infinite dilution chemical shifts obtained in the extensive studies by Popov's group.[149,153,154]

^{39}K chemical shifts have been studied for KI solutions in ethylenediamine and KOH solutions in MeOH[3] and recently Shih and Popov[151] have presented a rather extensive

TABLE 6.6 Chemical shifts/ppm at infinite dilution for $^{7}Li^{+}$, $^{23}Na^{+}$, $^{39}K^{+}$ and $^{133}Cs^{+}$ ions in various solvents according to refs 149–151

Solvent	Li[a]	Na[b]	K[c]	Cs[c]
Nitromethane	−0.36	−15.6	−21.10	−59.8
Acetonitrile	−2.80	−7	−0.41	+32.0
Dimethylsulphoxide	−1.01	−0.11	+7.77	+68.0
Propylene carbonate	−0.61	−9.4	−11.48	−35.2
Methanol	−0.54	−3.8	−10.05	−45.2
Dimethylformamide	0.45	−5.0	−2.77	−0.5
Acetone	1.34	−8.4	−10.48	−26.8
Pyridine	2.54	1.35	0.82	−31

[a] Reference: Aqueous 4.0 M $LiClO_4$. [b] Reference: Aqueous 3.0 M $NaClO_4$. [c] Reference: Infinitely dilute aqueous solutions.

report on the concentration and counterion dependence for a number of solvents. Their deduced infinite dilution shifts are included in Table 6.6.

As regards $^{133}Cs^{+}$ NMR, Richards and co-workers[38,39] have made a thorough attempt to rationalise the concentration dependence of the chemical shift in terms of the probability of interionic contact according to the theory of electrolyte solutions. Halliday *et al.*[118] reported an interesting difference in ^{133}Cs chemical shift between CH_3OD and CH_3OH solutions of CsCl appearing with increasing concentration. (This effect is confirmed in recent studies in our laboratory.) The solvent dependence of the $^{133}Cs^{+}$ chemical shift was briefly considered by Halliday *et al.*[38] and recently a more systematic approach to this problem has been reported.[150] Data from the latter work are given in Table 6.6.

Infinite dilution chemical shifts of the type summarized in Table 6.6 provide the starting point for discussions of the factors determining the solvation induced chemical shifts. According to expectation, the range of chemical shifts increases with the ion size, being (for common solvents) ca. 5–6 ppm for Li^{+}, ca. 15–20 ppm for Na^{+}, ca. 40–50 ppm for K^{+} and ca. 130 ppm for Cs^{+}. The chemical shift differences between different solvents are some tenths of the chemical shift differences between free and solvated ions demonstrating that the state of interaction of the ions may vary appreciably on transfer from one solvent to another. Bloor and Kidd[152] provided good arguments that the $^{23}Na^{+}$ shielding differences between solvents are predominantly due to the paramagnetic term and this certainly applies for the larger alkali ions as well. According to the overlap model of chemical shifts (see ch. 6C.5*b*) the overlap of the ion's outer *p* orbitals with outer orbitals of neighbouring solvent molecules should determine the shift. Attempts have therefore been made to correlate the chemical shifts with quantities expressing the electron-donating ability of the solvent. Bloor and Kidd[152] found a fair correlation with solvent Lewis basicities, while Popov and co-workers have considered the Gutmann solvent donor numbers, which are given by the enthalpy change for the reaction with $SbCl_5$ in 1,2-dichloroethane. As may be inferred from Fig. 6.7 a rather good linear relation between ^{23}Na infinite dilution chemical shift and solvent "donicity" is obtained[149,153] and by extrapolating the plot, donor numbers for basic solvents were estimated.[154] The correlation is good also for ^{39}K[151] but somewhat worse for ^{133}Cs,[150] while no correlation was found in the case of ^{7}Li.[148,149] The latter observation is understandable as for Li the diamagnetic and paramagnetic terms are probably of the same order of magnitude.

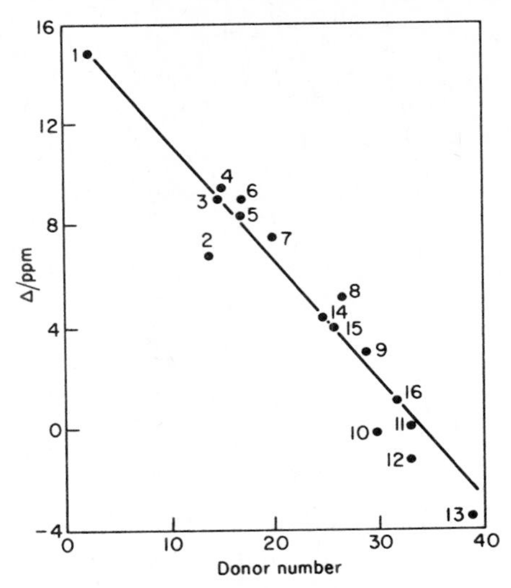

Fig. 6.7 Infinite dilution $^{23}Na^+$ chemical shifts plotted against the Gutmann donor number (reproduced with permission from ref. 149).

Therefore ring current and neighbour-anisotropy effects may influence the ^{7}Li chemical shift differences between different solvents, as discussed by Maciel *et al.*[144] *inter alia* in the case of pyridine and MeCN.

Most of the studies cited above have been concerned with the problem of ion pair formation. The monitoring of ion pair formation by chemical shift studies presupposes an appreciable chemical shift effect between ion-solvent and ion-ion contacts and according to the rather extensive experimental data available, the formation of contact ion pairs is generally easily discernible. Cahen *et al.*[148] postulated contact ion pairing of lithium ions with halide ions in several solvents, e.g. THF and $MeNO_2$, on the basis of appreciable positive chemical shifts relative to infinite dilution. The tendency to contact ion pair formation was much smaller with $[I_3]^-$, $[ClO_4]^-$ and $[BPh_4]^-$ as well as in some other solvents like DMSO and DMF. A corresponding difference between anions has been noted for Na^+ in many solvents.[135,146,153–156,158] The tendency of Na^+ contact ion pair formation was found to correlate well with solvent donicity while no good correlation with solvent dielectric constant was found.[149] Na^+ ion pairing in a number of ethers and other oxygenated solvents like THF, DME, diglyme and triglyme was studied by Canters[138] and Detellier and Laszlo.[157] Contact ion pairing was in the former study found to be much more important for $NaBH_4$ than for $NaBPh_4$ and in the latter for NaI than for $NaClO_4$. Finally, it may be mentioned that Cox *et al.*[147] studied Li^+ ion pairing with several planar aromatic anions and discussed the possibility of using the ring current shift effect to deduce information on geometrical features of an ion pair.

6D.4 Mixed Solvent Systems. Preferential Solvation Phenomena

a. *Quadrupole Relaxation Studies*

The quadrupole relaxation rate of the alkali ions has been seen to show a strong variation between different solvents and such a behaviour of an experimental parameter is a prerequisite if its value in a mixed solvent system should tell us about the composition of

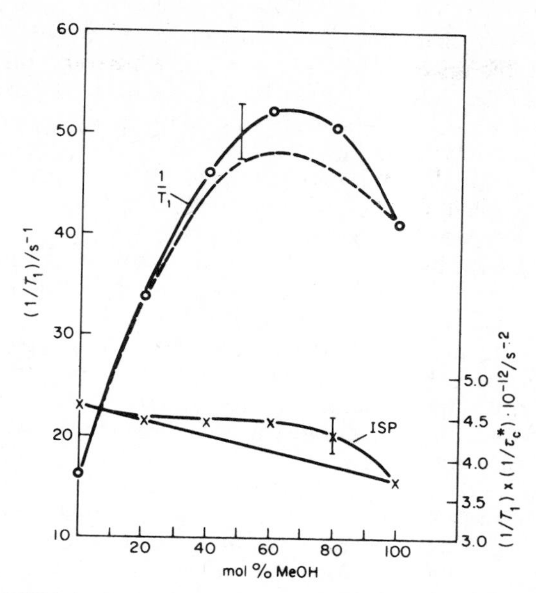

Fig. 6.8 Infinite dilution $^{23}Na^{+}$ relaxation rates in mixtures of MeOH and H_2O at 25°C. ○ denotes experimental data and the dashed line theoretical values (cf. text). × gives experimental data divided by the correlation time and ISP denotes the isosolvation point (reproduced with permission from ref. 164).

the solvation sphere. Studies of alkali ion relaxation in a large number of mixed solvent systems have been reported.[56,111,134,136,141,161–164] A notable feature of many of the investigations is the observation of a pronounced maximum in the alkali ion relaxation rate as a function of solvent composition. This applies for example to $^{7}Li^{+}$ relaxation in mixtures of water with DMF,[56,141] HMPT[141] or DMSO,[141] to $^{23}Na^{+}$ relaxation in EtCN—DMSO,[134] H_2O—THF[136] and H_2O—MeOH[164] and to $^{87}Rb^{+}$ relaxation in H_2O—MeOH.[161,164] In a number of cases, the relaxation rate maximum has been interpreted as an effect arising from asymmetric solvation, thus a field gradient effect.[134,136,162] Other possibilities are also plausible though, and for the H_2O—MeOH system the correlation time displays a marked maximum, thus rationalising the maximum in ion relaxation rate.[164] An important aim of the investigations cited has been to study preferential solvation phenomena but as mostly no independent information on the correlation time has been available the conclusions drawn may not be unambiguous. A further difficulty in most studies has been a discussion only of results obtained at finite electrolyte concentrations. Ion-ion contributions to relaxation may well vary appreciably with solvent composition.

Recently, a systematic approach to the applicability of ion quadrupole relaxation to the study of preferential solvation has been worked out by Holz *et al.*[164] Theoretically the study is based on Hertz' electrostatic theory and experimental quadrupole relaxation data are given for $^{23}Na^{+}$ and $^{87}Rb^{+}$ in mixtures of MeOH and H_2O. The relaxation rates used in the analysis of preferential solvation were obtained from the variable electrolyte concentration data by extrapolation to infinite dilution and are exemplified in Fig. 6.8 for $^{23}Na^{+}$. Schematically Holz *et al.*[164] express the quadrupole relaxation rate for an ion in a mixture of solvents A and B as

$$1/T_1 = n_A A_A \Lambda_A \tau_{cA} + n_B A_B \Lambda_B \tau_{cB} + \Lambda^* \qquad (6.9)$$

Here A is a factor characterizing the field gradients from one type of solvent molecules and depends on, *inter alia*, solvent dipole moment and ion-solvent distance. $\Lambda\,(=1-3^{-6\lambda})$ is the above-mentioned factor taking quenching of the field gradients due to symmetry into account, n is the number of solvent molecules in the first solvation sphere and τ_c is the correlation time. Λ^*, finally, characterises the non-additivity of the symmetry quenching effect. For estimating the field gradients, Holz *et al.*[164] use results obtained for the pure solvents while for the correlation time an average value is introduced and also, to make the analysis feasible, a constant total solvation number is assumed. If then also $\Lambda^* = 0$ is assumed, the dashed curve of Fig. 6.8 is obtained, estimating the correlation time from ^2H relaxation data and assuming non-preferential solvation. The difference between the experimental and theoretical curves then gives a measure of preferential solvation, i.e. the degree to which the composition of the ion's inner solvation sphere deviates from the bulk solvent composition. A more illustrative plot involves $1/(T_1 \cdot \tau_c)$ as this quantity under the assumptions used should vary linearly with bulk solvent composition for non-preferential solvation. As can be seen from Fig. 6.8 this analysis indicates preferential hydration of Na^+ in MeOH—H_2O mixtures while for Rb^+ the field gradients in the two pure solvents are not different enough to allow a safe conclusion. The paper of Holz *et al.*[164] suggests an interesting novel approach to study preferential solvation phenomena and gives also a discussion of the relevance of the different simplifying assumptions as well as of apparent discrepancies between quadrupole relaxation and chemical shift studies.

b. Chemical Shift Studies

As has been described above, species in direct contact have a dominating influence on the chemical shift of the alkali ions. Chemical shift studies therefore should be well suited for investigations of preferential solvation phenomena in mixed solvent systems and indeed this possibility has attracted great interest in the last few years. Thus a variety of solvent mixtures have been used in the study of $^7Li^+$,[86,144,166,170,171] $^{23}Na^+$,[136,140,152,159,165–169] $^{87}Rb^+$,[166] and $^{133}Cs^+$ shifts.[38,39,166,171] The $^{133}Cs^+$ ion is particularly attractive because of a wide chemical shift range and slow relaxation.

In most cases only qualitative deductions on preferential solvation have been made by noting at which solvent composition the chemical shift changes most rapidly; for example, if one solvent added up to a small mole fraction strongly affects the ion's chemical shift this solvent has been assumed to preferentially solvate the ion, An illustrative quantity in this type of investigation is the *isosolvation point* taken as the solvent composition where the chemical shift is the midpoint between the chemical shifts characterizing the pure solvents. Popov and co-workers[168,169] have determined isosolvation points for the sodium ion in a large number of mixed solvents and found solvation to increase in the sequence $MeNO_2 \ll$ MeCN $<$ pyridine $<$ TMU $\simeq$ DMSO $\simeq$ HMPT. In most cases, a good correlation with the Gutmann donor numbers is obtained and deviations in the case of TMU and DMSO are discussed in terms of solvent-solvent interactions.

As regards mixtures of water with an organic solvent it may be mentioned that both Li^+ and Na^+ have been found to be preferentially hydrated when the organic component is MeCN[144,159] while Na^+ is preferentially solvated by ethylenediamine.[152] For diacetin-water mixtures nonpreferential solvation of Na^+ was inferred.[165]

One important point in using the ion chemical shifts for predictions on preferential solvation phenomena is that the contribution from ion-ion effects, e.g. due to ion-pairing, must be eliminated since this contribution to the chemical shifts can often be assumed to vary strongly with solvent composition. The remedy to this problem is to make measurements at different salt concentrations and, at each solvent composition, extrapolate to infinite dilution. Preferably the result should be checked by using another counter-ion. Of course, on the basis of available information on ion-pairing an essentially inert counter-ion may often be found, thus reducing this problem.

A systematic approach to ion chemical shift studies of preferential solvation has recently been elaborated by Covington *et al.*[166,167,171,172] who examine the various simplifying assumptions inherent in this type of studies and also present a thermodynamic treatment of the problem. A general assumption is that the chemical shift varies in proportion to the composition of the first solvation sphere, i.e. that more long range effects are negligible as well as indirect solvent-solvent effects on the chemical shift. In their first paper, Covington *et al.*[166] analyse the situation with a constant total solvation number and with a constant equilibrium constant (except for a statistical factor) for the consecutive equilibria in the step-wise substitution of solvent A by solvent B.

$$MA_{n-x}B_x + B \rightleftarrows MA_{n-x-1}B_{x+1} + A$$

Under these conditions the chemical shift (relative to pure solvent A) is derived to be

$$\frac{\delta_{\text{obs}}}{\delta_B} = \frac{x_B K^{1/n}}{x_A + x_B K^{1/n}} \quad \text{or} \quad \frac{1}{\delta_{\text{obs}}} = \frac{1}{\delta_B}\left(1 + \frac{x_A}{x_B K^{1/n}}\right) \tag{6.10}$$

where δ_B is the chemical shift of the ion in pure solvent B, K the equilibrium constant for the process

$$MA_n + nB \rightleftarrows MB_n + nA$$

n the (constant) total solvation number and x_A and x_B the solvents' mole fractions. (For non-ideal cases, activities should be used.) From studies of the variation of the chemical shift with solvent composition, $K^{1/n}$ may be obtained. This simple model was found to apply with good approximation to several ions in H_2O_2 + H_2O mixtures (using infinite dilution chemical shifts), the result being that Rb^+ and Cs^+ are preferentially solvated by H_2O_2, Li^+ preferentially hydrated while for Na^+ essentially non-preferential solvation was found. Covington's treatment has recently been adopted by others, e.g. Greenberg and Popov[169] found the simple model to apply well for Na^+ in several non-aqueous binary mixtures, such as DMSO + pyridine (see Fig. 6.9).

In a number of cases, the simple model described is inappropriate and in ref. 171, Covington *et al.* investigate the effect of a changing solvation number. While for a constant total solvation number a symmetric plot of chemical shift vs. solvent composition obtains (and a linear change for non-preferential solvation), an asymmetric curve is obtained for both preferential and non-preferential solvation if the two solvents have different solvation numbers. For the special case of solvent B having half the solvation number (n) of solvent A the following equation is deduced

$$\delta_{\text{obs}}/\delta_B = x_B K^{2/n}/(x_A^2 + x_B K^{2/n})$$

This equation is found to apply well for Li^+ and Cs^+ ions in mixtures of DMSO and H_2O[171] as well as for the Na^+ ion in binary mixtures of a large number of monodentate and bidentate oxygenated mixtures.[140,159] The study of Detellier and Laszlo[140,159] led to important conclusions on coordination numbers, Na^+ being generally tetracoordinated, and mechanism for solvent exchange.

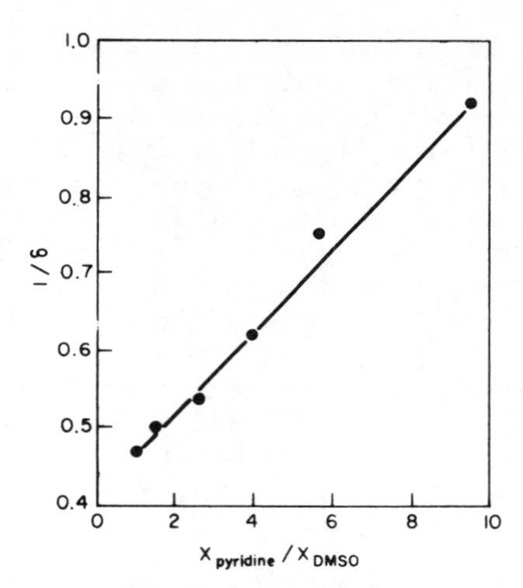

Fig. 6.9 The $^{23}Na^{+}$ chemical shift in mixtures of pyridine and DMSO plotted according to Eq. (6.10) (reproduced with permission from ref. 169).

In part 4 of the series, Covington and Thain[172] analyse the case where the consecutive solvent exchange equilibria are characterised by different equilibrium constants and in particular the situation with a regular free energy change is treated. Explicit expressions for this case are developed and used to analyse *inter alia* Li^{+} and Cs^{+} chemical shifts in mixtures of DMSO and H_2O, and Na^{+} shifts in mixtures of H_2O with MeCN or ethylenediamine. Among the results may be mentioned the finding of preferential hydration of Li^{+} and Na^{+} in mixtures of water with DMSO and MeCN, respectively, as well as the conclusion that more simplified analyses may easily lead to erroneous results.

The basic assumption that the ion's chemical shift varies in proportion to the composition of the first solvation sheath has probably to be critically examined. In fact the $^{133}Cs^{+}$ chemical shift data (extrapolated to zero concentration) of mixtures of H_2O and DMF show strong deviations from this simple behaviour in this particular case and may throw some doubt on the validity of the basic assumption also in other cases.

A neglected possibility, eliminating to a great extent these difficulties, would for aqueous systems be to study the water isotope effect in shielding (see ch. 6C.6), the idea being that this should measure the degree of ion-water contact in a mixed solvent system. Indeed exploratory studies in our laboratory of different mixed solvent systems have demonstrated the usefulness of this principle.

6E ALKALI ION COMPLEXES WITH LOW MOLECULAR WEIGHT COMPOUNDS

6E.1 Introduction and Survey of Systems Studied

The tendency of complex formation with other species is much less significant for the alkali ions than for other monoatomic metal ions. The complex chemistry of the alkali ions is therefore characterised by weak and rather non-specific associations with size effects being most important. Nevertheless, a number of alkali ion complexes is of considerable chemical and technical interest and in biology sodium and potassium complexes play a fundamental role and are the prerequisites for a number of life processes.

The potentials of NMR in the complex chemistry of the alkali ions are several: NMR is well suited for studies of weak complexes and it may provide unique information on complex structure and dynamics as well as on kinetics. From an NMR point of view, the considerations will be rather different for small complexes, like complexes with sugar molecules or amino acids, and for alkali ion binding in complex biological environments. The subdivision of the material in this treatment has been based on such considerations.

The present section deals with low molecular weight compounds and it may be useful to start with an enumeration of the systems studied. Aqueous $^{23}Na^+$ interactions with salts of a number of organic acids both of chemical interest, like EDTA, citric acid and NTA, and such of biological interest, like cystein and aspartic acid, were investigated by James and Noggle.[173,174] Jardetzky and Wertz[91,175,176] in their pioneering work obtained interesting qualitative results on the interaction of aqueous $^{23}Na^+$ with salts of a number of organic acids (e.g. pyruvic acid, malic acid, hydroxybenzoic acids) and these compounds were later reconsidered.[177] $^7Li^+$ complexing with NTA in aqueous solution was studied by Akitt and Parekh.[178]

$^{23}Na^+$ interactions with sugars have been examined both in aqueous solution[179,180] and in pyridine and Pr^iNH_2.[181,182]

Na^+ and K^+ interactions with many esters of phosphoric acid are important in biological systems and this has motivated studies on the interaction of aqueous $^{23}Na^+$ with phosphoserine and phosphoethanolamine,[183] of aqueous $^{39}K^+$ with ATP[184] and of aqueous $^{23}Na^+$ with nucleotides (guanosine and adenosine monophosphates.[185]

The penetration of alkali ions by ionophores through artificial and biological membranes is presently of great interest to chemists and biologists. Alkali ion NMR has developed into a very powerful tool for studies of alkali ion complexing with ionophores. These compounds are mostly cyclic, possessing several oxygens with which interaction mainly takes place. One example of a naturally occurring ionophore (valinomycin) and examples of two main groups of synthetic ionophores, i.e. crown ethers and cryptands, are illustrated in Fig. 6.10. $^{23}Na^+$ complexing with a variety of crown compounds has been investigated in different solvents,[137,186–189] and recently also studies of $^{39}K^+$,[191] $^{87}Rb^+$ [191] and $^{133}Cs^+$ [58,59,192] have been performed. The alkali ion binding to the C222 cryptand (see Fig. 6.10) as well as to some similar cryptands has been studied both by ^7Li,[193–196] and ^{23}Na[196–200] NMR. $^{23}Na^+$ interactions with several naturally occurring ionophores, i.e. monactin,[186] valinomycin,[186,201] enniatin B[186], monensin[186,202] and nigericin[186] were studied in MeOH and in the case of monension also $^7Li^+$ interactions were examined.[202]

It remains to mention a study of $^7Li^+$ interactions with bases like adenine and thymine and nucleosides like adenosine and inosine in DMSO[203] and studies of $^7Li^+$ and $^{23}Na^+$ interactions with polymethylenetetrazoles[204–206] and glutarimides[206] in $MeNO_2$ solutions. The latter compounds are of interest because of strong stimulating action on the central nervous system.

6E.2 Stability and Structure of Alkali Ion Complexes

The investigations listed cover a wide range of aspects of alkali ion complexation both such relating to the properties of the complexes (thermodynamic stability, structure, bonding and dynamics) and such relating to their rates of formation and dissociation. We will start by exemplifying the information alkali NMR has provided in the former respect.

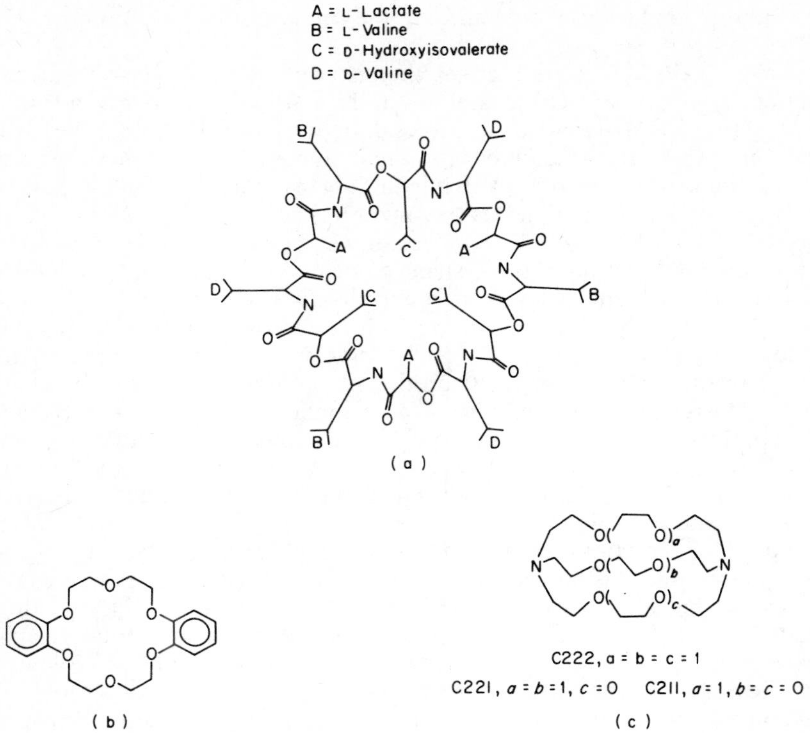

A = L-Lactate
B = L-Valine
C = D-Hydroxyisovalerate
D = D-Valine

(a)

(b)

C222, $a = b = c = 1$

C221, $a = b = 1$, $c = 0$ C211, $a = 1$, $b = c = 0$

(c)

Fig. 6.10 Structure of ionophores. (a) valinomycin (b) dibenzo-18-crown-6 and (c) C 222, C 221 and C 211 cryptands.

Qualitative information on complex stabilities is generally directly obtainable from studies of alkali ion chemical shift or relaxation rate as a function of concentration of the complexing agent or from comparisons between the effects of different compounds. For example, the ion specificity of cryptands and crown ethers has been examined in this way. The cryptands C 222 and C 221 thus complex Li^+ ions much more weakly than C 211 which has a cavity radius more appropriate to the unsolvated Li^+ ion.[193,195] Popov *et al.*[193,195] also demonstrated the great importance of the solvating power of the solvent; in nitromethane, for example, which is poorly solvating, a strong tendency to complex formation was observed. $^{133}Cs^+$ T_1 studies in the presence of K^+ or Rb^+ ions were used by Wehrli[59] to establish that the affinity sequence for binding to 18-crown-6 is $K^+ > Rb^+ > Cs^+$.

Quantitative information on complex stability may be obtained by fitting detailed variable composition signal intensity, relaxation or chemical shift data to a model of the equilibria occurring. For the case where exchange is slow enough (see ch. 6E.3) so that separate peaks of free and complexed alkali ions are registered[193–195,197] the task is, of course, facilitated since direct information on the concentrations of the different species is obtainable. The case where exchange is so rapid that only one weighted-average signal occurs has been treated by several authors.[173,185,204,205] One difficulty here is that the relaxation rate and chemical shift of the complexed state are generally unknown. Na^+

and Li^+ complexing with pentamethylenetetrazole (L) has been found to follow the simple equilibrium

$$M^+ + L \rightleftarrows ML^+$$

from chemical shift studies;[204,205] Li^+ was found to form a stronger complex than Na^+. For the case of the dilactam of cryptand C 222, strong complexing with Na^+ but considerably weaker with Li^+ and Cs^+ was found. This is according to expectation as the Na^+ size is close to the dimension of the cryptand cavity.[196] The 1 : 1 stoichiometry was found to apply also for the case of Na^+ complexing with two aminocarboxylic acids[173] but for other systems, less simple models[173,178,185,192] have to be used. An interesting example is provided by the self-association of 5′-guanosine monophosphate studied by ^{23}Na NMR by Paris and Laszlo.[185] The extensive aggregation and the Na^+ binding to the aggregates formed are evident from the finding of similar variable concentration behaviour of ^{23}Na relaxation as for micellar solutions (cf. ref. 119).

Concluding the discussion of complex stability it should be pointed out that, of course, information of formation constants may be obtained only if there is a significant dissociation; for stronger complexes only stoichiometric information is obtainable in combination with a lower limit of the formation constant. Popov and co-workers[196] in their studies of 1 : 1 complexes put this limit at $K \simeq 10^4\ \text{M}^{-1}$.

Both the chemical shift and the relaxation rate of an alkali ion in a complex can be assumed to contain significant information on structure and interactions. This possibility is far from being fully exploited yet but in a number of cases important information has been provided.

A most significant result is that the chemical shift of $^7Li^+$ in C 211 cryptate has been found to be the same in a number of solvents and also independent of counter-ion.[193,195] This strongly supports the basic assumption often made in chemical shift studies that only species in direct contact may influence the chemical shift.

The constant value of the $^7Li^+$ chemical shift of C 211 indicates that the Li^+ ion is completely encased. For the C 222, C 221 and C 222-dilactam complexes some solvent dependence is found which points to a looser structure allowing some Li^+-solvent contact.[193,195,196] For the case of different $^{133}Cs^+$ complexes, location of the ion at the outside of the ionophore has been inferred.[196] A progressively increasing Na^+ chemical shift with the number of oxygens in the cryptand was found by Kintzinger and Lehn[198] and this correlates well with the expected paramagnetic shielding contribution from Na^+—O orbital overlap. Another interesting work is due to Haynes *et al.*[186] who determined $^{23}Na^+$ chemical shifts in complexes with naturally occurring ionophores and correlated them with complex formation constants.

Chemical shift studies can be seen to be capable of adding significantly to our knowledge on alkali ion complexes. A further point is that, in giving information on geometrically well-defined states, these chemical shifts should prove useful in attempts to improve the theoretical understanding of alkali ion chemical shifts in general.

Information is available on the alkali ion relaxation of a large number of complexes mainly through linewidth studies. Quadrupole relaxation is the only significant contribution and for the case of a rigid complex with a reorientational correlation time τ_c, the relaxation rates are given by Eq. 1.17.

In the extreme narrowing limit, assumed to apply, a separation of the quadrupole coupling constant (χ) and the correlation time depends on independent information on either quantity. One way of doing this is to use simple hydrodynamic theory (e.g. the Debye equation) to estimate τ_c. While the magnitude of τ_c estimated from the Debye equation may be in error by an order of magnitude, this procedure is probably roughly appropriate for comparisons within a series of complexes similar in size and shape. A more satisfactory approach is, of course, to measure the correlation time experimentally, for example by ^{13}C, ^{14}N or ^{2}H NMR relaxation. There is though some danger in using such values: there may be internal mobility of the alkali ion or of some group of the complexing agent or reorientation may be appreciably anisotropic.

The first direct determination of alkali ion quadrupole coupling constants of ionophore complexes is due to Kintzinger and Lehn[198] who studied Na^+-cryptate complexes by means of ^{13}C (giving τ_c) and ^{23}Na NMR. The quadrupole coupling constants of C 211, C 221 and C 222 complexes, respectively, were determined to 2.20, 1.43 and 1.01 MHz. These values certainly reflect symmetry effects, the low χ value of the C 222 complex reflecting a fairly high symmetry in the case of 6 oxygens. The approach of Kintzinger and Lehn[198] is most promising and certainly deserves further consideration. In the same way, Paris and Laszlo[185] have recently obtained χ values of the Na^+ complexes with 5′-adenosinemonophosphate and 2′-guanosinemonophosphate to be 0.5 MHz.

Interpretation of relaxation data alone may, although with caution, also be made in terms of symmetry effects. For example, substitution in the aromatic rings of dibenzo-18-crown-6 was inferred to lead to larger field gradients for $^{23}Na^+$ in its complexes as a result of lowering of symmetry.[188] Haynes *et al.*[186] found correlation between χ values obtained using the Debye equation and symmetry for Na^+ complexes of some natural ionophores while Hourdakis and Popov[196] inferred a large asymmetry of $^{23}Na^+$ in complex with the cryptand C 222-dilactam.

6E.3 Studies of the Kinetics of Alkali Ion Complexes

Kinetic data on the complexation and decomplexation reactions have to a considerable extent been obtained through NMR studies. ^{1}H and ^{13}C NMR is applicable if the chemical shifts between the complex and the free ligand are sufficiently large to produce reasonable line broadenings in the intermediate exchange region.[207] NMR studies of the complexed cations offer an alternative approach. Schori *et al.*[187,188] studied the rate of Na^+ exchange for dibenzo-18-crown-6 (DBC) and other crown ether complexes in different solvents through ^{23}Na NMR. The ^{23}Na chemical shift difference ($rad{\cdot}s^{-1}$) between uncomplexed and complexed Na^+ ($\Delta\omega = 8 \pm 16$ $rad{\cdot}s^{-1}$ at 0.722T) was smaller than the observed linewidths. (At room temperature, T_2^{-1} in dilute DMF was ca. 140 s^{-1} and in the DBC complex ca. 2700 s^{-1}). In solutions containing both complexed and uncomplexed Na^+ the ^{23}Na signal will thus in the absence of chemical exchange essentially be a superposition of the two signals. An analysis of the Bloch-McConnell equations in the limit when the chemical shift between the sites, $\Delta\omega_{AB}$, equals zero, shows that in the region where the exchange rate becomes comparable to the relaxation rates, the lifetime, τ_A, of an uncomplexed Na^+ may be obtained from the following expression.[187,188]

$$\frac{1}{\tau_A} = \frac{(1/T_{iB} - 1/T_i)(1/T_i - 1/T_{iA})p_B}{(1/T_{iAv} - 1/T_i)}; \; i = 1, 2 \tag{6.11}$$

where

$$1/T_{iAv} = p_A/T_{iA} + p_B/T_{iB}, \tag{6.12}$$

p_A and p_B are the mole fractions and $1/T_{iA}$ and $1/T_{iB}$ the relaxation rates of uncomplexed and complexed sodium ion, respectively. $1/T_i$ is the relaxation rate of the *slowly* decaying component of the transverse or longitudinal ^{23}Na magnetization. In the fast exchange

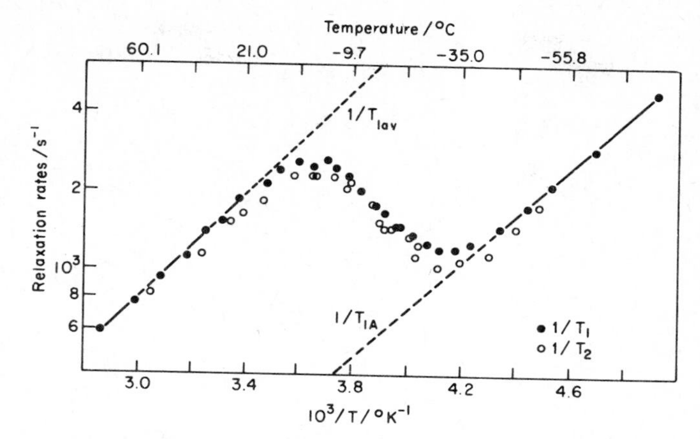

Fig. 6.11 Temperature dependence of $1/T_1$ (●) and $1/T_2$ (○) of ^{23}Na for a solution containing 0.57 M NaSCN and 0.2 M dibenzo-18-crown-6 (DBC) in DMF (reproduced with permission from ref. 188).

limit $1/T_i$ equals $1/T_{iAv}$ while in the slow exchange limit $1/T_i$ approaches $1/T_{iA}$. $1/T_{iB}$ may be calculated from $1/T_{iAv}$ using Eq. 6.12. Under the experimental conditions employed by Schori *et al.* p_B is close to the ratio of the total crown ether and sodium concentrations and $p_A \geq p_B$. Furthermore extreme narrowing conditions were found to apply and $T_1 = T_2$ for both complexed and uncomplexed $^{23}Na^+$.

Experimental results for a NaSCN solution in DMF containing DBC is shown in Fig. 6.11. $1/T_1$ was obtained from pulse experiments and $1/T_2$ from line width measurements. The good agreement between the two series of data in Fig. 6.11 confirms the validity of the assumption of $\Delta\omega_{AB} = 0$. The apparent Arrhenius activation energies, E_a, calculated from the temperature dependence of $1/\tau_A$ for sodium ions in different crown ether solutions are presented in Table 6.7. Included in this table is also a result obtained on K^+ through ^{39}K NMR studies[191] at a field of $6T$.

The two principal models for the exchange process

$$^*M^+(\text{solv.}) + M^+(\text{crown}) \rightleftarrows {}^*M^+(\text{crown}) + M^+(\text{solv.}) \quad (6.13)$$

$$M^+(\text{solv.}) + \text{crown} \rightleftarrows M^+(\text{crown}) + (\text{solv.}) \quad (6.14)$$

have been tested by Schori *et al.*[187,188] Through studies of $1/\tau_A$ at different crown ether and Na^+ concentrations it was concluded that the exchange takes place predominantly by the mechanism of Eq. 6.14 and the activation energies in Table 6.7 thus refer to the decomplexation step.

The equilibrium constants for complexation are generally strongly affected by the nature of the solvent—a result in line with the X-ray findings that in the complex the macrocyclic ring wraps around the cation and replaces most, if not all, of the molecules of the solvation shell.[208] By contrast, the activation energies for the decomplexation seem insensitive to the character of the solvent (Table 6.7). The number of solvents studied is, however, very limited and the apparent constancy of E_a may be accidental. It is nevertheless possible that the constancy of E_a implies that the rate limiting step in the decomplexation is a conformational change of the macrocylic ligand.[188] The difference of ca. 17 kJ/mole between the activation energy for the DBC and dicyclohexyl-18-crown-6 (DCC) complexes may along these lines be considered as the result of the greater flexibility of the macrocyclic ring in DCC.

TABLE 6.7 Kinetic parameters for alkali metal complexes

Complexing ligand	Cation studied	Solvent	E_a kJ/mole	$\Delta G\ddagger^a$ kJ/mole	$\Delta H\ddagger$ kJ/mole	$\Delta S\ddagger$ kJ/mole	Ref.
Dibenzo-18-crown-6	$^{23}Na^+$	DMF	52.8	—	—	—	187
Dibenzo-18-crown-6	$^{23}Na^+$	MeOH	54.9	—	—	—	188
Dibenzo-18-crown-6	$^{23}Na^+$	DME	55.7	—	—	—	188
Dibenzo-18-crown-6	$^{39}K^+$	MeOH	52.8	—	—	—	191
4,4-Dinitrodibenzo-18-crown-6	$^{23}Na^+$	DMF	52.4	—	—	—	188
4,4-Diaminodibenzo-18-crown-6	$^{23}Na^+$	DMF	54.9	—	—	—	188
Dicyclohexyl-18-crown-6	$^{23}Na^+$	MeOH	34.8	—	—	—	188
Crypt C222	$^{23}Na^+$	en	51.1 ± 4.6	62.0(50°C)	—	—	197
Crypt C221	$^{7}Li^+$	pyridine	56.6 ± 1.6	75.0 ± 0.4	54.0 ± 1.6	−62.4 ± 3.8	194
Crypt C211	$^{7}Li^+$	pyridine	82.1 ± 14.7	95.1 ± 4.6	79.6 ± 14.3	−52.3 ± 38.6	194
Crypt C211	$^{7}Li^+$	H_2O	89.2 ± 5.0	86.3 ± 1.3	86.7 ± 4.6	+1.7 ± 12.6	194
Crypt C211	$^{7}Li^+$	DMSO	67.5 ± 2.5	82.5 ± 0.4	64.9 ± 2.5	−57.8 ± 5.9	194
Crypt C211	$^{7}Li^+$	DMF	67.0 ± 2.5	83.8 ± 0.8	64.5 ± 2.5	−64.9 ± 5.9	194
Crypt C211	$^{7}Li^+$	formamide	59.1 ± 2.8	87.1 ± 0.8	56.6 ± 2.8	−95.5 ± 7.5	194
Crypt C211	$^{23}Na^+$	MeOH/D_2O (95/5)	—	64.5(58°C)	—	—	198
Crypt C222	$^{23}Na^+$	(95/5)					198
Crypt C222	$^{23}Na^+$	(95/5)	—	>67	—	—	198
Crypt C221	$^{23}Na^+$	(95/5)	—	>67	—	—	198
Crypt C222	$^{23}Na^+$	pyridine	—	76.69 ± 0.02	56.9 ± 0.8	−52.7 ± 2.5	210
Crypt C222	$^{23}Na^+$	THF	—	67.86 ± 0.08	57.7 ± 0.8	−33.9 ± 2.5	210
Crypt C222	$^{23}Na^+$	H_2O	—	60.62 ± 0.04	67.4 ± 0.8	+22.2 ± 3.3	210
Crypt C222	$^{23}Na^+$	EOA	—	60.42 ± 0.08	51.5 ± 0.8	−31.8 ± 2.5	210
Valinomycin	$^{23}Na^+$	MeOH	39.8				

[a] At 298 K if not otherwise specified

The rate of sodium ion exchange between valinomycin complexes and uncomplexed ion has been determined by ^{23}Na NMR in methanol solutions.[201] The ^{23}Na relaxation rate in the complex is very fast ($1/T_2 > 5000\ s^{-1}$ at $-80°C$) and only the signal from the uncomplexed ions could be detected. The same NMR technique as described above in the case of the crown ether complexes was used to obtain the rates. The result is summarised in Table 6.7.

Exchange rates of alkali ions in solutions of cryptate complexes have also been acquired through NMR studies of the cations. The effective symmetry of the cation environment in the cryptate complexes appears generally higher than in the crown ethers and relatively narrow $^{7}Li^{+}$ and $^{23}Na^{+}$ signals are obtained from the corresponding cryptate complexes.[194,197,198,210] The chemical shift difference between complexed and uncomplexed ions is such that in the slow exchange region two well resolved signals are obtained already at a field of 1.7T on a solution containing Na^{+} or Li^{+} in excess of the cryptate. (*Cf.* Fig. 6.12). The rate of ion exchange can accordingly be evaluated through ordinary line shape studies employing line shape equations valid for exchange in a two site system with different $T_2 : s$ in the sites.[194,197] The results are summarized in the lower half of Table 6.7. The exchange mechanism has not been thoroughly investigated in all cases but earlier ^{1}H NMR studies by Lehn *et al.*[209] favour the dissociation–complexation process analogous to that in Eq. 6.14 for the crown ethers.

The activation energy for decomplexation of Li^{+} is roughly correlated with the Gutmann donor number of the solvent; increasing donor number parallels increasing activation energy. This was taken to indicate that the transition state involves substantial ionic solvation.[194] On the other hand no similar correlation is observed for the decomplexation of Na^{+}. Possibly the systematic errors in the ^{7}Li studies were larger than originally estimated.

The positive entropy of activation for the H_2O solutions observed for the Na^{+} exchange is consistent with a solvent participation in the transition state.[210]

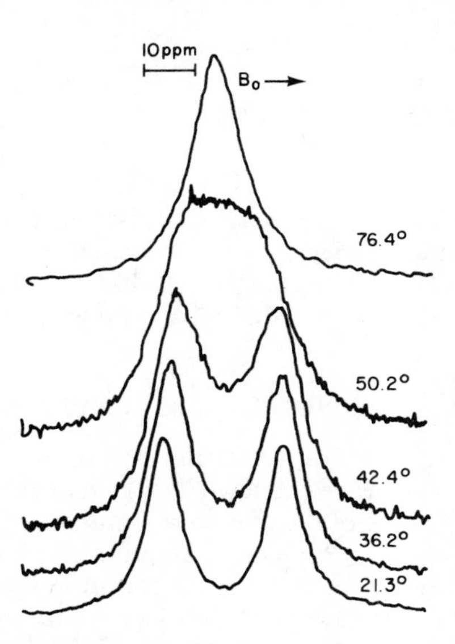

Fig. 6.12 ^{23}Na NMR spectra at different temperatures for an ethylene diamine solution containing 0.3 M C222 and 0.6 M NaBr. The spectra were obtained at 15.87 MHz (reproduced with permission from ref. 210).

6F ALKALI METAL ORGANIC COMPOUNDS

6F.1 Introduction

A variety of diamagnetic organic alkali metal compounds are known.[211] Some of these involve a decidedly covalent alkali metal bonding, like for example the C-Li bond in alkyllithium compounds, while others are more ionic in nature and best described as alkali metal salts of carbanions.

^{7}Li NMR studies of organic lithium compounds were initiated[212,213] in the early 1960s and have played an important role in the characterisation of the structure and dynamic properties of these compounds. This section will be devoted entirely to a discussion of NMR on lithium organic compounds. Other alkali metals give largely ionic organic compounds, the alkali metal NMR properties of which are almost exclusively determined by the solvents used. These types of compounds are dealt with in ch. 6G.

In solution, lithium organic compounds with the general structure RLi, can be monometric, dimeric, tetrameric or hexameric depending on temperature, solvent and the nature of the group R. When R is a primary saturated aliphatic group hexamers and tetramers may coexist in hydrocarbon solution[214] while in donating solvents like Et_2O or THF tetramers dominate. X-ray studies have shown the tetramers to have a tetrahedral structure, 6.I in the solid:

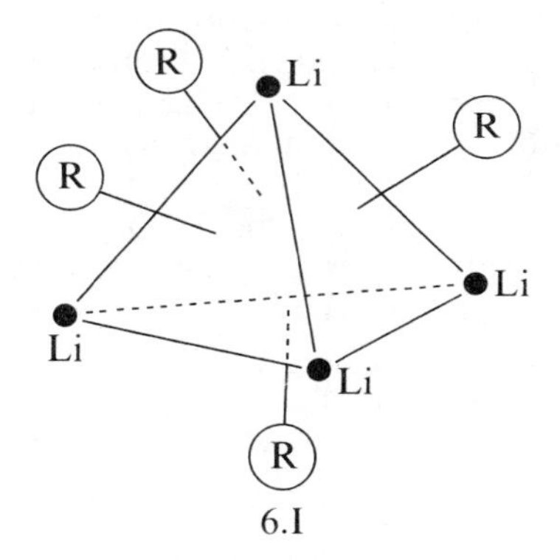

6.I

This structure is presumably retained in solution (see ch. 6F.2 and 6F.3).

Aryl lithium compounds are generally less associated than alkyl compounds. When R is an extensively resonance stabilised carbanion, like for example triphenylmethyl, the carbon-lithium bond is virtually ionised in strong donor solvents like HMPA.

6F.2 Chemical Shift Data of Lithium Organic Compounds

Although the quadrupole moment of ^{7}Li is relatively high, the ^{7}Li linewidths of lithium organic compounds are of the order of a few Hertz in hydrocarbon and other solvents at room temperature.[215–217] Some of the line broadening in alkyllithium compounds may in fact be due to unresolved $\underline{Li}$—C$\underline{H}$ spin coupling as shown by decoupling experiments,[218] T_1 relaxation studies[217] and by the use of deuterated alkyl groups.[215] ^{6}Li NMR signals in EtLi in toluene are even narrower than the ^{7}Li NMR signals.[215,219] The line width ratio is not as large as would be expected on the basis of a purely quadrupolar relaxation being responsible for the broadening.

There exists no generally accepted reference for lithium shifts. Most lithium shifts given in the literature are reported relative to an external standard, usually an aqueous or methanolic solution of a lithium salt. Susceptibility corrections have not always been applied, nor have the concentration dependent shifts in the reference solution been considered which makes a direct comparison of results from different laboratories difficult. A large number of ^{7}Li chemical shifts have however been remeasured relative to an internal proton lock.[220] The results are shown in Fig. 6.13. The lithium shift scale was chosen as to have its zero position at the ^{7}Li resonance of 70% LiBr in water.

The ^{7}Li shifts are seen to cover a fairly narrow range. Theoretical considerations indicate that the major factor determining the differences in chemical shifts for the alkyllithiums is substituent anisotropy effects.[220] Replacement of a proton on an α-carbon with a methyl group causes a change in chemical shift of about 0.4–0.6 ppm depending on the solvent, the change $Pr^iLi \rightarrow Bu^tLi$ being an exception. The steric crowding in Bu^tLi for one thing presumably prevents solvent molecules from reaching the Li core of the tetramer. Also almost all organolithium compounds can be seen to give an increased shielding on going from cyclopentane to ether.

Studies of the concentration dependence of the ^{7}Li shift of Pr^iLi in cyclopentane, where the aggregation number of the lithium complex is four at low concentrations (<0.03 M) and approaches six at higher concentrations, indicate the chemical shift of the tetramer and the hexamer to be the same within an experimental error of about 0.04 ppm.

It should be mentioned that in solutions of two (or more) different alkyllithium compounds, RLi and R′Li, mixed tetrameric and/or hexameric species may easily be formed. In these cases, and when *intra-* and *inter*molecular exchange processes are slow (see ch. 6F.5), the chemical shift of a Li nucleus is to a good approximation only dependent on the nature of the three neighbouring alkyl groups ("*the local environment assumption.*"[221])

The nature of the carbon-lithium bond in aryllithium compounds is still a matter of some controversy. In polar solvents the bond appears to be nearly ionic—for example in phenyllithium the proton-proton spin coupling pattern agrees much more with that of pyridine (which is isoelectronic with the phenyl carbanion) than with that of substituted benzenes.[222] On the other hand, the ^{7}Li shifts in *m*- and *p*-substituted phenyllithium compounds are clearly affected by the nature of the substituent and are well correlated with the H(2) proton shifts.[222]

As exemplified in Table 6.8 the ^{7}Li shifts in organolithiums in which the organic moiety may form a resonance stabilized carbanion can be markedly dependent on the solvent.[223] The variations have been taken to reflect changes in ion pair structure. The approximate constancy of the ^{7}Li shift of cyclopentadienyl Li in all solvents except HMPA is for example attributed to the existence of contact ion pairs. In HMPA, on the other hand, all compounds presumably form solvent separated ion pairs. The finer details in the shift changes may be due to ring currents from the organic moiety or to differences in aggregation of the ion pairs.[223]

^{7}Li shift data are available for a number of lithium organic systems other than those mentioned above: $LiBMe_4$, $LiAlMe_4$ and $LiTlMe_4$ in Et_2O, THF and DME;[224] MeLi together with Me_2M (where M=Mg, Zn and Cd) in THF;[225] MeLi together with LiX (where X=Br or I) in Et_2O or THF;[218,226,227] $LiMPh_3$ (where M=C, Si Ge, Sn and Pb) in THF;[224] $LiSnMe_3$ in THF.[224]

To round off this brief discussion of ^{7}Li shifts we would like to present a very nice experiment that illustrates the specificity of solvent—organolithium interactions and the shift changes produced. Lewis and Brown[214] studied the effect of added THF on the

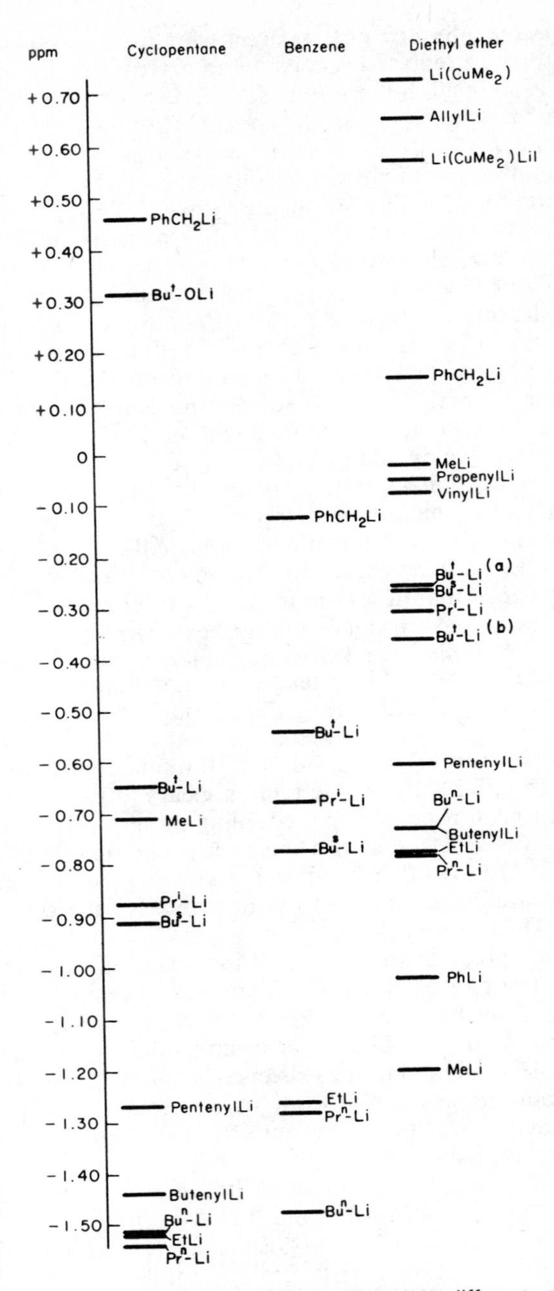

Fig. 6.13 ^{7}Li chemical shifts relative to an internal TMS lock in three different solvents. The shift scale is chosen as to have its zero position at the ^{7}Li resonance in 70% LiBr in water and a positive shift is upfield. (a) Data obtained on different samples of Bu^{t}Li (reproduced with permission from ref. 220).

TABLE 6.8 ^{7}Li chemical shifts in organolithium compounds where the organic moiety can form a resonance stabilised carbanion. Shifts for ca. 0.3 M solutions in ppm relative to external aqueous 1.0 M LiCl (Ref. 223)

Compound (R-Li)	Solvent				
R	Et_2O	THF	DME	Diglyme	HMPA
Cyclopentadienyl	−8.68[a]	−8.37	−8.67	−8.35	−0.88
Indenyl	−8.99	−6.12	−6.65	−3.11	−0.95
1-Phenylallyl	−0.66	−0.71	−0.96	−1.48	−0.66
Fluorenyl	−6.95	−2.08	−3.06	2.54	−0.73
1,3-Diphenylallyl	−1.44	−1.24	−2.42	—	−0.54
Triphenylmethyl	—	−1.11	−2.41	—	−0.66
Cyclononatetraenyl	—	−1.07	−3.68	—	−0.38

[a] This value is for dioxane solution.

^{7}Li shift for Bu^{n}Li in cyclopentane at 0°C and −75°C. The results are shown in Fig. 6.14 in which R is the molar ratio of THF to lithium.

Bu^{n}Li is known to be hexameric in cyclopentane and tetrameric in THF. The ^{7}Li shifts at R values greater than 1.0 are essentially the same as in neat THF and it appears that the transformation hexamer—tetramer is virtually complete already at R = 1.0 at −75°C.

6F.3 Spin-Spin Coupling

The existence of small but unresolved spin-spin couplings between C_α protons and lithium atoms in alkyllithiums has been indicated by linewidth studies and decoupling experiments.[215,218] The existence of $^1J(^{13}C, ^7Li)$ in MeLi was demonstrated in 1968 through the use of ^{13}C enriched compounds.[218,222] $^1J(^{13}C, ^7Li)$ were later found also in Bu^{t}Li and Bu^{n}Li.[216] The spin coupling pattern in the ^{7}Li NMR spectrum of ^{13}C enriched (57%) Bu^{t}Li in cyclohexane is shown in Fig. 6.15*a*. The pattern may be interpreted as a sum of subspectra of the four different local environments a lithium atom

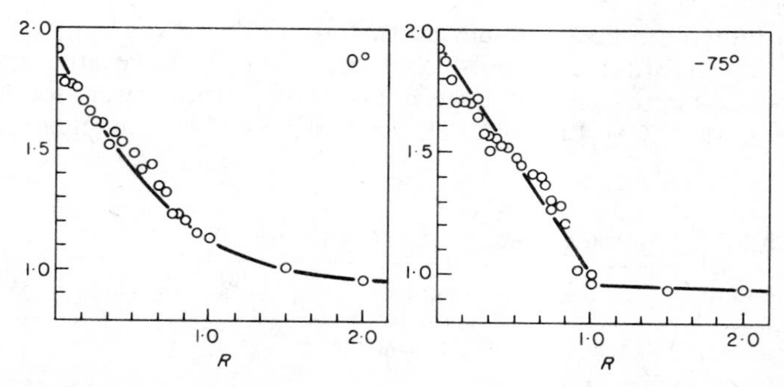

Fig. 6.14 The effect of added THF on the ^{7}Li chemical shift for Bu^{n}Li in cyclopentane at two temperatures. The chemical shift is given in ppm relative to an external aqueous LiBr solution. R is the molar ratio of THF to lithium (reproduced with permission from ref. 214).

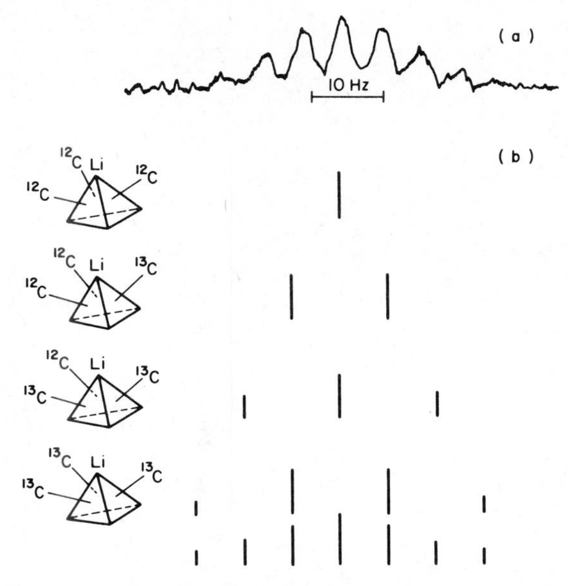

Fig. 6.15 (a) ^{7}Li NMR spectrum at room temperature of a ca. 0.1 M cyclohexane solution of Bu^{t}Li enriched with 57% ^{13}C in the α-carbons (reproduced with permission from ref. 216). (b) Tetrameric Bu^{t}Li illustrating the fourfold environment which determines the multiplicity of the above spectrum (reproduced with permission from ref. 228).

experiences in ^{13}C-enriched Bu^{t}Li (Fig. 6.15*b*). The observed spectrum supports the tetrameric structure of Bu^{t}Li. It also indicates that rapid intramolecular chemical exchange of the lithium atoms does not take place. The observed $^{1}J(^{13}C, ^{7}Li)$ are summarised below (Table 6.9).

Attempts have been made to observe $J(^{7}Li, ^{6}Li)$ in alkyllithiums enriched in ^{6}Li.[229] It was concluded that the coupling must be 0.3 Hz or less. This result has been taken to indicate that the metal-metal bond order in tetrameric alkyllithiums must be nearly zero.[229]

6F.4 Relaxation

An attempt to find the relaxation mechanism of ^{7}Li in alkyllithium compounds has been made.[217] T_1 was studied by pulse techniques over a range of temperatures. The "natural" linewidths in toluene solutions at room temperature were found to be about 1 Hz for EtLi and Me_3SiCH_2Li and 0.1 Hz for Bu^{t}Li. For MeLi in ether the value is also

TABLE 6.9 Some values of $^{1}J(^{13}C, ^{7}Li)$ for RLi

Compound	Solvent (temp.)	$^{1}J(^{13}C, ^{7}Li)$/Hz	Ref.
Li	THF, Et_2O, Et_3N (−60°?)	14.5–15	218, 228
Bu^{n}Li	Et_2O(−70°C)	14	216
Bu^{t}Li	Cyclohexane (room temp.)	11	216
Bu^{t}Li	Toluene (room temp.)	ca. 10	216

about 0.1 Hz. As mentioned above these values are considerably smaller than the actually observed linewidths pointing to the existence of unresolved proton spin couplings and possibly some exchange broadening.

Dipolar contributions to relaxation rates were estimated to be much smaller than the experimental values of T_1^{-1} and the dominant relaxation mechanism was assumed to be quadrupolar. In the absence of an experimental value of the rotational correlation times no attempt was made to calculate the quadrupole coupling constant.[217]

Through a study of the ^{7}Li linewidths for C_2D_5Li and Bu^{t}Li in toluene, Brown and Ladd calculated χ to be 0.57 MHz and 0.14 MHz, respectively.[215] The correlation times were calculated from the Debye-Stokes-Einstein equation and can not be considered reliable. From a study of the ^{7}Li quadrupolar splittings in solid EtLi Lucken has determined χ to be 0.098 MHz, a value considerably smaller than that obtained in toluene solution.[230]

Wehrli has recently carried out a detailed study of the ^{6}Li relaxation in MeLi, Bu^{n}Li and PhLi.[219] T_1 values were obtained through inversion recovery experiments and nuclear Overhauser enhancement (NOE) factors, $\eta_{^6\mathrm{Li}\{^1\mathrm{H}\}}$, were also determined. The T_1 value at 28°C of MeLi in Et_2O solution was 72 s, of Bu^{n}Li in *n*-hexane 19.8 s and of PhLi in a 70/30 mixture of C_6H_6 and Et_2O 12.8 s. The NOE values were found to be consistent with a 20, 35 and 17% ^{6}Li, ^{1}H dipole-dipole contribution to the relaxation rate in MeLi, Bu^{n}Li and PhLi, respectively. Under the assumption that the ^{7}Li relaxation in the same compounds is entirely quadrupolar, the quadrupolar contribution to the ^{6}Li relaxation rate was calculated to 3, 16.5 and 15%, respectively. The possibility of a ^{6}Li, ^{7}Li dipole-dipole contribution to the total relaxation rate was also considered. In MeLi a maximum contribution of 6% was estimated. The temperature dependence of T_1 for ^{6}Li in MeLi was compatible with a small spin-rotation contribution. It was finally concluded that due to the inefficient quadrupolar relaxation ^{6}Li NMR should be very advantageous in studies of larger and non-symmetric lithium organic compounds.

6F.5 Exchange Kinetics

The occurrence of rapid intermolecular exchange processes in mixtures of EtLi and MeLi in Et_2O was demonstrated by ^{7}Li NMR more than ten years ago.[221] At room temperature only one ^{7}Li signal is observed while at lower temperatures the line broadens and at $-80°$ the spectrum consists of several lines that can be attributed to mixed tetrameric species. The exchange is presumed to take place via a dissociation of the tetramer into dimers:

$$[\mathrm{RLi}]_4 \rightleftarrows 2[\mathrm{RLi}]_2$$

Further rapid dissociation of the dimers cannot be excluded on the basis of available experimental data.

Detailed NMR studies of the dissociation kinetics have been presented for two systems. The process

$$[\mathrm{MeLi}]_4 \rightarrow 2[\mathrm{MeLi}]_2$$

in ether proceeds with an enthalpy of activation, $\Delta H^{\neq}$, of 46 kJ/mole and an entropy of activation, $\Delta S^{\neq}$, of -26 J/mole, K.[231] Dissociation of the Bu^{t}Li tetramer into dimers is the rate limiting step in the formation of mixed complexes with $(CH_3)_3SiCH_2Li$ and can be conveniently studies by signal intensity measurements in cyclopentane. $\Delta H^{\neq}$

is calculated to be 100 ± 25 kJ/mole with $\Delta S^{\neq}$ approximately zero.[232] Solvent participation may possibly account for part of the large differences in $\Delta H^{\neq}$ for the $[CH_3Li]_4$ and $[Bu^tLi]_4$ dissociation. MeLi is unfortunately not very soluble in hydrocarbon solvents which precludes a direct comparison.

The rate of dissociation of $[Bu^tLi]_4$ in cyclopentane may be increased by a factor of at least 50 upon addition of Et_3N in low concentrations.[232,233] Since the bulky Bu^tLi tetramer is not likely to interact with the amine the rate increasing effect of the amine was ascribed to interactions with a transition state which more resembles the dissocated species.[233]

The intermolecular exchange of Bu^nLi dissolved in hydrocarbon solvents takes place much more rapidly than in Bu^tLi.[216] In hydrocarbon solvents, Bu^nLi is predominantly hexameric and it has been suggested that dissociation of the hexamer to form tetramer and dimer

$$[RLi]_6 \rightarrow [RLi]_4 + [RLi]_2$$

takes places much more readily than the dissociation of the tetramer.[216,233]

6G ALKALI METAL NMR OF AROMATIC RADICAL ION PAIRS

The radical nature of the molecular species formed by the action of alkali metals on solutions of aromatic hydrocarbons in THF and other polar solvents was clearly revealed by ESR spectroscopy in the 1950s. The ESR spectra were subsequently found to display hyperfine splittings due to interactions between the unpaired electron and magnetic alkali metal nuclei.[234] NMR spectroscopy on the magnetic alkali metal nuclei in these aromatic radical ion pairs has become a complementary technique in the study of their structure and dynamic properties.

6G.1 Theoretical Considerations

The observed chemical shift, δ_N, of a metal nucleus N in a radical ion pair may be considered the sum of two independent parts, δ_N^{Fc} and δ_N^D. δ_N^{Fc} is caused by the Fermi contact interaction and for a doublet radical δ_N^{Fc} is proportional to the hyperfine splitting constant (hfsc), a_N, (directly measurable in the ESR spectrum) and inversely proportional to the absolute temperature. A prerequisite for the observation of a single shifted alkali metal NMR signal is the fulfillment of the condition $|a_N|\tau_e \ll 1$ where τ_e is electron spin relaxation time of the radical ion pair. The value of the hfsc a_N is in turn directly proportional to the spin density at the nucleus, ρ_N. An important corollary is that the sign of the spin density may be inferred from the sign of δ_N^F. δ_N^D is simply the shift that should have been observed in the absence of the Fermi contact interaction. δ_N^D is clearly experimentally unobservable; limits on its value may however be obtained from diamagnetic model systems. In most cases, except when ρ_N is close to zero, δ_N^D may be neglected in comparison with δ_N^{Fc}.[235]

The transverse relaxation rate of alkali metal nuclei in radical ion pairs is generally considered to be the sum of three intramolecular contributions.[236–241]

$$T_2^{-1} = T_{2Fc}^{-1} + T_{2D}^{-1} + T_{2Q}^{-1} \tag{6.15}$$

The right-hand terms represent the Fermi contact interaction, the anisotropic electron-nucleus dipolar interaction and the quadrupolar interaction respectively.

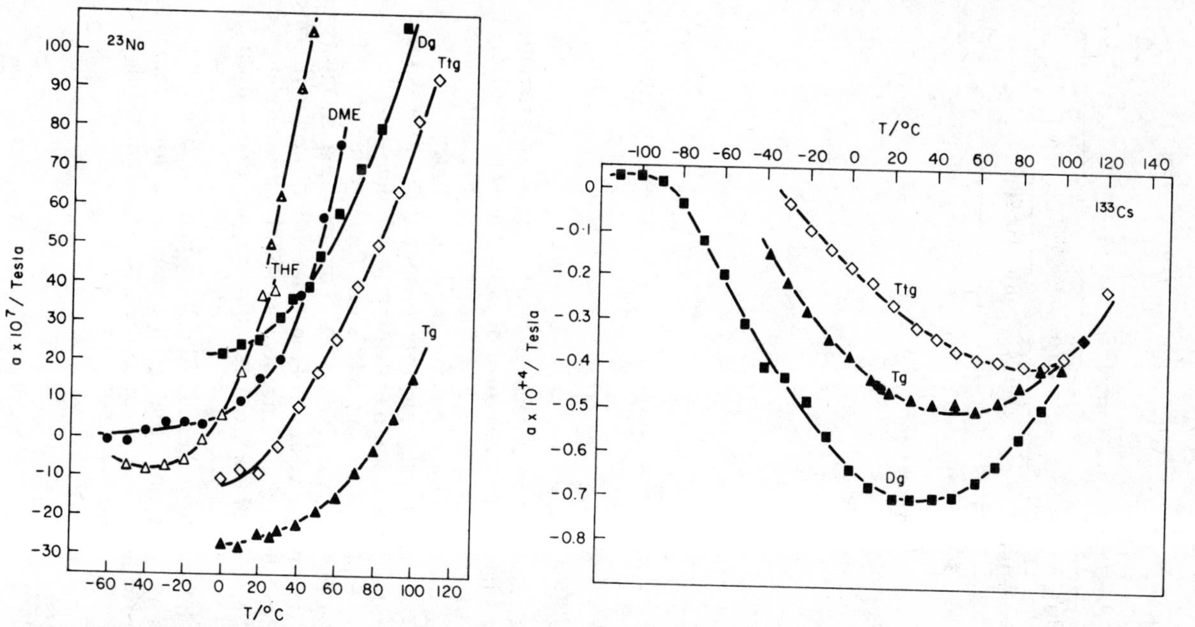

Figs 6.16 and 6.17 ^{23}Na and ^{133}Cs hyperfine splitting constants obtained from NMR studies as a function of temperature in solutions of (6.16) sodium biphenyl in THF (1.94 M), DME (1.0 M), diglyme (Dg) (1.7 M), triglyme (Tg) (1.0 M) and tetraglyme (Ttg) (0.77M). (Double triangles represent hsfc data for THF solutions obtained from ESR studies); (6.17) Caesium biphenyl in Dg (1.44 M), Tg (0.87 M) and Ttg (0.65 M) (reproduced with permission from ref. 235).

TABLE 6.10 NMR studies of aromatic radical ion pairs

	Nucleus studied							
Aromatic anion	^{6}Li	^{7}Li	^{23}Na	^{39}K	^{85}Rb	^{87}Rb	^{133}Cs	Ref.
Naphtalene	●	●	●	●	●	●	●	238, 239, 240
		●	●		●	●	●	243
		●				●	●	242
Biphenyl	●	●	●	●	●	●	●	235, 239, 241
			●		●	●		236
Fluorenone	●	●	●	●	●	●	●	235, 241
			●		●	●		236
		●						244
Anthracene		●	●		●	●	●	243
			●			●	●	242
2,2′-Dipyridyl		●	●		●	●	●	243
		●	●					242

6G.2 Experimental Data

Table 6.10 contains a summary of published NMR studies of aromatic radical ion pairs. The NMR studies have been performed on solutions with relatively high radical concentrations (0.1–1 M) in which τ_e is *ca* 10^{-10} to 10^{-11} s. Fig. 6.16 and 6.17 illustrate the type of information obtained for one aromatic anion: biphenyl.[235,241] The general trend is that the hfsc becomes smaller at lower temperatures. This is interpreted as a predominance of solvent separated ion pairs with low spin density at the metal nucleus at lower temperatures. Figures 6.16 and 6.17 show that the hfsc of ^{23}Na and ^{133}Cs becomes negative under certain conditions. MO studies of the spin density distribution show that this is dependent on the position of the metal with respect to the ring. Negative spin densities are predicted when the valence shell s-orbital of the metal interacts stronger with the highest bonding MO of the aromatic anion radical than with the lowest anti-bonding MO.[245]

Alkali metal NMR linewidth studies of radical ion pairs in different solvents and at different temperatures have indicated that quadrupole relaxation is important only for Na, K and Rb ion pairs while the Fermi contact relaxation is dominating in Cs ion pairs.

6H MISCELLANEOUS

6H.1 Alkali Metal Anions

The first NMR spectrum of an alkali metal anion, $^{23}Na^{-}$, was obtained in 1974 by Ceraso and Dye.[199] When excess sodium metal was added to a solution of the cryptand C222 in either $EtNH_2$ or THF the ionic species Na^{-} and $[Na(C222)]^{+}$ are formed. By cooling, crystals of Na^{-} $[Na(C222)]^{+}$ could be precipitated.

The ^{23}Na NMR spectrum of these solutions consists of two signals separated by about 50 ppm. The signal assigned to $^{23}Na^{-}$ was very narrow ($\Delta\nu_{1/2} < 3$ Hz in 0.15 M THF solution at -4°C), much narrower than the signal of $[^{23}Na(C222)]^{+}$ ($\Delta\nu_{1/2} \simeq 50$ Hz), and with a chemical shift that could be estimated to be within a few ppm of the shift of free $^{23}Na^-$ (see ch. 6C). The chemical shift of $^{23}Na^-$ was found to be nearly independent of the nature of the solvent. (By contrast the chemical shift of solvated $^{23}Na^+$ is markedly solvent dependent as already discussed in ch. 6D.) These results were taken to indicate that the two electrons in the 3s shell of the anion effectively shield the 2 p electrons from interaction with the solvent.

In a later paper, Dye *et al.*[189] have reported NMR studies also of Rb^- and Cs^- that have been prepared in much the same way as Na^-. In $EtNH_2$ and THF solutions containing Rb^- and Cs^- salts of $Rb(C222)^+$ and $Cs(C222)^+$, respectively, only the NMR signals of $^{87}Rb^-$ and $^{133}Cs^-$ are observed, the cation signals presumably being too broad to be detected. The resonance of $^{87}Rb^-$ in $EtNH_2$ and THF was found to be deshielded by 26 and 14 ppm, respectively, from the free rubidium atom and in THF the $^{133}Cs^-$ signal was about 52 ppm deshielded from the free caesium atom.

6H.2 Solutions of Alkali Metals in Liquid Ammonia

The nature of ammonia solutions of alkali metals has been the subject of an impressive number of theroretical and experimental studies (for recent reviews cf. Refs. 246–248). The most satisfactory models assume that in dilute ammonia solutions the main species are solvated metal ions, M_s^+, and solvated electrons, e_s^-. The electrons seem to be lodging in solvent cavities a few tenths of a nanometer in diameter—very much like electrons trapped in *F*-centres in ionic solids. NMR studies of ^7Li,[249,250] ^{23}Na,[249–251] ^{87}Rb,[249–250] and ^{133}Cs[249,250] in ammonia solutions of the corresponding metals show the resonances to be considerably deshielded compared to their position in ammonia solutions of diamagnetic salts. Also the linewidths are appreciably larger than in diamagnetic salt solutions.

By analogy with the deshielding observed for nuclei in metals[252] the deshielding in ammonia solutions have mostly been referred to as Knight shifts,* $K(N)$, and interpreted with the theoretical formalism developed for metals. The value of $K(N)$ depends on the electron spin density at the nucleus N, ρ_N. When several chemical species coexist in solution, each with its characteristic electron spin density, and if there is fast exchange of nuclei N between these species the observed value of $K(N)$ will be a weighted average of the values for the individual species.

A detailed NMR study of the Knight shifts of ^7Li, ^{23}Na, ^{87}Rb and ^{133}Cs in ammonia solutions of the metals at 300 K has been made by O'Reilly.[249] The results are shown in Fig. 6.18. The Knight shifts of ^7Li and ^{133}Cs differ by two orders of magnitude. Of particular interest is the finding that $K(^{23}Na)$ at the lowest concentrations ($\sim$3 mM, or about $R = 10\,000$) appears to decrease while the value at higher concentrations remains approximately constant. This was taken to indicate that at least two solute species involving unpaired electrons are present in solution at the lower concentrations. One of these is assumed to be the "free" solvated electron with virtually no hyperfine interaction with the metal ion and thus with a very small or zero contribution to the observed ^{23}Na

* The Knight shift of a nucleus N is defined as (at fixed frequency)

$$K(N) = \frac{B_0(N) - B_0^{\mathrm{Ref}}(N)}{B_0^{\mathrm{Ref}}(N)}$$

where $B_0(N)$ is the magnetic field at which the N-resonance occurs in the system under study and $B_0^{\mathrm{Ref}}(N)$ is the field at which the N-resonance occurs in a diamagnetic reference system. A weakness of this definition is that the value of $K(N)$ is dependent on the choice of reference system. The magnitude of the Knight shifts is however usually so large that this uncertainty becomes of secondary importance.

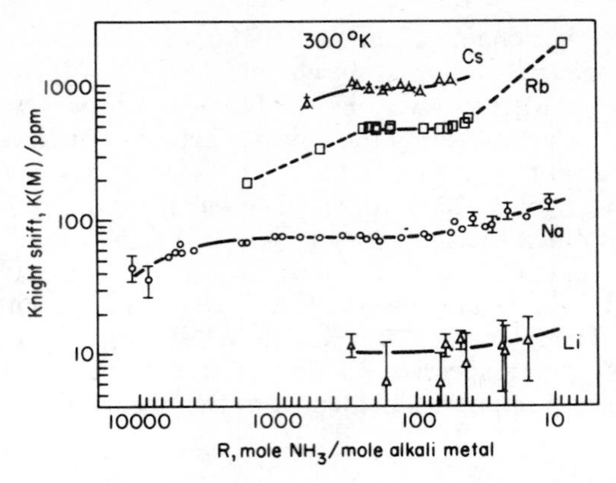

Fig. 6.18 Knight shifts of Li, Na, Rb and Cs in NH_3 solutions at 300 K plotted as a function of the mole ratio of NH_3 to alkali metal (reproduced with permission from ref. 249).

Knight shift. The second species is sometimes referred to as the solvated metal monomer (M_s) or sometimes as the solvent separated ion pair, $M^+\|e^-$, in which there is a considerable spin density of the electron at the metal. (For a theoretical calculation of the spin density in the monomer cf. ref. 253). At the highest metal concentrations spin-pairing begins to become important. The nature of the spin-paired species appears less well understood—it may be a solvated metal anion, M_s^-, a two-electron cavity species, $e_2^=$, or some other entity.

An interesting study of the self-diffusion coefficients of ^{7}Li, ^{23}Na and ^{1}H in lithium and sodium metal solutions in ammonia using the spin echo pulsed magnetic field gradient technique has been made.[254] The self-diffusion coefficient data suggest that the solvation number in the solutions is four both for lithium and sodium.

REFERENCES

1. B. Lindman and S. Forsén, "Methods of Biochemical Analysis" (D. Glick, ed.). Interscience, in press.
1. a. G. E. Hartwell and A. Allerhand, *J. Am. Chem. Soc.* **93**, 4415 (1971)
2. F. W. Wehrli, *J. Magn. Reson.* **23**, 527 (1976)
3. W. Sahm and A. Schwenk, *Z. Naturforsch.* **29a**, 1754 (1974)
4. A. Bolle, G. Puppi and G. Zantonelli, *Nuovo cimento* **3**, 412 (1946)
5. R. V. Pound, *Phys. Rev.* **72**, 1273 (1947)
6. A. Bolle and G. Zantonelli, *Ricerca Sci.* **18**, 847 (1948)
7. R. V. Pound, *Phys. Rev.* **73**, 1112 (1948)
8. F. Bitter, *Phys. Rev.* **75**, 1326 (1949)
9. J. R. Zimmerman and D. Williams, *Phys. Rev.* **76**, 163 (1949)
10. J. R. Zimmerman and D. Williams, *Phys. Rev.* **76**, 350 (1949)
11. W. H. Chambers and D. Williams, *Phys. Rev.* **76**, 638 (1949)
12. J. R. Zimmerman and D. Williams, *Phys. Rev.* **75**, 699 (1949)
13. K. Siegbahn and G. Lindström, *Nature* **163**, 211 (1949)
14. K. Siegbahn and G. Lindström, *Arkiv Fys.* **1**, 193 (1949)
15. T. L. Collins, *Phys. Rev.* **80**, 103 (1950)
16. G. D. Watkins and R. V. Pound, *Phys. Rev.* **82**, 343 (1951)

17. N. I. Adams, III, T. F. Wimett and F. Bitter, *Phys. Rev.* **82**, 343 (1951)
18. R. E. Sheriff and D. Williams, *Phys. Rev.* **82**, 651 (1951)
19. E. Yasaitis and B. Smaller, *Phys. Rev.* **82**, 750 (1951)
20. G. Lindström, *Arkiv Fys.* **4**, 1 (1951)
21. H. E. Walchli, W. E. Leyshan and F. M. Scheitlin, *Phys. Rev.* **85**, 922 (1952)
22. T. Kanda, Y. Masuda, R. Kusaka, Y. Yamagata and J. Itoh, *Phys. Rev.* **85**, 938 (1952)
23. E. Brun, J. Oeser, H. H. Staub and C. G. Telschow, *Phys. Rev.* **93**, 172 (1954)
24. L. C. Brown and D. Williams, *Phys. Rev.* **98**, 1537 (1955)
25. G. K. Yagola and E. E. Bogatirov, *Ukr. Fiz. Zh.* **7**, 145 (1962)
26. M. Eisenstadt and H. L. Friedman, *J. Chem. Phys.* **46**, 2182 (1967)
27. O. Lutz, *Phys. Lett.* **25A**, 440 (1967)
28. O. Lutz, *Z. Naturforsch.* **23a**, 1202 (1968)
29. D. Brinkman, *Phys. Lett.* **27A**, 466 (1968)
30. N. A. Schuster and G. E. Pake, *Phys. Rev.* **81**, 157 (1951)
31. R. V. Pound, *Phys. Rev.* **73**, 1247 (1948)
32. N. Bloembergen, E. M. Purcell and R. V. Pound, *Phys. Rev.* **73**, 679 (1948)
33. N. Bloembergen, Nuclear Magnetic Relaxation, Thesis, Leiden 1948; Reprinted by W. A. Benjamin, Inc. New York (1961)
34. V. A. Scherbakov and U. Eichoff, "Atlas of wide line NMR spectra of different nuclei", Khlopin Radium Institute Leningrad, 1978.
35. O. Lutz and A. Schwenk, *Phys. Lett.* **24A**, 122 (1967)
36. O. Lutz and A. Nolle, *Z. Naturforsch. A* **27a**, 1577 (1972)
37. H. Krüger, O. Lutz, A. Nolle, A. Schwenk and G. Stricker, *Z. Naturforsch.* **28a**, 484 (1973)
38. J. D. Halliday, R. E. Richards and R. R. Sharp, *Proc. Roy. Soc. A* **313**, 45 (1969)
39. C. Hall, R. E. Richards and R. R. Sharp, *Proc. Roy. Soc. Lond. A* **337**, 297 (1974)
40. A. H. Narten, F. Vaslov and H. A. Levy, *J. Chem. Phys.* **58**, 5017 (1973)
41. H. Kistenmacher, H. Popkie and E. Clementi, *J. Chem. Phys.* **61**, 799 (1974)
42. P. Kebarle, "Modern Aspects of Electrochemistry", Vol. 9, p. 1 (J. O'M. Bockris and B. E. Conway, eds). Butterworths, London (1974)
43. H. G. Hertz, M. Holz, R. Klute, G. Stalidis and H. Versmold, *Ber. Bunsenges. Phys. Chem.* **78**, 24 (1974)
44. B. Lindman and S. Forsén, "Chlorine, Bromine and Iodine NMR, Physico-chemical and Biological Applications" Springer Verlag, Heidelberg (1976)
45. M. Eisenstadt and H. L. Friedman, *J. Chem. Phys.* **44**, 1407 (1966)
46. V. I. Chizhik, *Proc. XVII Congr. Ampere 1973*, p. 339, V. Hovi, Ed., North-Holland
47. D. E. O'Reilly and E. M. Peterson, *J. Chem. Phys.* **51**, 4906 (1969)
48. H. G. Hertz, M. Holz, G. Keller, H. Versmold and C. Yoon, *Ber. Bunsenges. Phys. Chem.* **78**, 493 (1974)
49. A. Weiss, *Angew. Chem. Int. Ed.* **11**, 607 (1972)
50. L. Endom, H. G. Hertz, B. Thül and M. D. Zeidler, *Ber. Bunsenges. Phys. Chem.* **71**, 1008 (1967)
51. A. Geiger and H. G. Hertz, *Adv. Mol. Relaxation Proc.* **9**, 293 (1976)
52. C. Hall, R. E. Richards, G. N. Schulz and R. R. Sharp, *Mol. Phys.* **16**, 529 (1969)
53. H. G. Hertz, R. Tutsch and H. Versmold, *Ber. Bunsenges. Phys. Chem.* **75**, 1177 (1971)
54. E. J. Sutter and J. F. Harmon, *J. Phys. Chem.* **79**, 1958 (1975)
55. D. E. Woessner, B. S. Snowden, Jr. and A. G. Ostroff, *J. Chem. Phys.* **49**, 371 (1968)
56. R. A. Craig and R. E. Richards, *Trans. Far. Soc.* **59**, 1972 (1963)
57. R. G. Bryant, *J. Phys. Chem.* **73**, 1153 (1969)
58. F. W. Wehrli, *J. Magn. Reson.* in press.
59. F. W. Wehrli, *J. Magn. Reson.* **25**, 575 (1977).
60. H. G. Hertz, *Z. Elektrochemie* **65**, 20 (1961)
61. H. G. Hertz, G. Stalidis and H. Versmold, *J. Chim. Phys.* **66**, 177 (1969)
62. H. G. Hertz, *Ber. Bunsenges. Phys. Chem.* **77**, 531 (1973)
63. H. G. Hertz, *Ber. Bunsenges. Phys. Chem.* **77**, 688 (1973)
64. K. A. Valiev, *Soviet Phys. JETP* **10**, 77 (1960)
65. K. A. Valiev, *Soviet Phys. JETP* **11**, 883 (1960)

66. K. A. Valiev and B. M. Khabibullin, *Russ. J. Phys. Chem.* **35**, 1118 (1961)
67. K. A. Valiev, *J. Struct. Chem.* **3**, 630 (1962)
68. K. A. Valiev and M. M. Zaripov, *J. Struct. Chem.* **7**, 470 (1966)
69. R. M. Yul'met'ev, *J. Struct. Chem.* **7**, 783 (1966)
70. K. A. Valiev, *J. Struct. Chem.* **5**, 477 (1964)
71. V. I. Chizhik and V. I. Mikhailov, *Yad. Magn. Rezon.* **5**, 58 (1974)
72. C. A. Melendres and H. G. Hertz, *J. Chem. Phys.* **61**, 4156 (1974)
73. V. I. Ionov, R. K. Mazitov and I. I. Evdokimov, *J. Struct. Chem.* **10**, 197 (1969)
74. F. W. Cope and R. Damadian, *Physiol. Chem. Phys.* **6**, 17 (1974)
75. C. Deverell, *Progress in NMR Spectroscopy* **4**, 235 (1969).
76. C. Deverell, *Mol. Phys.* **16**, 491 (1969)
77. J. Itoh and Y. Yamagata, *J. Phys. Soc. Japan.* **13**, 1182 (1958)
78. A. Beckmann, K. D. Bökler and D. Elke, *Z. Phys.* **270**, 173 (1974)
79. S. J. Davis, J. J. Wright and L. C. Balling, *Phys. Rev. A* **9**, 1494 (1974)
80. C. W. White, W. M. Hughes, G. S. Hayne and H. G. Robinson, *Phys. Rev. A* **7**, 1178 (1973)
81. R. Huber, H. C. Weber, H. Nienstädt and G. Zu Putlitz, *Phys. Lett.* **55A**, 141 (1975)
82. R. Huber, F. König and H. G. Weber, *Z. Phys.* in press
83. V. I. Chizhik and Yu. A. Ermakov, *Yad. Magn. Rezonanz.* 60 (1971)
84. G. Laukien and K. Süss, *Physikalische Verhandlungen* 138 (1963)
85. V. I. Chizhik, *J. Struct. Chem.* (*USSR*) **8**, 303 (1967)
86. J. W. Akitt and A. J. Downs, *Chem. Soc. Spec. Publ. No. 22*, 199 (1967)
87. P. A. Speight and R. L. Armstrong, *Can. J. Phys.* **45**, 2493 (1967)
88. A. I. Mishustin and R. A. Sidorova, *Vestn. Mosk. Univ., Fiz. Astron.* **11**, 528 (1970)
89. A. I. Mishustin, *Vestn. Mosk. Univ., Fiz. Astron.* **13**, 503 (1972)
90. J. E. Wertz and O. Jardetzky, *J. Chem. Phys.* **25**, 357 (1956)
91. O. Jardetzky and J. E. Wertz, *J. Am. Chem. Soc.* **82**, 318 (1960)
92. V. S. Griffiths and G. Socrates, *J. Mol. Spectr.* **27**, 358 (1968)
93. F. W. Cope, *Biophys. J.* **10**, 843 (1970)
94. I. D. Campbell and D. J. Mackey, *Aust. J. Chem.* **24**, 45 (1971)
95. G. J. Templeman and A. L. Van Geet, *J. Am. Chem. Soc.* **94**, 5578 (1972)
96. H. G. Hertz and M. D. Zeidler, *Ber. Bunsenges. Phys. Chem.* **67**, 774 (1963)
97. R. K. Mazitov and L. S. Itkina, *Zh. Neorg. Khim.* **19**, 3203 (1974)
98. Yu. A. Buslaev, V. V. Evsikov, M. N. Buslaeva, R. V. Mazitov and V. N. Plakhotnik, *Proc. Acad. Sci. USSR, Phys. Chem. Sect.* **213**, 1120 (1973)
99. G. Z. Mal'tsev, G. V. Malinin, V. P. Mashovets and V. A. Shcherbakov, *Zh. Strukt. Khim.* **6**, 371 (1965)
100. V. I. Ionov, R. K. Mazitov and O. Ya. Samilov, *J. Struct. Chem. USSR.* **10**, 335 (1969)
101. G. Z. Mal'tsev, G. V. Malinin and V. P. Mashovets, *Zh. Strukt. Khim.* **6**, 378 (1965)
102. R. J. Moolenaar, J. C. Evans and L. D. McKeever, *J. Phys. Chem.* **74**, 3629 (1970)
103. R. A. Dwek, Z. Luz and M. Shporer, *J. Phys. Chem.* **74**, 2232 (1970)
104. H. G. Hertz, B. Lindman and V. Siepe, *Ber. Bunsenges. Phys. Chem.* **73**, 542 (1969)
105. J. W. Akitt and A. J. Downs, *Chem. Commun.* **1966**, 222
106. C. Deverell and R. E. Richards, *Mol. Phys.* **10**, 551 (1966)
107. G. J. Templeman and A. L. Van Geet, *J. Am. Chem. Soc.* **94**, 5578 (1972)
108. E. G. Bloor and R. G. Kidd, *Can. J. Chem.* **50**, 3926 (1972)
109. J. C. Hindman, *J. Chem. Phys.* **36**, 1000 (1962)
110. A. Carrington, F. Dravnicks and M. C. R. Symons, *Mol. Phys.* **4**, 174 (1960)
111. R. E. Richards and B. A. Yorke, *Mol. Phys.* **6**, 289 (1963)
112. O. Lutz, *Z. Naturforsch. A* **22**, 286 (1967)
113. D. Ikenberry and T. P. Das, *J. Chem. Phys.* **43**, 2199 (1965)
114. J. Kondo and J. Yamashita, *J. Phys. Chem. Solids* **10**, 245 (1959)
115. D. Ikenberry and T. P. Das, *J. Chem. Phys.* **45**, 1361 (1966)
116. C. Deverell, K. Schaumburg and H. J. Bernstein, *J. Chem. Phys.* **49**, 1276 (1968)
117. M. Halmann and I. Platzeus, *Proc. Chem. Soc.* 261 (1964)
118. J. Halliday, H. D. W. Hill and R. E. Richards, *Chem. Commun.* 219 (1969)

119. H. Gustavsson and B. Lindman, *J. Am. Chem. Soc.* **97**, 3923 (1975)
120. H. Gustavsson and B. Lindman, *Proc. Int. Conf. on Coll. and Surface Sci.*, Budapest 1975 (E. Wolfram, ed.). Vol. 1, p. 625. Akademia Kiado, Budapest (1976)
121. A. Loewenstein, M. Shporer, P. C. Lauterbur and J. E. Ramirez, *Chem. Commun.* **1968**, 214
122. A. Weiss and K. H. Nothnagel, *Ber. Bunsenges. Phys. Chem.* **75**, 216 (1971)
123. H. G. Hertz, M. Holz and R. Mills, *J. Chim. Phys.* **71**, 1355 (1974)
124. G. N. LaMar, W. Dew. Horrocks, Jr. and R. H. Holm (eds) "NMR of Paramagnetic Molecules". Academic Press, New York (1973)
125. F. Fister and H. G. Hertz, *Z. Phys. Chemie, N.F.* **40**, 253 (1964)
126. W. Hasenfratz, G. Heckmann, P. Ihlenburg and O. Lutz, *Z. Naturforsch. A* **22**, 583 (1967)
127. E. A. Nikoforov and A. A. Popel', *J. Struct. Chem. USSR* **11**, 1043 (1970)
128. G. M. Gusakov, I. I. Evdokimov and R. K. Mazitov, *J. Struct. Chem. USSR.* **12**, 642 (1971)
129. G. M. Gusakov, I. I. Evdokimov and R. K. Mazitov. *J. Struct. Chem. USSR.* **12**, 531 (1971)
130. M. Shporer, R. Poupko and Z. Luz, *Inorg. Chem.* **11**, 2441 (1972)
131. R. Goeller, H. G. Hertz and R. Tutsch, *Pure Appl. Chem.* **32**, 149 (1972)
132. Yu. M. Kessler, A. I. Mishustin and A. I. Podkovyrin, *J. Solution Chem.* **6**, 111 (1977).
133. E. Shchori, J. Jagur-Grodzinski, Z. Luz and M. Shporer, *J. Am. Chem. Soc.* **93**, 7133 (1971)
134. R. D. Green and J. S. Martin, *Can. J. Chem.* **50**, 3935 (1972)
135. M. S. Greenberg, R. L. Bodner and A. I. Popov, *J. Phys. Chem.* **77**, 2449 (1973)
136. A. L. Van Geet, *J. Am. Chem. Soc.* **94**, 5583 (1972)
137. A. M. Grotens, J. Smid and E. DeBoer, *Chem. Commun.* 759 (1971)
138. G. W. Canters, *J. Am. Chem. Soc.* **94**, 5230 (1972)
139. C. Detellier and P. Laszlo, *Bull. Soc. Chim. Belg.* **84**, 1087 (1975)
140. C. Detellier and P. Laszlo, in "Spectroscopic and Electrochemical Characterization of Solute Species in Non-Aqueous Solvents" (G. Momantov, Ed.) Plenum Press, New York, in press.
141. A. I. Mishustin and Y. M. Kessler, *J. Solution Chem.* **4**, 779 (1975)
142. H. G. Hertz, H. Weingärtner and B. M. Bode, *Ber. Bunsenges. Phys. Chem.* **79**, 1190 (1975)
143. H. G. Hertz and H. Weingärtner, *J. Solution Chem.* **4**, 790 (1975)
144. G. E. Maciel, J. K. Huncock, L. F. Lafferty, P. A. Mueller and W. K. Musker, *Inorg. Chem.* **5**, 554 (1966)
145. F. Koch, in *Proc. XV Colloque Ampere, Grenoble 1968*, p. 247 (P. Averbuch, ed.). North-Holland, Amsterdam (1969)
146. R. H. Erlich, E. Roach and A. I. Popov, *J. Am. Chem. Soc.* **92**, 4989 (1970)
147. R. H. Cox, H. W. Terry and L. W. Harrison, *J. Am. Chem. Soc.* **93**, 3297 (1971)
148. Y. M. Cahen, P. R. Handy, E. T. Roach and A. I. Popov, *J. Phys. Chem.* **79**, 80 (1975)
149. A. I. Popov, *Pure Appl. Chem.* **41**, 275 (1975)
150. W. J. DeWitte, R. C. Schoening and A. I. Popov, *Inorg. Nucl. Chem. Lett.* **12**, 251 (1976)
151. J. S. Shih and A. I. Popov, *Inorg. Nucl. Chem. Lett.* **13**, 105 (1977)
152. E. G. Bloor and R. G. Kidd, *Can. J. Chem.* **46**, 3425 (1968)
153. R. H. Erlich and A. I. Popov, *J. Am. Chem. Soc.* **93**, 5620 (1971)
154. M. Herlem and A. I. Popov, *J. Am. Chem. Soc.* **94**, 1431 (1972)
155. M. S. Greenberg, D. M. Wied and A. I. Popov, *Spectrochim Acta, Part A* **29**, 1927 (1973)
156. T. L. Buxton and J. A. Caruso, *J. Am. Chem. Soc.* **96**, 6033 (1974)
157. C. Detellier and P. Laszlo, *Bull. Soc. Chim. Belg.* **84**, 1081 (1975)
158. W. J. DeWitte and A. I. Popov, *J. Solution Chem.* **5**, 231 (1976)
159. C. Detellier and P. Laszlo, *Helv. Chim. Acta* **59**, 1333 (1976)
160. M. S. Greenberg, A. I. Popov, *J. Solution Chem.* **5**, 653 (1976)
161. P. Neggia, M. Holz and H. G. Hertz, *J. Chim. Phys.* **71**, 56 (1974)
162. C. Detellier and P. Laszlo, *Helv. Chim. Acta* **59**, 1346 (1976)
163. M. Goldsmith, D. Hor and R. Damadian, *J. Chem. Phys.* **65**, 1708 (1976)
164. M. Holz, H. Weingärtner and H. G. Hertz, *J. Chem. Soc. Farad. I* **73**, 71 (1977)
165. E. M. Arnett, H. C. Ko and R. J. Minasz, *J. Phys. Chem.* **76**, 2474 (1972)
166. A. K. Covington, T. H. Lilley, K. E. Newman and G. A. Porthouse, *J. Chem. Soc. Farad. I* **69**, 963 (1973)
167. A. K. Covington, K. E. Newman and T. H. Lilley, *J. Chem. Soc. Farad. I* **69**. 973 (1973)

168. R. H. Erlich, M. S. Greenberg and A. I. Popov, *Spectrochim. Acta, Part A* **29**, 543 (1973)
169. M. S. Greenberg and A. I. Popov, *Spectrochim. Acta, Part A* **31**, 697 (1975)
170. R. G. Baum and A. I. Popov, *J. Solution Chem.* **4**, 441 (1975)
171. A. K. Covington, I. R. Lantzke and J. M. Thain, *J. Chem. Soc. Farad. I* **70**, 1869 (1974)
172. A. K. Covington and J. M. Thain, *J. Chem. Soc. Farad. I* **70**, 1879 (1974)
173. T. L. James and J. H. Noggle, *J. Am. Chem. Soc.* **91**, 3424 (1969)
174. T. L. James and J. H. Noggle, *Bioinorg. Chem.* **2**, 69 (1972)
175. O. Jardetzky and J. E. Wertz, *Arch. Biochem. Biophys.* **65**, 569 (1956)
176. J. E. Wertz, *J. Phys. Chem.* **61**, 51 (1957)
177. G. A. Rechnitz and S. B. Zamochnick, *J. Amer. Chem. Soc.* **86**, 2953 (1964)
178. J. W. Akitt and M. Parekh, *J. Chem. Soc. A.* 2195 (1968)
179. J. Andrasko and S. Forsén, *Biochem. Biophys. Res. Commun.* **52**, 233 (1974)
180. D. O. Carpenter, M. M. Hovey and A. F. Bak, *Ann. N. Y. Acad. Sci.* **204**, 502 (1973)
181. C. Detellier, J. Grandjean and P. Laszlo, *J. Am. Chem. Soc.* **98**, 3375 (1976)
182. J. Grandjean and P. Laszlo, *Helv. Chim. Acta*, **60**, 259 (1977)
183. T. L. James and J. H. Noggle, *Anal. Biochem.* **49**, 208 (1972)
184. R. G. Bryant, *Biochem. Biophys. Res. Commun.* **40**, 1162 (1970)
185. A. Paris and P. Laszlo, *ACS Sympos. Series* **34**, 418 (1976)
186. D. H. Haynes, B. C. Pressman and A. Kowalsky, *Biochemistry* **10**, 852 (1971)
187. E. Shchori, J. Jagur-Grodzinski, Z. Luz and M. Shporer, *J. Am. Chem. Soc.* **93**, 7133 (1971)
188. E. Shchori, J. Jagur-Grodzinski and M. Shporer, *J. Am. Chem. Soc.* **95**, 3842 (1973)
189. J. L. Dye, C. W. Andrews and J. M. Ceraso, *J. Phys. Chem.* **79**, 3076 (1975)
190. H. Weingärtner and H. G. Hertz, *Ber. Bunsenges. Phys. Chem.* **81**, 1207 (1977)
191. M. Shporer and Z. Luz, *J. Am. Chem. Soc.* **97**, 665 (1975)
192. E. Mei, J. L. Dye and A. I. Popov, *J. Am. Chem. Soc.* **98**, 1619 (1976)
193. Y. M. Cahen, J. L. Dye and A. I. Popov, *Inorg. Nucl. Chem. Lett.* **10**, 899 (1974)
194. Y. M. Cahen, J. L. Dye and A. I. Popov, *J. Phys. Chem.* **79** 1292 (1975)
195. Y. M. Cahen, J. L. Dye and A. I. Popov, *J. Phys. Chem.* **79**, 1289 (1975)
196. A. Hourdakis and A. I. Popov, *J. Solution Chem.* **6**, 299 (1977)
197. J. M. Ceraso and J. L. Dye, *J. Am. Chem. Soc.* **95**, 4432 (1973)
198. J. P. Kintzinger and J. M. Lehn, *J. Am. Chem. Soc.* **96**, 3313 (1974)
199. J. M. Ceraso and J. L. Dye, *J. Chem. Phys.* **61**, 1585 (1974)
200. J. L. Dye, C. W. Andrews and J. M. Ceraso, *J. Phys. Chem.* **79**, 3076 (1975)
201. M. Shporer, H. Zemel and Z. Luz, *FEBS Lett.* **40**, 357 (1974)
202. P. G. Gertenbach and A. I. Popov, *J. Am. Chem. Soc.* **97**, 4738 (1975)
203. A. C. Plaush and R. R. Sharp, *J. Am. Chem. Soc.* **98**, 7973 (1976)
204. R. L. Bodner, M. S. Greenberg and A. I. Popov, *Spectroscopy Lett.* **5**, 489 (1972)
205. E. T. Roach, P. R. Handy and A. I. Popov, *Inorg. Nucl. Chem. Lett.* **9**, 359 (1973)
206. Y. M. Cahen, R. F. Beisel and A. I. Popov, *Inorg. Nucl. Chem.* H. H. Hyman Mem. Vol., 209 (1976)
207. J. M. Lehn, J. P. Sauvage and B. Dietrich, *J. Am. Chem. Soc.* **92**, 2916 (1970); K. H. Wong, G. Konizer and J. Smid, *J. Am. Chem. Soc.* **92**, 666 (1970); D. H. Haynes, *FEBS Lett.* **20**, 221 (1972)
208. M. R. Truter, *Structure Bonding* **16**, 71 (1973); M. Dobler, J. D. Dunitz and P. Seiler, *Acta Cryst.* **B30**, 2741 (1974); M. R. Truter, *Phil. Trans. Roy. Soc. Lond.* **B272**, 29 (1975)
209. J. M. Lehn, J. P. Sauvage and B. Dietrich, *J. Am. Chem. Soc.* **92**, 2916 (1970)
210. J. M. Ceraso, P. B. Smith, J. S. Landers and J. L. Dye, *J. Phys. Chem.* **81**, 760 (1977)
211. Cf. Chapter 3 by U. Schöllkopf in Vol. 13/1 of Houben-Weyl "Methoden der Organische Chemie". G. Thieme Verlag, Stuttgart (1970)
212. T. L. Brown, D. W. Dickerhoof and D. A. Bafur, *J. Am. Chem. Soc.* **84**, 1371 (1962)
213. M. A. Weiner and R. West, *J. Am. Chem. Soc.* **85**, 485 (1963)
214. H. L. Lewis and T. L. Brown, *J. Am. Chem. Soc.* **92**, 4664 (1970)
215. T. L. Brown and J. A. Ladd, *J. Organomet. Chem.* **2**, 373 (1964)
216. L. D. McKeever and R. Waack, *Chem. Commun.* 750 (1969)
217. G. E. Hartwell and A. Allerhand, *J. Am. Chem. Soc.* **93**, 4415 (1971)

218. L. D. McKeever, R. Waack, M. A. Doran and E. B. Baker, *J. Am. Chem. Soc.* **90**, 3244 (1968)
219. F. W. Wehrli, *Org. Magn. Reson.* **11**, 106 (1978)
220. P. A. Scherr, R. J. Hogan and J. P. Oliver, *J. Am. Chem. Soc.* **96**, 6055 (1974)
221. L. M. Seitz and T. L. Brown, *J. Am. Chem. Soc.* **88**, 2174 (1966)
222. J. A. Ladd and J. Parker, *J. C. S. Dalton Trans.* 930 (1972)
223. R. H. Cox and H. W. Terry, *J. Magn. Reson.* **14**, 317 (1974)
224. R. J. Hogan, P. A. Scherr, A. T. Weibel and J. P. Oliver, *J. Organomet. Chem.* **85**, 265 (1975)
225. L. M. Seitz and B. F. Little, *J. Organomet. Chem.* **18**, 227 (1969)
226. R. Waack, M. A. Doran and E. B. Baker, *Chem. Commun.* 1291 (1967)
227. D. P. Novak and T. L. Brown, *J. Am. Chem. Soc.* **94**, 3793 (1972)
228. L. D. McKeever, R. Waack, M. A. Doran and E. B. Baker, *J. Am. Chem. Soc.* **91**, 1057 (1969)
229. T. L. Brown, L. M. Seitz and B. Y. Kimura, *J. Am. Chem. Soc.* **90**, 3245 (1968)
230. E. A. C. Lucken, *J. Organomet. Chem.* **4**, 252 (1965)
231. K. C. Williams and T. L. Brown, *J. Am. Chem. Soc.* **88**, 4134 (1966)
232. M. Y. Darensbourg, B. Y. Kimura, G. E. Hartwell and T. L. Brown, *J. Am. Chem. Soc.* **92**, 1236 (1970)
233. T. L. Brown, *J. Pure Appl. Chem.* **23**, 447 (1970)
234. J. H. Sharp and M. C. R. Symons, in "Ions and Ion Pairs in Organic Reactions", Vol. 1, Ch. 5 (M. Szwarc ed.). J. Wiley-Interscience, New York (1972)
235. G. W. Canters and E. DeBoer, *Mol. Phys.* **26**, 1185 (1973)
236. G. W. Canters, E. DeBoer, B. M. P. Hendriks and H. van Willigen, *Chem. Phys. Lett.* **1**, 627 (1968)
237. G. W. Canters, E. DeBoer, B. M. P. Hendriks and A. A. K. Klaassen, *Proc. XV Colloque Ampère, Grenoble 1968*, p. 242 (P. Averbuch ,ed.). North-Holland, Amsterdam (1969)
238. B. M. Hendriks, G. W. Canters, C. Corvaja, J. W. M. DeBoer and E. DeBoer, *Mol. Phys.* **20**, 193 (1971)
239. E. DeBoer and J. L. Sommerdijk, in "Ions and Ion pairs in Organic Reactions", Vol. 1, p. 289 (M. Szwarc, ed.). John Wiley-Interscience, New York (1972)
240. E. DeBoer and B. M. P. Hendriks, *Pure Appl. Chem.* **40**, 259 (1974)
241. G. W. Canters and E. DeBoer, *Mol. Phys.* **27**, 665 (1974)
242. T. Takeshita and N. Hirota, *Chem. Phys. Lett.* **4**, 369 (1969)
243. T. Takeshita and N. Hirota, *J. Chem. Phys.* **58**, 3745 (1973)
244. G. W. Canters, H. van Willigen and E. DeBoer, *Chem. Commun.* 566 (1967)
245. G. W. Canters, C. Corvaja and E. DeBoer, *J. Chem. Phys.* **54**, 3026 (1971)
246. J. C. Thompson, "Electrons in Liquid Ammonia." Clarendon Press, Oxford (1976). (A comprehensive and critical review that can be strongly recommended as an introduction to the field)
247. J. Jortner and N. R. Kestner, eds, "Electrons in Fluids". Springer-Verlag, Berlin (1973)
248. M. C. R. Symons, *Chemical Soc. Reviews* **5**, 337 (1976)
249. D. E. O'Reilly, *J. Chem. Phys.* **41**, 3729 (1964)
250. D. E. O'Reilly, in "Metal-Ammonia Solutions, Physicochem. Properties," Colloq. Weyl, Lille, 1963, p. 215 (Publ. 1964)
251. K. S. Pitzer in "Metal-Ammonia Solutions, Physicochem. Properties," Colloq. Weyl, Lille 1963, p. 193. (Publ. 1964)
252. Cf. for example C. P. Slichter, "Principles of Magnetic Resonance." Harper & Row, New York (1963)
253. D. E. O'Reilly, *J. Chem. Phys.* **41**, 3736 (1964)
254. A. N. Garraway and R. M. Cotts, *Phys. Rev. A* **7**, 635 (1973)

7
THE ALKALINE EARTH METALS

BJÖRN LINDMAN, University of Lund, Sweden
STURE FORSÉN, University of Lund, Sweden

7A GENERAL PROPERTIES OF THE ALKALINE EARTHS

From an NMR point of view, the alkaline earths, i.e. Be, Mg, Ca, Sr and Ba, form a quite homogeneous series of elements and with the exception of Be (and to some extent also Mg) they have also quite similar chemical properties.[1] All these elements are widely distributed in minerals and in the sea, occurring in mineral for example as carbonates, chlorides and sulphates. For the higher elements, Ca, Sr and Ba, it is only the divalent ions which are of any greater interest to the chemist or biologist. The small sized beryllium, on the other hand, has a marked tendency to involve in covalent binding and organoberyllium compounds have attracted considerable interest. (Beryllium containing substances often show a pronounced toxicology.[2]) In aqueous solution beryllium occurs as the $[Be(H_2O)_4]^{2+}$ ion in the presence of acid but otherwise in the form of various complexes with hydroxide (or other species). Depending on the system, either the ionic or the covalent character of magnesium predominates. In aqueous solution it occurs as the hydrated Mg^{2+} ion, which is not acidic as the Be^{2+} hydrate. For the studies of aqueous Mg^{2+} solutions it is important to recall that the hydroxide is insoluble in water. Organomagnesium compounds, for example Grignard reagents, are among the most widely used of all organometallic compounds in organic synthetic work.

Both magnesium and calcium have a wide occurrence in biological systems and the relation of these elements to biological function forms the subject of intense research.[3–5] Calcium is found in bone and in the biological fluids as free Ca^{2+} ions as well as bound to various proteins where it can serve a structural role or as an activator of enzymic processes. Among important calcium requiring physiological processes may be mentioned blood clotting, visual excitation and muscle contraction. The presence of magnesium in chlorophyll is well known and magnesium also occurs in ionic form as the free ion or in complexes with other species like nucleotides and proteins. The complexing of Mg^{2+} with nucleic acids is of great significance, as is also the complexes between Mg^{2+} or Ca^{2+} and mucopolysaccharides in connective tissue.

In Tables 1.2 and 7.1 are summarized some nuclear properties of the stable alkaline earth isotopes and here are also given some other quantities of interest for the following discussion. All the nuclei can be seen to possess electric quadrupole moments and to have small magnetic moments. (It may be noted that no precise electric quadrupole moment seems to be available for ^{43}Ca.) The low NMR sensitivities, for ^{43}Ca accentuated by a very low natural abundance, constitute the main reason for the small number of studies performed. While for ^{9}Be a number of early NMR determinations of the magnetic

TABLE 7.1 Some properties of the alkaline earths

Element	Sternheimer antishielding factor of M^{2+} ion	Ionic radius of $M^{2+}/10^{-8}$ cm
Be	0.81[a]	0.31
Mg	4.32[a]	0.65
Ca	13.12[a]	0.99
Sr	41[a]	1.13
Ba	123[b]	1.35

[a] E. A. C. Lucken, "Nuclear Quadrupole Coupling Constants," Academic Press, London (1969); K. D. Sen and P. T. Narasimhan, "Advances in Nuclear Quadrupole Resonance" **1**, 277 (1974). [b] M. Weiden and A. Weiss, *Proc. 18th Ampere Congress*, Nottingham, 1974, p. 257.

moment were given,[6–12] the first observations of the magnetic resonances of ^{25}Mg,[10] ^{43}Ca,[13] ^{87}Sr,[14] ^{135}Ba[15] and ^{137}Ba[15] were reported relatively late. In recent years, very precise NMR determinations of the magnetic moments, which are of great importance for establishing absolute shielding scales (see below) have been performed by the Tübingen group for ^{25}Mg,[16] ^{43}Ca,[16,17] ^{87}Sr,[18,19] ^{135}Ba[20] and ^{137}Ba.[20] It may also be noted in this connection that the NMR signal of the radioactive ^{41}Ca isotope (half-life $1.1 \cdot 10^5$ years) has been observed for saturated aqueous solution of $Ca(NO_3)_2$.[21]

From the above outline of the chemistry and biology of the alkaline earths it is not unexpected to find that the majority of studies of these elements have been concerned with the aqueous ions. Our survey starts with the relaxation behaviour of the ions and thereby we follow a path of increasing complexity of the system. Thereafter the shielding of the ions will be discussed and then the few observations of the ^9Be NMR for organoberyllium compounds. Alkaline earth NMR studies of simple solid systems dealing with problems in solid state physics have been quite useful and these applications are mentioned, but not discussed, in the final section.

7B RELAXATION OF AQUEOUS ALKALINE EARTH IONS

7B.1 Relaxation Mechanisms

All the stable alkaline earth nuclei possess electric quadrupole moments but the magnitude of these is small for some isotopes. As, furthermore, appreciable quenching of the field gradients due to tetrahedrally or octahedrally symmetric solvation is probable for these rather strongly hydrated ions, it is evident that the contribution from other relaxation mechanisms than quadrupole relaxation must be carefully searched for. It is only for the Be^{2+} ion that this problem has been studied at depth. In this case, Wehrli[22,23] determined T_1 and the NOE as a function of temperature (−5 to 105°C) for 1 M $Be(NO_3)_2$. The temperature dependence reproduced in Fig. 7.1 demonstrates the predominance of spin-rotation relaxation at high temperatures.[22] At low temperatures, it was argued that mainly dipole-dipole and quadrupole relaxation are significant and a separation of these two contributions was achieved on the basis of the NOE measurements. (The use of the

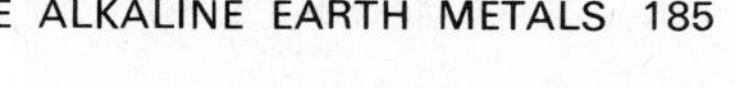

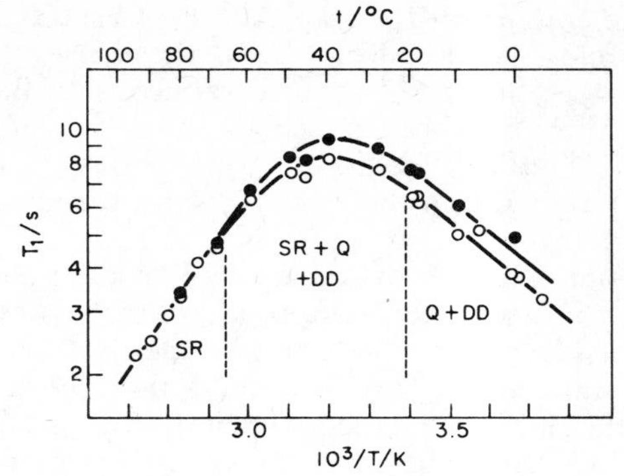

Fig. 7.1 Temperature dependence of ^9Be spin-lattice relaxation of 1 M aqueous $Be(NO_3)_2$. Lower curve gives observed relaxation time and upper curve the contribution from spin-rotation and quadrupole relaxation (reproduced with permission from ref. 22).

quadrupole relaxation rate to obtain the correlation time on the basis of the solid quadrupole coupling constant seems to presuppose strong complexation in solution.[22] An alternative interpretation of these data is given below.)

The considerably larger quadrupole moment (and smaller magnetic moment) for ^{25}Mg than for ^9Be hints to a marked predominance of quadrupolar effects in the relaxation of $^{25}Mg^{2+}$. This is supported by the relaxation rates for H_2O and D_2O solutions obtained by Simeral and Maciel,[24] the relaxation being more rapid in the latter case. For the Sr and Ba isotopes the quadrupole moments are so big that there is little doubt that non-quadrupolar effects are negligible for diamagnetic systems while the case of ^{43}Ca is worthy of careful consideration in this respect.

7B.2 Quadrupole Relaxation of the Infinitely Dilute Aqueous Ions

A prerequisite for a good interpretation of NMR data on chemical and biological systems is a detailed understanding of suitable simple reference states, in the present case the aquated alkaline earth ions. The relaxation behaviour at infinite dilution in water has recently been compared[25] with the electrostatic theory of ion quadrupole relaxation developed by Hertz.[26] From the data available in the literature, in combination with some recent direct relaxation time measurements in our laboratory, it is possible to establish rather reliable quadrupole relaxation rates of all the free hydrated ions. We will start by considering this matter and then turn to an interpretation in terms of Hertz' theory.

From his direct determination of T_1, Wehrli[22,23] obtains the quadrupolar contribution to $^9Be^{2+}$ relaxation to be ca. 0.07 s^{-1} for a 1 M $Be(NO_3)_2$ solution at 25°C. Due to some contribution from ion-ion effects, the infinite dilute value probably lies in the range 0.05–0.07 s^{-1}. For aqueous $^{25}Mg^{2+}$ a number of studies of the concentration dependence of the line width has been reported. The recent study by Simeral and Maciel,[24] probably the one least affected by instrumental broadening, gives an extrapolated infinite dilution value of $1/T_1^0 \simeq 8.5\ s^{-1}$. (Superscript 0 signifies infinite dilution situation.) Our own studies of T_1 suggest a slightly lower value, i.e. $1/T_1^0 = 6\ s^{-1}$ at 25°C. For $^{43}Ca^{2+}$, relaxation is slower and linewidth studies[16,17,24] not sufficient for establishing the relaxation rate. We have obtained $1/T_1 = 0.75\ s^{-1}$ for 0.2 M $CaCl_2$ and the infinite dilution value is probably close to this. For $^{87}Sr^{2+}$ the linewidth study of Banck and Schwenk[18] gave $1/T_1^0 = 205 \pm 40\ s^{-1}$. For the aqueous $^{135}Ba^{2+}$ and $^{137}Ba^{2+}$ ions the only information available is from the recent study of Lutz and co-workers[20] giving from the linewidths for a 1 M $BaCl_2$ solution $1/T_2 = 2580\ s^{-1}$ for $^{135}Ba^{2+}$ and $1/T_2 = 5470\ s^{-1}$ for $^{137}Ba^{2+}$. Judging from the $^{87}Sr^{2+}$ results, the infinite dilution values are up to 10–20% smaller.

This relaxation behaviour of the free hydrated ions provides a good basis for comparing with theoretical treatments. Essentially two approaches to the problem of ion quadrupole relaxation have been considered in the literature, i.e. the electronic distortion model[27] of Deverell and the electrostatic model[26] of Hertz and Valiev. (For a review see ref. 28.) We will here only consider the electrostatic theory of Hertz[26] as this has been very successful in accounting for the relaxation of alkali and halide ions. In this model the field gradients from the solvent molecules are computed considering them as point dipoles. As discussed in the treatment of alkali ions in this book the orientation and distribution of solvent molecules in the first solvation sheath are decisive for the magnitude of the relaxation rate; for example, for strongly solvated ions quenching due to symmetry of the field gradients of the inner solvation layer may occur. In the recent examination[25] of this problem it was found that very good agreement between Hertz' electrostatic model and experimental relaxation rates was obtained for Be^{2+}, Mg^{2+}, Sr^{2+} and Ba^{2+}. In all these cases, $1/T_1^0$ (or $1/T_2^0$) falls between the theoretical values based on solvation parameters suggested by Hertz[26] for Li^+ and Al^{3+}, but in view of the inaccuracy in some of the parameters entering in the theory, the relaxation rates can presently not be used to make safe deductions on solvation structure etc. For Ca^{2+} the calculated value is markedly larger than the experimental one and the reason for this is unclear. Possibly, the only reported value of the quadrupole moment is incorrect.

7B.3 Ion-Ion Effects on Quadrupole Relaxation

While thus the infinite dilution situation is well both experimentally established and theoretically understood, very little is known about the quadrupole relaxation behaviour of the alkaline earth ions at finite concentrations of inorganic salts. However, for $^{25}Mg^{2+}$ (refs 16 and 24) and $^{87}Sr^{2+}$ (ref. 18) rather detailed variable concentration studies have been reported for a number of salts and we give in Fig. 7.2 the results of Banck and Schwenk[18] for $^{87}Sr^{2+}$ as an illustrative example. Also for $^{25}Mg^{2+}$ the change with concentration of relaxation is slow and the difference between different anions small.[16,24] For the monovalent anions, the data of Simeral and Maciel[24] seem to give the sequence $[NO_3]^- > Br^- > [ClO_4]^- > Cl^-$, while according to Lutz *et al.*[16] $[ClO_4]^-$ and Br^- come in the reverse order. Both Banck and Schwenk[18] for $^{87}Sr^{2+}$ and Simeral and Maciel[24] for $^{25}Mg^{2+}$ attempted a correlation between observed linewidth and macroscopic viscosity giving in a number of cases a slower increase with increasing concentration of ion relaxation than of viscosity.

A thorough interpretation of these data, not yet attempted, will for several reasons be very difficult as the ion-ion contribution to ion quadrupole relaxation is composed of several terms some of which are quite difficult to estimate.[29–31] A detailed analysis has to involve information on the change of solvent mobility with concentration which may be obtained by studying solvent relaxation. A problematic matter arises for strongly solvated ions characterized by a partial or total quenching of the field gradient of the inner solvation shell. This quenching will be affected by other ions in a way which is very difficult to estimate. As regards the direct ion-ion effect it is important to consider

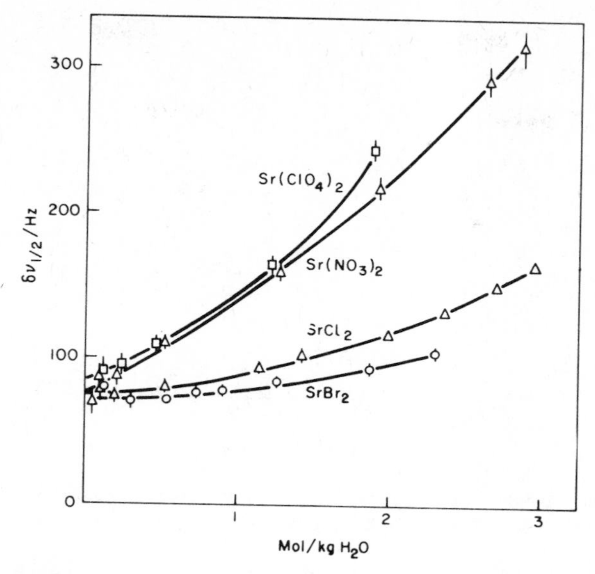

Fig. 7.2 $^{87}Sr^{2+}$ linewidths at half height of aqueous solutions of $SrCl_2$, $SrBr_2$, $Sr(NO_3)_2$ and $Sr(ClO_4)_2$ (reproduced with permission from ref. 18).

quenching of the field gradients from the ions, treated as point charges in Hertz' model, due to an ion cloud effect.[29]

7B.4 Quadrupole Relaxation Studies of Complex Formation in Aqueous Solution

With the exception of Ba^{2+}, all the alkaline earth ions have quadrupole relaxation rates in a range appropriate for studies of complexing with small molecules or macromolecules in much the same way as for the alkali and halide ions. Of course, the low sensitivities as well as a desire for lower ion concentrations in for example biological applications do provide serious obstacles. A further difficulty not analysed in detail yet, is that the stronger complexation of the alkaline earth ions may create unfavourable exchange conditions. Thus a condition for this type of studies is the use of a considerable excess of unliganded ion and that exchange proceeds reasonably rapidly compared to the relaxation of the free ion.

Four published studies[32–35] provide good insights into the possibilities of using ^{25}Mg and ^{43}Ca NMR to study complex formation of the ions with small molecules of ions. Magnusson and Bothner-By[32] studied the Mg^{2+} linewidth in the presence of a large family of compounds, for example organic carboxylic acids and biological phosphate compounds like nucleoside phosphates. By variable pH studies exemplified in Fig. 7.3 for ATP these authors estimated the complex formation constants as well as the relaxation of complexed Mg^{2+}. The latter was *ca.* 100 times or more than the relaxation of free Mg^{2+}. The contribution from exchange affects to relaxation was not investigated, while the failure of some compounds, like oxalate (known to complex Mg^{2+} strongly), to produce line-broadening was referred to slow exchange. A similar study was presented

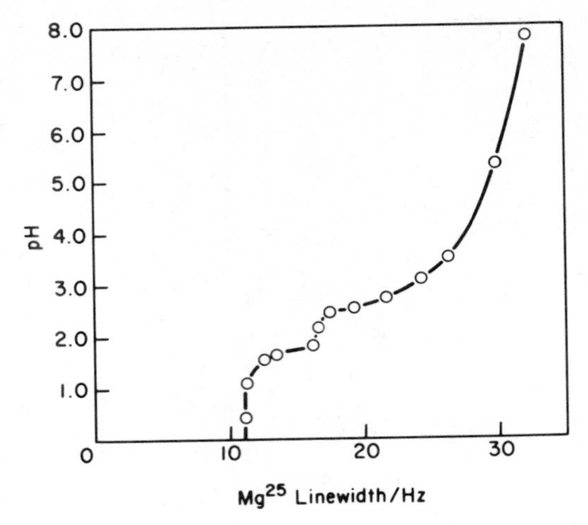

Fig. 7.3 pH dependence of the ^{25}Mg line width in a solution containing 0.041 M adenosine triphosphate and 2 M $MgCl_2$ (reproduced with permission from ref. 32).

by Bryant[34] and gave similar results and included also some variable temperature work. For the Mg^{2+}-ATP system, Bryant inferred firstly an appreciable exchange contribution to the linewidths and, secondly, a very large chemical shift difference between uncomplexed Mg^{2+} and the Mg-ATP complex. For some enzyme systems no effect was discernible which can be due either to slow exchange or a too small fraction of bound Mg^{2+}. Toma *et al.*[33] used ^{25}Mg NMR to study the complex formation between Mg^{2+} and EDTA and also to follow the displacement of Mg^{2+} by Cd^{2+}.

The only report considering the use of ^{43}Ca NMR to study complex formation is due to Bryant[35] who observed a linear reduction of T_2 on addition of ATP to a solution containing Ca^{2+}.

The studies cited demonstrate well the potential use of ^{25}Mg and ^{43}Ca NMR for investigation of complex formation with not too large molecules of biological interest. The high cation concentrations (about 1 M or more) used in these studies are of course a drawback from several points of view, for example the effect of high salt concentration on conformation and stability of biomolecules and the small fraction of cations involved in complex formation. However, these difficulties can presently to a large extent be overcome and according to our experience at least the concentration range 10–100 mM should cause no problems using Fourier transform technique and isotope-enriched samples.

From a biological point of view, the interest in ^{25}Mg and ^{43}Ca NMR is focussed on studying the ions' interactions with macromolecules. A recent ^{43}Ca NMR study[36] on a strongly Ca^{2+}-binding muscular protein, parvalbumin, reveals well both the possibilities and limitations of the technique. The $^{43}Ca^{2+}$ linewidth was studied as a function of protein concentration, pH and temperature over wide ranges. While at low temperature and neutral pH, no significant line-broadening was discernible, appreciable effects were noted at high temperatures or pH where the protein is slightly destabilized. These findings would suggest that while very strongly calcium-binding proteins generally not fall in the appropriate Ca^{2+} exchange range for this method to be applicable, the study of Ca^{2+}

binding in more weakly complexing systems like several other proteins as well as biological polyanions should be fruitful.

7C SHIELDING OF THE AQUEOUS ALKALINE EARTH IONS

The shielding of a monoatomic ion in solution is affected by near-by solvent molecules and ions and, except for the smallest ions, the paramagnetic shielding term can be assumed to dominate strongly. In contrast to the ions' quadrupole relaxation, the theoretical understanding of these shielding phenomena is far from complete. However, the overlap approach of Kondo and Yamashita,[37] originally put forward to explain ion shieldings in crystals, is currently preferred by most workers in the field and this theory may qualitatively account for shielding effects of the alkali and halide ions.[27,28] According to this model (which we discuss in more detail in the section on alkali ions, see ch. 6E.2, and elsewhere[28]) the shielding depends on three quantities, i.e. an average excitation energy, the expectation value of r^{-3} for an outer *p*-electron and sums of overlap integrals expressing the overlap between outer *p* orbitals of the studied ion and outer orbitals of other ions or solvent molecules.

A proper understanding of variable concentration effects on ion shielding presupposes information on the shielding of the free hydrated ion since the ion-solvent contribution to shielding may be modified with increasing electrolyte concentration. For the alkalis, absolute shielding scales have been well established but for the alkaline earths very limited information is available. While very precise measurements have been performed by the Tübingen group for the aqueous ions, the required data on the free atoms or ions is lacking or imprecise. For $^{9}Be^{2+}$ one obtains the shielding to be $(3 \pm 7) \cdot 10^{-4}$ for the aqueous ion compared to the free atom by comparing old NMR data[7,11] with a recent optical pumping study[38]. For $^{43}Ca^{2+}$ Lutz *et al.*[16,17] obtained in the same way $-(2 \pm 5) \cdot 10^{-4}$ while for $^{87}Sr^{2+}$ Sahm and Schwenk[19] obtained $\sigma_{M^{2+},aq} - \sigma_{M,free} = -(6.3 \pm 6.0) \cdot 10^{-4}$. (The uncertainties are due to uncertainties in the optical pumping data.) Apparently, at the present stage, not even the sign of the shielding can be unambiguously established, which is somewhat discouraging for the interpretation of chemical shift changes of aqueous solutions.

Chemical shift studies have been performed for the aqueous $^{25}Mg^{2+}$ (refs 16, 24 and 39), $^{43}Ca^{2+}$ (refs 16 and 17) and $^{87}Sr^{2+}$ (ref 18) ions and in addition the failure to observe $^{9}Be^{2+}$ chemical shift changes (within ca. 1 ppm) has been noted.[40] Mg^{2+} chemical shifts were first reported by Ellenberger and Villemin[39] but recent studies[16,24] have given results in conflict with these. Both Simeral and Maciel[24] and Lutz *et al.*[16] observed only small chemical shift changes (within ± 0.5 ppm up to 4 m) with concentration but these studies gave somewhat different anion ($[NO_3]^-$, $[ClO_4]^-$, Cl^-, Br^-) sequences. For $^{43}Ca^{2+}$, Lutz *et al.*[16,17] obtained a shift range of more than 20 ppm at the highest concentrations. Increasing chemical shift due to anion was found to follow the sequence $[NO_3]^- < [ClO_4]^- < H_2O$ (infinite dilution) $< Br^- < Cl^-$. For $^{87}Sr^{2+}$ the chemical shift changes more rapidly with concentration than for $^{43}Ca^{2+}$ but the different anions produce effects in the same order.[18] In analogy with the alkali ions, there is a steady increase in the chemical shift range with increasing alkaline earth atomic number and another resemblance is that oxyanions give increased shielding and halide ions a decreased one. In terms of the Kondo-Yamashita theory, increased cation-anion overlap compared to water occurs in the latter case while there is a decrease in the former case.

A notable feature of the chemical shifts of halide ions and the heavier alkali ions is the marked difference between H_2O and D_2O solutions and it is natural that attempts have been made to demonstrate a similar effect for the alkaline earths. Although Lutz *et al.*[16] found the chemical shift of $^{43}Ca^{2+}$ in all cases to be larger in H_2O than in D_2O, the experimental error (ca. 2 ppm) was comparable to the effect.

The only report[16] on non-aqueous solution gives a ca. 17 ppm higher shift of $^{43}Ca^{2+}$ in methanol than in water thus confirming the expectation that the chemical shifts should be useful for studies of solvation phenomena and ion-pairing in non-aqueous solutions and of preferential solvation in mixed solvents.

7D ORGANOMETALLIC AND OTHER COVALENT COMPOUNDS

The significance of organoberyllium and organomagnesium compounds has already been pointed out and both types of compounds have been considered in connection with alkaline earth NMR and in addition some other covalent beryllium compounds have been studied.

Kotz *et al.*[40] surveyed the possibilities of ^{9}Be NMR examining a range of substances. For the aqueous $[BeF_4]^{2-}$ ion these authors obtained, as shown in Fig. 7.4 for ^{9}Be, well resolved ^{9}Be-^{19}F spin coupling ($J = 33$ Hz) in both ^{9}Be and ^{19}F NMR. The ^{9}Be chemical shift of this ion is -2 ppm relative to the aqueous Be^{2+} ion. The sharp NMR signals of the $[BeF_4]^{2-}$ ion reflect the slow relaxation resulting from a tetrahedral symmetry leading to vanishing field gradients. ^{9}Be relaxation of this ion should respond very sensitively to interactions with other ions or molecules which produce some distortion of the symmetry.

An extended study of covalent beryllium compounds was presented by Kovar and Morgan[41] who reported chemical shifts, linewidths and spin-spin couplings in a number of adducts, e.g. $BeCl_2 \cdot 2SMe_2$, $MeBeCl \cdot 2SMe_2$, $Me_2Be \cdot NMe_3$ and $Me_2Be \cdot 2PMe_3$.

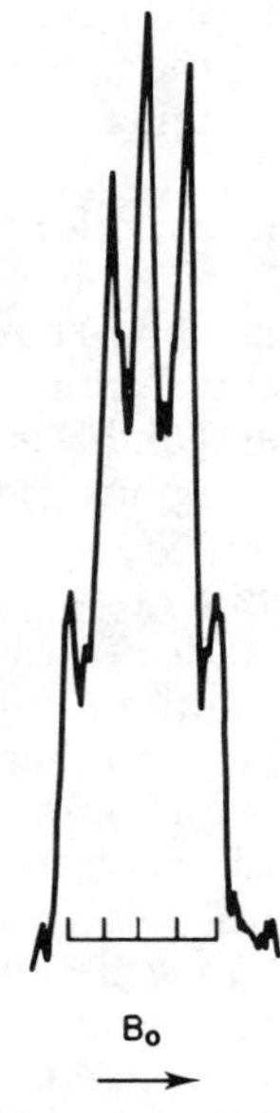

Fig. 7.4 ^{9}Be NMR spectrum of an aqueous solution of $(NH_4)2BeF_4$ (reproduced with permission from ref. 40).

The Be coordination number has a great effect both on chemical shift and on linewidth. Thus the chemical shift is higher for three-coordinate than for four-coordinate beryllium. Because of quadrupole relaxation effects, line-broadening may be much more important for three-coordinate than for four-coordinate beryllium as for the beryllium *t*-butoxide trimer (7.I), where a terminal Be has a half-height width of 64 Hz while the central Be

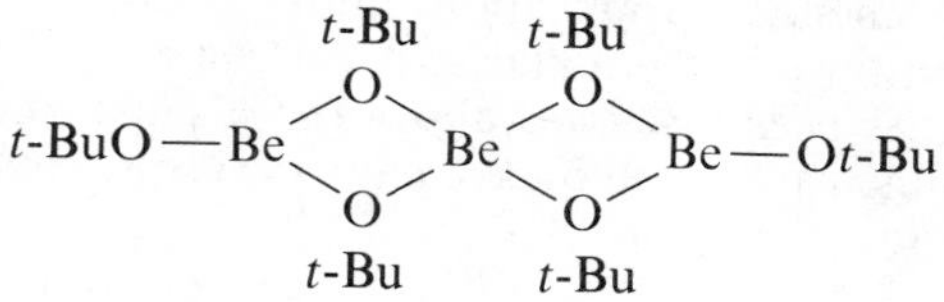

t-Bu *t*-Bu
O O
t-BuO—Be Be Be—O*t*-Bu
O O
t-Bu *t*-Bu

7.I

has a width of 5 Hz. Since for a number of compounds, increasing ^{9}Be chemical shift with increasing electronegativity of the groups bound to Be was noted it was suggested that the chemical shifts are strongly influenced by local diamagnetic factors determined by inductive effects. Spin-spin couplings were found to be small, $^{1}J(^{19}F, ^{9}Be)$ being observable but not $^{1}J(^{31}P, ^{9}Be)$ ($Me_2Be \cdot 2PMe_3$) or $J(^{9}Be, ^{1}H)$ (for example adducts of Me_2Be) couplings. The small spin-spin couplings were attributed to the strongly ionic character of the beryllium compounds.

Some interesting observations on the relaxation of covalent beryllium have been given by Wehrli[23], who *inter alia* determined the ^{9}Be relaxation of aqueous $[BeF_4]^{2-}$ and $[BeF_3OH_2]^{-}$ ions at different temperatures. For $Be(acac)_2$ (3 M in $CDCl_3$) a combination of ^{9}Be and ^{13}C relaxation of the olefinic carbon was used to determine the ^{9}Be quadrupole coupling constant to 350 kHz. This type of study will probably give most significant information on chemical bonds and symmetry effects and should be applicable also to the study of aqueous Ca^{2+}- and Mg^{2+}-complexes.

An unsuccessful attempt to observe the ^{25}Mg signal of some organomagnesium compounds has been reported by Toma *et al.*[33] but it is no doubt that the important field of covalent magnesium compounds deserves reconsideration using isotope-enriched samples and recent Fourier transform techniques.

7E SOLID STATE INVESTIGATIONS

NMR became a frequently employed method of studying the electronic state and dynamics of the alkaline earths in solids long before its usefulness to chemical problems was realized. It is especially ^{9}Be NMR which has been studied in the solid state but a number of studies of the other elements have also been reported.

^{9}Be NMR investigations have been concerned with both the relaxation times, the quadrupole coupling constant and the Knight shift (often over a wide temperature range) for a wide range of compounds: beryllium metal,[42–47] BeO,[48–50] intermetallic compounds[51–54] as well as beryllium basic acetate ($\nu_Q = 16.2$ kHz),[55] triglycine tetrafluoroberyllate,[56] Cr_2BeO_4,[57] $Be(NO_3)_2 \cdot 4H_2O$ ($\nu_Q = 9.8$ kHz),[22] $BeAl_2O_4$ (chrysoberyl)[12] and $Be_8Al_2Si_6O_{18}$ (beryl).[58] For ^{25}Mg, only studies of the Knight shift and quadrupole coupling constant of the metallic state seem to have been considered.[47,59–61]

All observations[62–65] of ^{43}Ca NMR in the solid state have concerned CaF_2 and double resonance techniques have been used to allow detection of the signal and the measurement of T_1. Two investigations[67,68] have used ^{87}Sr NMR to study the phase transition in $SrTiO_3$. Reported Ba NMR studies concern a determination[68] of the ^{137}Ba quadrupole coupling constant of $Ba(NO_3)_2$ and the observation[69] of both Ba resonances by a double resonance procedure for BaF_2.

As can be seen, NMR of the alkaline earths in the solid state remains almost wholly unexplored as concerns chemical applications and it is dfficult to judge if this picture will change in the near future. The popularity of for example ^{23}Na NMR in solid state chemistry and biology has been discussed above but it must be recalled that the NMR sensitivities of the alkaline earths will make many corresponding studies unrealistic in this case.

7F FUTURE PROSPECTS

Our survey of the NMR of the alkaline earths can be seen to have uncovered a field full of unanswered questions and unexplored possibilities both experimentally and theoretically and it may seem hazardous to prophetize about the developments in the few following years. A striking observation from the above survey is that attempts to examine potential uses mainly come from the pre-Fourier-transform era. Possibly this may have created undue pessimism. In fact, concentrations one or two orders of magnitude lower than those of the exploratory studies may now be studied in a routine way. Among the applications to follow in a near future the most important ones are in our opinion:

(*a*) Studies of Mg^{2+}- and Ca^{2+}-complexes of inorganic chemical or biochemical interest giving information on equilibria and exchange kinetics and by use of the double spin probe technique on binding aspects via the quadrupole coupling constant.

(*b*) While studies of Mg^{2+} and Ca^{2+} interactions to strongly binding proteins may prove impossible in many cases, more weakly binding macromolecules, like synthetic or biological polyelectrolytes (mucopolysaccharides, nucleic acids etc.) and transport proteins seem quite promising.[70]

(*c*) Alkaline earth ions in non-aqueous and mixed solvent systems (cf. ref. 71).

(*d*) Organomagnesium compounds.

(*e*) Mg^{2+} and Ca^{2+} ions in systems of surface-active agents and of model membranes.

REFERENCES

1. F. A. Cotton and G. Wilkinson, "Advanced Inorganic Chemistry," Ch. 7. Interscience, New York (1972)
2. G. Petzow and H. Zorn, *Chemiker-Zeitung* **98**, 236 (1974)
3. A. L. Lehninger, "Biochemistry". Worth Publishers, Inc., New York (1970)
4. R. H. Kretsinger, *Ann. Rev. Biochemistry* **45**, 239 (1976)
5. "Calcium in Biological Systems", Symposia of the Society for Experimental Biology, Symposium XXX. Cambridge University Press (1976)
6. J. R. Zimmerman and D. Williams, *Phys. Rev.* **75**, 699 (1949)
7. W. C. Dickinson and T. F. Wimett, *Phys. Rev.* **75**, 1769 (1949)
8. J. R. Zimmerman and D. Williams, *Phys. Rev.* **76**, 350 (1949)
9. W. H. Chambers and D. Williams, *Phys. Rev.* **76**, 638 (1949)
10. F. Alder and F. C. Yu, *Phys. Rev.* **82**, 105 (1951)

11. R. E. Sheriff and D. Williams, *Phys. Rev.* **82**, 651 (1951)
12. N. A. Schuster and G. E. Pake, *Phys. Rev.* **81**, 886 (1951)
13. C. D. Jeffries, *Phys. Rev.* **90**, 1130 (1953)
14. C. D. Jeffries and P. B. Sogo, *Phys. Rev.* **91**, 1286 (1953)
15. H. E. Walchli, *Phys. Rev.* **102**, 1334 (1956)
16. O. Lutz, A. Schwenk and A. Uhl, *Z. Naturforsch.* **30a**, 1122 (1975)
17. O. Lutz, A. Schwenk and A. Uhl, *Z. Naturforsch.* **28a**, 1534 (1973)
18. J. Banck and A. Schwenk, *Z. Physik* **265**, 165 (1973)
19. W. Sahm and A. Schwenk, *Z. Naturforsch.* **29a**, 1763 (1974)
20. O. Lutz, personal communication. H. Krüger, O. Lutz and H. Oehler, Verhandl. Deutschen Phys. Gesellschaft, p. 525 (1977); *Phys. Lett.* **62A**, 131 (1977)
21. E. Brun, J. J. Kraushaar, W. L. Pierce and W. J. Veigele, *Phys. Rev. Letters* **9**, 166 (1962)
22. F. W. Wehrli, *J. Magn. Reson.* **23**, 181 (1976)
23. F. W. Wehrli, *J. Magn. Reson.* in press
24. L. Simeral and G. E. Maciel, *J. Phys. Chem.* **80**, 552 (1976)
25. B. Lindman, S. Forsén and H. Lilja, *Chem. Scr.* **11**, 91 (1977)
26. H. G. Hertz, *Ber. Bunsenges. Phys. Chem.* **77**, 531 (1973)
27. C. Deverell, *Progr. NMR Spectroscopy* **4**, 235 (1969)
28. B. Lindman and S. Forsén, "Chlorine, Bromine and Iodine NMR. Physico-Chemical and Biological Applications", Springer-Verlag, Heidelberg (1976)
29. H. G. Hertz, *Ber. Bunsenges. Phys. Chem.* **77**, 688 (1973)
30. H. G. Hertz, M. Holz, R. Klute, G. Stalidis and H. Versmold, *Ber. Bunsenges. Phys. Chem.* **78**, 24 (1974)
31. H. G. Hertz, M. Holz, G. Keller, H. Versmold and C. Yoon, *Ber. Bunsenges. Phys. Chem.* **78**, 493 (1974)
32. J. A. Magnusson and A. A. Bothner-By, in "Magnetic Resonance in Biological Research", p. 365 (A. Franconi, ed.) Gordon and Breach, London (1969)
33. F. Toma, M. Villemin, M. Ellenberger and L. Brehamet, in "Magnetic Resonance and Related Phenomena", p. 317. Proc. XVIth Congress Ampere, Bucharest, 1970 (I. Ursu, ed.). Publ. House of the Academy of the Socialist Republic of Romania (1971)
34. R. G. Bryant, *J. Magn. Reson.* **6**, 159 (1972)
35. R. G. Bryant, *J. Amer. Chem. Soc.* **91**, 1870 (1969)
36. J. Parello, H. Lilja, A. Cavé and B. Lindman, *FEBS Lett.* **87**, 191 (1978)
37. J. Kondo and J. Yamashita, *J. Phys. Chem. Solids* **10**, 245 (1959)
38. E. W. Weber and J. Vetter, *Phys. Lett.* **56A**, 446 (1976)
39. M. Ellenberger and M. Villemin, *C. R. Acad. Sci. Paris* **B 266**, 1430 (1968)
40. J. C. Kotz, R. Schaeffer and A. Clouse, *Inorg. Chem.* **6**, 620 (1967)
41. R. A. Kovar and G. L. Morgan, *J. Amer. Chem. Soc.* **92**, 5067 (1970)
42. H. Alloul and C. Froidevaux, *J. Phys. Chem. Solids* **29**, 1623 (1968)
43. Y. Chabre and P. Segransan, *Solid State Commun.* **12**, 815 (1973)
44. Y. Chabre, *J. Phys. F, Metal Phys.* **4**, 626 (1974)
45. Y. Chabre and E. Geissler, *Scripta Metallurgica* **4**, 255 (1970)
46. W. D. Knight, *Phys. Rev.* **92**, 539 (1953)
47. T. J. Rowland, *Progress in Material Science* **9**, 30 (1961)
48. R. H. Thorland, A. K. Garrison and R. C. Du Varney, *Phys. Rev. B* **5**, 784 (1972)
49. J. F. Hon, *Phys. Rev.* **124**, 1368 (1961)
50. G. J. Troup, *Phys. Lett.* **2** (1962)
51. S. Downgang, *Compt. Rend. B* **276**, 701 (1973)
52. S. Downgang and R. Jesser, *J. Phys. F* **4**, 890 (1974)
53. F. Borsa and G. Olcese, *Phys. Status Solid A* **17**, 631 (1973)
54. H. Saji, T. Yamadaya and M. Asanuma, *J. Phys. Soc. Japan* **21**, 255 (1966)
55. D. W. Sheppard, *Dissertation Abstr.* **30B**, 332 (1969)
56. R. Blinc, J. Slak and J. Stepisnik, *J. Chem. Phys.* **55**, 4848 (1971)
57. H. Saji and T. Yamadaya, *Phys. Staus Solidi B* **63**, K103 (1974)
58. J. Hatton, B. V. Rollin and E. F. W. Seymour, *Phys. Rev.* **83**, 672 (1951)

59. P. D. Dougan, S. N. Sharma and D. L. Williams, *Can. J. Phys.* **47**, 1047 (1969)
60. E. M. Dickson and E. F. W. Seymour, *Proc. Phys. Soc.* (*London*), *J. Phys. Soc. C*, **3**, 666 (1970)
61. L. E. Drain, *Metal Mater.* **12**, 195 (1967)
62. J. F. Jacquinot, W. T. Wenckebach, M. Chapellier, M. Goldman and A. Abragam, *Compt. Rend. Ser. B* **278**, 93 (1974)
63. J. F. Jacquinot, W. T. Wenckebach, M. Goldman and A. Abragam, *Phys. Rev. Lett.* **32**, 1096 (1974)
64. D. A. McArthur, E. L. Hahn and R. E. Walstedt, *Phys. Rev.* **188**, 609 (1969)
65. E. L. Hahn, *Proc. Colloq. Ampere* (At. Mol. Etud. Radio Elec.) 199, **14**, 14 (1967)
66. G. Angelini, G. Bonera and A. Rigamonti, in "Magnetic Resonance and Related Phenomena", p. 346. Proc. 17th Congr. Ampere, 1972 (1973)
67. M. J. Weber and R. R. Allen, *J. Chem. Phys.* **38**, 726 (1963)
68. N. Weiden and A. Weiss, in "Magnetic Resonance and Related Phenomena", p. 257. Proc. 18th Congr. Ampere, 1974 (1975)
69. V. D. Shcheikin, D. I. Vainshtein, K. S. Saikin, V. M. Vinokurov and R. A. Dautov, *Fiz. Tverd. Tela* (*Leningrad*) **16**, 3469 (1974)
70. In this connection it may be mentioned that a general treatment of the relaxation of spin $\frac{5}{2}$(^{25}Mg) and $\frac{7}{2}$(^{43}Ca) nuclei undergoing fast chemical exchange between two sites—one of which is not in the extreme narrowing limit—has recently been worked out (T. Bull, S. Forsén and D. Turner, *J. Chem. Phys.* in press)
71. J.-J. Delpuech, A. Péguy, P. Rubini and J. Steinmetz, *Nouveau J. Chimie* **1**, 133 (1977)

8
THE TRANSITION METALS

R. GARTH KIDD,* University of Western Ontario, Canada
R. J. GOODFELLOW,† University of Bristol, England

8A INTRODUCTION

Although there are 61 stable isotopes in the transition metals and lanthanides which are NMR active, chemical use has only been made of the NMR spectra of 32 of these isotopes, and significant use has only been made of 8 of these isotopes. When the number of references for this chapter, 280, is compared with those for ch. 9–14, 848, where there are 32 stable isotopes, it is clear that the NMR spectra of the transition metals and the lanthanides has hardly been investigated. Indeed there is paucity of data on all nuclei apart from ^{51}V, ^{183}W, ^{55}Mn, ^{59}Co, ^{103}Rh, ^{195}Pt, ^{113}Cd, and ^{199}Hg. Some of the difficulties arise from the paramagnetism found for many of the compounds, and this in part accounts for the lack of information for the lanthanides. Nevertheless it is clear that considerable work is necessary on most of these nuclei to build up a data base on which future work can be done. At present for many elements it is difficult to estimate the chemical shift and linewidth in a particular compound.

There is a wealth of information and new discoveries awaiting all who venture into this uncharted area. Examples of the sort of information available may be taken from the explored areas. Thus for example there is a good correlation of chemical shift with oxidation state, enabling the oxidation state of a metal to be determined, and the relative stabilities of species in equilibrium, e.g., $[NbBr_nCl_{6-n}]^-$, have been determined. All the same even for the more extensively investigated nuclei only the initial investigations have been performed. When the few experiments that have been performed on ^{45}Sc or ^{67}Zn relaxation are compared with the very sophisticated treatments that have been applied to ^{23}Na relaxation, see ch. 6, it is clear that work in this area has hardly started, but most of the necessary tools are waiting to be used. No work has yet been done to use the nuclei in this chapter to investigate dynamic systems although the reaction is normally occurring at the metal. This would open up a powerful route to investigating homogeneous catalytic systems where the metal is the centre of the catalytic process.

It is hoped that this chapter will provide the reader with a guide to the work done to date and provide a basis on which future work can be built.

* Professor Kidd has written Sections 8B, 8C, 8D, 8E.1, 8E.2, 8F, 8G, 8H.1, and 8J.1
† Dr. Goodfellow has written Sections 8E.3, 8H.2, 8I, 8J.2, and 8K.

8B GROUP IIIA, THE LANTHANIDES AND THE ACTINIDES

Although these elements possess 23 stable isotopes between them capable of giving NMR signals, very little attention has been paid to the use of NMR spectroscopy to directly investigate the chemistry of these elements. This arises from three main causes. There has been only limited interest in the chemistry of scandium, yttrium, and lanthanum. Most of the lanthanides and many of the actinides only form paramagnetic compounds. Of the stable elements in this group, there is only cerium which does not have a stable isotope suitable for NMR spectroscopy.

A number of nuclei within this group offer possibilities for future work. The clearest example is ^{175}Lu which has $D^C = 173$ and forms many diamagnetic compounds, but with $I = \frac{7}{2}$ it does have a quadrupole moment of 5.68×10^{-28} cm^2. Relaxation investigations on ^{139}La would be complemented by also investigating ^{138}La, $D^C = 0.46$, $I = 5$, and quadrupole moment 2.7×10^{-28} cm^2 but perhaps the most intriguing nucleus is ^{141}Pr, $D^C = 1663$, $I = \frac{5}{2}$, and quadrupole moment -5.9×10^{-26} cm^2. Thus ^{141}Pr would appear to be ideal. Unfortunately all its compounds are paramagnetic but generally the electron relaxation is fast. Therefore there is still the possibility of obtaining signals with large Knight shifts. Similar reservations apply to many of the other lanthanides.

8B.1 Scandium, ^{45}Sc

Its 100% natural abundance and high receptivity of 1709 relative to ^{13}C make scandium-45 a favourable nucleus for NMR observation. The fact that only two research groups have reported any work in this area must reflect the limited interest in scandium chemistry rather than inherent difficulties in observation. With the exception of the $[ScF_6]^{3-}$ and $Sc(acac)_3$ complexes where scandium participates in normal covalent bonding, the interactions to come under NMR study have been the random collisions and ion pair formations occurring in aqueous solutions between the Sc^{3+} cation and various anions. In this respect, the NMR behaviour of scandium is more profitably compared with that of the alkali metal cations Na^+ and K^+ than with that of aluminium where static covalent bonds giving rise to discrete complexes dominate the NMR study of its structural chemistry.

The concentration dependence of various scandium salts in aqueous solution spans a chemical shift range of about 25 ppm, behaviour which is reminiscent of ^{23}Na and ^{39}K (see ch. 6). Because the ^{45}Sc nucleus has an electric quadrupole moment about twice that of ^{23}Na, equal field gradients at the metal nuclei give significantly broader resonance lines for scandium and as a result, the ^{45}Sc linewidth has been used to advantage in identifying cation-anion interactions in solution which cause departures from spherical symmetry about the cation and increases in the correlation time.

a. ^{45}Sc *Chemical Shifts*

The solution which has been adopted as an external reference by both sets of workers in this field is aqueous $Sc(ClO_4)_3$. Melson *et al.*[2] with the use of a superconducting solenoid and measurements at 55.1 MHz have achieved a level of precision in chemical shift measurement an order of magnitude better than that achieved with traditional instrumentation and they have adopted as external reference a capillary containing a 0.1 M aqueous $Sc(ClO_4)_3$ solution within a 5 mm sample tube. Buslaev *et al.*[3] use a 7.75% aqueous $Sc(ClO_4)_3$ solution as external reference. This solution will be about

TABLE 8.1 ^{45}Sc chemical shifts and linewidths

Compound	Solvent	Shift[a]/ppm	Linewidth/Hz	Ref.
$Sc(acac)_3$	Benzene	85		2
$ScCl_3$	H_2O; 0.1 M	−0.1	92	2
	H_2O; 1.0 M	0.5	80	2
	H_2O; 2.0 M	5.1	325	2
	H_2O	0	779	3
	H_2O; added HCl	−26	1038	3
	H_2O; added EtOH	−99	882	3
	H_2O; 2 molal		346	1a
	H_2O; 3 molal		432	1a
	H_2O; 4 molal		562	1a
	H_2O; 5 molal		952	1a
$ScBr_3$	H_2O; 0.1 M	0.0	161	2
	H_2O; 1.0 M	−2.5	303	2
	H_2O; 2.0 M	−5.4	317	2
	H_2O	0	407	3
	H_2O; added HBr	0	170	3
		−15	194	3
$Sc(ClO_4)_3$	H_2O; added KI	0	260	3
	H_2O; 0.1 M	0	149	2
	H_2O; 0.6 M	−2.2	140	2
	H_2O; 3.0 M	−14.8	182	2
				2
	H_2O; 1 molal		86	1b
	H_2O; 2 molal		173	1b
	H_2O; 3 molal		260	1b
	H_2O; 4 molal		389	1b
$(NH_4)_3ScF_6$	H_2O; added NH_4F	0	septet; $J_{Sc\text{-}F}$ = 180 Hz	3
$Sc(NO_3)_3$	H_2O; 0.1 M	4.3	789	2
	H_2O; 1.0 M	0.9	1737	2
	H_2O; 2.0 M	−0.1	2863	2
	H_2O; 2% solution	0	346	3
	H_2O; saturated	0	4550	3
	H_2O; added $(NH_2)_2CO$	16	1038	3
	H_2O; added NH_4SCN	27	865	3
	H_2O; added KI	0	260	3
	H_2O; added acac	0	3028	3
	H_2O; added HF:Sc = 1	0	1315	3
	H_2O; added HF:Sc = 2	0	1142	3
	H_2O; 2 molal		865	1b
	H_2O; 4 molal		1557	1b
	H_2O; 6 molal		3460	1b
	H_2O; 7 molal		4325	1b
$Sc_2(SO_4)_3$	H_2O; 0.1 M	9.1	341	2
	H_2O; 0.6 M	7.3	471	2
	H_2O; 1.0 M	6.7	785	2

[a] PPM relative to external 0.1 M aqueous $Sc(ClO_4)_3$ and using the δ scale where positive values represent higher frequencies than the reference.

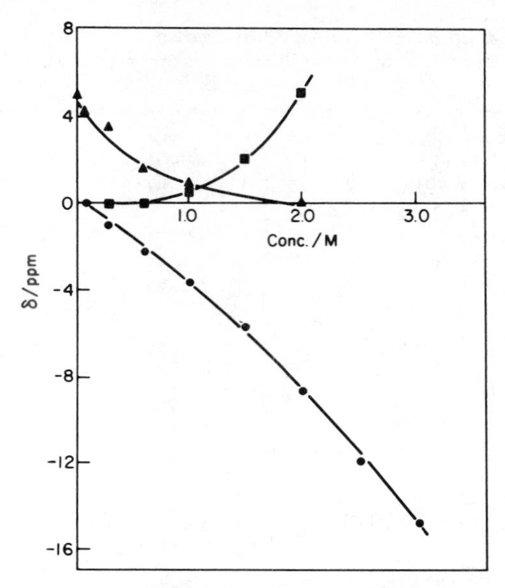

Fig. 8.1 Concentration and anion dependence of ^{45}Sc chemical shifts. ■, aqueous $ScCl_3$; ●, aqueous $Sc(ClO_4)_3$; ▲, aqueous $Sc(NO_3)_3$. Reproduced from G. A. Melson, D. J. Olszanski and E. T. Roach, *J. Chem. Soc., Chem. Commun.* 229 (1974), with permission.

0.24 M in $Sc(ClO_4)_3$ and Melson's data on the concentration dependence of the scandium shift shows that it will be displaced about 0.4 ppm to lower frequency of a 0.1 M solution. As this difference lies well within the experimental uncertainty of the Buslaev data, the chemical shift data in Table 8.1 quoted relative to either reference can be regarded as consistent. Where the exigencies of a particular experiment dictate the use of a more concentrated $Sc(ClO_4)_3$ solution as external reference, however, the concentration effect[2] upon δ of $-5.3\ M^{-1}$ should be corrected for in quoting chemical shift data.

The concentration dependence of the chemical shift of $Sc(ClO_4)_3$, and $ScBr_3$ follows a similar pattern to that already observed for the alkali metals (see ch. 6). The concentration plots shown in Fig. 8.1 converge on a single δ-value at infinite dilution, a value which is characteristic of some aquated form of the Sc^{3+} cation and which Melson *et al.*[2] attribute to $[Sc(OH)]^{2+}$. The decrease in linewidths for all three salts to values below 100 Hz at 0.1 M supports a picture of decreasing anion interaction as infinite dilution is approached. By contrast, $Sc(NO_3)_3$ and $Sc_2(SO_4)_3$ do not converge upon $\delta 0$ at low concentrations, but have values $\delta 4.3$ and $\delta 9.1$ respectively, suggesting nitrate ion and sulfate ion coordination even at low concentrations.[2] This view is reinforced by the fact that at 0.1 M concentration, these salts have linewidths of 789 Hz and 341 Hz, values which are 2–10 times greater than those for salts which appear to be well dissociated at 0.1 M concentration.

The ability of anions to increase scandium shielding, as determined by adding excess anion to the solution, is given by the sequence[3] $I^- < Br^- < Cl^-$, the same sequence as is observed[4] for ^{23}Na and ^{39}K. It should be noted, however, that the anion dependence of the cation shielding as defined by the slopes of concentration plots shows Br^- causing higher shielding of Sc^{3+} than Cl^-. This dichotomy requires rationalizing and indicates an area where further work is equired.

b. ^{45}Sc *Linewidths*

The narrowest ^{45}Sc resonances have been observed in dilute aqueous solutions of $Sc(ClO_4)_3$ where widths in the region of 90 Hz have been obtained.[1a,2] By extrapolating linewidths for these solutions to infinite dilution, a value of 70 Hz was obtained, from which an estimated value of 0.35×10^{-24} cm^2 for the electric quadrupole moment of the ^{45}Sc nucleus was calculated.[1a] This value is 2–3 times greater than that for ^{23}Na and ^{39}K, making ^{45}Sc more sensitive to electric field gradients than the alkali metals. Cation-anion interactions in solution, whether they be complexing, ion pairing, or just random collisions, will increase the electric field gradient at the cation nucleus above its near-zero value in the symmetrically aquated ion. This fact has been utilised to measure the relative complexing abilities of anions through their effects on the ^{45}Sc linewidth.[1a] The nitrate ion in particular, but also to a lesser extent the sulphate ion, gives rise to linewidths in the 4000 Hz region, indicating a marked lowering of the electrical symmetry about the scandium nucleus.

In the case of $[ScF_6]^{3-}$, the quadrupole coupling constant, χ, has been calculated from the ^{45}Sc linewidth and a value of 9.5 MHz obtained. This can be compared with the analogous values of 5.8 MHz for $[GeF_6]^{2-}$ and 4.5 MHz for $[TiF_6]^{2-}$ obtained from ^{73}Ge and ^{49}Ti linewidths by the same method.[1b]

c. ^{45}Sc *Coupling Constants*

An aqueous solution of $(NH_4)_3ScF_6$ with NH_4F added produces a well resolved ^{45}Sc septet having intensity ratios 1:6:15:20:15:6:1, reflecting coupling to the six equivalent $I = \frac{1}{2}$ fluorine nuclei.[2] The coupling constant $J(^{45}\mathrm{Sc}, ^{19}\mathrm{F})$ is 180 Hz. The fact that a septet is observed indicates that the electric field gradient at the scandium nucleus is sufficiently low that the scandium does not become decoupled from the fluorines through rapid quadrupolar relaxation.

8B.2 Yttrium, ^{89}Y

Yttrium-89 is 100% abundant in nature and has a relative receptivity, $D^C = 0.668$. It is a spin $I = \frac{1}{2}$ nucleus and does not suffer from signal deterioration due to quadrupole broadening. Lacking the quadrupole relaxation mechanism and having a low magnetic moment, however, its relaxation time is long, and NMR observation of the nucleus will be fraught with r.f. saturation difficulties. Kronenbitter and Schwenk,[5] using a pulsed spectrometer designed to cope with this difficulty, have obtained the ^{89}Y resonance from a 3 molal aqueous solution of $Y(NO_3)_3$ at pH = 1.0 and achieved a signal-to-noise ratio of 30 after 2^{16} pulse cycles. The average relaxation time $\frac{1}{2}(T_1 + T_2)$ for ^{89}Y in this solution is an incredibly long 60 seconds.

The concentration and anion dependence of the Y^{3+} chemical shift in aqueous solution has been studied[6] and concentration dependence similar to that observed for the alkali metals has been found. The ^{89}Y resonance position is roughly linear with concentration, and the slope towards lower frequencies increases in the order

$$Cl^- < [ClO_4]^- < [NO_3]^-.$$

The observation of coupling to ^{89}Y is unusual on account of rapid exchange but coupling has been observed in a few isolated cases. The first observation was for $(ButCH_2)_3Y \cdot 2THF$, where $^2J(^{89}\mathrm{Y}, ^1\mathrm{H}) = 2.5$ Hz.[7] $Cp_2Y(\mu\text{-}Me)_2AlMe_2$ is fluxional at room temperature, but at $-45°C$, $^{2}J(^{89}Y, ^{1}H) = 5$ Hz was observed[8] and subsequently $^{1}J(^{89}Y, ^{13}C) = 12.2$ Hz was determined[9] for the bridging methyl group.

8B.3 Lanthanum, ^{139}La, and the Lanthanides

The NMR study of lanthanum has been limited to three research groups at this stage in its development, and the results obtained show a general similarity with those of its congener scandium. It is the aqueous solutions of La^{3+} salts which have been the objects of most study, with information of chemical interest concerning interactions between the lanthanum cation and various counterions being obtained from linewidth and relaxation time data. The only non-aqueous solution study of lanthanum to appear combines NMR data with atomic radial distribution functions obtained from X-ray diffraction study of solutions in a powerful attack on the problem of ion-pair structure.

Lanthanum-139 is 99.9% abundant, has a spin $I = \frac{7}{2}$, has a natural abundance receptivity relative to ^{13}C of 336, and has a quadrupole moment of 0.21×10^{-28} cm^2, the same order of magnitude as those for the first-row transition metals and, along with gold and ^{201}Hg at the opposite end, significantly smaller than the quadrupole moments for other third-row transition metals. A 1971 study[10] of ^{139}La linewidths for aqueous LaX_3 solutions ($X = Cl^-$, Br^-, I^-, $[NO_3]^-$, $\frac{1}{2}[SO_4]^{2-}$, $[ClO_4]^-$) revealed, after correction for viscosity change, significantly greater concentration broadening with $[NO_3]^-$ and $[SO_4]^{2-}$ than with the other anions. This behaviour is similar to that observed for ^{45}Sc and has been taken as indicating formation of contact ion-pairs or inner-sphere complexes even at low concentrations. Subsequent investigation[11] of the lanthanum relaxation times on solutions with those anions yielding narrower ^{139}La resonances illustrates vividly the hazards associated with evaluating relaxation times from the linewidths of narrow lines when corrections for modulation broadening must be introduced.

Reuben[12,13] has given an elegant demonstration of how complex formation between a natural protein and a metal cation can be studied using NMR relaxation times. The introduction of bovine serum albumin (BSA), a protein molecule with about 100 free carboxylate groups, into aqueous $LaCl_3$ solution produced incremental increases in T_1^{-1} which were linear in protein concentration. This increase in ^{139}La relaxation rate results from the coordination of the La^{3+} cations by the carboxylate groups on BSA, thereby introducing a larger electric field gradient at lanthanum than is present in the aquated ion. The magnitude of the increase is proportional to n/T_1 where n is the number of carboxylate groups on each protein molecule participating in coordination. The coupling between the electric field gradient responsible for this increase and the nuclear quadrupole moment is modulated by rotational motion of the complex with time characteristic τ_c. For molecules with normal τ_c values of *ca.* 10^{-11} s, $\omega_0 \tau_c \ll 1$ and T_1 is independent of frequency. Due to the large size of the protein complex, however, $\tau_c \approx 10^{-8}$ s and $\omega_0 \tau_c \geq 1$, thereby making T_1 frequency dependent and permitting a separation of n from T_1. In this way, the number of complexing groups on the protein which are involved in complexing metal atoms can be independently assessed. In solving this problem, Reuben and Luz[13] have worked out a general theory for the longitudinal relaxation behaviour of a spin $\frac{7}{2}$ system which can be applied to any $I = \frac{7}{2}$ nucleus.

By combining the results of NMR linewidth studies with interatomic distance data obtained from atomic radial distribution functions Smith *et al.*[14] have deduced that while in aqueous solution, La^{3+} forms outer-sphere complexes with Cl^- counterions at an average distance of 4.7 Å, in methanol solution each lanthanum has three inner-sphere Cl^- neighbours in a binuclear $[La_2Cl_4(MeOH)_{10}]^{2-}$ complex with two bridging chlorines joining the two metal atoms. The methanol complex is chemically shifted by 293 ppm to high frequency of the aquo-complex and has a linewidth of 3800 Hz.

The use of direct NMR observation of the lanthanides has been restricted to solid state investigations which are marginal to the interests of this book. For example the ^{153}Eu spin-spin relaxation times for two spherical EuO single crystals have been investigated and their dependence on stocheiometry was found to be very marked.[15]

8B.4 The Actinides

NMR parameters have only been obtained in solution for ^{235}U in UF_6 where T_1 and $^1J(^{235}U, ^{19}F) = 4240$ Hz at 338 K and 4566 Hz at 356 K have been measured.[16] Considerable effort has been expended in trying to observe other actinide nuclei, with ^{239}Pu receiving the most study.[17,18,19]

Alloys and intermetallic compounds of the actinide elements have been studied to determine the temperature and field dependencies of their NMR linewidths, Knight shifts, and spin-lattice relaxation times.

The magnetic moment for ^{239}Pu has been determined both in the solid state and by atomic beam methods. The value of 0.200(4) μ_n[18] appears to be the most reliable value. The indirect observation of ^{239}Pu in a CaF_2 lattice has been reported[20] using the ENDOR technique in which the hyperfine coupling between the plutonium nucleus and its unpaired electrons is manifest in its EPR spectrum.

What NMR parameters are available for plutonium metal have been obtained by alloying with a few per cent of either ^{27}Al or ^{51}V whose NMR behaviour can be observed at a site characteristic of the host lattice.

8C GROUP IVA

Although the direct observation of NMR signals for titanium, zirconium and hafnium is feasible, the only solution studies reported are for titanium. Even for titanium there is only one publication which is on the halides. As the maximum linewidth found was only 200 Hz, and ^{49}Ti has a receptivity comparable to ^{13}C the lack of data must arise from the lack of interest in measuring rather than experimental difficulties.

Both zirconium and hafnium have suitable isotopes for NMR investigation. Zirconium-91 is 11.23%, $D^C = 6.0$, and $I = \frac{5}{2}$. The quadrupole moment is unknown, but is unlikely to be so large as to prevent the observation of a signal. The two isotopes of hafnium present a greater problem. Hafnium-177 is 18.5% abundant, $D^C = 0.67$, $I = \frac{7}{2}$ and the quadrupole moment is 3×10^{-28} cm^2 while ^{179}Hf is 13.75% abundant, $D^C = 0.19$, $I = \frac{9}{2}$ and the quadrupole moment is 3×10^{-28} cm^2. The effect of the larger spin of ^{179}Hf resulting in a smaller linewidth for ^{179}Hf. Consequently the observation of ^{179}Hf is generally to be preferred.

8C.1 Titanium, $^{47,49}Ti$

Titanium has received little attention from the NMR spectroscopist. The two magnetically active isotopes, ^{47}Ti and ^{49}Ti are about equally abundant at 7.3% and 5.5% respectively. Their natural abundance detection sensitivities relative to ^{13}C are

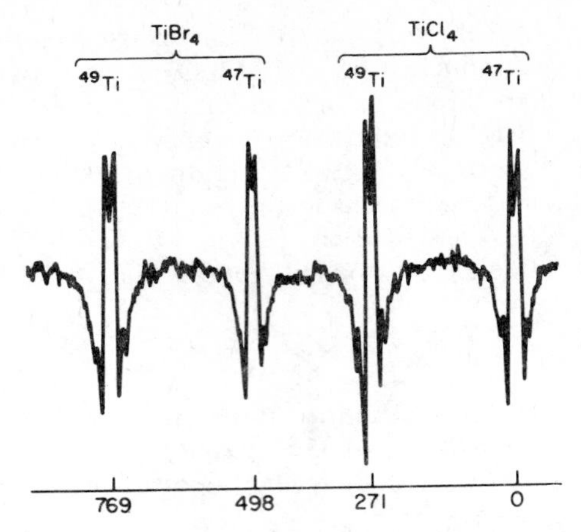

Fig. 8.2 The ^{47}Ti and ^{49}Ti spectra of $TiCl_4$ and $TiBr_4$ recorded as dispersion derivative signals. The splitting of each resonance is due to over modulation at 40 Hz. Reproduced from R. G. Kidd, R. W. Matthews and H. G. Spinney, *J. Amer. Chem. Soc.* **94**, 6686 (1973), by permission of the American Chemical Society.

0.864 and 1.18, and no information on their quadrupole moments is available. The most characteristic feature of titanium from an NMR viewpoint is the similarity of the magnetogyric ratios for the two isotopes, which results in their resonance frequencies differing by only 271 ppm. Each species contributing to a titanium spectrum generates a pair of lines of roughly equal intensity and separated by 271 ppm as shown in Fig. 8.2 where the ^{47}Ti spectrum and the ^{49}Ti spectrum overlap one another.

The limited chemical shift data available for titanium and given in Table 8.2 is sufficient to show that, with $TiBr_4$ absorbing at higher frequencies than $TiCl_4$, titanium shifts exhibit a halogen dependence which is the inverse of that for the MX_4 compounds of the main group elements. It has been suggested[21] that the larger δ-value for $TiBr_4$ arises from the low-lying electronic excited state which results in a larger paramagnetic shielding term for $TiBr_4$ than for $TiCl_4$. The inverse halogen dependence phenomenon is discussed more fully in Section 8D.2*a* with reference to ^{93}Nb chemical shifts.

TABLE 8.2 Titanium chemical shifts[21]

Compound	Solution	Shift^a/ppm	Linewidth/Hz	ΔE^b/cm^{-1}
$[TiF_6]^{2-}$	0.8 M TiF_4 in 48% aq. HF	0		
	3.4 M TiF_4 in 48% aq. HF	0	~220	
	TiF_4/NH_4F in H_2O	0		
	TiF_4 in H_2O	No signal		
$TiCl_4$	Neat liquid	1177	<25	34 840
	20 mol % in C_6H_{12}	1177		
	20 mol % in CCl_4	1177		
$TiBr_4$	Neat liquid	1675	<25	28 680

^{a} PPM relative to external $[TiF_6]^{2-}$ in 48% aqueous HF and using the δ scale where positive values represent higher resonance frequencies than the standard. ^{b} Lowest-energy electronic transition.

The titanium-fluorine coupling constant $J(^{49}Ti, ^{19}F)$ in $[TiF_6]^{2-}$ has been measured from the ^{19}F spectrum to be 33 Hz. Although the coupling is not resolved in the $^{47,49}Ti$ spectrum, the observed linewidth of 220 Hz would mask a septet with J(Ti, F) of 33 Hz. In common with the experience with other transition metal fluorides, this failure to resolve coupling to fluorine in highly symmetrical ions probably results from rapid exchange of F^- ions in the coordination shell of titanium.

8D GROUP VA

All three group VA elements possess NMR active nuclei in 100% or very nearly 100% abundance, all with high receptivity, an unusual event within the Periodic Table. All three nuclei are sensitive but possess quadrupole moments leading to broad signals which in the case of ^{181}Ta are so broad as to make ^{181}Ta very difficult to observe.

8D.1 Vanadium, ^{51}V

Vanadium is a favourable nucleus for observation. Because of its natural abundance of 100% and its high detection sensitivity, the securing of suitable signals is well within the capability of appropriate instruments. The ^{51}V nucleus has a quadrupole moment of 0.3×10^{-28} cm^2 putting it in the medium quadrupole category ($0.1 \leq Q < 1.0$) where signal widths are highly sensitive to electric field gradients at the nucleus, but where reduced signal intensity resulting from excessive linewidth is not generally a problem.

The aqueous chemistry of vanadium in the +5 oxidation state, including the rich variety of isopoly- and heteropolyanions formed in acidic media, has been studied by four groups of workers over a ten year period with a remarkable degree of unanimity regarding interpretations and assignments, given the sensitivity of anion structure to small changes in pH for these solutions. The spectral results for $[V_{10}O_{28}]^{6-}$ in particular provide an excellent example of how line position and linewidth data taken together can provide powerful insight into the solution structure of a very complex molecule.

The chemical shifts for V^I and V^{-I} carbonyl compounds have been measured by workers in Hamburg, who have also reported $J(^{51}V, ^{31}P)$ values for the phosphine-substituted carbonyls. These studies show the existence of discrete chemical shift ranges for vanadium in different oxidation states, and should be useful as a structural tool in future studies. The vanadium-fluorine coupling in vanadium(V) fluorides has been studied on a number of occasions, but vanadium NMR has not proven as fruitful as had been anticipated in characterising these compounds, due to intermediate rates of fluorine exchange and to asymmetric electric field gradients at vanadium, both of which decouple the vanadium from the fluorines.

a. ^{51}V *Chemical Shifts*

The choice of $VOCl_3$ as the external reference for vanadium chemical shifts has been adopted by most workers in the field, a choice which offers several advantages which weigh in favour of its continued use as the primary reference for reporting vanadium shieldings. Its use as a neat liquid precludes the possibility of solvent effects upon the resonance position. The occurrence of the $VOCl_3$ resonance at the high frequency end of the range for vanadium chemical shifts gives all other vanadium δ-values a single sign, in this case negative.

With the single exception of $\delta 5660$ reported for $[V(CO)_6]^-$ which will be discussed

TABLE 8.3 ^{51}V Chemical shifts, linewidths and coupling constants

Compound	Solvent	Shift[a]/ppm	Linewidth/Hz	Ref.
$[V(CO)_6]^-$	MeCN	5660		26
	$NH_3(-70°C)$	5590		26
$VOCl_3$	Neat liquid	0		
$[VO_2]^+$	H_2O; pH 1.2	−543.1	563	31
	H_2O; pH 2	−541	582	22
	H_2O; pH 0–2	−545 to −542	v. broad	28
	H_2O	−548	1386–1732	32
$[VO_4]^{3-}$	H_2O; pH 14	−536.2	36	31
	H_2O; pH 14	−536	38	22
	H_2O; pH 10–13	−536 to −530[b]	broad	28
			$J(^{51}V, ^{17}O) = 6.16 \pm 2.5$	33
$[VO_4H]^{2-}$	H_2O; pH 13.4–12.8	−533.0	87	31
$[V_2O_7]^{4-}$	H_2O; pH 13.4	−556.2	87	31
	H_2O; pH 10.5–13	−561 to −556[b]	narrow	28
V_3 and V_4 isopolyanions	H_2O; pH 9–7	−576		
		−573		31
	H_2O; pH 7	−580 to −564[b]	narrow	28
$[V_{10}O_{28}]^{6-}$	H_2O; pH 5.5	−426	485	22
		−495	291	
		−511	194	
$[V_{10}O_{28}]^{6-}$	H_2O; pH 5	−420.0	582	31
		−498.0	59	
		−510.3	194	
$[V_{10}O_{28}]^{6-}$	H_2O; pH 3–7	−409 to −421	medium	28
		−495 to −505[b]	broad	
		−525 to −509[b]	broad	
$[H_nV_{10}O_{28}]^{n-6}$	H_2O; pH 4	−427	504	22
		−513	329	
		−520	232	
$[H_2W_{11}VO_{40}]^{7-}$	H_2O; pH 4	−539	97	22
	H_2O	−539	104	32
$[PW_{11}VO_{40}]^{4-}$	H_2O; pH 1	−551	59	22, 34
$[PMo_{11}VO_{40}]^{4-}$	H_2O; pH 1.5	−532	38	22, 34
$[VW_5O_{19}]^{3-}$	H_2O; pH 1.7	−522	59	22
$[V_2W_4O_{19}]^{4-}$	H_2O; pH 4.5	−513	118	22
$[HV_2W_4O_{19}]^{3-}$	H_2O; pH 0.5	−527	175	22
$[V_2Mo_4O_{19}]^{4-}$	H_2O; pH 4.5	−502	135	22
$[H_8PV_{12}O_{40}]^{7-}$	H_2O: pH 1.3	−580	970	22
$[V(CO)_4(CN)_2]_2^{4-}$	MeCN	−760		26
$CpV(CO)_3AsPh_3$		−1250	348	25
$CpV(CO)_3PF_2Ph$		−1390	485	25
$CpV(CO)_3SbPh_3$		−1400	2214	25
$[CpV(CO)_3CN]^-$		−1440	175	25
$CpV(CO)_4$		−1520	156	25

TABLE 8.3 (*continued*)

Compound	Solvent	Shift[a]/ppm	Linewidth/Hz	$J(^{51}V, ^{31}P)$/Hz	Ref.
$CpV(CO)_3P(i\text{-}Bu)_3$		−1360	1976	660	25
$CpV(CO)_3P(OCH_2)_3CEt$		−1500	756	330	25
$CpV(CO)_3PH_2Ph$		−1390		180	23
$CpV(CO)_3PPh_3$		−1310	1841	370	25
$CpV(CO)_3P(n\text{-}Pr)_3$		−1400	523	160	25
$CpV(CO)_3PMe_3$		−1420	485	160	23, 25
$CpV(CO)_3P(NMe_2)_3$		−1340	582	230	23, 25
$CpV(CO)_3P(OEt)_3$		−1470	873	260	25
$CpV(CO)_3P(OMe)_3$		−1480		300	23
$CpV(CO)_3P(NEt_2)_3$		−1360	639	200	25
cis-$CpV(CO)_2[P(NMe_2)_3]_2$		−1230		240	23
cis-$CpV(CO)_2(PMe_3)_2$		−1250		160	23
cis-$CpV(CO)_2[P(OMe)_3]_2$		−1390		300	23
trans-$CpV(CO)_2(PMe_3)_2$		−990		210	23
trans-$CpV(CO)_2[P(OMe)_3]_2$		−1270		370	23
$[V(CO)_5NH_3]^-$	NH_3	−1670			26
$[V(CO)_5CN]^-$	NH_3	−1720			26
$[V(CO)_5P(OMe)_3]^-$		−1920		370	23
cis-$[V(CO)_4(PH_2Ph)_2]^-$		−1770		200	23
cis-$[V(CO)_4[P(OMe)_3]_2]^-$		−1880		350	23
$[V(CO)_5P(NMe_2)_3]^-$		−1840		230	23
$[V(CO)_5PPh_3]^-$	MeCN	−1850			26, 35
$[V(CO)_5AsPh_3]^-$		−1850			35
$[V(CO)_5PPh_2NMe_2]^-$		−1860			35
$[V(CO)_5PPh_2Cl]^-$		−1880			35
$[V(CO)_5PEt_3]^-$		−1890			35
$[V(CO)_5SbPh_3]^-$		−1920			35
$[V(CO)_5PPhF_2]^-$		−1960			35
$[V(CO)_5LL]^-$					
LL=$Ph_2P(CH_2)_2PPh_2$		−1830		190	36
$[V(CO)_4LL]^-$					
LL=$Ph_2P(CH_2)PPh_2$		−1590			36
$(CH_2)_2$		−1790		230	36
$(CH_2)_3$		−1710		190	36
$(CH_2)_4$		−1710			
$CpV(CO)_2LL$					
LL=$Ph_2P(CH_2)PPh_2$		−870		35	36
$(CH_2)_2$		−1110		51	36
$(CH_3)_4$		−1360		38	36

[a] PPM relative to external $VOCl_3$ and using the δ scale where positive values represent higher frequencies than the reference. Data quoted in ref. 28 relative to "aqueous NH_4VO_3" have been recalculated using $\delta_{VOCl_3} = \delta_{NH_4VO_3} - 576$. [b] Resonances which have been assigned or reassigned on the basis of more recent experimental evidence.

below, the vanadium(V) species exhibit the lowest vanadium shieldings as befits their high oxidation state. The lowest is for $VOCl_3$ whose yellow colour reveals the presence of low-lying electronic excited states signalling a large paramagnetic contribution to nuclear shielding. The vanadium in the aqueous oxyanions is more highly shielded by about 500 ppm, with all the isopolyvanadates and heteropolyvanadates lying in the δ-range -409 to -580. Within this 170 ppm range, no theoretical model has emerged which would relate a specific structural feature to a particular δ-value. What has been demonstrated,[22] however, is that the vanadium shieldings in the molybdenum and tungsten heteropolyvanadates correlate well with the average g-values measured from the ESR spectra of the analogous vanadium(IV) compounds which contain one unpaired electron. It is not unreasonable that these two parameters should be related to one another, since the vanadium shielding constant depends upon the orbital angular momentum of the electrons and the g-value provides a measure of spin-orbit coupling, but the opportunity of testing the correlation does not often present itself because one seldom encounters a series where diamagnetic and paramagnetic compounds of structurally similar forms are available.

The vanadium shieldings in about fifty V^{I} compounds of the types $CpV(CO)_3$ (phosphine) and $CpV(CO)_2$ (phosphine)$_2$ have been measured[23,24,25] and lie in the δ-range -990 to -1520. Within this range, the shielding for an individual compound can be used as a measure of phosphine ligand strength representing a combination of σ-donor and π-acceptor ability. It can be seen from the data in Table 7.3 that a change in phosphine ligand causes an increase in vanadium shielding in the order $P(NMe_2)_3 < PH_2Ph \leq PMe_3 < P(OMe)_3$. This is also the generally accepted order for increasing ligand strength which finds theoretical justification in the fact that $P(OMe)_3$ causes high ligand field splitting of metal atom d-orbital energies, ΔE, while $P(NMe_2)_3$ causes lower splitting. Because of its relationship to the ΔE factor appearing in the denominator of the analytical expression for the paramagnetic shielding term, a large ligand field splitting which moves the lowest electronic excited state away from the ground state will cause increased screening of the vanadium nucleus.

Vanadium ($-$I) compounds of the types $[V(CO)_5L]^-$ and $[V(CO)_4L_2]^-$ have chemical shifts in the more highly shielded region with δ-values in the range -1670 to -1920.[23,26] Again the δ-value provides a measure of the ligand strength as seen in Table 8.3.

Figure 8.3 shows that the compounds representing three different oxidation states of vanadium have ^{51}V shieldings in three distinct areas of the total range. Vanadium in its lowest oxidation state is the most highly shielded as one would expect from simple electron density considerations, and as the oxidation state increases, the vanadium becomes progressively deshielded.

The resonance at $\delta 5660$ reported[26] for $[V(CO)_6]^-$ calls for some comment at this point. This complex can be regarded as just another example of the general type $[V(CO)_5L]^-$, all of which lie within 130 ppm of $\delta = -1800$. When complexes with ligands having coordinating properties as different as NH_3, $[CN]^-$, and substituted phosphines all absorb within 250 ppm of each other, the suggestion that letting $L = CO$ will shift the vanadium resonance by 7400 ppm places too great a strain on the credibility of the experiment. This view is further reinforced by the fact that in the $CpV(CO)_3L$ series of complexes, the change in L from $[CN]^-$ to CO causes a ^{51}V shift difference of only 80 ppm.[25] The most probable explanation for a shift of 7400 ppm, given the range of 250 ppm for V^{-I} compounds and the range of 2000 ppm for diamagnetic vanadium compounds, is the presence of a paramagnetic decomposition product of $[V(CO)_6]^-$ in which a contact or pseudocontact interaction between the ^{51}V nucleus and an unpaired electron might possibly shift the vanadium resonance position by 7400 ppm.

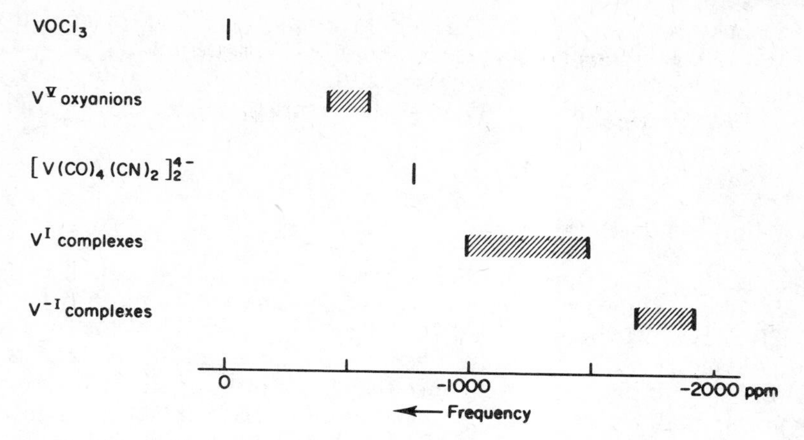

Fig. 8.3 ^{51}V chemical shift ranges for different chemical environments.

b. ^{51}V *Linewidths and Relaxation Times*

The linewidths for those ^{51}V resonances which have been observed are strongly dependent upon the electric field gradient at the nucleus as is to be expected for a nucleus having an intermediate quadrupole moment value of 0.3 barns. Values for the oxyanions in aqueous solution lie in the range 20–500 Hz with the narrowest ^{51}V resonance of 22 Hz that for $[VO_4]^{3-}$ in which the electric field gradient should be close to zero as a result of its tetrahedral symmetry. The polyvanadate anions which contain the tetrahedral VO_4 structural unit also yield ^{51}V resonances having widths less than 50 Hz. Where the effects of anion protonation resulting from pH change can be observed, it is useful to note that line broadening occurs in every case, indicating sufficient change in one of the oxygens to increase the electric field gradient experienced at the nucleus.

The linewidths observed in the lower oxidation state complexes of vanadium are generally larger than for the oxyanions, ranging up to about 1300 Hz. Although it is tempting to interpret these linewidths in terms of ligand bonding ability, no correlation between linewidth and σ-donor or π-acceptor properties of the ligands has been established.

In the course of studying the rotational diffusion of symmetric top molecules, the relaxation time T_1 for ^{51}V in $VOCl_3$ neat liquid was found to be 23.2×10^{-3} s at room temperature and shorter at reduced temperatures giving an Arrhenius activation energy for the relaxation time of $E_a = 8.23$ kJ/mol.[27]

c. ^{51}V *Coupling Constants*

Vanadium couplings to ^{31}P in phosphine complexes and to ^{19}F in fluoride complexes have been observed. Typical values for $J(^{51}V, ^{31}P)$ couplings are listed in Table 8.3. It has been shown[23] that these couplings are determined primarily by the nature of the phosphine ligand, increasing in the order $PMe_3 \sim PH_2Ph < P(NMe_2)_3 < P(OMe)_3$, and that they provide little information about the nature of the V-P bond in the complex.[25]

The ^{51}V spectrum of $[VOF_4]^-$ in 48% aqueous HF has been obtained, and while at 25°C it consists of a single line whose width is less than 100 Hz, at −10°C it becomes a quintet having $J(^{51}V, ^{19}F) = 116$ Hz.[28] A subsequent study[29] of $[VOF_4]^-$ at −20°C in 49% aqueous HF gives $J(^{51}V, ^{19}F) = 120$ Hz and at -80°C in anhydrous HF $J(^{51}V, ^{19}F) = 140$ Hz. While no fine structure in the ^{51}V spectrum of $[VF_6]^-$ could be observed, the ^{19}F spectrum in acetonitrile at -20°C yielded an octet with $J(^{51}V, ^{19}F) = 88$ Hz. The ^{13}C NMR spectrum of $[V(CO)_6]^-$ shows six lines attributable to $^{1}J(^{51}V, ^{13}C) = 116$ Hz.[30]

8D.2 Niobium, ^{93}Nb

Niobium-93 is 100% abundant and has a natural abundance receptivity relative to $^{13}C = 1$ of 2736. A comparison with the corresponding values of 5676 for ^{1}H and 4728 for ^{19}F shows that niobium ranks third among all the elements in receptivity. Its quadrupole moment of -0.2×10^{-28} cm^2 is the lowest of the first row transition metals, with the result that excessive linewidth is not a problem in ^{93}Nb spectroscopy. Although few linewidths have been reported, those that have are all less than 1000 Hz.

The study of vanadium and of manganese chemical shifts has shown that metal oxidation state is a primary determinant of metal atom shielding. Niobium(V) is the only oxidation state for which ^{93}Nb NMR data have been obtained. As a result, the range of chemical shifts observed is rather narrower than has been found with comparable metals.

Since the first niobium NMR data came out of W. G. Schneider's research laboratory at the National Research Council in 1965,[28] only two research groups have made significant inroads into the area. Out of this work has come the recognition that the halogen dependence of the niobium chemical shift is the inverse of that for most other metals. The other feature which has emerged from niobium spectra is one of interest to structural inorganic chemists. By adapting the pairwise additivity model worked out for tetrahedral complexes[37] to the octahedral coordination of niobium, it has been possible to identify in solution specific geometric isomers which are not identifiable by any other technique.

a. ^{93}Nb *Chemical Shifts*

Both $[NbCl_6]^-$ and $[NbF_6]^-$ have been used as reference compounds for the quoting of ^{93}Nb chemical shifts and both offer certain advantages. $[NbCl_6]^-$ gives rise to a sharp resonance and can be used as an internal reference in many niobium systems. As a reference for quoting purposes, it has the disadvantage of a resonance occurring in the middle of the chemical shift range for ^{93}Nb, thereby generating both positive and negative δ-values in about equal numbers. Although in some solvents $[NbF_6]^-$ gives a well-resolved septet due to coupling with the six fluorines, in 48% aqueous HF the fluorines are exchanging rapidly enough that the niobium resonance is a sharp singlet. Since this line occurs at the extreme low frequency end of the niobium spectral range, it has been adopted as the reference for the compendium given in Table 8.4. The question of whether $[NbF_6]^-$ displays a solvent effect of 60–70 ppm on going from aqueous to acetonitrile solution has not yet been resolved as two authors[38,39] report a solvent effect but one[40] reports no shift. When ^{93}Nb spectra for Nb^{I} and Nb^{-I} compounds are obtained, the resonances will likely occur at lower frequencies than $[NbF_6]^-$, again putting the reference in the middle of the range. At that time, the question of the most appropriate reference compound for ^{93}Nb chemical shifts should be reconsidered.

Virtually all the NMR data available for niobium concerns compounds in which the ligands are halide or pseudo-halide ions. In these compounds, iodine ligands shift niobium resonances to highest frequencies (lowest fields), with bromine, chlorine, and fluorine causing shifts to successively lower frequencies. This pattern is just the reverse of that which obtains for the metal resonances in the tetrahalides of ^{11}B, ^{27}Al, ^{69}Ga, ^{115}In, ^{13}C, ^{29}Si, ^{73}Ge, and ^{119}Sn. Another case* of this inverse halogen dependence is ^{47,49}Ti[21] where the bromides absorb at higher frequencies than the chlorides. It has been suggested[41] that the inverse halogen dependence results from *d*-orbital participation in the metal-halogen bond where transition metals are involved. The fact that ^{55}Mn and ^{59}Co exhibit normal halogen dependence, however, means that a satisfactory explanation for this phenomenon has yet to be given.

Niobium NMR has definitely characterised the nature of $NbCl_5$ and $NbBr_5$ in non-aqueous solvents. Both the $[NbX_6]^-$ anions and the NbX_5 MeCN solvent adducts have been identified by their ^{93}Nb shift and the experimental conditions appropriate to each have been determined.[42]

Mixtures of $NbCl_5$ and $NbBr_5$ in acetonitrile undergo a slow ligand redistribution reaction to give a mixture of products each of which gives a separate resonance in the niobium spectrum. In the case of the $[NbCl_nBr_{6-n}]^-$ system, random ligand distribution gives 7 different stoichiometries and 10 different structures allowing for geometrical isomerism where $n = 2, 3, 4$. Niobium chemical shifts can be measured with sufficient precision to show that coordinated ligands do *not* make an independent contribution to the niobium shielding and hence the substituent constant concept is not a viable one for niobium shieldings. Rather than being a theoretical disappointment, this situation has practical advantages in the case of six coordinate complexes for it means that geometrical isomers will have different niobium shieldings. The pairwise additivity model shows that shielding differences for geometrical isomers of the same molecule involving halide, pseudo-halide, and acetonitrile ligands lie in the range 5–25 ppm.[42,43] As shown in Table 8.4, specific geometric isomers of these niobium complexes have been identified on the basis of their niobium chemical shift.

b. ^{93}Nb *Linewidths*

There is little quantitative information available on niobium linewidths. The narrowest ^{93}Nb resonance to be reported is that for $NbCl_5$ in acetonitrile at 30 Hz which, on the basis of its narrow width is attributed to the octahedral $[NbCl_6]^-$ ion. In less symmetric environments, widths such as 700 Hz for $NbOCl_3$ have been observed.[38]

c. ^{93}Nb *Coupling Constants*

The anion $[NbF_6]^-$ is the only molecule in which coupling to niobium has been observed. There have been three[28,39,40] independent observations of the ^{93}Nb septet arising from this ion and values for $J(^{93}\mathrm{Nb}, {}^{19}\mathrm{F})$ from 334 Hz to 345 Hz have been reported. Of greater relevance to this ion is the fact that Aksnes *et al.*[44] have done a complete lineshape analysis of its ^{19}F decet as a function of fluorine exchange rate, and have obtained a two-parameter description of the spectrum which permits the separation of the quadrupole-induced transitions between spin states of ^{93}Nb from the fluorine exchange contribution to the line broadening.

* The shielding of alkali metal cations in aqueous solution is concentration dependent and counterion specific. The halide ions cause increased shielding of the hydrated metal cation in the order $I^- < Br^- < Cl^-$. While the mechanism for this anion effect is not fully understood at present, the nature of the interaction could not be described as covalent bonding and the phenomenon cannot usefully be regarded as another example of inverse halogen dependence.

TABLE 8.4 ^{93}Nb chemical shifts and linewidths

Compound	Solvent	Shift[a]/ppm	Linewidth/Hz	Ref.
$[NbBr_6]^-$	MeCN	2227		41
	MeCN	2221		42
$[NbBr_5Cl]^-$	MeCN	2113		41
	MeCN	2111		42
$[NbBr_4Cl_2]^-$	MeCN	1994		41
cis-$[NbBr_4Cl_2]^-$	MeCN	1991		42
$[NBr_3Cl_3]^-$	MeCN	1870		41
fac-$[NbBr_3Cl_3]^-$	MeCN	1870		42
$[NbBr_2Cl_4]^-$	MeCN	1743		41
cis-$[NbBr_2Cl_4]^-$	MeCN	1748		42
$[NbBrCl_5]^-$	MeCN	1619		41
	MeCN	1622		42
$[NbCl_6]^-$	MeCN	1489		41
	MeCN	1490		42
$NbBr_5 \cdot MeCN$	MeCN	2130	ca. 870	42
trans-$NbBr_4Cl \cdot MeCN$	MeCN	2019		42
cis-$NbBr_4Cl \cdot MeCN$	MeCN	1986		42
$NbBr_3Cl_2 \cdot MeCN$	MeCN	1844		42
$NbBr_2Cl_3 \cdot MeCN$	MeCN	1719		42
trans-$NbBrCl_4 \cdot MeCN$	MeCN	1580		42
$NbCl_5 \cdot MeCN$	MeCN	1441	ca. 870	42
trans-$[NbCl_4(SCN)_2]^-$	MeCN	1270		43
$[NbCl_5(NCS)]^-$	MeCN	1258		43
cis-$[NbCl_4(NCS)_2]^-$	MeCN	1033		43
trans-$[NbCl_2(SCN)_2]^-$	MeCN	1015		43
$[NbCl(SCN)_5]^-$	MeCN	857		43
$[NbCl_2(NCS)_2(SCN)_2]^-$	MeCN	820		43

fac-$[NbCl_3(NCS)_3]^-$	MeCN	809		43
$[Nb(SCN)_6]^-$	MeCN	710		43
cis-$[NbCl_2(NCS)_4]^-$	MeCN	588		43
trans-$[Nb(NCS)_2(SCN)_4]^-$	MeCN	550		43
cis-$[Nb(NCS)_2(SCN)_4]^-$	MeCN	543		43
trans-$[NbCl(NCS)_5]^-$	MeCN	372		43
trans-$[Nb(NCS)_4(SCN)_2]^-$	MeCN	365		43
$[Nb(NCS)_5(SCN)]^-$	MeCN	255		43
$[Nb(NCS)_6]^-$	MeCN	148		43
$[NbCl_5I]^-$	MeCN	1855		38
$[NbCl_5N_3]^-$	MeCN	1654		38
$[NbCl_5(SCN)]^-$	MeCN	1382 (calculated)		43
$[NbCl_5F]^-$	MeCN	1232		38
$[NbOBr_{3+n}]^{n-}$	($n \approx 2$) MeCN	1290	346	38
$[NbOCl_{3+n}]^{n-}$	($n \approx 2$) MeCN	1008	346	38
$NbOCl_3$	MeCN	1000	broad	39
	MeCN	987	1212	38
$NbF_5 \cdot NEt_3$	$EtOH/NEt_3$	155	v. broad singlet	28
$NbF_5 \cdot OEt_2$	Et_2O	70	broad singlet	28
$[NbF_6]^-$	EtOH; 0°C	5	septet; $J(^{93}Nb, ^{19}F) = 345$	28
	MeCN	0	septet; $J(^{93}Nb, ^{19}F) = 334$	40
	48% HF/H_2O; 20°C	0	sharp singlet	28
	48% HF/H_2O; −70°C		broad singlet	28
	100% HF; 20°C		199	45
	100% HF; reduced temp.		1502	45
	MeCN	−60	septet; $J(^{93}Nb, ^{19}F) = 342$	39
	MeCN	−68		38

[a] PPM relative to external $[NbF_6]^-$ singlet in 48% aqueous HF and using the δ scale where positive values represent higher frequencies than the reference. Data quoted relative to $[NbCl_6]^-$ have been recalculated using $\delta_{[NbF_6]^-} = \delta_{[NbCl_6]^-} + 1490$.

8D.3 Tantalum, ^{181}Ta

While tantalum-181 is 100% abundant and has a detection receptivity of 204 relative to ^{13}C, its very large quadrupole moment of 3×10^{-24} cm^2 has resulted in several unsuccessful attempts[28,46] to obtain its NMR spectrum in solution. The only successful report[47] has been that for a solution containing tantalum metal dissolved in a 1:1 HF/HNO_3 mixture for which the solute species is assumed to be the $[TaF_6]^-$ ion. In spite of the high symmetry and low electric field gradient which should obtain at the tantalum site, this solution yields a signal 36 kHz in width at room temperature, narrowing to 10 kHz at 80°C. The resonance frequency for this signal has been used to obtain the most precise measure of the tantalum magnetic moment yet achieved. After corrections for core diamagnetism and paramagnetic shift resulting from Ta-F covalency, a value of 2.361 ± 0.010 nuclear magnetons is obtained.

Tantalum-181 has the largest quadrupole moment of any nucleus whose NMR signal has been observed in solution.

8E GROUP VIA

Although NMR investigations have been carried out on all three elements they all have low receptivities making direct observation difficult. Of these nuclei only ^{183}W has attracted much attention, and this was by use of INDOR. Tungsten provides severe problems for direct observation, but it is to be expected that considerable use will be made of ^{53}Cr and ^{95}Mo NMR spectroscopies, especially for the organometallic compounds.

8E.1 Chromium, ^{53}Cr

Chromium-53, present in nature of the extent of 5.7%, has nuclear spin $I = \frac{3}{2}$ and low NMR detection sensitivity. Relative to ^{13}C in natural abundance its receptivity is 0.49, and although its quadrupole moment is not known, quadrupole broadening will produce lines whose signal-to-noise ratios are further diminished in comparison with carbon.

The only reports of ^{53}Cr NMR in solution have been those of Egozy and Loewenstein,[48] who chose $[CrO_4]^{2-}$ to minimise quadrupolar broadening and used a time-averaging technique to obtain chromium resonances from natural abundance samples, and of Lutz *et al.*[49] who have studied the concentration dependence, counter-ion dependence, and solvent dependence of $[CrO_4]^{2-}$ chemical shifts.

Egozy and Loewenstein[48] have used the broadening which is observed in the $[CrO_4]^{2-}$ signal as solution pH decreases to evaluate the bimolecular rate constant for the reaction of chromate ion with dichromate ion. There have been two previous studies of this reaction using ^{17}O NMR,[50,51] and the present work nicely complements that done previously.

Lutz *et al.*[49] have obtained half-height linewidths of about 11 Hz for ^{53}Cr in $[CrO_4]^{2-}$ and have observed resonance shifts to lower frequencies as the concentration in aqueous solution is increased. This concentration dependence is also cation dependent, with the slopes of the low frequency shifts increasing in the order $Na^+ < K^+ < Rb^+ < [NH_4]^+ < Cs^+$. These same authors have obtained a ^{53}Cr resonance for $Cr(CO)_6$ and find that it is 1795 ppm more highly shielded than $[CrO_4]^{2-}$.

This pattern of a large increase in metal shielding in the low oxidation state carbonyl compound is consistent with the pattern observed for ^{51}Mn and ^{59}Co.

When $[CrO_4]^{2-}$ is enriched to 35% in ^{17}O, broadening at the base of the ^{53}Cr signal occurs due to the complex pattern of satellites arising from the statistical distribution of ^{16}O and ^{17}O among the four oxygen sites. Computer analysis of the complex multiplet best fits the observed spectrum with a coupling constant $J(^{53}Cr, ^{17}O) = 10 \pm 2$ Hz.[33]

8E.2 Molybdenum, $^{95,97}Mo$

Molybdenum has two magnetically active isotopes, both with spin $I = \frac{5}{2}$ and both with low detection sensitivity. ^{95}Mo is the more readily detected of the two with natural abundance receptivity relative to ^{13}C of 2.9 while that for ^{97}Mo is 1.8. The quadrupole moment for ^{95}Mo falls in the intermediate category at 0.12×10^{-24} cm^2, while that for ^{97}Mo at 1.1×10^{-24} cm^2 is one of the larger quadrupole moments. It is unusual for two nuclei with the same spin and of about the same mass to have quadrupole moments that differ by such a large factor, and the quadrupole moment ratio has received close study.[52] This large difference in quadrupole moment has proven to be of some significance in several molybdenum NMR studies which have been reported.

By use of a Fourier transform spectrometer and data collection over an 18.6 hour period, a ^{17}O spectrum of $[MoO_4]^{2-}$ in which a well-resolved sextet due to coupling with ^{95}Mo has been obtained.[53] The coupling constant J (^{95}Mo, ^{17}O) is 40.3 ± 0.2 Hz. The fact that this coupling is resolved and the sharpness of the sextet components indicate a minimum relaxation time T_1 for ^{95}Mo of 0.7 ± 0.1 s. This yields a quadrupole coupling constant of ca. 400 kHz using a calculated value of 9×10^{-12} s for the rotational correlation time. A non-zero value for the electric field gradient at molybdenum probably arises from distortion of its tetrahedral symmetry through ion pair formation. Whatever the cause, the electric field gradient is sufficient to relax and completely decouple from ^{17}O the ^{97}Mo nuclei because of their much larger quadrupole moment. The oxygens bonded to ^{97}Mo give a single line at the approximate centre of the sextet, as shown in Fig. 8.4.

Following their success at resolving the coupling to molybdenum in the ^{17}O spectrum of the $[MoO_4]^{2-}$ ion, Vold and Vold[54] then obtained both the ^{95}Mo and ^{97}Mo spectra for the same ion and studied the two linewidths as a function of structure change which accompanies change in pH. In keeping with the difference in quadrupole moment for the two isotopes, the ^{97}Mo line is very much wider than the ^{95}Mo line as shown in Fig. 8.5. No coupling to ^{17}O is observed because of its extremely low natural abundance. In basic solution (pH 9–12) where protonation and polyanion formation are minimized, the narrowest signals are observed with relaxation times $T_1(^{95}Mo) = 840 \pm 20$ ms and $T_1(^{97}Mo) = 6.5 \pm 0.2$ ms. Protonation of $[MoO_4]^{2-}$ at lower pH broadens the ^{95}Mo line and T_1 is shortened to 200–300 ms. This line broadening provides structural information about polymolybdate formation in aqueous solution.

Recently Lutz and co-workers[55] have obtained chemical shift data for several molybdenum compounds and they find that the relationship between metal oxidation state and metal atom shielding which has been observed for other metals is preserved with molybdenum. Relative to $[MoO_4]^{2-}$, $[MoS_4]^{2-}$ absorbs at $\delta 496$, $[Mo(CN)_8]^{4-}$ absorbs at $\delta - 1309$ and $Mo(CO)_6$ in tetrahydrofuran absorbs at $\delta - 1856$. The cation, solvent, and pH dependence of the ^{95}Mo shielding in $[MoO_4]^{2-}$ have all been studied.[56] There is little information on coupling to molybdenum. As noted above, $^1J(^{95}Mo, ^{17}O) = 40.3$ Hz.[53] Examination of the ^{13}C NMR spectrum of $Mo(CO)_6$ shows a weak sextet

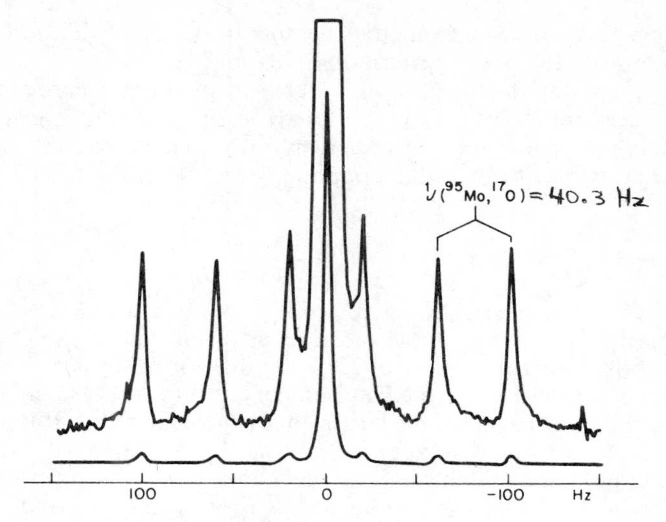

Fig. 8.4 The ^{17}O spectrum of aqueous $[MoO_4]^{2-}$ showing ^{95}Mo–^{17}O coupling but no ^{97}Mo–^{17}O coupling. Reproduced from R. R. Vold and R. L. Vold, *J. Chem. Phys.* **61**, 4360 (1974) by permission of the American Institute of Physics.

at the foot of the strong singlet, with $^1J(^{95}Mo, ^{13}C) = 68$ Hz.[57] Subsequently a similar observation has been made in the ^{31}P NMR spectrum of $(OC)_5MoL$, L = phosphorus compound, where $^1J(^{95}Mo, ^{31}P)$ in the range 133 to 284 Hz was found.[58]

8E.3 Tungsten, ^{183}W

While the only naturally occurring tungsten nucleus which possesses a nuclear spin has a desirable feature that $I = \frac{1}{2}$, this is counterbalanced by a low relative receptivity. The ^{183}W resonance of the finely powdered metal was reported by Sogo and Jeffries

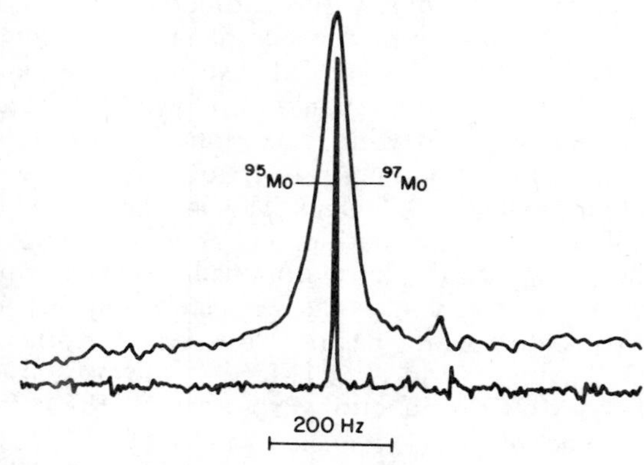

Fig. 8.5 Superimposed ^{95}Mo and ^{97}Mo resonances from aqueous $[MoO_4]^{2-}$. reproduced from R. R. Vold and R. L. Vold, *J. Magn. Reson.* **19**, 365 (1975), by permission of Academic Press.

in 1955.[59] Results have also been obtained for solid WO_3, some tungsten "bronzes"[60] and cobalt-tungsten alloys[61] but the direct observation of tungsten resonances for liquid samples was not reported until 1974.[62] Because of the difficulties of direct observation, the majority of shift data for ^{183}W comes from double resonance techniques.[63–66]

The recent direct observations by Schwenk and co-workers[62,67] were made with a Fourier transform spectrometer specially designed to detect weak NMR signals. The tungsten resonances were in the region of 3.21 MHz for the field of 1.8 T. Presumably partly as a consequence of the large diameter tubes used (i.d. 21 mm), T_2^* is rather short compared to T_2 and T_1 and a special observation procedure, "Quadriga"[68] was employed. The weakest sample examined, 0.88 molal WCl_6 solution gave a S/N ratio of 7:1 in 100 min. The spin-lattice relaxation time for ^{183}W in tungsten hexafluoride at 179 K was determined by a progressive saturation technique and the ratio of T_1 to T_2 by a procedure devised for the situation where T_1 and T_2 are long compared to T_2^*.[5] The results are: T_1, 4.2 ± 0.6 s; T_2, 2.1 ± 0.4 s.

The low magnetogyric ratio of ^{183}W can jeopardise double resonance experiments as spin-spin coupling to tungsten tends to be relatively small and the ^{183}W satellites (ca. 8% ea.) may appear only as shoulders on the main resonances or may not be distinguishable at all. McFarlane *et al.* were able to perform successful double resonance experiments provided $J(^{183}WH) > 1$ Hz.[66] Although no coupling was detectable for some trimethyl phosphite complexes, they were able to obtain ^{183}W data via ^1H-{^{183}W, ^{31}P} triple resonance experiments since $^1J(WP)$ and $^3J(PH)$ are substantial. An alternative approach in such situations is to use ^{31}P-{^{183}W} double resonance.[64]

a. ^{183}W *Chemical Shifts*

All reported ^{183}W chemical shifts have been quoted with respect to WF_6 despite its rather undesirable physical and chemical properties and uncertain purity. The value of $\Xi(^{183}W) = 4\,161\,780$ Hz. obtained by McFarlane *et al.* for WF_6 used CCl_3F as an internal reference and must involve uncertainties on the shift of the ^{19}F reference in this particular medium as well as its effect on the shift of WF_6 (e.g. the ^{183}W resonance is shifted ca. 12 ppm to high frequency in benzene solution).[65] Accordingly, the value $\Xi(^{183}W) = 4\,161\,733$ Hz derived from the result of Banck and Schwenk[67] is preferred and employed for the chemical shifts collected together in Tables 8.5 and 8.6. (A rather large correction is involved for the results of Narath and Wallace[60] and of Green and Brown[64] since they used the older value of Klein and Happe[63] which is considerably in error.[67])

To judge from their respective colours, the mean electronic excitation energy, ΔE is much less for WCl_6 than for WF_6 and the shielding is significantly less for the former in agreement with expressions for Ramsey's paramagnetic contribution to nuclear shielding. Thus the tungsten hexahalides provide one of the few instances where the relative shielding in complexes of different halides does not run counter to predictions based on ΔE (cf. $[PtF_6]^{2-}$ vs. $[PtCl_6]^{2-}$). It may be that in this case, the effects of ΔE override the shift to low frequency that usually accompanies an increase in the Period of the donor atom. However, the more common i.e. "anomalous" behaviour is shown by $W(CO)_3(\eta\text{-}C_5H_5)X$, X = Cl, Br and I and the differences have similar magnitudes to those observed for platinum complexes.

The range covered by the results in Tables 8.5 and 8.6 is some 6000 ppm which is about half the range of ^{195}Pt shifts. The shifts seem relatively indifferent to the substituents on the donor atom e.g. the very small range for $W(CO)_5L$ where L is a phosphite or most

TABLE 8.5 The ^{183}W chemical shifts in some tungsten(VI) compounds

Compound	$\delta(^{183}W)$/ppm	Medium	Ref.
WF_6	0^a	neat	67
$WF_5(OMe)$	52	C_6F_6	65
$WF_5(OPh)$	201	C_6F_6	65
$[WF_5O]^-$	616	CD_3CN	65
$[(WF_4O)_2F]^-$	628	$SO(OMe)_2$	65
cis-$WF_4(OMe)_2$	171	neat	65
cis-$WF_4(OPh)_2$	268	C_6F_6	65
trans-$WF_4O(OMe_2)$	577	neat	65
trans-$WF_4OPMe(OMe)_2$	544	CD_3CN	65
trans-$WF_4OS(OMe)_2$	540	$SO(OMe)_2$	65
mer-$WF_3(OMe)_3$	475	C_6F_6	65
cis-$WF_2(OMe)_4$	615	C_6F_6	65
$Na_2[WO_4]$	1121	H_2O	67
$K_2[WO_4]$	1118	H_2O	67
WO_3	840	solid	60
WCl_6	3302	CS_2	67

^{a} Reference $\Xi(^{183}W) = 4\,161\,733$ Hz.

notable $[PO(OMe)_2]^-$ which only extends 200 ppm if all phosphorus donors excepting $PCl_2(OMe)$ are included (the latter probably involves significant steric repulsion to judge from $PtX_3\{PCl_2(OMe)\}$[69]). Whilst this situation is found for ^{195}Pt chemical shifts, there the other ligands are essentially σ-donors whereas here they are carbon monoxide. The similarity of the shifts of $[WF_5O]^-$ and $[(WF_4O)_2F]^-$ also fits this pattern—the coordination of one fluorine to a second tungsten not greatly affecting the shift of the $[WF_5O]$ unit.

The change in ^{183}W chemical shift with temperature has been found by Banck and Schwenk[67] as 0.16(4) ppm/°K for $[WO_4]^{2-}$ and 0.34(16) ppm/°K for WCl_6. These are somewhat smaller than the general figure of ca. 1.5 ppm/°K quoted by McFarlane *et al.*[66]

The Knight shift in metallic tungsten is 10 460 ppm if WF_6 is taken as the origin or 13 380 ppm from the lowest frequency result in Table 8.6.[60] It is essentially independent of temperature which implies it is almost entirely due to *d*-band orbital paramagnetism.[70] The tungsten resonances in the bronzes, Na_xWO_3 are more shielded than in tungsten hexafluoride.[60]

b. ^{183}W *Coupling Constants*

One bond ^{183}W-^1H coupling constants in a variety of hydrides range from 28 to 80 Hz with most, including those for the bridging hydride in compounds such as $W_2H(CO)_{10}$, falling in the region 35–50 Hz.[71] In $WH_2Cp_2W(CO)_5$, $^2J(WH)$ is 19.2 Hz.[72] For WMe_6, $^2J(WH)$ is 3.0 Hz[73] and $^3J(WH)(Me)$ in $W(CO)_5(PMe_2Ph)$ is +2.2 Hz.[74] In a series of carbonyl complexes, $^1J(WC)$ varies from 125 to 186 Hz.[75] For the fluoride derivatives in Table 8.5, $^1J(WF)$ varies from 12 to 71 Hz. The sign of $^1J(WF)$ has not been determined but in those derivatives where there are two types of fluorine both have the same sign with the notable exceptions of $[WOF_5]^-$ and $[(WOF_4)_2F]^-$ where

TABLE 8.6 The ^{183}W chemical shifts in some low-valent tungsten complexes[a]

Compound	$\delta(^{183}W)^b$/ppm	Medium
$W(CO)_5\{PCl_2(OMe)\}$	−2018	c
$W(CO)_5\{P(OMe)_3\}$	−2354	neat
$W(CO)_5\{P(OMe)(OPh)_2\}$	−2312	$CDCl_3$
$W(CO)_5\{P(OMe)[(OCH_2)_2CMe_2]\}$	−2341	
$[NEt_3H][W(CO)_5\{PO(OMe)_2]$	−2316	c
$W(CO)_5\{P(NMe_2)_3\}$	−2270	d
$W(CO)_5\{PPh(OMe)_2\}$	−2271	c
$W(CO)_5\{PPh_2(OMe)\}$	−2225	d
$W(CO)_5(PMePh_2)$	−2192	d
$W(CO)_5(PMe_2Ph)$	−2205	d
$W(CO)_5(PH_2Ph)$	−2299	c
$W(CO)_5\{PPh_2(SnMe_3)\}$	−2160	c
$W(CO)_5\{PPh(SnMe_3)_2\}$	−2167	c
$W(CO)_5\{P(SnMe_3)_3\}$	−2180	c
trans-$W(CO)_4\{P(OMe)_3\}_2$	−2312	d
trans-$W(CO)_4(PBu^n_3)_2$	−2184	d
cis-$W(CO)_4\{P(OMe)_3\}_2$	−2286	d
cis-$W(CO)_4(PBu^n_3)_2$	−2128	d
cis-$W(CO)_4(PMe_2Ph)_2$	−1996	d
cis-$W(CO)_4\{P(OMe)_3\}$py	−1526	c
cis-$W(CO)_4\{P(OMe)_3\}$(4-Mepy)	−1539	c
cis-$W(CO)_4\{P(OMe)_3\}(NCPh)$	−1847	c
cis-$W(CO)_4\{P(OMe)_3\}\{S(SnMe_3)_2\}$	−1802	c
fac-$W(CO)_3(PMe_2Ph)_3$	−1697	d
$\{W(CO)_3Cp\}_2$	−2919	c
$W(CO)_3CpH$	−2896	c
$W(CO)_3CpD$	−2907	c
$W(CO)_3CpMe$	−2428	c
$W(CO)_3CpEt$	−2346	c
$W(CO)_3CpCl$	−1285	c
$W(CO)_3CpBr$	−1463	c
$W(CO)_3CpI$	−1875	c
$W(CO)_3Cp(SnMe_3)$	−2923	c
$W(CO)_2CpH\{P(OMe)_3\}$	−2772	c
$W(CO)_2CpH(PMe_2Ph)$	−2558	c
$W(CO)_2CpH(PMePh_2)$	−2560	c
WCp_2H_2	−3550	c

[a] From ref. 66 except for the two PBu^n_3 derivatives which are from ref. 64. [b] To high frequency of WF_6. [c] Benzene. [d] Dichloromethane.

surprisingly they have opposite signs despite similar magnitudes.[65] In fluorophosphine complexes of tungsten, $^2J(PF)$ is ca. 30 Hz.[76] Values of $^1J(WP)$ range from 200 to 500 Hz and have been reviewed by Verkade[77] and by Appleton *et al.*[78] The reduced coupling constant is about half that of $K(^{195}Pt, ^{31}P)$ but nevertheless McFarlane and Rycroft have shown that both have the same sign i.e. $^1J(WP)$ is positive.[74] A value of -150 ± 5 Hz has been obtained for $^1J(^{183}W, ^{119}Sn)$ in $W(CO)_3Cp(SnMe_3)$.[66]

8F GROUP VIIA

Within group VIIA, NMR spectroscopy is at present only really practical for ^{55}Mn although there have been a few isolated reports of investigations of ^{99}Tc, ^{185}Re and ^{187}Re nuclei. Even for manganese the largest proportion of manganese chemistry involves compounds which are paramagnetic as a result of unpaired electrons in their valence shells. Unfavourable relaxation times make these compounds unsuitable for study by NMR spectroscopy. With the single exception of the permanganate ion, NMR study in this area has been limited to the low oxidation states in which Mn^{I} and Mn^{-I} are stabilized by five carbonyl and/or phosphine ligands. The only structural variation which has been achieved in these compounds lies with the nature of the ligand occupying the sixth coordination site. Within this limitation, considerable diversity has been achieved and examples where the sixth ligand is hydrogen, a halogen, an alkyl group, an acyl group, a substituted Group IV metal, and cobalt carbonyl have all been studied.

8F.1 Manganese, ^{55}Mn

The ^{55}Mn nucleus is not difficult to observe and standard NMR techniques will yield favourable signal-to-noise ratios without the need for signal averaging or other signal enhancement methods. It is 100% abundant in nature and has natural abundance detection receptivity, $D^{C} = 994$, about the same as ^{27}Al but less than ^{51}V, ^{45}Sc, and ^{59}Co. The ^{55}Mn nucleus has a spin $I = \frac{5}{2}$ and a quadrupole moment of 0.55×10^{-28} cm^2. This Q is about twice that of ^{51}V whose diamagnetic complexes are structurally similar to those of manganese. Since linewidths are proportional to Q^2, one expects to encounter broader lines for ^{55}Mn and this expectation is confirmed by the data presented in Tables 8.3 and 8.7. This sensitivity of ^{55}Mn linewidths to variations in electric field gradient engendered by change in ligand has been exploited by chemists seeking to differentiate the σ-bonding and π-bonding properties of ligands. This subject is discussed in the section dealing with linewidths.

Since Calderazzo, Lucken and Williams[79] published the first compendium of ^{55}Mn data in 1967, four other research groups have added significant bodies of new experimental data to knowledge in this area. The theoretical interpretation of the data has been closely associated with isomer shift and quadrupole splitting values obtained from Mossbauer spectra and ^{55}Mn illustrates better than other nuclei the complementary nature of the information obtainable from these two spectroscopic sources.

a. ^{55}Mn *Chemical Shifts*

The reference compound which has been unanimously adopted by workers in this area for quoting ^{55}Mn chemical shifts is aqueous $KMnO_4$. Its resonance occurs at the extreme high frequency end of the range, and because of its tetrahedral symmetry, its linewidth is less than 10 Hz. The effects of different solvents on the manganese shielding in $[MnO_4]^-$ varies from $\delta 0.00$ for water to $\delta 17.79$ for hexamethylphosphoramide.[80] Although the shielding in pure acetone is $\delta 9.95$, the signal position does not vary linearly from $\delta 0.00$ to $\delta 9.95$ in water-acetone mixtures, but moves first to lower frequencies at low acetone concentrations with a minimum of $\delta -3.5$ at 30 volume % acetone, then moves to higher frequencies with increasing acetone concentration. A solvent effect of 25 ppm between the chemical shift of $Ph_3SnMn(CO)_5$ in tetrahydrofuran and that in dimethylformamide has also been reported,[81] confirming the fact that one can expect solvent effects of up to 30 or 40 ppm on ^{55}Mn shielding.

TABLE 8.7 ^{55}Mn Chemical shifts and linewidths

Compound	Solvent	$\delta(^{55}Mn)$/ppm[a]	$W_{1/2}$/Hz	Ref.
$KMnO_4$	HMPA	17.79	17	80
	DMF	11.93	17	80
	Trimethyl phosphate	11.47	17	80
	Acetone	9.95	17	80
	Propylene carbonate	5.84	17	80
	Water	0.00	17	80
$ClMn(CO)_5$	THF	−1004	331	79
$BrMn(CO)_5$	THF	−1160	688	79
$IMn(CO)_5$	THF	−1485	1013	79
$[I_2Mn(CO)_4]^-$	THF	−1020	4675	79
$CF_3COMn(CO)_5$	THF	−1850	3329	79
$CHF_2COMn(CO)_5$	THF	−1885	3729	79
$CH_2FCOMn(CO)_5$	THF	−2010	4675	79
$CH_3COMn(CO)_5$	THF	−1895	3838	79
$CF_3Mn(CO)_5$	THF	−1850	3201	79
$CHF_2Mn(CO)_5$	THF	−1970	4183	79
$CH_2FMn(CO)_5$	THF	−2130	4328	79
$CH_3Mn(CO)_5$	THF	−2265	3040	79
$ClCH_2COMn(CO)_5$	THF	−1855	6365	79
$Me_2CHCOMn(CO)_5$	THF	−1885	5184	79
$EtCH_2COMn(CO)_5$	THF	−1900	5820	79
$MeCH_2COMn(CO)_5$	THF	−1950	5002	79
$PhCH_2COMn(CO)_5$	THF	−2035	9331	79
p-$FC_6H_4COMn(CO)_5$	THF	−1950	8002	79
$HMn(CO)_5$	THF	−2630	4347	79
$Na[Mn(CO)_5]$	THF	−2780	10 585	79
$MeCOMn(CO)_4PPh_3$	THF	−1720	1609	79
$CpMn(CO)_3$	THF	−2225	10 039	79
$Ph(MeCO_2)_2C_5H_2Mn(CO)_3$		−2588	9945	92
$Mn_2(CO)_{10}$	THF	−2325	83	79
$MnCo(CO)_9$	Solution	−1840		93
	Solid	−1600		93
$Mn_2(CO)_8(PPh_3)_2$	THF	−2325	10 657	79
$Cl_3SnMn(CO)_5$	THF	−1967	v. sharp	81
$Cl_2PhSnMn(CO)_5$	THF	−2256	sharp	81
$ClPh_2SnMn(CO)_5$	THF	−2468	medium	81
$Ph_3SnMn(CO)_5$	THF	−2530	3000	81
$Cl_3SnMn(CO)_5$	CH_2Cl_2	−2024	327	83
$Cl_2PhSnMn(CO)_5$	CH_2Cl_2	−2278	546	83
$ClPh_2SnMn(CO)_5$	CH_2Cl_2	−2460	2638	83
$Ph_3SnMn(CO)_5$	CH_2Cl_2	−2610	6493	83
$Br_3SnMn(CO)_5$	CH_2Cl_2	−2044	310	83
$Br_2PhSnMn(CO)_5$	CH_2Cl_2	−2252	436	83
$BrPh_2SnMn(CO)_5$	CH_2Cl_2	−2468	3092	83
$Ph_3SnMn(CO)_5$	CH_2Cl_2	−2610	6493	83
$Cl_2MeSnMn(CO)_5$	CH_2Cl_2	−2312	509	83
$ClMe_2SnMn(CO)_5$	CH_2Cl_2	−2520	2000	83
$Me_3SnMn(CO)_5$	CH_2Cl_2	−2660	2873	83
$Br_3SnMn(CO)_5$	CH_2Cl_2	−2044	310	83

Continued overleaf

TABLE 8.7 (*Continued*)

Compound	Solvent	$\delta(^{55}Mn)$/ppm[a]	$W_{1/2}$/Hz	Ref.
$Br_2MeSnMn(CO)_5$	CH_2Cl_2	−2256	456	83
$BrMe_2SnMn(CO)_5$	CH_2Cl_2	−2485	1819	83
$Me_3SnMn(CO)_5$	CH_2Cl_2	−2660	2873	83
$(C_6F_5)_3SnMn(CO)_5$	THF	−2280	600	81
$(C_6F_5)_2PhSnMn(CO)_5$	THF	−2380	1400	81
$(C_6F_5)Ph_2SnMn(CO)_5$	THF	−2470	2000	81
$Cl_3SiMn(CO)_5$	Acetone	−2200		81
$Cl_3GeMn(CO)_5$	Acetone	−1870		81
$Cl_3SnMn(CO)_5$	Acetone	−1910		81
$(C_6F_5)_3SiMn(CO)_5$	Acetone	−2290		81
$(C_6F_5)_3GeMn(CO)_5$	Acetone	−2140		81
$(C_6F_5)_3SnMn(CO)_5$	Acetone	−2260		81
$Ph_3SiMn(CO)_5$	Acetone	−2470		81
$Ph_3GeMn(CO)_5$	Acetone	−2430		81
$Ph_3SnMn(CO)_5$	Acetone	−2510		81
$Ph_3PbMn(CO)_5$	Acetone	−2350		81
$HMn(CO)_5$	neat	−2578	4147	94
$HMn(CO)_4PF_3$	neat	−2673	4275	94
$HMn(CO)_3(PF_3)_2$	neat	−2742	4020	94
$HMn(CO)_2(PF_3)_3$	neat	−2813	4275	94
$HMn(CO)(PF_3)_4$	neat	−2888	5820	94
$HMn(PF_3)_5$	neat	−2953	10 785	94
$Ph_3SnMn(CO)_5$	DMF	−2505		81
	Benzene	−2510		81

[a] PPM relative to external aqueous K MnO_4 and using the δ scale where positive values represent higher frequencies than the reference.

The total variation in manganese shielding covers a range of 3000 ppm. This range is 50% greater than that for ^{51}V where similar variety in the compounds studied has been achieved. The pattern of increased shielding with decrease in oxidation state is maintained, with Mn^{VII} at $\delta 0$, Mn^{I} in the region δ −1000 to δ −1500, and Mn^{-I} in the region δ −1700 to δ −3000.

Since the nuclear shieldings of some transition metals ($^{47,49}Ti$, ^{93}Nb) show an "inverse" halogen dependence, it is worth noting that in the $XMn(CO)_5$ compounds, the shielding of the metal increases in the order $Cl < Br < I$. This dependence is the same as that for compounds containing halogens bonded to ^{11}B, ^{27}Al, ^{69}Ga. ^{115}In, ^{13}C, ^{29}Si, ^{73}Ge and ^{119}Sn, and can be regarded as the normal behaviour pattern. A theoretical explanation has been provided[82] for normal behaviour in terms of the nephelauxetic effect exerted by the halogen upon the orbitals of the central atom and the resultant variation in $\langle r^{-3} \rangle$.

There is general agreement[81,83,84] that for the extensive $LMn(CO)_5$ series of compounds, the ^{55}Mn chemical shift provides a reliable measure of the polarity of the L-Mn bond and hence a measure of the σ-donor properties of the ligand L. The compounds studied span the polarity range from Cl^{-}-$^{+}Mn(CO)_5$ (δ − 1004) through the non-polar $(CO)_5$ Mn-$Mn(CO)_5$ (δ − 2325) to Me_3Sn^{+}-$^{-}Mn(CO)_5$ (δ − 2660) and as the character of the ligand changes in such a way as to increase the electron density on Mn a steady increase in the NMR shielding of the manganese is observed. Two general correlations have been drawn, both of which support this picture. The ^{55}Mn chemical shifts for twelve $R_{3-n}X_nSn\text{-}Mn(CO)_5$ compounds have been measured and shown to correlate with the ^{119}Sn isomer shifts obtained from the Mossbauer spectra for the same compounds.[83] As the Mn δ-value becomes more negative, the isomer shift for Sn decreases, indicating a depletion of the $s\sigma$-electron density on Sn consistent with the polarity change revealed by the ^{55}Mn chemical shift. The other correlation which bears upon the L-Mn bond polarity involves Graham's σ-parameter.[85] This parameter based upon carbonyl stretching force constants in $LMn(CO)_5$ compounds provides a measure of the σ-donor/σ-acceptor properties of L. A comparison of the ^{55}Mn σ-values with the σ-parameters for 21 compounds[81,84] shows a linear correlation in which the strongest σ-donor ligands having the most negative σ-parameters give the most highly shielded manganese atoms. This correlation further supports the view that manganese shielding constants measured by ^{55}Mn NMR reflect the total electron density on manganese and the extent to which the polarity of the M-L bond affects this electron density. The magnitude of the shielding change, which spans a range of 1600 ppm for these compounds, indicates its origin in the paramagnetic term of the Ramsey screening equation. The fact that simple changes in electron density, which could be accommodated in the diamagnetic term, are generating such large shifts is due to the C_{4v} symmetry of all the examples under consideration. Changes in bond polarity along the z-axis cause perturbations of the manganese $3d$-electrons which introduce changes in the orbital angular momentum factor in the paramagnetic term.

b. ^{55}Mn *Linewidths*

The ^{55}Mn quadrupole moment of 0.55×10^{-28} cm^2 is the largest of the first row transition metals and this fact is reflected in the linewidths of the NMR signals observed. Because its tetrahedral symmetry generates a near-zero electric field gradient at manganese, the permanganate ion exhibits an extremely sharp line of less than 10 Hz in width,[80] a value which probably reflects magnetic field inhomogeneity rather than intrinsic linewidth. In spite of significant quadrupole moments possessed by both nuclear species, spin-spin coupling between ^{55}Mn and ^{17}O has been partially resolved[86] in the ^{17}O spectrum of $[MnO_4]^-$ and shows a relaxation time T_1 for ^{55}Mn of 0.84 s, a value which converts to a linewidth of 3.5 Hz. The axial symmetry for $Mn_2(CO)_{10}$ also gives rise to a fairly narrow line 48 Hz in width.[79] While a few others lie in the range 100–1000 Hz, most ^{55}Mn lines exceed 1000 Hz in width and range up to 10 657 Hz for $Mn_2(CO)_8(PPh_3)_2$ in solution. This large variation offers the prospect of determining electric field gradients from linewidths, provided the rotational correlation time for the molecule in solution is either known or remains constant. T. L. Brown's group at the University of Illinois has shown[87] that manganese linewidths determined in solution will correlate well with χ^2 determined by NQR spectroscopy done on the solid compounds. As the linewidth and hence the relaxation time are related to the quadrupole coupling constant by the Eq. 1.17 the straight line relationship between $\Delta\nu$ and χ^2 indicates that τ_c is constant and can be evaluated from the slope. Once this slope has been ascertained, the linear relationship can be used to obtain quadrupole coupling constants from NMR linewidths in cases where the NQR data are not available.

There are a number of nuclei for which a structurally related series of compounds shows a correlation between the width and the chemical shift of the same NMR line. One example of this phenomenon is the ^{35}Cl resonances in the chloromethanes,[88] and another is the ^{55}Mn resonances in $R_3Sn\text{-}Mn(CO)_5$ compounds.[81,84] The origin of the correlation lies in the $\langle r^{-3} \rangle$ factor for the atom under investigation, a factor which is common to both the quadrupole coupling constant expression[89]

$$e^2qQ = e^2Q\,(\text{constant})\langle r^{-3} \rangle R^1 \tag{8.1}$$

and the paramagnetic shielding constant expression[90]

$$\sigma_p = -\frac{\text{constant}}{\Delta E}\langle r^{-3} \rangle \tag{8.2}$$

In the ^{55}Mn case it has been possible, using the *linewidth* vs. *chemical shift* correlation, to obtain approximate values for the manganese quadrupole coupling constant from the manganese chemical shift, and show that in the $LM(CO)_5$ series of compounds, the quadrupole coupling constant changes sign from being positive for weak σ-donors such as Cl, to being negative for strong σ-donors such as Ph_3Sn, with the crossover point occurring at a chemical shift value of about δ -2150 ± 200. It is significant that, within the accuracy limits of this analysis, the quadrupole coupling constant crossover point coincides with the $LMn(CO)_5$ example in which the L-Mn bond is not polarised in either direction [$(CO)_5Mn$-$Mn(CO)_5$ for which δ -2325] and also coincides with the H_3C-$Mn(CO)_5$ example having δ -2265 which Graham selected as the "neutral" reference for ligand σ-donor/acceptor properties, assigning the CH_3 ligand a σ-parameter of zero.

The widths of manganese-55 NMR lines have been used[91] to study the kinetics of the electron transfer reaction $[(RNC)_6Mn]^{2+} + \varepsilon \rightarrow [(RNC)_6Mn]^{+}$. Since the Mn^{II} complex exists as a diamagnetic dimer in equilibrium with the paramagnetic Mn^{I} complex, the electron transfer reaction manifests itself as broadening of the ^{55}Mn resonance of the diamagnetic complex, for which minimum linewidths of about 70 Hz were observed. Rate constants dependent upon the nature of R and lying in the range 10^4 to 10^5 l mol^{-1} s^{-1} were evaluated using the line broadening technique.

c. ^{55}Mn *Coupling Constants*

No cases of ^{55}Mn spectra in which coupling to other nuclei is exhibited have been reported. Broze and Luz,[86] however, have obtained a ^{17}O spectrum for $[MnO_4]^-$ enriched in ^{17}O in which partial splitting into a broad sextet is observed. The coupling constant obtained is J (^{55}Mn, ^{17}O) $= 30 \pm 3$ Hz. Lutz and coworkers[33] obtain a comparable value of 28.9 ± 2.8 Hz using the same method.

8F.2 Technetium, ^{99}Tc

Although technetium occupies a central position among the transition metals in the Periodic Table, no stable isotope of the element exists. Technetium-99 is a 0.29 Mev β^- emitter with a half-life of 5×10^5 years, which is made by neutron bombardment of molybdenum. It is the most stable isotope of the element, and the one used in studying technetium chemistry.

Once fabricated, ^{99}Tc comprises 100% of the available technetium and has a spin $I = \frac{9}{2}$, a high detection receptivity relative to natural abundance ^{13}C of 2134, and a quadrupole moment of 0.3×10^{-28} cm^2.

A number of years ago, the ^{17}O resonance in an aqueous solution of $NaTcO_4$ was reported.[95] A strong ^{99}Tc resonance was also observed[96] at 6.000 MHz for a 0.63 M aqueous solution of this $[TcO_4]^-$, the half-height linewidth for which was 29 Hz measured under high-resolution conditions. This represents a peak-to-peak linewidth for the absorption derivative signal of 16.8 Hz. This linewidth can be compared with that of 8280 Hz for ^{185}Re in $[ReO_4]^-$ to provide confirmation that the quadrupole moment value of 0.3 barns for ^{99}Tc is the right order of magnitude. The linewidth ratio $\Delta\nu(Re)/\Delta\nu(Tc)$ is 493. Then, by the linewidth Eq. 1.17 if the electric field gradient is assumed to be the same at the two metals by virtue of their identical symmetries and iso-electronic valence shells, the linewidth ratio should be the ratio of the spin factors times the ratio Q^2 values times the ratio of τ values. The spin factor ratio Re/Tc is 4.3 and the Q^2 ratio Re/Tc is 87 for a product ratio of 374 if the correlation time ratio is unity. Since τ is proportional to the cube of the molecular radius, the τ ratio Re/Tc will be slightly greater than unity, bringing the calculated linewidth ratio into good agreement with that observed. Due to the Q^2 factor in the equation, the linewidth is very sensitive to variations in Q and this agreement confirms the quadrupole moment of 0.3×10^{-24} cm^2 for technetium-99.

8F.3 Rhenium, $^{185,187}Re$

Rhenium has two magnetically active isotopes, both with spin $I = \frac{5}{2}$, which together comprise all of the naturally occurring rhenium. Rhenium-185 comprises 37.07% of the element and has a natural abundance detection sensitivity relative to ^{13}C of 280. Rhenium-187 comprises 62.93% of the element and has a receptivity, $D^C = 489$. Both isotopes have large quadrupole moments, with the more detectable isotope, ^{187}Re, having the smaller Q at 2.6×10^{-28} cm^2 and ^{185}Re having $Q = 2.8 \times 10^{-28}$ cm^2. Since linewidths are proportional to Q^2, nuclei with large quadrupole moments will have extremely broad lines except in environments where the electric field gradient is vanishingly small. The only nucleus having a comparable or larger quadrupole moment than rhenium to yield a solution phase NMR signal is ^{181}Ta in $[TaF_6]^-$, and there the linewidth is 35.8 kHz.

The ^{185}Re and ^{187}Re resonances in a 0.8 M aqueous solution of $NaReO_4$ have both been observed,[97] and their linewidths are 8280 Hz ($T_2 = 22$ μs) and 7820 Hz ($T_2 = 23.5$ μs) respectively. The fact that the linewidths are proportional to Q^2 for the respective nuclei is consistent with a quadrupolar relaxation mechanism. The possibility that chemical exchange of oxygen between solvent and $[ReO_4]^-$ might contribute to relaxation of rhenium has been ruled out by the fact that the linewidths *narrow* with increasing temperature. Where quadrupole relaxation is the dominant mechanism, the linewidth is given by Eq. 1.17. The temperature dependence of the rhenium linewidths is clearly attributable to variation in the rotational correlation time τ which will decrease with increasing temperature. A comparison of the ^{185}Re linewidth (8280 Hz) with that of ^{181}Ta (35,800 Hz) gives a ratio $\Delta\nu(Ta)/\Delta\nu(Re) = 4.3$ which, on an order of magnitude basis, is consistent with the square of the quadrupole moment ratio $Q^2(Ta)/Q^2(Re) = 5.4$. While the spin-related coefficient in equation 1.17 for ^{185}Re ($I = \frac{5}{2}$) is larger at 0.024 than that for ^{181}Ta($I = \frac{7}{2}$) at 0.010, the size-related τ for $[TaF_6]^-$ will be larger than that for $[ReO_4]^-$ and these two differences will tend to cancel one another. We are left with the fact that the linewidths are roughly proportional to Q^2 and therefore *the electric field gradients at the two metals are roughly the same.*

This fact raises some fundamental questions about the structures of these ions in solution. Since both have cubic symmetry, we expect the electric field gradient in each case to be zero. The presence of non-vanishing field gradients in ions of this type will shed new light on the subject of ion-ion and ion-solvent interactions in solution.

8G GROUP VIIIa

Group VIIIa elements present a problem to the NMR spectroscopist. Direct observation of these nuclei is very difficult. Although both ^{57}Fe and ^{187}Os have $I = \frac{1}{2}$ they both have very low receptivities. Although ^{99}Ru, ^{101}Ru, and ^{189}Os have receptivities comparable with ^{13}C, they also possess quadrupole moments making observation difficult. As a consequence very little use has been made of direct observation but by use of ^{57}Fe enrichment, INDOR has proved to be a useful technique for ^{57}Fe investigation.

8G.1 Iron, ^{57}Fe

Iron-57 is present to the extent of 2.2% in all samples and has a natural abundance detection sensitivity relative to ^{13}C of 0.004. Its NMR detection is further hindered by a nuclear spin of $I = \frac{1}{2}$. Lacking a quadrupolar mechanism for relaxation, the upper spin state for ^{57}Fe will tend to saturate under the RF power levels necessary for its observation. These two factors taken together make iron an unlikely element for NMR study.

By using a pulse technique involving four different irradiation frequencies and hence known as Quadruple Fourier Transform, Schwenk[68] has obtained from $Fe(CO)_5$ a ^{57}Fe resonance having signal-to-noise ratio of greater than 100. An observation time of 20 hours and 6.5×10^6 pulses were expended on this achievement. These experimental conditions will deter all but the most persistent and persuade most chemists to seek alternative avenues to structural information about iron. Even if the necessary instrumentation were readily available, the number of diamagnetic iron compounds is strictly limited, and methods for which paramagnetism is not a drawback will prove more fruitful for characterisation.

Coupling of ^{57}Fe to ^{13}C and to ^{31}P has been observed in iron carbonyl and in iron phosphine carbonyls. Mann[98] has reported $J(^{57}\text{Fe}, ^{31}\text{P})$ values of 25.9 Hz, 26.5 Hz, and 27.4 Hz for $Fe(CO)_4PEt_3$, $Fe(CO)_4PEt_2Ph$, and $Fe(CO)_4PEtPh_2$ respectively, and $J(^{57}\text{Fe}, ^{13}\text{C}) = 23.4$ for $Fe(CO)_5$. The coupling constant with ^{13}C is significant in that it yields the series of reduced coupling constants K (M, ^{13}C) = 1.46×10^{20} cm^{-3}, 238×10^{20} cm^{-3}, and 402×10^{20} cm^{-3} for $[V(CO)_6]^-$, $Fe(CO)_5$, and $[Co(CO)_4]^-$ respectively, as is expected from the Fermi contact equation.

By using samples enriched to 82% in ^{57}Fe and a triple resonance technique involving observation of ^{13}C with successive decoupling of ^{1}H and then ^{57}Fe, Koridize *et al.*[99] have obtained ^{57}Fe parameters for a number of substituted ferrocenes. Relative to a solution of ferrocene (Fc) in CS_2 at $\delta 0$, ^{57}Fe shieldings in substituted ferrocenes are: Fc-H$^+$ $\delta = -1098.85$, Fc-C(O)Me $\delta 215.55$, Fc-CH(OH)Me $\delta 0$, Fc-CH_2Me $\delta 36.6$. Coupling between iron and carbons in both the substituted and unsubstituted cyclopentadienyl rings have been measured and lie in the range 4.73–5.25 Hz. No coupling to exocyclic carbons was observed.

8G.2 Osmium, $^{187, 189}$Os

Osmium has received scant attention from NMR spectroscopists. Although OsO_4 is the only compound to have been studied in the liquid phase, the osmium relaxation times obtained are of some interest.

Osmium-187 has spin $I = \frac{1}{2}$, is 1.64% abundant, and lacking a quadrupole moment, its relaxation results from a magnetic dipole-dipole mechanism. Osmium-189 has spin $I = \frac{3}{2}$, is 16.1% abundant, has a natural detection sensitivity relative to ^{13}C of 2, and has a quadrupole moment of 0.8 barns. In the course of making high precision measurements of the magnetic moments for nuclei which yield weak NMR signals, Schwenk in Tubingen has obtained resonances for both osmium isotopes in molten OsO_4.

Both the transverse and the longitudinal relaxation times for ^{187}Os with $I = \frac{1}{2}$ have been determined[100] using a pulse technique which yields the value ranges

$$\begin{aligned} 1\text{ s} &\leq T_1 \leq 26\text{ s} \\ 0.7\text{ s} &\leq T_2 \leq 12\text{ s} \\ 1.5 &\leq T_1/T_2 \leq 2.2 \end{aligned} \tag{8.3}$$

For ^{189}Os which has a quadrupole moment and hence a potential quadrupolar relaxation mechanism if an electric field gradient is present, the transverse relaxation time in molten OsO_4 at 70°C evaluated both from the pulse decay pattern and from the linewidth is very much shorter at $T_2 = 230 \pm 10\ \mu s$.[101] This difference in relaxation time by a factor of 10^4 indicates the presence of a quadrupolar relaxation mechanism which is absent for ^{187}Os, and hence the presence of a residual electric field gradient at the osmium nucleus in spite of the tetrahedral shape of the osmium tetroxide molecule. The unexpected presence of this electric field gradient is consistent with similar results obtained from other tetrahedral MO_4 species and discussed in the section of this chapter dealing with the appropriate metal.

8H GROUP VIIIb

Cobalt-59 has received considerable attention from the early days of NMR spectroscopy and is now a very useful.method of investigating diamagnetic cobalt compounds. Rhodium-103, although it has $I = \frac{1}{2}$ and is 100% abundant has proved to be very insensitive for direct observation and most results have been obtained by use of INDOR. In contrast, the observation of iridium has proven to be extremely difficult. There are two possible nuclei, ^{191}Ir and ^{193}Ir to observe. Both isotopes have $I = \frac{3}{2}$, quadrupole moments 1.5×10^{-28} cm^2 and $D^C = 0.054$ and 0.12 respectively. From these properties coupled with the low observation frequencies of 1.72 MHz and 1.87 MHz, respectively, at 2.35 T, it would appear that $^{191,193}Ir$ NMR spectroscopy is feasible for only very symmetric complexes. Thus by use of Eq. 1.17 assuming that $[TaF_6]^-$ and $[IrF_6]^{3-}$ have similar correlation times and field gradients, then $[IrF_6]^{3-}$ will have a $^{191,193}Ir$ linewidth of ca. 20 kHz. It is therefore unlikely that iridium NMR spectroscopy will receive much attention from chemists.

8H.1 Cobalt, ^{59}Co

Cobalt-59 occupies a unique place in nuclear magnetic resonance history. Since a radiofrequency is one of the most accurately measurable physical quantities known, the discovery of the NMR phenomenon in 1945 provided physicists with a highly accurate tool for the determination of nuclear magnetogyric ratios. Several years of research activity produced for each of the common nuclei a γ-value having an absolute accuracy of better than one part in 10^6. It was with considerable consternation that Proctor and Yu[102] discovered in 1951 while measuring the magnetogyric ratio of cobalt-59 using a number of different cobalt compounds, that the resonance frequency for cobalt could vary by as much as one part in 10^2 depending upon the compound selected for measurement. Unless the factor giving rise to the compound specificity could be evaluated, this chemical effect placed a limit of 1% on the accuracy with which the cobalt-59 magnetogyric ratio could be determined. The term "chemical shift" was the most disparaging designation the physicists could find to label an effect which was vitiating the last four decimal places in their highly accurate magnetogyric ratio determinations. The development of nuclear magnetic resonance in the succeeding 25 years has converted the physicists annoyance into the most versatile tool for structure determination available to the contemporary chemist.

Under ordinary chemical conditions, cobalt is most commonly encountered in the oxidation state Co^{II}. These compounds are paramagnetic and are therefore beyond the scope of NMR investigation. When the cobalt atom is coordinated by six donor ligands, however, cobalt(III) becomes relatively more stable, and it is these low-spin d^6 complexes of cobalt(III) that have provided most of the samples for ^{59}Co NMR investigation. The smaller group of cobalt compounds which has also been investigated contains the d^{10} electron configuration with the lower oxidation state cobalt(O) and cobalt(−I) carbonyl compounds as examples.

a. General Features of Cobalt NMR

Cobalt-59 is 100% abundant in nature and has a natural abundance detection sensitivity relative to ^{13}C of 1572, putting it among the top six nuclear species for ease of detection. It has nuclear spin $I = \frac{7}{2}$ and a quadrupole moment of 0.4×10^{-28} cm^2. A quadrupole moment of this intermediate magnitude makes ^{59}Co linewidths sensitive to electric field gradients at cobalt and hence to the symmetry about the cobalt atom. The structure for most cobalt compounds is based on an octahedral array of six ligands surrounding the cobalt, and this results in lower electric field gradients than would be the case for structures having lower symmetry. Departures from strict octahedral symmetry caused by ligand variation within the six-fold array cause line broadening which is useful for detecting subtle variations in structure.

There have been about twenty studies conducted in which the cobalt-59 NMR spectra for groups of related compounds have been obtained and their NMR parameters compared. An additional fifteen studies in which kinetic, structural, or equilibrium information about particular compounds has been obtained from cobalt spectra have been reported. Our present knowledge about cobalt-59 NMR rests on these works.

Chemical shifts for the cobalt nucleus in different chemical environments span a range of 18 000 ppm or 1.8% with $[Co(CO_3)_3]^{3-}$ showing the highest resonance frequency yet observed for cobalt and $[Co(PF_3)_4]^-$ exhibiting the lowest. The factors which determine cobalt shielding within this chemical shift range are discussed below in the section dealing with chemical shifts.

Cobalt linewidths in excess of 30 kHz have been observed in some solution spectra. Linewidths for other compounds cover the complete range between this high value and the narrow resonance for $[Co(CN)_6]^{3-}$ in the region of 40 Hz. A number of studies discussed below in the section on cobalt linewidths deal with the relationships between linewidth, electric field gradient, and molecular structure.

In contrast with other nuclei which have received the same degree of study, the story on ^{59}Co coupling constants is almost nonexistent; but not quite. Couplings to ^{13}C, ^{19}F, and ^{31}P have been reported and are discussed below.

b. ^{59}Co Chemical Shifts

(*i*) *Reference Compounds.* A number of different reference compounds have been used for reporting ^{59}Co shielding data. While $[Co(CN)_6]^{3-}$ is the reference which has been adopted by the greatest number of workers in this field, $[Co(en)_3]^{3+}$, $[Co(NH_3)_6]^{3+}$, and $Co(acac)_3$ have all been employed by various investigators either as primary or as secondary references. It has been found convenient in the reporting of ^{1}H and ^{13}C shielding data to select a reference whose NMR signal lies near one end of the chemical shift range experienced by that nucleus. By using this factor as a criterion in selecting a reference, one achieves a tabulation of δ-values most of which have the same sign. In addition to satisfying this criterion, $[Co(CN)_6]^{3-}$ is reasonably soluble in water and gives one of the narrowest ^{59}Co NMR signals.

In the tabulations of ^{59}Co data which follow, those chemical shifts which are reported relative to a different reference have been converted to the $[Co(CN)_6]^{3-}$ scale using the following conversion factors:

$$\delta_{[Co(CN)_6]^{3-}} = \delta_{[Co(en)_3]^{3+}} + 7120 \quad \text{(ref. 103)} \tag{8.4}$$

$$\delta_{[Co(CN)_6]^{3-}} = \delta_{Co(acac)_3} + 12\,500 \quad \text{(ref. 104)} \tag{8.5}$$

$$\delta_{[Co(CN)_6]^{3-}} = \delta_{[Co(NH_3)_6]^{3+}} + 8150 \quad \text{(ref. 104)} \tag{8.6}$$

Use of the bare ^{59}Co nucleus as an "absolute" shielding reference has been eschewed since our knowledge of this resonance position depends either upon a long extrapolation of a σ vs. ΔE^{-1} plot, or else upon a theoretically calculated value of the paramagnetic shielding term in a particular compound.

Due to the ever-present possibility of ligand exchange when dealing with cobalt complexes, the use of $[Co(CN)_6]^{3-}$ as an external reference physically separated from the solution under study, rather than as an internal reference added directly to the test solution, has been adopted. This has meant that, in theory, a correction factor proportional to the magnetic susceptibility difference between reference and sample should be applied to all δ-values. In practice, it is found that this correction factor is less than the experimental uncertainty in the measurement and it is therefore ignored. In a study where $[Co(CN)_6]^{3-}$, $[Co(en)_3]^{3+}$, and $Co(acac)_3$ were used as internal, then as external, references, chemical shift differences of less than 5 ppm were observed.[105]

(*ii*) *Error Limits on Chemical Shifts.* Although pulsed Fourier transform instrumentation offers some sensitivity and precision enhancement, virtually all cobalt-59

TABLE 8.8 Cobalt chemical shifts in $Co^{III}C_5X$-type complexes[103]

Complex	Solvent	$\delta(^{59}Co)$/ppm[a]	$W_{1/2}$/Hz
$[Co(CN)_5(OH)]^{3-}$	0.4 M, H_2O	1840 ± 60	5250 ± 175
$[Co(CN)_5(H_2O)]^{2-}$	0.4 M, H_2O	1822 ± 60	6120 ± 175
$[Co(CN)_5(NO_2)]^{3-}$	H_2O	1400[b]	17 500[b]
$[Co(CN)_5Br]^{3-}$	0.4 M, H_2O	1220 ± 60	11 370 ± 350
$[Co(CN)_5I]^{3-}$	0.4 M, H_2O	780 ± 60	9970 ± 350
$[Co(CN)_6]^{3-}$		0	175

[a] PPM relative to external saturated aqueous $K_3Co(CN)_6$ and using the δ scale where positive values represent higher resonance frequencies than the reference. [b] Ref. 104.

TABLE 8.9 Cobalt chemical shifts in $Co^{III}N_6$-type complexes

Complex	Solvent	$\delta(^{59}Co)$/ppm[a]	$W_{1/2}$/Hz	$\Delta E^{b}/cm^{-1}$	Ref.
fac-$Co(NH_3)_3(N_3)_3$	DMSO	9970	297		105
mer-$Co(NH_3)_3(N_3)_3$	DMSO	9900	1242		105
cis-$[Co(NH_3)_4(N_3)_2]^+$	H_2O	9400	525		104
cis-$[Co(NH_3)_4(N_3)_2]^+$		9260	367		105
trans-$Co(NH_3)_4(N_3)_2^+$	H_2O	9170	875		104
trans-$Co(NH_3)_4(N_3)_2^+$		9220	527		105
$[Co(NH_3)_5(N_3)]^{2+}$	H_2O	9100	525		104
$[Co(NH_3)_5(N_3)]^{2+}$		8680	350		105
$[Co(en)_2(NH_3)_2(N_3)_2]^+$		8820	1050		105
cis-$[Co(en)_2(N_3)_2]^+$		8400	875		105
fac-$[Co(en)(NH_3)_3(N_3)]^{2+}$		8240	647		105
$[Co(NH_3)_6]^{3+}$		8300			102
$[Co(NH_3)_6]^{3+}$		8218		21 050	108
$[Co(NH_3)_6]^{3+}$		8080		21 000	121
$[Co(NH_3)_6]^{3+}$	H_2O	8100	350	21 000	123
$[Co(NH_3)_6]^{3+}$		8100			104
$[Co(NH_3)_6]^{3+}$		8175		21 100	130
$[Co(NH_3)_6]^{3+}$		8100		21 050	124
$[Co(NH_3)_6]^{3+}$		8150	87		105
$[Co(en)_2(NH_3)(N_3)]^{2+}$		7895	1225		105
$[Co(en)(NH_3)_4]^{3+}$		7790	402		105
$[Co(NH_3)_5NO_2]^{2+}$		7460		21 840	121
$[Co(NH_3)_5NO_2]^{2+}$		7500		21 050	
$[Co(NH_3)_5NO_2]^{2+}$		7490		20 750	122
$[Co(NH_3)_5NO_2]^{2+}$	H_2O	7440			104
$[Co(NH_3)_5NO_2]^{2+}$	H_2O	7630 ± 5	175		103
$[Co(NH_3)_5NO_2]^{2+}$		7420[c]		21 830	125
$[Co(NH_3)_5NO_2]^{2+}$		7625	262		105
cis-$[Co(en)_2(NH_3)_2]^{3+}$	H_2O	7300	350		104
		7440	472		105
trans-$[Co(en)_2(NH_3)_2]^{3+}$	H_2O	7510	700		104
		7390	752		105
$[Co(NO_2)_6]^{3-}$		{7400 {8100			102
$[Co(NO_2)_6]^{3-}$		{7490 {8060 (weak)		20 830	108
$[Co(NO_2)_6]^{3-}$		7350		20 670	121
$[Co(NO_2)_6]^{3-}$		7440		20 700	130
$[Co(NO_2)_6]^{3-}$	H_2O	{7440 {8160 (weak)	175 700		104
fac-$[Co(en)(NH_3)_3NO_2]^{2+}$		7255	297		105
mer-$[Co(en)(NH_3)_3NO_2]^{2+}$		7235	490		105
mer-$Co(NH_3)_3(NO_2)_3$		7200	875		104
cis-$[Co(NH_3)_4(NO_2)_2]^+$		7290		22 222	108
cis-$[Co(NH_3)_4(NO_2)_2]^+$		6880		22 510	121
cis-$[Co(NH_3)_4(NO_2)_2]^+$	H_2O	7080	875		104
cis-$[Co(NH_3)_4(NO_2)_2]^+$		7255	262		105
trans-$[Co(NH_3)_4(NO_2)_2]^+$		7199		22 470	108
trans-$[Co(NH_3)_4(NO_2)_2]^+$		7150		22 630	121

TABLE 8.9 (*continued*)

Complex	Solvent	$\delta(^{59}Co)$/ppm[a]	$W_{1/2}$/Hz	$\Delta E^b/cm^{-1}$	Ref.
trans-$[Co(NH_3)_4(NO_2)_2]^+$		7080	875		104
trans-$[Co(NH_3)_4(NO_2)_2]^+$		7190	490		105
$[Co(pn)_3]^{3+}$		7220		21 280	108
Λ-$[Co(pn)_3]^{3+}$		7100	175	21 370	134a
Δ-$[Co(pn)_3]^{3+}$		7170	525	21 370	134a
$[Co(en)_3]^{3+}$		7300			102
$[Co(en)_3]^{3+}$		7177		21 280	108
$[Co(en)_3]^{3+}$		7010		21 400	121
$[Co(en)_3]^{3+}$	H_2O		210		104
$[Co(en)_3]^{3+}$		7120	1750		103
$[Co(en)_3]^{3+}$	0.3 M, H_2O	7010			106
$[Co(en)_3]^{3+}$	0.1 M, H_2O	6990			106
$[Co(en)_3]^{3+}$		7150		21 300	130
$[Co(en)_3]^{3+}$		7120			105
$[Co(en)_3]^{3+}$	deprotonated H_2O; ph >10.	6980			134b
$[Co(phen)_3]^{3+}$		7080	1400		104
$Co(NH_3)_3(NO_2)_3$		6940		23 310	121
$[Co(NH_3)_2(NO_2)_4]^-$	H_2O	6950	1750	23 470	108
$[Co(NH_3)_2(NO_2)_4]^-$		6860			104
$[Co(en)(NH_3)_2(NO_2)_2]^+$		6890	595		105
$[Co(en)(NH_3)_2(NO_2)_2]^+$		6875	875		105
cis-$[Co(en)_2(NH_3)_2(NO_2)]^{2+}$		6885	507		105
$[Co(dipy)_3]^{3+}$		6620		22 230	121
$[Co(NH_2OH)_6]^{3+}$		6500	350		104
cis-$[Co(en)_2(NO_2)_2]^+$		6470		23 000	121
$[Co(en)_2(NO_2)_2]^+$		6450	1225		105
trans-$[Co(en)_2(NO_2)_2]^+$		6350		23 300	121
trans-$[Co(en)_2(NO_2)_2]^+$		6260	1925		105
$[Co(bmip)_2(NH_3)_2]^+$	H_2O	5510 ± 50			135
$[Co(dmgH)_2(NH_3)_2]^+$	H_2O	5120 ± 50			135

[a] PPM relative to external aqueous $K_3Co(CN)_6$ and using the δ scale where positive values represent higher resonance frequencies than the reference. [b] Lowest energy electronic absorption band. [c] Resonance line assigned by authors to $[Co(CN)_5NO_2]^{3+}$. Both δ and ΔE values favour this assignment.

spectra obtained to date have been secured using a continuous wave spectrometer. Spectral calibration is accomplished in these cases either by interpolation between two signals of known δ-value, or through the generation of audio-frequency sideband signals which provide calibration points in the spectrum. For the narrower cobalt lines having widths in the region less than 500 Hz relative line positions can be determined within $\pm 0.5\%$. By choosing a secondary reference substance lying within 2000 ppm of the sample, the accuracy of the resulting shielding data should be within ± 10 ppm. Since the centre of a spectral line is only locateable to within $\pm 5\%$ of its width, the broader cobalt signals which can be as wide as 10–20 kHz even for solution spectra will have error limits of ± 50 to ± 100 ppm on the shielding data derived therefrom.

TABLE 8.10 Cobalt chemical shifts in $Co^{III}N_5X$-type complexes

Complex	Solvent	$\delta(^{59}Co)$/ppm[a]	$W_{1/2}$/Hz	$\Delta E^b/cm^{-1}$	Ref.
$[Co(NH_3)_5F]^{2+}$		8710		19 610	122
$[Co(NH_3)_5F]^{2+}$	H_2O	9520 ± 5	1750		103
$[Co(NH_3)_5OH_2]^{3+}$		8310		20 410	122
$[Co(NH_3)_5OH_2]^{3+}$		8300		20 400	123
$[Co(NH_3)_5OH_2]^{3+}$	H_2O	9170	5248		104
$[Co(NH_3)_5OH_2]^{3+}$	H_2O	9060 ± 5	2974		103
$[Co(NH_3)_5OH_2]^{3+}$	H_2O/H^+		4898		136
$[Co(NH_3)_5OH_2]^{3+}$		9080	4898		105
$[Co(NH_3)_5OH]^{2+}$	H_2O	9117 ± 5	1924		103
$[Co(NH_3)_5OH]^{2+}$	H_2O/OH^-		315		136
$[Co(NH_3)_5OH]^{2+}$		9140	402		105
$[(NH_3)_4Co(NH_2)(OH)Co(NH_3)_4]^{4+}$		9388	8834		137
$[Co(NH_3)_5SO_4H]^{2+}$	H_2O	8720		18 870	122
$[Co(NH_3)_5SO_4]^{+}$		8750		19 050	122
$[Co(NH_3)_5SCN]^{2+}$		8760		19 610	122
$[Co(NH_3)_5CH_3COO]^{2+}$		8910		19 920	122
$[Co(NH_3)_5NO_3]^{2+}$		8930		19 610	122
$[Co(NH_3)_5C_2O_4]^{+}$		9130		19 800	124
$[Co(NH_3)_5CO_3H]^{2+}$		9000		19 680	122
$[Co(NH_3)_5NCS]^{2+}$		8200		19 840	122
$[Co(NH_3)_5(CO_3)]^{+}$		9000		19 670	121
$[Co(NH_3)_5(CO_3)]^{+}$		9000		20 410	126
$[Co(NH_3)_5Cl]^{2+}$		9070		18 720	121
$[Co(NH_3)_5Cl]^{2+}$		8750		18 870	122
$[Co(NH_3)_5Cl]^{2+}$	H_2O	8850 ± 5	1750		103
$[Co(NH_3)_5Cl]^{2+}$		8850	1750		105
$[Co(NH_3)_5Br]^{2+}$	H_2O	8710		18 350	122
$[Co(NH_3)_5Br]^{2+}$	H_2O	8820 ± 5	1750		103
$[Co(NH_3)_5Br]^{2+}$		8820	1750		105
$[Co(NH_3)_5I]^{2+}$	H_2O	8940		17 240	122
$[Co(NH_3)_5I]^{2+}$	H_2O	8760 ± 5	2100		103
mer-$[Co(en)(NH_3)_3(OH)]^{2+}$		8785	750		105
fac-$[Co(en)(NH_3)_3(OH)]^{2+}$		8710	315		105
mer-$[Co(en)(NH_3)_3(H_2O)]^{3+}$		8700	14 000		105
fac-$[Co(en)(NH_3)_3(H_2O)]^{3+}$		8700	8750		105
$[Co(NH_3)_5(N_3)]^{2+}$		8680	350		105
$[Co(en)(NH_3)_3Cl]^{2+}$		8420	4375		105
$[Co(en)(NH_3)_3Br]^{2+}$		8390	4375		105
cis-$[Co(en)_2(NH_3)(OH)]^{2+}$		8370	507		105
cis-$[Co(en)_2(NH_3)(H_2O)]^{3+}$		8350	12 250		105
trans-$[Co(en)_2(NH_3)(OH)]^{2+}$		8260	315		105
trans-$[Co(en)_2(NH_3)(H_2O)]^{3+}$		8250	12 800		105
cis-$[Co(en)_2(NH_3)Cl]^{2+}$		8050	4375		105
cis-$[Co(en)_2(NH_3)Br]^{2+}$		8040	3500		105
$[Co(NH_3)_5(NO_2)]^{2+}$		7625	262		105

[a] PPM relative to external saturated aqueous $K_3Co(CN)_6$ and using the δ scale where positive values represent higher resonance frequencies than the reference. [b] Lowest energy electronic absorption band.

TABLE 8.11 Cobalt chemical shifts in $Co^{III}N_4XY$-type complexes

Complex	Solvent	$\delta(^{59}Co)$/ppm[a]	$W_{1/2}$/Hz	ΔE[b]/cm^{-1}	Ref.
$[Co(NH_3)_4(OH)_2]^+$		10 090	245		105
cis-$[Co(NH_3)_4(H_2O)_2]^{3+}$					123
cis-$[Co(NH_3)_4(H_2O)_2]^{3+}$	H_2O	9820	5250		104
cis-$[Co(NH_3)_4(H_2O)_2]^{3+}$		9960	5250		105
c,c-$[Co(en)(NH_3)_2(OH)_2]^+$		9820	400		105
c,t-$[Co(en)(NH_3)_2(OH)_2]^+$		9810	1030		105
trans-$[Co(en)(NH_3)_2(OH)_2]^+$		9730	400		105
cis-$[Co(NH_3)_4Cl_2]^+$		9810	3500		105
trans-$[Co(NH_3)_4Cl_2]^+$		9810	7000		105
$[Co(NH_3)_4(CO_3)]^+$		9146			108
$[Co(NH_3)_4(CO_3)]^+$		9735			
$[Co(NH_3)_4(CO_3)]^+$		9070		19 060	121
$[Co(NH_3)_4(CO_3)]^+$		9680	5250		104
$[Co(NH_3)_4(CO_3)]^+$		9620		19 080	126
$[Co(NH_3)_4(CO_3)]^+$		9730	1925		105
$[Co(NH_3)_4(H_2O)Cl]^{2+}$		9600		19 200	123
$[Co(en)(NH_3)_2(CO_3)]^+$		9330	3325		105
trans-$[Co(en)(NH_3)_2Br_2]^+$		9310	8400		105
$[Co(en)(NH_3)_2Cl_2]^+$		9320	5250		105
$[Co(en)(NH_3)_2Cl_2]^+$		9300	7000		105
cis-$[Co(en)_2(OH)_2]^+$		9300	400		105
trans-$[Co(en)_2(OH)_2]^+$		9280	260		105
$[Co(NH_3)_4(C_2O_4)]^+$		9190		19 530	124
$[Co(NH_3)_4(C_2O_4)]^+$		10 000[c]			105
cis-$[Co(en)_2(H_2O)_2]^{3+}$		9150	11 370		105
trans-$[Co(en)_2(H_2O)_2]^{3+}$		9100			105
cis-$[Co(en)_2Cl_2]^+$		8970	7870		105
trans-$[Co(en)_2Cl_2]^+$		7109		19 800	108
trans-$[Co(en)_2Cl_2]^+$		8970	14 000		105
cis-$[Co(en)_2Br_2]^+$		8960	5250		105
trans-$[Co(en)_2Br_2]^+$		8960	9620		105
$[Co(en)_2(CO_3)]^+$		7131		19 530	108
$[Co(en)_2(CO_3)]^+$	H_2O	9181	5250		104
$[Co(en)_2(CO_3)]^+$		8880	3325		105
$[Co(en)_2(C_2O_4)]^+$	H_2O	8960	5250		104
$[Co(en)_2(C_2O_4)]^+$		8610	3500		105
$[Co(acacen)(NH_3)_2]^+$	H_2O	8300 ± 50			135
$[Co(salen)(NH_3)_2]^+$	H_2O	8060 ± 50			135
$[Co(NH_3)(NO_2)_3(C_2O_4)]^{2-}$		8350		21 830	125
$[Co(NH_3)(NO_2)_3(CO_3)]^{2-}$		7400		21 740	126
$[Co(NH_3)_4(CN)(H_2O)]^{2+}$		6950		22 170	125
$Co(NH_3)_4(CN)(SO_3)$		5910		23 310	125
$Co(dmgH)_2(Me)(H_2O)$	MeOH	4020 ± 50			135
$Co(dmgH)_2(Bu)(py)$	$CHCl_3$	3420 ± 50			135
$Co(dmgH)_2(Me)(py)$	$CHCl_3$	3370 ± 50			135
$Co(dmgH)_2(Me)(ba)$	$CHCl_3$	3320 ± 50			135
$Co(dmgH)_2(Ph_3Sn)(ba)$	$CHCl_3$	3150 ± 50			135
$Co(dmgH)_2(Ph_3Sn)(py)$	$CHCl_3$	3090 ± 50			135

[a] PPM relative to external saturated aqueous $K_3Co(CN)_6$ and using the δ scale where positive values represent higher resonance frequencies than the reference. [b] Lowest-energy electronic transition. [c] This signal may be ascribed to $Co(NH_3)_4(H_2O)_2^{3+}$.

TABLE 8.12 Cobalt chemical shifts in $Co^{III}N_3XYZ$-type complexes

Compound	Solvent	$\delta(^{59}Co)$/ppm[a]	$W_{1/2}$/Hz	ΔE^b/cm^{-1}	Ref.
$Co(NH_3)_3(CO_3)Cl$		11 650		18 150	126
$Co(en)(NH_3)(OH)_3$		10 860	350		105
$[\{Co(NH_3)_3OH\}_2O_2CR]^{3+}$		10 525 ± 20	varies with R		138
fac-$[Co(NH_3)_3(H_2O)_3]^{3+}$	H_2O	10 500	350		104
fac-$[Co(NH_3)_3(H_2O)_3]^{3+}$		10 780	350		105
$[Co(NH_3)_3(H_2O)_2Cl]^{2+}$		10 600		18 900	123
$Co(NH_3)_3(C_2O_4)Cl$		10 080		18 690	124
Co(acacen)(Me)(py)	$CHCl_3$	6960 ± 50			135
Co(acacen)(Me)(ba)	$CHCl_3$	6900 ± 200			135
Co(Salen)(Me)(ba)	$CHCl_3$	6800 ± 50			135
$Co(inaa)_3$	$CHCl_3$	6650			139

[a] PPM relative to external saturated aqueous $K_3Co(CN)_6$ and using the δ scale where positive values represent higher resonance frequencies than the reference. [b] Lowest energy electronic absorption band.

Because much of the cobalt-59 shielding data reported in the Tables were measured relative to $[Co(CN)_6]^{3-}$ without the intervention of a secondary reference, the line separations lie in the range 8,000–14,000 ppm, where precision of ±40 to ±70 ppm is the best that can be achieved. For this reason, the use of $[Co(en)_3]^{3+}$ (δ12 500) as secondary reference lying within 2000 ppm of most ^{59}Co resonances is much to be recommended. Although it would be nice to have another secondary reference around δ4500, no compound with the appropriate features of ready availability, narrow linewidth and reasonable solubility is apparent. This situation is mitigated, however, by the fact that few cobalt resonances occur in this region, and most of those that do are broad lines where signal width is the limiting factor in locating the line position.

As with any physical measurement, the level of reproducibility achievable within one laboratory is considerably greater than that which is achieved among different labs. Those shielding constants listed in the tables which have been studied by a number of different workers show degrees of scatter represented by the following standard deviations:

	Number of determinations	Average δ/ppm	Standard deviation	
$[Co(NH_3)_6]^{3+}$	7	8160	73	0.9%
$[Co(NH_3)_5(NO_2)]^{2+}$	7	7495	80	1.1%
$[Co(en)_3]^{3+}$	8	7110	91	1.3%
$[Co(NH_3)_5(H_2O)]^{3+}$	5	8784	393	4.5%
$[Co(CO_3)_3]^{3-}$	4	13975	91	0.7%
$[Co(C_2O_4)_3]^{3-}$	4	12973	71	0.5%
$Co(acac)_3$	4	12500	135	1.1%

TABLE 8.13 Cobalt chemical shifts in $Co^{III}O_6$-type complexes

Complex	Solvent	$\delta(^{59}Co)$/ppm[a]	$W_{1/2}$/Hz	$\Delta E^b/cm^{-1}$	Ref.
$[Co(CO_3)_3]^{3-}$		14 130		15 500	108
$[Co(CO_3)_3]^{3-}$		13 900	437		104
$[Co(CO_3)_3]^{3-}$		13 950		15 870	126
$[Co(CO_3)_3]^{3-}$		13 920	437		105
$[Co(C_2O_4)_3]^{3-}$		13 000			102
$[Co(C_2O_4)_3]^{3-}$		13 040		16 390	108
$[Co(C_2O_4)_3]^{3-}$		13 000	437		104
$[Co(C_2O_4)_3]^{3-}$		7610		16 560	124
$[Co(C_2O_4)_3]^{3-}$		12 850	437		105
$Co(acac)_3$	C_6H_6	12 680		16 750	108
$Co(acac)_3$		12 300		16 900	121
$Co(acac)_3$		12 500	175		104
$Co(acac)_3$		12 520		16 800	130

Substituted acetylacetonates

Ligand X,Y	Complex	$\delta(^{59}Co)$/ppm	$W_{1/2}$/Hz	Ref.
CF_3, S	$Co(tfthbd)_3$	12 795	440	134a
CF_3, CH_3	*trans*-$Co(tfac)_3$	12 580	330	134a
	cis-$Co(tfac)_3$	12 576	280	134a
	$Co(tfac)_3$	12 500	280	104
	$Co(tfac)_3$	12 500	280	104
CH_3, C_6H_5	*trans*-$Co(bzac)_3$	12 512	470	134a
	cis-$Co(bzac)_3$	12 550	440	
	$Co(bzac)_3$	12 400	870	104
CH_3, CH_3	$Co(acac)_3$	12 500	210	104
CH_3, Br, CH_3	$Co(acac\text{-}Br)_3$	12 500	870	104
H, C_6H_5	*trans*-$Co(ppd)_3$	12 441	560	134a
	cis-$Co(ppd)_3$	12 441	230	134a
C_6H_5, C_6H_5	$Co(dbzm)_3$	12 400	870	104

[a] PPM relative to external saturated aqueous $K_3Co(CN)_6$ and using the δ scale where positive values represent higher resonance frequencies than the reference. [b] Lowest-energy electronic transition.

TABLE 8.14 Cobalt chemical shifts in $Co^{III}O_4XY$-type and $Co^{III}O_3XYZ$-type complexes

Complex	Solvent	$\delta(^{59}Co)/ppm^a$	$W_{1/2}/Hz$	$\Delta E^b/cm^{-1}$	Ref.
$Co^{III}O_4XY$-type complexes					
$[Co(NH_3)_2(CO_3)_2]^-$		11 500	5250		104
		11 740		17 270	126
		11 700	2800		105
$[Co(en)(CO_3)_2]^-$		11 090	4200		105
$[Co(NH_3)_2(C_2O_4)_2]^-$		10 400		17 950	124
		10 930	2600		105
$[Co(en)(C_2O_4)_2]^-$		10 400	5600		105
$[Co(edta)]^-$		10 300	7000		104
$Co^{III}O_3XYZ$-type complexes					
$Co(NH_3)_2(H_2O)_3Cl$		11 800		16 700	123
$Co(acacen)(Me)(H_2O)$		7150 ± 50			135
$Co(salen)(Me)(H_2O)$	MeOH	6820 ± 50			135

[a] PPM relative to external saturated aqueous $K_3Co(CN)_6$ and using the δ scale where positive values represent higher resonance frequencies than the reference. [b] Lowest energy electronic absorption band.

TABLE 8.15 Cobalt chemical shifts in $Co^{III}S_6$-type, $Co^{III}Se_6$-type and $Co^{III}As_6$-type complexes

Complex	$\delta(^{59}Co)/ppm^a$	$W_{1/2}/Hz$	$\Delta E^b/cm^{-1}$	Ref.
$Co\{(EtO)_2PS_2\}_3$	8880		13 500	128
$Co(H_2NCS_2)_3$	7231		15 300	130
$Co\{(CH_2)_4NCS_2\}_3$	7074		15 200	130
$Co\{Me_2NCS_2\}_3$	7074		15 200	130
$Co\{Et_2NCS_2\}_3$	6450	520		104
$Co\{Et_2NCS_2\}_3$	6630		15 300	128
$Co(EtSCS_2)_3$	6460		16 000	128
$Co(Pr^iOCS_2)_3$	6360	870		104
$Co(EtOCS_2)_3$	6100		16 100	128
$Co(EtOCS_2)_3$	6360	520		104
$Co(EtOCS_2)_3$	6208		15 400	130
$Co\{Me_2NCSe_2\}_3$	6759		15 000	130
$[Co(triarsine)_2]^{3+}$	4902		15 200	130
	5398 (weak)			

[a] PPM relative to external saturated aqueous $K_3Co(CN)_6$ and using the δ scale where positive values represent higher resonance frequencies than the reference. [b] Lowest energy electronic absorption band.

TABLE 8.16 Cobalt chemical shifts in Co^0 and Co^{-1} complexes

Complex	Solvent	$\delta(^{59}Co)$/ppm[a]	$W_{1/2}$/Hz	Ref.
$[Co(NO)_2Br]_2$		3390		117
$Co(NO)_2BrPPh_3$		1920	30 000	117
$Co(NO)_2BrP(OPh)_3$		1420	23 000	117
$Ph_3PbCo(CO)_4$		−2950		140
$Ph_3SnCo(CO)_4$		−3170		140
$Cl_3SnCo(CO)_4$		−2950		140
$Br_3SnCo(CO)_4$		−2710		140
$I_3SnCo(CO)_4$		−2570		140
$Ph_3GeCo(CO)_4$		−3050		140
$Cl_3GeCo(CO)_4$		−2640		140
$Br_3GeCo(CO)_4$		−2230		140
$I_3GeCo(CO)_4$		−2160		140
$Ph_3SiCo(CO)_4$		−3270		140
$Cl_3SiCo(CO)_4$		−2920		140
$Co_4(CO)_{12}$	*n*-Hexane	−370	25 700	141
		−1700	22 900	
$Co_4(CO)_{12}$	Pentane	−814 (1)		109
		−1961 (3)		
$Co_4(CO)_{12}$	Hexane	−250 (1)		142
		−1520 (3)		
$Co(CO)_3NO$	Heptane	−1365	960	109
$Co_2(CO)_8$	Solid	−2200		143
$Co_2(CO)_8$	Solution	−2200		143
$Co_2(CO)_8$	Pentane	−2101	10 930	109
$Co_2(CO)_8$	Hexane	−1860	27 100	141
Cp_2CoCl	H_2O	−2410	25 500	109
$HgCo_2(CO)_8$	Benzene	−2520	40 000	109
$CpCo(CO)_2$	Heptane	−2675	11 900	109
$CoMn(CO)_9$	Solid	−3000		93
$CoMn(CO)_9$	Solution	−2900		93
$Na[Co(CO)_4]$	H_2O	−3100	v. sharp	109
$HCo(CO)_4$	Pentane	−3720	5670	109
$HCo(PF_3)_4$	Heptane	−3910	6720	109
$K[Co(PF_3)_4]$	H_2O	−4220	v. sharp	109
$Co(NO)_2Br(PZ_3)$ complexes				
PZ_3				
PPh_2CN		1610	15 700	117
PEt_2Cl		1600	14 000	117
PMe_2NMe_2		1590	12 200	117
PEt_3		1590	12 200	117
PPh_2OEt		1570	26 200	117
$PPh_2C_6F_5$		1580	26 200	117
$P(OEt)_3$		1490	17 500	117
$PPh(OEt)_2$		1460	19 200	117
$P(OPh)_3$		1420	22 700	117
$PPh(OPh)_2$		1380	24 500	117

[a] PPM relative to external saturated aqueous $K_3Co(CN)_6$ and using the δ scale where positive values represent higher resonance frequencies than the reference.

TABLE 8.17 Halogen dependence of cobalt shifts[a,b] in $Co(NO)_2X(PZ_3)$ complexes

PZ_3 \ X	Cl	Br	I
PPh_2Cl	1990	1810	1040
$PPhCl_2$	2040	1750	1020
$PPhMe_2$	1950	1740	1200
PPh_2NMe_2	1970	1700	1100
PPhMeCl	1910	1670	1020
PMe_2NMe_2	1850	1590	1030
PEt_3	1770	1590	1030
$PPh(OEt)_2$	1710	1460	850
$P(OPh)_3$	1610	1420	760

[a] PPM relative to external saturated aqueous $K_3Co(CN)_6$ and using the δ scale where positive values represent higher resonance frequencies than the reference. [b] Data obtained from ref. 117.

Even within the same laboratory, the degree of scatter over a period of time can be considerable as is illustrated by a comparison of Yamasaki *et al.* (1968)[104] with Yajima, Koike, Yamasaki, and Fujiwara (1974).[105] In these two studies carried out 6 years apart, 15 of the complexes reported were common to both studies. Since the 1968 study reported chemical shifts relative to $[Co(CN)_6]^{3-}$ while the 1974 study reported chemical shifts relative to $[Co(NH_3)_6]^{3+}$, the difference between the two reported shift values for each of these 15 complexes represents the shift of $[Co(NH_3)_6]^{3+}$ relative to $[Co(CN)_6]^{3-}$ and should be constant. The average difference is 8170 ppm, with a standard deviation of 198 ppm or 2.4%.

(*iii*) *Medium Effects on Cobalt Shieldings.* In studying the NMR spectrum of a compound, the chemist is usually concerned with the determination of molecular structure. It is because the NMR spectrum is extremely sensitive to subtle changes in structure that the technique has become an indispensible part of the laboratory equipment. This same sensitivity, however, makes the NMR parameters subject to medium effects other than those specific covalent binding interactions included under the general designation of structure. The medium effects whose influence upon cobalt shieldings must be considered are:

(*a*) variation with change in concentration,
(*b*) variation with change in solvent or change in phase, and
(*c*) variation with change in temperature and pressure.

Only if the medium effects cause variations in the cobalt shieldings which are an order of magnitude less than the variations caused by the structural effects can the shieldings be regarded as diagnostic of structure. The following sections provide the evidence to show that this favourable situation does in fact obtain.

Concentration effects. Cobalt shieldings are relatively insensitive to solution concentration. The largest concentration effect to be observed occurs in $[Co(en)_3]Cl_3$, where the cobalt in a 0.1 M aqueous solution is more highly shielded by 20 ppm than in a 0.3 M solution.[106] This pattern of increased shielding with decrease in concentration is also observed for $K_3[Co(CN)_6]$, where the cobalt shielding extrapolated to infinite dilution in water is 1.5 ppm greater than that for a 1 M solution. The larger concentration effect of 20 ppm observed with $[Co(en)_3]^{3+}$ is about the same magnitude as the experimental uncertainty in all but the most precise measurements of cobalt shielding, and indicates that even fairly subtle structural changes in cobalt complexes are unlikely to be masked by concentration effects when studied by cobalt NMR. A theoretical understanding of the observed concentration effect can be gained from the recognition that the magnitudes of both the paramagnetic shielding term σ and the electric field gradient eq at the cobalt nucleus are directly proportional to $\langle 1/r^3 \rangle$.[89,107] The electric field gradient present at a cobalt nucleus situated in an octahedrally symmetric ligand environment must be small and will result from interactions with the counter-ion in solution. Since these interactions increase with increasing concentration, the observed deshielding of cobalt with increasing concentration is rationalized.

Solvent and phase change effects. Due to the fact that most six-coordinate cobalt complexes carry an ionic charge, water is the solvent of choice for measuring their solution spectra, and most cobalt spectra have been taken from aqueous solutions. Cobalt trisacetylacetonate is sufficiently soluble in a range of solvents to enable the solvent effect upon the cobalt shielding to be assessed.[108] In acetone, the complex absorbs at δ12 741 and moves progressively downward through δ12 726 in dimethyl formamide, δ12 678 in benzene, δ12 673 in toluene, δ12 660 in carbon disulphide, δ12 646 in ether, δ12 644 in ethanol, to a low of δ12 565 in chloroform. Although this 200 ppm spread is small relative to the total range of cobalt shieldings, the solvent effect can be a significant influence which must be recognised when comparing compounds having similar δ-values.

The influence of a phase change on spectral parameters can be regarded as a medium effect and its mechanism of operation is similar to that of a solvent effect. Both $Co_2(CO)_8$ and $CoMn(CO)_9$ have been investigated as solids and in solution,[93,109] and within the experimental error of ± 200 ppm which apply to the solid spectra because of their broad lines, the cobalt shieldings in the solid and in solution are the same.

Temperature and pressure effects. Cobalt-59 shieldings exhibit a linear dependence upon temperature, with an increase in temperature yielding a decrease in shielding which moves the resonance position to higher δ-values. Two independent and very precise studies on temperature dependence have been carried out, and the agreement for the compound studied by both workers is excellent as shown in Table 8.18. The temperature

TABLE 8.18 Temperature dependence of cobalt shieldings

Complex	Solvent	$\dfrac{d\sigma}{dT}\Big/$ppm., K^{-1}	Ref.
$K_3Co(CN)_6$	H_2O	−1.38	102
$K_3Co(CN)_6$	H_2O	-1.38 ± 0.01	110
$Co(NH_3)_6Cl_3$	H_2O	-1.42 ± 0.06	110
$Co(acac)_3$	Toluene	−2.29	102
$Na_3Co(NO_2)_6$	H_2O	-2.85 ± 0.05	110
$Co(acac)_3$	$CHCl_3$	−2.96	102

has been shown to affect the shielding through the ΔE factor in the shielding equation.[108] Temperature variations alter the distance between the mean thermally excited ground level and the effective excited state to a degree that becomes manifest in the temperature coefficient observed.

Increasing pressure causes an increase in cobalt shielding which moves the resonance position to lower δ-values.[110] This increased shielding also operates through the ΔE factor and can be understood in terms of the crystal field model for cobalt complexes. Since an increase in pressure is going to reduce the metal-ligand distance in the complex, the crystal field splitting parameter, Δ, will increase. The related ΔE factor must also increase, thereby reducing the magnitude of the paramagnetic term and increasing the overall screening.

(*iv*) *Structural Correlations.* Cobalt chemical shifts cover a range of 18 000 ppm and within this span a number of structural correlations are clearly recognisable. Figure 8.6 presents chemical shift values for cobalt complexes grouped according to structure type. With the exception of the cyano-complexes, cobalt(III) compounds absorb in the high frequency region from δ3000 to δ14 000. Cobalt(II) compounds are paramagnetic and unobservable by normal NMR methods. Lower oxidation state cobalt(O) and cobalt($-$I) compounds are found in the lower frequency region between δ-values 2000 and $-$4000. The pentacyano and hexacyano-cobalt(III) complexes also absorb in this region between δ1800 and δ0. This pattern of metal in a lower oxidation state

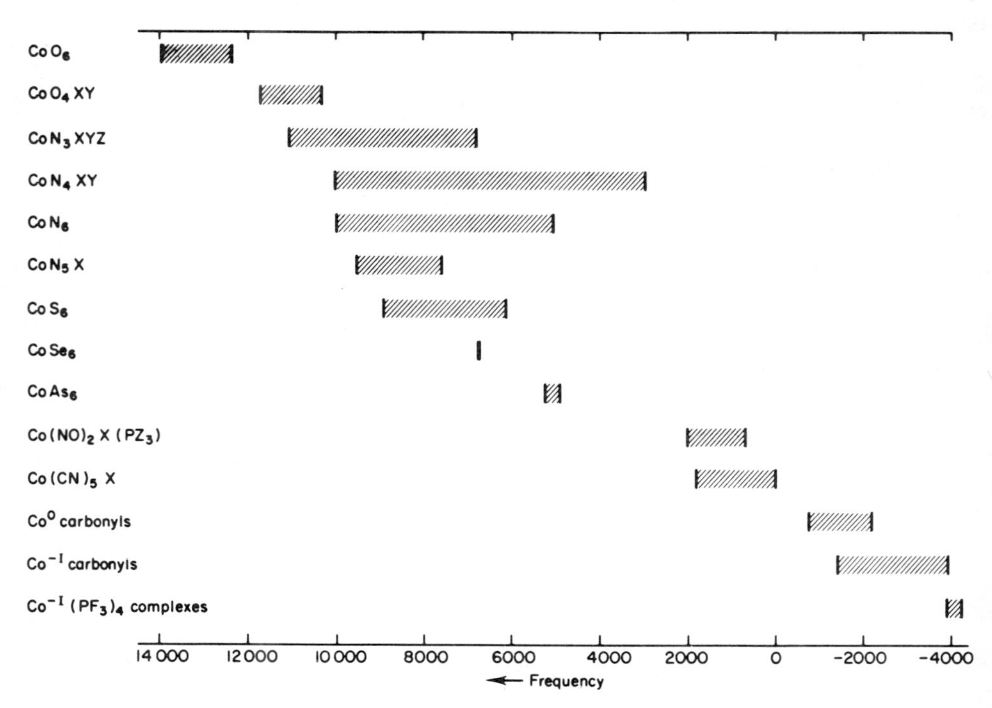

Fig. 8.6 ^{59}Co chemical shift ranges for different chemical environments.

absorbing at a lower frequency than does the metal in a higher oxidation state is consistent with the ^{51}V and ^{55}Mn picture where the same situation is observed. General shielding considerations can be used to explain the higher screening of a metal nucleus by the greater electron density present in a lower oxidation state.

Cobalt(III) complexes. Complexes of cobalt(III) all contain six ligands bonded octahedrally about the cobalt atom. Indeed it is the coordination by these six ligands which stabilises the higher oxidation state since in their absence Co^{II} is the more stable form. The common ligands have C, N, or O donor atoms from among first row elements or S, Se, As, Cl, Br, or I from among heavier elements. While each ligand does not contribute independently of the other ligands present to the shielding of the cobalt atom, the nature of the donor atom determines to a first approximation the shielding produced by the ligand. Oxygen-donor ligands provide the least shielded environment for cobalt because of the large paramagnetic contribution which they make to its chemical shift, and the δ-values for $Co^{III}O_6$ type complexes lie in the region 14 000 to 12 000. Nitrogen-donor ligands occasion higher shielding, and δ-values for $Co^{III}N_6$ type complexes occur in the region 10 000 to 5000. Carbon-donor ligands cause the highest shielding of cobalt(III), with the shift for $[Co(CN)_6]^{3-}$ at $\delta 0$. Ligands with sulphur, selenium, and arsenic donor atoms cause intermediate shielding levels giving δ-values in the range 7000 to 5000. This demarcation according to cobalt environment has been utilised in characterising polynuclear cobalt(III) complexes containing more than one type of cobalt coordination.[111]

The halogen dependence of ^{59}Co shieldings can be determined from the shifts for $[Co(CN)_5X]^{3-}$ complexes given in Table 8.8. The order of increasing cobalt shielding is seen to be $Br^- < I^- < [CN]^-$ which is the normal halogen-dependence order obtaining with the MX_4 compounds of the main group metals. Although the chemical shift for the analogous Cl^- complex in this series has not been reported, both the chloro- and fluoro-complexes in the series $[Co(NH_3)_5X]^{2-}$ have been studied, and again the normal halogen dependence with cobalt shielding increasing in the order $F^- < Cl^- < Br^- < I^-$ can be deduced from the data in Table 8.10. It is also evident from this series of complexes that the azide ion can be added to the series in the order $F^- < Cl^- < Br^- < I^- < [N_3]^-$.

Under favourable conditions, it is possible to distinguish geometric isomers on the basis of ^{59}Co shielding. Yajima *et al.*[105] report that *trans*-isomers are always more highly shielded than the corresponding *cis*-isomers. In the case of $[Co(en)_2(NO_2)_2]^+$ the difference is 190 ppm, but in other cases the difference is less than 50 ppm. Where linewidths are narrow and positions can be determined to within ± 10 ppm, differences of this magnitude are sufficient to permit an assignment. With broader lines whose positions are less accurately determined, it may not be possible to distinguish *cis* from *trans* on the basis of chemical shift alone. Under these circumstances, linewidth differences will generally prove useful. With one exception,* the *trans*-isomer has a broader line than the *cis*-isomer[105] as is expected from a simple point-charge model which gives the electric field gradient for a *trans*-isomer as twice that of the corresponding *cis*-isomer.[104]

^{59}Co NMR parameters have been used by several workers to study the problem of second-sphere coordination of Co^{III} complexes in aqueous solution. Both cobalt shieldings[112,113] and linewidths[113] provide evidence for second sphere coordination of $[SO_4]^{2-}$, $[PO_4]^{3-}$, Cl^-, and Br^- to $[Co(en)_3]^{3+}$, and in several cases association

* The linewidth for *cis*-$[Co(en)_2(OH)_2]^+$ is 402 Hz while that for *trans*-$[Co(en)_2(OH)_2]^+$ is 262 Hz.

constants have been determined. In the case of second-sphere phosphate coordination, it has been shown[114] that the increase in ^{59}Co linewidth results not from an increase in the electric field gradient experienced by cobalt, but from the increased rotational correlation time for the larger complex.

A study of optical and geometrical isomerism in tris [(±)propylenediamine] cobalt^{III} complexes using ^{59}Co NMR yielded a number of spectral multiplets which have been assigned to specific structures from among the 24 possibilities.[115]

Both the ^{59}Co shieldings and quadrupole coupling constants have been measured for a series of bisdimethylglyoxime complexes of cobalt(III) having a variety of axial ligands.[116] A bonding model is developed which attempts to relate the shielding and quadrupole coupling data. The relationship is not, however, a simple one.

Cobalt(O) and cobalt(−I) complexes. The phosphine complexes of cobalt have been extensively studied by Rehder and Schmidt[117] who find normal halogen dependency for $Co^{(-I)}$ shift with shielding in the $Co(NO)_2X(PZ_3)$ compounds increasing in the order $Cl^- < Br^- < I^-$. The cobalt shift in these compounds provides a measure of the phosphine π-acceptor ability. If Z is varied for the bromide compounds, one finds that δ varies from 1920 for the poor π-acceptor PPh_3 to 1380 for the good π-acceptor $P(OPh)_3$.

Lucken *et al.*[109] have studied the cobalt carbonyls and find that their resonances all lie to low frequency of $[Co(CN)_6]^{3-}$. Since in these compounds cobalt has a formal $3d^{10}$ electron configuration, there are no low-lying excited states created by ligand field splitting of the *d*-orbitals and the paramagnetic contribution to the cobalt shielding will be small as is confirmed by the chemical shift range observed for $Co^{(-I)}$ compounds.

Wuyts and Van der Kelen[118] have measured the cobalt shieldings in 22 compounds of the $R_3Sn\text{-}Co(CO)_4$, $R_2Sn[Co(CO)_4]_2$, and $RSn[Co(CO)_4]_3$ types in an attempt to differentiate the σ and π contributions to the Sn-Co bond. While the data did not provide an unambiguous answer to this problem, the cobalt shieldings were found to be linearly related to the ^{59}Co quadrupole coupling constants in a manner similar to that already discussed for ^{55}Mn resonances in $R_3Sn\text{-}Mn(CO)_5$ compounds.

(*v*) *Theoretical Basis for Shieldings.* All discussions of nuclear shielding which go beyond the establishment of substituent effects or structural correlations ultimately end up using the theoretical model introduced by Ramsey[119] in which the screening constant is divided into diamagnetic and paramagnetic terms. Because their electronic excitation and the nature of the excited state were both well understood, the octahedral complexes of cobalt(III) provided an excellent system with which to test Ramsey's screening theory and calculated values of cobalt screening for $[Co(C_2O_4)_3]^{3-}$, $[Co(NH_3)_6]^{3+}$, and $[Co(en)_3]^{3+}$ relative to $[Co(CN)_6]^{3-}$ were found to be in good agreement with the experimental values obtained by Proctor and Yu.[102] By choosing a series of 14 such complexes having a high degree of structural similarity, Freeman *et al.*[108] demonstrated a linear relationship between σ_p and ΔE^{-1}, thereby showing that the angular momentum factor B remains essentially constant for this group of complexes, and that there was some experimental justification for Ramsey's theoretical model.

This model is fully described in Ch. 3 and will be represented for the present discussion in the form

$$\sigma = A - B/\Delta E \tag{8.7}$$

where A is the diamagnetic term and $B/\Delta E$ is the paramagnetic term. As its designation indicates, an increase in the magnitude of the paramagnetic term causes decreased shielding of the nucleus and results in its resonance position occurring at lower fields or higher frequencies and increased values on the δ-scale. An increase in the magnitude of the diamagnetic term causes increased shielding of the nucleus and results in its resonance position occurring at higher fields or lower frequencies and decreased values on the δ-scale. While the absolute magnitude of the diamagnetic term is substantial, the *variation* in A from compound to compound is a negligibly small proportion of the large variation in total screening observed. This arises from the fact that its value is largely determined by the core electrons whose wave functions are practically uninfluenced by chemical bonding, and calculations for atoms beyond the first row of the Periodic Table where the core electrons comprise a large fraction of the total show this to be the case.[108,120]

In a classic paper which has served as a model for calculating the magnitude of the paramagnetic term for all nuclei, Griffiths and Orgel[90] showed that for octahedral complexes of cobalt(III), the paramagnetic term in the screening constant could be calculated semi-empirically using wave functions derived from a crystal-field description of the complex to evaluate the orbital angular momentum induced by the magnetic field (B), and spectroscopically determined values of the electronic excitation energy for ΔE in Eq. 8.2. This expression gives σ numerically as $-(9.3 \times 10^7/\Delta E)\langle r^{-3}\rangle$ when ΔE is in wave numbers and $\langle r^{-3}\rangle$ is in atomic units.

To obtain a theoretical insight into cobalt shieldings, two simplified treatments of the Ramsey equation have been used, both of which attribute the chemical shifts to variations in the paramagnetic term.

Linear dependence on $1/\Delta E$. Griffith and Orgel,[90] Freeman, Murray, and Richards,[108] Kanekar,[121,122] and Biradar[123–126] have all analysed their cobalt chemical shifts by establishing a linear dependence upon ΔE^{-1}. Six-coordinate complexes of cobalt(III) with oxygen-donor and nitrogen-donor ligands were the first to be studied by ^{59}Co NMR and their chemical shifts showed a remarkably good correlation with the wavelengths of their characteristic *d-d* bands in the visible region of the spectrum. The plot of this correlation is given in Fig. 8.7 and a value for the angular momentum factor B is obtained directly from the slope of this line. Since $B/\Delta E$ is a dimensionless constant, and since ΔE is normally expressed in cm^{-1} units, B is most conveniently reported in cm^{-1} units. Because the ΔE^{-1} axis in chemical shift correlation plots is frequently calibrated in mμ or nm wavelength units, the slope conversion factor 1(ppm/nm) = 10 cm^{-1} should be recognized. The y-intercept at $\Delta E^{-1} = 0$ represents the shielding of a hypothetical Co^{3+} ion with spherically symmetric charge distribution for which the σ_p-term is zero. The resonance frequency for this point on the shielding axis differs from that for the bare ^{59}Co nucleus by the magnitude of the diamagnetic term A which Dickinson[127] has estimated to be 2140 ppm. Those authors[124–126,128,129] who report "absolute" shieldings for different cobalt environments are using this extrapolated y-intercept as their shielding reference. While there is some theoretical justification for relating paramagnetic chemical shifts to a reference having $\sigma_p = 0$, the fact that its value depends upon the extrapolation of a line which is not uniquely defined makes it a poor primary reference. As a result of slope variations the intercept value relative to the $[Co(CN)_6]^{3-}$ resonance position can vary from a high of 15 395 ppm to a low of 11 000 ppm.

Effects of orbital reduction. Kanekar,[128] Fujiwara, Yajima, and Yamasaki,[129] and Martin and White[130] all observe departures from the above correlation for complexes with S-donor, Se-donor, and As-donor ligands. Although the complexes containing ligands having first row donor atoms C, N, and O all lie fairly close to the same correlation line, complexes with S-, Se-, and As-donor ligands taken together with the y-intercept value obtained previously which must be the same for this new group of complexes, establish

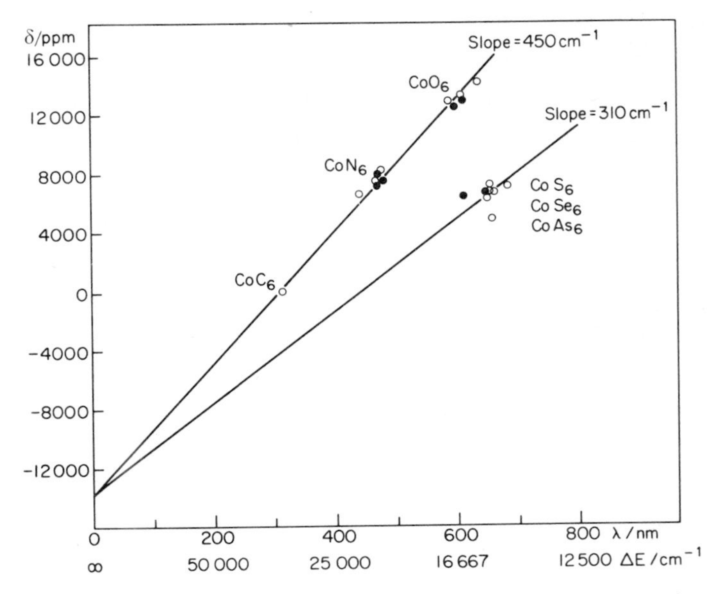

Fig. 8.7 ^{59}Co chemical shift dependence upon $1/\Delta E$ for octahedral cobalt (III) complexes. ○ represents data from R. L. Martin and A. H. White, *Nature* **223**, 394 (1969). ● represents data from S. Fujiwara, F. Yajima and A. Yamasaki *J. Magn. Reson.* **1**, 203 (1969).

a new correlation line having a lower slope than previously. The lower slope defines a reduced B factor of 310 cm^{-1} and can be explained in terms of the varying effects which different ligands have upon $\langle r^{-3}\rangle$ for the $3d$-orbitals of the cobalt atom. Variations in the inter-electron repulsion (as measured by spectroscopic term separations) and spin-orbit coupling constant (as measured by magnetic moments of paramagnetic complexes) of metal ions as the ligand is changed, both indicate that all ligands increase the radial size of the valence orbitals of a metal beyond that which obtains in the uncomplexed ion, and that some ligands cause a greater cloud expansion or "nephelauxetic effect" than others. Any ligand-induced expansion of the cobalt $3d$-orbitals causes a reduction in $\langle r^{-3}\rangle$ which is reflected in a lower slope for the σ vs. ΔE^{-1} plot. Along with this nephelauxetic effect on the B factor, a number of authors have included the "orbital reduction factor," k, into the B factor, thereby providing another mechanism for explaining a reduced slope. The orbital reduction factor[131,132] is an empirical constant which describes the extent to which the orbital angular momentum for a metal atom in a complex is reduced below its free-ion value as a result of covalent bond formation. Its value as determined from magnetic suceptibility measurements lies in the range $0.7 \leq k \leq 1$. It is problematic whether there is any justification for regarding the nephelauxetic phenomenon and the reduced orbital angular momentum phenomenon as two separate effects. There has been no experimental demonstration of the one effect operating independently of the other and since any structural change which affects one is expected to alter the other in the same direction, the two effects will be regarded as synonymous, and variations in B will be attributed to variations in the product $k^2\langle r^{-3}\rangle$ without any attempt being made to further differentiate the cause. Thus, the different slopes in Fig. 8.7 show that B is the same for oxygen-donor and nitrogen-donor ligands, at a value 0.65 of that for the free Co^{3+} ion, while the value for the cyanide complex is slightly lower at 0.57 of the free-ion value. These data of Fujiwara *et al.*[129] confirm the claim made by Martin and White[130] that complexes with ligands having first row donor atoms all lie on one regression line whereas those with ligands having second and third row donor atoms give a different regression line having lower slope. Two independent sets of data confirm that the slope for the S-donor, Se-donor, and As-donor ligands lies 0.7 below that for the first row donor line and 0.45 below the free ion value.

While it is very satisfying for the NMR spectroscopist to have established a general theoretical relationship between the large paramagnetic chemical shifts observed in cobalt(III) complexes and the presence in these complexes of the low-lying excited states responsible for their colours, it is with a sense of *deja vu* that we recognise just another manifestation of the temperature-independent paramagnetism phenomenon explained by Van Vleck in 1932. During the past 40 years, chemists engaged in the characterisation of transition metal compounds by magnetic susceptibility measurements have routinely included a TIP correction factor in their susceptibility calculations for strongly coloured ions. This TIP factor and our $B/\Delta E$ paramagnetic term have a common origin and the NMR manifestation provides the more sensitive technique for its evaluation.

c. ^{59}Co *Linewidths and Relaxation Times*

The widths of ^{59}Co resonance lines span an extremely large range from the narrowest at 175 Hz for $[Co(CN)_6]^{3-}$ to the widest at 40 kHz for $HgCo_2(CO)_8$. With a quadrupole moment of 0.4×10^{-28} cm^2, cobalt will couple effectively with electric field gradients present at the nucleus and there is general agreement that the relaxation of ^{59}Co among different nuclear spin states is dominated by the quadrupolar mechanism. Given the extent to which cobalt shieldings have been studied via the chemical shift parameter it is surprising that there has not been more interest in the study of cobalt relaxation times and the electric field gradients which give rise to them. While one or two studies have devoted major attention to the cobalt relaxation process, there is still much that remains to be learned in this area.

Four possible relaxation mechanisms which could contribute to cobalt linewidths have been considered.[104] Dipolar coupling, scalar coupling, quadrupolar coupling, and chemical exchange have been assessed, and all but quadrupolar coupling have been eliminated using order-of-magnitude arguments. A point-charge model has been used to determine the relative electric field gradients (EFG's) at the centers of CoA_xB_{6-x} complexes. This model predicts that CoA_6, CoB_6, and *fac*-CoA_3B_3 will have EFG's of O; CoA_5B, $CoAB_5$, *cis*-CoA_4B_2, and *cis*-CoA_2B_4 will have EFG's of 2 units; *mer*-CoA_3B_3 will have an EFG of 3 units; *trans*-CoA_4B_2 and *trans*-CoA_2B_4 will have EFG's of 4 units. Although there are some exceptions, the predictions of this model tend to be confirmed experimentally, with *trans*-isomers usually giving broader lines than *cis*-isomers, and *mer*-isomers broader than *fac*-isomers.

Ader and Loewenstein[107] have studied the ^{59}Co relaxation times for $[Co(CN)_6]^{3-}$, $[Co(NH_3)_6]^{3+}$, and $[Co(en)_3]^{3+}$ in both aqueous and non-aqueous solvents. The influence of viscosity/temperature, concentration, solvent composition, and counter-ion on the longitudinal relaxation time T_1 were determined and used to estimate the effective radii for the ions in solution using the Debye model for the rotational correlation time τ_c.

d. ^{59}Co *Coupling Constants*

Despite the extensive literature dealing with cobalt shieldings in a broad range of cobalt(III) complexes, very few instances of spin-spin coupling involving cobalt have been observed. The first place one looks for coupling to a quadrupole nucleus is where the atom is either octahedrally or tetrahedrally coordinated by $I = \frac{1}{2}$ nuclei providing an environment of cubic symmetry. The most readily available compounds of this type are the hexafluoroanions of various metals, and $J(\mathrm{M}, {}^{19}\mathrm{F})$ values have been obtained for $[\mathrm{ScF_6}]^{3-}$, $[\mathrm{TiF_6}]^{2-}$, and $[\mathrm{NbF_6}]^{-}$. In the case of cobalt, however, the ligand field created at the cobalt(III) ion by six fluoride ions is so weak that $[\mathrm{CoF_6}]^{3-}$ is one of the few examples of a high-spin $\mathrm{Co^{III}}$ complex and as such is paramagnetic. It is unlikely that an NMR spectrum will be observed under this circumstance.

In $[\mathrm{Co(CN)_6}]^{3-}$ the ^{59}Co-^{13}C coupling has been observed[133] as two satellite peaks in the ^{59}Co spectrum, each of which constitutes 3% of the main peak area. The coupling constant is 126 Hz. The midpoint of the two satellite peaks is displaced from the position of the main peak giving an isotope shift of 0.9 ppm. Coupling to ^{13}C has also been observed[109] in $\mathrm{Na[Co(CO)_4]}$ where $^{1}J(^{59}\mathrm{Co}, {}^{13}\mathrm{C}) = 287 \pm 13$ Hz.

Both ^{59}Co-^{31}P and ^{59}Co-^{19}F coupling have been observed[109] in $\mathrm{K[Co(PF_3)_4]}$ yielding coupling constants $^{1}J(^{59}\mathrm{Co}, {}^{31}\mathrm{P}) = 1222 \pm 25$ Hz and $^{2}J(^{59}\mathrm{Co}, {}^{19}\mathrm{F}) = 57 \pm 2$ Hz.

8H.2 Rhodium, ^{103}Rh

Rhodium is one of the few elements that occur solely as an isotope which has $I = \frac{1}{2}$. This desirable feature is counterbalanced by a very low magnetogyric ratio. Sogo and Jeffries,[59] using wideline techniques, obtained a ^{103}Rh signal from the finely powdered metal by 1955 and confirmed that the magnetic moment was negative. They were unable to observe any resonances for solutions. Although there have been several reports on rhodium metal and alloys since,[143a–145] there are as yet still no reported direct observations for solutions. Despite being an element of major chemical interest, its low resonance frequency and relative receptivity seem to have deterred spectroscopists e.g. ^{103}Rh frequencies are usually below the range quoted for multinuclear modifications of Fourier transform spectrometers.

Thus all reported rhodium resonances in complexes have been obtained by double resonance methods. The spin-spin couplings to ^{103}Rh are often small on account of the magnetogyric ratio but this is less of a problem than for nuclei which are of low abundance (e.g. ^{183}W). Possible alternatives for phosphine complexes are ^{31}P-{^{103}Rh} double resonance as used by Brown and Green[146] for some triphenyl phosphine complexes or ^{1}H-{^{31}P, ^{103}Rh} triple resonance cf. McFarlane *et al.*[66] The ^{1}H-{^{103}Rh} INDOR bands observed for complexes such as *mer*-$[\mathrm{RhCl_3(SMe_2)_3}]$ are only slightly broader than the width of the envelope due to Rh-H coupling but such couplings (ca. 1 Hz) could not be resolved in the rhodium spectrum.[147] An exception is $[\mathrm{Rh}(\eta\text{-}\mathrm{C_5H_4CO_2Me})(\mathrm{C_2H_4})_2]$[148] where five of the nine lines due to coupling to the ethylene protons are resolved with a linewidth of <1 Hz in the rhodium INDOR spectrum (observed via the cyclopentadienyl protons).[149] Interestingly, this coupling is not observable in the proton spectrum under these conditions because the ethylene resonances are broadened by slow rotation. In cases where $J(\mathrm{RhH}) < 1$ Hz, better INDOR spectra were obtained if the centre of the doublet due to coupling to ^{103}Rh was observed i.e. the INDOR spectrum is positive-going, corresponding to a decoupling experiment rather than "tickling." Because of the small magnitude of the coupling constant, the power required is low and the linewidth is only marginally broader than the envelope due to coupling to the protons i.e. similar to the more normal "tickling" INDOR linewidths.[147]

a. ^{103}Rh *Chemical Shifts*

The reference used by Brown and Green[146] was a figure based on the result of Sogo and Jeffries[59] for rhodium metal. This does not seem a very suitable reference in view of its poor accuracy and the difficulties involved in its redetermination. McFarlane and co-workers[150,151] have used *mer*-$RhCl_3(SMe_2)_3$ (saturated in dichloromethane, at 24°C) as their reference. Since this is a separate sample rather than an external reference observed at the same time as the peaks to be referenced and the value is not likely to be accurately reproducible, the real reference is the value of $\Xi(^{103}Rh)$, 3 172 310 Hz. If an essentially arbitrary value of $\Xi(^{103}Rh)$ is to be used as reference, then 3.16 MHz would be more desirable as it is simpler for the mathematical calculations and nearly all shifts are then positive. The latter has been employed for all the ^{103}Rh shifts collected together in Tables 8.19–8.22. When adjustment of reported results to this scale was necessary, Eq. 8.8 (which takes account of the change in the reference frequency in the denominator of the formula for chemical shift) was used. In addition to the data in

$$\delta = \delta' \frac{\Xi(\text{ref})'}{\Xi(\text{ref})} + \frac{\Xi(\text{ref})' - \Xi(\text{ref})}{\Xi(\text{ref})} \cdot 10^6 \tag{8.8}$$

Table 8.20, reference 150 reports the shifts for the mixed halide species formed by equilibration of *mer*-$RhCl_3L_3$ and *mer*-$RhBr_3L_3$ (L = SMe_2, SMePh and $SeMe_2$), some mixed neutral ligand complexes, *mer*-$RhCl_3LL'_2$ (L and L′ = SMe_2, SMePh or $SeMe_2$) of unidentified stereochemistry and seven isomers of $[RhCl_2(MeSCH_2CH_2SMe)_2]Cl$. (In this reference (e) to (h) in the Table should be associated with footnotes (d) to (g) respectively.)

Since the present volume of ^{103}Rh chemical shift data does not justify analysis in isolation it seems more practicable to make a comparison with the features observed for ^{195}Pt chemical shifts especially in view of the similarity of the chemistry of rhodium and platinum. Although the shifts in platinum(IV) compounds overlap the range for

TABLE 8.19 The ^{103}Rh chemical shifts in some low valent rhodium derivatives

Compound	$\delta(^{103}Rh)^a$	Ref.
$RhCl(PPh_3)_3$	−81	146
$RhBr(PPh_3)_3$	−142	146
$RhI(PPh_3)_3$	−268	146
trans-$RhCl(CO)(PMe_3)_2$	−415	155
trans-$RhCl(CO)(PMe_2Ph)_2$	−405	151
trans-$RhCl(CO)\{PMe_2(o\text{-}MeOC_6H_4)\}_2$	−407	151
trans-$RhCl(CO)(PPh_3)_2$	−368	146
trans-$RhBr(CO)(PMe_3)_2$	−448	155
trans-$RhBr(CO)(PMe_2Ph)_2$	−441	151
trans-$RhI(CO)(PMe_2Ph)_2$	−523	151
$Rh(\eta\text{-}C_5H_4CO_2Me)(C_2H_4)_2$	−818	149
$[Rh_{13}(CO)_{24}H_2]^{3-}$	4554,[b] −408, −522	152
$[Rh_{13}(CO)_{24}H_3]^{2-}$	3547,[b] −532, −600	152

[a] In ppm to high frequency of $\Xi(^{103}Rh)$ = 3.16 MHz. [b] Central metal atom.

TABLE 8.20 The ^{103}Rh chemical shifts in some rhodium(III) complexes with phosphorus, arsenic and chalcogen ligands

Complex	$\delta(^{103}Rh)^a$			Ref.
	X = Cl	Br	I	
$[RhX_5(SOMe_2)]^{2-}$	5728			147
$Rh_2Cl_6(PBu^n_3)_4$	2978			146
cis-$[RhX_4(SMe_2)_2]^-$	4882	4339	3070	147
cis-$[RhX_4(SOMe_2)_2]^-$	4145	3637		147
trans-$[RhX_4(NMe_3)_2]^-$	7942			147
trans-$[RhX_4(SMe_2)_2]^-$	5226	4532	2958	147
trans-$[RhX_4(SOMe_2)_2]^-$	4506	3836		147
mer-$RhX_3(PMe_3)_3$	2202	1746	809	147, 155
mer-$RhX_3(PBu^n_3)_3$	2770	2334		146
mer-$RhX_3(AsMe_3)_3$	2806	2276	1191	146
mer-$RhX_3(SMe_2)_3$	3897	3437	2448	147, 150
mer-$RhX_3(SMePh)_3$	4121	3646		150
mer-$RhX_3(SeMe_2)_3$	3904	3385		150
mer-$RhX_3(TeMe_2)_3$	3179	2567	1352	147
mer-$RhX_3(SOMe_2)_3$	3478	3061		147
mer cis-$RhX_3(OSMe_2)(SOMe_2)_2$	4133	3739	2903	147
mer trans-$RhX_3(OSMe_2)(SOMe_2)_2$	4918	4437		147
fac-$RhX_3(PMe_3)_3$	1437	1229		155
fac-$RhX_3(AsMe_3)_3$	2125	1798		147
fac-$RhX_3(SMe_2)_3$	3658			150
fac-$RhX_3(SMePh)_3$	3904	3529		150, 147
fac-$RhX_3(SEt_2)_3$	3599			150
cis-$[RhX_2(AsMe_3)_4]^+$	1457	1222		147
trans-$[RhX_2(PMe_3)_4]^+$	1564	1154	398	147
trans-$[RhX_2(AsMe_3)_4]^+$	1946	1486		147
trans-$[RhX_2(SMe_2)_4]^+$	3081	2756		147
cis cis cis-$[RhX_2(OSMe_2)_2(SOMe_2)_2]^+$	4661	4424		147
trans cis cis-$[RhX_2(OSMe_2)_2(SOMe_2)_2]^+$	4340	4049		147
fac-$[RhX(OSMe_2)_3(SOMe_2)_2]^{2+}$	4760	4591		147

^{a} In ppm to high frequency of $\Xi(^{103}Rh) = 3.16$ MHz.

platinum(II) complexes, they are generally to higher frequencies. The shifts in rhodium(III) compounds are likewise to high frequency and at present there is no overlap with the shifts for rhodium(I) species (Table 8.19). The span of the latter is rather small, largely due to the close similarity of the complexes as yet studied. Although the present range of shifts for rhodium(III) is less than for platinum, this is due to the absence of data for rhodium species analogous to $[PtF_6]^{2-}$ or $[PtI_6]^{2-}$. Comparison of the differences in chemical shift of related complexes of Rh(III) and Pt(II) suggests that the sensitivity of ^{103}Rh shifts to structure is at least as great as that of ^{195}Pt shifts. The central rhodium atom in the cluster compounds $[Rh_{13}(CO)_{24}H_2]^{3-}$ and $[Rh_{13}(CO)_{24}H_3]^{2-}$ is remarkable in having a shift clearly in the Rh(III) region whereas the peripheral rhodium atoms have shifts similar to Rh(I) compounds.[152]

Increasing the atomic number of the halides in a complex usually causes the metal resonance to move to low frequency. This is contrary to the decrease in ΔE which should

TABLE 8.21 The ^{103}Rh chemical shifts in some rhodium(III) complexes with carbon monoxide and tertiary phosphines

Complex	$\delta(^{103}Rh)^a$	Ref.
mer trans-$RhCl_3(CO)(PMe_3)_2$	1477	155
mer trans-$RhCl_3(CO)(PMe_2Ph)_2$	1588	151
mer trans-$RhBr_3(CO)(PMe_3)_2$	983	155
mer trans-$RhBr_3(CO)(PMe_2Ph)_2$	1074	151
cis trans-$RhCl_2Me(CO)(PMe_2Ph)_2$	499	151
cis trans-$RhBr_2Me(CO)(PMe_2Ph)_2$	351	151
cis trans-$RhI_2Me(CO)(PMe_2Ph)_2$	52	151
trans-$RhBrIMe(CO)(PMe_2Ph)_2{}^b$	268	151
trans-$RhClIMe(CO)(PMe_2Ph)_2{}^b$	194	151
trans-$RhClIMe(CO)\{PMe_2(o\text{-}MeOC_6H_4)\}_2{}^b$	254	151
trans-$RhClI(CF_3)(CO)(PMe_2Ph)_2{}^b$	707	151
trans-$RhBrI(CF_3)(CO)(PMe_2Ph)_2{}^b$	625	151
cis trans-$RhI_2(CF_3)(CO)(PMe_2Ph)_2$	453	151
trans trans-$RhCl_2H(CO)(PMe_2Ph)_2$	240	151

[a] In ppm to high frequency of $\Xi(^{103}Rh) = 3.16$ MHz. [b] I *trans* to CH_3 or CF_3

TABLE 8.22 The ^{103}Rh chemical shifts in some isonitrile complexes of rhodium(III)[a]

Complex	$\delta(^{103}Rh)^b$
$[RhMe(CNBu^t)_5]^{2+}$	82
trans-$[RhMe(CNBu^t)_4py]^{2+}$	526
trans-$[RhMe(CNBu^t)_4(4\text{-}Mepy)]^{2+}$	525
trans-$[RhMe(CNBu^t)_4(PMe_3)]^{2+}$	156
trans-$[RhMe(CNBu^t)_4(PMe_2Ph)]^{2+}$	105
trans-$[RhMe(CNBu^t)_4(PPh_3)]^{2+}$	39
trans-$[RhMe(CNBu^t)_4\{P(OMe)_3\}]^{2+}$	151
trans-$[RhMe(CNBu^t)_4\{P(OPh)_3\}]^{2+}$	95
trans-$[RhMe(CNBu^t)_4(AsPh_3)]^{2+}$	57
trans-$[RhMe(CNBu^t)_4(SbPh_3)]^{2+}$	231
trans-$[RhMe(CNBu^t)_4(SMe_2)]^{2+}$	289
trans-$[RhClMe(CNBu^t)_4]^+$	526
trans-$[RhBrMe(CNBu^t)_4]^+$	367
trans-$[RhIMe(CNBu^t)_4]^+$	107
trans-$[RhIEt(CNBu^t)_4]^+$	183
trans-$[RhIPr^n(CNBu^t)_4]^+$	174
trans-$[RhIBu^n(CNBu^t)_4]^+$	175
trans-$[RhI(CH_2Ph)(CNBu^t)_4]^+$	338

[a] From ref. 155. [b] In ppm to high frequency of $\Xi(^{103}Rh) = 3.16$ MHz.

increase the paramagnetic contribution to the shielding. Both rhodium(I) and rhodium(III) halide complexes show the same behaviour. The data for mixed bromide-chloride complexes of *mer*-RhX_3L_3 where L is SMe_2, SMePh and $SeMe_2$ [150] allow detailed comparison of the effects of halides with those reported for $\delta(^{195}Pt)$ in $[PtX_3L]^-$.[153] The effect on $\delta(^{103}Rh)$ of changing a bromide *cis* to SMe_2 into chloride (ca. + 170 ppm) is about two thirds the equivalent change of $\delta(^{195}Pt)$ in $[PtX_3(SMe_2)]^-$ and the same relationship holds for the smaller difference when the halide *trans* to SMe_2 is replaced. In contrast to the platinum case, these differences are almost identical between cases when the remaining halides are chloride or bromide. Similarly the effect of the *cis* neutral ligands being $SeMe_2$ rather than SMe_2 is much less for rhodium if allowance is made for the increased number of them. The effect on the metal chemical shift of altering the neutral ligands seems at least as great as for platinum. It appears that the competing effects of ΔE and increasing atomic number of the donor are balanced more in favour of the former for $\delta(^{103}Rh)$ than in the case of $\delta(^{195}Pt)$.

For platinum(II) complexes, the metal chemical shift was not greatly affected by changes of the substituents on the donor atom. This seems true for the rhodium(I) compounds *trans*-$RhCl(CO)(PR_3)_2$ (Table 8.19) but for rhodium(III) (Tables 8.20, 8.21 and 8.22) the differences are much greater. The particularly high value of *mer*-$RhCl_3(PBu^n_3)_3$ compared to its PMe_3 analogue (Table 8.20) suggests the shift may be affected by steric crowding in the octahedral complexes. Increasing the alkyl chain length in HgR_2 has a very marked effect on the ^{199}Hg chemical shift.[154] The changes in $\delta(^{103}Rh)$ in the series *trans*-$[RhIR(CNBu^t)_4]^+$, R = Me, Et, Pr^n, and Bu^n (Table 8.22) are also quite large[155] but are in exactly the opposite sense to the mercury compounds ($Me \ll Et > Pr^n \simeq Bu^n$ compared to $Me \gg Et < Pr^n \simeq Bu^n$).

At the present level of understanding it seems unlikely that ^{103}Rh chemical shifts will give much information on the bonding in rhodium compounds. The usefulness is more likely to be in the identification of complexes and the elucidation of their structures especially when there is the additional information inherent in a double resonance experiment. For example, samples analysing as $RhX_3(AsMe_3)_3$ showed two proton resonances in the ratio 2:1 as expected for a meridional isomer. However ^1H-$\{^{103}Rh\}$ INDOR proved these to be related to two different rhodium species with $\delta(^{103}Rh)$ 1946 and 4735 ppm for X = Cl and 1486 and 3950 ppm for X = Br from which the complexes can be identified as $[RhX_2(AsMe_3)_4][RhX_4(AsMe_3)_2]$.[147]

The temperature dependence of $\delta(^{103}Rh)$ seems similar to that of other transition metals viz: *trans*-$RhCl(CO)(PMe_2Ph)_2$, 0.26;[151] $Rh(\eta\text{-}C_5H_4CO_2Me)(C_2H_4)_2$, 0.25 in $(CD_3)_2CO$, 0.32 in $CDCl_3$;[149] $RhH(CO)(PPh_3)_3$, 0.44;[156] *mer*-$RhCl_3(SMe_2)_3$ and isomers of $[RhCl_2(MeSCH_2CH_2SMe)_2]Cl$, 1.0 ± 0.1 ppm/°K.[150]

The ratio $\nu(^{103}Rh)/\nu(^2H)$ measured by Sogo and Jeffries[59] for rhodium metal leads to $\Xi(^{103}Rh)$, 3 155 700(100) Hz or −1360 ppm. There are problems with theoretical estimations of the Knight shift in the metal[143] but the figures of 0.22%[59] or 0.36%[143] lead to −3500 or −5000 ppm as the origin for Knight shifts for rhodium. Alternatively, if $RhSn_2$ for which more reliable estimates are possible[143] is used, the origin is −4400 ppm.

b. ^{103}Rh *Coupling Constants*

The ^{103}Rh-^1H coupling in rhodium(III) hydrides is usually in the range 15 to 30 Hz but in rhodium(I) complexes of the type $RhHL_4$ is much less.[71] Hyde *et al.*[151] have shown that for *trans trans*-$RhCl_2H(CO)(PMe_2Ph)_2$, $^1J(RhH)$ is negative, i.e. the reduced coupling constant is positive as expected. For a number of methyl derivatives, $^2J(RhH)$

is *ca.* 2 Hz and positive.[71,151] With ligands such as PMe_3, ^3J(RhH) is 0.3 to 1.5 Hz and negative.[151,157] Spin-spin coupling between ^{103}Rh and ^{13}C has been reviewed by Mann.[75] Cairns *et al.*[158] report ^1J(RhF) for *trans*-$RhF(CO)(PPh_3)_2$ as 55 Hz. For the trifluoromethyl complexes in Table 8.21, ^2J(RhF) is −14 Hz[151] and in fluorophosphine complexes is ca. 30 Hz.[76] Rhodium phosphorus coupling has been reviewed by several authors.[78,159,160] The values are normally in the region 81–150 Hz, have negative sign (*K* positive)[151,157] and show similar behaviour to $^1J(^{195}Pt, ^{31}P)$ e.g. in respect of *trans* influence and oxidation. In *mer*-$RhX_3(TeMe_2)_3$, $^1J(^{125}Te, ^{103}Rh)$ is +71, +70 and +66 Hz for ^{125}Te *trans* to tellurium and +94, +93 and +69 Hz for ^{125}Te *trans* to halide where *X* is Cl, Br and I respectively.[147] The sign was established by comparison (via ^{1}H-{^{125}Te} and ^{1}H-{^{103}Rh} experiments) with ^3J(RhH) which was assumed to be negative as in complexes of PMe_3 etc. and with ^2J(TeH) which was assumed to be negative as in platinum complexes of $TeMe_2$.[161] The values are far less sensitive to change of halide than the analogous coupling in $[PtX_3(TeMe_2)]^-$[162] or to being *trans* to tellurium rather than X[161] but the reduced couplings constants still have the same sign for ^{195}Pt and ^{103}Rh.

On the basis of the ratios of their magnetogyric ratios, couplings to ^{103}Rh would be expected to be about a seventh of those to ^{195}Pt. In practice, they seem to be reduced by a further factor of three.

8I GROUP VIIIc

Within the triad nickel, palladium and platinum there is only ^{195}Pt which presents the NMR spectroscopist with an observable nucleus in practice. Nickel has only one isotope, ^{61}Ni, of use for NMR spectroscopy, but it is only 1.19% abundant, $I = \frac{3}{2}$ and D^C = 0.24. Consequently ^{61}Ni NMR spectroscopy has received no attention from chemists. The situation is somewhat better for palladium where ^{105}Pd is 22.23% abundant, $I = \frac{5}{2}$ and D^C = 1.41. The increased sensitivity has enabled some solid state investigations to be made on the Pd/H_2 system but to date no investigations of direct chemical interest have been made. Even for ^{195}Pt there have been only a few cases of direct observation, even though it is 33.8% abundant $I = \frac{1}{2}$ and D^C = 19.1. This hesitance probably arises from the ease of using INDOR to observe ^{195}Pt rather than inherent difficulties in direct observation.

8I.1 Platinum, ^{195}Pt

The only naturally occurring isotope of platinum with non-zero nuclear spin, ^{195}Pt has $I = \frac{1}{2}$ and reasonable abundance and sensitivity. In 1951, Proctor and Yu[102] reported the ^{195}Pt resonance of a molar solution of chloroplatinic acid doped with $MnCl_2$ and confirmed that the moment was positive. In addition to measurements on the metal (Rowland (1958)[163] and Drain (1963)[164]) there were two further studies on solutions using broadline CW instrumentation which are of particular interest to chemists, namely concerning complex halogeno-anions[165] and a series of neutral complexes.[166] In the same year, 1968, there were two separate reports[167,168] illustrating the ability to determine platinum chemical shifts by double resonance methods. As coupling constants to ^{195}Pt are quite large and ligand exchange generally slow, ^{195}Pt "satellites" are visible for a wide range of types of complexes and double resonance still continues to produce the majority of the results for $\delta(^{195}Pt)$. Since the sensitivity of ^{195}Pt is $2\frac{1}{2}$ times that of ^{113}Cd or $3\frac{1}{2}$ times that of ^{199}Hg, there should be no difficulty in observing it directly with the multinuclear Fourier transform systems used for the latter two metals, yet such reports for ^{195}Pt have been much slower to appear.[169–171]

a. ^{195}Pt *Chemical Shifts*

Before discussing the problem of the reference for ^{195}Pt chemical shifts, we should consider the effects of the medium on $\delta(^{195}Pt)$. Both Freeman *et al.*[170] and Pesek and Mason[171] have investigated the solvent dependence of the resonance of $[PtCl_4]^{2-}$. Their results span some 240 ppm which is however less than for $[PtCl_6]^{2-}$ (400 ppm[171]) so that the possibility of the solvent approaching along the z-axis of a platinum(II) complex does not seem important. Whilst polar solvents such as DMSO and DMF give shifts to high frequency, water results in by far the lowest frequency observed [cf. the effect of solvent on $\delta(^{199}Hg)$, Table 8.30]. The difference between $[PPh_4]_2[PtCl_4]$ and $[NBu^n_4]_2[PtCl_4]$ (24 ppm) is notably less than on going from dichloromethane to chloroform as solvent (61 ppm).[170] A possible conclusion is that increasing hydrogen bonding ability of the solvent (presumably towards the coordinated chloride ions) results in shifts to low frequency. A similar order was obtained for $[NPr^n_4][PtCl_3(PMe_3)]$ but the spread is much less (24 ppm),[153] in keeping with the smaller formal negative charge on the complex. For the neutral complexes, *trans*-$PtCl_2(AsBu^n_3)_2$,[166] *trans*-$PtH(m\text{-}OCOC_6H_4Me)(PEt_3)_2$[168] and *trans*-$PtBr_2(SMe_2)_2$,[153] the order of $\delta(^{195}Pt)$ *vs.* solvent is quite different with the more polar solvents in the middle of the range. Varying the concentration of $Na_2[PtCl_4]$ in water resulted in a change of up to 24 ppm[170] and Hackbusch[172] comments that the resonance of $[PtCl_6]^{2-}$ in water is concentration dependent although Freeman *et al.*[170] say this is not so for $Na_2[PtCl_6]$ in the range 0.03 to 0.3 M. For $[NPr^n_4][PtCl_3(PMe_3)]$ and *trans*-$PtBr_2(SMe_2)_2$ in CH_2Cl_2, the variation is less than ± 1 ppm for 0.01 to 0.2 M concentrations.[153]

The temperature dependence of the platinum chemical shift of aqueous $[PtCl_6]^{2-}$ is 1.1 ppm/°K[172] and that of $[PtCl_4]^{2-}$ is similar.[170] For the singly charged ion $[PtCl_3(PMe_3)]^-$ in CH_2Cl_2 it was 0.5 ppm/°K. Results have been obtained for a number of neutral complexes viz. *trans*-$PtCl_2(PBu^n_3)_2$ (0.34), *cis*-$PtCl_2(PEtPh_2)_2$ (0.54), *cis*-$PtCl_4(PPr^n_3)_2$ (0.84), *trans*-$PtBr_2(PPr^n_3)_2$ (0.33), *trans*-$PtBrH(PBu^n_3)_2$ (0.12),[174] *cis*- and *trans*-$PtCl_2(SMePh)_2$ (0.35 in $CDCl_3$, 0.47 in PhCN),[175] *trans*-$PtBr_2(SMe_2)_2$ and *trans*-$PtCl_4(SMe_2)_2$ (0.25).[155] The larger values for the *cis*-isomers and anions suggest that interaction with the solvent (which should be temperature dependent) is adding to the effect of temperature on the distribution amongst vibrational levels.

One requirement of a reference for chemical shifts is that its position should be accurately determined since all errors and uncertainties are automatically transferred to the shifts referred to it. The results of direct measurements have mainly been referred to $[PtCl_6]^{2-}$. In addition to the unresolved matter of its concentration dependence, it has the disadvantages of an especially large temperature dependence, an 11 ppm shift between H_2O and D_2O solutions,[171] a tendency to decompose when left in solution for long periods and that nearly all shifts are negative with respect to it. For double resonance determinations, several different complexes have been used as reference—*trans*-$PtH(ONO_2)(PEt_3)_2$,[168] *trans*-$PtClH(PEt_3)_2$,[176] and *cis*-$PtCl_2(SMe_2)_2$[175]— not necessarily under well-defined conditions (e.g. $PtCl_2(SMe_2)_2$ is a mixture of isomers in solution). By their nature, these references are merely frequencies as observed on one particular occasion. The number of digits specified in the divisor can result in significant differences when evaluating chemical shifts e.g. if Freeman *et al.*[170] were actually working at a field corresponding to 90 MHz for TMS, then $\nu(Na_2[PtCl_6])$ was 19.347

MHz rather than 19.34 MHz making the shift of $K_2[PtBr_4]$ 2663 instead of 2664 ppm. Thus it is desirable to have as few non-zero significant figures as possible in the reference frequency and 21.4 MHz seems a good choice since the vast majority of results then have positive sign whilst none are unnecessarily large. For direct measurements, Hackbusch[172] recommends aqueous $[Pt(CN)_6]^{2-}$ because of its insensitivity to concentration and counter-ion, and its temperature coefficient of only 0.51 ppm/°K [$\Xi(^{195}Pt)$ = 21 414 376(20) Hz]. However as the effects of the medium are clearly greater for charged species, a neutral complex with no dipole moment which is also suitable for double resonance determination would be preferable. There are some uncertainties in correlating $[PtCl_6]^{2-}$ with a scale based on 21.4 MHz but using such comparisons as are available in refs 170–172, a mean value of 4522 ppm [$\Xi(^{195}Pt)$ = 21 496 770] is derived and used in Eq. 8.8 to make conversions. (The formula given in ref. 153 for converting the results of Pidcock *et al.*[166] is incorrect; it should be $[32955 - (\nu_r - \nu_s)]/12.782$, and the results in ref. 171 are clearly to low frequency despite the footnotes to the tables.)

The recent Fourier transform observations[170–172] greatly extend the number of

TABLE 8.23 The ^{195}Pt chemical shifts of some complex ions of platinum

Compound	Medium	$\delta(^{195}Pt)^a$	Ref.
$[PPh_3CH_2Ph]_2[PtF_6]$	CH_2Cl_2(0.1 M)	11 847	183
fac-$[PtCl_3F_3]^{2-}$	H_2O	7605	183
$[PtCl_6]^{2-}$	aq. HCl/HBr (total Pt 2 M)	4521	172
$[PtBrCl_5]^{2-}$	aq. HCl/HBr (total Pt 2 M)	4236	172
trans-$[PtBr_2Cl_4]^{2-}$	aq. HCl/HBr (total Pt 2 M)	3942	172
cis-$[PtBr_2Cl_4]^{2-}$	aq. HCl/HBr (total Pt 2 M)	3941	172
mer-$[PtBr_3Cl_3]^{2-}$	aq. HCl/HBr (total Pt 2 M)	3636	172
fac-$[PtBr_3Cl_3]^{2-}$	aq. HCl/HBr (total Pt 2 M)	3635	172
trans-$[PtBr_4Cl_2]^{2-}$	aq. HCl/HBr (total Pt 2 M)	3320	172
cis-$[PtBr_4Cl_2]^{2-}$	aq. HCl/HBr (total Pt 2 M)	3319	172
$[PtBr_5Cl]^{2-}$	aq. HCl/HBr (total Pt 2 M)	2990	172
$[PtBr_6]^{2-}$	aq. HCl/HBr (total Pt 2 M)	2651	172
$[NBu^n_4]_2[PtCl_6]$	CH_2Cl_2	4783	171
$H_2[PtI_6]$	aq. HI	−1528	171
$Na_2[Pt(CN)_6]$	H_2O	672	172
$[Pt(en)_3][ClO_4]_4$	H_2O (sat'd)	3579	171
$[PtCl_4]^{2-}$	H_2O (total Pt 0.5 M)	2887	172
$[PtBrCl_3]^{2-}$	H_2O (total Pt 0.5 M)	2661	172
trans-$[PtBr_2Cl_2]^{2-}$	H_2O (total Pt 0.5 M)	2418	172
cis-$[PtBr_2Cl_2]^{2-}$	H_2O (total Pt 0.5 M)	2408	172
$[PtBr_3Cl]^{2-}$	H_2O (total Pt 0.5 M)	2138	172
$[PtBr_4]^{2-}$	H_2O (total Pt 0.5 M)	1843	172
$[NBu^n_4]_2[PtCl_4]$	CH_2Cl_2(0.3 M)	3055	170
$[NBu^n_4]_2[Pt_2Cl_6]$	CH_2Cl_2	3325	b
$[NBu^n_4]_2[PtBr_4]$	CH_2Cl_2(0.3 M)	2066	170
$[NBu^n_4]_2[Pt_2Br_6]$	CH_2Cl_2	2233	b
$H_2[PtI_4]$	aq. HI	−993	171
$[NBu^n_4]_2[Pt_2I_6]$	CH_2Cl_2	−600	b
$Na_2[Pt(CN)_4]$	H_2O	−204	172
$K_2[Pt(SCN)_4]$	H_2O(2M)	604	172
$Na_2[Pt(NO_2)_4]$	H_2O(sat'd)	2353	171
$[Pt(NH_3)_4][ClO_4]_2$	H_2O(ca. 310K)	1900	171
$[Pt(en)_2]Cl_2$	H_2O(sat'd)	1493	171

[a] In ppm to high frequency of $\Xi(^{195}Pt)$ = 21.4 MHz. [b] P. S. Pregosin, unpublished results.

results for the simpler complex ions with halides etc. (see Table 8.23). With modern instrumentation, the resonances of individual isomers of $[PtBr_nCl_{(6-n)}]^{2-}$ can now be resolved (cf. ref. 165) and are assigned on the basis of their intensities which approximated to the statistical distribution.[172] The CW investigation of Pidcock *et al.*[166] concentrated on neutral complexes of the type PtX_2L_2 where L was chosen to give good solubility in the organic solvents used. The results of McFarlane[175] also cover a range of complexes, PtX_2L_2 although chalcogenide ligands predominate. The ligands usually include a methyl group to facilitate the double resonance measurements and data are included on some mixed halide complexes. Another general survey of ^{195}Pt chemical shifts,[153] obtained by ^1H-$\{^{195}Pt\}$ INDOR, gives results for complexes of the types $[PtX_3L]^-$ and $[PtXL_3]^+$ as well as *cis*- and *trans*-PtX_2L_2 with ligands involving a wide selection of donor atoms, and again includes mixed halide complexes. The early double resonance investigation of Dean and Green[168] was concerned solely with complexes of the type *trans*-$PtHX(PEt_3)_2$ where X is a coordinated anion. For a series of substituted benzoates, the range of $\delta(^{195}Pt)$ was only 6 ppm. The study of Anderson *et al.*[176] is

TABLE 8.24 Variation of ^{195}Pt chemical shift[a] with the neutral ligand in $[NR_4][PtX_3L]^b$

L	X = Cl	Br	I
C≡O[c]	1264	563	−953
C≡NMe	1487	806	−730
$H_2C{=}CH_2$	1748	1060	
$NH_2(p\text{-}C_6H_4CH_3)$	2717	2099	
$NHMe_2$	2670	2114	529
NMe_3	2818	2282	
$MeN{=}CMe_2$			526
4-Etpy	2772	2181	522
N≡CMe	2511	1827	
PMe_3	1033	414	−973
PPh_3	1020		
$P(OMe)_3$	1037	371	−1060
PF_3	907	183	−1370
$AsMe_3$	1360	664	−913
$SbMe_3$	1390	605	−1109
OH_2[d,e]	3330		
SMe_2	1776	1118	−440
$SeMe_2$	1764	1057	−596
$TeMe_2$	1474	707	−995
$SOMe_2$	1535	892	
X[d]	3055	2066	

[a] In ppm to high frequency of $\Xi(^{195}Pt) = 21.4$ MHz. Results from CH_2Cl_2 solutions ($CDCl_3$ for C_2H_4), are due to R. J. Goodfellow and co-workers except for [c]P. S. Pregosin, unpublished and [d] ref. 170. [b] R = Pr[n] or Bu[n]. [e] In aqueous solution of

$Na_2[PtCl_4]$.

mainly directed at the use of ^{1}H-{^{195}Pt} and ^{1}H-{^{31}P} double resonance to establish the identities of the unstable platinum(IV) hydride complexes formed by addition of HX to *trans*-$PtX_2(PEt_3)_2$ and *trans*-$PtHX(PEt_3)_2$ but data are also given (and discussed) for the platinum(II) precursors and a range of silyl derivatives *trans*-PtX(silyl) $(PEt_3)_2$. This has been followed recently by results for the complexes where (silyl) is -SiH_2YSiH_3 or (-$SiH_2)_2Y$ in which Y is O, S, Se or Te.[177] Kennedy *et al.*[178] have investigated a wide variety of organometallic derivatives of platinum(II) and (IV). Some data on PtX_2L_2 are also presented including X = SPh and SePh. (There are surprising discrepancies between the results here and those previously reported by Hall *et al.*[179] for five of the complexes). Some pentafluorophenyl derivatives of platinum(II) and (IV) have been studied by Crocker *et al.*[180] The observation of coupling to ^{14}N on the ^{195}Pt resonance provides a convenient means of distinguishing *N*-bonded thiocyanate from *S*-bonded.[181] Anderson *et al.* have used this approach to study the isomerism of $Pt(SCN)_2L_2$ and $[Pt(SCN)_3L]^-$ for a range of neutral ligands L and data are also included on some cyanate complexes.[182] Crocker *et al.* have reported data for $[NBu^n_4][PtX_3L]$ where L is PF_3 or $P(OMe)_nX_{(3-n)}$, $n = 1, 2$ or 3. (X = F, Cl or Br).[69]

The number of ^{195}Pt results is thus already quite large and too great for a comprehensive listing to be given here. Instead, following the complexes with anions in Table 8.23, we illustrate in Table 8.24 the effect of different neutral ligands on $\delta(^{195}Pt)$ with results for $[PtX_3L]^-$, as this type of complex can be obtained with the widest range of ligands, and in Table 8.25, the relationship between different types of platinum(II) and (IV) complexes for some typical ligands. For data on a particular type of system, the reader is directed to the papers outlined above but some ^{195}Pt chemical shifts which have appeared piecemeal in papers not primarily concerned with this subject are brought together in Table 8.26.

A striking feature of Table 8.23 is the spread of values for $[PtX_6]^{2-}$, which encompasses the complete range of ^{195}Pt chemical shifts—some 13 000 ppm. The shifts

TABLE 8.25 Variation of ^{195}Pt chemical shift[a] with the structure of the complexes

L	*cis*-$PtCl_2L_2$	*trans*-$PtCl_2L_2$	$[PtClL_3]^+$	$[PtL_4]^{2+}$
CNMe	665		203	−318
NMe_3		2647		
PMe_3	125	583	−146	−358
$AsMe_3$	242	753	−211	
$SbMe_3$	−79		−646	
SMe_2	982	1109	523	
$SeMe_2$	798	1029	241	
	$[PtCl_5L]^-$	*cis*-$PtCl_4L_2$	*trans*-$PtCl_4L_2$	*mer*-$[PtCl_3L_3]^+$
CNMe	3323	2447		
PMe_3	3060	1921	2367	1348
$AsMe_3$	1359	2126	2502	1449
SMe_2	3656	2788	2850	

[a] In ppm to high frequency of $\Xi(^{195}Pt)$ = 21.4 MHz; data by R. J. Goodfellow and co-workers.

TABLE 8.26 Some miscellaneous ^{195}Pt chemical shifts

Compound	$\delta(^{195}Pt)^a$	Ref.
$PtMe(ONO_2)(Ph_2PCH_2CH_2PPh_2)$	28	191
$PtClMe(o\text{-}PPh_2C_6H_4CH{=}CH_2)$	372	191
$PtMe_2(o\text{-}PPh_2C_6H_4CH{=}CH_2)$	582	191
$[PtCl_3(\mu\text{-}SMe_2)PtCl_2(SMe_2)]^-$	1 109, 983	b
fac-$PtMe_3(acac)(PPh_3)$	2152	192
$Pt(Me_2C_2B_7H_7)(PEt_3)_2$	−747	c
Pt(*trans*-stilbene)$(PEt_3)_2$	−589	d
$Pt(CF_3C{\equiv}CCF_3)(PEt_3)_2$	−179	d
$Pt\{1,2\text{-}\eta\text{-}C_6(CF_3)_6\}(PEt_3)_2$	−105	d
$Pt\{1,3\text{-}4,6\text{-}\eta\text{-}C_6(CF_3)_6\}(PEt_3)_2$	1275	d
$Pt\{1,3\text{-}4,6\text{-}\eta\text{-}C_6(CF_3)_6\}(Ph_2PC_2H_4PPh_2)$	901	d
$(cod)Pt\{\mu\text{-}(CF_3)_2CO\}Pt(cod)$	824, 190	e
trans-$PtCl_2(C_5H_5NO)(C_2H_4)$	1657	169
trans-$PtCl_2$(piperidene)(*cis*-2-butene)	1567	169
trans-$PtCl_2$(piperidine)(*trans*-2-butene)	1569	169
$K_2[Pt_3(CO)_6]$	33	201
$[NBu_4]_2[Pt_6(CO)_{12}]$	−5	201
$[NBu_4]_2[Pt_9(CO)_{18}]$	70, −513[f]	201

[a] In ppm to high frequency of $\Xi(^{195}Pt) = 21.4$ MHz. [b] P. L. Goggin, R. J. Goodfellow and F. J. S. Reed, *J. Chem. Soc. Dalton Trans.* 576 (1974). [c] M. Green, J. L. Spencer, F. G. A. Stone and A. J. Welch, *J. Chem. Soc. Chem. Commun.* 571 (1974). [d] J. Browning, M. Green, J. L. Spencer and F. G. A. Stone, *J. Chem. Soc. Dalton Trans.* 97 (1974). [e] M. Green, J. A. K. Howard, A. Laguna, L. E. Smart, J. L. Spencer and F. G. A. Stone, *J. Chem. Soc. Dalton Trans.* 278 (1977). [f] Middle Pt_3 triangle.

of the fluoro-anions[183] are especially notable and it is a pity that so far these are the only two data in the high frequency half of the range. Obviously there is no clear division of the shifts according to oxidation state but those of divalent platinum tend to come at lower frequency, particularly for the complexes with neutral ligands. This trend is continued for Pt° e.g. $Pt^\circ L_4$, L = $PMe_2Ph(-195)$, $P(OMe)_3(-1297)$,[153] $PF(OPh)_2(-1057)$, and $F\overline{POC_6H_4O}(-1096)$.[184].

The regularity of the spacings between the shifts of $[PtBr_nCl_{(6-n)}]^{2-}$ (Table 8.23) has prompted Goggin *et al.*[153] to investigate the differences in $\delta(^{195}Pt)$ resulting from replacement of one halide by another in platinum(II) complexes. They conclude that such differences depend not only on the ligand *trans* to the halide, as might be anticipated on the basis of *trans*-influence, but also on the *cis* ligands. However, this dependence is systematic and by means of a table of corrections, the difference in shift on changing a halide in a particular complex can be calculated with fair accuracy. Successive substitutions involving neutral ligands are much less regular than for halides (Table 8.25) precluding a simple analysis. However, there is one feature which is likely to be of use in the identification of unknown complexes, namely the relatively small effect that substituents on the donor atom have on the shift especially compared to the effect of change of the donor atom, e.g. complexes of different phosphines have much the same ^{195}Pt chemical shifts.[153]

The large shifts to low frequency on increasing the atomic number of the halides in a complex are contrary to predictions on the basis of the paramagnetic term derived by Ramsey, since ΔE clearly decreases. However, Goggin *et al.*[153] found that for variation of L in $[PtCl_3L]^-$, a moderately good correlation with ΔE did exist. This correlation can be improved and extended to $[PtBr_3L]^-$ and $[PtI_3L]^-$ if coefficients corresponding to the period of each donor atom are introduced. The results for several complexes of the type *cis*- and *trans*-PtX_2L_2 and $[PtXL_3]^+$ also fit quite well when these same coefficients are used—the major difficulty being the identification of the electronic absorption bands from which to derive ΔE.

Extrapolation of this plot leads to the shift when the paramagnetic contribution is zero as -5906 ± 106 ppm (Ξ, 21,274(2) MHz). This is significantly lower than the earlier prediction of Drain (-2500 ppm)[164] or the similar figure Brown and Cohen[185] derived from the temperature dependence of some complexes *via* rather dubious assumptions. Even with this lower origin, the resonance of platinum metal is to low frequency ($K \simeq -2.48\%$ at 300 K).[164]

^{195}Pt nmr has also been used in the study of the 'one-dimensional metal,' $K_2[Pt(CN)_4]Br_{0.3}$, 2.3 H_2O.[186] Below 120 K, the shift is not very different from that of $K_2[Pt(CN)_4]$ but on raising the temperature, the resonance moves to high frequency as the metallic state develops and has altered by some 3000 ppm at 300 K. From measurements on oriented single crystals,[187] $\sigma_{\parallel}$ and $\sigma_{\perp}$ have been obtained and compared with the values for $K_2[Pt(CN)_4]$, 3 H_2O. The value of $\sigma_{\perp}$ is notably larger than for the non-metallic complexes studied, where it does not vary greatly.[188] In the latter, the $\sigma_{\parallel} - \sigma_{\perp}$ ranges from 2500 to 10 500 ppm in keeping for the results for ^{199}Hg.[189]

b. ^{195}Pt *Relaxation Times*

Freeman *et al.*[170] have determined the spin-lattice relaxation times of $Na_2[PtCl_4]$ and $Na_2[PtCl_6]$ in aqueous solution (not degassed) as 0.7 and 0.3 s respectively which are similar to the values reported by Pesek and Mason[171] for a range of complex ions (0.32–1.28 s). For both $[PtCl_6]^{2-}$ and $[PtCl_4]^{2-}$, T_1 decreased with increasing temperature implying that spin-rotation is the dominant mechanism.[171] The latter authors also report values for T_2 of the order of 5 ms in keeping with a linewidth of ca. 100 Hz. However, their reported widths at half height for $H_2[PtCl_6]$ and $H_2[PtBr_6]$ are equivalent to 6 ppm yet Hackbusch[172] has resolved resonances due to the isomers of $[PtBr_nCl_{(6-n)}]^{2-}$ which differ by only one ppm (Table 8.23). Clearly the linewidth observed by Pesek and Mason must be due to inhomogeneity of the field (the measurements were done at 9.75 MHz on a spectrometer whose usual field is equivalent to 21.4 MHz, and without lock). The very short values for T_2 must stem from the same cause and should be disregarded.

Freeman *et al.*[170] suggest that spectra may be obtained using concentrations of the order of 10^{-2} M and quote results for some 0.03 M solutions. By comparison with ^{113}Cd and ^{199}Hg, rather lower concentrations than this ought to be feasible. Of course, broad resonances will give much poorer S/N ratios and this is likely to be the case[170] when platinum is bonded to ^{14}N because of the unresolved coupling or to ^{75}As and $^{121,123}Sb$ which cause relaxation by scalar coupling of the second kind.[153,173] For double resonance (INDOR) determinations, the sensitivity depends on the spectral complexity of the ligand but for favourable situations such as with SMe_2, 0.1 M solutions are routine and 0.01 M perfectly possible.[153]

c. ^{195}Pt *Coupling Constants*

Jesson[71] and Appleton *et al.*[78] have surveyed ^1J(PtH) in platinum(II) hydrides. Values range from 700 to over 1300 Hz, largely according to the *trans* influence of the *trans* group. The sign is positive.[190] Apparently, ^1J(PtH) does not show the same proportionality to the coordination number of the metal as ^1J(PtP) since a similar range is observed for the unstable platinum(IV) hydrides.[176] Appleton *et al.*[78] also considered ^2J(PtH) of methyl derivatives of platinum(II) and (IV) which have values of 40–90 Hz and are negative.[178,191,192] For complexes of ligands such as PMe_3, ^3J(PtH) is positive with values in the range 15–50 Hz.[157] One bond ^{195}Pt-^{13}C couplings have been reviewed by Mann.[75] Values are ca. 600 Hz for methyl groups but are larger for CN^- and CO reaching 2155 Hz in *cis*-[$PtCl_2Et(CO)$].[193] Anderson *et al.*[176] have obtained the value −1600(100) Hz for $^1J(^{195}Pt, ^{29}Si)$ in *trans*-$PtCl(SiH_2Cl)(PEt_3)_2$. Despite its quadrupole moment, the relaxation of ^{14}N is relatively slow in some platinum complexes of [NCS]$^-$,

TABLE 8.27 Some ^{195}Pt-^{195}Pt spin-spin coupling constants

Compound		J(PtPt)/Hz
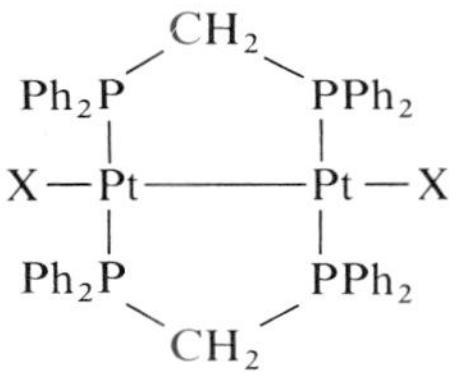 $X{-}Pt(\mu\text{-}Ph_2PCH_2PPh_2)_2Pt{-}X$ (Ph$_2$P–CH$_2$–PPh$_2$ bridges; X–Pt——Pt–X)	X = Cl	8197
	X = I	9007
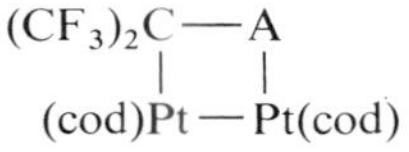 $(CF_3)_2C{-}A$ / (cod)Pt—Pt(cod)	A = O	+5355
	A = NH	+6235
$Pt_3(\mu\text{-}CO)_3(PBu^t_2Me)_3$		1770
$Pt_3(\mu\text{-}CNBu^t)_3(CNBu^t)_3$		188
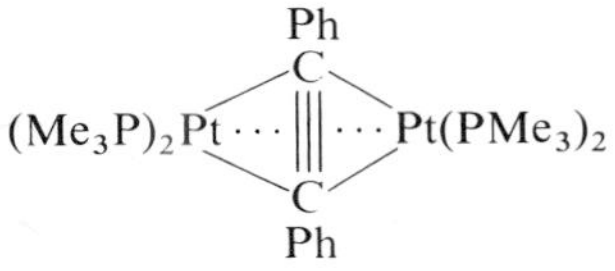 $(Me_3P)_2Pt\cdots(PhC{\equiv}CPh)\cdots Pt(PMe_3)_2$		470
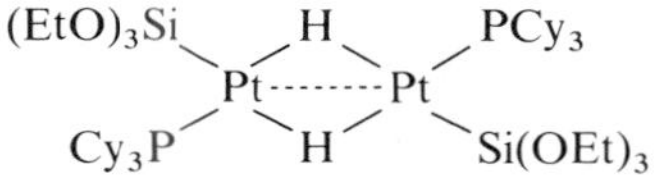 $(EtO)_3Si(Cy_3P)Pt(\mu\text{-}H)_2Pt(PCy_3)Si(OEt)_3$ (Pt- - - - - - - -Pt)		92
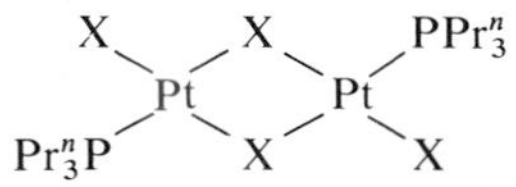 $X(Pr^n_3P)Pt(\mu\text{-}X)_2Pt(PPr^n_3)X$	X = Cl	200
	X = Br	239
	X = I	391

[a] C. Crocker, R. J. Goodfellow and M. Murray unpublished results.

$[NCO]^-$ and CH_3NC permitting the evaluation of $J(^{195}Pt, ^{14}N)$. For *N*-bonded cyanates and thiocyanates, $^1J(^{195}Pt, ^{14}N)$ varies from 200 to 500 Hz according to the *trans* influence.[182] In platinum(II) complexes of methyl isocyanide, $^2J(^{195}Pt, ^{14}N)$ is 90–120 Hz and is approximately halved on oxidation of the complex to Pt(IV).[194] For a few complexes of trimethylamine, coupling to ^{14}N is partially resolved[195] but in general for amine complexes, it is better to observe ^{15}N. The results of Pregosin *et al.*[196] for a selection of complexes of *n*-dodecylamine show that $^1J(^{195}Pt, ^{15}N)$ behaves in a similar manner to $^1J(^{195}Pt, ^{31}P)$ with respect to oxidation state and *trans*-influence. There is now a wealth of information on the latter coupling constant which has been reviewed by Nixon and Pidcock[160] and Appleton *et al.*[78] The sign was established as positive for *cis*- and *trans*-$PtCl_2(PEt_3)_2$ by McFarlane[197] and as, in view of its magnitude, it is unlikely to reverse in any complex, is the most common basis for the determination of the signs of other couplings to ^{195}Pt. The values of $^1J(^{195}Pt, ^{77}Se)$ and $^1J(^{195}Pt, ^{125}Te)$ which have been determined for a number of complexes of $SeMe_2$ and $TeMe_2$[161,162,198] are exceptionally sensitive to the other ligands present and $^1J(^{195}Pt, ^{77}Se)$ reverses sign on occasions. This sensitivity is probably related to the presence of a lone pair on the donor atom and the same applies to the very low values of $^1J(PtF)$ (200–250 Hz) found in $[PtF(PR_3)_3][BF_4]$[158] compared to ca. 2000 Hz in $[PtF_6]^{2-}$.[199] $^1J(^{195}Pt, ^{125}Te)$ is much larger for bridging $TeMe_2$ groups.[162] The variation in $^2J(PtF)$ of CF_3 derivatives of platinum has been discussed by Appleton *et al.*[78] The sign is positive for platinum(II) and (IV).[200]

A number of ^{195}Pt-^{195}Pt coupling constants have now been obtained from the analysis of ^{31}P spectra or ^{19}F-$\{^{195}Pt\}$ INDOR measurements (Table 8.27). The values display a remarkable range and seem to be largest for ligands of low *trans*-influence. Unfortunately, the signs have only been determined (positive) in the cases of two of the bigger values. The directly observed ^{195}Pt signals of $[Pt_9(CO)_{18}]^{2-}$ are multiplets due to coupling between the ^{195}Pt nuclei in different triangular layers. Only an average value of 274 Hz is seen due to rapid rotation of the triangles with respect to one another.[201]

8J GROUP IB

The NMR possibilities within group IB have hardly been exploited. However examination shows that attention has to be paid to both the NMR and chemical properties. Thus both ^{63}Cu and ^{65}Cu are readily detected but chemically there are problems. Copper(I) is diamagnetic, but only a limited number of compounds are known, the majority of which are readily oxidised to copper(II). Copper(II) is paramagnetic with a long T_{1e} leading to severe broadening with $[Cu(aq)]^{2+}$ being one of the most efficient relaxation agents.

Silver presents a two-fold problem. It is insensitive with receptivities of 0.195 (^{107}Ag) and 0.276 (^{109}Ag) with long T_1 values leading to difficulties in observation. This problem is exacerbated by the lack of stable complexes. In general equilibria occur requiring concentration dependent studies to identify individual complexes. It is therefore to be anticipated that silver NMR spectroscopy will be a very restricted technique.

Gold also presents problems, but these are mainly for the NMR spectroscopists as there are many well-defined compounds, especially for gold(III). Gold-197 is 100% abundant, $I = \frac{3}{2}$, but its D^C is only 0.142. The quadrupole moment of 0.59×10^{-28} cm^2 means that the resonances will be broad but they should be detectable in a range of gold compounds. The major difficulty will be the very restricted number of symmetric compounds.

8J.1 Copper, $^{63,65}Cu$

Both ^{63}Cu and ^{65}Cu have spin $I = \frac{3}{2}$ and intermediately-valued quadrupole moments of 0.16×10^{-28} cm^2 and 0.15×10^{-28} cm^2 respectively. Their natural abundance detection receptivities relative to ^{13}C are 365 and 201, making ^{63}Cu more readily detectable than ^{39}K and ^{35}Cl, and only slightly less so than ^{23}Na. Nevertheless copper has received little attention from NMR spectroscopists. While there have been several reports of copper resonances being observed in solid samples[202–207] only three studies of copper NMR in solution have been reported, two dealing with the $[Cu(CN)_4]^{3-}$ ion and one with $[Cu\{P(OMe)_3\}_4]^+$.

Ellis *et al.*[208] describe a method for observing a number of different nuclear species including ^{63}Cu in either continuous-wave or Fourier-Transform modes by making minor adaptations to a standard Varian XL-100 spectrometer. The method is illustrated with the ^{63}Cu signal from a 1 M solution of $K_3Cu(CN)_4$ in D_2O. Some shielding data for copper have been obtained by Yamamoto *et al.*[209] who find that the ^{63}Cu resonance in $[Cu(CN)_4]^{3-}$ occurs at a frequency 820 ppm higher than that for powdered CuCl used as an external reference. On the reasonable assumption that the copper shielding in solid CuCl where Cu^I is tetrahedrally coordinated will be about the same as that for $[CuCl_4]^{3-}$ in solution, this chemical shift indicates an increasing shielding order $CN^- < Cl^-$. Since the increasing shielding order for ^{59}Co which exhibits normal halogen dependence is $Cl^- < Br^- < I^- < [CN]^-$, copper shieldings appear to follow the inverse halogen dependence. Further evidence for this tendency is found in the fact that solid CuBr absorbs 57 ppm to low frequency of solid CuCl,[205] although this evidence is somewhat attenuated by the fact in the same study, solid CuI absorbs 10 ppm to high frequency of CuCl. Because of the extremely broad lines encountered in solids, the error limits on line positions and hence shieldings obtained from solid samples are necessarily large. The conclusions one draws from such data must be qualified accordingly.

The tetrahedral symmetry of the phosphine complex $[Cu\{P(OMe)_3\}_4]^+$ provides a system with ^{63}Cu-^{31}P coupling. Using a triple resonance technique involving observation of ^{31}P with successive decoupling of ^1H and then ^{65}Cu, McFarlane and Rycroft[210] have obtained the spectral parameters for ^{65}Cu. The temperature dependence of the line position, at 0.055 ppm K^{-1}, is 1–2 orders of magnitude less than those for ^{195}Pt, ^{103}Rh, and ^{59}Co, as is to be expected for a d^{10} ion with no ligand-field transitions to affect σ_p. Couplings observed were $^1J(^{63}Cu, ^{31}P) = 1210 \pm 10$Hz and $^4J(^{63}Cu, ^1H) = 0$.

8J.2 Silver, $^{107,109}Ag$

Naturally occurring silver consists of almost equal amounts of the isotopes ^{107}Ag and ^{109}Ag, both with $I = \frac{1}{2}$. Because of a 15% larger magnetogyric ratio, the less abundant ^{109}Ag is the more receptive to NMR observation but only by a factor of 1.4. Despite the low magnetogyric ratio, there were two separate reports of the resonances of both isotopes for metallic silver and strong aqueous solutions of $Ag[NO_3]$ in 1954.[211,212] The solutions were "doped" with $Mn[NO_3]_2$ to reduce the relaxation times so as to be compatible with the broadline technique used. Apart from further studies on the metal,[213,214] silver NMR was in abeyance until the application of Fourier transform methods in the last few years. In particular, Schwenk and co-workers,[215,216] using large sample volumes (up to 21 mm i.d.) and the "Quadriga" pulsed Fourier transform technique[68] which is especially suited to the observation of weak signals when T_1, $T_2 \gg T_2^*$, have been able to study silver resonances in the absence of paramagnetic relaxing agents so avoiding the effects the latter can have on chemical shifts. Thus a S/N ratio of better than 100:1 could be achieved for a 9 molal silver nitrate solution in 12 min.

a. ^{107,109}Ag *Chemical Shifts*

The silver resonances of $Ag[ClO_4]$, $Ag[NO_3]$ and AgF are highly concentration dependent[215,217] e.g. for 5 molal solutions the shifts are ca. -55, -40 and 10 ppm from the reference respectively. However, at high dilution they approach the same value i.e. that of the aquated Ag^+ ion and Schwenk and co-workers use this as the reference for silver shifts $\Xi(^{107}Ag)$ 4 047 897(6) Hz, $\Xi(^{109}Ag)$ 4 653 623(8).[218] Comparison of the value of the apparent magnetic moment in this reference with the value measured for silver atoms by the atomic beam magnetic resonance technique leads to an estimate of the shielding in aquated Ag^+ as -940 ppm with respect to the free atom. For the measurements of chemical shifts by alternate measurement of the resonances in the sample and a reference, a reference which gives a satisfactory signal in a short time is desirable. Burges *et al.* favour a solution 9.1 molal in $Ag[NO_3]$ and 0.24 molal in $Fe[NO_3]_3$ since this amount of the paramagnetic additive reduces T_1/T_2 by a factor of three and corresponds to the maximum in the plot of shift against added $Fe[NO_3]_3$.[215] The shift of this sample is -47 ppm.

The ratio $\nu(^{109}Ag)/\nu(^{107}Ag)$ was determined for several systems and did not vary outside the errors of measurement (0.9 ppm) although the chemical shifts span 270 ppm.[216] Accordingly, within these limits of error there is no primary isotope effect and at least for the present range of chemical shifts (< 1000 ppm) there is little point in measuring ^{107}Ag as well as the more sensitive ^{109}Ag.

Jucker *et al.*[216] have studied silver salts in a number of coordinating solvents. In all cases only one resonance was observed which must represent the average when more than one species is present i.e. chemical exchange is rapid. The shifts of $Ag[NO_3]$ in acetonitrile and propionitrile are very concentration dependent (ca. -10 ppm M^{-1}). At low concentrations, $[Ag(NCR)_2]^+$ predominates and extrapolation gives the shifts $[Ag(NCMe)_2]^+$, 335(2) ppm and $[Ag(NCEt)_2]^+$, 303(5) ppm. The silver resonances in these solutions have a negative temperature coefficient (ca. -0.5 ppm/°K) because increasing the temperature promotes dissociation. The isosolvation point for mixtures of acetonitrile and water occurs at 7.5 mole % CH_3CN in keeping with a strong preference for solvation by the nitrile.[217] The concentration dependence is much less for Ag[OOCMe] in pyridine since the formation constant for $[Ag(py)_2]^+$ is much higher. Extrapolation gives $\delta(^{109}Ag)$, 350(3) ppm. The formation of $[Ag(en)_2]^+$ is even more favoured and the resonances showed no detectable concentration effect: $\delta(^{109}Ag)$, 554(2) ppm. Several species occur in solutions of silver salts in the presence of thiosulphates and the resonance of AgCl in $Na_2[S_2O_3]$(3M) is by far the most concentration dependent of those reported. Extrapolation to minimal concentration gives the highest silver shift so far, 841(7) ppm for $[Ag(S_2O_3)_3]^{5-}$. Since the silver chemical shifts of AgF, AgCl, AgBr and AgI in 70 % aqueous ethylamine do not converge to the same point on dilution, complexes involving coordinated halide must be present.[216]

The recent accurate determination of the Knight shift in metallic silver by Sahm and Schwenk[219] ($K^{107} = -6130(170)$, $K^{109} = -6150(170)$ ppm *re* free atom) compares well with those of earlier reports.[211,212] However, the difference between the values for the two isotopes is significantly different from the hyperfine structure anomaly although theories for Knight shifts predict that they should be the same. The metal has also been studied up to and above its melting point[220] and at low temperature[213] including the measurement of relaxation times.

b. ^{107,109}Ag *Relaxation Times*

Schwenk and co-workers used a steady state procedure to evaluate T_1/T_2 and a method which observes the approach to steady state magnetisation to obtain $(T_1 + T_2)$.[5] For a 12 molal solution of $Ag[NO_3]$ in acetonitrile, T_1/T_2 was 5.0 at 298 K whilst $(T_1 + T_2)$ are 320 s and 292 s at 303 K and 283 K respectively. Extrapolation leads to ca. 4 min for T_1 at 298 K. For an 8.3 molal aqueous silver fluoride solution, T_1 was 49 ± 20 s and T_2 8.2 ± 2 s.[215]

c. ^{107,109}Ag *Coupling Constants*

Probably as a consequence of the lability of most silver complexes and of the low magnetogyric ratios, there are relatively few reports of spin-spin coupling to silver isotopes. Thus the first report of coupling to ^{1}H was in 1973 by Schmidbaur *et al.* for the unusually stable phospha-alkyl $[Ag_2\{(CH_2)_2P(CH_3)_2\}_2]$ where $^{2}J(^{107,109}Ag^{1}H)$ is 9.8 Hz.[221] The couplings ^{2}J(AgF) and ^{3}J(AgF) have been observed in several fluorocarbon derivatives.[222] Coupling to ^{31}P has been observed in phosphine complexes of the types $[AgXL_n]$ where n is 2, 3 or 4 but normally cooling is required to slow intermolecular exchange.[223] The separate couplings to each isotope can be observed with values from 220 to 760 Hz for ^{107}Ag according to the phosphine and the hybridisation of the silver atom.

8K GROUP IIB

Zinc, cadmium, and mercury can all be observed by NMR spectroscopy with a difficulty similar to that of ^{13}C nuclei. The observation of well-defined samples is relatively easy but problems can arise. In many cases, an averaged signal over many species is observed. This averaging may not be fast on the NMR time scale leading to broadened signals. Consequently broad signals which are concentration, solvent and temperature dependent are often found. In spite of these difficulties there is considerable interest in these nuclei arising in part from the possible biological applications and it is to be anticipated that considerable work will be done.

8K.1 Zinc, ^{67}Zn

Only one naturally occurring isotope of zinc possesses a nuclear magnetic moment, ^{67}Zn (4.11 %). Weaver[224] observed its resonance in aqueous solution in 1953 from which he obtained a value for the magnetic moment and confirmed that its sign was positive and that the nuclear spin was $\frac{5}{2}$. Despite the biological significance of zinc, ^{67}Zn has received very little attention since that date, presumably because of its low receptivity and quadrupole moment. Recently, Epperlein *et al.* have observed zinc resonances in the region of 4.8 MHz for a number of systems using a Fourier transform spectrometer operating at a magnetic field of 1.807 T. Their results are brought together and discussed in ref. 225. As an example, a 2.3 molal solution of $Zn[ClO_4]_2$ in a 10 mm diam. tube gave a S/N ratio of 35:1 in five minutes. Linewidths range from 20 Hz to 600 Hz, increasing with concentration and on lowering the temperature. The addition of 5% ethanol to an aqueous solution of $ZnBr_2$ increased the linewidth from 75 Hz to 115 Hz and no resonances could be obtained for $ZnCl_2$ or $ZnBr_2$ in ethanol, dimethyl ether or acetone. This suggests that excessive linewidth will preclude the measurement of zinc resonances in a large variety of situations.

The resonances of aqueous zinc perchlorate, nitrate and sulphate showed only a minimal variation with concentration and that of the perchlorate was independent of temperature. Accordingly, the value for $Zn[ClO_4]_2$, extrapolated to infinite dilution seems a reasonable choice for the standard for ^{67}Zn chemical shifts. The ratio of the zinc frequency to that of ^{37}Cl in the perchlorate counteranion was used to obtain the ratio ν_{Zn}/ν_H. Comparison of this with the value for free atomic zinc in the $^1S_0$ state obtained by optical pumping techniques leads to the shielding constant for hydrated zinc compared to the free atom as -690 ppm. For the zinc perchlorate reference, Ξ is ca. 6 256 753 Hz.

The resonances of aqueous solutions of zinc chloride, bromide and iodide showed large and non-linear dependencies on concentration which is understandable since they represent the average for the various complexes involving water and halide that are present in such solutions. For 1.1 molal solutions, the shifts are 70, 26 and 10 ppm to high frequency of the reference for $ZnCl_2$, $ZnBr_2$ and ZnI_2 respectively. Although this follows the usual order for halide-complexes, it is also the order of decreasing complexing ability of the halide ions towards zinc, $Cl^- > Br^- > I^-$. The degree of association with halide is no doubt significant in the temperature dependence of the resonances ($+0.5$ to $+0.7$ ppm/°K) and is responsible for the shifts to higher frequencies in D_2O solutions.

The only other zinc resonance to have been reported appears to be of the zintl alloy, LiZn.[226] With adjustment for the shift of $Zn[NO_3]_2$, the reference used, the shielding constant is -2690 ppm w.r.t. the free zinc atom.

8K.2 Cadmium, $^{111,113}Cd$

A quarter of cadmium nuclei have a spin of one half but this is split almost equally between two isotopes. The fractionally less abundant ^{113}Cd has ca. 10% more sensitivity but neither is especially difficult to observe. Thus Proctor in 1950 reported[227] the ratio of their resonance frequencies and their magnetic moments from measurements on saturated cadmium chloride and also confirmed that I was one half and μ negative for both. During the next twenty years, there are several reports of cadmium resonances in metals and alloys, often with measurements of aqueous salts to provide a reference for Knight shifts. Presumably because the likely returns of cadmium NMR were felt not to justify the outlay of a dedicated system, investigation by chemists did not begin until the development of multinuclear observation Fourier transform systems.[228–230] Such measurements have concentrated on ^{113}Cd, no doubt on the grounds of sensitivity although ^{111}Cd has the point that its resonance frequency is so close to that of ^{195}Pt that there would be no need to readjust the system to measure both. From measurements of $\nu(^{111}Cd)$ and $\nu(^{113}Cd)$ for $CdCl_2$ and $Cd[NO_3]_2$ solutions, Kruger *et al.*[231] have shown that their ratio is, within experimental error, the same as for the free atom despite a shift of about a thousand ppm with respect to the latter so there is no significant primary isotope effect and results for either isotope can be simply interconverted. Subsequent to the measurements of Hildenbrand and Dreeskamp in 1970,[232] reports of double resonance studies of cadmium have been rather sparse especially when compared to ^{199}Hg. Suitable satellites are frequently not observable because of the lability of cadmium complexes and the present results are largely confined to organometallic derivatives.[233,234]

Using 12 mm i.d. sample tubes, a satisfactory signal can be obtained for solutions 15–30 mM in cadmium (with natural abundance of ^{113}Cd) in a few hours with a Fourier transform spectrometer.[236] Recently, Byrd and Ellis[240] have achieved a ca. tenfold sensitivity enhancement by the use of 18 mm i.d. tubes. Further gains (up to ×8) are possible by the use of ^{113}Cd enriched materials which are likely to be most relevant for work with large molecular weight natural products. However, loss of sensitivity as a result of signal broadening often occurs with such systems e.g. Armitage *et al.*[239] required 10 h to observe the signals in 25 mM 96% ^{113}Cd solutions in the presence of carbonic anhydrases.

a. ^{111,113}Cd *Chemical Shifts*

Workers making direct measurements have used $Cd[ClO_4]_2$ as their reference for chemical shifts either at 0.1 M concentration[230,236,240,241] or extrapolated to infinite dilution[242,243] the latter being 0.15 ppm to low frequency.[242] Of course, this reference is not observable by double resonance techniques and Turner and White[233] and Kennedy and McFarlane[234] have used dimethyl cadmium. Although $Cd[ClO_4]_2$ at infinite dilution i.e. $[Cd(H_2O)_6]^{2+}$ seems a more fundamental reference, its Ξ value has not been determined directly and must be evaluated by the shift from $CdMe_2$ involving some uncertainty. Thus $CdMe_2$ seems a preferable reference at least for the shifts of organometallic compounds and it has been used for the results of such compounds collected in Table 8.28. The value $\Xi(^{113}Cd) = 22\,193\,173$ Hz is that of Kennedy and McFarlane[234] for the neat liquid at 24°C although the authors do not explicitly say that no internal reference was added. $\Xi(^{111}Cd)$ is 21 215 478.[234]

The chemical shift of dimethyl cadmium in solution is significantly solvent dependent. Presumably this has the same origin as the solvent dependence of $\delta(^{199}Hg)$ in $HgMe_2$ since they have similar ranges and sense although the ranking of solvents vs. shift is not identical.[244] Polar solvents cause shifts to low frequency e.g. tetrahydrofuran(1M), −67 ppm whilst chlorocarbons and hydrocarbons have the opposite effect e.g. dichloromethane(1 M), +3 and cyclohexane(1 M), +35 ppm.[236] However, this sensitivity to the medium is insufficient to explain the large differences between the results of different workers for $CdPr^n_2$ and $CdBu^n_2$ (see Table 8.28). Turner and White have checked their results and are confident that their samples are not contaminated with CdRX or CdX_2 which, by analogy with the ^{199}Hg shifts of the related compounds, might explain the low values of the other authors. It is of some importance to establish the truth of the situation since Turner and White's shifts follow the same pattern as $\delta(^{199}Hg)$ in HgR_2[245] $Me \gg Et < Pr^n < Bu^n$ whereas the alternative values show successive shifts to low frequency with increasing chain length. The methyl cadmium alkoxides are believed to be tetrameric[234] and for several of these complexes listed in Table 8.28, a second species appeared after a day in solution which, since they have very similar NMR parameters including δ(Cd) to those of the original species, the authors suggest might be hexamers. The shift of these compounds showed a temperature dependence of $+0.10 \pm 0.03$ ppm/K while for $CdMe_2$ it was -0.17 ± 0.05 ppm/K under the same conditions.

Cadmium complexes tend to be rather labile especially in aqueous solution so that although many of these solutions must contain more than one species, only one resonance is observed. One exception is the ^{113}Cd resonance observed in the presence of bovine carbonic anhydrase which is unaffected by the addition of further cadmium ions but rather a second resonance appears corresponding to a different site.[239] Solutions of $[Cd_{10}(SCH_2CH_2OH)_{16}]^{4+}$ also show two resonances corresponding to different environments but these coalesce on raising the temperature.[241]

TABLE 8.28 The cadmium chemical shifts for some organic cadmium derivatives

Compound	δ(Cd)	Medium	Ref.
$CdMe_2$	0[a]	neat	234
CdEtMe	−50	$CdMe_2/CdEt_2$	236
$CdEt_2$	−94	CH_2Cl_2(20 %)	233
$CdPr^n_2$	−48	CH_2Cl_2(20 %)	233
	−139	neat	236
$CdPr^i_2$	−207	CH_2Cl_2(20 %)	233
$CdBu^n_2$	−45	CH_2Cl_2(20 %)	233
	−154	neat	230
$CdBu^i_2$	−24	CH_2Cl_2(20 %)	233
$CdPh_2$	−314	dioxan (1 M)	236
CdMe(OMe)	−323	benzene (sat'd, 351 K)	234
CdMe(OEt)	−293	benzene (0.8 M)	234
$CdMe(OPr^n)$	−300	benzene (0.8 M)	234
$CdMe(OPr^i)$	−272	benzene (0.8 M)	234
$CdMe(OBu^n)$	−299	benzene (0.8 M)	234
$CdMe(OBu^i)$	−307	benzene (sat'd)	234
$CdMe(OBu^s)$	−275	benzene (0.8 M)	234
$CdMe(OBu^t)$	−259	benzene (0.8 M)	234
$CdMe(Oneo\text{-}C_5H_{11})$	−314	benzene (0.8 M)	234
$(CdMeOCH_2)_2CH_2$	−302	benzene (0.8 M)	234
$CdMe(OCH_2Ph)$	−312	benzene (0.8 M)	234
$CdMe(OCHPh_2)$	−322	benzene (0.8 M)	234
$CdMe(OCPh_3)$	−329	benzene (0.8 M)	234
CdMe(OPh)	−383	benzene (0.8 M)	234
$CdMe(OC_6H_3Me_2\text{-}2,6)$	−381	benzene (0.8 M)	234
$\{CdMe\mu\text{-}(OPh)py\}_2$	−298	benzene/pyridine	234
$CdMe(SPr^i)$	−31	benzene (0.8 M)	234
$CdMe(SBu^t)$	−44	benzene (0.8 M)	234

[a] Reference, $\Xi(^{111}Cd)$ = 21 215 478 Hz $\Xi(^{113}Cd)$ = 22 193 173 Hz.[234]

The resonances for solutions of $Cd[ClO_4]_2$ and $Cd[SO_4]$ are a little concentration dependent but tend to the same point at low concentrations as do those of the more concentration dependent $Cd[NO_3]_2$.[242,243] Krüger *et al.* have evaluated $\nu(^{111}Cd)/\nu(^1H)$ for this extrapolation and hence, by comparison with the ratio for the free cadmium atom from optical pumping techniques, derived the shielding in $[Cd(H_2O)_6]^{2+}$ as −1106(4) ppm from the free atom.[243] Because the association constants involved are much greater, the shift of cadmium halides show a large, nonlinear concentration dependence which actually turn in the cases of bromide and iodide.[242] A selection of results for some cadmium salts and complexes is given in Table 8.29 (referenced to $Cd[ClO_4]_2$ since this seems a more relevant comparison than $CdMe_2$ for most cases). Many of these solutions will contain several species so that although for specific conditions accurate chemical shifts can be found, it is not possible to associate these with particular complexes e.g. compare the results for glycine and imidazole systems under different conditions. Sometimes there is a surprising lack of consistency between

TABLE 8.29 The ^{113}Cd chemical shifts for some solutions of cadmium salts and complexes

Compound	Medium (Molarity)	$\delta(^{113}Cd)$	Ref.
$Cd[ClO_4]_2$	$H_2O(0.1)$	0^a	236
	$H_2O(3.0)$	−13	242
	MeOH(1.0)	−29	241
	DMSO(0.5)	−26	241
	DMF(1.0)	−46	241
	pyridine(0.2)	80	241
$Cd[SO_4]$	$H_2O(3.0)$	−5	242
$Cd[NO_3]_2$	$H_2O(3.0)$	−49	242
$Cd[CF_3SO_3]_2$	$H_2O(1.0)$	−3	241
$CdCl_2$	$D_2O(1.0)$	98	236
	12M HCl(1.0)	265	241
$CdBr_2$	$D_2O(1.0)$	109	236
	9M HBr(1.0)	167	241
CdI_2	$D_2O(1.0)$	55	236
$Cd[SCN]_2$	$D_2O(0.2)$	60	236
$[Cd(CN)_4]^{2-}$	$D_2O(1.0)$	510	236
$[Cd(NH_3)_6]Cl_2$	$1:1 NH_3/D_2O(1.0)$	287	236
$[Cd(EDTA)]^{2-}$	H_2O pH5(0.5)	85	241
Cd^{2+}-glutathione(1:2)	H_2O pH10(0.5)	318	241
$CdCl_2$-glycine(1:2)	H_2O pH7	161	239
(1:3)	H_2O pH7	177	239
(1:2)	H_2O pH9	211	239
$Cd[SO_4]$-glycine(1:2)	H_2O pH8	108	239
Cd^{2+}-imidazole(1:6)	H_2O pH8.3	133	239
(1:25)	H_2O pH5(0.5)	74	241
$Cd(SCH_2CH_2OH)_2$	DMSO(0.2)	655	241
$[Cd_{10}(SCH_2CH_2OH)_{16}][ClO_4]_4$	DMF(0.2)	520, 387	241
$Cd\{SPS(OPr^i)_2\}_2$	toluene(0.5)	401	241
$[NEt_4]_2[Cd\{S_2C_2(CN)_2\}_2]$	DMSO(0.5)	813	241
$Cd(S_2CNEt_2)_2$	DMSO(0.08)	215	241
$[NEt_4][Cd(S_2COEt)_3]$	$CHCl_3(0.5)$	277	241
$Cd[Mn(CO)_5]_2$	MeOH(0.8)	552	241

[a] Reference for this table, $\Xi(^{113}Cd) = 22\,178\,900$, shift from $CdMe_2$, −643.1 ppm.

results e.g. the approximately linear dependence of the shift of $Cd[NO_3]_2$ with concentration is estimated as 16.5, 14 or 11 ppm M^{-1} from refs. 242, 243 and 230 respectively. Kostelnik and Bothner-By[242] have examined the effects on δ(Cd) of additions of small amounts of a series of ligands to aqueous $Cd[SO_4]$. Although the extrapolation to a shift for a fully complexed species that they suggest seems doubtful in view of the several species likely to be present in equilibrium, the trends observed agree with the results in Table 8.29, viz: ligands coordinating through oxygen give shifts to low frequency whereas those coordinating through nitrogen or sulphur give high frequency shifts which are notably larger in the latter case. The not insignificant changes in $\delta(^{111}Cd)$ of $Cd[NO_3]_2$ and $CdCl_2$ that result on replacing H_2O by D_2O as solvent parallel the shifts these anions produce[243] and so presumably are a consequence of the effect of the solvent on the degree of association between the metal and the anions.

Maciel has suggested[230] that ^{113}Cd chemical shifts provide a means of studying the equilibrium and dynamics of complexation. However the number of different species, sometimes of different stereochemistry, likely to be present and the uncertainty as to the effect of every component on the chemical shift of each one [cf. $\delta(^{199}Hg)$] suggest that such studies are not likely to be reliable, at least at our present state of knowledge. Similarly, the significant contribution from chemical shift anisotropy is likely to limit the usefulness of relaxation time measurements.

The ^{113}Cd resonances of solid cadmium sulphide, selenide and telluride have been measured by Look[246] and the first of these also by Krüger *et al.*[243] The approximate shifts from $CdMe_2$ are +60, −70 and −320 ppm.

The Knight shift of metallic cadmium has a large anisotropy (ca. 10% of the shift)[247] which inverts on cooling to 1.2 K.[248] The asymmetric broadening of the resonance signal may be considerably narrowed by spinning at the "magic angle" leading to a more accurate value for the shift[249] of -5380 ± 10 ppm from the free atom.

b. $^{111,113}Cd$ *Relaxation Times*

Holz *et al.*[235] have reported some investigations into the spin-lattice relaxation of ^{113}Cd in aqueous $Cd[ClO_4]_2$ (unfortunately incomplete as a result of a perchloric acid explosion and consequent damage to one of the samples and an experimentalist!). The increase in the relaxation rate at higher temperatures in H_2O implies a spin-rotation contribution. In D_2O solution, T_1 (63 s) is over twice as long as in H_2O (29 s) so there must be a significant contribution from dipole-dipole interactions with the protons of water. In fact, with ^1H irradiation, an NOE of −1.2 is observed which compared to the maximum of −2.5 indicates a ca. 50% contribution from the dipolar mechanism.[236] The rate is also greater when the D_2O is ^{17}O enriched.[235] Dipolar interactions with ^{17}O should only contribute ca. 4% so scalar coupling of the first kind is the most likely process from which Holz *et al.* calculate $J(^{113}Cd^{17}O)$ as 248 Hz. Schwartz[237] has proposed that transient chemical shift anisotropy could make a significant contribution to the relaxation of heavy metal nuclei and this is presumably the predominant mechanism for ^{113}Cd in D_2O. The spin lattice relaxation times of cadmium halides in D_2O decrease in the order Cl > Br > I (viz. 40.5, 7.1, and 6.5 s respectively) and the dipolar contribution in H_2O decreases similarly.[236] The suggestion of Cardin *et al.* that relaxation for CdI_2 may be due to scalar coupling of the second kind does not seem justified in view of the similarity of the rates for $CdBr_2$ and CdI_2—chemical shift anisotropy seems more probable. Increasing alkyl chain length in CdR_2 lowers T_1 from 3.5 s for $[CdMe_2]$[236] to 0.2 s for $[CdBu^n_2]$.[230] The likely mechanism is chemical shift anisotropy with spin-rotation contributing above 25°C in the case of $CdMe_2$[236] for which there was no observable NOE.[228] The quite long (ca. 1 min) relaxation times that can occur for ^{113}Cd in the more symmetric environments are a potential hazard to the sensitivity of cadmium observations. They may be counteracted by the addition of paramagnetic salts[236] (but with as yet undetermined effects on the chemical shift cf. ref. 238) or the use of the "Quadriga" technique.[68] A further hazard is the possibility of a partial negative NOE, e.g. see ref. 239.

c. $^{111,113}Cd$ *Coupling Constants*

In dialkyl cadmiums, $^2J(^{113}Cd, ^1H)$ is ca. 50 Hz[233] and is positive[232] (K negative as usual) but the value is much larger in the methyl cadmium alkoxides[234] in keeping with the lower *trans* influence of alkoxide. Values of $^3J(^{113}Cd, ^1H)$ for CdR_2 are larger than the two bond coupling but with opposite sign.[233] In these compounds $^1J(^{113}Cd, ^{13}C)$ is ca. 500 Hz[236] and is negative (K positive).[232] Cadmium-phosphorus coupling has been observed for CdI_2L_2 where L is an alkyl or alkyl-aryl phosphine,[250,251] but cooling is normally required to reduce the rate of phosphine exchange. $^1J(^{113}Cd, ^{31}P)$ decreases from nearly 1400 to below 1200 Hz on increasing the number of phenyl groups on the phosphine so paralleling the behaviour of $^1J(^{199}Hg, ^{31}P)$ but in contrast to the couplings to other metals.[160] In $Cd_2I_4(PEt_3)_2$, the coupling is larger (1710 Hz) and has the same sign as $^2J(PH)$[252] which may reasonably be presumed to be negative in such complexes.[197]

8K.3 Mercury, $^{199,201}Hg$

Mercury has two isotopes with non-zero nuclear spins, both occurring in reasonable proportions. The more receptive, ^{199}Hg with $I = \frac{1}{2}$ can be observed without too much difficulty and its resonance in an aqueous solution of $Hg_2[NO_3]_2$ was reported by Proctor and Yu[102] as early as 1951. Although of similar receptivity to ^{13}C and ^{67}Zn in natural abundance, the other isotope, ^{201}Hg has proved much more difficult to observe presumably as a consequence of the quadrupole moment associated with a value of $\frac{3}{2}$ for I. Proctor and Yu[102] were unable to detect its resonance and the first report was in 1965 and then for metallic mercury.[253] As yet there seem to be no observations for solutions.

Following Proctor and Yu, there were several ^{199}Hg measurements on aqueous solutions of mercury salts and organo-mercurials using wideline instrumentation.[120,254–256] McLauchlan *et al.* reported a double resonance study of dimethyl mercury in 1966[257] since when this technique esp. the INDOR mode has played an important role in the determination of ^{199}Hg chemical shifts as well as the signs of coupling constants. In 1973, Maciel and Borzo[258] reported the observation of ^{199}Hg resonances using a Fourier transform spectrometer adapted for multinuclear operation and this approach is now providing important information, particularly on systems not amenable to double resonance.

Maciel and Borzo[258] estimate that measurements on mercury concentrations as low as 0.01 M would be practicable using large tubes and long accumulation times and Sens *et al.*[244] obtained an S/N ratio of 9:1 after 10 h for a 25 mM solution of $HgCl_2$. However, they concluded that observation at the concentrations involved for biological or pesticide work was not feasible with their existing instrumentation, at least without isotope enrichment. For comparison, INDOR measurements on 0.2 M solutions of methyl mercury halides are routine and satisfactory results were obtained on a ca. 50 mM aqueous solution of [HgClMe].[262] With current generation spectrometers and wide tubes, an improvement in the INDOR sensitivity of up to five-fold ought to be possible.

a. ^{149,201}Hg *Chemical Shifts*

The generally accepted reference for ^{199}Hg chemical shifts is dimethyl mercury. It gives a strong signal even in a capillary as an external reference, is very convenient for double resonance determination and being a liquid, problems of solvent and concentration dependence can be avoided. Unfortunately almost all other shifts are then negative and it is, of course, a serious health hazard. For neat dimethyl mercury, $\Xi(^{199}Hg) =$ 17 910 841 Hz[261] and agrees with the value of Johannesen and Duerst[263] when TMS was added as a reference; this is some 60 Hz greater than the values when benzene is added (ref. 245 and probably 189) which correspond to the solvent effects reported by

Sens *et al.*[244] The ratio of the Larmor frequency of the free mercury atom to that of the proton in water has been measured by optical pumping techniques[264] from which $\Xi(^{199}Hg)$ is 17 827 145(30) Hz, equivalent to a shift from $HgMe_2$ of $-4673(2)$ ppm or, alternatively, a shielding of $-4695(2)$ ppm for dimethyl mercury *re* the free atom.

Although some mixing with *d*-functions may occur, the valence shell in mercury compounds is mainly of *s* and *p* origin. The equation of Jameson and Gutowsky[265] for the paramagnetic contribution to the shielding is considerably simplified if *d*-orbital terms are neglected and then (Eq. 8.10) becomes relatively easy to apply. The

$$\sigma_p = -(2e^2\hbar^2\mu_0/12\pi\Delta Em^2)\langle r^{-3}\rangle P_u = -\tfrac{2}{3}A \cdot P_u \tag{8.10}$$

term P_u involves the coefficients of the *p*-electrons and Jameson and Gutowsky have described it as representing the "unbalance" in the valence electrons presumably because the maximum value of 2 is attained when one *p* orbital is filled and the other two empty and vice versa (note however this refers to unshaired electrons i.e. lone pairs). This label is unfortunate and has lead to misconceptions e.g. P_u for a tetrahedrally bonded carbon has a maximum value of $\frac{3}{2}$ whereas it is only one for fluorine[265] despite the latter being more "unbalanced." For linearly bonded mercury, P_u has a maximum value of 1 whilst when three coordinate or four coordinate it is $\frac{3}{2}$.

These equations also predict that $\sigma_{\parallel} - \sigma_{\perp}$ should be $+A$ for a linear mercury compound. Combining the values of $\sigma_{\parallel} - \sigma_{\perp}$ measured by Kennedy and McFarlane[189] for $HgMe_2$ and HgMeX, X = Cl, Br and I, with the shielding which should be $-\frac{2}{3}A$, the origin for the paramagnetic contribution comes out as -5000, -4500, -4500 and -4700 respectively. These agree remarkably well with the value for the free atom in view of the approximations made. However, using the value 221 nm[266] to calculate ΔE for $HgMe_2$, $\langle r^{-3}\rangle$ is 27 a.u. which seems rather high.

Mercury chemical shifts are very medium dependent. This is not so surprising for salts which may solvate or dissociate but is more remarkable for the relatively inert organic derivatives. There is a lot of data on the shifts of some compounds in different solvents and some representative values for $HgMe_2$ and HgMeCl are collected in Table 8.30. In general, the more polar solvents cause shifts to low frequency[244] which are

TABLE 8.30 Effect of medium on the ^{199}Hg chemical shifts[a] of dimethyl mercury and methyl mercury chloride

Medium	$HgMe_2$	HgClMe
DMSO	-108^b	-848^b
Pyridine	-94^b	-780^c
Acetonitrile	-78^b	-862^c
THF	-76^b	-861^c
Benzene	-50^b	-813^c
CH_2Cl_2	-46^c	-814^c
Hexane	$+5^b$	
CH_2Cl_2 + $[NBu^n_4]Cl^d$	-46^c	-702^c

[a] From neat $HgMe_2$. [b] 1.0 M conc. from ref. 244. [c] 0.2 M conc. from ref. 262. [d] One equivalent weight per Hg.

accompanied by increases in ${}^1J$(HgC) and ${}^2J$(HgH).[267] These trends have led various workers to suggest that the solvent is coordinating to mercury[267] and Kennedy and McFarlane[189] surmised that the "imbalance" would be less on coordination of a solvent molecule leading to a smaller value of P_u. However, as explained above, P_u should increase, causing a shift to high frequency. Although HgMeCl has a much greater affinity for coordinating additional ligands than $HgMe_2$, the range of shifts is similar for both. The situations of pyridine and added chloride ion are very different for the two compounds. Pyridine is the most likely of the solvents in Table 8.30 to coordinate to mercury and vibrational spectra[262] show that in low polarity solvents there is partial formation of $[HgMeCl_2]^-$ on addition of $[NBu^n_4]Cl$. Thus, the especially high frequencies of ^{199}Hg (the direction predicted by changes in P_u) in HgMeCl in the presence of pyridine and chloride ion are to be associated with coordination (in the usual meaning) and the effect of the other polar solvents on HgMeCl and all media on $HgMe_2$ are something quite distinct. Presumably the latter involves interactions of the quadrupolar $HgMe_2$ with the fields produced by the polar solvents in a solvation shell but there is no simple correlation with dielectric constant or dipole moment. The addition of halide to HgX_2 and $Hg(CN)_2$ to produce $[HgX_3]^-$ and $[Hg(CN)_2X]^-$ is accompanied by high frequency shift of the mercury resonance (Table 8.31) of similar magnitude to that for HgMeCl.[262]

TABLE 8.31 The ^{199}Hg chemical shifts for some solutions of mercury salts and complexes

Compound	Medium	$\delta(^{199}Hg)^a$	Ref.
Hg^{2+} aquated	$Hg[ClO_4]_2$ inf.dil.	−2253	260
$Hg[NO_3]_2$	0.84 M HNO_3(2 M)	−2361	258
$Hg(OCOCH_3)_2$	1 M AcOH(0.5 M)	−2389	269
$HgCl_2$	EtOH(0.5 M)	−1497	244, 261
$HgBr_2$	EtOH(0.5 M)	−2152	261
HgI_2	DMSO(1 M)	−3106	261
$Hg(CN)_2$	pyridine(2 M)	−1070	258
$[NBu^n_4][HgCl_3]$	CH_2Cl_2(0.5 M)	−1183	261
$[NBu^n_4][HgBr_3]$	CH_2Cl_2(0.5 M)	−1935	261
$[NBu^n_4][HgI_3]$	CH_2Cl_2(0.5 M)	−2822	261
$[NBu^n_4][Hg(CN)_2Cl]$	CH_2Cl_2(0.5 M)	−878	261
$[NBu^n_4][Hg(CN)_2Br]$	CH_2Cl_2(0.5 M)	−898	261
$Hg_2Cl_4(PBu^n_3)_2$	$CDCl_3$(0.2 M)	−741	261
$Hg_2Br_4(PBu^n_3)_2$	$CDCl_3$(0.2 M)	−1020	261
$Hg_2I_4(PBu^n_3)_2$	$CDCl_3$(0.2 M)	−1760	261
$HgCl_2(PBu^n_3)_2$	$CDCl_3$(0.5 M)	−404	261
$HgBr_2(PBu^n_3)_2$	$CDCl_3$(0.5 M)	−471	261
$HgI_2(PBu^n_3)_2$	$CDCl_3$(0.5 M)	−716	261
$[HgCl(PMe_3)][BF_4]$	D_2O(0.1 M)	−1299	279
$[HgBr(PMe_3)][BF_4]$	D_2O(0.1 M)	−1490	279
$[HgI(PMe_3)][BF_4]$	D_2O(0.1 M)	−1856	279
$[Hg(CN)(PMe_3)][BF_4]$	D_2O(0.1 M)	−1267	279
$[Hg(PMe_3)_2][BF_4]_2$	D_2O(0.1 M)	−1159	279

[a] In ppm to high frequency of $[HgMe_2]$.

The sensitivity of mercury chemical shifts to the medium makes the presentation of a representative set of values very difficult. It is impracticable to list all the recorded shifts. There is no one solvent in which all have been or can be measured and, in any case, the effect of a particular solvent is likely to differ between compounds. Thus, somewhat arbitrarily, we have selected the least polar medium where this does not involve unreasonable additional errors or break up a series.

The resonance of aqueous $Hg[ClO_4]_2$ shows a small linear dependence on concentration (4.19 ppm per mole %)[260] and extrapolation to infinite dilution leads to a value for the aquated Hg^{2+} ion. The shielding compared to the free atom is −2432 ppm.[260] The shift of aqueous $Hg[NO_3]_2$ is not far from this point but is shifted to low frequency in nitric acid.[258,260] Extrapolation of the results for aqueous solutions of $[Hg_2][ClO_4]_2$ gives a shift of ca. −1450 ppm or a shielding of ca. −3200 ppm *re* the free atom.[260] Not surprisingly, the resonances of aqueous $HgCl_2$ solutions are very concentration dependent.[260] Halide exchange seems to be fast compared to the NMR timescale so that mixtures of $HgCl_2$ and $HgBr_2$ in ethanol gave only a single peak at an intermediate position.[261] Similarly, the values of the shifts for $[HgX_4]^{2-}$ reported by Dharmatti[254] must represent averages for the various species formed by partial dissociation.[256] As with the resonances of other metal nuclei, there is a trend to low frequency for halide complexes in the order Cl > Br > I. The differences are much less for $[HgX(PMe_3)]^+$ and HgMeX (Table 8.33) than for HgX_2 (Table 8.31) in keeping with the *trans* influence of the ligand opposite the halide. In contrast, the resonances of some silyl and germyl derivatives[268] are to high frequency of $HgMe_2$ (Table 8.32). Dessy *et al.* in 1959[255] drew attention to the large shifts to lower frequencies that are produced by replacing a β-H in $HgMe_2$ by a methyl group i.e. for the series $HgMe_2$, $HgEt_2$, $HgPr^i_2$ etc. but as yet the effect seems unexplained. Extending the chain beyond $HgEt_2$ results in small changes to high frequency giving the order Me ≫ Et < Pr^n < Bu^n (Table 8.32) which also occurs for HgRX (Table 8.33) but with ca. half the magnitude i.e. the same per alkyl group present. Another example of the sensitivity of $\delta(^{199}Hg)$ to the structure of the ligands is the effect of *para* substituents in $Hg(p\text{-}C_6H_4X)_2$[269] and $Hg(p\text{-}C_6H_4X)Cl$[270] for both of which a range of over 100 ppm is reported. Hudson and Smith[271] quote the frequencies of ^{199}Hg in two porphyrin derivatives but unfortunately there is no reference from which to evaluate the shift.

The measurements by Blumberg *et al.*[253] on metallic mercury dispersed in vaseline give the Knight shift as −27,240 ppm from the free atom. Weinert and Schumacher[272] have observed the resonance for a single crystal or mercury at 1.2 K and found an anisotropy in the Knight shift of 7.8% of the isotropic shift.

b. $^{199,201}Hg$ *Relaxation Times*

The spin-lattice relaxation time for $HgCl_2$ (1.4 M ethanolic solution) is ca. 1.4 s[258] while for dimethyl- and divinyl-mercury, Sens *et al.*[244] report 0.87 s and 0.25 s respectively. The temperature variation (−35 to +40°C) was only marginally greater than the errors but it was possible to identify a maximum for T_1 at ca. 18°C. There was no observable NOE, so it is concluded that the two competing mechanisms are spin-rotation (at higher temperatures) and chemical shift anisotropy (at lower temperatures). This appears contrary to the results of Lassigne and Wells[259] who have obtained spin-lattice relaxation rate data for ^{199}Hg in $HgMe_2$ by comparison of the relaxation rates of the protons in molecules containing ^{199}Hg with those containing the other isotopes. They conclude that SR is the major contribution even at 234 K for which they estimate CSA as only 20%. However, their measurements are at 1.41 T which would reduce the

TABLE 8.32 The ^{199}Hg chemical shift in some organo-mercurials

Compound	Medium	$\delta(^{199}Hg)$	Ref.[a]
$HgMe_2$	neat	0^b	262
$HgEt_2$	neat	−289	244, 245
$HgPr^n_2$	neat	−240	245, 255
$HgPr^i_2$	neat	−640	245, 255
$HgBu^n_2$	CCl_4(2 M)	−208	245
$HgBu^t_2$	toluene (50%)	−829	189
$Hg(CH_2CMe_3)_2$	neat	−153	189
$Hg(CH_2C_6H_5)_2$	CH_2Cl_2(1 M)	−703	245
$Hg(CH_2CH_2C_6H_5)_2$	neat	−305	258
$Hg(CH_2COOCH_3)_2$	CH_2Cl_2(2 M)	−772	245
$Hg(CH{=}CH_2)_2$	neat	−642	244, 245
$Hg(CCl{=}CH_2)_2$	CH_2Cl_2(2 M)	−1145	245
$Hg(CH{=}CHCl)_2$		−840	270
$Hg(CF{=}CF_2)_2$	C_6F_6 (90%)	−957	263
$Hg(CF{=}CF_2)Me$	C_6F_6 (86%)	−482	263
$Hg(C_5H_5)Me$	$CDCl_3$ (20%)	−707	280
$Hg(C_6H_5)_2$	CH_2Cl_2(1 M)	−745	244, 245, 276
$Hg(C_6F_5)_2$	acetone (sat'd)	−920	276
$Hg(C_6H_5)(C_6F_5)$	CH_2Cl_2(1 M)	−832	245
$Hg(C_6H_5)(CBr_2Cl)$	DMSO(1 M)	−1187	244
$Hg(SiH_3)_2$		196	268
$Hg(SiH_3)(SiMe_3)$	benzene	327	268
$Hg(SiMe_3)_2$		481	268
$Hg(GeH_3)_2$		−147	268
$Hg(GeH_3)(SiMe_3)$		159	268

[a] Including information on the shift under different conditions. [b] Reference, $\Xi(^{199}Hg)$ = 17 910 841 Hz.

CSA rate to less than one third of that at 2.35 T, the field used by Sens *et al.*[244] and further their deduced value of $\sigma_{\parallel} - \sigma_{\perp}$ (4600 ppm) is much less than that recently reported by Kennedy and McFarlane[189] (the latter value of 7475 ppm would increase the CSA rate by × 2.6). These T_1 values present no problems for pulse Fourier transform observation nor do the linewidths of 3 to 37 Hz reported by Maciel and Borzo.[258] However, Krüger *et al.*[260] have found that the linewidths for aqueous $Hg[ClO_4]_2$ increase with concentration up to 90 Hz and for $Hg_2[NO_3]_2$ (1 M in DNO_3) was 700 Hz! Large linewidths could result from exchange between inequivalent sites or scalar coupling of the second kind e.g. when bonded to ^{127}I or ^{75}As.[173] The large range of chemical shifts compared to the frequency range of the spectrometer is a possible hazard and seems the most reasonable explanation of the inability of Sens *et al.*[244] to find resonances for $HgBr_2$ and HgI_2 in DMSO since Goodfellow and Taylor observed them without difficulty.[261]

c. $^{199,201}Hg$ *Coupling Constants*

The spin-spin coupling of ^{199}Hg to ^1H or ^{19}F in organo mercury derivatives has received considerable attention. Values of $^2J(HgH)$ are normally in the range 100 to 300 Hz

TABLE 8.33 The ^{199}Hg chemical shifts of some organo-mercury derivatives

Compound	Medium	$\delta(^{199}Hg)$[a]	Ref.[b]
HgMeCl	CH_2Cl_2(0.2 M)	−814	244, 262[c]
HgMeBr	CH_2Cl_2(0.2 M)	−915	244, 262[c]
HgMeI	CH_2Cl_2(0.2 M)	−1097	244, 262
HgMe(CN)	C_6H_6(0.1 M)	−719	279
HgMe(SCN)	CH_2Cl_2(0.2 M)	−705	262
$HgMe(OCOCH_3)$	3.5 M AcOH(0.5 M)	−1136	269
$HgMe(ONO_2)$	C_6H_6(0.1 M)	−1211	[d]
HgMe(SMe)	CH_2Cl_2(0.1 M)	−497	[d]
HgMe(SPh)	CH_2Cl_2(0.1 M)	−587	[d]
$(HgMe)_2S$	CH_2Cl_2(0.1 M)	−316	[d]
$[HgMe(OD_2)][NO_3]$	D_2O(ca. 1 M)	−1146	[d]
$[HgMe(NH_3)][BF_4]$	CD_3OD(0.1 M)	−943	[d]
$[HgMe(PMe_3)][NO_3]$	D_2O(0.1 M)	−655	279
HgEtCl	C_6H_6(0.2 M)	−978	244, 262
HgEtBr	C_6H_6(0.2 M)	−1071	262
$HgPr^nCl$	C_6H_6(0.2 M)	−947	262
$HgPr^nBr$	C_6H_6(0.2 M)	−1039	262
$HgBu^nCl$	C_6H_6(0.2 M)	−944	262
$HgBu^nBr$	C_6H_6(0.2 M)	−1036	262
$Hg(CH_2CMe_3)Cl$	THF(0.5 M)	−990	269
$Hg\{CH_2C(OMe)Me_2\}Cl$	$CHCl_3$(40 mol %)	−1145	[e]
$Hg\{CH_2C(OMe)Me_2\}Br$	$CHCl_3$(40 mol %)	−1234	[e]
$Hg\{CH_2C(OMe)Me_2\}I$	$CHCl_3$(40 mol %)	−1386	[e]
$Hg\{CH_2C(OMe)Me_2\}(CN)$	$CHCl_3$(40 mol %)	−927	[e]
$Hg\{CH_2C(OMe)Me_2\}(SCN)$	$CHCl_3$(40 mol %)	−939	[e]
$Hg(C_6H_{11})(OCOPr^n)$	DMSO(0.5 M)	−1176	244
Hg(CH=CHCl)Cl		−1150	270
$Hg(C_5H_5)Cl$	$CDCl_3$(5 %)	−1204	280
HgPhCl	DMSO(0.5 M)	−1187	244
$HgPh(OCOCH_3)$	DMSO(1 M)	−1437	244, 276
$HgPh(OCOC_2H_5)$	DMSO(0.5 M)	−1425	269
$HgPh(OCOC_6H_5)$	DMSO(1 M)	−1428	244
$Hg_2Ph_2\{o\text{-}(OCO)_2C_6H_4\}$	DMSO(0.5 M)	−1443	244
$HgPh(o\text{-}C_6H_4OH)$	DMSO(0.5 M)	−1440	269
$[HgPh(PMe_3)][NO_3]$	D_2O(0.1 M)	−1044	279
$Hg(p\text{-}C_6H_4CH_3)Cl$	DMSO(1 M)	−1151	244
$Hg(C_6F_5)(OCOCH_3)$	DMSO (sat'd)	−1426	276

[a] In ppm to high frequency of $HgMe_2$. [b] Including information on the shift under different conditions. [c] V. Lucchini and P. R. Wells, *J. Organometallic Chem.* **92**, 283 (1975). [d] R. P. Carlile, P. L. Goggin and R. J. Goodfellow, unpublished results. [e] T. Ibusiki and Y. Saito, *Chem. Lett.* 311 (1974).

according to the *trans* group and medium.[73,78,267] The sign of ^2J(HgH) in $HgMe_2$ was shown to be negative by double resonance,[257] and the results of further sign determinations are listed by Anet and Sudmeier.[274] Fedorov *et al.*[275] give ^{199}Hg-^{19}F couplings for a range of fluoroalkyl derivatives and Johannesen and Duerst[263] have determined the signs in two perfluorovinyl compounds—^2J(HgF) is positive. For CF_3 derivatives, ^2J(HgF) varies from ca. 1200 to nearly 2000 Hz and is positive.[252] The signs of the couplings in phenyl and pentafluoro-phenyl mercury derivatives have been established by McFarlane.[276] Values of $^1J(^{199}$Hg, ^{13}C) vary from ca. 600 to 1800 Hz and are positive as expected.[257,274] In $HgX_2(PR_3)_2$, ^1J(HgP) is 3700-5000 Hz but can be considerably larger in $Hg_2X_4(PR_3)_2$[277] (up to 7500 Hz) or in some phosphinate complexes, HgX{PO(OEt$_2$)}.[278] The sensitivity to the *trans* ligand in the linear species $[HgX(PMe_3)]^+$ is much greater than for $^1J(^{195}$Pt, ^{31}P) in comparable situations viz. Cl(7852) > Br(7308) > CN(6966) > I(6357) > PMe_3(5173) ≫ Ph(2641) > Me(1747 Hz) but remains positive throughout.[279] The directly observed mercury spectra of the phosphine complexes in Table 8.31 show the appropriate multiplicity confirming the number of phosphines coordinated,[261] while for $Hg(CH_3)(C_5H_5)$, the intensities of the lines of the multiplet observed by ^{1}H-{^{199}Hg} INDOR agree with those predicted if there is equal coupling to all five cyclopentadienyl protons as a result of fast exchange of the Hg-C bond.[280]

REFERENCES

1. a. Y. A. Buslaev, V. P. Tarasov, M. N. Buslaeva and S. P. Petrosyants, *Doklady Akad. Nauk. S.S.R.* **209**, 882 (1973); *Doklady Phys. Chem.* **209**, 290 (1973)
   b. V. P. Tarasov and Y. A. Buslaev, *J. Magn. Reson.* **25**, 197 (1977)
2. G. A. Melson, D. J. Olszanski and E. T. Roach, *J. Chem. Soc., Chem. Commun.* 229 (1974)
3. Y. A. Buslaev, S. P. Petrosyants, V. P. Tarasov and V. I. Chagin, *Zhur. neorg. Khim.* **19**, 1790 (1974); *Russ. J. Inorg. Chem.* **19**, 975 (1974)
4. C. Deverell and R. E. Richards, *Mol. Phys.* **6**, 289 (1966)
5. J. Kronenbitter and A. Schwenk, *J. Magn. Reson.* **25**, 147 (1977)
6. C. Hassler, J. Kronenbitter and A. Schwenk, *Z. Physik A*, **280**, 117 (1977)
7. M. F. Lappert and R. Pearce, *J. C. S. Chem. Comm.* 126 (1973)
8. D. G. H. Ballard and R. Pearce, *J. C. S. Chem. Comm.* 621 (1975)
9. J. Holton, M. F. Lappert, G. R. Scollary, D. G. H. Ballard, R. Pearce, J. L. Atwood and W. E. Hunter, *J. C. S. Chem. Comm.* 425 (1976)
10. K. Nakamura and K. Kawamura, *Bull. Chem. Soc. Japan* **44**, 330 (1971)
11. J. Reuben, *J. Chem. Phys.* **79**, 2154 (1975)
12. J. Reuben, *J. Amer. Chem. Soc.* **97**, 3822 (1975)
13. J. Reuben and Z. Luz, *J. Chem. Phys.* **80**, 1357 (1976)
14. L. S. Smith, D. C. McCain and D. L. Wertz, *J. Amer. Chem. Soc.* **98**, 5125 (1976)
15. H. G. Bohn, R. R. Arons, H. Luetgemeier and K. J. Fischer, *J. Magn. Magn. Mater.* **2**, 67 (1976), *Chem. Abs.* **85**, 151484 (1976); R. R. Arons, Report AED-CONF-75-449-001 (1975), *Chem. Abs.* **85**, 85154 (1976)
16. I. Ursu, D. E. Demco, V. Simplaceanu, N. Valcu and N. Ilie, *Magn. Resonance and Related Phenomena, Proc. Congr. Ampere, 18th* (1974), **2**, 533 (1975), *Chem. Abs.* **83**, 170511 (1975)
17. J. Butterworth, *Phil. Mag.* **3**, 1053 (1958)
18. J. Faust, R. Marrus and W. A. Nierenberg, *Phys. Letters* **16**, 71 (1965)
19. F. Y. Fradin and M. Brodsky, *Int. J. Magnetism* **1**, 89 (1970)
20. W. Kolbe and N. Edelstein, *Phys. Rev.* **B4**, 2869 (1971)
21. R. G. Kidd, R. W. Matthews and H. G. Spinney, *J. Amer. Chem. Soc.* **94**, 6686 (1973)

22. L. P. Kazanskii and V. I. Spitsyn, *Doklady Akad. Nauk. S.S.R.* **223**, 381 (1975); *Doklady Phys. Chem.* **223**, 721 (1975)
23. D. Rehder, *J. Magn. Reson.* **25**, 177 (1977)
24. D. Rehder and W. L. Dorn, *Transition Metal Chem.* **1**, 233 (1976)
25. D. Rehder and W. L. Dorn, *Transition Metal Chem.* **1**, 74 (1976)
26. D. Rehder, *J. Organometal. Chem.* **37**, 303 (1972)
27. K. T. Gillen and J. H. Noggle, *J. Chem. Phys.* **53**, 801 (1970)
28. J. V. Hatton, Y. Saito and W. G. Schneider, *Can. J. Chem.* **43**, 47 (1965)
29. J. A. S. Howell and K. C. Moss, *J. Chem. Soc.* (*A*), 270 (1971)
30. P. C. Lauterbur and R. B. King, *J. Amer. Chem. Soc.* **87**, 3266 (1966)
31. O. W. Howarth and R. E. Richards, *J. Chem. Soc.* 864 (1965)
32. C. M. Flynn, Jr, M. T. Pope and S. O'Donnell, *Inorg. Chem.* **13**, 831 (1974)
33. O. Lutz, W. Nepple and A. Nolle, *Z. Naturforsch.* **31a**, 1046 (1976)
34. L. P. Kazanskii, M. A. Fedotov, M. N. Ptushkina and V. I. Spitsyn, *Dokl. Akad. Nauk. SSSR*, **224**, 866 (1975); *Doklady Phys. Chem.* **224**, 1029 (1975)
35. D. Rehder and J. Schmidt, *J. Inorg. Nucl. Chem.* **36**, 333 (1974)
36. D. Rehder, L. Dahlenberg, and I. Muller, *J. Organometal. Chem.* **122**, 53 (1976)
37. T. Vladimiroff and E. R. Malinowski, *J. Chem. Phys.* **45**, 1830 (1967)
38. H. G. Spinney, Thesis submitted to University of Western Ontario, 1972
39. Y. A. Buslaev, V. D. Kopanev, S. M. Sinitsyna and V. G. Khlebodarov, *Zhur. neorg. Khim.* **18**, 2567 (1973); *Russ. J. Inorg. Chem.* **18**, 1362 (1973)
40. K. J. Packer and E. L. Muetterties, *J. Amer. Chem. Soc.* **85**, 3035 (1963)
41. Y. A. Buslaev, V. D. Kopanev and V. P. Tarasov, *Chem. Comm.* 1175 (1971)
42. R. G. Kidd and H. G. Spinney, *Inorg. Chem.* **12**, 1967 (1973)
43. R. G. Kidd and H. G. Spinney, submitted to *J. Amer. Chem. Soc.*
44. D. W. Aksnes, S. M. Hutchison and K. J. Packer, *Mol. Phys.* **14**, 301 (1968)
45. J. A. S. Howell and K. C. Moss, *J. Chem. Soc.* 2481 (1971)
46. J. A. S. Howell and K. C. Moss, *J. Chem. Soc.* (*A*), 2483 (1971)
47. L. C. Erich, A. C. Gossard and R. L. Hartless, *J. Chem. Phys.* **59**, 3911 (1973)
48. Y. Egozy and A. Loewenstein, *J. Magn. Reson.* **1**, 494 (1969)
49. B. W. Epperlein, H. Krüger, O. Lutz, A. Nolle and W. Mayr, *Z. Naturforsch.* **30a**, 1237 (1975)
50. B. N. Figgis, R. G. Kidd and R. S. Nyholm, *Can. J. Chem.* **43**, 145 (1965)
51. J. A. Jackson and H. Taube, *J. Phys. Chem.* **69**, 1844 (1965)
52. J. Kaufmann, J. Kronenbitter and A. Schwenk, *Z. Physik. A.* **274**, 87 (1975)
53. R. R. Vold and R. L. Vold, *J. Chem. Phys.* **61**, 4360 (1974)
54. R. R. Vold and R. L. Vold, *J. Magn. Reson.* **19**, 365 (1975)
55. O. Lutz, A. Nolle and P. Kroneck, *Z. Naturforsch.* **31a**, 454 (1976)
56. W. D. Kautt, H. Krüger, O. Lutz, H. Maier and A. Nolle, *Z. Naturforsch.* **31a**, 351 (1976)
57. B. E. Mann, *J. C. S. Dalton* 2012 (1973)
58. D. S. Milbrath, J. G. Verkade and R. J. Clark, *Inorg. Nucl. Chem. Letters* **12**, 921 (1976)
59. P. B. Sogo and C. D. Jeffries, *Phys. Rev.* **98**, 1316 (1955)
60. A. Narath and D. C. Wallace, *Phys. Rev.* **127**, 724 (1962)
61. A. Narath, K. C. Brog and W. H. Jones, *Phys. Rev.* **B2**, 2618 (1970)
62. W. Sahm and A. Schwenk, *Z. Naturforsch.* **29a**, 1763 (1974)
63. M. P. Klein and J. Happe, *Bull. Amer. Phys. Soc.* **6**, 104 (1961)
64. P. J. Green and T. H. Brown, *Inorg. Chem.* **10**, 206 (1971)
65. W. McFarlane, A. M. Noble and J. M. Winfield, *J. Chem. Soc. A*, 948 (1971)
66. H. C. E. McFarlane, W. McFarlane and D. S. Rycroft, *J. Chem. Soc. Dalton Trans.* 1616 (1976)
67. J. Banck and A. Schwenk, *Z. Physik* **20b**, 75 (1975)
68. A. Schwenk, *J. Magn. Reson.* **5**, 376 (1971)
69. C. Crocker, P. L. Goggin and R. J. Goodfellow, *J. Chem. Soc. Dalton Trans.* 2494 (1976)
70. A. Narath and A. T. Fromhold, *Phys. Rev.*, **139A**, 794 (1965)
71. J. P. Jesson, in "Transition Metal Hydrides", p. 75, (E. L. Muetterties, ed.). Dekker, New York (1971)

72. B. Deubzer and H. D. Kaesz, *J. Amer. Chem. Soc.* **90**, 3276 (1968)
73. A. Shortland and G. Wilkinson, *J. Chem. Soc. Chem. Commun.* 318 (1972)
74. W. McFarlane and D. S. Rycroft, *J. Chem. Soc. Chem. Commun.* 336 (1973)
75. B. E. Mann, *Adv. Organometallic Chem.* **12**, 135 (1974)
76. J. W. Emsley, L. Phillips and V. Wray, *Progr. N. M. R. Spectr.* **10**, 85 (1976)
77. J. G. Verkade, *Coord. Chem. Rev.* **9**, 1 (1972)
78. T. G. Appleton, H. C. Clark and L. E. Manzer, *Coord. Chem. Rev.* **10**, 335 (1973)
79. F. Calderazzo, E. A. C. Lucken and D. F. Williams, *J. Chem. Soc.* (*A*), 154 (1967)
80. D. Gudlin and H. Schneider, *J. Magn. Reson.* **17**, 268 (1975)
81. G. M. Bancroft, H. C. Clark, R. G. Kidd, A. T. Rake and H. G. Spinney, *Inorg. Chem.* **12**, 728 (1973)
82. R. G. Kidd and D. R. Truax, *J. Amer. Chem. Soc.* **90**, 6867 (1968)
83. S. Onaka, Y. Sasaki and H. Sano, *Bull. Chem. Soc. Japan* **44**, 726 (1971)
84. S. Onaka, T. Miyamoto and Y. Sasaki, *Bull. Chem. Soc. Japan* **44**, 1851 (1971)
85. W. A. G. Graham, *Inorg. Chem.* **7**, 315 (1968)
86. M. Broze and Z. Luz, *J. Phys. Chem.* **73**, 1600 (1969)
87. P. S. Ireland, C. A. Deckert and T. L. Brown, *J. Magn. Reson.* **23**, 485 (1976)
88. Y. Saito, *Can. J. Chem.* **43**, 2530 (1965)
89. T. P. Das and E. L. Hahn. Nuclear quadrupole resonance spectroscopy, in "Solid State Physics". Suppl. 1, p. 120. Academic Press, New York (1958)
90. J. S. Griffith and L. E. Orgel, *Trans. Faraday Soc.* **53**, 601 (1957)
91. D. S. Matteson and R. A. Bailey, *J. Amer. Chem. Soc.* **91**, 1975 (1969)
92. J. P. Williams and A. Wojcicki, *Inorg. Chim. Acta* **15**, L19 (1975)
93. E. S. Mooberry and R. K. Sheline, *J. Chem. Phys.* **56**, 1852 (1972)
94. R. J. Miles, Jr, B. B. Garrett and R. J. Clark, *Inorg. Chem.* **8**, 2817 (1969)
95. B. N. Figgis, R. G. Kidd and R. S. Nyholm, *Proc. Roy. Soc.* **A269**, 469 (1962)
96. R. G. Kidd, unpublished results.
97. R. A. Dwek, Z. Luz and M. Shporer, *J. Phys. Chem.* **74**, 2232 (1970)
98. B. E. Mann, *J. Chem. Soc., Chem. Commun.* 1173 (1971)
99. A. A. Koridze, P. V. Petrovskii, S. P. Gubin and E. I. Fedin, *J. Organometal. Chem.* **93**, C26 (1975)
100. W. Sahm and A. Schwenk, *Z. Naturforsch.* **29a**, 1763 (1974)
101. A. Schwenk and G. Zimmermann, *Phys. Letters* **26A**, 258 (1968)
102. W. G. Proctor and F. C. Yu, *Phys. Rev.* **81**, 20 (1951)
103. N. A. Matwiyoff and W. E. Wageman, *Inorg. Chim. Acta* **4**, 460 (1970)
104. A. Yamasaki, F. Yajima and S. Fujiwara, *Inorg. Chim. Acta* **2**, 39 (1968)
105. F. Yajima, Y. Koike, A. Yamasaki and S. Fujiwara, *Bull. Chem. Soc. Japan* **47**, 1442 (1974)
106. T. H. Martin and B. M. Fung, *J. Phys. Chem.* **77**, 637 (1973)
107. R. Ader and A. Loewenstein, *J. Magn. Reson.* **5**, 248 (1971)
108. R. Freeman, G. R. Murray and R. E. Richards, *Proc. Roy. Soc.* **A242**, 455 (1957)
109. E. A. C. Lucken, K. Noack and D. F. Williams, *J. Chem. Soc.* (*A*) 148 (1967)
110. G. R. Benedek, R. Englman and J. A. Armstrong, *J. Chem. Phys.* **39**, 3349 (1963)
111. W. Hackbush, H. H. Rupp and K. Weighardt, *J. Chem. Soc. Dalton* 1015 (1975)
112. B. M. Fung, S. C. Wei, T. M. Martin and I. Wei, *Inorg. Chem.* **12**, 1203 (1973)
113. K. L. Craighead, P. Jones and R. G. Bryant, *J. Phys. Chem.* **79**, 1868 (1975)
114. K. L. Craighead and R. G. Bryant, *J. Phys. Chem.* **79**, 1602 (1975)
115. K. L. Craighead, *J. Amer. Chem. Soc.* **95**, 4434 (1973)
116. R. A. LaRossa and T. L. Brown, *J. Amer. Chem. Soc.* **96**, 2072 (1974)
117. D. Rehder and J. Schmidt, *Z. Naturforsch.* **27b**, 625 (1972)
118. L. F. Wuyts and G. P. Van der Kelen, *J. Mol. Structure* **23**, 73 (1974)
119. N. F. Ramsey, *Phys. Rev.* **78**, 699 (1950)
120. W. G. Schneider and A. D. Buckingham, *Discuss. Faraday Soc.* **34**, 147 (1962)
121. S. S. Dharmatti and C. R. Kanekar, *J. Chem. Phys.* **31**, 1436 (1959)
122. C. R. Kanekar and N. S. Biradar, *Current Sci.* **35**, 37 (1966)
123. N. S. Biradar and M. A. Pujar, *Current Sci.* **35**, 385 (1966)

124. N. S. Biradar and M. A. Pujar, *Z. anorg. allg. Chem.* **379**, 88 (1970)
125. N. S. Biradar and M. A. Pujar, *Inorg. Nucl. Chem. Letters* **7**, 269 (1971)
126. N. S. Biradar and M. A. Pujar, *Z. anorg. allg. Chem.* **391**, 54 (1972)
127. W. C. Dickinson, *Phys. Rev.* **80**, 563 (1950)
128. C. R. Kanekar, M. M. Dhingra, V. R. Marathe and R. Nagarajan, *J. Chem. Phys.* **46**, 2009 (1967)
129. S. Fujiwara, F. Yajima and A. Yamasaki, *J. Magn. Reson.* **1**, 203 (1969)
130. R. L. Martin and A. H. White, *Nature* **223**, 394 (1969)
131. M. Gerlock and J. R. Miller, *Prog. Inorg. Chem.* **10**, 1 (1968)
132. J. S. Griffith, "The Theory of Transition Metal Ions", p. 284. Cambridge University Press (1964)
133. D. D. Traficante and J. A. Simms, *Rev. Sci. Instr.* **43**, 1122 (1972)
134. a. A. Johnson and G. W. Everett, Jr, *Inorg. Chem.* **12**, 2801 (1973)
b. J. Wilinski and R. J. Kurland, *Inorg. Chem.* **12**, 2202 (1973)
135. B. E. Reichert and B. O. West, *J. Chem. Soc. Chem. Commun.* 177 (1974)
136. F. Yajima, A. Yamasaki and S. Fujiwara, *Inorg. Chem.* **10**, 2350 (1971)
137. M. R. Hyde and A. G. Sykes, *J. Chem. Soc. Dalton* 1583 (1974)
138. W. Hackbusch, H. H. Rupp and K. Wieghardt, *J. Chem. Soc. Dalton* 2364 (1975)
139. N. J. Patel and B. C. Haldar, *J. Inorg. Nucl. Chem.* **29**, 1037 (1967)
140. H. W. Speiss and R. K. Sheline, *J. Chem. Phys.* **53**, 3036 (1970)
141. a. H. Haas and R. K. Sheline, *J. Inorg. Nucl. Chem.* **29**, 693 (1967)
141. b. M. A. Cohen, D. R. Kidd, and T. L. Brown, *J. Amer. Chem. Soc.* **97**, 4408 (1975)
142. E. S. Mooberry, M. Pupp, J. L. Slater and R. K. Sheline, *J. Chem. Phys.* **55**, 3655 (1971)
143. J. A. Seitchik, V. Jaccarino and J. H. Wernick, *Phys. Rev.* **138A**, 148 (1965)
144. A. Narath, A. T. Frombold and E. D. Jones, *Phys. Rev.* **144**, 428 (1966)
145. H. T. Weaver and R. K. Quinn, *Phys. Rev.* **B10**, 1816 (1974) and references therein.
146. T. H. Brown and P. J. Green, *J. Amer. Chem. Soc.* **92**, 2359 (1970)
147. J. R. Barnes, P. L. Goggin and R. J. Goodfellow, unpublished results.
148. M. Arthurs, S. M. Nelson and M. G. B. Drew, *J. Chem. Soc. Dalton Trans.* 779 (1977)
149. R. J. Goodfellow, unpublished results
150. H. C. E. McFarlane, W. McFarlane and R. J. Wood, *Bull. Soc. Chim. Belg.* **85**, 864 (1976)
151. E. M. Hyde, J. D. Kennedy, B. L. Shaw and W. McFarlane, *J. Chem. Soc. Dalton Trans.* 1571 (1977)
152. S. Martinengo, B. T. Heaton, R. J. Goodfellow and P. Chini, *J. Chem. Soc. Chem. Comm.* 39 (1977)
153. P. L. Goggin, R. J. Goodfellow, S. R. Haddock, B. F. Taylor and I. R. H. Marshall, *J. Chem. Soc. Dalton Trans.* 459 (1976)
154. A. P. Tupčiauskas, N. M. Sergeyev, Yu. A. Ustynyuk and A. N. Kashin, *J. Magn. Reson.* **7**, 124 (1972)
155. B. F. Taylor, Ph.D. Thesis, University of Bristol (1973)
156. T. H. Brown and P. J. Green, *Phys. Lett.* **31A**, 148 (1970)
157. P. L. Goggin, R. J. Goodfellow, J. R. Knight, M. G. Norton and B. F. Taylor, *J. Chem. Soc. Dalton Trans.* 2220 (1973)
158. M. A. Cairns, K. R. Dixon and J. J. McFarland, *J. Chem. Soc. Dalton Trans.* 1159 (1975)
159. G. Mavel, *Ann. Rev. N. M. R. Spectr.* **5B** (1973)
160. J. F. Nixon and A. Pidcock, *Ann. Rev. N. M. R. Spectr.* **2**, 345 (1969)
161. S. R. Haddock, Ph.D. Thesis, University of Bristol (1975)
162. P. L. Goggin, R. J. Goodfellow and S. R. Haddock, *J. Chem. Soc. Chem. Commun.* 176 (1975)
163. T. J. Rowland, *J. Phys. Chem. Solids* **7**, 95 (1958)
164. L. E. Drain, *J. Phys. Chem. Solids* **24**, 379 (1963)
165. A. von Zelewsky, *Helv. Chim. Acta* **51**, 803 (1968)
166. A. Pidcock, R. E. Richards and L. M. Venanzi, *J. Chem. Soc. A*. 1970 (1968)
167. W. McFarlane, *Chem. Commun.* 393 (1968)
168. R. R. Dean and J. C. Green, *J. Chem. Soc. A*, 3047 (1968)
169. P. S. Pregosin and L. M. Venanzi, *Helv. Chim. Acta* **58**, 1548 (1975)

170. W. Freeman, P. S. Pregosin, S. N. Sze and L. M. Venanzi, *J. Magn. Reson.* **22**, 473 (1976)
171. J. J. Pesek and W. R. Mason, *J. Magn. Reson.* **25**, 519 (1977)
172. W. Hackbusch, unpublished results.
173. P. L. Goggin, R. J. Goodfellow, S. R. Haddock and J. G. Eary, *J. Chem. Soc. Dalton Trans.* 647 (1972)
174. S. M. Cohen and T. H. Brown, *J. Chem. Phys.* **61**, 2985 (1974)
175. W. McFarlane, *J. Chem. Soc. Dalton Trans.* 324 (1974)
176. D. W. W. Anderson, E. A. V. Ebsworth and D. W. H. Rankin, *J. Chem. Soc. Dalton Trans.* 2370 (1973)
177. E. A. V. Ebsworth, J. M. Edward and D. W. H. Rankin, *J. Chem. Soc. Dalton Trans.* 1667 (1976)
178. J. D. Kennedy, W. McFarlane, R. J. Puddephatt and P. J. Thompson, *J. Chem. Soc. Dalton Trans.* 874 (1976)
179. P. W. Hall, R. J. Puddephatt and C. F. H. Tipper, *J. Organometallic Chem.* **71**, 145 (1974)
180. C. Crocker, R. J. Goodfellow, J. Gimeno and R. Uson, *J. Chem. Soc. Dalton Trans.* 1448 (1977)
181. S. J. Anderson and R. J. Goodfellow, *J. Chem. Soc. Chem. Commun.* 443 (1975)
182. S. J. Anderson, P. L. Goggin and R. J. Goodfellow, *J. Chem. Soc. Dalton Trans.* 1959 (1976); S. J. Anderson and R. J. Goodfellow, *ibid.* 1683 (1977)
183. P. L. Goggin and R. J. Goodfellow, unpublished results
184. C. Crocker and R. J. Goodfellow, *J. Chem. Soc. Dalton Trans.* 1687 (1977)
185. T. H. Brown and S. M. Cohen, *J. Chem. Phys.* **58**, 395 (1973)
186. H. Niedoba, H. Launois, D. Brinkmann and H. U. Keller, *J. Phys.* (*Paris*) *Lett.* **35**, L251 (1974)
187. H. H. Rupp, *Z. Naturforsch.* **26a**, 1937 (1971)
188. H. J. Keller and H. H. Rupp, *Z. Naturforsch.* **26a**, 785 (1971)
189. J. D. Kennedy and W. McFarlane, *J. Chem. Soc. Faraday Trans II*, **72**, 1653 (1976)
190. W. McFarlane, *Chem. Commun.* 772 (1967)
191. M. A. Bennett, R. Bramley and I. B. Tomkins, *J. Chem. Soc. Dalton Trans.* 166 (1973)
192. R. Bramley, J. R. Hall, G. A. Swile and I. B. Tomkins, *Aust. J. Chem.* **27**, 2491 (1974)
193. J. Browning, P. L. Goggin, R. J. Goodfellow, N. W. Hurst, M. Murray and L. G. Mallinson, *J. Chem. Soc. Dalton Trans.* 872 (1978)
194. J. Browning, P. L. Goggin, and R. J. Goodfellow, unpublished results
195. P. L. Goggin, R. J. Goodfellow, S. R. Haddock, J. R. Knight, F. J. S. Reed and B. F. Taylor, *J. Chem. Soc. Dalton Trans.* 523 (1974)
196. P. S. Pregosin, H. Omura, and L. M. Venanzi, *J. Amer. Chem. Soc.* **95**, 2047 (1973)
197. W. McFarlane, *J. Chem. Soc. A*, 1922 (1967)
198. W. McFarlane, *Chem. Commun.* 755 (1968)
199. D. F. Evans and G. K. Turner, *J. Chem. Soc. Dalton Trans.* 1238 (1975)
200. J. D. Kennedy, W. McFarlane and R. J. Puddephatt, *J. Chem. Soc. Dalton Trans.* 745 (1976)
201. C. Brown, B. T. Heaton, P. Chini, A. Fumagalli and G. Longoni, *J. Chem. Soc. Chem. Commun.* 309 (1977)
202. E. R. Andrew, W. S. Hinshaw and R. S. Tiffen, *J. Phys.* (C), 6, 2217 (1973)
203. E. R. Andrew and W. S. Hinshaw, *Phys. Letters* (A), **43**, 113 (1973)
204. A. Kawamori and R. Takagi, *Mol. Phys.* **25**, 489 (1973)
205. B. D. Guenther and R. A. Hultsch, *J. Magn. Reson.* **1**, 609 (1969)
206. K. D. Becker and G. W. Herzog, *Ber. Bunsenges. Physik. Chem.* **78**, 45 (1974)
207. A. Kawamori and G. Soda, *Mol. Phys.* **29**, 1085 (1975)
208. P. D. Ellis, H. C. Walsh and C. S. Peters, *J. Magn. Reson.* **11**, 426 (1973)
209. Y. Yamamoto, H. Haraguchi and S. Fujiwara, *J. Phys. Chem.* **74**, 4369 (1970)
210. W. McFarlane and D. S. Rycroft, *J. Magn. Reson.* **24**, 95 (1976)
211. E. Brun, J. Oeser, H. H. Staub and C. G. Telschow, *Phys. Rev.* **93**, 172 (1954)
212. P. B. Sogo and C. D. Jeffries, *Phys. Rev.* **93**, 174 (1954)
213. J. Poitrenaud and J. M. Winter, *J. Phys. Chem. Solids* **25**, 123 (1964)
214. A. Narath, A. T. Fromhold and E. D. Jones, *Phys. Rev.* **144**, 428 (1966)
215. C. W. Burges, R. Koschmieder, W. Sahm and A. Schwenk, *Z. Naturforsch.* **28a**, 1753 (1973)
216. K. Jucker, W. Sahm and A. Schwenk, *Z. Naturforsch.* **31a**, 1532 (1976)
217. A. K. Rahimi and A. I. Popov, *Inorg. Nucl. Chem. Letters* **12**, 703 (1976)

218. A. Sahm and A. Schwenk, *Z. Naturforsch.* **29a**, 1763 (1974)
219. A. Sahm and A. Schwenk, *Phys. Lett.* **51A**, 357 (1975)
220. U. El-Hanany, M. Shaham and D. Zamir, *Phys. Rev.* **B10**, 2343 (1974)
221. H. Schmidbaur, J. Adlkofer and W. Buchner, *Angew. Chem. Internat. Edn.* **12**, 415 (1973)
222. B. L. Dyatkin, B. I. Martynov, L. G. Martynova, N. G. Kizim, S. R. Sterlin, Z. A. Stumbrevichute and L. A. Federov, *J. Organometal. Chem.* **57**, 423 (1973) and references therein
223. E. L. Muetterties and C. W. Alegranti, *J. Amer. Chem. Soc.* **94**, 6386 (1972)
224. H. E. Weaver, *Phys. Rev.* **89**, 923 (1953)
225. B. W. Epperlein, H. Krüger, O. Lutz, and A. Schwenk, *Z. Naturforsch.* **29a**, 1553 (1974)
226. L. H. Bennett, *Phys. Rev.* **150**, 418 (1966)
227. W. G. Proctor, *Phys. Rev.* **79**, 35 (1950)
228. C. S. Peters, R. Codrington, H. C. Walsh and P. D. Ellis, *J. Magn. Reson.* **11**, 431 (1973)
229. G. E. Maciel, "N.M.R. Spectroscopy of Nuclei other than Protons", p. 347 (T. Axenrod and G. A. Webb, eds). Wiley, New York (1974)
230. G. E. Maciel and M. Borzo, *J. Chem. Soc. Chem. Commun.* 394 (1973)
231. H. Krüger, O. Lutz, A. Nolle, A. Schwenk and G. Stricker, *Z. Naturforsch.* **28a**, 484 (1973)
232. K. Hildenbrand and H. Dreeskamp, *Z. Phys. Chem.* **69**, 171 (1970)
233. C. J. Turner, R. F. M. White, *J. Magn. Reson.* **26**, 1 (1977)
234. J. D. Kennedy and W. McFarlane, *J. Chem. Soc. Perkin Trans. II.* 1187 (1977)
235. M. Holz, R. B. Jordan and M. D. Zeidler, *J. Magn. Reson.* **22**, 47 (1976)
236. A. D. Cardin, P. D. Ellis, J. D. Odom and J. W. Howard, *J. Amer. Chem. Soc.* **97**, 1672 (1975)
237. R. N. Schwartz, *J. Magn. Reson.* **24**, 205 (1976)
238. C. W. Burges, R. Koschmieder, W. Sahm and A. Schwenk, *Z. Naturforsch.* **28a**, 1753 (1973)
239. I. M. Armitage, R. T. Pajer, A. J. M. S. Uiterkamp, J. F. Chlebowski and J. E. Coleman, *J. Amer. Chem. Soc.* **98**, 5710 (1976)
240. R. A. Byrd and P. D. Ellis, *J. Magn. Reson.* **26**, 169 (1977)
241. R. A. Haberkorn, L. Que, W. O. Gillum, R. H. Holm, C. S. Liu and R. C. Lord, *Inorg. Chem.* **15**, 2408 (1976)
242. R. J. Kostelnik and A. A. Bothner-By, *J. Magn. Reson.* **14**, 141 (1974)
243. H. Krüger, O. Lutz, A. Schwenk and G. Stricker, *Z. Physik* **266**, 233 (1974)
244. M. A. Sens, N. K. Wilson, P. D. Ellis and J. D. Odom, *J. Magn. Reson.* **19**, 323 (1975)
245. A. P. Tupčiauskas, N. M. Sergeyev, Yu. A. Ustynyuk and A. N. Kashin, *J. Magn. Reson.* **7**, 124 (1972)
246. D. C. Look, *Phys. Stat. Sol.* **50B**, K97 (1972)
247. T. J. Rowland, *Phys. Rev.* **103**, 1670 (1956)
248. S. N. Sharma and D. L. Williams, *Phys. Lett.* **25A**, 738 (1967)
249. E. R. Andrew, W. S. Hinshaw and R. S. Tiffen, *J. Magn. Reson.* **15**, 191 (1974)
250. B. E. Mann, *Inorg. Nucl. Chem. Letters* **7**, 595 (1971)
251. A. Yamasaki and E. Fluck, *Z. anorg. Chem.* **396**, 297 (1973)
252. P. L. Goggin, R. J. Goodfellow, and K. Kessler, unpublished results.
253. W. E. Blumberg, J. Eisinger and R. G. Shulman, *J. Phys. Chem. Solids*, **26**, 1187 (1965)
254. S. S. Dharmatti, C. R. Kanekar and C. S. Mathur, *Proc. 2nd U.N. Intern. Conf. Peaceful Uses At. Energy, Geneva* **28**, 644 (1958)
255. R. E. Dessy, T. J. Flautt, H. H. Jaffe and G. F. Reynolds, *J. Chem. Phys.* **30**, 1422 (1959)
256. P. D. Godfrey, M. L. Heffernan and D. F. Kerr, *Aust. J. Chem.* **17**, 701 (1964)
257. K. A. McLauchlan, D. H. Whiffen and L. W. Reeves, *Mol. Phys.* **10**, 131 (1966)
258. G. E. Maciel and M. Borzo, *J. Magn. Reson.* **10**, 388 (1973)
259. C. R. Lassigne and E. J. Wells, *Canad. J. Chem.* **55**, 1303 (1977)
260. H. Krüger, O. Lutz, A. Nolle and A. Schwenk, *Z. Physik* A273, 325 (1975)
261. R. J. Goodfellow and B. F. Taylor, unpublished results
262. P. L. Goggin, R. J. Goodfellow and N. W. Hurst, unpublished results
263. R. B. Johannesen and R. W. Duerst, *J. Magn. Reson.* **5**, 355 (1971)
264. B. Cagnac and J. Brossel, *Compt. Rend.* **249**, 77 (1959)
265. C. J. Jameson and H. S. Gutowsky, *J. Chem. Phys.* **40**, 1714 (1964)
266. N. W. Hurst, Ph.D. Thesis, University of Bristol (1975)

267. A. J. Brown, O. W. Howarth and P. Moore, *J. Chem. Soc. Dalton Trans.* 1589 (1976)
268. S. Cradock, E. A. V. Ebsworth, N. S. Hosmane and K. M. Mackay, *Angew. Chem. Internat. Edn.* **14**, 167 (1975)
269. M. Borzo and G. E. Maciel, *J. Magn. Reson.* **19**, 279 (1975)
270. P. J. Banney, M.Sc. Thesis, University of Queensland (1969) quoted by P. R. Wells, in "Determination of Organic Structures by Physical Methods", Vol. 4, Ch. 5 (F. C. Nachod and J. J. Zuckerman, eds). Academic Press, New York (1971)
271. M. F. Hudson and K. M. Smith, *Tetrahedron Lett.* 2223 (1974)
272. R. W. Weinert and R. T. Schumacher, *Phys. Rev.* **172**, 711 (1968)
273. L. A. Federov, *Zh. Strukt. Khim.* **17**, 236, 247 (1976) [*J. Struct. Chem.* **17**, 207, 216 (1976)]
274. F. A. L. Anet and J. L. Sudmeier, *J. Magn. Reson.* **1**, 124 (1969)
275. L. A. Fedorov, Z. A. Stumbrevichyute and E. I. Fedin, *Zh. Strukt. Khim.* **16**, 976 (1975) [*J. Strukt. Chem.* **16**, 899 (1975)]
276. W. McFarlane, *J. Chem. Soc. A*, 2280 (1968)
277. S. O. Grim, P. J. Lui and R. L. Keiter, *Inorg. Chem.* **13**, 342 (1974)
278. J. Bennett, A. Pidcock, C. R. Waterhouse, P. Coggon and A. T. McPhail, *J. Chem. Soc. A*, 2094 (1970)
279. P. L. Goggin, R. J. Goodfellow and A. J. Griffiths, unpublished results
280. R. J. Goodfellow and S. R. Stobart, *J. Magn. Reson.* **27**, 143 (1977)

9

GROUP III—ALUMINIUM, GALLIUM INDIUM AND THALLIUM

J. F. HINTON, University of Arkansas, U.S.A.

R. W. BRIGGS, University of Arkansas, U.S.A.

Boron-11 has been a well-used quadrupolar nucleus for many years and is briefly discussed (along with boron-10) in ch. 4. The other group III elements have been rather less studied by NMR, but the advent of multinuclear, pulsed-Fourier transform techniques has made the direct observation of ^{27}Al, ^{69}Ga, ^{71}Ga, ^{113}In, ^{115}In, ^{203}Tl and ^{205}Tl relatively easy. The nuclear properties of these elements associated with the NMR experiment are shown in Tables 1.1 and 1.2. The receptivity of the group III nuclei relative to that of the ^{13}C nucleus indicates a high sensitivity to detection. All of the nuclei within this group possess a quadrupole moment, except the two thallium isotopes; however, the quadrupolar properties of Al, Ga and In can be a source of significant information rather than a detriment to the experiment, as is frequently assumed. The linewidths expected for nuclei in identical environments (based on Eq. 1.18) are in the ratio 1.0:5.9:2.3:13.5:14.0 for ^{27}Al, ^{69}Ga, ^{71}Ga, ^{113}In, and ^{115}In. The nuclei not only have a relatively high sensitivity to detection, but they also exhibit very large chemical shift ranges. This combination of circumstances makes the group III nuclei very sensitive probes for chemical studies and for the investigation of solution properties such as complexation, ion-solvent interaction and preferential solvation. The NMR literature involving the direct observation of Al, Ga, In and Tl may be divided into two basic areas of investigation of interest to the chemist: (1) organometallic chemistry, and (2) solvation and complexation of the elements in the ionic form.

9A ALUMINIUM-27

The data in Table 1.2 suggest that ^{27}Al is a very favorable nucleus for direct NMR investigation. Although ^{27}Al possesses a quadrupole moment, one can obtain narrow lines in the NMR spectra. The linewidth depends upon the geometry of the substituents about the ^{27}Al nucleus. Octahedrally solvated Al^{3+}, tetrahedral AlX_4^-, and dimers of the type Al_2X_6 and Al_2R_6 exhibit high symmetry about the ^{27}Al nucleus; consequently, the field gradient at the nucleus will be small, and relatively narrow lines will be observed. For trigonal species, linewidths are much larger because of the large field gradient about the ^{27}Al nucleus. Thus linewidths vary from 3 to 6000 Hz. The chemical shift range of the ^{27}Al nucleus is approximately 450 ppm.

Table 9.1 contains ^{27}Al chemical shifts and linewidths taken from the review by Akitt[1] and supplemented with more recent data. Discussion of the data in this table will be abridged here, in view of the existence of the earlier review.[1] In general, ^{27}Al chemical shifts may be categorized according to the ligand symmetry about the aluminium atom. In the low frequency range one finds aluminium species with octahedral symmetry covering approximately 70 ppm (i.e. $Al(PhCN)_6^{3+}$ at -46 ppm to $Al(EtNCS)_6^{3+}$ at $+20$ ppm). Aluminium species with tetrahedral symmetry are to be found in an intermediate region on the chemical shift scale. This symmetry group is normally found at high frequency, relative to $[Al(H_2O)_6]^{3+}$, and extends from -28 ppm for $[AlI_4]^-$ to $\sim +108$ ppm for $AlCl_3 \cdot Me_3P$. One also finds within this region dimers such as Al_2X_6 (where X is a halogen) and adducts of the type $Al(BH_4)_3 \cdot Me_3P$ whose symmetry about the aluminium atom is nearly tetrahedral. At the highest frequency one finds the alkyl alanes, Al_2Me_6 at $+156$ ppm, and their adducts, such as $Et_2S \cdot AlEt_3$ at $+221$ ppm. Akitt[1] has suggested that since many of these molecules maintain tetrahedral symmetry, the shifts are a function of substitution instead of stereochemistry. The very large linewidths associated with these species could arise from dissociation and exchange mechanisms rather than symmetry distortion.

Since the ^{27}Al nucleus possesses a quadrupole moment, the linewidth of the resonance signal will be a sensitive function of the symmetry and arrangement of ligands about the nucleus. Consequently, the combination of linewidth and chemical shift determination can provide an extremely powerful probe for studying the chemical and physical environment about an aluminium atom.

Aluminium-27 NMR spectroscopy has been used to investigate the physical and chemical properties of aqueous solutions of aluminium salts. These studies involved the determination of the hydration number of the Al^{3+} ion,[2] hydrolysis and polymerization[2,3] and complexation.[4–16] Solvation and complexation in nonaqueous and mixed aquo-organic solvent systems have also been studied using ^{27}Al NMR spectroscopy.[9,16–32]

Because the Al^{3+} ion has a relatively large ratio of charge to ionic radius, it can interact quite strongly with basic solvent molecules. Under favorable conditions, one can observe spin-spin coupling between ^{27}Al and the nuclei of the solvent molecules; furthermore, in some mixed solvent systems the ligand exchange rate is slow enough to allow the observation of a variety of solvated species.[23,25,31] Equilibrium constants, rate constants and the kinetic parameters, $\Delta H^{\ddagger}$ and $\Delta S^{\ddagger}$, may also be obtained for these systems.

Using ^{27}Al NMR, Delpuech[23] has studied octahedral and tetrahedral solvates of the Al^{3+} ion, viz. the solid solvates of $AlA_6^{3+} \cdot 3ClO_4^-$, where A is trimethylphosphate, triethylphosphate, dimethylmethylphosphonate, diethylethylphosphonate, dimethyl hydrogenphosphite and hexamethylphosphorotriamide, in nonaqueous nitromethane at 0°. The spin-spin coupling patterns of ^{27}Al gave definite evidence for octahedral and tetrahedral coordination. The ^{27}Al chemical shift in the coordination species with the same coordination number was found to correlate with the basicity of the ligand molecule. The chemical shift was also found to be very dependent upon the number of coordinating ligands; for example, the ^{27}Al chemical shift difference between the tetrahedral hexamethylphosphorotriamide-Al species and the octahedral trimethylphosphate-Al species is approximately 54 ppm. In aqueous-organic solutions separate ^{27}Al NMR signals were observed for the different aqueous-organic solvates. It was found that the ^{27}Al chemical shift is determined mainly by the number of organic ligands and that the effect of substituting a water molecule by an organic ligand is almost additive (see Table 9.1). An additivity relationship is not always observed, however, between the ^{27}Al chemical shift and the number of ligands. In a study of mixed tetrahalo aluminate ions (e.g. $[AlCl_nI_{4-n}]^-$ where $0 \leq n \leq 4$) the ^{27}Al NMR signal of all five

TABLE 9.1 ^{27}Al chemical shifts and linewidths

Compound	Solvent	Chemical Shift δ_{Al}/ppm[a]	Linewidth $W_{1/2}$/Hz	Ref.
$Et_2S \cdot AlEt_3$	C_6H_{12}	+221	1890	36
$Al_2Bu^i_6$	C_6H_{12}	+220	6000	37
$MePhS \cdot AlEt_3$	C_6H_{12}	+204	2200	36
		+178	1160	36
Anisole$\cdot AlEt_3$	C_6H_{12}	+176	2100	36
$THF \cdot AlEt_3$	C_6H_{12}	+176	1280	36
Pyrrolidine$\cdot AlEt_3$	C_6H_{12}	+176	1220	36
Pyridine$\cdot AlEt_3$	C_6H_{12}	+167	1440	36
$Et_3N \cdot AlEt_3$	C_6H_{12}	+165	1100	36
Quinoline$\cdot AlEt_3$	C_6H_{12}	+164	~2500	36
$[AlHBu^i_2]_n$		+162	10 000	37
Al_2Et_6	C_7H_{16}	< +150	2750	17
	C_6H_{12}	+174	1000	36
	C_6H_{12}	+142	960	38
		+171	1700	37
Thiophene$\cdot AlEt_3$	C_6H_{12}	+154	1050	36
Al_2Me_6	neat	+156	450	37
$MeAl(BH_4)_2$	neat	+141		1
$AlCl_3 \cdot PMe_3$	CH_2Cl_2	+108.2		60
AlH_4^-	Et_2O[b]	+100	420	17
	Et_2O	+103	70	18
$LiAlH_4$[c]	glycol dimethyl ether	+100		24
$NaAlH_4$[c]	glycol dimethyl ether	+98.7		24
$LiAlH_4$[c]	THF	+97.7		24
$NaAlH_4$[c]	THF	+96.7		24
$LiAlH_4$[c]	Et_2O	+101		24
$KAlH_4$	diglyme	+100		61
$AlCl_4^-$	MeCN	+110		19
	CH_2Cl_2	+102.4	15.1	20
	$POCl_3$	+101.3 to +102.5	10 to 30	18
	MeCN	+108	33	17
	acrylonitrile	+105	32	17
	benzonitrile	+103	~30	17, 62
	nitromethane	+107	3	25
$AlBr_3 \cdot PMe_3$	CH_2Cl_2	+100.8		63
Al_2Cl_6	Et_2O (2 M)[d]	+105	126	17
	toluene	+91	300	37, 39
	Et_2O	+101	110	4, 37
$[AlBrCl_3]^-$	CH_2Cl_2	+99		20, 62
$AlBr_3 \cdot PEt_3$	CH_2Cl_2	+99		63
$MeCO \cdot AlCl_4$	MeCOCl[b]	+97	166	17
$Al(BH_4)_3$	C_6H_6	+97.4	180	64
Al_2Br_6	Et_2O(1.5 M)	+95	104	17
	Et_2O	+96	100	37
	Br_2	+101	300	37

Continued overleaf

TABLE 9.1 *(Continued)*

Compound	Solvent	Chemical Shift δ_{Al}/ppm[a]	Linewidth $W_{1/2}$/Hz	Ref.
$[AlBr_2Cl_2]^-$	CH_2Cl_2	+93		20
		+94		62
$Al_2Me_3Cl_3$		+93		37
$Al_2(OMe)_2Cl_4$	neat	+90		37
$[AlBr_3Cl]^-$	CH_2Cl_2	+87.2		20
		+87.5		62
$[AlICl_3]^-$	CH_2Cl_2	+86.2	42	20
$Al_2Cl_4Bu^i_2$	neat	+86	4000	37
$Co(AlI_4)_2$	C_6H_6	+81		39
$[Al(OH_4)]^-$	KOH/H_2O (0.5 M)	+80	60 to 100	17
		+80	90	37
		+80	23	5
$AlBr_4^-$	MeCN[e]	+80	35	17
$AlIBrCl_2^-$	CH_2Cl_2	+79		20
$PhCOCl \cdot AlCl_3$	PhCOCl[b]	+77	2000	17
$(C_6H_6)AlBr_3$	C_6H_6[b]	+75	510	17
$Me_3N \cdot Al(BH_4)_3$	C_6H_6	+71.1	9 to 20	64
$[AlIBr_2Cl]^-$	CH_2Cl_2	+69.3		20
$Al(BH_4)_3 \cdot PMe_3$	toluene	+63.2		1
$Al_{13}(OH)_{24}$	H_2O	+62.5	50	2
$[AlIBr_3]^-$	CH_2Cl_2	+60.6		20
$Et_2O \cdot AlCl_3$	Et_2O	+50	160	17
$Et_2O \cdot AlBr_3$	Et_2O	+47	210	17
$[AlI_2BrCl]^-$	CH_2Cl_2	+47.7		20
$[AlI_2Br_2]^-$	CH_2Cl_2	+37		20
$[AlI_2Cl_2]^-$	CH_2Cl_2	+59.4	57	20
Al_2I_6	Et_2O(1.5 M)	+40	81	17
	Et_2O	+39	90	37
	EtNCS	+36	83	17
$[Al(HMPT)_4]^{3+}$	nitromethane	+34		23
$Al(OOCCH_2)_3N$	H_2O	+24	180	1
$[AlI_3Cl]^-$	CH_2Cl_2	+21.7	46	20
$[Al(EtNCS)_6]^{3+}$	EtNCS	+20	60	17
$Al_2I_6 + Al_2Cl_6$	toluene	+18		39
$[AlI_3Br]^-$	CH_2Cl_2	+8		20
$Al_2(OBu^i)_6$		+7	100	37
$[Al_2(OH)_2(H_2O)_n]^{4+}$	H_2O	+3	450	2, 3
$[AlF_n(H_2O)_{6-n}]^{(3-n)+}$	H_2O	+2 to +15	550	37, 6
$[Al(C_2O_4)_3]^{3-}$	H_2O[b]	0	125	17
$[Al(Pr^iO)_3]_4$	*n*-hexane	0	90	17
$[Al(EtOH)_6]^{3+}$	EtOH(0.3-1 M)	0	70	17
$[Al(Pr^iOH)_6]^{3+}$	Pr^iOH (0.3-1 M)	0	70	17
$Al(acac)_3$	C_6H_6[b]	0	93	17
		+2.5	120	65
$[Al(H_2O)_6]^{3+}$	H_2O	0	3	7
$[Al(DMF)_6]^{3+}$	DMF	−1.7	39	21
	nitromethane	~1.7	5.6	25
$[Al(H_2O)_5(SO_4)]^+$	H_2O-H_2SO_4	−3.3	5	7
$[Al(H_2O)_4(SO_4)_2]^-$	H_2O-H_2SO_4	−6.8	80	1

TABLE 9.1 *(Continued)*

Compound	Solvent	Chemical Shift δ_{Al}/ppm[a]	Linewidth $W_{1/2}$/Hz	Ref.
Al^{3+}	0.1 M tartaric acid[f]			
	pH = 2.5	~ +2		10
	pH = 11.5	~ −83		10
H_3PO-Al complexes	H_2O/H_3PO_4	−7.7	100	8
$[AlL_i(H_2O)_{6-i}]^{3+}$ L = trimethyl phosphate[g]	nitromethane			
$i = 1$		−3.7		9
= 2		−6.7		9
= 3		−10.0		9
= 4		−14.0		22
= 5		−17.5		22
= 6		−20.5		23
$Al(POCl_3)_6^{3+}$	$POCl_3$	−21.2 to −22.4	55 to 77	18
$AlI_3 \cdot C_6H_6$	C_6H_6[b]	−23	800	17
AlI_4^-	MeCN[h]	−28	58	17
	CH_2Cl_2	−27	23.9	20
$Al(ViCN)_6^{3+}$	acrylonitrile	−33	83	17
$Al(MeCN)_6^{3+}$	MeCN	−34	73	17
	MeCN	−14		19
$Al(PhCN)_6^{3+}$	benzonitrile	−46	100	17
$Co(AlBr_4)_2$	C_6H_6	−144		39
$Co(AlCl_4)_2$	C_6H_6	−225		39

[a] Reference: $[Al(H_2O)]_6^{3+}$. [b] Saturated solution. [c] Data are also given[24] for the corresponding deuterated compounds. [d] Identical data are reported[17] for solutions in THF and EtNCS. [e] Identical data are reported[17,20,62] for solutions in acrylonitrile, benzonitrile, EtNCS and CH_2Cl_2. [f] Data are also given[10] for solutions in mandelic, 5-sulphosalicylic, malic, citric, thiomalic, gluconic, saccharic and lactobionic acids. [g] Similar data are listed[22,23] for L = triethylphosphate, dimethylmethylphosphonate, diethylethylphosphonate, and dimethylhydrogenphosphite. [h] Identical data are reported[17] for solutions in acrylonitrile and benzonitrile.

TABLE 9.2 ^{27}Al coupling constants

Compound	Solvent	Coupling constant/Hz	Ref.
$LiAlH_4$[a]	glycol dimethyl ether	$^1J(Al, H) = 172$	24
AlH_4^-	Et_2O	$^1J(Al, H) = 110$	18
$NaAlEt_4$	DMSO	$^1J(Al, C) = 73.0$	66
		$^2J(Al, C) \approx 1.0$	66
	DME	$^3J(Al, H) = 5.8$	67
		$^2J(Al, H) = 7.3$	
$NaAlBu^n_4$	DMSO	$^1J(Al, C) = 76.0$	66
		$^2J(Al, C) \approx 4.0$	66
$LiAlMe_4$		$^1J(Al, C) = 71.5$	68
$LiAlMe_4$	DME	$^2J(Al, H) = 6.34$	69
Al_2Me_6	toluene	$^1J(Al, C) = 110$	56
		$^2J(Al, C) = 19$	
$[AlMe_2Cl]_2$	toluene	$^1J(Al, C) = 105$	56
$[AlMe_2Br]_2$	toluene	$^1J(Al, C) = 105$	56
$[AlEt_2Cl]_2$	toluene	$^1J(Al, C) = 105$	56
Al_2Me_6	Et_2O	$^1J(Al, C) = 91$	56
$AlCl_3 \cdot PMe_3$	CH_2Cl_2	$^1J(Al, P) = 263, 280$	63, 70
$AlCl_3 \cdot PMe_2Ph$		$^1J(Al, P) = 290$	70
$AlBr_3 \cdot PMe_3$	CH_2Cl_2	$^1J(Al, P) = 248$	63
$AlBr_3 \cdot PEt_3$	CH_2Cl_2	$^1J(Al, P) = 240$	63
$Al(BH_4)_3 \cdot NMe_3$	C_6H_6	$^2J(Al, H) = 45, 44, 46$	1, 55, 64
		$^1J(Al, B) \approx 9$	64
$Al(BH_4)_3 \cdot PMe_3$	toluene	$^1J(Al, P) = 265$	1
$[Al(HMPT)_4]^{3+}$	nitromethane	$^2J(Al, P) = 30$	23
$\{Al[(MeO)_3PO]_6\}^{3+}$	nitromethane	$^2J(Al, P) = 19.5$	23

[a] Closely similar results are reported[24,61,71] for $NaAlH_4$ and $KAlH_4$ (solvents including THF and diglyme).

TABLE 9.3 T_1 data[a] for ^{27}Al

Compound		T_1/ms	Compound	T_1/ms
$AlCl_3$[b]	2.4 M	22.7	Al_2Me_6[c,d]	0.34
	2.0 M	32.6	$[AlMe_2Cl]_2$[c]	0.15
	1.5 M	50.5	$[AlMe_2Br]_2$[c]	0.16
	1.0 M	71.4[e]	$[AlEt_2Cl]_2$[c]	0.084

[a] Data taken from refs. 56 and 58. [b] In H_2O. [c] In toluene. [d] 0.18 ms for a Et_2O solution. [e] Value extrapolated to infinite dilution: 133.3 ms.

theoretically possible ions was observed.[20] Although substitution of I^- for Cl^- in $AlCl_4^-$ causes a monotonic increase in the shielding of the ^{27}Al nucleus, the increase is not a linear function of the number of iodine atoms attached to aluminium.

The coalescence of the ^{27}Al NMR spectra with increasing temperature was used to determine the rate constant for the exchange of a ligand molecule between bulk and bound solvent sites.[23] The rate laws were found to be zero and first order in free ligand for octahedral and tetrahedral solvates, respectively. These data are consistent with a dissociative and an associative substitution mechanism, respectively. This mechanistic change is accompanied by a dramatic decrease in the activation enthalpies and entropies. Typical values are $\Delta H^\ddagger = 82.8$ and 32.2 kJ mol^{-1} and $\Delta S^\ddagger = 28.9$ and -42.7 J mol^{-1} K^{-1} for Al (dimethylmethylphosphonate)$_6^{3+}$ and Al (hexamethylphosphorotriamide)$_4^{3+}$, respectively.

A number of studies of trisubstituted, and adducts of trisubstituted, aluminium compounds have been made utilizing ^{27}Al NMR spectroscopy. Compounds of the type AlR_3(R = Me, Et, Pr, Bu, and hexyl) have been examined,[33–35] as well as Al_2R_6,[17,36–38] $(AlXR_2)_2$[38] and $Co(AlX_4)_2$.[39] Aluminium alkoxy compounds[40–44] and adducts of the trisubstituted aluminium compounds[36,45–48] (including adducts of aluminium borohydride[49–55]) have also been studied.

Because of the relatively fast relaxation of ^{27}Al, spin-spin coupling is not generally observed in ^{27}Al spectra; however, a number of examples of aluminium coupling have been reported, as shown in Table 9.2.

A few investigations have been made of the spin-lattice relaxation time, T_1, of the ^{27}Al nucleus.[56–59] Table 9.3 contains T_1 values obtained for several aluminium species.

The chemical shift of the ^{27}Al nucleus in AlH_4^- has been calculated by a variation procedure using both VB- and MO-type wave functions. The results were found to be in reasonable agreement with the experimental value.[37]

Although the subject will not be discussed here, the ^{27}Al nucleus has proved to be an attractive nucleus for solid state NMR studies. A detailed list of references on this topic is available from the authors upon request.

9B GALLIUM

Gallium has two isotopes that possess spin ($I = \frac{3}{2}$), ^{69}Ga and ^{71}Ga. The latter has the higher receptivity and narrower relative linewidth. This indicates that ^{71}Ga is the more favourable isotope for direct NMR observation in spite of the adverse natural abundance; however, resonances of both isotopes have been studied.

The chemical shift range of the gallium nucleus for species observed to the present time is approximately 1400 ppm and, as discussed by Akitt,[1] the chemical shifts appear to follow the same trends as found with the corresponding aluminium compounds, although there are some individual differences. The types of compounds studied by ^{69}Ga and/or ^{71}Ga NMR are more limited than those studied by ^{27}Al NMR. Table 9.4 contains gallium chemical shifts relative to $[Ga(H_2O)_6]^{3+}$ and linewidths, taken from the review by Akitt[1] and supplemented with more recent data.

Complexation studies of the gallium ion have been performed using ^{69}Ga and ^{71}Ga NMR as the experimental probes.[72–77] The relaxation properties of the gallium nuclei and the relatively small number of compounds studied have resulted in little information being available on spin-coupling. The coupling constants for ^{69}Ga and ^{71}Ga with hydrogen have been determined, however, in the GaH_4^- ion.[61] The values obtained

TABLE 9.4 ^{71}Gallium chemical shifts and linewidths

Compound	Solvent	δ_{Ga}/ppm	$W_{1/2}$/Hz	Ref.
GaH_4^-	Et_2O	+682		82
$GaCl_4^-$	Ga_2Cl_4 melt	+197	300	82
	Ga_2Cl_4/benzene	+223	300	82
	H_2O/HCl	+257[a]	100	82
		+239	90	79
		+243	870(^{69}Ga)	78
$[Ga(OH)_4]^-$	H_2O/NaOH	+192	600	82
$Et_2O \cdot GaCl_3$	neat	+137		82
$GaBr_4^-$	H_2O/HBr	+69	100	82
		+64	80	79
	Ga_2Br_4 melt	+127		82
$[Ga(H_2O)_6]^{3+}$	H_2O	0	300	82
			200	79
			1300(^{69}Ga)	78
$[Ga(DMF)_6]^{3+}$	DMF	−25	120	79
$[Ga(MeCN)_6]^{3+}$	MeCN	−75		83
GaI_4^-	H_2O/HI	−450	100	82
		−479	540(^{69}Ga)	78
Ga^+ in Ga_2Br_2	benzene	−652	150	82
	melt	−413		82
Ga^+ in Ga_2Cl_2	benzene	−685	150	82
	melt	−493	150	82

[a] Other reported values are +250 (ref. 83), +240 (ref. 84) and +200 (ref. 85).

were $|J(^{69}Ga, ^1H)| = 515$ Hz and $|J(^{71}Ga, ^1H)| = 650$ Hz. The differences in linewidth between ^{69}Ga and ^{71}Ga species predicted by Table 1.2 have been proven experimentally to be correct;[78–80] however, the predicted values represent a lower limit. The effect of solvent, complexation and concentration on the relaxation rates of the ^{69}Ga and ^{71}Ga nuclei has been studied.[74,81]

Gallium NMR has proved to be a very useful technique for studying the solid state properties of gallium compounds. References are available upon request from the authors.

9C INDIUM

NMR studies of ^{115}In and ^{113}In have been almost exclusively physics-, rather than chemistry-, oriented. A wide range of investigations of solids or melts of indium metals, alloys, and salts has been undertaken. Several theoretical studies of indium NMR properties have been reported. Neither the purely theoretical nor the physics-oriented papers will be referenced or discussed here. The references are, however, available from the authors upon request.

The large quadrupole moments of ^{115}In and ^{113}In, and the low receptivity of the latter, probably account for the scarcity of chemistry-oriented NMR studies of these isotopes. Moreover, ^{115}In is radioactive (β^- decay), though it has a very long half-life

TABLE 9.5 ^{115}In chemical shifts and linewidths

Species	Conc.[a]	Solvent[b]	δ_{In}/ppm[c]	$W_{1/2}$/Hz	Ref.
$InCl_4^-$ [d]		Et_2O	+440	675 ± 25	93
$InCl_4^-$ [d]	<1.5 M	Et_2O [e]	+420	1255	91
$InCl_4^-$ [d]	<1.5 M	ethyl acetoacetate [e]	+415	2070	91
$InCl_4^-$ [d]	<1.5 M	Pr^i_2O [e]	+420	675	91
$InCl_4^-$ [d]	<1.5 M	$MeBu^iCO$ [e]	+420	745	91
$InCl_4^-$ [d]	<1.5 M	Bu^nOAc [e]		2030	91
$InCl_4^-$ [d]	1:4:12	4:12 H_2O:acetone	+430	1950	92
$InCl_4^-$ [d]	1:4:20	4:20 H_2O:acetone	+480	1120	92
$InCl_4^-$ [d]	1:4:28	4:28 H_2O:acetone	+430	840	92
$[InCl_x(H_2O)_{6-x}]^{(3-x)+}$ [d]		conc. HCl	+180	8100 ± 750	93
$InBr_4^-$ [f]		Et_2O	+180	375 ± 10	93
$InBr_4^-$ [f]	<1.0 M	Et_2O [g,h]	+180[h]	540[h]	91
$[In(H_2O)_6]^{3+}$ [i]		H_2O	0	375 ± 25[j]	93
$[In(H_2O)_6]^{3+}$ [k]	<0.6 M	1 M HNO_3	0	1470	91
$[In(H_2O)_6]^{3+}$ [i]	1:16	H_2O	0	2240	92
$[In(H_2O)_6]^{3+}$ [i]	1:1:20	1:20 $HClO_4$:H_2O	+10	1780	92
$[In(H_2O)_6]^{3+}$ [i]	1:1:20:20	1:20:20 $HClO_4$:H_2O:acetone	+10	3920	92
$[In(H_2O)_6]^{3+}$ [i]	1:2:15:15	2:15:15 $HClO_4$:H_2O:acetone		18 000	92
$[In(SO_4)_2]^-$ [l]		H_2O	−18	300 ± 20[j]	93
$[InBr_x(H_2O)_{6-x}]^{(3-x)+}$ [i]		conc. HBr	−300	11 300 + 1500	93
InI_4^- [m]		conc. HI	−578	1075 ± 50	93
InI_4^- [m]	<0.5 M	conc. HI	−440	850	91
InI_4^- [m]		Et_2O [h]	−583	247 ± 5	93
InI_4^-; $[In(H_2O)_6]^{3+}$ [i]	1:1:15:150	1 : 15 : 150 HI : H_2O:acetone	−620; −7	210; 5000	92
InI_4^-; $[In(H_2O)_6]^{3+}$ [i]	1:1:15:75	1:15:75 HI:H_2O:acetone	−575; −4	220; 5000	92
InI_4^-; $[In(H_2O)_6]^{3+}$ [i]	1:1:15:60	1:15:60 HI:H_2O:acetone	−555; −4	190; 5000	92
InI_4^-; $[In(H_2O)_6]^{3+}$ [i]	1:1:15:45	1:15:45 HI:H_2O:acetone	−600; −30	190; 5000	92
InI_4^-; $[In(H_2O)_6]^{3+}$ [i]	1:1:15:30	1:15:30 HI:H_2O:acetone	−560; −15	140; 5000	92
InI_4^-; $[In(H_2O)_6]^{3+}$ [i]	1:1:15:15	1:15:15 HI:H_2O:acetone	−555; −2	470; 5000	92

[a] Listed as molarity or mole fraction ratios of indium salt and solvent components. [b] Temperature either 25°C or ambient and not specified. [c] Reference: $[In(H_2O)_6]^{3+}$. [d] As $InCl_3$. [e] In equilibrium with conc. HCl. [f] As $InBr_3$. [g] In equilibrium with conc. HBr. [h] For data on the effects on the linewidth of solvent and concentration variation see refs. 91 and 92. The chemical shifts reported vary only in the range 180 to 195 ppm for $InBr_4^-$ and in the range −470 to −540 ppm for InI_4^-. [i] As $In(ClO_4)_3$. [j] $(W_{1/2}/\eta)$/Hz cP^{-1}. [k] As $In(NO_3)_3$. [l] As $In_2(SO_4)_3$. [m] As InI_3.

(5×10^{14} yr). Four papers appeared in the 1950s, in which aqueous solutions of In^{III} salts were studied.[86–89] The NMR frequencies (Ξ/MHz) of ^{115}In and ^{113}In in a saturated $In(NO_3)_3$ solution in 30% HNO_3 doped with $Mn(NO_3)_2$ were found to be 21.910 and 21.863, respectively.[87,88] In what is apparently the only study of the relaxation of ^{115}In in solution, a T_1 of 0.33 ms and a T_2 of 0.24 ms were obtained[89] for 1.0 M $In(ClO_4)_3$ in 1.3 M aqueous $HClO_4$.

Four other papers reporting ^{115}In chemical shifts and linewidths in aqueous and nonaqueous indium salt solutions have been published in the last decade.[90–93] The trivalent indium ion tends to form complexes with anions, particularly in nonaqueous solvents, with exchange rates slow enough in some cases to cause broadening. In the $In(ClO_4)_3$:HI:H_2O:acetone system, the rate is sufficiently slow to allow resolution of resonances for both the hydrated In^{3+} ion and the complex InI_4^- ion.[92]

Indium chemical shifts can be categorized according to the anion participating in complex formation, the shifts of each complex being only slightly and unpredictably dependent on solvent and concentration. With respect to the hydrated In^{3+} ion, $InCl_4^-$ species resonate to high frequency by 420–440 ppm; $InBr_4^-$ to high frequency by 180–190 ppm; mixed aquo-chloro complexes to high frequency by 180 ppm; $[In(SO_4)_2]^-$ to low frequency by 18 ppm; mixed aquo-bromo complexes to low frequency by 300 ppm (rather unusual, in that they do *not* resonate in the range between $[In(H_2O)]^{3+}$ and $InBr_4^-$); and InI_4^- to low frequency by 470–600 ppm. The total shift range is thus ca. 1100 ppm (see Table 9.5).

The solution species and their shifts are most easily assigned by linewidth criteria. Due to their large quadrupole moments, ^{115}In and ^{113}In possess linewidths very sensitive to the environmental symmetry of the indium nucleus, a fact which can be used to advantage to monitor formation of ion pairing and mixed complex formation. The sensitivity of ^{115}In to this type of study may be appreciated when it is realised that its linewidth in nonaqueous solutions of tetrahaloindate anions obtained by extraction from acidified aqueous salts is determined by perturbation of tetrahedral symmetry by ion-pairing with hydrogen ions.[91–93] In some cases ligand exchange among quadrupole-broadened species can give rise to a weighted-average signal too broad to be detected.[92]

It seems that no coupling constants involving ^{115}In or ^{113}In have been experimentally determined, but a value of 968 Hz for the one-bond (^{115}In, ^{1}H) coupling in InH_4^- has been calculated.[61]

9D THALLIUM

9D.1 General

The isotopes ^{205}Tl and ^{203}Tl are the only non-quadrupolar NMR-active nuclei in the Group III family, and have high receptivities—^{203}Tl being slightly less receptive than ^{31}P, while ^{205}Tl, the third most receptive spin-1/2 nuclide, is twice as receptive as ^{31}P. Thallium-205 is the most receptive heavy metal spin-1/2 nucleus (by a factor of thirty over ^{119}Sn, which the second most receptive).

Solids or melts of thallium metal, alloys and salts have received much attention with thallium NMR. Some theoretical papers concerning the NMR parameters of thallium have been published. The interested reader can request a list of these solid-state and theoretical papers from the authors. Several investigators have reported the use of thallium NMR to study thallous ions adsorbed on zeolites.[94–97]

9D.2 Tl^+ Solutions

a. Biological Studies

Because of its similarities to the alkali metal cations, the thallous ion has potential as a probe for studies of Na^+ and K^+ functions in biological systems. Arnold and Scholl[98] monitored the ^{205}Tl chemical shift of 0.1 M $TlNO_3$ as a function of pH in 0.0125 M and 0.025 M phosphate solutions, finding high-frequency shifts of 13.5 ppm and 27 ppm at the second H_3PO_4 dissociation point. They also observed a 5.75 ppm high-frequency shift for a D_2O solution of $TlNO_3$ and dipalmitoylphosphatidylcholine (DPPC) as the Tl^+:DPPC ratio decreased and as the temperature was lowered through DPPC's thermal phase transition point. Reuben and Kayne[99,100] noted ^{205}Tl chemical shifts, together with T_1's and T_2's of Tl^+, in the presence of pyruvate kinase and its substrates. Grisham, Mildvan, and co-workers[101,102] deduced from ^{205}Tl longitudinal relaxation rates that the monovalent cation binding site in (Na^+, K^+)-ATPase is very near the divalent cation binding site.

b. Chemical Shifts

The concentration and anion dependence of chemical shifts for thallous salts in aqueous solutions[103–116] and in nonaqueous solvents[117–123] have been investigated (see Table 9.6). There have also been studies of preferential solvation in mixed solvents.[117,118,120,123] Solvent isotope shifts for D_2O and H_2O were found to vary with anion and concentration by as much as 10 ppm, increasing to as much as 100 ppm upon complexation of Tl^+ with ethylenediamine.[124] An iterative method for determining shifts of thallium salts in solvents in which they are insoluble[117] and in complexes with crown ethers[121] was devised. The solvent-dependent shift range exceeds 2400 ppm; the anion and concentration dependences are in the range 10–100 ppm.

c. Relaxation

The first investigators of $^{205}Tl^+$ relaxation in aqueous solution noted a linear increase in longitudinal relaxation rate with added paramagnetic $Fe(CN)_6^{3-}$ up to a 0.01 M concentration.[103] Reeves' group found that ^{203}Tl and ^{205}Tl relaxation rates were equal, that $T_1 \geq T_2$, and that the relaxation rates were independent of resonance frequency, solvent isotopic substitution (H_2O, D_2O), concentration (0.03 M–2.0 M), and anion[124,125] (see Table 9.7). Degassed Tl^+ samples in water and in methanol gave T_1's of 1.85 s and 0.91 s, respectively.[124] A large, linear increase of T_1^{-1} with dissolved oxygen was found,[124,125] up to an oxygen partial pressure of five atmospheres. Relaxation rates also increased with the addition of paramagnetic Cu^{2+} and $Fe(CN)_6^{3-}$ ions, and less markedly increased with addition of the chelates ethylenediamine and *o*-phenanthroline.[124]

Reeves suggested that spin-rotation relaxation was dominant in aqueous Tl^+ solutions.[124,125] A recent determination of the temperature-dependence of T_1(^{205}Tl) in H_2O and in HMPT has demonstrated that although spin-rotation is indeed the dominant relaxation mechanism for aqueous solutions, it is not[122] for HMPT as solvent. The solvent-dependence of the ^{205}Tl T_1 was found to vary by over an order of magnitude for a series of nine solvents, ranging from 0.08 s in *n*-butylamine to 1.80 s in water.[122] A recent theoretical paper has predicted the importance of transient spin-rotation and chemical shift anisotropy mechanisms for thallium relaxation.[126] The same group earlier determined the increases of T_1^{-1} and T_2^{-1} with increasing concentration of the radical 4-hydroxy-2,2,6,6-tetramethylpiperidine-1-oxyl (TANOL),

T_2^{-1} being much the more strongly affected.[127] The nuclear-electron Tl^+-TANOL dipolar interaction dominates longitudinal relaxation, while the scalar hyperfine interaction dominates transverse relaxation.[127]

d. *Coupling Constants*

A proton NMR study of the thallium-cryptate complex of

$$N[(CH_2)_2O(CH_2)_2O(CH_2)_2]_3N$$

yielded (^{205}Tl, ^1H) couplings of 11.5 Hz to the *N*-methylene protons and 14.5 Hz to the *O*-methylene protons.[128] A ^{13}C-{^1H} investigation of the Tl^+/valinomycin complex in $CDCl_3$ showed thallium-carbon couplings of 96 Hz and 101 Hz, which were assigned to the carbonyl carbons of the D- and L-valine residues.[129] No coupling was observed in CD_3OD and other polar solvents, presumably due to fast exchange.

9D.3 Tl^{3+} Solutions

a. *Chemical Shifts*

Thallic salts tend to hydrolyze in aqueous solution; only five thallium NMR studies of such systems have been reported.[110,116,130–132] Gutowsky and McGarvey reported[110] concentration-dependent shifts for $Tl(NO_3)_3$ in HNO_3 and for dilute $TlCl_3$ in varying ratios of $HCl:HNO_3$. The most extensive study of Tl^{3+} solutions was carried out by Figgis, who reported shifts of $TlCl_3$, $TlBr_3$, and $Tl(NO_3)_3$ with increasing concentrations of their respective acids.[131] He also reported ^{205}Tl shifts of $Tl(NO_3)_3$ with added H_2SO_4, HF, $HClO_4$, HBr, and HCl.[131] The shift range for the concentration-, anion-, and pH-dependence of the $^{205}Tl^{3+}$ chemical shift is nearly 2000 ppm. Its solvent-dependence has not been adequately examined.

b. *Relaxation*

No relaxation studies, *per se*, of Tl^{3+} in solution have been reported, although Figgis measured linewidths of non-degassed aqueous solutions.[131] These varied from about 10 Hz for $TlCl_3$ in HCl to about 5000 Hz for $Tl(NO_3)_3$ in HBr at $Br^-:Tl^{3+}$ ratios near 1.5, with a linewidth range of 75–350 Hz for most of the solutions.

c. *Coupling Constants*

A proton NMR study of a Tl^{III} complex with nitrilotriacetate ion, $N(CH_2CO_2^-)_3$, in D_2O, at $pH \leq 3$, showed coupling of the methylene protons to ^{205}Tl of 387 Hz and to ^{203}Tl of 384 Hz.[133]

9D.4 Alkylthallium (III) Compounds

a. *Chemical Shifts*

Chemical shifts (Table 9.6) of the alkylthallium(III) compounds Me_3Tl,[116,130,134] Et_3Tl,[116] Me_2Tl^+,[116,124,135–138] and $MeTl^{2+}$,[139] as well as of the complex bis-(*cis*-1,2-bithioethene)thallate anion[137] and of the ethoxide, *N,N*-dimethylamine, pyrazole, and *N,N*-diethyldithiocarbamate derivatives of dimethylthallium[116,137] have been determined. The solvent-, anion-, and concentration-dependences of the shifts for Me_2Tl^+ and $MeTl^{2+}$, which can be as much as 350 ppm, were investigated.[135,139] The trialkyls resonate at highest frequency, with the dialkyls giving signals about 1000–1500 ppm to low frequency, the monoalkyls about 1500–2000 ppm to low frequency, and the *cis*-1,2-dithioethene, ethoxide, *N,N*-dimethylamine, and *N,N*-diethyldithiocarbamate dialkyls resonating at about 5000–1000 ppm to low frequency. Solvent isotope shifts for Me_2Tl^+ in H_2O and D_2O were concentration-dependent and amounted to about 5 ppm.[124]

b. Relaxation

Chan and Reeves reported the T_1 of aqueous Me_2Tl^+ to be 0.56 s for a degassed solution, independent of anion, concentration, resonance frequency, solvent isotope composition (H_2O, D_2O), and thallium isotope (203 or 205) observed.[124] The times T_1 and T_2 were nearly the same, and the increase in relaxation rate with increased oxygen partial pressure, although large ($T_1^{-1}/s = 1.8 + 1.1p(O_2)/atm$ for a 0.8 M solution at 26°C for $p(O_2)$ in the range 0–5 atm*), was an order of magnitude smaller than for the Tl^+ ion.[124] The ^{205}Tl dynamic nuclear polarization of Me_3Tl in solution with an organic radical was measured.[140]

c. Coupling Constants

Nearly all the reported coupling constants (Tables 9.8 to 9.13) have been determined without direct observation of ^{205}Tl or ^{203}Tl. Couplings to ^1H in trialkyls,[134,141–149] mixed trialkyls,[137,141,149–156] mixed alkyl-vinyl,[143] alkyl-alkynyl,[153] and mixed alkyl-cyclopentadienyl[153,157,158] compounds have been reported (Table 9.8). Couplings of ^{205}Tl with ^{13}C in Me_3Tl[134] ($^1J = +1930$ Hz) and in $[(Me_3Si)_2N]_3Tl$[149] ($^3J = 83$ Hz) and with ^{19}F in $Tl(O_2CCF_3)_3$[146] ($^4J = 85.2$ Hz for a solution in THF at -110°C) have also been reported. In addition, it was found that $|^2J(^{205}Tl, ^{205}Tl)|$ is 536.2 Hz in $(Me_2TlNMe_2)_2$[137,155] and 1037 Hz and 971 Hz in the *cis* and *trans* conformers of $(Me_2TlNHMe)_2$,[116] respectively. The value of $|^2J(^{205}Tl, ^{203}Tl)|$ in $(Me_2TlOEt)_2$[137] is 1200 ± 20 Hz.

There is rather extensive literature concerning ($^{203,205}Tl$, ^1H) coupling in dialkylthallium(III) compounds (Tables 9.9 and 9.10), with a plethora of studies[116,124,134,135,137,139,142,152,153,159–178] of the effect of solvent and anion on the coupling in Me_2Tl^+. Coupling constants for diethyl,[116,152,164,179] dipropyl (*n*-, *i*-),[116,142,180] and dibutyl[142] compounds have been observed, as have those for $[(Me_3SiCH_2)_2Tl]^+$[181] and $(Me_3Si)_2Tl^+$.[182] Thallium coupling with ^{13}C in the dimethylthallium ion has been reported (Table 9.13) as a function of anion and solvent.[134,135,137,139] Couplings ($^{203,205}Tl$, ^1H) in mixed dialkylthallium,[142,174,181,183–185] alkyl-vinyl,[186] alkyl-aryl,[174,183,186] alkyl-alkynyl,[174,183] and alkyl-cyclopentadienyl[186] organothallium compounds have also been recorded.

Monoalkylthallium(III) couplings in the literature (Table 9.11) include ($^{203,205}Tl$, ^1H) values for $MeTl^{2+}$,[139,165,168,187,188] $EtTl^{2+}$[168,188] and others.[181,189–191] For $MeTl^{2+} \cdot 2[O_2CMe]^-$, $^1J(^{205}Tl, ^{13}C)$ has been reported as $+5631$ Hz in $CDCl_3$ and $+5976$ Hz in CH_3OH.[139]

Thallium-proton couplings have been measured in a variety of substituted vinyl[142,180,192–196] and divinyl[142,180,197] thallium(III) derivatives (Tables 9.8 to 9.11).

* The slope is nonlinear in the 0–1 atm range of $p(O_2)$, being 3.5 up to $p(O_2) = 0.2$ atm.

9D.5 Arylthallium (III) Compounds

a. Chemical Shifts

Thallium chemical shifts (Table 9.6) have been determined only for phenylthallium(III) dichloride and its complexes with PPh_3 and dipyridine in methanol and pyridine,[116] diphenylthallium(III) chloride in liquid ammonia,[134] diphenylthallium(III) bromide in DMSO,[116] a series of substituted arylthallium(III) bis(trifluoroacetates) in several solvents,[136] and triphenylthallium(III) in ether.[116] Diaryls appear to resonate about 200 ppm to high frequency of monoaryls (this is a crude estimate because the same anion and solvent were not used for the two cases) and 600 ppm to low-frequency of triaryls. Ring substituent-induced shifts are approximately 100 ppm; solvent-induced shifts are also about 100 ppm.

b. Coupling Constants

Representative values are listed in Tables 9.8 to 9.12. Thallium-205 coupling constants with protons in Ph_3Tl have been measured.[142,143,180] Mixed aryl-alkyl diorganothallium(III) cation $J(^{205}Tl, ^{1}H)$ values have also been determined.[174,183,186,198] Several diarylthallium thallium-proton coupling constants may be found in the literature,[116,134,142,180,199–203] and a few thallium-fluorine couplings;[204,205] for instance, the compound $(C_6F_5)_2TlBr$ has $(^{205}Tl, ^{19}F)$ coupling constants $^3J = 799$ Hz, $^4J = 343$ Hz and $^5J = 99$ Hz. A host of monoarylthallium $(^{205}Tl, ^{1}H)$ coupling constants have been measured,[136,142,180,199–203,206–212] a few $(^{205}Tl, ^{19}F)$ values,[205] and several $(^{205}Tl, ^{13}C)$ couplings.[205,213–216] Of note are coupling constants for one di-thallated aromatic compound,[212] 3-[$Tl(TFA)_2$]-4-(MeO)$C_6H_3Tl(TFA)_2$.

9D.6 Miscellaneous Compounds

a. Thallous Ethoxide

This compound exists as a tetramer both in solution and as the neat liquid. Its chemical shift is 2914 ppm to high-frequency of the ^{205}Tl signal of aqueous $TlNO_3$;[116,130] $^2J(^{205}Tl, ^{203}Tl)$ for the tetramer was found to be 2560 Hz.[130]

b. Coupling Constants for Various Thallium(III) Compounds

The value $^1J(^{205}Tl, ^{1}H) = 6144$ Hz has been estimated[61] for the unstable hydride TlH_4^- using data for XH_4^- with X = ^{11}B, ^{27}Al, ^{69}Ga and ^{71}Ga. Values of $J(^{205}Tl, ^{1}H)$ have been determined for a series of thallated naphthyl, phenanthryl, and other fused-ring aromatic systems,[142,202] for a series of dichloropyridiomethylthallium(III) and chloro-bis (pyridiomethyl) thallium(III) cations,[217] and for the bis (*cis*-1,2-dithioethene) thallate anion[218] (see I). Both $(^{205}Tl, ^{1}H)$ coupling constants[219–223] and $(^{205}Tl, ^{13}C)$ couplings[224–226] for Tl(III) metalloporphyrins and for the related four-coordinate dimethylthallium(III) dipyrromethane[227] (see II) have been detected. Large $(^{205}Tl, ^{1}H)$ coupling constants have been found for norbornyl- and norbornenyl-thallium compounds[228] and thallium norbornyl lactones.[229] Weibel and Oliver[144,145,230] measured thallium-proton couplings for four-coordinate Tl in the series $Li[(Me_3Sn)_xTlMe_{4-x}]$, with $x = 0, 1, 2, 3$, and 4. Values of $^2J(^{205}Tl, ^{1}H)$ are in the range -214 to -228 Hz, while $^3J(^{205}TlSnC^{1}H)$ are 32 to 48 Hz. Proton spectra of several methyl-[186] and dimethyl-[153,157,158] thallium(III) cyclopentadienyls, for the bimetallic[231] species

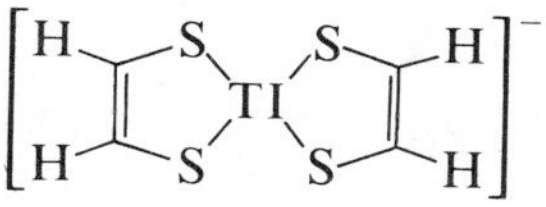

I

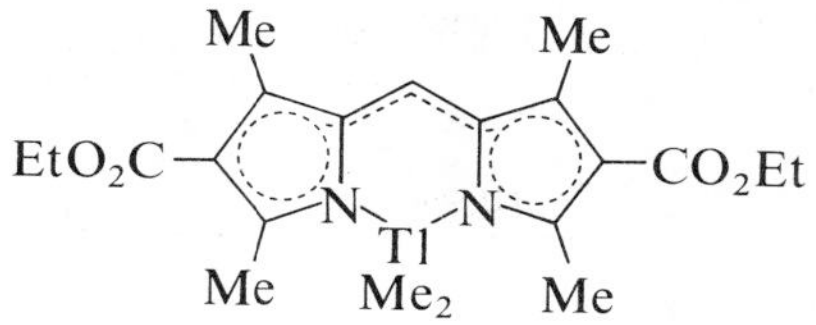

II

$[Me_2Tl][Mo\text{-}(\pi\text{-}Cp)(CO)_3]$ and $[Me_2Tl][W\text{-}(\pi\text{-}Cp)(CO)_3]$[231] and for dicyclopentadienylthallium(III)[142] have been recorded, in which a coupling of 214–225 Hz for thallium to the cyclopentadienyl (Cp) protons was observed for only the methylcyclopentadienyl[186] and dicyclopentadienyl[142] species, although the thallium coupling constants to the methyl protons for the dimethyl[153,157,158] and bimetallic[231] species were suggestive of some covalent Tl-(Cp)interaction. Thallium-proton couplings for the bimetallic $Tl[Co(CO)_3P(CH_2SiMe_3)_3]$ have been measured,[232] as well as (^{205}Tl, ^{1}H), (^{205}Tl, ^{31}P), and (^{205}Tl, ^{13}C) couplings for $Me_3PCH_2TlMe_3$[233] and for $(Me_2TlCH_2PMe_2CH_2)_2$,[234] which contains an eight-membered ring.

Finally, (^{205}Tl, ^{1}H) and (^{205}Tl, ^{11}B) couplings for the metallocarboranes

$$[Me_2Tl]^+[B_{10}H_{12}TlMe_2]^- \quad \text{and} \quad [Ph_3PMe]^+[B_{10}H_{12}TlMe_2]^-$$

have been obtained by $^{1}H\text{-}\{^{11}B\}$ and $^{11}B\text{-}\{^{1}H\}$ NMR.[172]

9D.7 Concluding Remarks

a. Chemical Shifts

The chemical shift range for thallium is extremely large, covering nearly 4000 ppm for Tl^{III} species and about 3400 ppm for Tl^{I} species, in compounds measured to date, with a total shift range of over 5500 ppm. There is some overlap of the two regions, e.g. $(TlOEt)_4$ resonates in the same region as arylthallium(II) dications and at a higher frequency (by ca. 500 ppm!) than aqueous Tl^{3+} solutions. Resonances of Tl^{+} in highly basic, electron-donating solvents are higher than those of Tl^{3+} in aqueous solution (see Table 9.6).

b. Relaxation

Thallium relaxation processes, in which the spin-rotation and chemical shift anisotropy mechanisms are expected to be instrumental, are very efficient, expecially in the presence of oxygen or other paramagnetics, making FT work very advantageous (see Table 9.7).

c. Coupling Constants

Representative coupling constants, listed* in Tables 9.8 to 9.13, are often large and can vary significantly with changes in intervening atoms, solvents, temperature, anions, and

* Only ^{205}Tl couplings are normally listed in this book (or mean values when spectral splittings due to ^{203}Tl and ^{205} Tl cannot be resolved). The original literature often contains separate data for ^{203}Tl coupling. Comprehensive tabulations of coupling constants to thallium are available from the authors of the present chapter.

TABLE 9.6 Thallium-205 chemical shifts

Compound	Conc./M[a]	Solvent	Temp./°C	δ_{Tl}[b]/ppm	Ref.
Me_3Tl		n-pentane	24	+5093	116
Me_3Tl		Et_2O	24	+4753	116
Me_3Tl	20%	Et_2O		+4755 ± 20	130
Me_3Tl	10%	acetone	−70	+4504[c]	134
Et_3Tl		n-pentane	24	+4963	116
Ph_3Tl		Et_2O	24	+3814	116
$[Et_4N][Tl(S_2C_2H_2)_2]$		acetone-d_6	56	+4075[c]	137
$Me_2TlSSCNEt_2$		$CDCl_3$	−50	+3728[c]	137
$(Me_2TlNMe_2)_2$		C_6D_6	37	+4025[c,n]	137
$(Me_2TlNMe_2)_2$		C_6H_6	24	+3988	116
$(Me_2TlNMe_2)_2$		Et_2O	24	+3841	116
$(Me_2TlOEt)_2$		C_6D_6	37	+3846[c,n]	137
$(Me_2TlOEt)_2$		toluene-d_8	37	+3839[c,n]	137
$(Me_2TlOEt)_2$		toluene-d_8	−60	+3869[c]	137
$(Me_2TlOEt)_2$		n-hexane	24	+3753	116
Me_2Tl-NNCHCHCH		toluene	54	+3734	116
Me_2Tl-NNCHCHCH		neat (melt)	54	+3730	116
Me_2Tl-NNCHCHCH		toluene	24	+3619	116
Me_2TlOH		H_2O	24	+3636	116
Me_2TlOPh[d]	0.2	CH_2Cl_2	29	+3743	135
Me_2TlOPh[d]	0.2	pyridine	29	+3595	135
Me_2TlOPh[d]	0.2	DMSO	29	+3561	135
Me_2TlI	0.1	pyridine	29	+3689	135
Me_2TlI	1.2	DMSO		+3594	116
Me_2TlI[d]	0.2	DMSO	29	+3570	135
Me_2TlBr	sat.	DMSO		+3575	116
Me_2TlBr	5%	NH_3(liq.)	−30	+3544[c]	134
Me_2TlF	sat.	H_2O		+3499	116
Me_2TlF	sat.	DMSO		+3413	116
Me_2TlNO_3[e]	1.2	50% en		+3650	116
Me_2TlNO_3	∞-dil.(0.026)	n-butylamine	23	+3586	138
Me_2TlNO_3	1.0	D_2O		+3478	116
Me_2TlNO_3	1.0	H_2O		+3473	116
Me_2TlNO_3	∞-dil.(0.0005)	H_2O	23	+3514	138
Me_2TlNO_3	∞-dil.(0.001)	formamide	23	+3477	138
Me_2TlNO_3	0.6	MeOH		+3435	116
Me_2TlNO_3	∞-dil.(0.008)	DMSO	23	+3450	138
Me_2TlNO_3	∞-dil.(0.001)	NMF	23	+3440	138
Me_2TlNO_3	∞-dil.(0.0025)	NEF	23	+3438	138
Me_2TlNO_3	1.0	pyridine	29	+3569	135
Me_2TlNO_3	0.2	pyridine	29	+3409	135
Me_2TlNO_3	∞-dil.(0.006)	pyridine	23	+3424	138
Me_2TlNO_3	∞-dil.(0.005)	DMF	23	+3443	138
Me_2TlNO_3	∞-dil.(0.011)	HMPT	23	+3346	138
Me_2TlOAc	0.4	H_2O	29	+3519	139
Et_2TlBr		DMSO		+3429	116
Et_2TlNO_3		DMSO		+3176	116

TABLE 9.6 *(Continued)*

Compound	Conc./M[a]	Solvent	Temp./°C	δ_{Tl}[b]/ppm	Ref.
Pr^n_2TlBr		DMSO		+3468	116
$Pr^n_2TlNO_3$		DMSO		+3223	116
Ph_2TlBr		DMSO		+3185	116
Ph_2TlCl	5%	NH_3(liq.)	−20	+3184	134
$PhTlCl_2 \cdot PPh_3$		pyridine	24	+3278	116
$PhTlCl_2 \cdot$dipy		pyridine	24	+3093	116
$PhTlCl_2$		pyridine	24	+3087	116
$PhTlCl_2$		MeOH	24	+2989	116
$MeTl(OAc)_2$	1.0	MeOH	29	+3135	139
$MeTl(OAc)_2$	0.8	$CHCl_3$	29	+3075	139
2,4,6-$Me_3C_6H_2Tl(tfa)_2$[f]	0.4	MeCN	25	+2930	136
2,4,6-$Me_3C_6H_2Tl(tfa)_2$[f]	0.4	MeOH	25	+2901	136
2,4,6-$Me_3C_6H_2Tl(tfa)_2$[f]	0.4	DMSO	25	+2887	136
2,4-$Me_2C_6H_3Tl(tfa)_2$	0.4	DMSO	25	+2855	136
2,5-$Me_2C_6H_3Tl(tfa)_2$	0.4	DMSO	25	+2833	136
3,4-$Me_2C_6H_3Tl(tfa)_2$[f]	0.4	DMSO	25	+2828	136
4-$MeC_6H_4Tl(tfa)_2$[f]	0.4	DMSO	25	+2817	136
$C_6H_5Tl(tfa)_2$[f]	0.4	DMSO	25	+2790	136
4-$Bu^tC_6H_4Tl(tfa)_2$	0.4	DMSO	25	+2818	136
4-$Pr^iC_6H_4Tl(tfa)_2$	0.4	DMSO	25	+2823	136
4-$Pr^nC_6H_4Tl(tfa)_2$	0.4	DMSO	25	+2816	136
4-$EtC_6H_4Tl(tfa)_2$	0.4	DMSO	25	+2817	136
4-$BrC_6H_4Tl(tfa)_2$	0.4	DMSO	25	+2779	136
4-$FC_6H_4Tl(tfa)_2$	0.4	DMSO	25	+2776	136
4-$ClC_6H_4Tl(tfa)_2$	0.4	DMSO	25	+2769	136
$TlCl_3$	—	neat	25	+2585	130
$TlCl_3$	—	neat	25	+2515	235
$TlCl_3$	8.0	H_2O	24	+2553	116
$TlCl_3$[g]	∞-dil.(0.1)	H_2O		+2307	131
$Tl(NO_3)_3$[h]	1.5	H_2O	25 ± 3	+1850	110
$Tl(NO_3)_3$	0.69	1.5 M HNO_3		+1900	131
$Tl(NO_3)_3$	∞-dil.(0.25)	∞-dil. HNO_3(0.125)		+2592	131
$TlBr_3$	0.29	H_2O		+1235	131
$(TlOEt)_4$		neat		+2915	130
$(TlOEt)_4$		neat	24	+2893	116
$(TlOEt)_4$		C_6H_6	37	+2917[n]	137
$TlBF_4$	∞-dil.(0.2)[j]	*n*-butylamine		+1958	117
$TlNO_3$	∞-dil.(0.2)[j]	*n*-butylamine		+1708	117
$TlNO_3$	∞-dil.(0.002)	*n*-butylamine	25	+1848	122
$TlNO_3$	∞-dil.(0.002)	pyridine	25	+783	120
$TlNO_3$	∞-dil.(0.002)	DEA	25	+531	120
$TlNO_3$	∞-dil.(0.02)	HMPT	25	+502	122
TlOAc	*ca.* 0.3	MeOH		+512	144
$TlClO_4$[i]	∞-dil.(0.2)[j]	MeOH[k]		+35	117
$TlClO_4$	∞-dil.(0.2)[j]	DMSO		+369	118
$TlNO_3$	∞-dil.(0.002)	DMSO	25	+360	120
$TlNO_3$	∞-dil.(0.002)	NEF	25	+306	120
$TlNO_3$	∞-dil.(0.002)	NEA	25	+280	120
$TlNO_3$	∞-dil.(0.002)	NMF	25	+161	120

Continued overleaf.

TABLE 9.6 *(Continued)*

Compound	Conc./M[a]	Solvent	Temp./°C	$\delta_{Tl}{}^{b}$/ppm	Ref.
$TlNO_3$	∞-dil.(0.002)	DMF	25	+145	122
$TlNO_3$; $TlClO_4$	∞-dil.(0.2)[j]	formamide		+96	118
$TlNO_3$	∞-dil.(0.002)	formamide	25	+77	120
$TlNO_3$	∞-dil.(0.002)	H_2O	25	0	120
TlF; TlOAc; TlO_2CH; $TlClO_4$; $TlBF_4$	∞-dil.(0.2)[j]	H_2O		0	118
$TlNO_3$	∞-dil.(0.002)	pyrrole	25	−389	122
TlOAc	ca. 0.3	MeOH (and dibenzo-18-crown-6)		−35[l]	121
TlOAc	ca. 0.3	MeOH (and 18-crown-6)		−70[l]	121
$TlClO_4$	ca. 0.3	DMF (and dibenzo-18-crown-6)		−110[l]	121
$TlClO_4$	ca. 0.3	DMF (and 18-crown-6)		−180[l]	121
$TlNO_3$	0.3	H_2O		−7	103
$TlNO_3$	0.3	H_2O (and 0.15 M OAc)		+18	103
$TlNO_3$	0.3	H_2O (and 0.15 M citrate)		+168	103
$TlNO_3$	0.3	H_2O (and 0.15 M $Fe(CN)_6^{3-}$)		+918	103
$TlNO_3$[m]	0.5	H_2O	26	−11	124
$TlNO_3$	0.5	H_2O (and 0.25 M en)	26	+117	124
$TlNO_3$	0.5	H_2O (and 0.5 M en)	26	+263	124
$TlNO_3$	0.5	D_2O	26	−19	124
TlO_2CH[m]	1.0	H_2O	26	+34	124

[a] As molarity unless otherwise indicated. Values in parentheses for ∞-dilution extrapolations denote lowest concentration measured. [b] Shifts are reported with respect to infinite-dilution aqueous $TlNO_3$, for which $\Xi = 57.683833$ MHz.[120] Values for compounds from each literature reference are internally consistent; uncertainties propagated in calculating shifts with respect to a single reference are ±(5 to 15) ppm, occasionally as great as ±50 ppm, for comparison of shifts from different laboratories. Data on Ξ for various compounds are reported directly in refs 120, 122, 134–136, 138 and 139. [c] Measured by ^1H-{^{205}Tl} INDOR. [d] Ref. 135 contains data for other concentrations. [e] Data for other solution conditions is contained in refs 116, 124, 135 and 137 [f] Ref. 136 also contains data for THF solutions. [g] See also refs. 110 and 112. [h] There is additional data in ref. 131. The shift reported for $Tl(NO_3)_3$ in ref. 130 indicates the sample was probably $TlNO_3$. [i] Many additional data for $TlClO_4$ are contained in refs 117 and 118. Shifts reported range from +2147 to −506 ppm. [j] Extrapolated to infinite-dilution anion concentration. [k] Thallium salt insoluble in this solvent; shift calculated from an iterative extrapolation of mixed solvent data. [l] Calculated shift of the Tl^+/crown complex; obtained by an iterative fitting procedure. [m] There is additional information on aqueous solutions in ref. 124. [n] Reference 137a incorrectly[137b] lists the shift of $(TlOEt)_4$ in C_6H_6 as +111 ppm, rather than −599 ppm, w.r.t. aqueous Me_2TlNO_3. Its listing of chemical shifts w.r.t. $MeTlNO_3$ for Me_3Tl in acetone at −70° and Me_2TlBr in liquid NH_3 at −30°, calculated from the (correct) frequencies of ref. 134, are also erroneous[137b] and should be +986 ppm and +26 ppm, respectively. In ref. 137a, the frequency ratios for $(Me_2TlNMe_2)_2$ in C_6H_6, and $(Me_2TlOEt)_2$ in C_6H_6 and in toluene (all at 37°), are in error[137b] and should be 0.5791611, 0.5790578, and 0.5790538, respectively.

Compound	Conc./M	Solvent	Temp./°C	T_1^{-1}/s^{-1}	T_2^{-1}/s^{-1}	Added paramagnetic	Ref.
TlO_2CH	2.0	H_2O	26	$0.54 + 37.0\,p(O_2)$		$p(O_2) = 0\text{-}5$ atm.	124
TlO_2CH	2.0	H_2O (and 0.5 M phen)	26	$2.0 + 37.2\,p(O_2)$		$p(O_2) = 0\text{-}5$ atm.	124
TlO_2CH	2.0	H_2O (and 0.5 M en)	26	$5.1 + 38.5\,p(O_2)$		$p(O_2) = 0\text{-}5$ atm.	124
TlO_2CH	1.0	MeOH	26	$1.1 + 148.6\,p(O_2)$		$p(O_2) = 0\text{-}5$ atm.	124
$TlNO_3$	0.5	H_2O	26	0.54 + 2.9 [en]		none	124
TlO_2CH	1.0	H_2O	26	0.53 + 9.7 [en]		none	124
TlO_2CH	1.0	H_2O	26	0.53 + 3.1 [phen]		none	124
$TlNO_3$	0.0803	H_2O	26	0.54	0.83	none	125
$TlNO_3$	0.0803	H_2O	26	8.3	9.6	$p(O_2) = 0.2$	125
$TlNO_3$	0.0803	H_2O	26	38.0		$p(O_2) = 1.0$	125
$TlNO_3$	0.080	H_2O	26	15.3	13	8×10^{-5} M $Fe(CN)_6^{3-}$	125
$TlNO_3$	0.15	H_2O	26	9.9	19	7×10^{-3} M Cu^{2+}	125
$TlNO_3$	0.0839	D_2O	26	0.44	0.8	none	125
$TlNO_3$	0.0839	D_2O	26	8.3	9.6	$p(O_2) = 0.2$ atm.	125
$TlNO_3$	0.0839	D_2O	26	41.0		$p(O_2) = 1.0$ atm.	125
$TlNO_3$	0.3	H_2O		$8 + 54\,000[Fe(CN)_6^{3-}]$		$[Fe(CN)_6^{3-}] = 0\text{-}10^{-2}$ M	103
$TlNO_3$	0.2	H_2O	22	0.57 + 105 000 [TANOL][a]	$0.75 + (2 \times 10^6)$[b] [TANOL]	[TANOL][a] = $0\text{-}10^{-3}$ M	127
$TlNO_3$	0.2	H_2O	10	5.3	161.0	[TANOL][a] = 5×10^{-5} M	127
$TlNO_3$	0.2	H_2O	18	4.8	139.0	[TANOL][a] = 5×10^{-5} M	127
$TlNO_3$	0.2	H_2O	40	5.2	88.0	[TANOL][a] = 5×10^{-5} M	127
$TlNO_3$	0.2	H_2O	62	5.4	54.0	[TANOL][a] = 5×10^{-5} M	127
$TlNO_3$	0.2	H_2O	83	5.4	45.0	[TANOL][a] = 5×10^{-5} M	127
$TlNO_3$	ca. 0.2	H_2O	25	0.56		none	122
$TlNO_3$	ca. 0.2	pyrrole	25	0.95		none	122
$TlNO_3$	ca. 0.2	DMF	25	1.52		none	122
$TlNO_3$	ca. 0.2	formamide	25	1.47		none	122
$TlNO_3$	ca. 0.2	NEF	25	2.50		none	122
$TlNO_3$	ca. 0.2	DMSO	25	0.56		none	122
$TlNO_3$	ca. 0.2	HMPT	25	1.64		none	122
$TlNO_3$	ca. 0.1	pyridine	25	11.1		none	122
$TlNO_3$	ca. 0.2	*n*-butylamine	25	12.5		none	122
$TlNO_3$	ca. 0.2	H_2O	70	0.69		none	122
$TlNO_3$	ca. 0.2	HMPT	110	0.90		none	122

[a] TANOL is the radical 4-hydroxy-2,2,6,6-tetramethylpiperidine-1-oxyl. [b] The slope is not constant, but increases with TANOL concentration.

TABLE 9.8 Representative (^{205}Tl, ^{1}H) coupling constants for triorganothallium (III) compounds[a]

Compound	Solvent	Temp./°C	^2J/Hz	^3J/Hz	^4J/Hz	Ref.
Me_3Tl	acetone	−70	−268.8			134
Et_3Tl	CH_2Cl_2	−85	−198.2	+396.1		141, 142
Ph_3Tl[b]	NMe_3			+259.4	+80 ± 5	142, 143, 180
Me_2EtTl	CH_2Cl_2	−85	−223.0 (Me)			141
			−242.4 (Et)	+472.4		
Et_2MeTl	CH_2Cl_2	−85	−186.9 (Me)			141
			−218.8 (Et)	+441.5		
$Me_2(Cp)Tl$	pyridine	39	−379 (Me)			153
$Me_2(SeCF_3)Tl$	pyridine		−383			154
$Me_2(MeS)Tl$	pyridine	39	−372			153
$Me_2(PhC{\equiv}C)Tl$	pyridine	39	−393			153
Me_2ViTl	NMe_3	−60	−295.5 (Me)			143
Vi_2MeTl	NMe_3	−60	−317.4 (Me)			143
$(Me_2TlNMe_2)_2$	C_6D_6	37	−343.8	94.5	−2.2	137,155
$(Et_2TlNMe_2)_2$	C_6H_6		−381.8	81.1 (NMe)		155
				+612.4 (Et)		
$(Pr^n_2TlNEt_2)_2$	C_6D_6		−382	+518 (Pr^n)	8.5 (Et)	150
				117 (Et)		
$(Me_2TlNC_4H_4)_2$	pyridine-d_5		−404			150
$(Me_2TlOEt)_2$	toluene	−60	−371			137
$Tl(O_2CMe)_3$	CD_3OD	−85			26.3	146
$(Me_3Si)_3Tl$	CH_2Cl_2	−60		102.1		156
$(Me_3Si)_3Tl$	NMe_3[c]	−60		112.9		156
$[(Me_3Si)_2N]_3Tl$	toluene				5	149
$Me_2TlN(SiMe_3)_2$	C_6D_6		−324			149

[a] The signs given in this and the following five tables (and in the text) are those determined experimentally, plus those which can be confidently predicted by analogy. [b] ^5J = +35 ± 5 Hz. [c] 1:1 $(Me_3Si)_3Tl$:NMe_3 adduct formed.

degree of ion pairing or association (for organothallium cations). For example, two-bond thallium-proton couplings range from −187 Hz to −393 Hz for trialkylthallium compounds, from −259 Hz to −828 Hz for dialkylthallium species, and from −630 Hz to −1120 Hz for monoalkylthallium ions; three-bond coupling ranges for the same types of molecules are, respectively, +81 Hz to +612 Hz, ±143 Hz to +639 Hz, and +739 Hz to +1626 Hz; and one-bond carbon-thallium couplings are ca. +1930 Hz, +2460 Hz to +3456 Hz, and +5630 Hz to +5976 Hz, respectively.

Di- and mono-arylthallium cations exhibit coupling of thallium to ring protons of ca. +455 Hz and +812 Hz to +1143 Hz (three-bond, *ortho* hydrogens), +117 Hz to +208 Hz and +280 Hz to +570 Hz (four-bond, *meta* hydrogens), and +40 Hz to +55 Hz and +54 Hz to +115 Hz (five-bond, *para* hydrogens). Monoarylthallium cations are characterized by carbon-thallium coupling ranges of 8841 Hz to 10718 Hz (one-bond), +500 Hz to +756 Hz (two-bond), +793 Hz to +1101 Hz (three-bond), and −140 Hz to −217 Hz (four-bond).

These large couplings could in some cases cause experimental problems because of an insufficiently large spectral window, aliassed peaks, and computer-limited resolution

TABLE 9.9 Representative (^{205}Tl, ^{1}H) coupling constants for diorganothallium(III) cations,[a] R_2Tl^+.

R	Solvent	^2J/Hz	^3J/Hz	^4J/Hz	Ref.
Me	D_2O	$-408 \pm 1.5\%$, -439[b]			135, 139, 124, 142, 152, 116, 164–170
Me	liq. NH_3[c,d]	-448.9			134
Me	HMPT	-461,[e] -473,[e,f] -445[g,h]			173, 116, 152, 171
Me	$SbCl_5/SO_2$[i]	-336			173
Et	D_2O	-338 ± 2	$+623$ to $+634$		142, 152, 164, 179
Et	$CDCl_3$[j]	-306	$+612$		168
Et	DMSO	-376[d], -393[e]	$+639$,[d] $+633$[e]		116
Pr^n	D_2O[k]	-341	$+469$	$+20.5$	142, 180
Pr^i	D_2O[k]	-259	$+574$		142
Bu^n	D_2O[k]	-320	$+452$		142
Me_3Si	monoglyme[i]		143.4		182
Me_3SiCH_2	$CDCl_3$	-537,[d,i] -566[j]			181
Vi	D_2O[f]	$+842$[l]	$+1618$ (*trans*), $+805$[l](*cis*)		142, 180
cis-MeCH=CH	D_2O[f]	$+637$[l]	$+1485$[l]	-94.0	142
trans-MeCH=CH	D_2O[f]	$+640 \pm 10$	$+809 \pm 10$	-47.1	142
MeCH=C(Me)[m]	D_2O[f]		398 (Me), $+1348$[l](CH)	-98.0	142
Ph[n]	DMSO-d_6[o]		$+455$	$+117$	199
4-MeC_6H_4[p]	DMSO-d_6[o]		$+448$	$+128$	199
2,5-$Me_2C_6H_3$[q]	DMSO-d_6[o]		$+454$	$+208$(H-3), 48(Me)	199

[a] The anions are given in the footnotes. The terms anion and cation are used somewhat loosely in this and the following four tables, since the extent of ion association and pairing is seldom known; thus the anions may sometimes be acting effectively as ligands. [b] F^-. [c] -30°C. [d] Br^-. [e] NO_3^-. [f] ClO_4^-. [g] acac. [h] ox. [i] Cl^-. [j] $Pr^iCO_2^-$. [k] SO_4^{2-}. [l] ± 5 Hz. [m] Methyl groups mutually *trans*. [n] $^5J = +55$ Hz. [o] $CF_3CO_2^-$. [p] $^6J = 24$ Hz. [q] $^5J = +40$ Hz(H-4), 17.6 Hz(Me).

TABLE 9.10 (^{205}Tl, ^{1}H) Coupling constants for selected diorganothallium(III) cations of the type $R^1R^2Tl^{+a}$

R^1	R^2	Solvent	^2J/Hz	^3J/Hz	^4J/Hz	Ref.
Me	Et	D_2O	$-356 \pm 1\%^{b,c}$ (Me) -399^c (Et)	$+680^c$		142, 174 183
Me	Me_3SiCH_2	$CDCl_3$	$-342,^d$ -376^e (Me) $-539,^d$ -562^e ($-CH_2Si-$)			181
Me	Vi		-412^e (Me)			186
Me	$ViCH_2$	$CDCl_3$	$-364 \pm 1.5\%^{b,e,f,g}$ (Me) $-533 \pm 1^{b,e,h}$ ($-CH_2-$)	$196 \pm 2^{b,e,f,h}$	$224 \pm 2^{b,e,h}$ (*cis*) $202 \pm 4^{b,e,h}$ (*trans*)	185
Me	Ph	$CDCl_3$	-426^e			186
Me	Ph	D_2O	-455^b			174, 183
Me	PhC≡C	D_2O	-672^b			174, 183
Me	N≡C	D_2O	-828^b			174, 183
Me	Cp	$CDCl_3$	$-450 \pm 6^{b,e,h,i}$ (Me) $218 \pm 2^{b,e,h,i}$ (Cp)			186
Ph	$Me_2NCS_2CH_2$	$CDCl_3$	$-360,^b$ $-318 \pm 2^{j,k}$			198^l

[a] The anions are given in the footnotes. [b] AcO^-. [c] SO_4^{2-}. [d] Br^-. [e] $Pr^iCO_2^-$. [f] tropolonate. [g] salicylaldehydate. [h] $EtCO_2^-$. [i] 4-Pr^i-tropolonate. [j] $S_2CNMe_2^-$. [k] $S_2CNPh_2^-$. [l] Values of ^6J to the non-equivalent methyl protons for different counter-ions are: both 7.5 Hz[b]; 6.4 and 7.9 Hz[k]; 6.8 and 9.7 Hz[j].

TABLE 9.11 (^{205}Tl, ^{1}H) Coupling constants for selected mono-organothallium(III) cations, RTl^{2+} [a]

R	Solvent	^2J/Hz	^3J/Hz	^4J/Hz	Ref.
Me	D_2O	-936[b]			165, 168, 187
Me	$CDCl_3$	-911,[b] -676 ± 2,[c,d]			139, 168, 174,
		-790[e]			187, 188
Et	$CDCl_3$	-822,[f] -769,[e]	$+1626$[f], $+1398$[e]		168, 188
		-631[c,d]	$+1370$,[c] $+1355$[d]		
Me_3SiCH_2	$CDCl_3$	-1121[f]			181
$(HO)Me_2CCH_2$	DMSO-d_6[b]	-871.5,[g] -864.0[g]		101	191
$PhMe(EtO)CCH_2$	CD_3OD,[b] $CDCl_3$[b]	-822,[g] -890[g]		29	189
$Ph(OBu^i)CHCH_2$	CD_3OD,[b] $CDCl_3$[b]	-772,[g] -917[g]	$+828$		189
Vi	D_2O[h]	$+2004 \pm 10$	$+3750 \pm 10$(*trans*)		142, 180
			$+1806 \pm 10$(*cis*)		
$Me_2C{=}CCMe_2(OMe)$[i]	D_2O[b]			-107(*trans*), -187(*cis*)	192
				80(CMe_2)	
$MeO_2CC(Et){=}CEt$[j]	$CDCl_3$[b]		1338	-126(CH_2)	195
$MeO_2CC(Me){=}CPh$[k,l]	CD_3OD[b]			-142(Me), 125.5(Ph)	194, 196
Ph[m]	DMSO-d_6[n]		$+1035$[o]	$+396$[o]	199, 207, 208
	pyridine[o]		$+812$	$+306$	
4-MeC_6H_4[p]	DMSO-d_6[n]		$+1025$	$+376$	199, 207, 208
2,5-$Me_2C_6H_3$[q]	DMSO-d_6[n]		$+1109$	$+559$(H-3), 113(Me)	199, 207, 208

[a] The anions are given in the footnotes. [b] $CH_3CO_2^-$. [c] $S_2CNMe_2^-$. [d] $S_2CNEt_2^-$. [e] ox. [f] $Pr^iCO_2^-$. [g] The CH_2 protons are non-equivalent. [h] ClO_4^-. [i] ^5J(COMe) = 4 Hz. [j] Et groups mutually *trans*. [k] MeO_2C *trans* to Tl. [l] ^5J(Ph) = 64 Hz; ^6J(Me) = 13 Hz. [m] $^5J = +115$ Hz[n], $+105.5$ Hz[o]. [n] $CF_3CO_2^-$. [o] Cl^-. [p] $^6J = 66$ Hz. [q] ^5J(Me) = 55 Hz, ^5J(H-4) = $+115$ Hz.

TABLE 9.12 Thallium, carbon coupling constants for selected monoarylthallium(III) cations,[a] RTl^{2+}

R	Solvent	$^{1}J(^{205}Tl, ^{13}C\text{-}1)$/Hz	$^{2}J(^{205}Tl, o\text{-}C)$/Hz	$^{3}J(^{205}Tl, m\text{-}C)$/Hz	$^{4}J(^{205}Tl, p\text{-}C)$/Hz	Ref.
Ph	DMSO-d_6		+500	+1010	−185	213, 214
		10 718	+527.3	+1047.4	−202.2	215
Ph	THF	9902	+676	+878	−183	205, 216
4-MeC_6H_4[b]	THF	9756	+756	+895	−195	205, 216
4-$Pr^nC_6H_4$[c]	DMSO-d_6		+562	+1075	−204	213, 214
2,4-$Me_2C_6H_3$[d]	THF	9329	+646[e](C-2), +549[e](C-6)	+793(C-3), +849(C-5)	ca. −140	205, 216
3,4-$Me_2C_6H_3$	THF	9634	+565(C-2), ca. +500(C-6)	+800(C-3), +820(C-5)	−183	205, 216
3,4-$Me_2C_6H_3$[f]	DMSO-d_6		+552(C-2), +515(C-6)	+1071(C-3), +1101(C-5)	−217	213, 214
2,4,6-$Me_3C_6H_2$[g]	THF	8841	+512	+968	−166	205, 216

[a] The anion is trifluoroacetate. [b] $^{5}J = +110$ Hz. [c] $^{5}J = +94$ Hz; $^{6}J = 47$ Hz; $^{7}J < 1$ Hz. [d] $^{3}J(Me) = 402$ Hz. [e] Assignments incorrectly switched in Ref. 205; correctly listed in Ref. 216. [f] $^{4}J(Me) = 76$ Hz; $^{5}J = +92$ Hz. [g] $^{3}J(Me) = 459$ Hz; $^{5}J = +107$ Hz.

TABLE 9.13 Some (^{205}Tl, ^{13}C) coupling constants for the dimethylthallium(III) cation, Me_2Tl^{+} [a].

Solvent	D_2O	Pyridine	NH_3(liq.)[b]	DMSO
^1J/Hz	+2459,[c] +2478[d] (NO_3^-) +2513[f] (AcO^-)	+3018,[e] +3080[d] (NO_3^-) +3012[g] (I^-) +2897,[h] +2918[d] (PhO^-)	+3456(Br^-)	+2905,[e] +2903[d] (NO_3^-) +2928,[h] +2934[d] (I^-)
Ref.	135, 139	135, 139	134	135

[a] The anion is listed in brackets. [b] At −30°C. [c] 0.8 M. [d] 0.2 M. [e] 1.0 M. [f] 0.4 M. [g] 0.1 M. [h] 0.9 M.

(in FT work), or large sweep widths and times (for CW). In FT work, intensity anomalies due to insufficient transmitter power[124] or unnecessarily long accumulation times (if pulses are shortened to eliminate intensity errors) could be a problem.

For observation of nuclei to which thallium is coupled, it must be remembered that there are two thallium isotopes, and that decoupling of one might still leave a fairly complicated spectrum (decoupling ^{205}Tl should cause the greatest simplification). Also, for ^{203,205}Tl-{X} and X-{^{203,205}Tl} experiments, very high-power broadband irradiation might be required for efficient decoupling.

REFERENCES

1. J. W. Akitt, *Ann. Reports NMR Spectry.* **5A**, 465 (1972)
2. J. W. Akitt, N. N. Greenwood, B. Khandelwal and G. D. Lester, *J.C.S. Dalton* 604 (1972)
3. J. W. Akitt, N. N. Greenwood and G. D. Lester, *J.C.S. Chem. Comm.* 988 (1969)
4. H. Ohtaki, *Bull. Chem. Soc. Japan* **43**, 2463 (1970)
5. R. J. Molenaar, J. C. Evans and L. D. McKeever, *J. Phys. Chem.* **74**, 3629 (1970).
6. N. A. Matwiyoff and W. E. Wageman, *Inorg. Chem.* **9**, 1031 (1970)
7. B. W. Epperlein and O. Lutz, *Z. Naturforsch.* (*A*) **23**, 1413 (1968)
8. J. W. Akitt, N. N. Greenwood and G. D. Lester, *J. Chem. Soc.* (*A*), 2450 (1971)
9. D. Canet, J. J. Delpuech, M. R. Khaddar and P. Rubini, *J. Magn. Reson.* **9**, 329 (1973)
10. A. D. Toy, T. D. Smith and J. R. Pilbrow, *Austral. J. Chem.* **26**, 1889 (1973)
11. J. W. Akitt, N. N. Greenwood and G. D. Lester, *J. Chem. Soc.* 803 (1969)
12. J. W. Akitt, "Nuclear Magnetic Resonance Spectroscopy of Nuclei Other Than Protons" (T. Axenrod and G. Webb, eds.) Wiley, New York (1974)
13. A. P. Shut'ko, I. Y. Mulih and B. A. Geller, *Ukr. Khim. Zhur.* **42**, 993 (1976)
14. J. W. Akitt, N. N. Greenwood and B. L. Khandelwal, *J.C.S. Dalton* 1226 (1972)
15. V. S. Lyubernov and S. P. Ionov, *Zhur. fiz. Khim.* **46**, 838 (1972)
16. E. R. Malinowski, *J. Amer. Chem. Soc.* **91**, 4701 (1969)
17. H. Haraguchi and S. Fujiwara, *J. Phys. Chem.* **73**, 3467 (1969)
18. R. G. Kidd and B. R. Truax, *J.C.S. Chem. Comm.* 160 (1969)
19. J. F. Hon, *Mol. Phys.* **15**, 57 (1968)
20. R. G. Kidd and D. R. Truax, *J. Amer. Chem. Soc.* **90**, 6867 (1968)
21. W. G. Movius and N. A. Matwiyoff, *Inorg. Chem.* **6**, 847 (1967)
22. J. J. Delpuech, M. R. Khaddar, A. Peguy and P. Rubini, *J.C.S. Chem. Comm.* 154 (1974)
23. J. J. Delpuech, M. R. Khaddar, A. Peguy and P. Rubini, *J. Amer. Chem. Soc.* **97**, 3373 (1975)
24. S. Hermanek, O. Kriz, J. Plesek and T. Hanslik, *Chem. Ind.* (*London*) 42 (1975)
25. E. Schippert, *Adv. Mol. Relaxation Processes* **9**, 167 (1976)
26. L. D. Supran and N. Sheppard, *J.C.S. Chem. Comm.* 832 (1967)

27. J. P. Oliver and C. A. Wilkie, *J. Amer. Chem. Soc.* **89**, 163 (1967)
28. J. F. Ross and J. P. Oliver, *J. Organometal Chem.* **22**, 503 (1970)
29. E. S. Gore and H. S. Gutowsky, *J. Phys. Chem.* **73**, 2515 (1969)
30. J. W. Akitt and R. H. Duncan, *J. Magn. Reson.* **25**, 391 (1977)
31. D. Gudlin and H. Schneider, *J. Magn. Reson.* **16**, 362 (1974)
32. J. Huet, J. Durand, and Y. Infarnet, *Org. Magn. Reson.* **8**, 382 (1976)
33. L. Petrakis, *J. Phys. Chem.* **72**, 4182 (1968)
34. C. P. Poole, H. E. Swift and J. F. Itzel, *J. Phys. Chem.* **69**, 3663 (1965)
35. C. P. Poole, H. E. Swift and J. F. Itzel, *J. Chem. Phys.* **42**, 2576 (1965)
36. H. E. Swift, C. P. Poole and J. F. Itzel, *J. Phys. Chem.* **68**, 2509 (1964)
37. D. E. O'Reilly, *J. Chem. Phys.* **32**, 1007 (1960)
38. E. N. DiCarlo and H. E. Swift, *J. Phys. Chem.* **68**, 551 (1964)
39. D. E. O'Reilly, C. P. Poole, R. F. Belt and H. Scott, *J. Polymer Sci.* **A2**, 3257 (1964)
40. J. W. Akitt and R. H. Duncan, *J. Magn. Reson.* **15**, 162 (1974)
41. L. Maijs and V. A. Shcherbakov, *Latv. PSR Zinat. Akad. Vestis* (*Kim. Ser.*) **6**, 751 (1974)
42. L. Maijs and V. S. Shcherbakov, *Latv. PSR Zinat. Akad. Vestis* (*Kim. Ser.*) **5**, 548 (1975)
43. J. W. Akitt and R. H. Duncan, *J.C.S. Dalton* 1119 (1976)
44. L. S. Bresler, I. Y. Poddubnyi, T. K. Smirnova, A. S. Khachaturov and I. Y. Tsereteli, *Doklady Akad. Nauk SSSR* **210**, 847 (1973)
45. H. E. Swift and J. F. Itzel, *Inorg. Chem.* **5**, 2048 (1966)
46. S. I. Vinogradova, V. M. Denisov and A. I. Kol'tosov, *Zhur. obsch. Khim.* **42**, 1031 (1972)
47. V. M. Denisov, E. B. Toporkova, L. V. Alferova and A. I. Kol'tosov, *Izv. Akad. Nauk SSSR* (*Ser. Khim.*) **1**, 157 (1975)
48. L. P. Kazanskii, A. I. Gasanov, V. F. Chuvaev and V. I. Spitsyn, *fiz. Mat. Metody Koord. Khim.* (*Tezisy Doklady, Vses. Soveshch.*) 98 (1974)
49. G. N. Boiko, Y. I. Malov and K. N. Semenenko, *fiz. Mat. Metody Koord. Khim.* (*Tezisy Doklady, Vses, Soveshch.*) 104 (1974)
50. G. N. Boiko, Y. I. Malov, K. N. Semenenko and S. P. Shilkin, *Koord. Khim.* **2**, 686 (1976)
51. L. V. Titov, E. R. Eremin, L. N. Erofeev and V. Y. Rosolovskii, *Izv. Akad. Nauk SSSR* (*Ser. Khim.*) **1**, 7 (1974)
52. G. N. Boiko, *Izv. Akad. Nauk SSSR* (*Ser. Khim.*) **6**, 1215 (1976)
53. G. N. Boiko, Y. I. Malov, K. N. Semenenko and S. P. Shilkin, *Zhur. neorg. Khim.* **19**, 1485 1974)
54. G. N. Bioko, Y. I. Malov and K. N. Semenenko, *Izv. Akad. Nauk SSSR* (*Ser. Khim.*) **5**, 1143 (1973)
55. P. H. Bird and M. G. H. Wallbridge, *J. Chem. Soc.*, 3923 (1965)
56. O. Yamamota, *J. Chem. Phys.* **63**, 2988 (1975)
57. A. Takahashi, *J. Phys. Soc. Japan* **38**, 1225 (1975)
58. H. G. Hertz, R. Tutsch and H. Versmold, *Ber. Bunsenges. Phys. Chem.* **75**, 1171 (1971).
59. T. T. Ang, *Diss. Abstr. Int. B* **33**, 5750 (1973)
60. B. M. Cohen, A. R. Cullingworth and J. D. Smith, *J. Chem. Soc.* (*A*), 2193 (1969)
61. V. P. Tarasov and S. I. Bakum, *J. Magn. Reson.* **18**, 64 (1975)
62. D. E. H. Jones, *J.C.S. Dalton* 567 (1972)
63. N. H. N Vriezen and F. Jellinek, *Chem. Phys. Letters* **1**, 284 (1967)
64. P. C. Lauterbur, R. C. Hopkins, R. W. King, U. V. Ziebarth and C. W. Heitsch, *Inorg. Chem.* **7**, 1025 (1968)
65. J. L. Atwood and G. D. Stucky, *J. Amer. Chem. Soc.* **91**, 2538 (1969)
66. T. D. Westmoreland, N. S. Bhacca, J. D. Wander and M. C. Day, *J. Amer. Chem. Soc.* **95**, 2019 (1973)
67. T. D. Westmoreland, N. S. Bhacca, J. D. Wander and M. C. Day, *J. Organometal. Chem.* **38**, 1 (1972)
68. O. Yamamoto, *Chem. Letters* **6**, 511 (1975)
69. A. T. Weibel and J. P. Oliver, *J. Organometal. Chem.* **74**, 155 (1974)
70. J. P. Laussac, J. P. Laurent and G. Commenges, *Org. Magn. Reson.* **7**, 72 (1975)
71. R. E. Santini and J. B. Grutzner, *J. Magn. Reson.* **22**, 155 (1976)

72. B. G. Gribov, G. M. Gusakov, B. I. Kozyrkin and E. N. Zorina, *Doklady Akad. Nauk SSSR* **210**, 1350 (1973)
73. Y. A. Buslaev, V. P. Tarasov, N. N. Mel'nikov and S. P. Petrosyants, *Doklady Akad. Nauk SSSR* **216**, 796 (1974)
74. Y. A. Buslaev, V. P. Tarasov, S. P. Petrosyants and N. N. Mel'nikov, *Zhur. Strukt. Khim.* **15**, 617 (1974)
75. J. D. Glickson, T. P. Pitner, J. Webb and R. A. Gams, *J. Amer. Chem. Soc.* **97**, 1679 (1975)
76. Y. A. Buslaev, S. P. Petrosyants and V. P. Tarasov, *Zhur. Strukt. Khim.* **15**, 200 (1974)
77. Y. A. Buslaev, V. P. Tarasov, S. P. Petrosyants and N. N. Mel'nikov, *Koord. Khim.* **1**, 1435 (1975)
78. A. Fratiello, R. E. Lee and R. E. Schuster, *Inorg. Chem.* **9**, 82 (1970)
79. W. G. Movius and N. A. Matwiyoff, *Inorg. Chem.* **8**, 925 (1969)
80. M. Rice and R. V. Pound, *Phys. Rev.* **99**, 1036 (1955)
81. V. P. Tarasov and Y. A. Buslaev, *Mol. Phys.* **24**, 664 (1972)
82. J. W. Akitt, N. N. Greenwood and A. Storr, *J. Chem. Soc.*, 4410 (1965)
83. S. F. Lincoln, *Austral. J. Chem.* **25**, 2705 (1972)
84. S. F. Lincoln, A. C. Sandercock and D. R. Stranks, *J.C.S. Dalton* 669 (1975).
85. Y. A. Buslaev, V. P. Tarasov, M. N. Buslaeva and S. P. Petrosyants, *Doklady Akad. Nauk SSSR*, **209**, 882 (1973)
86. W. G. Procter and F. C. Yu, *Phys. Rev.* **81**, 20 (1951)
87. Y. Ting and D. Williams, *Phys. Rev.* **89**, 595 (1953)
88. Y. Ting, F. K. Biard, and D. Williams, *Phys. Rev.* **86**, 618 (1951)
89. M. Rice and R. V. Pound, *Phys. Rev.* **106**, 953 (1957)
90. Y. A. Buslaev, V. P. Tarasov, M. N. Buslaeva, and S. P. Petrosyants, *Doklady Akad. Nauk SSSR* **209**, 882 (1973)
91. H. Haraguchi, K. Fuwa and S. Fujiwara, *J. Phys. Chem.* **17**, 1497 (1973)
92. A. Fratiello, D. D. Davis, S. Peak and R. E. Schuster, *Inorg. Chem.* **10**, 1627 (1971)
93. T. H. Cannon and R. E. Richards, *Trans. Faraday Soc.* **62**, 1378 (1966)
94. D. Freude, U. Lohse, H. Pfeifer, W. Schirmer, H. Schmiedel and H. Stach, *Z. phys. Chem. (Lpzg)* **255**, 443 (1974)
95. D. Freude, A. Hauser, H. Pankau and H. Schmiedel, *Z. Phys. Chem. (Lpzg)* **251**, 13 (1972)
96. H. Bittner and E. Hauser, *Monatsh.* **101**, 1471 (1970)
97. T. Wada and J. P. Cohen-Addad, *Bull. Soc. Fr. Miner. Cryst.* **92**, 238 (1969)
98. K. Arnold and R. Scholl, *Stud. Biophys.* **59**, 47 (1976)
99. J. Reuben and F. J. Kayne, *J. Biol. Chem.* **246**, 6227 (1971)
100. F. J. Kayne and J. Reuben, *J. Amer. Chem. Soc.* **92**, 220 (1970)
101. C. M. Grisham, R. K. Gupta, R. E. Barnett and A. S. Mildvan, *J. Biol. Chem.* **249**, 6738 (1974)
102. C. M. Grisham and A. S. Mildvan, *Fed. Proc.* **23**, 1331 (1974)
103. R. P. H. Gasser and R. E. Richards, *Mol. Phys.* **2**, 357 (1959)
104. R. Freeman, R. P. H. Gasser, R. E. Richards, and D. H. Wheeler, *Mol. Phys.* **2**, 75 (1959)
105. E. B. Baker and L. W. Burd, *Rev. Sci. Instr.* **34**, 238 (1963)
106. R. E. Sheriff and D. Williams, *Phys. Rev.* **82**, 651 (1951)
107. H. L. Poss, *Phys. Rev.* **72**, 637 (1947)
108. H. L. Poss, *Phys. Rev.* **75**, 600 (1949)
109. W. G. Proctor, *Phys. Rev.* **75**, 522 (1949)
110. H. S. Gutowsky and B. R. McGarvey, *Phys. Rev.* **91**, 81 (1953)
111. D. Herbison-Evans, P. B. P. Phipps and R. J. P. Williams, *J. Chem. Soc.* 6170 (1965)
112. R. Freeman, R. P. H. Gasser and R. E. Richards, *Mol. Phys.* **2**, 301 (1959)
113. R. W. Vaughan and D. H. Anderson, *J. Chem. Phys.* **52**, 5287 (1970)
114. N. H. Nachtrieb and R. K. Momii, "Reactiv. Solids, Proc. Int. Symp. (6th., 1968)," p. 675, (J. W. Mitchell, ed.). Wiley-Interscience, New York (1969)
115. S. Hafner and N. H. Nachtrieb, *J. Chem. Phys.* **42**, 631 (1965)
116. H. Köppel, J. Dallorso, G. Hoffman and B. Walther, *Z. anorg. allg. Chem.* **427**, 24 (1976)
117. J. J. Dechter and J. I. Zink, *Inorg. Chem.* **15**, 1690 (1976)
118. J. J. Dechter and J. I. Zink, *J. Amer. Chem. Soc.* **97**, 2937 (1975)

119. J. J. Dechter and J. I. Zink, *J.C.S. Chem. Comm.* 96 (1974)
120. J. F. Hinton and R. W. Briggs, *J. Magn. Reson.* **19**, 393 (1975)
121. J. J. Dechter and J. I. Zink, *J. Amer. Chem. Soc.* **98**, 845 (1976)
122. J. F. Hinton and R. W. Briggs, *J. Magn. Reson.* **25**, 379 (1977)
123. J. J. Dechter, *Diss. Abstr. Int. B* **36**, 3944 (1976)
124. S. O. Chan and L. W. Reeves, *J. Amer. Chem. Soc.* **96**, 404 (1974)
125. M. Bacon and L. W. Reeves, *J. Amer. Chem. Soc.* **95**, 272 (1973)
126. R. N. Schwartz, *J. Magn. Reson.* **24**, 205 (1976)
127. B. W. Bangerter and R. N. Schwartz, *J. Chem. Phys.* **60**, 333 (1974)
128. J. M. Lehn, J. P. Sauvage, and B. Dietrich, *J. Amer. Chem. Soc.* **92**, 2916 (1970)
129. V. F. Bystrov, Y. D. Gavrilov, V. T. Ivanov and Y. A. Ovchinnikov *Eur. J. Biochem.* **78**, 63 (1977) cited by B. F. Bystrov, *Progr. NMR Spectr.* **10**, 41 (1976)
130. W. G. Schneider and A. D. Buckingham, *Disc. Faraday Soc.* **34**, 147 (1962)
131. B. N. Figgis, *Trans. Faraday Soc.* **55**, 1075 (1959)
132. T. Sidei, S. Yano, and S. Sasaki, *Ôyô Butsuri* **27**, 513 (1958)
133. M. G. Voronkov, V. A. Pestunovich, S. V. Mikhailova, J. Popelis and E. Leipins, *Khim. Geterotsikl. Soedin* **3**, 312 (1973)
134. K. Hildenbrand and H. Dreeskamp, *Z. phys. Chem. (N.F.)* **69**, 171 (1970)
135. P. J. Burke, R. W. Matthews, and D. G. Gillies, *J. Organometal. Chem.* **118**, 129 (1976)
136. J. F. Hinton and R. W. Briggs, *J. Magn. Reson.* **22**, 447 (1976)
137. (a) G. M. Sheldrick and J. P. Yesinowski, *J.C.S. Dalton*, 870 (1975)
(b) J. P. Yesinowski, private communication.
138. J. F. Hinton and R. W. Briggs, *J. Magn. Reson.* **25**, 555 (1977)
139. C. S. Hoad, R. W. Matthews, M. M. Thakur, and D. G. Gillies, *J. Organometal. Chem.* **124**, C31 (1977)
140. W. Mueller-Warmuth and R. Vilhjalmsson, *Z. phys. Chem.* (*N.F.*) **93**, 283 (1974)
141. J. P. Maher and D. F. Evans, *Proc. Chem. Soc.* 208 (1961)
142. J. P. Maher and D. F. Evans, *J. Chem. Soc.* 637 (1965)
143. J. P. Maher and D. F. Evans, *J. Chem. Soc.* 5534 (1963)
144. A. T. Weibel and J. P. Oliver, *J. Amer. Chem. Soc.* **94**, 8590 (1972)
145. A. T. Weibel and J. P. Oliver, *J. Organometal. Chem.* **74**, 155 (1974)
146. R. J. Abraham, G. E. Hawkes, and K. M. Smith, *Tetrahedron Letters* 1999 (1975)
147. M. L. Maddox, S. L. Stafford, and H. D. Kaesz, *Adv. Organometal Chem.* **3**, 1 (1966)
148. F. Schindler and H. Schmidbaur, *Chem. Ber.* **100**, 3655 (1967)
149. P. Krommes and J. Lorberth, *J. Organometal. Chem.* **131**, 415 (1977)
150. B. Walther, A. Zschunke, B. Adler, A. Kolbe and S. Bauer, *Z. anorg. allg. Chem.* **427**, 137 (1976)
151. A. G. Lee, *J. C. S. Chem. Comm.* 1614 (1968)
152. G. D. Shier and R. S. Drago, *J. Organometal. Chem.* **5**, 330 (1966)
153. A. G. Lee and G. M. Sheldrick, *Trans. Faraday Soc.* **67**, 7 (1971)
154. C. J. Marsden, *J. Fluorine Chem.* **5**, 423 (1975)
155. B. Walther and K. Thiede, *J. Organometal. Chem.* **32**, C7 (1971)
156. A. G. Lee and G. M. Sheldrick, *J. Chem. Soc.* (*A*), 1055 (1969)
157. A. G. Lee and G. M. Sheldrick, *J.C.S. Chem. Comm.* 441 (1969)
158. A. G. Lee, *J. Chem. Soc.* (*A*), 2157 (1970)
159. B. Schaible, W. Haubold and J. Weidlein, *Z. anorg. allg. Chem.* **403**, 289 (1974)
160. Y. Lee and L. W. Reeves, *Canad. J. Chem.* **53**, 161 (1975)
161. W. Schwarz, G. Mann and J. Weidlein, *J. Organometal. Chem.* **122**, 303 (1976)
162. R. I. Jolly and J. M. Titman, *J. Phys.* (*C*) **5**, 1284 (1972)
163. B. Walther, R. Mahrwald, C. Jahn, and W. Klar, *Z. anorg. allg. Chem.* **423**, 144 (1976)
164. J. V. Hatton, *J. Chem. Phys.* **40**, 933 (1964)
165. H. Kurosawa and R. Okawara, *Inorg. Nucl. Chem. Letters* **3**, 21 (1967)
166. H. Kurosawa, K. Yasuda and R. Okawara, *Bull. Chem. Soc. Japan* **40**, 861 (1967)
167. H. Kurosawa, K. Yasuda and R. Okawara, *Inorg. Nucl. Chem. Letters* **1**, 131 (1965)
168. H. Kurosawa and R. Okawara, *J. Organometal. Chem.* **10**, 211 (1967)
169. U. Knips and F. Huber, *J. Organometal. Chem.* **107**, 9 (1976)

170. P. Krommes and J. Lorberth, *J. Organometal. Chem.* **120**, 131 (1976)
171. M. Aritomi and Y. Kawasaki, *J. Organometal. Chem.* **90**, 185 (1975)
172. N. N. Greenwood, B. S. Thomas and D. W. Waite, *J.C.S. Dalton*, 299 (1975)
173. Y. Kawasaki and M. Aritomi, *J. Organometal. Chem.* **104**, 39 (1976)
174. H. Kurosawa and R. Okawara, *Trans. N.Y. Acad. Sci., Ser. II* **30**, 962 (1968)
175. H. Schmidbaur and F. Schindler, *Angew. Chem.* (*Int.*) **4**, 876 (1965)
176. A. G. Lee, *J. Chem. Soc.* (*A*), 467 (1970)
177. L. Pellerito, R. Cefalu and G. Ruisi, *J. Organometal. Chem.* **63**, 41 (1973)
178. H.-U. Schwering and J. Weidlein, *J. Organometal. Chem.* **99**, 223 (1975)
179. D. W. Turner, *J. Chem. Soc.* 847 (1962)
180. J. P. Maher and D. F. Evans, *Proc. Chem. Soc.* 176 (1963)
181. S. Numata, H. Kurosawa and R. Okawara, *J. Organometal. Chem.* **70**, C21 (1974)
182. E. A. V. Ebsworth, A. G. Lee and G. M. Sheldrick, *J. Chem. Soc.* (*A*), 1052 (1969)
183. H. Kurosawa, M. Tanaka and R. Okawara, *J. Organometal. Chem.* **12**, 241 (1968)
184. M. Tanaka, H. Kurosawa and R. Okawara, *J. Organometal. Chem.* **21**, 41 (1970)
185. T. Abe, H. Kurosawa and R. Okawara, *J. Organometal. Chem.* **25**, 353 (1970)
186. T. Abe and R. Okawara, *J. Organometal. Chem.* **35**, 27 (1972)
187. H. Kurosawa and R. Okawara, *Inorg. Nucl. Chem. Letters* **3**, 93 (1967)
188. H. Kurosawa and R. Okawara, *J. Organometal. Chem.* **14**, 225 (1968)
189. S. Uemura, K. Zushi, A. Tabata, A. Toshimitsu, and M. Okano, *Bull. Chem. Soc. Japan* **47**, 920 (1974)
190. S. Uemura, H. Miyoshi, A. Toshimitsu and M. Okano, *Bull. Chem. Soc. Japan* **49**, 3285 (1976)
191. W. Kruse and T. M. Bednarski, *J. Org. Chem.* **36**, 1154 (1971)
192. R. K. Sharma and E. D. Martinez, *J.C.S. Chem. Comm.*, 1129 (1972)
193. J. P. Maher, Ph.D. Thesis, University of London (1963)
194. S. Uemura, K. Sohma, H. Tara and M. Okano, *Chem. Letters*, 545 (1973)
195. R. K. Sharma and N. H. Fellers, *J. Organometal. Chem.* **49**, C69 (1973)
196. S. Uemura, H. Tara, M. Okano and K. Ichikawa, *Bull. Chem. Soc. Japan* **47**, 2663 (1974)
197. A. N. Nesmeyanov, A. E. Borisov, N. V. Novikova and E. I. Fedin, *Doklady Akad. Nauk SSSR* **183**, 118 (1968)
198. T. Abe, S. Numata and R. Okawara, *Inorg. Nucl. Chem. Letters* **8**, 909 (1972)
199. A. McKillop, J. D. Hunt and E. C. Taylor, *J. Organometal. Chem.* **24**, 77 (1970)
200. B. Walther, *Z. anorg. allg. Chem.* **395**, 112 (1973)
201. H. B. Stegmann, K. B. Ulmschneider, and K. Scheffler, *J. Organometal. Chem.* **72**, 41 (1974)
202. J. P. Maher, M. Evans, and M. Harrison, *J.C.S. Dalton*, 188 (1972)
203. G. B. Deacon and I. K. Johnson, *J. Organometal. Chem.* **112**, 123 (1976).
204. D. E. Fenton, D. G. Gillies, A. G. Massey and E. W. Randall, *Nature* **201**, 818 (1964)
205. W. Kitching, D. Praeger, C. J. Moore, D. Doddrell and W. Adcock, *J. Organometal. Chem.* **70**, 339 (1974)
206. K. Ichikawa, S. Uemura, T. Nakano and E. Uegaki, *Bull. Chem. Soc. Japan* **44**, 545 (1971)
207. A. McKillop, J. S. Fowler, M. J. Zelesko, J. D. Hunt, E. C. Taylor and G. McGillivray, *Tetrahedron Letters* 2423 (1969)
208. E. C. Taylor and A. McKillop, *Acc. Chem. Res.* **3**, 338 (1970)
209. A. G. Lee, *J. Organometal. Chem.* **22**, 537 (1970)
210. H. C. Bell, J. R. Kalman, J. T. Pinhey and S. Sternhell, *Tetrahedron Letters* 3391 (1974)
211. B. Davies and C. B. Thomas, *J.C.S. Perkin I*, 65 (1975)
212. G. B. Deacon, R. N. M. Smith and D. Tunaley, *J. Organometal. Chem.* **114**, Cl (1976)
213. L. Ernst, *J. Organometal. Chem.* **82**, 319 (1974)
214. L. Ernst, *Org. Magn. Reson.* **6**, 540 (1974)
215. J. Y. Lallemand and M. Duteil, *Org. Magn. Reson.* **8**, 328 (1976)
216. W. Kitching, C. J. Moore, D. Doddrell and W. Adcock, *J. Organometal. Chem.* **94**, 469 (1975)
217. R. G. Coombes, M. D. Johnson and D. Vamplew, *J. Chem. Soc.* (*A*), 1055 (1969)
218. E. Hoyer, W. Dietzsch, H. Müller, A. Zschunke and W. Schroth, *Inorg. Nucl. Chem. Letters* **3**, 457 (1967)
219. R. J. Abraham, G. H. Barnett, E. S. Bretschneider and K. M. Smith, *Tetrahedron* **29**, 553 (1973)

220. R. J. Abraham, G. H. Barnett and K. M. Smith, *J.C.S. Perkin I.* 2142 (1973)
221. R. J. Abraham and K. M. Smith, *Tetrahedron Letters* 3335 (1971)
222. K. M. Smith, *J.C.S. Chem. Comm.* 540 (1971)
223. C. A. Busby and D. Dolphin, *J. Magn. Reson.* **23**, 211 (1976)
224. R. J. Abraham, G. E. Hawkes and K. M. Smith, *J.C.S. Chem. Comm.* 401 (1973)
225. R. J. Abraham, G. E. Hawkes and K. M. Smith, *J.C.S. Perkin II* 627 (1974)
226. R. J. Abraham, G. E. Hawkes, M. F. Hudson and K. M. Smith, *J.C.S. Perkin II*, 204 (1975)
227. A. T. T. Hsieh, C. A. Rogers and B. O. West, *Austral. J. Chem.* **29**, 49 (1976)
228. F. A. L. Anet, *Tetrahedron Letters* 3399 (1964)
229. A. McKillop, M. E. Ford, and E. C. Taylor, *J. Org. Chem.* **39**, 2434 (1974)
230. A. T. Weibel and J. P. Oliver, *J. Organometal. Chem.* **57**, 313 (1973)
231. B. Walther and C. Rockstroh, *J. Organometal. Chem.* **44**, C4 (1972)
232. A. T. T. Hsieh and G. Wilkinson, *J.C.S. Dalton* 867 (1973)
233. H. Schmidbaur, H.-J. Füller and F. H. Köhler, *J. Organometal. Chem.* **99**, 353 (1975).
234. H. Schmidbaur and H.-J. Füller, *Chem. Ber.* **107**, 3674 (1974)
235. S. Hafner and N. H. Nachtrieb, *J. Chem. Phys.* **40**, 2891 (1964)

10

GROUP IV–SILICON, GERMANIUM, TIN AND LEAD

ROBIN K. HARRIS,* University of East Anglia, England
JOHN D. KENNEDY,† University of Leeds, England
WILLIAM McFARLANE,† City of London Polytechnic, England

Group IV is unique in the Periodic Table in that four of the five elements have spin-$\frac{1}{2}$ nuclei that are suitable for NMR, viz. ^{13}C, ^{29}Si, ^{119}Sn and ^{207}Pb. All four have useful natural abundances, ranging from 1.11% for ^{13}C to 22.6% for ^{207}Pb. The magnetic moments are moderate, giving rise to standardised NMR frequencies between $\Xi = 19.87$ MHz (^{29}Si) and $\Xi = 37.29$ MHz (^{119}Sn). Receptivities relative to the proton lie between $D^p = 1.76 \times 10^{-4}$ (^{13}C) and $D^p = 4.44 \times 10^{-3}$ (for ^{119}Sn). Consequently NMR is able to yield much chemically useful information for these elements. The introduction of FT methods resulted immediately in a boom in ^{13}C NMR work, for chemical reasons, and ^{13}C now rivals ^{1}H as the most popular nucleus for study (see ch. 4). For reasons to be given below, together with the obviously diminished significance of silicon (as opposed to carbon) chemistry, the study of ^{29}Si has lagged behind that of ^{13}C and ^{15}N, so that it cannot yet be counted as a "common NMR nuclide." However, it is rapidly approaching that status, and silicon has a rich and important chemistry, so that information on ^{29}Si parameters in the literature is quite extensive, though it is not yet sufficiently comprehensive (or well-understood from a theoretical viewpoint!). The ^{119}Sn nucleus has also been studied a great deal; in fact much information has been available from ^{1}H-{^{119}Sn} double resonance studies. This is currently being supplemented by results from direct observation of ^{119}Sn resonances using FT techniques. The ^{207}Pb nucleus has been studied rather less. In consequence of the above facts the present survey is able to discuss most of the published work on ^{207}Pb, but only a proportion of that on ^{119}Sn and ^{29}Si. Even so, considerable space is devoted here to studies of ^{29}Si, since it is arguably the most important individual nucleus for NMR which is not considered in ch. 4. The balance of topics for the three nuclei also differs, much more space being devoted to relaxation for ^{29}Si than for the other two nuclei, reflecting the information available. One final comparison of the properties of the spin-$\frac{1}{2}$ group IV nuclei is worth mentioning at this point, viz. the magnetic moments of ^{13}C and ^{207}Pb are positive whereas those of ^{29}Si and ^{119}Sn are negative. It is clear that the future is bright for NMR work on all four nuclei, with the order of importance remaining as $^{13}C > {}^{29}Si > {}^{119}Sn > {}^{207}Pb$, reflecting the chemical interest in these elements.

* Dr Harris has written Section 10A.
† Drs Kennedy and McFarlane have written Sections 10B, 10C and 10D.

The situation for germanium is very different. The ^{73}Ge nucleus has $I = \frac{9}{2}$, a 7.8% natural abundance, and a receptivity that is slightly lower than that of ^{13}C. The other naturally-occurring isotopes of Ge have $I = 0$. Thus germanium has been scarcely studied by NMR, probably partly also because the chemistry of this element is not as important as those of the other group IV elements. It is doubtful whether ^{73}Ge NMR will ever be of great significance for structural studies.

10A SILICON-29

10A.1 General

Silicon has only one naturally-occurring isotope with non-zero spin, namely ^{29}Si, which has $I = \frac{1}{2}$ and a natural abundance of 4.70%. This places it among the category of "rare" spins, and implies that all the modern techniques can be applied in a similar fashion to their use for ^{13}C. However, isotopic enrichment of ^{29}Si is much less common than for ^{13}C and, indeed, no NMR studies of enriched ^{29}Si have yet been reported. The magnetogyric ratio of ^{29}Si is roughly one fifth that of the proton, resulting, for instance, in a resonance frequency of ca. 20 MHz in a magnetic field such that protons resonate at 100 MHz. At constant applied magnetic field the natural abundance and magnetogyric ratio lead to an estimation that the receptivity of ^{29}Si is 3.69×10^{-4} that of ^{1}H or 2.10 that of ^{13}C. However, this is over-optimistic as a basis for ^{29}Si work since (a) ^{29}Si tends to have a relatively long spin-lattice relaxation time (see below), and (b) γ (^{29}Si) is negative. The latter factor causes the nuclear Overhauser effect to be generally negative, giving reduced signal intensity or even negative or null signals (see below).

There is only a little work on the direct observation of ^{29}Si NMR for liquids in the early literature, and this was summarized by Lauterbur[1] in 1962. Interest in ^{29}Si NMR was renewed in the late 1960s and several papers were published[2–5] using CW hetero-nuclear double resonance techniques before Fourier transform spectrometers took over in the early 1970s. A review by Wells[6] surveys the situation as at 1970, and Marsmann[7] reviewed ^{29}Si chemical shifts in 1973. The greatly increased rate of appearance of papers in the period 1970–5 prompted a comprehensive review by Schraml and Bellama.[8] It is in no way the purpose of the present book to repeat material adequately surveyed elsewhere, so the reader is referred to the review by Schraml and Bellama[8] for much of the basic data on ^{29}Si NMR. In particular data on coupling constants involving ^{29}Si, most of them obtained from the resonances of nuclei such as ^{1}H, ^{19}F, ^{13}C and ^{31}P, are covered fully by Schraml and Bellama,[8] and will therefore be dealt with very scantily here. However, those authors did not refer to ^{29}Si relaxation data very much, so this area will be treated fully here, as will certain aspects of chemical shift information, particularly on siloxanes.

10A.2 Experimental

The older ^{29}Si chemical shift data in the literature was obtained using CW spectrometers, either observing ^{29}Si resonances directly or via ^{1}H-{^{29}Si} or ^{19}F-{^{29}Si} double resonance. Where measurements have been repeated using FT equipment the earlier values will be ignored. As in the case of ^{13}C NMR, the use of ^{1}H noise decoupling has greatly increased the accuracy of ^{29}Si measurements, as have the other modern advances in instrumentation.

In collating ^{29}Si chemical shift data, two features that affect comparability must be borne in mind. Firstly, there are medium effects to be considered, including those on the reference signal.[9,10] There have been few investigations of such influences, mainly because the relatively low S/N of ^{29}Si NMR necessitates the use of highly concentrated solutions (neat liquids if possible). However, Williams, Cargioli and LaRochelle have shown[11] that substantial effects exist for silanols of the type $HO[Me_2SiO]_nH$; these effects depend on the electron-pair donor ability of the solvent (increasing donation causing low frequency shifts). This is clearly a special case, but there is plenty of less well-defined evidence for substantial solvent effects on other ^{29}Si shifts. Secondly the literature contains many discrepancies in data for the same compound. This is due to several factors, viz. (*i*) the solvent effects mentioned above, (*ii*) variations in referencing compounds and techniques, and (*iii*) the inferior instrumentation available for the earlier work. Consequently data can only be rigorously compared when taken from the same source, though this is not always possible.

Nearly all reports of ^{29}Si shifts now use TMS as the reference compound, in spite of a relatively recent suggestion[9] that $(EtO)_4Si$ might be preferable. In certain specialised cases secondary references are desirable.[12] Conversion factors between various references have been quoted by Schraml and Bellama.[8] The indirect calculation of ^{29}Si shifts via measurement of absolute resonance frequencies has been urged and used;[13] this has the merit of dispensing with the necessity of adding TMS, thus giving better S/N, but it is essentially an external referencing system. The TMS silicon resonance frequency in a field such that the protons resonate at 100 MHz has been given variously as 19 867 220 Hz,[14] 19 867 184 Hz,[13] and 19 867 185 Hz.[15] Data presented here will be quoted on the TMS scale. To save space, solution conditions will not always be quoted and the reader is urged to consult the original literature on this point if necessary. As noted above, even TMS has been shown[9,10] to have a solvent-dependent ^{29}Si resonance frequency.

The only "special" techniques frequently required for ^{29}Si are those connected with the "null signal" problem. The nuclear Overhauser enhancement, η, for ^{29}Si-{^{1}H} studies (in the absence of the SC relaxation mechanism) is $(\gamma_H/2\gamma_{Si})(T_1/T_{1dd}) = -2.52\ T_1/T_{1dd}$, where T_{1dd} is the (^{29}Si, ^{1}H) dipole-dipole contribution to $T_1(^{29}Si)$. Thus if the dipolar mechanism is dominant, a negative signal is obtained. Unfortunately if $T_{1dd} \approx 1.52\ T_{1o}$ signals are very weak or absent and it is necessary to circumvent this problem. Since molecular oxygen often gives a significant contribution to $T_1(^{29}Si)$, substantial improvements in S/N can sometimes be made[16,17] by bubbling the solution with O_2 or N_2, or by degassing. A more certain effect is obtained[17,18] by the addition of a shiftless relaxation reagent, such as $Cr(acac)_3$; usually ca. 0.03 M is the optimum concentration. However, when that is undesirable, the gated decoupling technique may be employed,[18] though of course this takes more spectrometer time. Gated decoupling is also usually used for NOE measurements.[19]

When signals in the region −80 to −130 ppm are to be observed the broad resonance given by the glass of the insert and NMR tube may be a nuisance. In such circumstances it may be desirable to use NMR tubes made of teflon, kel-F or FEP (and even, perhaps, to use a teflon insert[20]). Alternatively the broad glass resonance may be removed using difference spectroscopy, i.e. subtracting (using the computer attached to the spectrometer) the FID of a blank run (from an NMR tube devoid of silicon-containing species) from that of the sample of interest.[21]

It has been shown[22] that substantial gains in S/N can be achieved in the solution state for molecules containing a direct Si-H bond by using the ^{29}Si-{^{1}H} cross-polarization technique. The sample used for these experiments was the polysiloxane $Me_3SiO[SiMeHO]_{35}SiMe_3$.

10A.3 Chemical Shifts

a. General Description and Empirical Relations

(*i*) *Introduction.* Silicon-29 chemical shifts cover a range of ca. 400 ppm, known values being from $\delta_{Si} = +60.2$ or $+64.0$ ppm[23,24] for $(Me_3Si)_2Hg$ to -346.2 or -351.7 ppm[25,26] for SiI_4 (all values being quoted with respect to Me_4Si). This range is smaller than that for ^{13}C. Nevertheless the dispersion arising from structural effects is considerable, and ^{29}Si chemical shift measurements thus provide a valuable tool for the determination of molecular structure. This is particularly the case for siloxanes, which are discussed separately below. Naturally the factors affecting δ_{Si} are many and various. Full theoretical understanding is not yet available, but this aspect is discussed further on p. 326–7. It is the aim of the present Section to relate the shifts to structural factors in an empirical way. The following influences, among others, may be expected to operate, viz. (*a*) substituent electronegativity, (*b*) steric interactions (*c*) variations from a tetrahedral arrangement around silicon, (*d*) coordination number at silicon, and (*e*) pi bonding involving the silicon d orbitals. Other factors, such as the magnetic anisotropy of neighbouring groups, have been invoked qualitatively.[27,28] The above influences are not, of course, independent, and this renders prediction of shifts problematic.

(*ii*) *Direct Substituents* The nature of the atoms bonded directly to silicon is, of course, of very great importance. However, their electronic effects can only be judged properly by studying molecules of the type SiX_4, since otherwise departures from tetrahedral environment at silicon introduce complications. Data for species SiX_4 are given in Table 10.1, and it can be seen that dependence of substituent electronegativity is by no means linear. Indeed, it appears there may be[8,47] a maximum in δ_{Si} as a function of electronegativity. The compounds $SiBr_4$ and SiI_4 have anomalously low resonance frequencies; their efficient shielding is not fully understood.[48,49]

When data for species SiX_3Y are examined, it is found that the shielding change from SiX_4 is frequently in the opposite direction to that for the total substitution $SiX_4 \rightarrow SiY_4$. A plot of $\sigma_i - \sigma_{ref}$ data versus n for series of molecules SiX_nY_{4-n} reveals typically (Fig. 10.1) a "sagging" pattern (a "hump" pattern for a plot of δ_{Si} vs. n), such as is also found for ^{31}P chemical shifts in tricoordinated and tetracoordinated phosphorus species. Relevant ^{29}Si chemical shifts are given in Table 10.1. For a few of the series (e.g. $SiMe_4 \rightarrow SiH_4$ and $SiCl_4 \rightarrow SiBr_4$) the shifts deviate little from additivity. Only a handful (e.g. $SiMe_4 \rightarrow SiPh_4$ and $SiMe_4 \rightarrow SiVi_4$) show the opposite behaviour (a hump in $\sigma_i - \sigma_{ref}$ vs. n), and in these cases the deviations from additivity are probably within experimental variation when medium influences are taken into account. Vongehr and Marsmann[40] have tried, with some success, to fit the experimental data by using empirical pairwise interaction parameters between the substituents, which include the "heavy atoms," Br and I. Attempts[47] to correlate relative shielding constants, $\sigma_i - \sigma_{ref}$, for compounds of the type $SiR^1R^2R^3R^4$ with the sum of substituent electronegativities have not been very successful, but once again a general sagging pattern is indicated.

Probably the trimethylsilyl group is the commonest in occurrence in silicon chemistry, and there are many reports of ^{29}Si chemical shifts for this group. Some typical values are given in Table 10.2; they cover a range of ca. 120 ppm. There is a rough trend with substituent electronegativity but, as point out above, other factors such as geometry and hybridisation changes complicate the situation. A similar pattern is found for silyl compounds, which are listed in Table 10.3.

Nguyen-Duc-Chuy et al.[46] have shown that Si shifts for related compounds correlate well, e.g. for the series $Me_{4-n}X_nSi$, $Me_{3-n}X_nSiPh$, and $Me_{3-n}X_nSiCH_2Ph$.

TABLE 10.1 Silicon-29 chemical shifts,[a] δ_{Si}/ppm, for series of compounds,[b] SiX_nY_{4-n}

Y	$n = 1$	2	3	4	X
Me	−15.5(29) −16.34(13)	−37.3(29) −41.5(3)	−65.2(29)	−93.1(29) −91.9(26)	H
Me	1.6(30)	4.6(30) 5.0(4)	6.5(30)	8.4(23)	Et
Me	−5.1(31) −4.41(13)	−9.4(31)	−11.9(31)	−13.98(32)	Ph
Me	−6.80(4,8)	−13.67(4,8)	−20.55(4,8)	−22.5(3)	Vi
Me	5.9(33) 6.52(28)	−1.7(33) −1.85(28)	−17.5(33) −16.80(28)	−28.1(33) −28.60(28)	NMe_2
Me	17.2(34) 17.75(28)	−2.5(34) −1.62(28)	−41.4(34) −39.8(28)	−79.2(9) −79.15(28)	OMe
Me	14.53(19) 13.5(35)	−5.74(19) −6.1(35)	−44.21(19) −44.5(35)	−82.40(19)[c]	OEt
Me	17.2(33) 17.72(36)	−6.1(33)	−54.0(33)	−101.1(33)	OPh
Me	16.9(37)	−3.2(37)	−41.5(37)		$OCH_2CH_2NH_2$
Me	15.6(37)	−4.4(37)	−43.8(37)		$OCH_2CH_2NMe_2$
Me	22.3(38)	4.4(38)	−42.7(38)	−74.5(3)	OAc
Me	6.87(8,9) 6.83(13)	−21.5(39)	−65.0(34)	−105.2(34) −104.4(13)	$OSiMe_3$
Me	35.4(2,8) 32.01(13)	8.8(2,8)	−51.8(2,8) −55.7(2,40)	−113.6(2,8)[d] −109.0(25)	F
Me	−19.78(13)	−48.7(42) −48.45(4)		−135.5(42) −135.2(23)	$SiMe_3$
Me	16.46(28) 15.92(43)	28.14(28)	34.00(28)	38.59(28)	SMe
Me	30.21(27) 30.27(13)	32.17(27)	12.47(27)	−18.5(33) −20.0(26)	Cl
Me	26.41(27) 26.24(13)	19.86(27)	−18.18(27)	−93.6(25) −92.7(26)	Br
Me	8.72(27) 8.89(13)	−33.68(27)	−144.0(40)[e]	−346.2(25) −351.7(26)	I
H	−60.1(32) −61.5(3)	−33.24(32) −34.5(3)	−17.75(32) −22.5(41)	−13.98(32)	Ph
H	−17.4(40)	−28.5(40)	−77.8(40)	−109.0(25) −113.6(2,8)[d]	F
H	−36.1(29)	−11.0(29)	−9.6(29)	−18.5(33) −20.0(26)	Cl
H	−48.54(40) −49.0(44)	−30.36(40)	−43.30(40)	−93.6(25) −92.7(26)	Br
H	−83.25(40)	−99.60(40)	−175.90(40)	−346.2(25) −351.7(26)	I
Et	35.9(45)	36.2(45)		−18.5(33) −20.0(26)	Cl
Ph	−12.6 to −25.4(11)	−30.2 to −37.5(11)		−73.3(26)	OH
Ph	−4.7(33)	−31.0(41)	−76.2(41) −73.2(46)	−109.0(25) −113.6(2,8)[d]	F

Continued overleaf

TABLE 10.1 *(Continued)*

Y	$n = 1$	2	3	4	X
Ph	1.2 to 1.5(11)	6.3(41) 6.19(32)	5.3(41)	−18.5(33) −20.0(26)	Cl
F	−81.7(2,8) −82.6(2,7)	−55.0(2,8)	−32.1(2,8)	−18.5(33) −20.0(26)	Cl
F	−82.4(2,8)	−67.4(2,8)	−67.0(2,8)	−93.6(25) −92.7(26)	Br
Cl	−8.0(26)	−9.8(26) −7.2(42)	−34.2(26)	−81.1(26) −80.0(42)	$SiCl_3$
Cl	−34.3(26)	−50.7(26)	−69.8(26)	−92.7(26) −93.6(25)	Br
Cl	−75.4(26)	−151.5(26)	−245.9(26)	−351.7(26) −346.2(25)	I
Br	−149.5(26)	−212.3(26)	−280.1(26)	−351.7(26) −346.2(25)	I

[a] In the various tables of ^{29}Si chemical shifts herein the data listed for a given compound will not necessarily be exhaustive. In particular, results of older work will tend to be omitted in cases where the same compound has been studied later. The reference numbers are given in brackets after the shift values. [b] Listed only if data for $n = 1, 2$ and 3 are all available, or else at least two of these plus the values for SiX_4 and SiY_4. [c] Reported as solvent-dependent, $\delta_{Si} = -81.2$ to -82.5 ppm, in ref. 9. [d] The value −117.4 given in ref. 41 is in error. [e] The value −17.96 in ref. 27 appears to be in error.

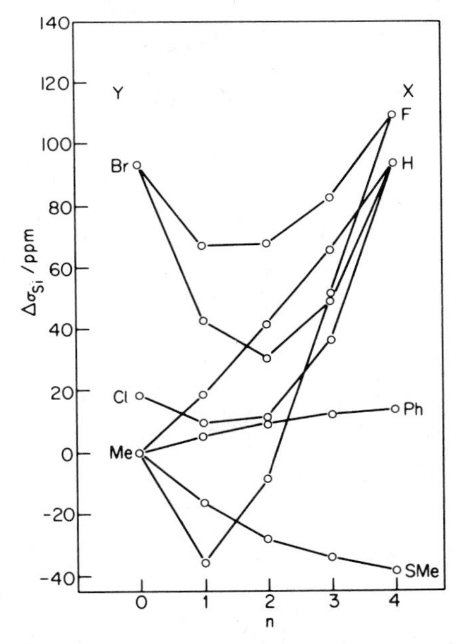

Fig. 10.1 The effect of substituent changes on ^{29}Si shielding for compounds of the type SiX_nY_{4-n}, illustrating the generally "sagging" pattern. The data are from Table 10.1. The y axis gives the shielding change in ppm from $SiMe_4$

TABLE 10.2 ^{29}Si Chemical shifts for trimethylsilyl groups[a] in some representative compounds Me_3SiX.

X	CN	$N(SiMe_3)_2$	NCS	NCO	N_3	
δ_{Si}/ppm	−12.3(41)	2.37(43)	5.4(41)	7.4(41)	16.7(41)[b]	
X	$NPN(SiMe_3)_2$	NSO	$P(SiMe_3)_2$	$SSiMe_3$	$SnMe_3$	$HgSiMe_3$
δ_{Si}/ppm	2.2(50)	−62(51)	2.73(52) 0.4(41)	12.8(53)	−11.0(54)	60.2(23) 64.0(24)

[a] See Tables 10.1 and 10.9 for data on X = Me, H, F, Cl, Br, I, Vi, Ph, $SiMe_3$, OMe, $OSiMe_3$, OAc, SMe, NMe_2 and $Si(SiMe_3)_3$ [b] This appears as 15.2 in ref. 8, but attributed to ref. 41.

(*iii*) *Effects of Remote Substituents.* Naturally substitution at a point remote from the silicon has, in general, a smaller effect on δ_{Si} than substitution at the α-position, and this leads to smaller non-additivity. However, effects of remote substituents can be quite substantial, at least in the case of SiX_4 with X = —O—, which gives a range from $\delta_{Si} = -70$ to $\delta_{Si} = -115$ ppm. A number of publications have concentrated on the effect of substituting Y in compounds of the type $Me_{3-n}X_nSi(CH_2)_mY$, and the results have been reviewed by Schraml and Bellama.[8] Later work by Schraml and his colleagues[56] shows that δ_{Si} is more sensitive to remote substituents acting through oxygen than through carbon. Some of the evidence for this statement can be seen in Table 10.4. In addition, it has been shown[59] that for meta- and para-substituted phenoxytrimethylsilanes, $XC_6H_4OSiM_3$ the dependence of δ_{Si} on the Hammett σ constant of X is roughly twice as great as for the corresponding phenyltrimethylsilanes, in spite of the closer proximity (through the bonds) of X to Si in the latter series. The sensitivity of δ_{Si} to substituent in Me_3SiO-compounds is potentially of considerable value to organic chemists in general and to carbohydrate chemists in particular, because of the commonness of the trimethylsilation procedure. There is good dispersion in the ^{29}Si spectra of trimethylsilated sugars[56,60,61] but the effects have not yet been adequately explained. However, the sensitivity to substituent of $\delta_{Si}(Me_3SiO$-) has been related to the Taft polar substituent constant and to the occurrence of multiple bonding between silicon and oxygen.[62]

Schraml *et al.*[63] have contrasted the effects of a remote substituent Y on δ_{Si} for compounds containing the linkages Si—X···Y and X—Si···Y where X is capable of $(p\text{-}d)_\pi$ bonding to silicon (typical examples are the series $Me_{4-n}Si(OCH_2CH_2NH_2)_n$ and

TABLE 10.3 Silicon chemical shifts of some silyl compounds[a]

Compound	$(H_3Si)_2O$	$(H_3Si)_2S$	$(H_3Si)_2Se$	$(H_3Si)_2Te$
δ_{Si}/ppm	−37.5(44)	−48.0(44)	−57.1(44)	−88.9(44)
Compound	$(H_3Si)_3N$	$(H_3Si)_2Hg$	H_3SiSiH_3	$LiOSiH_3$
δ_{Si}/ppm	−39.9(55a)	−10.3(24)	−104.8(29)	−48(55b)
Compound	$LiSSiH_3$	$LiSeSiH_3$	$LiP(SiH_3)_2$	$LiAs(SiH_3)_2$
δ_{Si}/ppm	−58(55b)	−71(55b)	−62(55b)	−75(55b)

[a] See Table 10.1 for data on SiH_4, H_3SiMe, H_3SiPh, SiH_3F, SiH_3Cl, SiH_3Br and SiH_3I.

TABLE 10.4 ^{29}Si Chemical shift comparisons for Me_3SiOR and Me_3SiCH_2R

R	$\delta_{Si}(Me_3SiOR)$	Δ_1	$\delta_{Si}(Me_3SiCH_2R)$	Δ_2
Me	17.2(34)	0.0	1.6(58)	0.0
Et	13.5(56)	−3.7	0.7(56)	−0.9
CH_2CH_2Br	18.29(56)	1.1	2.0(4)	0.4
CH_2CH_2OMe	16.25(56)	−0.9	0.4(58)	−1.2
CH_2Cl	6.81(56)	−10.4	−0.4(58)	−2.0
CH_2Vi	16.59(56)	−0.6	1.0(58)	−6.0
CH_2Ph	17.40(56)	0.2	1.1(58)	−0.5
Ph	17.72(36)	0.5	0.4(58)	−1.2

$\Delta_1 = \delta_{Si}(Me_3SiOR) - \delta_{Si}(Me_3SiOMe)$;
$\Delta_2 = \delta_{Si}(Me_3SiCH_2R) - \delta_{Si}(Me_3SiEt)$

$(EtO)_nMe_{3-n}SiCH_2NH_2$). In the former case deshielding of Si is observed as Y increases in electronegativity whereas for the latter shielding is seen. The authors explain this as a decrease in $(p\text{-}d)_\pi$ back-donation to Si for the Si—X bond in Si—X···Y under the influence of Y, added to the direct inductive effect of Y on Si, whereas for the X—Si···Y case the X—Si back-donation is enhanced, over-compensating for the direct influence of Y on Si.

Examination of data[36,64] for the series $C_6H_4(OSiMe_3)_2$ and $C_6H_4(SiMe_3)_2$ (see Table 10.5) reveals ^{29}Si deshielding for the ortho isomers; the effect is small (0.6 ppm) for the latter series but somewhat larger (ca. 1.3 ppm) for the former. The shift has been explained as due to steric interactions (the ^{13}C shifts of the Me_3Si and Me_3SiO groups are similarly affected). Steric effects have been more strikingly illustrated by results[65] for trimethylsiloxy derivatives of aliphatic hydrocarbons, for which a γ-shielding influence of a methyl group on δ_{Si} is ca. 3 ppm (see Table 10.6).

Following the work[67] of Grant and Paul ^{13}C data for the alkanes has been examined in terms of the carbon skeleton arrangement according to Eq. 1, which expresses δ_C for the kth carbon in terms of a constant B (which will be the chemical shift of methane),

$$\delta_C(k) = B + \sum_{l} A_l + \sum_{l=\alpha} C_{n(k),\, m(l)} \tag{10.1}$$

additive shift parameters A_l for carbons at the lth position relative to k, and correction terms C which depend on the number of carbon atoms directly bonded to the kth and lth carbons, $n(k)$ and $m(l)$ respectively (the summation for the C terms runs over the

TABLE 10.5 ^{29}Si Chemical shifts, δ_{Si}/ppm, for some benzene derivatives

Compound	*ortho*	*meta*	*para*
$C_6H_4(SiMe_3)_2$	−4.1(64)	−4.7(64)	−4.7(64)
$C_6H_4(OSiMe_3)_2$	18.90(36)	17.87(36)	17.25(36)

TABLE 10.6 ^{29}Si chemical shifts for some trimethylsiloxy aliphatic compounds, Me_3SiOR

R	Me	Et	Pr^i	Bu^t	1-adamantyl	2-adamantyl
δ_{Si}/ppm	17.2(34)	14.53(13)	12.12(56)	6.20(65)	6.18(66)	12.58(66)

TABLE 10.7 Silicon-29 chemical shift data for alkylsilanes[a]

Compound	Et_3SiH	Pr^n_3SiH	Bu^n_3SiH	$(n\text{-hexyl})_3SiH$	Bu^i_3SiH
δ_{Si}/ppm	0.15(68)	−8.11(68); −8.5(33)	−6.58(68); −6.7(33)	−6.71(68)	−14.7(33)

Compound	Bu^t_3SiH	Et_3SiBu^n	Et_3SiBu^s	$EtSiPr^n_3$	Me_3SiPr^n	$Me_3Siheptyl^n$	$Me_3SiC_6H_{11}$
δ_{Si}/ppm	17.8(45)	6.91(4)	8.16(4)	3.75(4)	0.7(57)	1.41(4)	2.38(4)

[a] See also Table 10.1 for data on SiH_4, $SiMe_4$, $SiEt_4$, $MeSiH_3$, Me_2SiH_2, Me_3SiH, Me_3SiEt, Me_2SiEt_2 and $MeSiEt_3$.

carbons directly bonded to the kth carbons only). Scholl *et al.*[4] were the first to apply the analogous equation (without the C terms) to δ_{Si}, while Harris and Kimber[68] gave a further discussion. Collation of 18 of the pieces of data for alkylsilanes (plus SiH_4) listed or referred to in Tables 10.1 and 10.7 can be shown[69] to yield $B = -92.5$ ppm (the average of the two reported shifts for SiH_4) plus the parameters listed in Table 10.8. These can be seen to bear some similarity to the corresponding ^{13}C parameters; the well-known γ-shielding effect is included. The remaining ^{29}Si data on alkylsilanes involve other parameters which cannot be well-determined, and the accuracy of the shifts does not make more sophisticated attempts at correlation feasible at the present time.

(*iv*) *Further Comments and Data.* Results are now available for ^{29}Si chemical shifts of such a wide variety of compounds that it is not feasible to supply a comprehensive list here. For earlier work the reader is referred to the reviews and compilations already cited.[1,4,6–8] For completeness, however, references are now listed[70–86] for every known

TABLE 10.8 Chemical shift parameters[a] for ^{29}Si (from alkylsilanes[b]) and for ^{13}C (from alkanes)

Parameter[c]	A_α	A_β	A_γ	A_δ	C_{41}	C_{42}
^{29}Si value[d,e]/ppm	26.27[16]	4.24[15]	−2.24[9]	0.29[4]	−3.11[7]	−5.15[9]
^{13}C value[d,f]/ppm	9.09[52]	9.40[48]	−2.49[36]	0.31[23]	−1.50[2]	−8.36[1]

[a] These refer to contributions to δ rather than σ. [b] From 18 measurements involving SiH_4 and 13 alkylsilanes. [c] See the text. [d] The number of observations affected by the parameter is given in square brackets. [e] The value for C_{32} cannot be distinguished from zero. [f] From the work of Grant and Paul.[67]

TABLE 10.9 Silicon-29 chemical shifts, δ_{Si}/ppm, for some polysilanes

(a) General formula Si_nX_{2n+2}[a,b]

X	$n = 2$	$n = 3$[c] A	$n = 3$[c] B	*neo*[c] $n = 5$ A	*neo*[c] $n = 5$ B
Me	−19.58(4)	−16.1(42) −15.93(4)	−48.7(42) −48.45(4)	−9.8(42) −8.7(23)	−135.5(42) −135.2(23)
Cl	−8.0(26)	−3.5(42) −6.0(26)	−7.2(42) −9.8(26)	3.5(42) 3.2(26) 3.7(89)	−80.0(42,89) −81.1(26)
F	−73.5(2,8)	−75.5(2,8)	−13.3(2,8)		
OMe	−52.5(3)				
Et	−9.4(45)				

$n = 4$

normal: $Cl_3Si\ SiCl_2\ SiCl_2\ SiCl_3$ — $SiCl_3$: −6.1(26); $SiCl_2$: −6.6(26)

iso: $(Cl_3Si)_3\ SiCl$ — $(Cl_3Si)_3$: −2.3(26); SiCl: −34.2(26)

(b) Some simple disilanes

compound		Cl_3Si (A) $SiFCl_2$ (B)	Me_3Si (A) $SiMe_2H$ (B)	Me_3Si (A) $SiMe_2Ph$ (B)	Me_3Si (A) $SiPh_3$ (B)
δ_{Si}/ppm	A	−2.5(2)	−18.9(42)	−19.3(42)	−18.4(42)
	B	−18.7(2)	−39.1(42)	−21.7(42)	−20.2(42)

compound		$Me_3Si\ SiMe_2Cl$	$Me_3Si\ SiMe_2F$	$(MeO)_3Si\ SiPh_3$	$[(MeO)_2MeSi]_2$
δ_{Si}/ppm	A	−18.2(42)	−22.5(42)	−45.9(42)	−7.5(3)
	B	+22.8(42)	+34.0(42)	−29.2(42)	

[a] For $n = 1$ see Table 10.1. [b] For Si_2H_6 see Table 10.3. The remaining data on Si_nH_{2n+2} in ref. 29 are subject to some uncertainty (H. C. Marsmann, private communication). [c] The silicon skeletons are labelled as follows (group A is always of the type SiX_3):

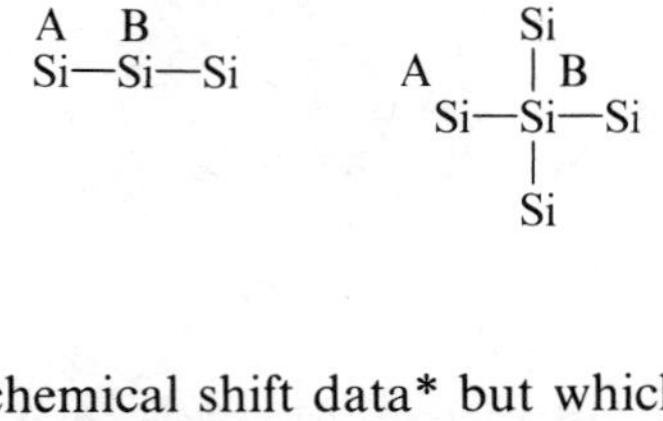

```
A   B
Si—Si—Si
```

```
       Si
A      | B
Si—Si—Si
       |
       Si
```

article which contains ^{29}Si chemical shift data* but which is neither given in Schraml and Bellama's review[8] nor mentioned elsewhere in the present work.

Certain classes of compound, however, merit somewhat more comment here. Schraml and Bellama[8] have quoted data for silicon hydrides, polysilanes and silacyclic compounds, all of which have special interest. Rather more chemical shift values have now been reported for hydrides,[11,29,32,42,44,68,87,88] but many of these results have been mentioned above. The polysilanes have also received more attention, and data are given in Table 10.9. Additional information has also been reported[90–93] for cyclic systems† with only C and Si as ring atoms, and the rather anomalous range −49 to −60 ppm

* The groups of G. Fritz, E. Niecke, and A. Meller use the outmoded convention of reporting shifts with the positive direction to low frequency.

TABLE 10.10 ^{29}Si chemical shifts for species with silicon bonded to a metal

Species	$F_3SiCo(CO)_4$	*trans*-$[PtCl(SiH_2Cl)(PEt_3)]_2$	$Me_3SiHg\underline{Si}H_3$[a]	$Me_3SiSnMe_3$
δ_{Si}/ppm	−29.3	−25.0	−22.1	−11.0
ref.	2,3,8	101	24	54

Species	$Hg(SiH_3)_2$	$Me_3SiHgGeH_3$	$(Bu^t_3Si)_2Cd$	$(Et_3Si)_2Hg$	$(Me_3Si)_2Hg$	$Me_3\underline{Si}HgSiH_3$[a]
δ_{Si}/ppm	−10.3	41.1	47.5	50.2	60.2; 64.0	63.7
ref.	24	24	45	45	23 24	24

[a] The silicon involved is underlined.

has been established[94,95] for δ_{Si} of $SiMe_2$ groups in silacyclopropanes, with −106.2 ppm found[96] for a silacyclopropene.

The silazanes, which contain the Si-N(R)-Si bond, have not received much more attention since the work of Jancke *et al.*[97] The range of ^{29}Si shifts for these compounds appears to be much smaller than that of the analogous siloxanes, which are discussed fully below. There are many reports[11,98–100] for δ_{Si} in systems involving Si-N(sp^3) or Si-N(sp^2) bonds.

Table 10.10 gives some data on δ_{Si} for systems in which silicon is bonded to a metal. Since there are relatively few such results in the literature the range observed may be atypical. However, the highest shift (lowest shielding) known for ^{29}Si involves bonds to Hg.

It is worth mentioning the data for compounds with co-ordination number differing from the normal four. There are three classes of such systems. Thus the anion SiH_3^-, with co-ordination three, has[102] $\delta_{Si} = -165$ ppm. Secondly, results for six-fold co-ordination have been reported;[103,104] such compounds also have high shielding (e.g. SiF_6^- with $\delta_{Si} = -185.3$). Finally there are the interesting series of compounds of type I,

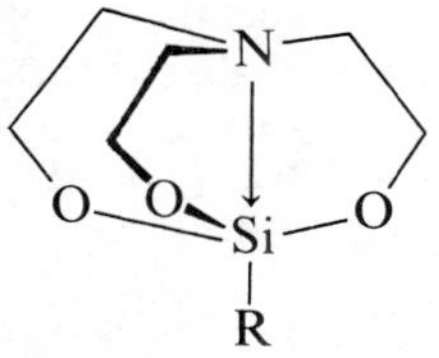

I

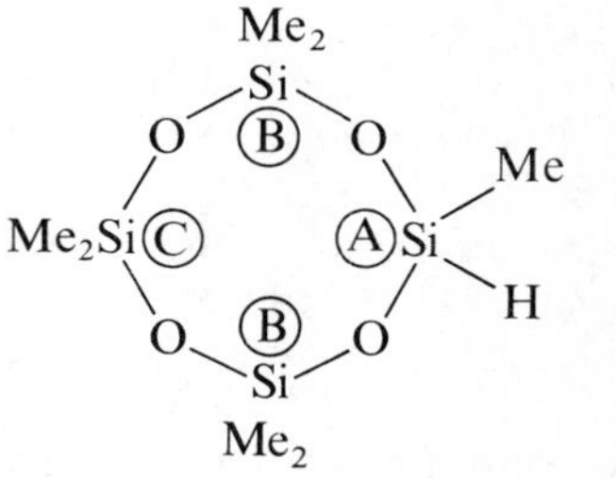

III

known as the silatranes, which apparently have five-fold co-ordination. The data in Table 10.11 show that in comparison with the corresponding triethoxysilanes, $(EtO)_3SiR$(II), the silatranes are additionally shielded by ca. 20 ppm. This is in the expected direction, by analogy with co-ordination effects on ^{11}B and ^{31}P resonances.

TABLE 10.11 Silicon chemical shifts[105] (δ_{Si}/ppm) of silatranes (I) and analogous triethoxysilanes(II)

R	H	Me	Vi	Ph
I	−83.6	−64.8	−81.6	−81.7
II	−59.5	−44.2	−59.5	−58.4

b. Siloxanes and Related Compounds

(*i*) *Introduction.* Silicon has a well-developed propensity to form macromolecules involving the Si-O-Si linkage. Such molecules are often referred to as silicones, the systematic name being siloxanes. They are of great importance commercially, but structure determination has not always been easy and it became obvious that ^{29}Si NMR could play a big role in this area. The subject has been reviewed,[17] but developments have been rapid. In order to discuss the results it is necessary to introduce the notation normally used for the units constituting a siloxane. Formally these are defined as containing split oxygen atoms, and the nomenclature is based on methylated units in four groups, as follows. The M, D, T, Q notation refers to the number of oxygen atoms bonded to the silicon (mono-, di-, tri-oxo and quaternary). Substitution of methyl

$Me_3SiO_{\frac{1}{2}}$-	$-O_{\frac{1}{2}}-Si(Me)_2-O_{\frac{1}{2}}-$	$-O_{\frac{1}{2}}-Si(Me)(O_{\frac{1}{2}})-O_{\frac{1}{2}}$	$-O_{\frac{1}{2}}-Si(O_{\frac{1}{2}})_2-O_{\frac{1}{2}}-$
end units	chain units	branch units	quaternary units
M	D	T	Q

groups is indicated by a superscript suffix, e.g. M^{Ph} implies the unit $PhMe_2SiO_{\frac{1}{2}}$—. In this article the usage will be such as to imply that the bonding of a unit is always to another siloxane unit, and groups such as —OH, —OMe etc. will be specified separately, e.g. M^{OH} implies $HOMe_2SiO_{\frac{1}{2}}$—, except when silicate ions are under discussion.

The dispersion in the ^{29}Si NMR spectrum for siloxanes is large (see Fig. 10.2)—observed shifts range from $\delta_{Si} = 12.4$ to $\delta_{Si} = -114.0$ ppm. Moreover, the ranges for unsubstituted M, D, T and Q units are very well separated at ~6, ~ −21, ~ −67 and ~ −106 ppm. Substitution makes big differences; such changes are in general, of course, largest for electronegative substituents, but even units such as M^H and M^{Ph} give chemical shift ranges which do not overlap with each other or with that for unsubstituted M. The shift difference between M^{OH} and D resonances is ca. 10 ppm, more than adequate for integration to give[106] acceptable data for the average degree of polymerisation of low molecular weight disiloxanol fluids. Figure 10.2 gives an overview of the substituent effects and shows that silicon-29 NMR is clearly established as a powerful tool for the structure determination of siloxanes. A number of factors affecting the shifts will be described below. Tables 10.12, 10.13 and 10.14 give ^{29}Si shift data for a number of siloxane

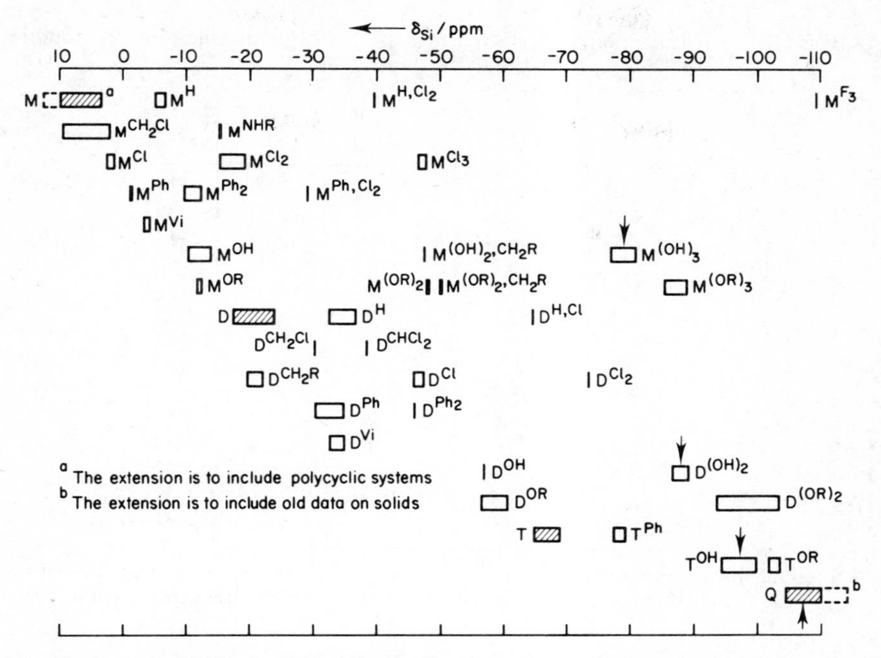

Fig. 10.2 The ranges of ^{29}Si chemical shifts for siloxane units (excluding trimeric rings and polycyclic systems). The ranges for the unsubstituted M, D, T and Q units are shaded. The ranges for silicate anions (see the text) $Q^1 \equiv M^{(OH)_3}$, $Q^2 \equiv D^{(OH)_2}$, $Q^3 \equiv T^{OH}$ and $Q^4 \equiv Q$ are indicated by vertical arrows.

oligomers which are not quoted elsewhere in this section. These results illustrate the wide range of shifts observed but they will not be discussed in detail here.

(*ii*) *Cyclisation*. Ring-size affects ^{29}Si chemical shifts (Table 10.15). This is particularly noticeable for three-membered rings, which are deshielded by ca. 10 ppm compared to four-membered rings for the compounds with D and D^{Ph} units. This shift is smaller for true Q units, being only ca. 8 ppm. The effect of cyclisation is also seen in exocyclic M groups for compounds of the type $M_{2n}Q_n$, which give δ_{Si} = 6.4 and 9.2 ppm for n = 3

TABLE 10.12 ^{29}Si chemical shifts of siloxane dimers[a] M_2^X

X^b	—		H		CH_2Cl	Ph	Vi	$NHPr^i$
δ_{Si}/ppm	6.8 to 8.0,[c]	6.83[d]	−5.25,	−5.27	3.5	−1.8	−3.53	−15.2 to −16.7[c]
ref.	9	13	107	68	34	31	108	11

X	OMe	OH	Cl_2	$(OMe)_2$	$(OEt)_2$	Et_3	F_3	Cl_3
δ_{Si}/ppm	−12.0	−10.5 to −16.9[c]	−15.2	−48.04	−88.89	9.11	−113.0	−46.8
ref.	34	11	109	110	110	4	2, 8	5

[a] Excluding silicate ions. [b] All Me except where otherwise stated. [c] Solvent dependent. [d] Data for this compound may also be found in refs 31, 33 and 34.

TABLE 10.13 ^{29}Si chemical shifts, δ_{Si}/ppm, of linear siloxane oligomers[a] $M^X D^Y_n M^X$

X[b]	Y[b]	$n = 1$ M	$n = 1$ D	$n = 2$ M	$n = 2$ D	Ref.	Comments
—	—	6.61	−21.46	6.78	−22.08	111	[c](39,111)
		6.70	−21.5	6.80	−22.0	39	
—	H	8.72	−36.86	9.32	−36.41	111	[c](39)
—	Cl	9.8	−46.2			5	
—	Ph	7.2	−34.9			31	
—	OH	5.9	−57.1			5	
—	Vi	d	−35.1			112	
H	—	−7.27	−18.59	−7.35	−20.41	107	[c](107)
Cl	—	2.3	−18.3			5	[c](5)
CH_2Cl	—	2.5	−19.6	2.1	—20.7	34	
OMe	—	−12.5	−22.0	−12.0	−20.4	34	[c](34)
Ph_2	—	−12.5	−19.4			31	
Ph_2	Ph	−11.1	−32.4			31	
Ph_2	Ph_2	−10.5	−46.1			31	

[a] Excluding silicate ions and $M^{OH}D_nM^{OH}$ (see Fig. 10.3). [b] All Me except where otherwise stated. [c] Data on higher linear oligomers are known (ref. cited in brackets). [d] Not reported.

TABLE 10.14 ^{29}Si NMR chemical shifts for some siloxane oligomers

	M	T	Q	Ref.
M_4Q	8.21, 7.6, 8.6 to 8.8		−104.4, −105.1, −104.0 to −104.2, −105.2	13, 21, 34, 114
M_6Q_2	8.0, 8.92		−107.5, −106.54	21, 114
M_8Q_3	8.0, 8.99(M_3Q)		−107.7, −106.67(M_3Q)	21, 114
	8.3, 9.31(M_2Q)		−110.1, −109.09(M_2Q)	21, 114
$M_{10}Q_4$	7.6(M_3Q)		−108.6(M_3Q)	114
	7.7(M_2Q)		−110.4(M_2Q)	114
M_3T	6.7	−65.0		31, 34
M_4T_2	7.3	−66.2		31
M_3T^{Ph}	8.1	−77.5		31
$M_3^{Ph_2}T$	−11.5	−64.7		31
$M_3^{Ph_2}T^{Ph}$	−10.3	−78.0		31
$M_4T_2^{Ph}$	8.8	−7.92		31

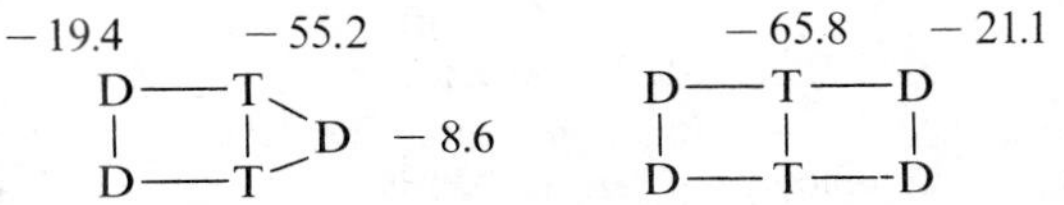

```
 −19.4    −55.2                 −65.8    −21.1
   D——T                       D——T——D
   |    |  >D  −8.6           |    |    |          115
   D——T                       D——T——D
```

TABLE 10.15 Silicon-29 chemical shifts, δ_{Si}/ppm, for some cyclic siloxanes[a]

X	D_3^X	D_4^X	D_5^X	D_6^X	Ref.
Me	−9.2	−20.0	−22.8	−23.0	31
		−19.69	−22.04	−22.66	107
H		−32.39 to −33.02	−34.45 to −34.85		116
Ph	−20.62 to −20.70	−30.11 to −30.42	−32.39 to −32.77		116
Vi		−32.65 to −32.75			116
$(OSiMe_3)_2$[b]	−100.7	−108.8			114
		−107.76		−107.76	21, 117
Ph_2	−33.8				31
Et		−19.7			1

[a] For the four isomers of $D_2D_2^{Vi}$ the shifts of the D units are[113] in the range −18.28 to −19.04, and the shifts of the D^{Vi} units are in the range −33.11 to −33.80. For $D_3D^{CHCl_2}$ the D^{CHCl_2} resonance is at −38.4 ppm and the D peaks are at −17.8 ppm.[34] For $D_3D^{CH_2Cl}$ the D^{CH_2Cl} resonance is at −30.6 ppm and the D peaks are at −18.5 and 19.5 ppm.[34] [b] These compounds have the general formula $M_{2x}Q_x$. Their M units resonate[21,114] at 6.4 for $x = 3$; 9.2, 10.20 for $x = 4$ and[117] 10.33 for $x = 6$.

and 4 respectively;[114] in such cases cyclisation causes shielding. Trimethylsilylation of minerals has produced a range of exotic polycyclic systems, some of which have cage structures. The shifts are given in Table 10.16, but the observed effects are not, at this stage, understood.

(*iii*) *Polymerisation—Effect of End Groups.* For long-chain polymers of the type $M^XD_n^YM^X$ it is of interest to examine how the shifts of the D^Y units depend on their proximity to the end groups. Such effects can be detected for up to six units from the end. Data are given in Table 10.17. It can be seen that the end-group effects depend greatly on the nature of X and Y. Thus for X = Me, Y = H there is a monotonic shielding effect on D^Y as the unit gets nearer the end-group. For X = H, Y = Me the reverse is true—there is a monotonic *de*shielding effect; this is also shown for X = OH, Y = Me, but for this system the shifts are heavily dependent on the solvent because of the influence of hydrogen bonding.[11] Indeed, for solutions of $M^{OH}D_{20}M^{OH}$ in Et_3N all resolvable peaks due to D units near chain ends resonate to *low* frequency of the main-chain units.[120] In the case of X = Y = Me, the influence of the end-group is to cause shielding up to and including the penultimate D unit, but the final D unit is substantially *de*shielded compared to the main chain. This type of information becomes important if NMR is to be used to measure average chain lengths, since clearly it is often necessary to look for rather weak peaks due to groups near the chain ends; for chain-lengths of the order of 50 this presents no real problems.[119]

(*iv*) *Polymerisation—Effect of Chain Length.* As chain lengths increase, both the end-groups and the chain units show shifts which are basically due to replacement of the end-group, M^X say, by an end-group plus a chain unit, M^XD^Y, at the relevant distance from

TABLE 10.16 Silicon-29 chemical shifts,[a] δ_{Si}/ppm, for some polycyclic siloxanes[b] of general formula M_xQ_y

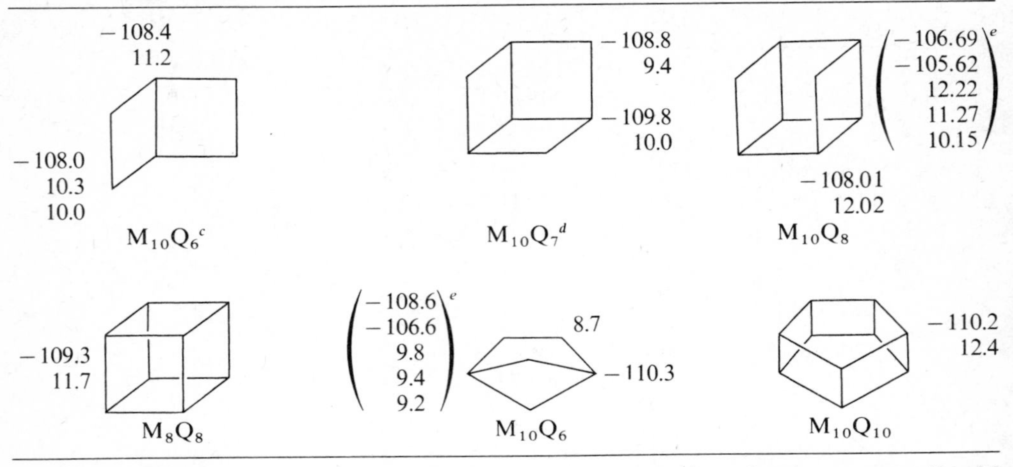

[a] The numbers near −105 are shifts for the Q units; those near 10 are for the corresponding M units. [b] The structural diagrams are drawn such that Q-type Si atoms are at each corner, connected by siloxane bridges, and with additional M units as required by the formulae. Data are from ref. 114, except for $M_{10}Q_{10}$, first quoted in ref. 118, and $M_{10}Q_8$, which is reported in ref. 117. It should be noted that the values in ref. 114 are systematically ca. 1 ppm lower than those of ref. 117. [c] The M units on QM_2 groups are non-equivalent, but the detailed assignment is not known. [d] In principle there should be more ^{29}Si signals, but only two in each region could be resolved. [e] The assignment of these signals is not known.

the site of interest. If X and Y are the same, this replacement is effectively that of Me by M^X, and the shift conferred may be classified according to the distance of the site observed from the site substituted. Thus the change MM → M$\underline{D}$M (the site observed is underlined) is an *a* effect, whereas $\underline{M}$M → $\underline{M}$DM or $\underline{M}$$\underline{D}$M → M$\underline{D}$DM or M$\underline{D}$DM → MD$\underline{D}$DM is a *b* effect. Unfortunately most data in the literature were not recorded under sufficiently standard conditions to test the consistency of the shift effects. The *a*-shifts are, of course, frequently large, so the inconsistencies and variations are relatively unimportant. For the MD_nM series[107] the *b*-effect on δ_{Si} is negative (ca. −0.6 ppm, except for the change

TABLE 10.17 Silicon-29 chemical shifts,[a] δ_{Si}/ppm, for linear polysiloxanes, $M^XD^Y_nM^X$

X	Y	n	$D^Y(1)$	$D^Y(2)$	$D^Y(3)$	$D^Y(4)$	$D^Y(5)$	Ref.
Me	Me	8	−21.86	−22.45	−22.30	−22.20		39
Me	H	≈50	−35.82	−35.33	−34.95	−34.88	−34.88[c]	119
OH[b]	Me	10	−21.34	−21.63	−21.68	−21.80	−21.90	107
H	Me	7	−20.51	−22.31	−22.34	−22.35		107

[a] The units are numbered from the chain ends thus: $M^X D^Y(1) D^Y(2) D^Y(3)$.... [b] The shifts for this type of system are strongly dependent on solvent and concentration.[11] [c] The resonances of the remaining D units are also, indistinguishably, at δ_{Si} = −34.88 ppm.

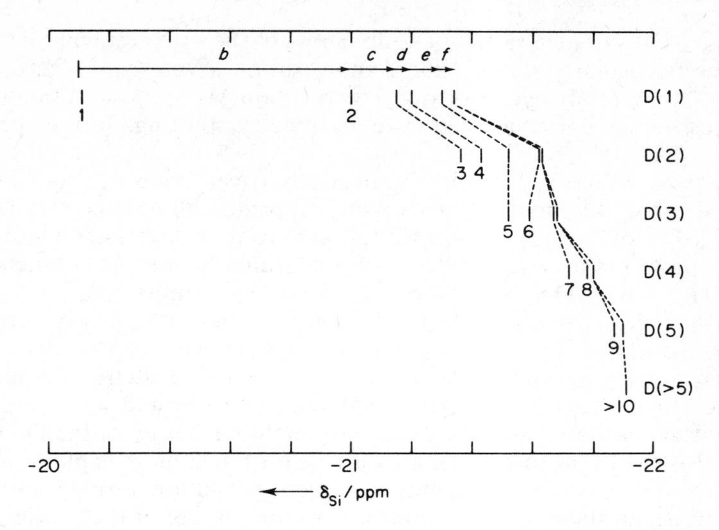

Fig. 10.3 Silicon-29 chemical shifts for the D units in $M^{OH}D_nM^{OH}$ oligomers (average n ~ 4), present together in a single solution in benzene. The data are from ref. 107 and were measured by the absolute frequency method. The unit referred to is listed at the right. The dashed lines link resonances for a given oligomer, with the value of n given underneath. The substituent shift effects *b*, *c*, *d*, *e* and *f* are indicated for D(1) only.

$\underline{M}M \rightarrow \underline{M}DM$, which is much smaller), but the longer-range shifts are positive (the *c*-effect is ca. +0.17 ppm). This situation is consistent with the end-group effects discussed above, since for a long-chain polymer D(1) → D(2) is a *b* change, D(2) → D(3) is a *c*-change etc. The data on the MD_n^HM series[107,119] show that the induced shifts from *b* onwards are positive, also in agreement with observations on long-chain polymers.

The data in Fig. 10.3, for the $M^{OH}D_nM^{OH}$ series, are instructive because they were measured[107] on a single solution containing a mixture of the various oligomers; certain medium effects are thereby minimised. The results show (*a*) the high degree of dispersion in the ^{29}Si spectrum, allowing the study of complex mixtures of species to be productive, (*b*) the sensitivity of ^{29}Si shifts to long-range substitution effects (the *f* shift is due to substitution 11 bonds from the observed nucleus), and (*c*) the subtlety of the factors influencing shifts due to substitution. As far as the last point is concerned, the data show non-additivity of the substitution shifts—the effects vary with the species under examination. In particular, the *b* shift in δ_{Si} drops from ca. −0.9 ppm to ca. −0.3 ppm as the species get larger than $M^{OH}D_2M^{OH}$; this results in a greater bunching than expected of the resonances to low frequency. Explanations of these subtleties are not yet available, but it should be recalled that the —OH-ended oligomers are particularly prone to large solvent effects[11] and also probably involve hydrogen-bonding between each other and even intramolecularly.

(*v*) *Asymmetry Effects.* Monosubstituted D units provide asymmetric centres, and therefore suitable oligomers and polymers should give evidence of stereoisomerism, leading, in some cases to tacticity studies. For cyclic oligomers such effects lead to the existence of configurational isomers, and silicon-29 NMR has shown the existence of such species for the systems[116] D_4^H, D_5^H, D_3^{Ph}, D_4^{Ph}, D_5^{Ph} and D_4^{Vi}, and also[113] for $D_2D_2^{Vi}$ (this case also gives rise to positional isomers). The shift effects involved are nearly additive for D_4^H but are markedly not so for most of the other systems. The D^H group appears to be particularly suitable for demonstrating asymmetry effects. Thus it has been shown[119] that a polymer of average composition MD_{50}^HM has a random arrangement as far as tacticity is concerned. The asymmetry splittings in the spectra are ca. 0.4 ppm.

(*vi*) *Copolymers.* Silicon-29 NMR is potentially a very valuable tool for the study of chemical structure of copolymers involving siloxane units. So far, remarkably little has been published on this topic. Suitable copolymers may be classified into three types, viz. (*a*) linear polymers involving differently-substituted D units, (*b*) branched or cross-linked copolymers with D/T or D/Q or T/Q mixtures of units, and (*c*) copolymers of siloxanes with purely organic units (which may or may not involve cross-linking). Some rather qualitative work has been carried out on type (*b*) systems, but the only really detailed study so far reported[17] has involved type (*a*) systems with mixtures of D and D^H units. Fine structure has been observed at the heptad level, combined with asymmetry effects of the D^H groups, and a random distribution of the D and D^H units were found. It is clear that this whole area is ripe for much more exploitation.

(*vii*) *Silicate solutions.* Ionic silicates in aqueous solution form a special class of "siloxanes" in which there is much interest since the nature of the species present has been obscure until recently. Silicon-29 NMR has already proved of great value in this area.[12,20,121–124] For dilute solutions of molecular ratios corresponding to the "metasilicate" composition (alkali hydroxide: silica ratio ca. 2:1) six bands can be seen in the spectrum,[20] labelled A to F from high to low frequency, which can be linked to the resonance (G) due to solid silica (e.g. glass). Bands A, B, D, F and G form a regular sequence at $\delta_{Si} \simeq -70, -77, -86, -95$ and -107 ppm respectively, and are attributed to "monomer" (SiO_4^{4-} and related protonated species), end-groups, middle-groups, branching positions and "quaternary" silica respectively. All groups have silicon tetrahedrally co-ordinated to oxygen, and have been variously labelled (one notation[122] uses the symbols Q^0, Q^1, Q^2, Q^3 and Q^4 to indicate with a superscript the number of siloxy units bonded to the silicon designated, though this notation is not entirely compatible with the general M, D, T, Q usage). Thus there is a reasonably additive shift of ca. 7–12 ppm for the replacement $—O^-$ (or $—OH$) $\rightarrow$ $—OSiO^{3-}$. There is some dispute as to the origin of bands C and E, which are found ca. -9.7 and -17.5 ppm from the Q^0 peak, respectively, but it appears to be most likely that these arise from Q^2 and Q^3 peaks of cyclic "trimers," respectively, with cyclisation shifts of ca. $+7$ ppm from bands D and F respectively. In more concentrated solutions the spectra become very complex, but it may be anticipated that a detailed assignment will become feasible; already there exists a suggested assignment for a spectrum of a solution of 8.0 mol dm^{-3} KOH plus 3.0 mol dm^{-3} SiO_2 in terms of six chemical species. The rate of condensation of the "monomeric" species in acid solution has been followed[121] using ^{29}Si NMR.

For details of the ^{29}Si chemical shifts for silicate solutions the reader is referred to the original literature.[12,20,121–124]

c. Theoretical Considerations

The theory of ^{29}Si chemical shifts is not well-advanced. Early attempts to explain the data were mostly at a qualitative level. Recent semi-empirical calculations involve the use of the Jameson–Gutowsky formulation, Eq. 3.36, for the "local" paramagnetic term, σ_p, involving the 3p and 3d electrons on silicon. This approach, which is discussed (together with other early theoretical work on ^{29}Si shifts) in the review by Schraml and Bellama,[8] contains a number of approximations, especially (*a*) the use of an "average excitation energy," ΔE, and, in most cases, (*b*) neglect of d-electron term. Nevertheless, Engelhardt, Radeglia *et al.*[33] succeeded in reproducing the typical "sagging pattern" for Si shifts as a function of substitution, though it was found necessary to introduce an empirical correction factor. A similar approach, involving CNDO/2 calculations, has been described by Roelandt, van de Vondel and van den Berghe.[125] Moreover, Radeglia *et al.* have reformulated[48] and extended[126–129] their work. They note[48] that the results tend to be poor for molecules containing bromine and iodine. Such a heavy atom effect, which may be described[40] by pairwise interaction parameters, is also found for ^{13}C NMR and has been variously ascribed[49] to dispersion forces (presumably affecting the electric field contribution to σ) and steric influences on the local paramagnetic contribution to σ, though there is some doubt about both the sign of such effects and about the exponent of the interatomic distance required. Radeglia[48] has fitted some ^{29}Si shift data involving I or Br to an empirical equation involving a term proportional to $(\sum_i Z_i^m)^n$, where Z_i is the atomic number of atom i bonded to Si, and the best fit is obtained when the exponents m and n are 1 and 2 respectively (though this result is not properly linked to either the dispersion effect or the steric effect). Radeglia *et al.* have also calculated[127] the effect of bond-angle changes on the shielding, and find that these introduce considerable corrections in some cases, notably Si_2F_6, $Si_2(OMe)_6$ and the central silicon of Si_3F_8, giving better agreement with experiment. Wolff and Radeglia[128] have also applied CNDO/2 calculations *without* the average energy approximation to the problem of Si shielding. The improvements are not spectacular, but it seems clear that the use of a constant ΔE for all silicon compounds is a gross oversimplification.

It should be noted that these calculations neglect d-orbitals, and therefore implicitly assume $(d\text{-}p)\pi$-bonding is unimportant. Wolff and Radeglia[129] have, in fact, carried out CNDO/2 calculations with inclusion of silicon d-orbitals, but find that this makes agreement between experimental and calculated data worse! However, other evidence[63] suggests that finer details of ^{29}Si shifts may indeed involve $(d\text{-}p)\pi$-bonding.

It remains clear that further work on the theory of ^{29}Si chemical shifts is required.

10A.4 Coupling Constants to Silicon

a. General Discussion

Since data on coupling to silicon may be rather readily obtained by observing spectra of the other nucleus involved, it is not surprising that there were many investigations in this area during the 1960s using CW spectrometers. This was particularly true for J_{SiH} and J_{SiF}. However, determinations with ^{29}Si in natural abundance required observation of so-called "^{29}Si satellites" which were relatively weak compared to peaks due to per-^{28}Si-isotopomers; thus it was difficult to obtain values for small coupling constants, and frequently only one-bond couplings were determined. Recently, the FT-inspired boom in ^{13}C NMR has led to information about (Si, C) coupling. To date, splittings in ^{29}Si spectra themselves have not been greatly studied, with the possible exceptions of those giving (Si, P) data. Naturally, ^{29}Si spectra without proton noise decoupling frequently present both S/N problems and difficulties of spectral analysis, so they have not often been observed and interpreted. Values of J_{SiC} are usually more readily obtained by observing ^{29}Si satellites in ^{13}C resonance than from ^{13}C satellites in ^{29}Si resonance.[13,130] A few papers have reported values of J_{SiSi} from observation of ^{29}Si satellites in ^{29}Si resonance.[13,42]

Because early studies established the general trends adequately, information about (Si, H), (Si, F) and to some extent (Si, C) coupling contained in the review by Schraml and Bellama[8] still gives an excellent survey of the area. Consequently only a brief summary of the general features will be given here, emphasizing more recent work. Some general ranges for the reported values are given in Table 10.18.

Many signs of coupling constants have been determined, and these are included in Table 10.18, but in some cases uncertainty remains. There may be sign changes in cases where magnitudes near zero are known. Frequently signs are only known for specific situations, e.g. for $^2J_{SiH}$ only when the intervening atom is carbon. Most of the known signs were determined using heteronuclear double resonance with CW spectrometers. However, pulse techniques can also be used, as has been done[140] by the selective population transfer method to confirm that $^2J_{SiH}$ is positive for $(Me_2HSi)_2O$. It is important to realise that because the magnetogyric ratio of ^{29}Si is negative coupling constants to nuclei ^1H, ^{19}F and ^{13}C, which have positive values of γ, will involve a negative sign that is wholly irrelevant to the mechanism of coupling. Thus, for purposes of comparison around the Periodic Table, signs of *reduced* coupling constants, K, should be used. Moreover it is relative signs of K rather than J that are directly produced by double resonance experiments. When the negative γ_{Si} is borne in mind, it becomes clear that (Si, H) and (Si, C) couplings have much in common with (C, H) and (C, C) couplings respectively. Thus $|^1J_{SiH}|$ in silane[8] is 202.5 Hz, and it increases markedly with halogen substitution (see ref. 8), rising to 381.7 Hz for $SiHF_3$. Values of $|^1J_{SiC}|$ depend on the s-character of the carbon atom:[130,141] 50.3 Hz for Me_4Si, 64 Hz for $Me_3Si\underline{C}H{=}CH_2$ and 83.5 Hz for $Me_3Si\underline{C}{\equiv}CPh$ (coupling to the underlined carbon in the second and third cases). Substituent electronegativity, χ, has been shown to have a general correlation with $^1J_{SiC}$. The relationship $|^1J_{SiC}|/Hz = 7.90\chi_X + 31.5$ has been given for trimethylsilyl compounds.[13] However, the importance of such relations should not be over-emphasised since other factors are clearly of importance for coupling to silicon (as in corresponding cases for other nuclei). Thus $|^1J_{SiH}|$ values for SiH_2F_2 (282 Hz) and SiH_2I_2 (280.5 Hz) differ little[142] in spite of big differences in substituent electronegativity. Moreover, in the series $Me_nSi(OEt)_{4-n}$, $|^1J_{SiC}|$ varies[19] regularly with n, being 50.3, 59.0, 73.0 and 96.2 Hz for $n = 4$, 3, 2 and 1 respectively. This is as expected on electronegativity grounds, but the variation is larger than would be anticipated if simply linked to s-character changes (compare 96.2 Hz for $MeSi(OEt)_3$ with 83.6 Hz for $Me_3SiC{\equiv}CPh$). However, the data for $|^1J_{SiC}|$ in $Me_nSi(OEt)_{4-n}$ correlate very well[19] with values for $|^1J_{SiH}|$ in the corresponding series $H_nSi(OEt)_{4-n}$: $|^1J_{SiH}|/Hz = 2.20\,|^1J_{SiC}|/Hz + 86.6$. A correlation of $^1J_{SiC}$ with s-characters of the orbitals of the Si-C bond has recently been reported[131] for a number of chlorinated organosilicon compounds. An excellent correlation also exists[13] between $|^1J_{SiC}|$ for trimethylsilyl compounds and $|^1J_{CC}|$ in corresponding compounds Me_3CX. Expressed in reduced form this is $|^1K_{SiC}|/NA^2m^{-3} = 1.72|^1K_{CC}|/NA^2m^{-3} + 0.80$. Similar remarks to those for $^1J_{SiC}$ may be made for $^1J_{SiSi}$. Thus Sharp *et al.* report[42] a reasonably good correlation between $^1J_{SiSi}$ and the sum of substituent electronegativities. They also find a relationship between $^1J_{SiSi}$ and $^1J_{SiC}$ for molecules of the type $Me_3Si\underline{Si}Me_2X$ (the silicon involved in the $^1J_{SiC}$ discussed is underlined).

Values of $|^1J_{SiF}|$ are usually somewhat greater than those of $|^1J_{SiH}|$ in analogous compounds, though for SiF_4 the coupling constant (169–178 Hz, dependent on medium) is actually less than $^1J_{SiH}$ for SiH_4. The substituent electronegativity effect on $^1J_{SiF}$ is not very clear.[8] A small isotope effect on $^1J_{SiF}$ has been found[143] by measurement of values for $HSiF_3$ (276.6 Hz) and $DSiF_3$ (274.6 Hz). An isotope effect of the H → D substitution has also been reported[133] for $^1J_{SiH}$.

TABLE 10.18 Ranges and signs of coupling constants between silicon and elements of atomic number less than 18[a]

Coupling[b]	^{1}J/Hz	$^{2}J^{c}$/Hz	$^{3}J^{c}$/Hz	$^{4}J^{c}$/Hz
Si, ^{1}H	-147 to -382[d] (8)	1 to 13 (8, 28, 131, 132)[e]	1 to 8 (8, 19, 32, 43, 28, 133, 134)[f]	1.0 to 1.5 (32)
Si, C	-37 to -113 (8, 13, 42)	4.0 to 5.8 (13, 32), 16.1 (130)	1.0 to 3.6 (32)	
Si, ^{15}N	$+6$ (55)[g], $+12$ (88)			
Si, F	$+167$[h] to $+488$ (8)	-14 to -91 (8)	-15.7 (8), 2.5 (8)	
Si, Si	$+53$ to $+186$ (42)	1.0 (134), 4.0 (13)		
P, Si	$+7$ to 50 (8, 88, 136)[i]	1 to 27 (50, 137, 138)[i]	-34 (139)	

[a] Refs given in brackets. [b] Signs are omitted where they are uncertain. [c] Through elements of atomic number less than 18. [d] Exceptionally -420 for $H_2Si[(Mn(CO)_5]_2 \cdot 4$ pyridines.[8] [e] Positive where determined in cases with an intervening sp^3 carbon, but a negative value (-6.37) is suggested[132] for Vi_4Si. [f] Negative where determined. [g] Incorrectly quoted as -4.2 in ref. 8, Table XXV. [h] Lower for hexacoordinated anions,[8,135] e.g. $+110$ Hz for SiF_6^{2-}. [i] Some data for $^{2}J_{PSi}$ are erroneously listed as $^{1}J_{PSi}$ in ref. 8, Table XXV.

TABLE 10.19 Coupling constants[a] to silicon involving elements of atomic number greater than 18

Coupling type	Molecule	J/Hz[b]
^1J(SeSi)	$(SiH_3)_2Se$	+110.6
^1J(^{119}SnSi)	$Ph_3SiSnMe_3$	+650[c]
^1J(^{195}PtSi)	*trans*-$PtCl(SiCH_2Cl)(PEt_3)_2$	−1600
^2J(SiFe^{1}H)	$(Cl_3Si)_2FeH(CO)(Cp)$	20
^2J(SiFeC)	*cis*-$(Me_3Si)_2Fe(CO)_4$	5.3
^2J(PPtSi)	*trans*-$PtCl(SiCH_2Cl)(PEt_3)_2$	+18
^2J(^{119}SnNSi)	$[(Me_3Si)_2N]_2Sn$	16.2
^3J(SiSeCH)	$Me_nSi(SeMe)_{4-n}$	3.6 to 5.7
^3J(SiSeSiH)	$(SiH_3)_2Se$	−4.6
	$SiH_3SeSiMe_3$	2.1
^3J(SiAsSiH)	$(SiH_3)_3As$	5.3
^3J(SiSbSiH)	$(SiH_3)_3Sb$	4.5
^1J(^{119}SnSi)	$Me_3SiSnMe_3$	+656 ± 10[d]

[a] *Loc. cit.* in ref. 8, except where otherwise stated. [b] Signs only given where known. [c] Wrongly quoted in ref. 8. [d] Ref. 54.

The reader is referred to the work of Schraml and Bellama[8] for further details of one-bond coupling of silicon to H, C and F, and also for some data on longer-range coupling, particularly $^2J_{SiH}$. In contrast to the situation for C and Sn, there is as yet little information on coupling to Si over more than two-bonds except for $^3J_{SiH}$. Tables 10.18 and 10.19 include some information on silicon coupling to nuclei other than H, C and F.

b. Theoretical Work

The brief discussion above makes it clear that (Si, H), (Si, C) and (Si, Si) coupling constants present no features which severely differentiate them from the well-understood (C, H) and (C, C) couplings. The implication is that the Fermi contact term dominates all their couplings. A number of theoretical papers have discussed the situation on this basis. Thus Kovačevič and Maksič have used[144] the maximum overlap approximation to give hybridization parameters for use in the Fermi contact term, with the average energy approximation, and have derived values of $^1J_{SiC}$ which compare well with the experimental data of Levy *et al.*[130] The relationship could be expressed as:

$$|^1J_{SiC}|/\mathrm{Hz} = 5.554 \times 10^{-2}\alpha_{Si}^2\alpha_C^2 + 18.2$$

where α_{Si}^2 and α_C^2 are the % *s*-characters of the relevant hybrid orbitals describing the C-Si bond. Summerhays and Deprez[145] turned their attention to data of Harris and Kimber[13] for $^1J_{SiC}$ in trimethylsilyl compounds. They used a finite perturbation treatment with an INDO scheme applied to the Fermi contact term only and obtained good agreement with experiment. They speculate, with some evidence, that deviations for certain substituents may be traced to effects of d-orbital participation. They also give a clear discussion in terms of parameters broadly describable as hybridization and substituent electronegativity. Beer and Grinter[146] have also used the Finite perturbation INDO method to calculate $^1J_{SiH}$ (Fermi contact term only), and find reasonable agreement with experiment, though the larger coupling constants for $SiHCl_3$ and SiH_2Cl_2 were underestimated. In a later paper[147] they extend their work to $^1J_{SiC}$, also calculating

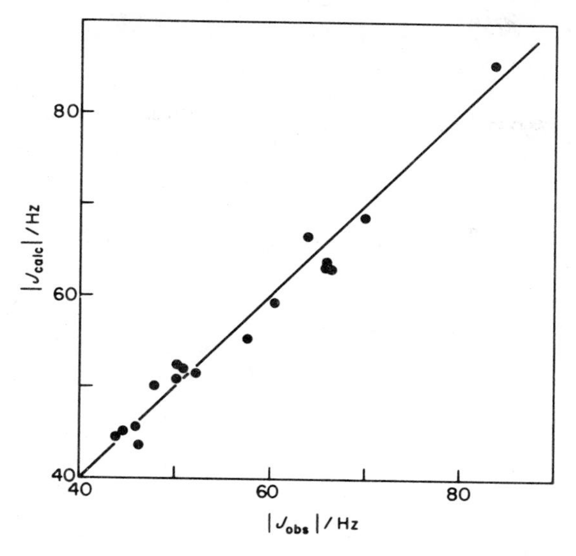

Fig. 10.4 Plot of |$^1J_{SiC}$|, as calculated by Beer and Grinter,[147] vs. the experimentally observed values.

the orbital and spin-dipolar contributions. Their best set of results (see Fig. 10.4) conformed to the equation:

$$^1J_{SiC}(\text{obs})/\text{Hz} = 0.983\ ^1J_{SiC}(\text{calc})/\text{Hz} + 1.24$$

with a correlation coefficient of 0.985 for 17 pieces of data. The orbital and spin-dipolar terms (particularly their variation) can be neglected, especially as they tend to be mutually cancelling. The authors[147] also comment that there is no need to introduce silicon *d*-orbitals. It should be noted that none of the compounds discussed involved multiple substitution by electronegative substituents.

Kovačevič and co-workers[148] have extended their maximum overlap approximation calculations to cases involving $^1J_{SiH}$, with some success, but the range of experimental values considered was rather small. For earlier theoretical work, particularly on $^1J_{SiH}$, the reader is referred to the review by Schraml and Bellama.[8] Much of this work was directed to explaining or parameterising the non-additivity of substituent effects.

10A.5 Spin-lattice Relaxation

(*i*) *Introduction.* Silicon-29 relaxation times have been measured for a number of systems. However, most of the work has been on small molecules, so that the results may not be entirely representative. Pioneering studies in this area were carried[16,39,149,150] out in the General Electric Company's laboratories and the results have been summarised (as at 1973) by Levy *et al.*[150]

The results discussed below show that the dominant relaxation mechanisms for ^{29}Si nuclei in solution (at least for small molecules) are intramolecular (Si, H) dipole-dipole and spin-rotation, together with effects from paramagnetic species where appropriate. There is also evidence[150] for small contributions from the intermolecular DD mechanism and from SA. No definitive examples of relaxation by the SC mechanism have been reported. Since $\gamma(^{29}Si) \sim \frac{4}{5}\gamma(^{13}C)$ and the covalent radius of Si is appreciably greater than that of C, values of $T_{1dd}(^{29}Si)$ may be expected to be ca. a factor of 10 longer than those of $T_{1dd}(^{13}C)$ in cases where there are direct Si—H or C—H bonds. Other factors, such as molecular mobility, will also be important. Moreover, direct Si—H bonds are less common in silicon chemistry than direct C—H bonds for organic chemistry. On the other hand, when comparing analogous systems the increased size of Si over C will have less effect in the direct bonds to hydrogen. Nevertheless relatively long values of $T_{1dd}(^{29}Si)$ are to be expected, and this helps to explain the greater importance of T_{1sr} for ^{29}Si than for ^{13}C.

The influence of the medium (other than the effect of paramagnetic additives) on $T_1(^{29}Si)$ has not been adequately investigated, and it may be that some of the anomalies in the results discussed below arise because of medium effects. Harris and Kimber[13] have noted a 25 % change in $T_1(^{29}Si)$ for hexamethyldisiloxane on halving its concentration in benzene.

(ii) Techniques. Most of the results in the literature have been obtained using the inversion-recovery method, with or without the Freeman-Hill modification, under conditions of proton noise decoupling. However, the null-signal problem, arising from the fact that γ_{Si} is negative, renders this method inapplicable in some cases, so that gated decoupling techniques must be used. Levy gated the decoupler "on" only during data acquisition but this method may suffer from non-exponential effects (though in practice these appear to be minimal). The method of NOE growth (see ch. 2), applied by Kimber and Harris[151] to this problem, does not suffer from this disadvantage.

As for other nuclei, determination of the mechanism of relaxation relies heavily on the measurement of the nuclear Overhauser effect to separate (usually) (^{29}Si, ^{1}H) dipolar contributions from those of other mechanisms. Since $\gamma_H/\gamma_{Si} = -5.04$, such measurements will only give accurate results for both T_{1dd} and T_{1o} when the nuclear Overhauser enhancement, η, lies in the range $2.0 > |\eta| > 0.5$. Outside this range, results for one or other of the two parameters must be treated with caution, as must all values of T_1 greater than ca. 80 s. Relaxation rates due to (Si, H) dipolar interactions are exactly in balance with those due to other mechanisms when $\eta = -1.26$. Other information about mechanisms of silicon-29 relaxation has been obtained by variation of temperature[150,152] and by deuteration studies.[150,153] In addition to the direct measurement of $T_1(^{29}Si)$, some results have been obtained[154,155] by observation of proton relaxation for isotopomers containing ^{28}Si and ^{29}Si.

(iii) Effect of Molecular Size. Spin-lattice relaxation times depend heavily on the correlation time for the relevant molecular motion. Dipole-dipole contributions to

TABLE 10.20 Silicon-29 relaxation data[68] for trialkylsilanes, R_3SiH

R	T_1/s	η	T_{1dd}/s	T_{1o}/s
Me	15.9	−0.21	[191][a]	17.4
Et	42.5	−0.67	160	57.9
Pr^n	45.4	−1.87	61.2	176
Bu^n	32.3	−2.34	34.8	[452][a]
n-hexyl	11.7	−2.42	12.2	[295][a]

[a] Values in square brackets in this and the following seven tables are liable to large errors.

TABLE 10.21 Silicon-29 relaxation data[150] for cyclic polydimethylsiloxanes, $[Me_2SiO]_n$

n	T_1/s	η	T_{1dd}/s	T_{1o}/s
3	80	1.1	183	142
4	100	1.83	138	365
6	91	2.24	102	[820]

T_1^{-1} therefore generally increase as molecular size (or, more strictly, local mobility) increases. However, the reverse is generally true for the spin-rotation mechanism. For silicon-29 this means that size change can readily cause the dominant mechanism to change. This is well illustrated for trialkylsilanes (Table 10.20) and cyclosiloxanes (Table 10.21). For each series it can be seen that for the larger molecules the DD mechanism predominates whereas for the smaller ones "other mechanisms" (presumably mainly SR) are more important. This is indicated by the steady increase of $|\eta|$ with molecular size. The changeover is rather abrupt, so that for ambient probe temperature few molecules have nuclear Overhauser enhancements of ca. -1.26 (the condition for $T_{1dd} = T_{1o}$). This also implies that the null signal situation is relatively rare, though substantially diminished intensities are not uncommon. The values for T_1(total) for the series of compounds listed in Tables 10.20 and 10.21 show maxima, in qualitative agreement with Eq. 3.65, the Hubbard relationship.[156] However, the data cannot be used to check Eq. 3.65 quantitatively for several reasons, including the fact that only in few cases are both T_{1dd} and T_{1sr} well-defined.

(*iv*) *Effect of Temperature*. As with the effect of molecular size, temperature variation changes T_{1dd} and T_{1sr} in opposing directions (the basic influence being that of molecular mobility in the region of the ^{29}Si nucleus). This has been demonstrated for tetramethylsilane (Table 10.22) and for tri-*n*-propylsilane (Fig. 10.5). In both cases the Hubbard relationship[156] seems to be well obeyed, with the slopes of the graphs of ln T_{1dd} and ln T_{1sr} vs. T^{-1} being approximately equal in magnitude but opposite in sign. The "null signal" situation is reached[149,150] for Me_4Si at -62.5°C. Levy *et al.* derive 7.7 kJ mol^{-1} for the "activation energy" for molecular rotation of Me_4Si. The experiments on Me_4Si and Pr^n_3SiH provide the best evidence for the importance of the SR mechanism for ^{29}Si. Variable temperature $T_1(^{29}Si)$ data are also available[150] for $(EtO)_4Si$, $(Me_2SiO)_6$ and for

TABLE 10.22 Variable temperature silicon-29 relaxation data[150] for tetramethylsilane

T/°C	T_1/s	η	T_{1dd}/s	T_{1o}/s
+25	19	−0.09	[522]	19.7
0	23.3	−0.205	286	25.4
−20	32.5	−0.35	234	37.8
−50	42.0	−0.495	213.8	52
−62.5	55	−1.0	139	91
−83	37	−1.59	58.6	100

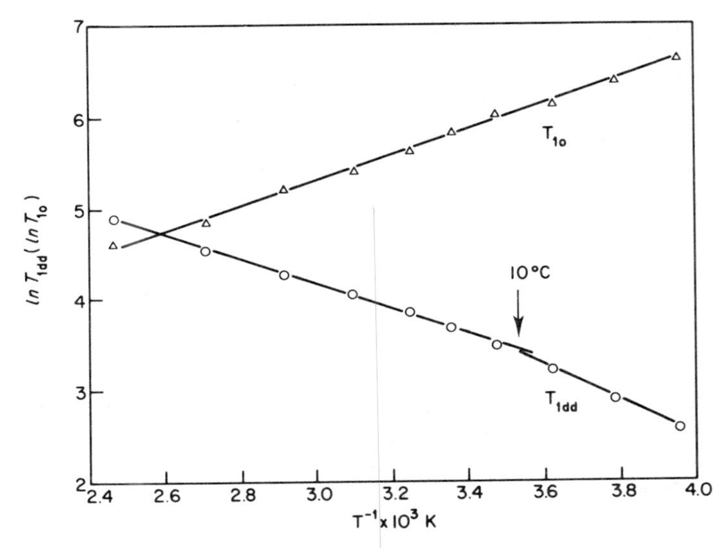

Fig. 10.5 The variation of T_{1dd} and T_{1o} with temperature for the ^{29}Si nucleus in tri-*n*-propylsilane (data for a sample containing 30% $(CDCl_2)_2$ at 19.87 MHz from ref. 152). Values of T_{1o} below 40°C are unlikely to be accurate. Activation energies of 11–12 kJ mol^{-1} are indicated by both T_{1o} and T_{1dd}, though there may be an unexplained deviation in the latter case below about 10°C.

the polysiloxane of average composition $MD_{50}M$. In these cases, however, T_{1o} exhibits a complex dependence on temperature, which is consistent with the shielding anisotropy relaxation mechanism becoming more important than SR at temperatures below ambient (though the dominant mechanism at such temperatures is DD).

(*v*) *The Influence of Directly-bonded Hydrogens.* Since the (Si, H) dipolar interactions provide a very important relaxation mechanism for most cases, the existence or otherwise of one or more direct Si-H bonds is likely to have a profound effect on $T_1(^{29}Si)$. The magnitude of T_1 will be smaller in the presence of directly-bonded hydrogens, and domination of the DD mechanism is more likely. Such effects are illustrated by the data in Table 10.23 for MM^H, the cyclic compound $III(D_3D^H)$ and for the polysiloxane $MH^H_{50}M$ (compare $MD_{50}M$, with[150] $T_1 = 46s, \eta = -2.16, T_{1dd} = 53.7s$ and $T_{1o} = [322]s$ for the main-chain D units). However, comparisons between molecules are complicated by the effects of changes in molecular size, e.g. for the series $M^H_2 \rightarrow MM^H \rightarrow M_2$. In the case of the phenylsilanes[32] replacement of SiH by SiPh (between Ph_2SiH_2 and Ph_3SiH) actually shortens T_1 substantially because of the lengthening of τ_c, and although $PhSiH_3$ has a high T_1^{-1}, the contribution of T_{1dd}^{-1} is very small (see below). The effect of incremental mass change is not so apparent for larger molecules, so that $T_1(^{29}Si)$ for silatrane (I, R=H) is substantially shorter[105] than that for phenylsilatrane ($T_1 \approx 70$–100 s). Consideration of all the known data for compounds containing Si-H bonds (Tables 10.20, 10.23 and the values for MD^H_5M given below) shows that, due to the generally compensatory effects of the DD and SR mechanisms, the total $T_1(^{29}Si)$ lies between 5 s and 50 s at ambient probe temperature.

(*vi*) *Long-range and Intermolecular* (*Si, H*) *Interactions.* Systems without direct Si-H bonds are clearly likely to have longer values for $T_1(^{29}Si)$. This can be seen from Tables 10.20 and 10.21, which show that the maximum T_1 for the D_n series is over twice as long as

TABLE 10.23 Silicon-29 relaxation data for some compounds containing a direct Si-H bond

Compound		T_1/s	η	T_{1dd}/s	T_{1o}/s	Ref.
M_2^H		26.0	−0.35	[187]	30.2	68
MM^H	M^H	34	−0.56	150	50	150
	M	43	−0.35	[310]	44	
M_2		42.5	−0.34	[314]	49.2	13
D_3D^H(III)	D^H A	31	−1.8	43.4	108	150
	D B	66	−1.45	115	155	
	D C	67	−1.40	121	151	
$MD^H_{50}M$[a]		26.1	−2.16	30.5	[181]	17
$PhSiH_3$		5.5	−0.07	[200]	5.6	32
Ph_2SiH_2		26.5	−2.13	31	[170]	
Ph_3SiH		12.0	−2.41	12.5	[275]	
$(PhCH_2)_3SiH$		19.3	−2.43	20.0	[540]	68
silatrane(I,R=H)		23.2	−2.21	26.5	[190]	105
$PhMe_2SiH$		46	−0.62	187	61	150
Ph_2MeSiH		35.4	−2.27	39.3	[341]	32

[a] The data are for the main-chain D^H units.

that for the series R_3SiH. However, such effects do not necessarily mean that the DD mechanism loses its dominance, as can be seen by the data in Table 10.24, which shows that $|\eta|$ actually increases[19] as n decreases for the series $Me_nSi(OEt)_{4-n}$ in spite of the progressive replacement of two-bond (Si, H) interactions by three-bond ones (which causes T_1 itself to progressively lengthen). This is, in fact, further evidence that for small molecules it is the SR mechanism which competes with DD interactions. For systems such as $Si(OEt)_4$, however, with $T_1 > 100$ s, it is possible that *inter*molecular (Si, H) dipolar interactions become relatively important. Levy *et al.* have demonstrated[150] this to be true for Si_2Cl_6($T_1 \approx 115$ s) by measuring the Si-{^{1}H} intermolecular NOE for a solution in benzene ($\eta = -0.36$ at 10°C) and showing that it is absent for a solution in benzene-d_6. These authors suggest[150] that T_{1dd} (intermolecular) may frequently be in the range 700–1000 s. If this is true, such a contribution would still be

TABLE 10.24 Silicon-29 relaxation data[19] for methylethoxysilanes $Me_nSi(OEt)_{4-n}$

n	T_1/s	η	T_{1dd}/s	T_{1o}/s
4	19.8	0.0	[a]	19.8
3	37.1	−0.23	[406]	40.8
2	70.5	−0.62	287	93.5
1	92.6	−1.00	233	153.5
0	145.0	−1.49	245	345

[a] Indeterminately large.

TABLE 10.25 Silicon-29 relaxation data[13, 43] for trimethylsilyl compounds,[a] Me_3SiX

X	H	F	Me	Cl	OEt	Br	$OSiMe_3$
T_{1dd}/s	[190]	[798]	[b]	[571]	[406]	[486]	[314]
T_{1o}/s	17.5	19.5	19.8	28.6	40.8	33.1	49.2

X	I	SMe	$SiMe_3$	Ph	$N(SiMe_3)_2$	$OSi(OSiMe_3)_3$
T_{1dd}/s	[439]	[360]	[434]	179	187	173
T_{1o}/s	55.2	51	50.2	55	123	64.0

[a] Listed in the order of their boiling points. [b] Indeterminate.

unimportant even for Ph_2SiCl_2 ($T_1 = 126.2$ s, $\eta = -1.28$, $T_{1dd} = 248$ s, $T_{1o} = 255$ s),[32] which has no hydrogen atoms nearer Si than three bonds.

(*vii*) *Trimethylsilyl Compounds.* A substantial series of measurements has been made[13,43] on compounds of the type Me_3SiX (see Table 10.25). The results show that for reasonably small molecules of this type T_{1dd} is >150 s, and that relaxation is dominated by "other" mechanisms. Since T_{1o} shows a pronounced inverse dependence on the molecular volatility, it is reasonable to suppose that it is dominated by SR, as is confirmed by the variable temperature measurements mentioned above. Within closely-related series of compounds, e.g. the trimethylsilyl halides, the variations can be seen as a smooth dependence on substituent mass. It may be noted that, with the exception of $N(SiMe_3)_3$ ($T_{1o} = 123$ s), the values for T_{1o} lie in the range 17–64 s, and T_1(total) is in the range 15–50 s (74.3 s for $N(SiMe_3)_3$).

Relaxation measurements have also been reported[17,150,157] for trimethylsiloxy groups in siloxane oligomers (see below), with $T_1(^{29}Si)$ ranging from 35 to 56 s (dominated by SR interactions), and for trimethylsilated sugars (pyranosides).[158] In the latter case T_1 ranges from 23 to 50 s and η from -1.7 to -2.3 for the instances measured so far. Thus the DD mechanism dominates for such compounds, in contrast to the results discussed above, presumably because of molecular size and relative rigidity. There is no clear distinction between results for CH_2OSiMe_3 sidechains and for $OSiMe_3$ groups bonded directly to ring-carbon atoms of the pyranosides.

(*viii*) *Segmental Motion in Silicones.* The selectivity inherent in the FT method may be used to derive separate T_1 values for different silicons in the same molecule, as for the corresponding ^{13}C case. This is particularly valuable for siloxane (i.e. silicone) oligomers. Levy *et al.* have reported[150] data for a number of systems of the type MD_nM. Work has also been carried out[157] on MD_5^HM. All the results for T_{1dd} and T_{1o} are given in Table 10.26. For the dimethylsiloxy (D) units T_1(total) varies from 54 to 82 s and $|\eta|$ lies between 0.55 and 2.0 in the cases reported. The data in Table 10.26 show that, as expected, increasing the length of the oligomer chain causes a decrease in T_{1dd} and an increase in T_{1o} for a given unit. The value of $|\eta|$ therefore increases and the dominant mechanism changes for D(1) and D(2) units, in the manner discussed above. For the end-stopper units (M) and for D(1) units, T_1 itself goes through a maximum value. There are some indications that it increases again beyond $n = 6$, but the variation is within experimental error. The fact that for the M units the maximum value of T_1 is only ca. 50 s indicates the efficiency of the spin-rotation contribution for trimethylsiloxy groups

TABLE 10.26 Silicon-29 relaxation times[150,157] for siloxane oligomers

Compound	T_{1dd}/s					T_{1o}/s				
	M	D(1)	D(2)	D(3)	D(4)	M	D(1)	D(2)	D(3)	D(4)
MDM	[245]	247				41	69			
MD_2M	181	164				43	149			
MD_3M	203	188	162			59	146	147		
MD_6M	151	101	73	78		58	175	224	240	
MD_8M	158	112	84	69	69	61	166	210	266	266
MD^H_5M	265	65.4	57.7	47.7		70.8	209	217	[226]	

(see above). The behaviour of T_{1dd} and T_{1o} for the different silicon atoms of the same oligomer is of considerably greater interest. The value of T_{1dd} increases towards the end-groups, whereas T_{1o} (presumably dominated by T_{1sr}) decreases. It is clear that relaxation is dominated by segmental motion of the chain, and the T_1 data therefore give information about the mobility at different points in the chain. Since all the values of T_1 are relatively high, it is clear that the systems are very mobile. This is confirmed by the data for the main chain units of $MD_{50}M$ and $MD^H_{50}M$ (see above). This implies that the microviscosity bears little relation to the macroscopic viscosity. The T_{1dd} values for the D^H units of MD^H_5M are shorter than the corresponding results for MD_3M or MD_6M, presumably due to the direct Si-H bond of the D^H unit (only partially compensated by increased mobility). However, the data for the M unit and all the T_{1o} values for MD^H_5M are not so readily rationalised in relation to the MD_nM results.

No ^{29}Si relaxation data appears to have been reported yet for cross-linked polysiloxanes, which might be expected to have relatively short $T_1(^{29}Si)$ values. However, some values of η have been published[159] for silicone oligomers, including a few with T and Q units.

(ix) *Effects of Paramagnetic Additives.* Since ^{29}Si relaxation times are often longer than desirable from an experimental FT point of view it is often necessary to reduce them in order to improve S/N. The easiest way to do this is to add small amounts of paramagnetic substances. Such action also affects the NOE by decreasing T_1/T_{1dd}, which may be helpful if a null-signal situation is encountered. Consequently, some effort has been put[150,160] into quantitatively examining the effect of paramagnetic substances on $T_1(^{29}Si)$.

Thus the effect of dissolved oxygen has been examined.[150,160] Levy *et al.*[150] give data on T_{1e} at the contribution to T_1 from interactions with the unpaired electrons on O_2, defined for a partial pressure of 1 atm over the solution. The results are in the range 30–40 s (with one at 52 s). These are, as expected, somewhat longer than the corresponding data for ^{13}C (if only the nuclear dipole-electron dipole is effective $T_{1e}(^{13}C)$ should be shorter than $T_{1e}(^{29}Si)$ by a factor of 1.6). However, they suffice to materially affect observed ^{29}Si relaxation times for undegassed solutions, though it is clear that O_2 will not usually be efficient enough as a relaxation agent.

The commonest shiftless relaxation agents for use in ^{29}Si NMR are the acetylacetonates of Cr^{III} and Fe^{III}. Kimber has reported[160] detailed $T_1(^{29}Si)$ measurements for the effects of $Cr(acac)_3$ and $Fe(acac)_3$ on M_2 and D_4 with the paramagnetics in the concentration range up to 0.06 mol l^{-1}. He also gives data for $Cr(acac)_3$ acting on a siloxane polymer of average composition $MD^H_{50}M$. Since $\eta \rightarrow 0$ at high paramagnetic concentrations, he concludes there is no "three spin effect," i.e. relaxation by the paramagnetic via the protons in the system. The Fe^{III} ion induces relaxation more efficiently than Cr^{III}, as expected since it has a larger magnetic moment. However, the broadening effects are similar, so that $Fe(acac)_3$ can be considered as a better shiftless relaxation reagent than $Cr(acac)_3$. Kimber has presented[160] approximate data, derived from the T_1 measurements, for the distances between the paramagnetic centre and the relevant silicon atom in the weak complexes; these are in the range 0.45–0.65 nm. Information about the effects of $Cr(acac)_3$ on $T_1(^{29}Si)$ is also given by Levy *et al.*[150] It should be noted[160,161] that values of T_1 for different silicon nuclei are not necessarily reduced to a common level, i.e. interactions with the relaxation reagent can occur preferentially even for different sites in the same compound.

(*x*) *Silicate Solutions.* Engelhardt[153] has reported ^{29}Si relaxation times for some of the anionic species present in solutions of sodium silicate in the presence of excess NaOH. These times are very short, ca. 0.5 to 5 s. More detailed studies have been carried out by Harris and Newman[12] on solutions of sodium and potassium metasilicate (i.e. alkali:silica ratios of 2:1). Short values of T_1 were again found, but it was remarked that the results depended on the history of the solution; in particular T_1's decrease with ageing of the solution in the pyrex NMR tube. Moreover values of $|\eta|$ also decrease with ageing. Consequently it appears that leaching of paramagnetic ions from the pyrex may have a pronounced effect. Data have been presented[12] for solutions to which Mn^{II} or Cr^{III} have been deliberately added, but the exact causes of relaxation in the solutions without such additives have yet to be determined. However, it is clear that the different silicate species and environments in the solutions differ in their relaxation times. Moreover it is shown that DD contributions are significant (though, so far, in no case dominant) and have been found to be in the range 4–9 s when the gegenion ion is Na^+ but appreciably longer (ca. 25 s) for the less viscous K^+ solutions.

(*xi*) *Special Situations.* Two of the compounds listed in Table 10.25 merit additional mention because in principal there are other relaxation mechanisms that could come into play. The data,[13] however, suggest that (Si, F) dipolar interactions are not of great importance for Me_3SiF, and a direct measurement of η for a ^{29}Si-$\{^{19}F\}$ experiment confirmed this (η was found to be only -0.06). In the case of Me_3SiI (^{29}Si, ^{127}I) scalar interactions could contribute to ^{29}Si relaxation, since the resonance frequencies for these nuclei differ by only ca. 150 kHz at the magnetic field used. However, the data for this molecule[13] do not appear to be unusual, so this mechanism is probably not important for Me_3SiI.

Phenylsilane is certainly a special case. Apart from the silicate solutions, this molecule[32,162] has the lowest ^{29}Si T_1 yet measured in the absence of paramagnetic species. Moreover, the DD contribution to T_1^{-1} is negligible. The low value of T_1 is in fact believed to be due to interactions between the nuclear spin and *internal* rotation about the C-Si bond. The effect is large because the barrier (which is necessarily six-fold) to internal rotation is low (variously estimated[163,164] to be 146* J mol^{-1} or 74 J mol^{-1}), as for the analogous molecule toluene. Measurements of ^{13}C relaxation times[32] for $PhSiH_3$ serve to reinforce the above conclusions.

10A.6 Physical States other than Solutions

Since it appears to be likely that chemists will increasingly extend their usage of NMR beyond the confines of the isotropic solution state, it is worth summarizing some of the

* Incorrectly quoted as 8.5 *J* mol^{-1} in ref. 32.

experiments already reported for ^{29}Si resonance. These experiments do not yet form a coherent pattern.

(*i*) *Gas-Phase Work*. Johannesen, Brinckman and Coyle[2] have reported that the silicon-29 nucleus in SiF_4 is 1.1 ppm less shielded for the gas at 30 atm. pressure than for a 15 mol % solution in CCl_3F. Moreover, studies using ^{19}F and ^{1}H NMR have shown[165] that $^{1}J_{SiF}$ in SiF_4 and $^{1}J_{SiH}$ in SiH_4 are pressure-dependent, the former being an order of magnitude more sensitive than the latter.

(*ii*) *Glasses, Gums and Gels*. Silicon chemistry is replete with non-crystalline "solid" states, particularly for siloxanes, and it is therefore of some considerable interest to discover to what extent these are amenable to investigation by normal FT NMR instrumentation. The first point of note is that ordinary pyrex glass, as present in NMR inserts and tubes, gives[15] a broad (linewidth ≈ 700 Hz) ^{29}Si signal at ca. $\delta \approx -110$ ppm. The spin-lattice relaxation time can be quite short,[15] e.g. ca. 16 s, presumably because of the presence of paramagnetic impurities (probably involving spin diffusion among the ^{129}Si—selective irradiation experiments show[107] the line broadening is homogeneous). A typical NMR tube appears to contain ca. 180 ppm of Fe (as determined by atomic absorption techniques).

Sodium silicate glasses were studied in some detail, using CW ^{29}Si NMR (absorption, dispersion and adiabatic rapid passage methods) by Mosel *et al.*,[166] who found that spin-lattice relaxation times were ca. 10–30 ms when the glass contained 0.1 mol % of Fe_2O_3, but were 10–85 s in other cases for which Fe_2O_3 was essentially absent. Linewidth data were also listed.

Amorphous silica, of the low-density and high surface area type ("Aerosil") which finds many useful applications in chemical industry, also gives rise[15] to an observable ^{29}Si signal (see Fig. 10.6), which is barely distinguishable from that of pyrex glass. The band of the aerosil, corrected for that of the glass, appears to be ca. 350 Hz wide.

High molecular weight polydimethylsiloxanes form gum rubbers, and it is of interest to monitor the molecular motion in such systems by measuring the spin-lattice relaxation time. Levy and co-workers report[150] $T_1 = 27$ s and $|\eta| = 1.7$ for the D units of a sample

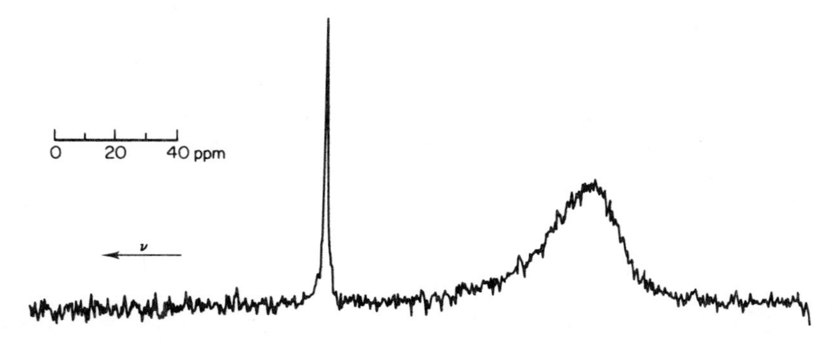

Fig. 10.6 Silicon-29 NMR spectrum of 50 cSt polydimethylsiloxane (PMS) (average number of D units 33 molecule^{-1}) adsorbed onto Aerosil 200 silica. This shows a sharp peak at high frequency due to the main-chain D units of the adsorbed PMS, and a broad band at lower frequency due to the silica of the NMR insert, tube and Aerosil together. Benzene was added to the sample to improve the magnetic field homogeneity. Gated noise decoupling of the protons (eliminating the NOE) was employed. Spectrometer conditions: 19.87 MHz, $T_{ac} = 0.8$ s $T_d = 19.2$ s, 2386 pulses at 90°. The total time taken was ca. 13h.

with chain-length of ca. 3400, indicating negligible restriction of motion. Even cross-linking does not ncessarily reduce T_1 for the observable resonances. Thus curing an OH-ended dimethylsiloxane (ca. 550 units, viscosity 3000 cSt) with ca. 6% of its weight of poly(ethoxysilane) of approximate composition $EtO[Si(OEt)_2O]_4Et$ gave[69] no observable change in $T_{1dd}(42 \pm 4$ s) for the D-unit resonance, though the resulting material was a solid rubber. It is probable that the extreme narrowing situation still holds for the observable ^{29}Si sites in these systems.

(*iii*) *Crystalline Solids.* One of the earliest papers on ^{29}Si NMR, by Holzman *et al.*[167] gave, inter alia, chemical shifts for cristobalite, quartz (from three sources), hyper-silicon and both green and black silicon carbides. The dispersion mode (CW) was used for these experiments, usually with high r.f. field intensity, and this technique offers some advantages, even over TF operation, so that the work has not yet been reproduced using modern equipment. The most interest result was the discovery of a 48 ppm difference in chemical shift between the green and black SiC.

The dipolar decoupling and cross-polarization techniques pioneered by Waugh and his colleagues[168] (see ref. 169) show promise of revolutionising NMR as applied to solids. One paper has appeared[170] using ^{29} Si for such studies of polycrystalline materials. This reports the three principal components of the shielding tensor (relative to the shielding for solid Me_4Si, which is isotropic) for 8 compounds containing Me_3Si—, $Me_2Si<$ or $MeSi\lessgtr$ groups. The differences between the largest and smallest components vary from 26 to 70 ppm. The deviation of the mean of the shielding components from the value for the melt ranges up to 5.5 ppm. A number of other interesting points emerge, and clearly there is a need for more detailed investigations. So far no analogous work on single crystals using ^{29}Si has been reported, nor has the auxiliary use of the powerful magic-angle-spinning method.[169,171] Of course, the cross-polarisation method requires the proximity to the ^{29}Si of abundant spin species such as ^1H, so that it would not be readily applicable to the solid materials studied by Holzman *et al.*[167] However, amorphous silica of high surface area contains enough protons (in water of hydration) to give successful cross-polarization experiments.[22]

(*iv*) *Adsorbed Species.* It has been shown[15] that when polydimethylsiloxane (PMS) is absorbed on to Aerosil 200 silica a relatively sharp ^{29}Si peak can be observed for the main-chain D units (see Fig. 10.6). The linewidth is 10 to 40 Hz, depending on circumstances, but the spin-lattice relaxation time (measured for the case of 1000 cSt PMS) is high (20 s for the dry material, 38 s when benzene is added), being of the same order as for the neat liquid siloxane. The nuclear Overhauser enhancement, $\eta = -1.7$, is unchanged (within experimental error) by adsorption. Clearly, adsorption does not prevent flexing of the main-chain units, probably in loops between sites of adsorption. The observed signal presumably does not involve the silicon atoms near the point or points of attachment of the PMS to the Aerosil.

10B GERMANIUM-73

The majority of germanium compounds are expected to give very broad ^{73}Ge resonance lines as a result of quadrupolar broadening, and this, together with the low receptivity, has so far restricted studies to species in which the electric field gradient might be expected to be zero or small. With one exception,[172] all chemical shifts for ^{73}Ge have been obtained by direct observation using either CW[173] or pulsed FT techniques,[174] and the available results are collected in Table 10.27. Linewidths are *ca.* 2 Hz for the tetrahalides, but only 0.58 Hz in Me_4Ge (for which fine structure due to coupling to the protons was

TABLE 10.27 Germanium chemical shifts[a]

No.	Species	$\delta(^{73}Ge)^b$/ppm	Ref.	No.	Species	$\delta(^{73}Ge)^b$/ppm	Ref.
1	Me_4Ge	0	174	11	$GeCl_2Br_2$	-130 ± 4^e	173
2	Et_4Ge	$+18.1 \pm 0.7$	174	12	$GeClBr_3$	-219 ± 4^e	173
3	Pr^n_4Ge	$+2.1 \pm 0.8$	174	13	$GeCl_3I$	-213 ± 4^f	173
4	Bu^n_4Ge	$+5.5 \pm 0.8$	174	14	$GeCl_2I_2$	-518 ± 4^f	173
5	$Ge(OMe)_4$	-36.0 ± 0.7	174	15	$GeClI_3$	-808 ± 4^f	173
6	$Ge(SMe)_4$	$+153 \pm 3^c$	172	16	$GeBr_3I$	-513 ± 4^g	173
7	$GeCl_4$	$+30.9 \pm 0.5^d$	173, 174	17	$GeBr_2I_2$	-708 ± 4^g	173
8	$GeBr_4$	-312.1 ± 0.8^d	173, 174	18	$GeBrI_3$	-901 ± 4^g	173
9a	GeI_4	-1080.7 ± 1	174	19	$GeCl_2BrI$	-316 ± 4^h	173
9b	GeI_4	-1108.1 ± 1	174	20	$GeClBr_2I$	-407 ± 4^h	173
10	$GeCl_3Br$	-47 ± 4^e	173	21	$GeClBrI_2$	-601 ± 4^h	173

[a] Solution conditions: Entries 1–5, 7, 8, 10–12—neat liquid; entry 6—50% v/v in in CH_2Cl_2; entries 9a, 13–21—CS_2 solution; entry 9b—benzene solution. [b] To high frequency of Me_4Ge. [c] $\Xi(^{73}Ge) = 3.488850$ MHz by $^1H - \{^{73}Ge\}$ double resonance. δ derived assuming $\Xi(^{73}Ge) = 3.488423$ MHz in $GeCl_4$. [d] No change on dilution in CS_2 (ref. 173). [e] Present in 1:1 mixture of $GeCl_4$ and $GeBr_4$. [f] Present in 1:1 mixture of $GeCl_4$ and GeI_4. [g] Present in 1:1 mixture of $GeBr_4$ and GeI_4. [h] Present in 1:1:1 mixture of $GeCl_4$, $GeBr_4$ and GeI_4.

clearly resolved[174]). In other cases the linewidths are less than 25 Hz, being dominated by unresolved coupling to protons in the tetraalkyl derivatives[174] and (probably) by magnetic field inhomogeneity[173] for the mixed halides. The failure[174] of an attempt to observe a resonance from $MeGeCl_3$ was attributed to rapid quadrupolar relaxation.

The trends in ^{73}Ge chemical shifts are similar to those found for ^{119}Sn (*vide infra*). Thus an approximately additive relation is found[173] for the tetrahalides and mixed halides although this is improved by the inclusion of "pairwise" additivity parameters. CNDO/2 calculations have shown[175] that charge densities at the germanium atom change in a parallel manner in these species, and it has therefore been suggested that there is a direct relation between charge density and the germanium shielding. However, this clearly does not explain the chemical shift of Me_4Ge, and it should be noted that the increased shielding of many nuclei (and especially tin) brought about by the heavier halogens has also been attributed to other causes. The other compounds examined gave the results expected from a comparison with data for ^{119}Sn, although the value of $\delta(^{73}Ge)$ for Et_4Ge is perhaps larger than could be expected from comparison with those for Pr^n_4Ge and Bu^n_4Ge. The ^{73}Ge resonance frequency in $GeCl_4$ has been compared with those of a number of other nuclei,[176–179] and from the results it is possible to deduce that $\Xi(^{73}Ge) = 3488315 \pm 10$ Hz in Me_4Ge, which we propose as a suitable reference material for ^{73}Ge chemical shifts.

Spin coupling involving ^{73}Ge has been observed only in species in which the germanium atom is in an environment of tetrahedral symmetry, the results being:

$^1J(^{73}GeH) = -97.6$ Hz in GeH_4;[180] $^1J(^{73}Ge^{13}C) = -18.7$ Hz in Me_4Ge;[181]

$^1J(^{73}Ge^{19}F) = +178.5$ Hz in GeF_4;[182] $^2J(^{73}Ge, H) = +2.99$ Hz in Me_4Ge;[174]

$^3J(^{73}Ge, H) = -1.9$ Hz* in $Ge(OMe)_4$;[174] and $^3J(^{73}Ge, H) = -2.5$ Hz* in $Ge(SMe)_4$.[172]

* Estimated by bandshape fitting.

In all cases we have based the signs of these coupling constants upon comparison with those in compounds of other group IV elements rather than upon double resonance experiments, and it should be noted that since $\gamma(^{73}Ge) < 0$ the corresponding reduced coupling constants will have the opposite signs.

Nothing seems to have been done on T_1 for ^{73}Ge, but T_2 has been measured both from linewidths and by spin-echo methods[174] for compounds [1]–[5] and [7]–[9] of Table 10.27. Values ranged from 0.03 s (in [5]) to 0.74 s (in [1]).

Thus whilst a chemical shift range of at least 1200 ppm makes ^{73}Ge attractive for chemical purposes the difficulties associated with its study will probably militate against its extensive use.

10C TIN

10C.1 General

Three of the naturally occurring isotopes of tin have $I = \frac{1}{2}$ (the others all have $I = 0$) and ^{119}Sn is the one usually studied since it is somewhat superior to ^{117}Sn with regard to abundance (8.6 against 7.6 %) and sensitivity. The remaining isotope, ^{115}Sn (abundance 0.35 %) is very seldom used.

Most values of coupling constants involving tin and ^{1}H, ^{13}C or ^{19}F have been obtained from the spectra of the latter nuclei which often exhibit a very characteristic appearance, as shown in Fig. 10.7. Each resonance is usually flanked by a double pair of $^{117/119}Sn$ satellites in which the ^{117}Sn–^{119}Sn spacing is 2.3 % of the relevant coupling to ^{119}Sn; this feature is specially valuable for the identification of such satellites under conditions of poor S/N. In second order proton and ^{19}F spectra the presence of $^{117/119}Sn$ may introduce sufficient magnetic asymmetry to permit partial first-order analysis.[183]

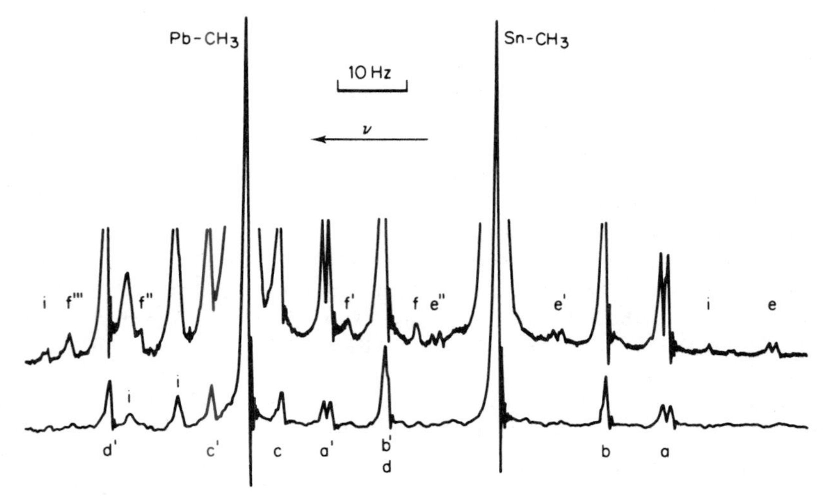

Fig. 10.7 60 MHz frequency-sweep proton spectrum of $Me_3SnPbMe_3$ in benzene solution, the upper trace being recorded at higher spectrometer gain. Lines marked i are due either to impurities or are ^{13}C satellites, and the other satellites arise from the following species: (a) and (c) $Me_3{}^{117/119}SnPbMe_3$; (b) and (d) $Me_3Sn^{207}PbMe_3$; (e) and (f) $Me_3{}^{117/119}Sn^{207}PbMe_3$. Reproduced with permission from ref. 224.

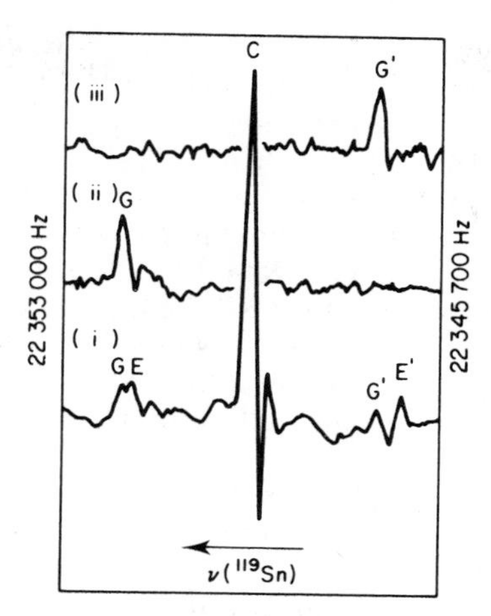

Fig. 10.8 ^{119}Sn INDOR spectra produced by collapsing the coupling $^{3}J(^{119}Sn, H)$ = ca. 2 Hz in $(Me_3Sn)_3Sn^-Li^+ \cdot 3THF$. The traces were obtained by accumulating 32 scans, each of 25 s duration: (*i*) central line in proton spectrum monitored, G′-G = $^{1}J(^{119}Sn, ^{117}Sn)$, E′-E = $^{1}J(^{119}Sn, ^{119}Sn)$, and the asymmetry is due to second order features; (*ii*) high-frequency ^{117}Sn satellite due to $^{2}J(^{117}Sn, H)$ monitored; (*iii*) low-frequency ^{117}Sn satellite monitored. Reproduced with permission from ref. 186.

The first studies of ^{119}Sn resonances were by direct observation using CW and (fairly) rapid passage techniques;[3,184] subsequently the ^{1}H-{^{119}Sn} indirect double resonance method[14,185] has proved valuable to the extent that the majority of over 700 tin chemical shifts now in the literature were measured in this way. In species with at least one methyl or *t*-butyl group attached to tin it is easy by this method to get results from solutions at 0.1 M concentration, and with the use[186] of spectral accumulation this sensitivity can be improved by two orders of magnitude or more. Fig. 10.8 shows a ^{119}Sn INDOR spectrum obtained in this way. Recently, proton decoupling allied to pulsed FT methods has been used[187,188] to give dramatic improvements in sensitivity for direct ^{119}Sn observation and this approach is especially advantageous when the ^{119}Sn spectrum is very complex as a result of coupling to many inequivalent protons, as commonly occurs when there are several organic groups bound to tin. The negative magnetogyric ratio of ^{119}Sn leads to nuclear Overhauser enhancement factors of from $\eta = 0$ to -1.34 in ^{119}Sn-{^{1}H} experiments, depending upon the relative importance of (^{1}H, ^{119}Sn) dipole-dipole relaxation compared to other ^{119}Sn relaxation contributions, and so the effect can never be beneficial with regard to signal intensities, although it may be useful for assignments. It is therefore desirable to use gated decoupling to suppress the NOE (although this may impose excessive time intervals between pulses) or to use a paramagnetic relaxation time reagent such as $[Cr(acac)_3]$.[188] These mild difficulties, however, are generally outweighed by the advantage of collapsing a multiplet (which might easily contain more than one hundred lines) into a singlet by proton decoupling. In the case of molecules with several tin atoms the ^{119}Sn resonance will be flanked by satellites arising from coupling between ^{119}Sn and $^{117/119}$Sn in doubly substituted species which have natural abundances of $<1\%$. It is important to realise that whilst species containing one ^{119}Sn and one ^{117}Sn nucleus will give first order ^{119}Sn spectra, those containing two ^{119}Sn nuclei will not, since $^{n}J(^{119}Sn, ^{119}Sn)$ will in general not be small compared with the chemical shift differences, especially if n is small. Indeed, in

$Me_3{}^{119}Sn^{119}SnMe_3$ for example, the two ^{119}Sn nuclei are magnetically equivalent under conditions of proton decoupling and $^1J(^{119}Sn^{119}Sn)$ will have no effect upon the appearance of the spectrum. This coupling constant can however be derived from $^1J(^{119}Sn^{117}Sn)$ in this molecule[187] or from $^1H\text{-}\{^{119}Sn\}$ double resonance experiments.[189] It is perhaps worth commenting that in view of the superior sensitivity to NMR detection of ^{119}Sn compared with ^{13}C it might prove more effective to measure values of $J(^{119}Sn, {}^{13}C)$ from ^{13}C satellites in ^{119}Sn FT spectra rather than vice versa, although the computer problems of dynamic range will be more serious. Studies of ^{119}Sn NMR have been essentially confined to chemical shifts and coupling constants and their interpretation in (*quasi*) theoretical terms, and their use in structural work. It is convenient to consider the chemical shift and coupling constant aspects separately (see below).

In $SnCl_4$ the dominant longitudinal relaxation mechanism of ^{119}Sn over a wide range of temperature is spin-rotation, and this is confirmed by an essential field-independence of T_1. In $SnBr_4$ and SnI_4, however, T_1 is strongly field-dependent and exhibits a minimum with increasing temperature owing to competing scalar (modulated by halogen relaxation) and spin-rotation interactions. On the other hand T_2 in all three halides is governed by scalar coupling to the halogen, and analysis of the data gives $|^1J(^{119}Sn^{35}Cl)| = 470$ Hz, $|^1J(^{119}Sn^{81}Br)| = 920$ Hz, and $|^1J(^{127}I^{119}Sn)| = 940$ Hz.[190,191] Both scalar and spin-rotation mechanisms also contribute[192] to liquid-state $T_1(^{119}Sn)$ values of SnI_3Cl and $SnCl_3I$, and evidence was obtained for a maximum in T_{1sc}^{-1} as a function of temperature, which arises when $2\pi T_1(^{127}I) \approx [\nu(^{119}Sn) - \nu(^{127}I)]^{-1}$. The investigation[192] yielded $|^1J(^{127}I, {}^{119}Sn)| = 1638$ Hz and 1097 Hz for $SnCl_3I$ and SnI_3Cl respectively, with $|J(^{119}Sn, {}^{35}Cl)| = 421$ Hz for SnI_3Cl. It also appears that spin-rotation is an important T_1 mechanism in organotin species, although the observation of a $^{119}Sn\text{-}\{^1H\}$ nuclear Overhauser enhancement of $\eta = -1.0$ in $Bu^n_6Sn_2$ indicates a ca. 75% contribution from dipole-dipole coupling to protons in this molecule at least.[188]

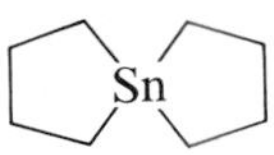

Sn IV

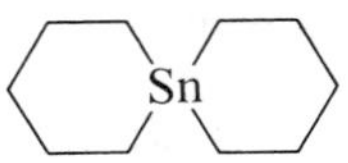

Sn V

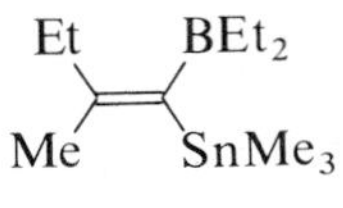

Et BEt$_2$ Me SnMe$_3$ VI

10C.2 Chemical Shifts

These are summarised in Tables 10.28–10.34, in which the entries have been selected to provide as wide a view as possible in the space available of the total data. Comprehensive lists of ^{119}Sn chemical shifts up to 1971[193] and 1973[194] have been published. With certain exceptions to be noted below, solvent and concentration effects upon the chemical shifts do not exceed a few ppm provided that there is no chemical interaction. Thus results obtained in such solvents as benzene, carbon tetrachloride, chloroform, or dichloromethane are directly comparable with those obtained on neat liquids, which is fortunate in view of the range and in some cases the uncertainty of the conditions used for many of the literature measurements. The exceptions arise when the tin atom has one or more electronegative substituents, and donor atoms are present (either in the solvent or the tin-containing molecule itself), for then reversible association or auto-association can occur to extents which will depend upon conditions. The tin atom thereby becomes five- or six-coordinate and there is a concomitant increase[3,184,195] in shielding, which may be as much as several hundred ppm and which will depend upon the position of the (generally rapid) dynamic equilibrium. This behaviour can be used to study these processes and is discussed in this light in a later section; solvents likely to coordinate to tin include the hard bases dimethyl sulphoxide (DMSO), hexamethylphosphortriamide (HMPT), and tertiary amines,[196,197] while among tin compounds the alkoxides are especially prone to auto-association.[33,198,199]

It has been claimed[200,201] that there is a significant $^{117/119}$Sn primary isotope shielding effect. That is, it has been suggested[200,201] that in general ^{117}Sn and ^{119}Sn chemical shifts when expressed in ppm can differ by up to several ppm, and the results of certain ^{1}H-{^{117}Sn} and ^{1}H{^{119}Sn} double resonance experiments are said to support this. It has not proved possible to reproduce these results in our own and other laboratories,[201] and it is noteworthy that the molecules which display the largest discrepancies are those for which proton coupling makes the tin spectrum very complex and so aggravates the difficulty of the task of determining $\nu(^{117/119}\mathrm{Sn})$ with sufficient precision.[202] The mass difference is less than 2%, and comparison with results for $^{10/11}$B and $^{14/15}$N makes effects of this magnitude appear improbable;[203,204] in any case they would be negligible for most chemical purposes in view of the wide range of tin chemical shifts of over 2000 ppm. Consequently, in what follows discussion is limited to ^{119}Sn.

Almost all workers, whether using the ^{1}H-{^{119}Sn} method or direct observation have adopted Me_4Sn [1]* as the reference compound, even though its resonance occurs rather near the middle of the chemical shift range and it is correspondingly more important that readers be left in no doubt as to the sign convention adopted. For this compound $\Xi(^{119}\mathrm{Sn})$ has been reported as 37 290 665 ± 3[195] and 37 290 662 ± 2[205] Hz; we use the former value [$\Xi(^{117}\mathrm{Sn})$ is 35 632 295 ± 3 Hz for Me_4Sn].

Figs. 10.9 and 10.10 illustrate the patterns of tin chemical shifts for a variety of classes of tin compound, and it is convenient to organize the discussion of the data on the basis of the factors which have been suggested as affecting tin shielding.

a. *Effective Electronegativity of the Substituents on Tin*

An increase in the electron-withdrawing ability of groups bound to tin generally leads to a decrease in the tin shielding. This is especially well exemplified by series of related molecules in which the point of variation is not adjacent to the tin atom. For example, (*a*) in $Me_3SnCH_nCl_{3-n}$ (compounds [1] and [20] to [22]) in which the ^{119}Sn shielding decreases by 85 ppm as n changes from 3 to 0;[195] (*b*) in organotin carboxylates in which changes in $\delta(^{119}\mathrm{Sn})$ parallel those of the pK_a of the parent carboxylic acid[207] (compounds [127] to [129]); (*c*) in substituted aromatic derivatives of tin, for which there are correlations between $\delta(^{119}\mathrm{Sn})$ and the Hammett σ-constant of the substituents[208] (compounds [62]–[77]); and (*d*) in organotin mercaptides, for which the tin shielding increases as the Taft σ-constant of the S-organo group becomes more positive[197] (compounds [172]–[174]). When the atom attached to tin is changed other effects (see below) may predominate, and it can be difficult to recognise the effect of substituent electronegativity.

In principle this dependence upon substituent electronegativity could be attributed to a change in the electron density on tin and hence in the diamagnetic contribution to the shielding. However, it is generally held that the observed changes are too large to be due to this and that rather a change in the paramagnetic term is responsible. A useful

* Numbers in square brackets refer to the tin compounds as listed in Tables 10.28 to 10.34.

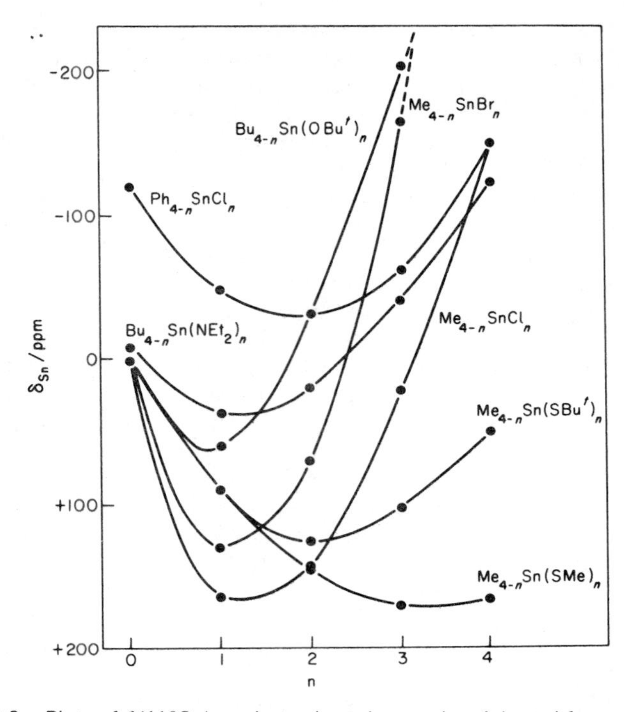

Fig. 10.9 Plots of $\delta(^{119}Sn)$ against n in various series. Adapted from ref. 194.

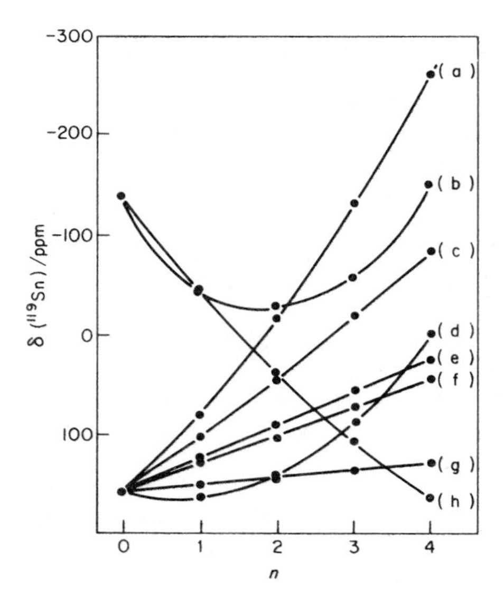

Fig. 10.10 Plots of $\delta(^{119}Sn)$ against n in mixed species $SnX_{4-n}Y_n$. (a) X = MeS, Y = Bu^{t}Se; (b) Ph, Cl; (c) MeS, MeSe; (d) MeS, Me; (e) MeS, Bu^{t}S; (f) MeS, PhS; (g) MeS, $PhCH_2S$; (h) Ph, MeS. Reproduced with permission from ref. 206.

approximation to this is given by Eq. 3.36, which for Sn should involve the 5*p* and 5*d* electrons. For four-coordinate tin compounds the *d*-electron term may be ignored (if there is no π-bonding) and the variations in $\delta(^{119}Sn)$ can be discussed in terms of changes of P_{5p} and/or $\langle r^{-3} \rangle_{5p}$. The first of these factors will be a maximum when two valence *p*-orbitals of tin are filled and one is vacant or vice versa,[209] and thus the observed decreases in tin shielding when one or two substituents are more electron-withdrawing than the others are to be expected. Plots of $-\delta(^{119}Sn)$ against n in series $R_{4-n}SnX_n$ often exhibit a characteristic 'U' shaped appearance (e.g. Fig. 10.9), as is also the case for ^{29}Si (see Fig. 10.1) in which minimum shielding occurs at or near $n = 2$, which is the point expected to give maximum *p*-electron imbalance[210] and hence maximum paramagnetic contribution to the tin shielding. However, there are clearly additional factors involved since the values of $\delta(^{119}Sn)$ for $n = 0$ and $n = 4$ (for both of which P_{5p} should be zero) are often very different.

b. Variations of Interbond Angles at Tin

The chemical shifts of mixed tetra-alkyl tins show wide and apparently random variations which it is tempting to suppose arise from deviations of the interbond angles at tin from the normal tetrahedral value. This view is supported by the fact that the deviations appear to be largest when groups such as *iso*propyl or *tert*-butyl are present, and by measurements on cyclic species.[211] Thus the ^{119}Sn nucleus in dimethylstannacyclopentane [14] is 50 ppm less shielded than that in dimethyldiethyltin, which should be comparable in most respects. Inspection of a model indicates that in the cyclic molecule one of the C—Sn—C interbond angles is significantly reduced from the "normal" value by the constraints of the 5-membered ring system, and this apparently affects the tin shielding. A similar effect is apparent in diphenylstannacyclopentane, and in the stannaspirane(IV) [17], for which two C—Sn—C interbond angles are reduced and four are enlarged (the reduction in tin shielding from that in the acyclic analogue is 120 ppm[211]). The same kind of behaviour is shown by cyclic tin derivatives of dithiols [182], and the phenomenon appears to be a general one.[211] Several other molecules have been prepared recently which have tin in a five-membered ring, and these too have abnormally low tin shielding,[212] (e.g. [60]), when allowance is made for the adjacent double bonds. By contrast, stannacyclohexanes (e.g. [15] and (V) [18]) have tin shieldings which are abnormally high for reasons which are not yet clear, while 2,6-dithio analogues (e.g. [183] and [185]), and also species in which the tin atom is in a larger ring (e.g. [16]), have essentially "normal" values of $\delta(^{119}Sn)$.[211]

c. Effects Due to Ring Currents

Suggestions that these can be of significance in understanding tin shielding in known molecules appear to be rooted in a misconception of their nature.[193,213]

d. Variations in ΔE

In discussing nuclear shielding it has often been found convenient to assume that ΔE remains essentially constant in series of related compounds, and this approach has usually been adopted with tin. An exception to this generalisation is provided by the chemical shifts in species in which tin is bound to one or more transition elements.[214] For such elements in the first series this leads to remarkably low shielding, and indeed the value of $\delta(^{119}Sn) = +483$ ppm in $MeSn[Co(CO)_4]_3$ [248] is the most positive yet recorded.[214] Normally the attachment of an electropositive element to tin (e.g. H, Li, B) leads to increased shielding, so the behaviour of the transition elements cannot readily be accounted for in terms of simple electron withdrawal. Rather it has been suggested[214] that dπ-dπ overlap involving tin and the transition metal leads to an effective reduction in ΔE (or more probably vitiates the whole basis of Eq. 3.36) and so increases the paramagnetic term. For heavier transition metals there may be less π-interaction of this type and/or the "bulky atom effect" may supervene. An alternative possibility is that the high frequency shifts arise from electric field effects of the carbonyl groups in [244]–[248], but this does not seem probable.

e. Electric Field Effects

The local electric fields associated with groups such as CF_3 and C=O are known to have significant effects (up to several ppm) upon ^{13}C shielding[215] and since their mode of action is to distort the electron cloud of the atom studied they may be expected to have effects upon tin shielding which are an order of magnitude or more greater. In fact the only experimental evidence to support this is provided by work on some benzotrifluoride derivatives (e.g. [62], [73] and [74]) from which it appears that a CF_3 group *ortho* to tin can cause reduced tin shielding.[205,216]

f. π-bonding

There has long been a school of thought which holds that dπ-pπ overlap is significant in bonds between tin and elements in groups V, VI and VII of the periodic table.[217] However, direct evidence for this kind of interaction is hard to find, and it is additionally far from clear just what effect it might have upon tin shielding, although a reasonable view would be that it should reduce the electronic imbalance in the bonds to tin and so reduce the paramagnetic term.[195,205,218] Certainly, many species with several electronegative substituents at tin have abnormally high shielding which it is tempting to explain by such π-bonding. For example, it is difficult to attribute the difference[197] of over 100 ppm in $\delta(^{119}Sn)$ between $(Bu^tS)_4Sn$[197] and $(MeS)_4Sn$ [186] solely to the difference in effective electronegativity of the two mercapto groups involved, and the shifts to low frequency of unassociated organotin di- and tri-alkoxides cannot be understood in terms of inductive electron withdrawal,[198,199] although it should be pointed out that similar trends in ^{29}Si chemical shifts have been rationalised with specific exclusion of *d*-orbital participation.[33]

Similarly, the systematic linear increases in shielding when saturated groups attached to tin are replaced by unsaturated ones in series such as $Et_{4-n}SnZ_n$ ($n = 0$–4; Z = Ph, Vi, CH_2Ph, C≡CH) suggest that π-interaction involving the tin 5*d* orbitals and the π-system of Z may be important.[219]

g. Change in the Coordination Number of Tin

When the coordination number of tin increases from four to five or six the tin shielding usually increases by several hundred ppm.[195] This kind of change can be brought about by the addition of donor molecules (e.g. DMSO, pyridine) to four-coordinate species in which the tin atom has at least one electronegative substituent, and such behaviour is exhibited by many of the compounds in Tables 10.29 to 10.31 and 10.34. It can also arise as a result of the auto-association of tin compounds with electronegative substituents on tin, and, in cases where a definite adduct cannot be isolated but there is a rapid reversible equilibrium in solution, measurements of $\delta(^{119}Sn)$ can provide valuable structural[220,221] and thermodynamic information. The origin of the effect is not yet fully understood: the additional ligand(s) should increase the neighbouring atom diamagnetic term (see Sec. h) but this will not generally be large enough to account for all of the effect. It could therefore arise from the reduction in effective nuclear charge, which will decrease the paramagnetic contribution (Eq. 3.36) via the $\langle r^{-3} \rangle$ terms, and it is also possible that the use of the tin 5d orbitals could be a major factor, although this has not been accounted for in detail.

h. "Bulky Atom Effects"

It is well established that whilst replacement of one group R in R_4Sn by an electron-withdrawing group X will generally decrease the tin shielding, further replacement will reverse this trend, and $\delta(^{119}Sn)$ in SnX_4 may well be much more negative than in R_4Sn in spite of the relative effective electronegativities of R and X.[3,184,195,218] This behaviour is most extreme when X is bulky and polarizable, (e.g. I) but there is no consensus as to the mechanism involved. One treatment[222] supposes that magnetic fields associated with the diamagnetic circulation of all the electrons of X are responsible and hence the effect depends upon the atomic number of X. However, this leads to comparable predicted effects for all observed nuclei (e.g. Si, Sn, Pb, etc.) which experimentally is definitely not the case, and it also predicts a linear relation between $\delta(^{119}Sn)$ and n in series $R_{4-n}SnX_n$, which again is by no means so. In the case of the mixed tin halides $\delta(^{119}Sn)$ can be predicted[223] accurately from the sum of the electronegativities of the appropriate halogens with little need for cross-terms, as the linearity of plots of $\delta(^{119}Sn)$ against n would suggest, but for many other series (e.g. $Me_{4-n}SnCl_n$) this is not so.

A typical plot (Fig. 10.10) can be characterised either by the use of cross terms taken pair-wise, or by the equivalent process[206] of specifying the difference in $\delta(^{119}Sn)$ between the points for $n = 0$ and 4 and the extent of the departure from linearity. When this latter approach is adopted it turns out that when the atoms attached to tin are chemically similar (especially in respect of potential π-bonding ability) the departure from linearity is small whereas when the atoms are dissimilar (e.g. C and S) then there will be a significant departure from linearity.

Recently it has been found that in general, shifts to low frequency are not produced by bulky atoms when these do not have available electrons in orbitals of suitable symmetry to form π-bonds with tin, as for example in the series Me_3SnEMe_3 (E = Si, Sn, Pb), compounds [227], [228], and [233], and similar behaviour has been noted for ^{195}Pt and ^{207}Pb shielding.[54,224] The current position is that more theoretical work is needed to determine the extent to which the "bulky atom effect" upon tin shielding depends upon polarizability, π-bonding and other factors.

10C.3 Spin-Spin Coupling Involving ^{119}Sn

The ratio $^nJ(^{119}Sn, X)/^nJ(^{117}Sn, X)$ should be $\gamma(^{119}Sn)/\gamma(^{117}Sn) = 1.0465$, and there is no evidence that this does not always hold; reports in much of the earlier proton literature[242] of apparently erratic variations in the ratio of the two couplings appear merely to reflect instrumental inadequacies, and the present discussion will largely ignore ^{117}Sn (and ^{115}Sn). It should further be noted that since $\gamma(^{119}Sn)$ is negative $^nJ(^{119}Sn, X)$ will be of opposite sign to $^nK(Sn, X)$ if $\gamma(X)$ is positive.

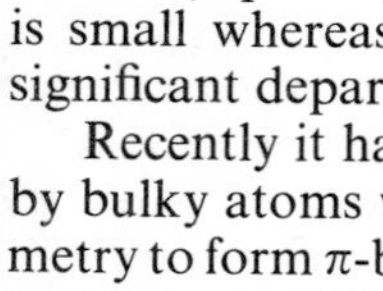

VII 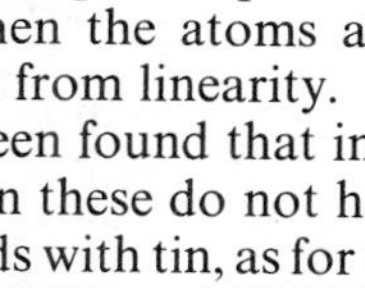 VIII 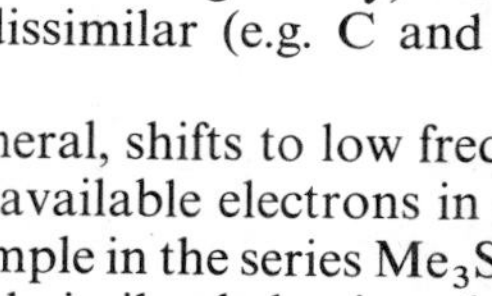 IX 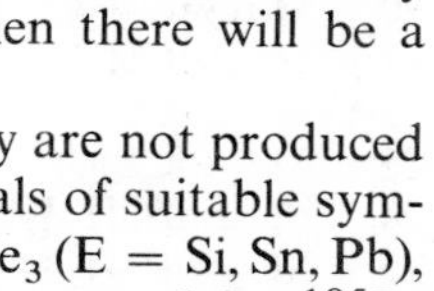 X

TABLE 10.28 Tin chemical shifts in compounds with only carbon bound to tin

No.	Compound	$\delta(^{119}Sn)$/ppm[a]	Conditions	Ref.
1	Me_4Sn	0.0	5% soln. in CH_2Cl_2	195
2	Et_4Sn	+1.4[b]	20% in CCl_4	219, 225
3	Pr^n_4Sn	−16.8	neat	3
4	Pr^i_4Sn	−43.9	neat	3
5	Bu^n_4Sn	−6.6	in CCl_4	225, 226
6	Me_3Bu^tSn	+17.5	neat	3
7	$Me_2Bu^t_2Sn$	+11.5 ± 2	neat	227
8	$MeBu^t_3Sn$	−25.4 ± 4	neat	227
9	7-$SnMe_3$-norbornane	−9.2 ± 0.2	25% v/v in CH_2Cl_2	208
10	1-$SnMe_3$-norbornane	+4.4 ± 0.2	25% v/v in CH_2Cl_2	208
11	1-$SnMe_3$-adamantane	−6.9 ± 0.2	25% v/v in CH_2Cl_2	208
12	2-$SnMe_3$-adamantane	−13.7 ± 0.2	25% v/v in CH_2Cl_2	208
13	Me_3Sn *cyclo* C_3H_5	+14.0 ± 0.2	25% v/v in CH_2Cl_2	208
14	$Me_2\overline{Sn[CH_2]_3CH_2}$	+54 ± 2	neat liquid	211
15	$Me_2\overline{Sn[CH_2]_4CH_2}$	−42 ± 2	30% soln. in CH_2Cl_2	211
16	$Me_2\overline{Sn[CH_2]_5CH_2}$	−2 ± 1	30% soln. in CH_2Cl_2	227
17	structure IV	+121 ± 2	neat liquid	227
18	structure V	−80 ± 2	neat liquid	227
19	$[Et_3Sn[CH_2]_2]_2$	−1.1 ± 0.1	95% in C_6D_6	188
20	Me_3SnCH_2Cl	+4	in C_6H_6	195
21	$Me_3SnCHCl_2$	+33	in C_6H_6	195
22	Me_3SnCCl_3	+85	+10% C_6H_6	195
23	Me_3SnCBr_3	+101	+10% C_6H_6	195
24	$Me_3Sn[CH_2]_2CN$	+6 ± 2	neat liquid	227
25	$Me_3Sn[CH_2]_2COMe$	+1.5 ± 2	neat liquid	227
26	$Me_3SnCH_2Ti(Cp)_2Cl$	+14 ± 1	sat. soln. in C_6H_6	214
27	$Me_3SnCH_2Zr(Cp)_2Cl$	+15 ± 1	sat. soln. in C_6H_6	214
28	Me_3SnCH_2Ph	−4	50% in CH_2Cl_2	205
29	$(PhCH_2)_4Sn$	−36		213
30	$(Cp)_4Sn$	−25.89 ± 0.15[c]	in CCl_4	200
31	$MeSn(Cp)_3$	−7.0	in CCl_4	228
32	$Me_2Sn(Cp)_2$	+23.2	in CCl_4	228
33	$Me_3Sn(Cp)$	+26.0[d]	in CCl_4	228
34	$(ViCH_2)_4Sn$	−47.9[e]	neat liquid	226
35	*syn*-7-$Me_3SnC_7H_9$[f]	−13.2 ± 0.2	in C_6H_6	229
36	*anti*-7-$Me_3SnC_7H_9$[f]	−25.6 ± 0.2	25% v/v in CH_2Cl_2	229
37	*endo*-2-$Me_3SnC_7H_9$[f]	−1.2 ± 0.6	in C_6H_6	229
38	*exo*-2-$Me_3SnC_7H_9$[f]	+7.8 ± 0.6	in C_6H_6	229
39	$(Me_2Sn)_2\overline{CCH{:}CHC(SiMe_3)CH}$	+7.0	in CCl_4	228
40	$(Me_3Sn')_2\overline{CCH{:}CHC(SnMe_3)CH}$	+11.7[g]	in CCl_4, 20°C	228
41	$Me_3Sn[CH_2]_nVi$	+5.4 ± 0.4 ($n = 1$)[R]	in CH_2Cl_2	230
42	$(HC{\equiv}C)_4Sn$	−279	20% in Et_2O	219
43	$(HC{\equiv}C)_2SnEt_2$	−141	30% in CCl_4	219
44	Vi_4Sn	−157.4	in CCl_4	226
45	Vi_2SnEt_2	−81	20% in CCl_4	219

TABLE 10.28 *(Continued)*

No.	Compound	$\delta(^{119}Sn)$/ppm[a]	Conditions	Ref.
46	Ph_4Sn	-137 ± 2	30% in $C_2H_3Cl_3$ at 100°C	206
47	Ph_3SnMe	-98	30% in CCl_4	219
48	Ph_2SnMe_2	-60	neat liquid	3
49	$PhSnMe_3$	-30[i]	neat liquid	3, 206, 208
50	$Ph_2\overline{Sn[CH_2]_3CH_2}$	0 ± 2.5	20% v/v in CH_2Cl_2	227
51	$Ph_2\overline{Sn[CH_2]_4CH_2}$	-113 ± 2	40% v/v in CH_2Cl_2	227
52	$Ph_2\overline{Sn[CH_2]_5CH_2}$	-57.4 ± 4	40% v/v in CH_2Cl_2	227
53	1-$SnMe_3$-naphthalene	-31.8 ± 0.2	25% v/v in CH_2Cl_2	208
54	2-$SnMe_3$-pyridine	-56.7	+5% C_6D_6	70
55	3-$SnMe_3$-pyridine	-27.4	+5% C_6D_6	70
56	4-$SnMe_3$-pyridine	-27.8	+5% C_6D_6	70
57	4-$XC_6H_4SCH_2SnPh_3$[j]	-119.3 ± 1.5	20% in CH_2Cl_2	206
58	structure VI	-51.4 ± 1	in CH_2Cl_2	212
59	$Me_2\overline{SnCH{:}CHB(NEt_2)CHCH}$	-160 ± 1	in CH_2Cl_2	212
60	$Me_2\overline{SnCEtCH{:}CHCBEt_2}$	$+19.5 \pm 1$	in CH_2Cl_2	212
61	$Me_2\overline{SnCHCHSiMe_2CHCH}$	-170 ± 1	Neat liquid	212
62	2-$CF_3C_6H_4SnMe_3$[k]	-16.6 ± 0.2	50% v/v in CCl_4	208
63	3-$CF_3C_6H_4SnMe_3$[k]	-23.9 ± 0.2	50% v/v in CCl_4	208
64	2-$ClC_6H_4SnMe_3$[k]	-24.2 ± 0.2	50% v/v in CCl_4	208
65	3-$ClC_6H_4SnMe_3$[k]	-24.5 ± 0.2	50% v/v in CCl_4	208
66	4-$ClC_6H_4SnMe_3$[k]	-25.6 ± 0.2	50% v/v in CCl_4	208
67	2-$MeOC_6H_4SnMe_3$[k]	-34.6 ± 0.2	50% v/v in CCl_4	208
68	3-$MeOC_6H_4SnMe_3$[k]	-28.0 ± 0.2	50% v/v in CCl_4	208
69	4-$MeOC_6H_4SnMe_3$[k]	-28.8 ± 0.2	50% v/v in CCl_4	208
78	$Me_3SnPh \cdot Cr(CO)_3$	$+3$	5% in CH_2Cl_2	205
79	$(Me_3Sn)_2C_2B_{10}H_{10}$[l]	$+40.3$	in CCl_4	226
80	$(Cp)_2Sn$[m]	-2199 ± 10	in cyclohexane	227

[a] For the data in this and the following six tables the reference used is Me_4Sn. [b] Values ranging from -7 to $+1.4$ ppm have been reported.[3,200,205,206] [c] This measurement was performed without proton decoupling, and the claimed accuracy is therefore surprisingly high in view of the complexity of the tin spectrum. See discussion in text (p. 345) on 117/119-tin isotope effects. [d] A range of 5 ppm has been obtained by different workers (refs. 227, 228). [e] Changes to -147.5 ppm on dilution with CCl_4. Ref. 226. [f] C_7H_9 = norborn-2-enyl. [g] At $-80°$ in toluene $\delta(^{119}Sn) = -49.8$ ppm, $\delta(^{119}Sn') = +11.1$ ppm. [h] $\delta(^{119}Sn) = -1.6$ ppm ($n = 2$); $+0.5$ ppm ($n = 3$).[230] [i] Hardly affected by dilution with CCl_4 or CH_2Cl_2 ref. 206. [j] X = Cl, Br, NH_2, OMe, Bu^t. [k] Ref. 208 also contains data for 2,6-Me_2Y [70]; 2,4-Me_2Y [71]; 2,4-Me_2Y [72]; 2,6-$(CF_3)_2Y$ [73]; 2,4-$(CF_3)_2Y$ [74]; 3,5-$(CF_3)_2Y$[75]; 2,6-$(MeO)_2Y$[76], where Y = $C_6H_3SnMe_3$, and for 2,4,6-$(MeO)_3$ $C_6H_2SnMe_3$ [77]. [l] Icosahedral carborane derivative. [m] Tin (II) compound.

TABLE 10.29 Tin chemical shifts in compounds with tin-halogen bonds

No.	Compound	$\delta(^{119}Sn)$/ppm	Conditions	Ref.
81	$SnCl_4$	-150 ± 2^a	neat liquid	184
82	$SnBr_4$	-638 ± 1^a	neat liquid	184
83	SnI_4	-1701 ± 2	in CS_2	184
84	$SnCl_3Br$	-263 ± 3	mixture of [81] and [82]	184
85	$SnCl_2Br_2$	-385 ± 1	mixture of [81] and [82]	184
86	$SnClBr_3$	-508 ± 1	mixture of [81] and [82]	184
87	$SnCl_2$	-236 ± 1	in tetrahydrofuran	184
88	$(PhMe_2CCH_2)_3SnF$	$+139$	in $CDCl_3$	207
89	$MeSnCl_3$	$+6.3 \pm 1^b$	neat liquid at 74°C	231
90	$(n\text{-hexyl})SnCl_3$	$+2.0 \pm 0.1$	neat liquid	188
91	$PhSnCl_3$	-63 ± 0.5	in CH_2Cl_2	195
92	Me_2SnCl_2	$+137 \pm 0.2^c$	20% in CH_2Cl_2	195
93	$Bu^t_2SnCl_2$	$+52 \pm 5$	neat liquid at 50°C	227
94	$Me(n\text{-hexyl})SnCl_2$	$+124.4 \pm 0.1$	neat liquid	188
95	Ph_2SnCl_2	-32 ± 0.8	in CH_2Cl_2	195
96	Me_3SnCl	$+154\text{-}+166^d$	various	3, 200, 218, 225, 226
97	Bu^t_3SnCl	$+50 \pm 1$	neat liquid	227
98	$Me_2(n\text{-hexyl})SnCl$	$+156.9 \pm 0.1$	neat liquid	188
99	Ph_3SnCl	-48 ± 1	in CH_2Cl_2	195
100	$Me_2[MeC(O)CH_2CH_2]SnCl$	$+69.5 \pm 2^e$	in CH_2Cl_2	227
101	$Me_2(Cp)SnCl$	$+101.6$	in CCl_4	228
102	$MeSnBr_3$	-170 ± 0.1	in CH_2Cl_2	195
103	$EtSnBr_3$	-141 ± 0.2	50% in CCl_4	219
104	Me_2SnBr_2	$+70$	in C_6H_6	218
105	Et_2SnBr_2	$+99 \pm 2.5$	in CH_2Cl_2	195
106	$Bu^t_2SnBr_2$	$+76.5 \pm 1$	sat. soln. in C_6H_6	227
107	$\overline{CH_2[CH_2]_4Sn}Br_2$	$+43.5 \pm 3$	10% v/v in CH_2Cl_2	227
108	Me_3SnBr	$+128 \pm 0.3$	in C_6H_6	195
109	Et_3SnBr	$+148 \pm 0.5$	neat liquid	219
110	Bu^t_3SnBr	$+74.8 \pm 1$	in C_6H_6	227
111	$MeSnI_3$	-699.5 ± 1^c	sat. in CCl_4	206
112	Me_2SnI_2	-157 ± 0.15	20% in CH_2Cl_2	205
113	$\overline{CH_2[CH_2]_4Sn}I_2$	-169 ± 3	in CH_2Cl_2	227
114	Me_3SnI	$+38.6$	in C_6H_6	218
115	Bu^t_3SnI	$+82.7 \pm 1$	in C_6H_6	227
116	Ph_3SnI	-114.5	in CCl_4	226
117	$MeSnCl_2I$	-202 ± 5	in C_6H_6	206
118	$MeSnClI_2$	-436 ± 3	in C_6H_6	206

[a] In the presence of water or other donor solvent different results are obtained. Refs 3 and 184. [b] Changes monotonically to +21 ppm on dilution with C_6H_6 or $CHCl_3$. Ref. 231. [c] Sensitive to solvent effects. [d] Depends on nature and amount of "inert solvent;" cyclohexane, CCl_4, CH_2Cl_2, C_6H_6 or $CHCl_3$. [e] Thought to contain five-coordinate tin, see text.

TABLE 10.30 Tin chemical shifts in compounds with tin-oxygen bonds

No.	Compound	$\delta(^{119}Sn)/ppm$[a]	Conditions	Ref.
119	Me_3SnOH	$+118 \pm 1$	sat. in CH_2Cl_2	232
120	$(Me_2PhCCH_2)_3SnOH$	$+161$	in CH_2Cl_2	207
121	Ph_3SnOH	-86	in CH_2Cl_2	207
122	$(Me_3Sn)_2O$	$+109.5 \pm 1$[b]	neat liquid	232
123	$(Bu^t_3Sn)_2O$	-39.3	sat. in C_6H_6	227
124	$(Ph_3Sn)_2O$	-80.6	in $CHCl_3$	3
125	$Me_3SnO{\cdot}CO{\cdot}H$	$+150$[c]	0.05 M in $CDCl_3$	207
126	$Me_3SnO{\cdot}CO{\cdot}Me$	$+129$	sat. in CH_2Cl_2	193
127	$(Me_2PhCCH_2)_3SnO{\cdot}CO{\cdot}Me$	$+79$[d]	0.4 M in $CDCl_3$	207
128	$(Me_2PhCCH_2)_3SnO{\cdot}CO{\cdot}CF_3$	$+146$[d]	0.4 M in $CDCl_3$	207
129	$Ph_3SnO{\cdot}CO{\cdot}R$	-80 to -121[d]	0.4 M in $CDCl_3$	207
130	$Bu^n_2Sn(O.CO.Me)_2$	-164 ± 4	neat liquid	202
131	$Bu^nSn(O{\cdot}CO{\cdot}Me)_3$	-535 ± 5	50% in CCl_4	202
132	Me_3SnOMe	$+129 \pm 2$[e]	sat. in C_6H_6	233
133	Me_3SnOBu^t	$+91 \pm 2$[e]	50% v/v in C_6H_6	233
134	Bu^n_3SnOMe	$+83 \pm 7$[e]	neat liquid	198
135	$Bu^n_3SnOBu^t$	$+60 \pm 2$[e]	neat liquid	198
136	Bu^n_3SnOMe	-38.7 ± 2[e]	50% v/v in CCl_4	227
137	$Me_2Sn(OMe)_2$	-126	in C_6H_6	221
138	$Me_2Sn(OBu^t)_2$	-1.8 ± 1[e]	50% v/v in C_6H_6	234
139	$Bu^n_2Sn(OMe)_2$	-165 ± 2[f]	neat liquid	198
140	$Bu^n_2Sn(OBu^t)_2$	-34 ± 5[e]	neat liquid	198
141	$Bu^t_2Sn(OMe)_2$	-114.7 ± 0.1[e]	neat liquid	227
142	$Bu^t_2Sn(OBu^t)_2$	-123.4 ± 0.4[e]	neat liquid at 60°C	227
143	$MeSn(OEt)_3$	-434 ± 10[g]	25% v/v in mesitylene	220, 233
144	$MeSn(OBu^t)_3$	-117 ± 1[e]	neat liquid	233
145	$Bu^nSn(OEt)_3$	-428 ± 14[g]	neat liquid	199
146	$Bu^nSn(OBu^s)_3$	-321 ± 2[f]	neat liquid	199
147	$Bu^nSn(OBu^t)_3$	-200 ± 5[e]	neat liquid	199
148	$Bu^n_3Sn(ox)$	$+29.4 \pm 5$[f]	neat liquid	202
149	$Bu^t_2Sn(ox)_2$	-341 ± 2[g]	sat. in C_6H_6	227
150	$Me_2Sn(acac)_2$	-366[g]	10% in $CDCl_3$	227
151	$Bu^t_2Sn(acac)_2$	-480 ± 4[g]	in C_6H_6	227
152	$Me_2Sn(OMe)Cl$	$+126$,[e] -90[f]	in CH_2Cl_2	235
153	$(Bu^n_2SnF)_2O$	-168 ± 14	in CCl_4	236
154	$(Bu^n_2SnCl)_2O$	-145 ± 5, -94 ± 5	in CCl_4	236
155	$Bu^n_3SnO{\cdot}OBu^t$	$+105 \pm 1$	in CH_2Cl_2	195
156	$Ph_3SnO{\cdot}OBu^t$	-95 ± 1	in CH_2Cl_2	195
157	$Bu^n_3SnOSiMe_3$	$+71 \pm 2$	neat liquid	198
158	$Bu^n_2SnOCH_2CH_2O$ (cyclic)	-189 ± 5	sat. in $CDCl_3$	198
159	$Bu^n_2SnSCH_2CH_2O$ (cyclic)	-32 ± 2	sat. in $CDCl_3$	198
160	$Bu^n_2SnOCH_2CMe_2CH_2O$ (cyclic)	-213 ± 5	neat liquid at 120°C	198
161	$Bu^t_2SnOCH_2CH_2O$ (cyclic)	-223 ± 4	sat. in C_6H_6	227
162	$Bu^t_2SnOCH_2CH_2CH_2O$ (cyclic)	-281 ± 3	sat. in C_6H_6	227
163	$Me_3SnON{=}C[CH_2]_4CH_2$ (cyclic)	$+132 \pm 0.8$	20% in CH_2Cl_2	205
164	$Me_3SnON{=}C(n\text{-}C_5H_{11})_2$	$+20 \pm 0.1$	50% in CH_2Cl_2	205
165	$(Me_3SnO)_3P{=}O$	$+24.9$	neat liquid	3

[a] The chemical shifts of many of these species are sensitive to concentration and/or presence of donor solvents. [b] $\delta(^{119}Sn) = +117.1 \pm 1$ ppm, 50% soln. in CH_2Cl_2. [c] $\delta(^{119}Sn) = +2.5$ ppm for 3 M solution in $CDCl_3$. [d] Approximately linear algebraic increase in $\delta(^{119}Sn)$ with decreasing p*Ka* of parent carboxylic acid between these extremes. [e] Four-coordinate monomeric species. [f] Five-coordinate dimeric species. [g] Six-coordinate species.

TABLE 10.31 Tin chemical shifts in compounds with tin-sulphur, selenium or tellurium bonds

No.	Compounds	$\delta(^{119}Sn)$/ppm	Conditions	Ref.
166	$(Me_3Sn)_2S$	$+86.5 \pm 0.3$	20% in CH_2Cl_2	232
167	$(Me_3Sn)_2Se$	$+44.5 \pm 1$	20% in CH_2Cl_2	232
168	$(Me_3Sn)_2Te$	-66.8 ± 1	20% in CH_2Cl_2	232
169	$(Ph_3Sn)_2S$	-48.7	neat liquid	3
170	structure XIII	$+128$	in C_6H_6	195
171	$SnPh_2$ analogue of XIII	$+19.5$	in $CHCl_3$	3
172	Me_3SnSMe	$+90$[a]	10% in C_6H_6	218
173	Me_3SnSBu^t	$+55.5$[a]	neat liquid	197
174	Ph_3SnSMe	$+90.5$[a]	neat liquid	197
175	$Me_3SnS(4\text{-}XC_6H_4)$	$+89$ to $+104$[b]	in CH_2Cl_2	206
176	$Ph_3SnS(4\text{-}XC_6H_4)$	-70 to -61[b]	in CH_2Cl_2	206
177	$Me_3SnS[CH_2]_{2,3,4}SSnMe_3$	$+82.3$, $+81.5$, $+80.5$	neat liquid	211
178	$Me_2Sn(SMe)_2$	$+144$[a]	10% in C_6H_6	218
179	$Ph_2Sn(SMe)_2$	$+38.5 \pm 1$[a]	30% in CH_2Cl_2	206
180	$MeSn(SMe)_3$	$+167$[a]	10% in C_6H_6	218
181	$PhSn(SMe)_3$	$+107 \pm 2$	30% in CH_2Cl_2	206
182	$Me_2\overline{SnSCH_2CH_2S}$	$+194$	20% in CH_2Cl_2	205
183	$Me_2\overline{SnS(CH_2)_3S}$	$+148.8$	30% in CH_2Cl_2	202
184	$Ph_2\overline{SnSCH_2CH_2S}$	$+78 \pm 2$	in CH_2Cl_2	211
185	$Ph_2\overline{SnS[CH_2]_3S}$	$+28.5 \pm 2$	in CH_2Cl_2	211
186	$(MeS)_4Sn$	$+160$	50% in CH_2Cl_2	197
187	$(MeS)_3SnSeMe$[c]	$+101 \pm 2$	mixture of [186] + [190]	206
188	$(MeS)_2Sn(SeMe)_2$[c]	$+43 \pm 2$	mixture of [186] + [190]	206
189	$MeSSn(SeMe)_3$[c]	-21 ± 3	mixture of [186] + [190]	206
190	$Sn(SeMe)_4$	-80.5	neat liquid	237
191	$MeSn(SeMe)_3$	$+14.8$	neat liquid	237
192	$Me_2Sn(SeMe)_2$	$+57.1$	neat liquid	237
193	$Me_3SnSeMe$	$+45.6$	neat liquid	237
194	$[Me_2SnCl]_2S$	$+144$	20% in CH_2Cl_2	205
195	$Me_2SnCl\cdot S\cdot CS\cdot NEt_2$[d]	-204	in CH_2Cl_2	195, 234
196	$Me_2Sn(S\cdot CS\cdot NEt_2)_2$[e]	-336	in CH_2Cl_2	195, 234

[a] Marked variations if *S*-alkyl group is changed—see text p. 348. [b] Varies according to nature of X (=Me, Bu^t, OMe, Cl, Br, NO_2, NH_2), see text. [c] Many other mixed species $SnX_{4-n}Y_n$ (X, Y = SMe, SeMe, SPh, SePh, SBu^t, I) can be made similarly. See text. [d] Probably contains five-coordinate tin. [e] Probably contains six-coordinate tin.

TABLE 10.32 Tin chemical shifts in compounds with tin-nitrogen or phosphorus bonds

No.	Compound	$\delta(^{119}Sn)$/ppm	Conditions	Ref.
197	Me_3SnNMe_2	+75	25% in C_6H_6	238
198	$Me_3SnNMePh$	+73 ± 1	30% in C_6H_6	227
199	$Me_2Sn(NMe_2)_2$	+58.8	25% in C_6H_6	238
200	$Me_2Sn(NMePh)_2$	+30.5 ± 1	25% in C_6H_6	227
201	$MeSn(NMe_2)_3$	−15.1	25% in C_6H_6	238
202	$MeSn(NMePh)_3$	−66 ± 1	25% in C_6H_6	227
203	$(Me_2N)_4Sn$	−118	25% in C_6H_6	238
204	$(MePhN)_4Sn$	−175.6 ± 1	25% in C_6H_6	227
205	$Me_3SnNHPh$	+47 ± 2	neat liquid	227
206	$(Me_3Sn)_2NPh$	+61 ± 2	neat liquid	227
207	$Me_2SnNMeCH_2CH_2NMe$ (ring: Sn–N bond)	−78 ± 1.5	sat. in C_6H_6	227
208	$(Me_2SnNMe)_3$	+92.1 ± 1.5	sat. in C_6H_6	227
209	$Me_2Sn(NEt_2)Cl$	+75 ± 2[a]	30% in C_6H_6	202
210	Me_3SnPPh_2	−2.3 ± 0.1	65% in C_6H_6	239
211	$Me_2Sn(PPh_2)_2$	−11.5 ± 2	20% in C_6H_6	202
212	$(Me_3Sn)_2PPh$	+14.2 ± 0.1	neat liquid	239
213	$(Me_3Sn)_3P$	+36.3 ± 0.1	65% in C_6H_6	239
214	$(Me_3Sn)_3P\cdot W(CO)_5$	+63.4 ± 0.1	in C_6H_6	239

[a] Shifts to high frequency on dilution.

TABLE 10.33 Tin chemical shifts in compounds with tin-hydrogen, metal or metalloid bonds

No.	Compound	$\delta(^{119}Sn)$/ppm	Conditions	Ref.
215	Me_3SnH	−104.5		226
216	Me_2SnH_2	−224.6		133
217	$MeSnH_3$	−346 ± 2	60% in toluene at −35°C	229
218	Ph_3SnH	−164.5 ± 5	neat liquid	227
219	Ph_2SnH_2	−233.6		133
220	SnH_3^+	−186	in FSO_3H at −80°C	240
221	Me_3SnLi	−183	20% in THF	205
222	$Me_3SnLi\cdot 3HMPT$	−162	20% in HMPT	186, 241
223	$(Me_3Sn)_3Sn'Li\cdot 3THF$	−107(Sn); −1030(Sn′)	in C_6H_6	186
224	$Me_3SnB(NMe_2)_2$	−150 ± 1	neat liquid	54
225	$Me_3SnBClNMe_2$	−139 ± 1	neat liquid	54
226	$[Me_3SnTlMe_3]^-Li^+$	−376.5 ± 1	in dimethoxyethane	202
227	$Me_3SnSiMe_3$	−126.7 ± 1.5	neat liquid	54
228	$Me_3SnSnMe_3$	−108.7	$+C_6D_6$	187
229	$Me_3SnSn'Et_3$	−108.1(Sn); −61.8(Sn′)	$+C_6D_6$	187
230	$Et_3SnSnEt_3$	−59.9	$+C_6D_6$	187
231	$Me_3SnSn'Ph_3$	−91.5(Sn), −153(Sn′)	sat. in CH_2Cl_2	54
232	$(Me_3Sn)_2Sn'Me_2$	−101(Sn), −263.5(Sn′)	$+C_6H_6$	186

TABLE 10.33 (*Continued*)

No.	Compounds	$\delta(^{119}Sn)$/ppm	Conditions	Ref.
233	$Me_3SnPbMe_3$	-57.0	in C_6H_6	224
234	$Me_3SnTaH_2(Cp)_2$	$+53 \pm 2$	sat. in C_6H_6	214
235	$Me_3SnCr(CO)_3(Cp)$	$+161 \pm 0.5$	in C_6H_6	214
236	$Me_3SnMo(CO)_3(Cp)$	$+121 \pm 0.5$	in C_6H_6	214
237	$Me_3SnW(CO)_3(CP)$	$+42.0 \pm 0.5$	in C_6H_6	214
238	$Me_3SnMoH(Cp)_2$	$+123 \pm 2$	in C_6H_6	214
239	$Me_3SnMoCl(Cp)_2$	$+90 \pm 2$	in C_6H_6	214
240	$Me_3SnMn(CO)_5$	$+63 \pm 1$	in C_6H_6	214
241	$Me_3SnRe(CO)_5$	-89 ± 1	in C_6H_6	214
242	$Me_2Sn[Mn(CO)_5]_2$	$+150 \pm 1$	in C_6H_6	214
243	$Me_2Sn[Re(CO)_5]_2$	-223 ± 2	in C_6H_6	214
244	$MeSn[Mn(CO)_5]_3$	$+284 \pm 2$	in $CDCl_3$	214
245	$(Me_3Sn)_2Fe(CO)_4$	$+79 \pm 1$	in $CDCl_3$	214
246	$Me_3SnCo(CO)_4$	$+151.0 \pm 0.2$	in C_6H_6	214
247	$Me_2Sn[Co(CO)_4]_2$	$+293 \pm 1$	in C_6H_6	214
248	$MeSn[Co(CO)_4]_3$	$+483 \pm 1$	in C_6H_6	214
249	$Me_3SnRh(C{\equiv}CPh)_2(PPh_3)_2$	$+172 \pm 2$	in CH_2Cl_2	214
250	$Me_3SnIr(CO)(C{\equiv}CPh)_2(PPh_3)_2$	-106 ± 2	in CH_2Cl_2	214
251	*cis*-$Me_3SnPt(C{\equiv}CPh)(PMe_2Ph)_2$	-48 ± 1	in $C_6D_5CD_3$	214

TABLE 10.34 Tin chemical shifts in complexed species[a]

No.	Species	$\delta(^{119}Sn)$/ppm	Conditions	Ref.
252	$Me_3SnCl \cdot C_5H_5N$	$+25.4$	in $CHCl_3$	3
253	$Me_3SnCl \cdot DMSO$	$+3$	in DMSO	197
254	$Me_3SnCl \cdot HMPT$	-47.5 ± 2	10% in HMPT	202
255	$Me_3SnCl_2^- NEt_4^+$	$+53 \pm 0.2$	in acetone	205
256	$(Me_3Sn \cdot bipy)^+ BPh_4^-$	-18 ± 0.4	in CH_2Cl_2	205
257	$Me_3SnBr \cdot PhNH_2$	$+149.9$	in $CHCl_3$	3
258	$Me_3SnI \cdot DMSO$	$+6.5 \pm 2$	10% in DMSO	202
259	$Me_3SnI \cdot HMPT$	-48.5 ± 2	10% in HMPT	202
260	$Me_2SnCl_2 \cdot 2DMSO$	-246	in DMSO	197
261	$Me_2SnI_2 \cdot 2DMSO$	-315 ± 2	in DMSO	202
262	$MeSnI_3 \cdot 2DMSO$	-795 ± 4	in DMSO	202
263	$Sn^{II} \cdot EDTA$	-713 ± 2	in NH_4OH aq.	202
264	$SnF_6^{2-} \cdot 2[Pr^i_2NH_2]^+$	-888	in H_2O	195
265	$SnF_5Cl^{2-} \cdot 2Pr^i_2NH_2]^+$	-826	in H_2O	227
266	$Sn(OH)_6^{2-} \cdot 2K^+$	-590 ± 2	in H_2O	184

[a] Other tables also contain species in this category. It should be noted that many of the complexes have a tendency to dissociate in solution, and thus the results should be viewed with caution as it is by no means always clear exactly what species is being measured. For example, widely different results have been quoted for [253] depending on whether the solvent was CH_2Cl_2 (which permits dissociation) or DMSO which inhibits it. In general it is desirable for these species that a careful study of solvent effects be made before a chemical shift is quoted.

Tables 10.35 to 10.40 list coupling constants involving ^{119}Sn. Most of these have been measured from the spectrum of the other nucleus, especially when this is ^{1}H, ^{13}C or ^{19}F, and a large proportion of the remainder have been determined by ^{1}H-{^{119}Sn} and ^{1}H-{X} heteronuclear double resonance techniques,[185] which have the advantage of usually giving the sign in addition to the magnitude of the coupling constant. A simple approach using Pople and Santry MO theory[243] might lead one to expect general parallels in the behaviour of coupling constants involving ^{119}Sn and the corresponding ones involving ^{13}C, except that the former should be of opposite sign and numerically much larger owing to the different γ's and greater value of $S^2(0)$ for tin. The data[54] in Table 10.40 confirm this view for the one-bond couplings, and show that for X with lone pair electrons ^{1}K(SnX) is negative, as is ^{1}K(CX), but otherwise ^{1}K(SnX) tends to be positive, as again is the case with ^{1}K(CX). Apparent exceptions to this generalisation are the values of ^{1}J(^{119}Sn, ^{31}P) in Table 10.38, which are all positive[239] [corresponding to negative ^{1}K(SnP)], whereas in derivatives of four-coordinate phosphorus ^{1}K(PC) is usually positive.[244] However, it will be noticed that upon quaternisation of phosphorus the reduced tin-phosphorus coupling becomes substantially less negative, a change in the same direction as that of the analogous carbon-phosphorus coupling constant.[244]

When the tin atom acquires an electron lone pair, as in $Me_3Sn^-Li^+$ [221] and $(Me_3Sn)_3Sn^-Li^+$, it ceases to be comparable with carbon, and in accord with this ^{1}K(SnC) and ^{1}K(SnSn) become negative in these species.[241,245] In terms of Pople and Santry MO theory[243] this implies a change in sign of the mutual polarisability of the valence *s* orbitals of tin on the one hand and carbon or tin on the other. Other changes in ^{1}K(SnX) brought about by varying the substituents at tin are less spectacular, and are as expected from the MO treatment. Thus factors such as replacement of alkyl by more electronegative groups which increase the effective nuclear charge of the tin atom and/or the *s*-character of the relevant tin hybrid orbitals generally increase $|^{1}K(SnX)|$ provided that this is not small enough for changes in the mutual polarisability to dominate. This behaviour is exemplified by such couplings as ^{1}J(^{119}Sn, ^{13}C) in series $Me_{4-n}SnX_n$[246] and by ^{1}J(^{119}Sn, ^{77}Se), which changes from +1015 Hz in $Me_3SnSeMe$ to +1520 Hz in $Sn(SeMe)_4$. [See fig. 10.12.][237] The range of variation of ^{1}J(^{119}Sn, ^{119}Sn) is very large indeed,[187,245] and although the trends are in the right direction it is difficult to see how replacing methyl groups by *iso*propyl or *tert*-butyl can sufficiently affect $S^2_{Sn}(0)$ to bring about the observed effects as has been proposed.[187] In fact extrapolation suggests that in $Bu^t_6Sn_2$ this coupling constant will be very close to zero or even negative, so it seems clear that a major part of the variation must arise from alteration of the s-overlap integral β between the two tin atoms, which will affect tin-tin mutual polarizability directly.

Figure 10.11 shows a plot[241] of ^{1}J(^{119}Sn, ^{13}C) in methyltin species against the corresponding ^{2}J(^{119}Sn, H) in species in which there are no special conformational relationships. The two coupling constants are normally of opposite sign (i.e. ^{2}J(^{119}Sn, H) is positive, this corresponding to negative *K*, in species in which the tin atom has no electron lone pairs), and one can normally use the plot to deduce one from the other. However, it is inadvisable to attempt to make precise estimates of the tin hybridization from either of these two coupling constants. Fairly detailed studies have been undertaken of ^{2}J(^{119}SnCH) in methyltin compounds by several groups, and it is found that the coupling constant can often be predicted by the use of additivity parameters which depend upon the substituents of the tin atom, more electronegative groups leading to a larger magnitude.[247,248] When some of the methyl groups are replaced by other substituents the behaviour is more complex. In addition it is known that stereochemical

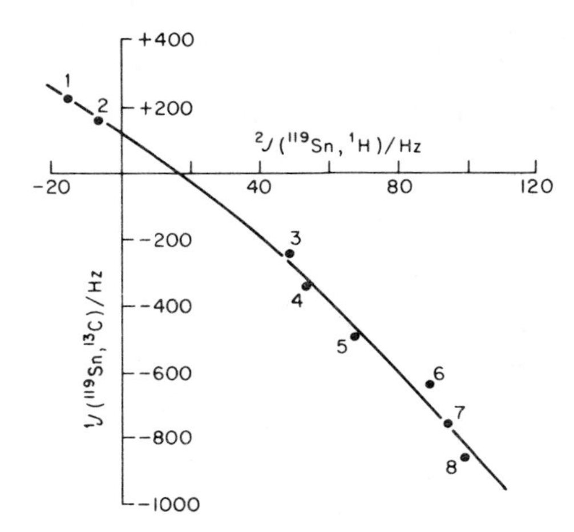

Fig. 10.11 Plot of $^1J(^{119}Sn, ^{13}C)$ against $^2J(^{119}Sn, H)$ in representative organotin compounds. (1) Me_3SnLi in HMPT; (2) Me_3SnLi in tetrahydrofuran; (3) Me_6Sn_2; (4) Me_4Sn; (5) Me_3SnBr in H_2O; (6) $MeSnBr_3$ in C_6H_6; (7) Me_2SnCl_2 in H_2O/Me_2CO; (8) Me_2SnCl_2 in H_2O. Reproduced with permission from ref. 241.

factors can be important since in (XI) different values (56 and 36 Hz) are found for the coupling between ^{119}Sn and the two inequivalent CH_2 protons.[249] This is an area which would repay more thorough investigation. Geminal couplings between ^{119}Sn and ^{13}C are usually quite small (20–30 Hz), apparently of positive sign,[250] and appear to be controlled by factors similar[251,252] to the ones which affect $^2J(^{119}Sn, H)$. In various cyclic species where there are specific fixed steric relationships between tin and the relevant carbon nuclei good evidence has been obtained for angular dependence of the coupling constant, and this should prove valuable in structural work.[253] Other geminal couplings involving ^{119}Sn can be quite large (e.g. $^2J(^{119}Sn, ^{31}P) = 1650$ Hz in $Me_3SnPt(PPh_2Me)_2C{\equiv}CPh$[227]) and presumably will show wide variations according to the nature of the intervening atom, although little work has been done in this area yet. Figure 10.12 shows the pattern of variation of some couplings involving ^{119}Sn in related compounds.

Three-bond tin-proton coupling constants are often numerically larger than two-bond ones and of opposite sign[254] (i.e. ^3J negative, ^3K positive); their magnitudes depend upon factors similar to those which determine $^1J_{CH}$, i.e. effective nuclear charge and *s*-character of orbitals of the tin atom.[255] It also seems likely that there will be a Karplus type of dependence upon dihedral angle, although there is little direct evidence for this. On the other hand there is evidence of such a relationship in the case of

$Ph{-}C(Me)_2{-}CH_2{-}Sn(Bu^t)(Ph){-}I$

XI

$(Me_2SnZ)_3$ ring: Me_2Sn, Z, $SnMe_2$, Z, Me_2Sn, Z

XII Z = NEt_2
XIII Z = S

Me_2Sn(S)–S–$SnMe_2$ ring with N–Et: Me_2Sn, S, $SnMe_2$, S, Me_2Sn, N, Et

XIV

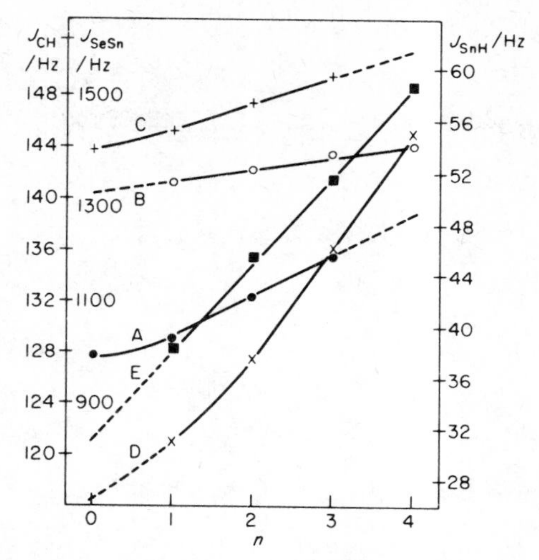

Fig. 10.12 Coupling constants in Hz as functions of n in series $Me_{4-n}Sn(SeMe)_n$ showing characteristic dependence upon the effective nuclear charge and/or hybridization of the central tin atom. (A) $^1J(^{13}CH)$ (Sn); (B) $^1J(^{13}CH)$ (Se); (C) $^2J(^{119}SnH)$; (D) $-^3J(^{119}SnH)$; (E) $^1J(^{119}Sn, ^{77}Se)$. Reproduced with permission from ref. 237.

$^3J(^{119}Sn, ^{13}C)$, which typically has values of 5 to 70 Hz according to the stereochemical relationship.[251–253] Longer range couplings to tin have been relatively little studied.

10C.4 Applications of Tin NMR to Chemical Problems

Although most work on ^{119}Sn NMR has been aimed at understanding the factors which determine coupling constants and shielding there have been some studies which have given results of direct chemical value. The applications of NMR (including ^{119}Sn work) to structural aspects of organotin compounds have been reviewed.[289] In mixtures of XII and XIII the fine structure of ^{119}Sn INDOR spectra (obtained by monitoring appropriate satellites in the proton spectra) gave unequivocal evidence of the presence of intermediate species such as XIV with lifetimes long on an NMR timescale.[195] Similarly, the direct ^{119}Sn spectra of mixtures of tin tetrahalides provided evidence for the existence of all possible intermediate species, again with long lifetimes,[184] and other mixtures have been investigated by INDOR techniques.[206,290] Samples of $[Ph_2Sn(CH_2)_4]_n$, which may have $n = 1$ or 2 (corresponding to tin in a five or ten membered ring) according to method of preparation, can be distinguished by their different values of $\delta(^{119}Sn)$ arising from different interbond angles at tin.[291] In XV separate ^{119}Sn resonances have been detected,[236] with a chemical shift difference of 20–50 ppm corresponding to the two types of tin environment, and INDOR experiments have established a structural analogy[235] with the Grignard reagent for the dialkyltinhalidemethoxides, $R_2Sn(X)OMe$. Tin chemical shifts have been used to distinguish species with Sn—N and Sn—C bonds in certain reaction mixtures.[292]

Undoubtedly the greatest contribution made by ^{119}Sn NMR has been to the study of the formation and stability of complex species in which the tin atom is more than four-coordinate.[198,199,207,220,225,233] As pointed out earlier[195] this can lead to an increase in the tin shielding of several hundred ppm, and in cases such as $Me_2Sn(acac)_2$ [150], for which the stability is high, $\delta(^{119}Sn)$ is essentially unaffected by dilution by inert

TABLE 10.35 Coupling constants between tin and protons, $J(^{119}Sn, ^{1}H)$[a]

Compound	^{1}J/Hz	^{2}J/Hz	Ref.	Compound	^{2}J/Hz	^{3}J/Hz	Ref.
SnH_3^+ (−80°C)	2960	—	240	Vi_4Sn	96.0	183.1	266
SnH_3^- (−78°C)	109.4	—	255	Me_3SnLi[b]	−18 to −5.2	—	256, 267, 268
SnH_4	(−) 1930	—	133	$(Me_3Sn)_3SnLi$[b]	+36 to +39	2 to 8	267, 269
$MeSnH_3$	(−) 1852	+63.5	256	Et_4Sn	+32.2	−71.2	270
Me_2SnH_2	(−) 1797	+58.0	133, 256	Et_3SnCl	+40.6	−92	270
Me_3SnH	(−) 1744	+56.5	256	Et_2SnCl_2	+51.7	−130.8	270
Ph_3SnH	(−) 1935.8	—	257	$EtSnCl_3$	+83.4	−242	270
H_3SnCl	(−) 2448	—	258	$Ph_3SnC{\equiv}C\underline{H}$	—	34.2	271
$MeSnH_2Cl$	(−) 2228	—	259	$PhSnCl_3$	—	122	272
Me_4Sn	—	+53.9	256	(4-MeO-$C_6H_4)_4Sn$	—	45.8[c]	264
$Me_3SnW(CO)_3(Cp)$	—	+48.7	260	$Me_3SnSi\underline{H}_3$	51.6	—	273
$Me_2Sn[Co(CO)_4]_2$	—	+45	261	$(Me_3Si)_4Sn$		22.2	274
$MeSn[Co(CO)_4]_3$	—	+33	261	$Me_3SnSnMe_3$	+49	−17.3	189
Me_3SnCCl_3	—	+59	262	$Me_3SnSeMe$	+55.5	−31.1	237, 268
Me_3SnCl	—	+58.2	263	Me_3SnSMe	+56.9	(−)37.5	218, 268
Me_2SnCl_2	—	+68.9	263	$(MeS)_4Sn$	—	(−)66.0	218, 268
$MeSnCl_3$	—	+96.9	263	$Me_3SnOCH_2CCl_3$	—	−31.0	275
$Me_3SnN\underline{H}Ph$	—	20	264	$(Me_2N)_4Sn$	—	−51.0	276
$Me_3SnP\underline{H}_2$	—	62.5	265				

[a] Signs in parentheses have been inferred by comparison with similar molecules. Note that $\gamma(^{119}Sn)$ is negative, and therefore the reduced coupling constants $^{n}K(^{119}Sn, H)$ are of opposite sign to the corresponding $^{n}J(^{119}Sn, H)$. [b] Solution conditions affect the coupling constant. [c] $^{4}J(^{119}Sn, H) = 8.5$ Hz.

TABLE 10.36 Coupling constants between tin and carbon, $J(^{119}Sn, ^{13}C)$[a]

Compound	^1J/Hz[b]	^2J/Hz	^3J/Hz	^4J/Hz	Ref.
Me_4Sn	−340	—	—	—	246
Me_3SnCl	−380[c]	—	—	—	246
Me_2SnCl_2	−566[d]	—	—	—	246
Me_3SnBr	−380	—	—	—	246
$MeSnBr_3$	−640	—	—	—	246
Et_4Sn	−307	+23.5	—	—	250
Pr^n_4Sn	(−)313.4	20.1	51.4	—	251
Bu^n_4Sn	(−)313.7	19.3	52.0	"0"	251
Bu^nSnCl_3	(−)645	40	120	—	277
Vi_4Sn	(−)519.3	<6	—	—	251
$(ViCH_2)_4Sn$	(−)250	49.5	50.6	—	252
$Me_3Sn(cyclo\text{-}C_3H_5)$	341.6, 502.8	18.8	—	—	251
$Me_3Sn(cyclo\text{-}C_4H_7)$	313.5, 389.7	23.3	57.6	—	251
$Me_3Sn(cyclo\text{-}C_5H_9)$	311.8, 405.6	<6	51.2	—	251
$Me_3Sn(cyclo\text{-}C_6H_{11})$	303.9, 407.4	14.4	57.5	6.0	251
$Me_2Sn[CH_2]_3CH_2$ (ring)	302, 335.1	17.5	—	—	211
$Me_2Sn[CH_2]_4CH_2$ (ring)	312, 321.5	30.5	46.4	—	211
structure VII	314.2, 405.6	[f]	(x) 67.5	—	253
			(y) 11.9	—	253
1-$SnMe_3$-adamantane	290.4, 451.7	12.2	51.1	7.3	253
structure VIII	313.9, 411.8	(x) 12.4	(r) 57.8		253
			(s) 53.6		
		(y) 6.0	(t) 9.7		
			(w) 8.0		
structure IX	[f],416	(x) 12.0	(r) 23.0		253
		(y) 9.0	(s) 0		
			(w) 69[g]		
structure X	[f],432	(x) [f]	(r) 22.0		253
		(y) 10.0	(s) 59		
			(w) 36[g]		
Me_3SnPh[e]	347.5, 474.4	36.6	47.4	10.8	278
Ph_2SnCl_2	(−)785	63	90	16	277
$Me_2Sn(acac)_2$	(−)966	—	—	—	277
$Me_2Sn(NEt_2)_2$	(−)472	12	13	—	277
$(Me_3Sn)_2$	−240	−56	—	—	189
$Me_3SnLi \cdot 3THF$	+155	—	—	—	241
$Me_3SnLi \cdot 3HMPT$	+220	—	—	—	241

[a] Signs are only specified when known. Note that $\gamma(^{119}Sn)$ is negative. [b] The first figure refers to J in the Sn-methyl group, the second to J in the other organo-groups. [c] $^1J(^{119}Sn, ^{13}C) = -494$ Hz in 50% aqueous solution. [d] In acetone, addition of water increases the magnitude of the coupling. [e] Variations are noted in the various couplings in compounds with substituted phenyl groups. [f] Not observed or not determined. [g] Coupling to carbon t not reported.

TABLE 10.37 Coupling constants between tin and fluorine, $J(^{119}Sn, ^{19}F)$

Species	$^nJ/Hz^a$	n	Ref.
$[SnF_6]^{2-}$	1550	1	279
$[Sn(OH)F_5]^{2-}$	1776(b–e),[b] 127.8(f)[b]	1	280
cis-$[Sn(OH)_2F_4]^{2-}$	1820(c,e)[b] 1518(d,f)[b]	1	280
trans-$[Sn(OH)_2F_4]^{2-}$	1956	1	280
fac-$[Sn(OH)_3F_3]^{2-}$	1605	1	280
mer-$[Sn(OH)_3F_3]^{2-}$	1909(c, e)[b], 1670(d)[b]	1	280
$(Me_2PhCCH_2)_3SnF$	+2298[a]	1	207
$Me_3SnCF_2CF_2H$	249.5; 10.0	2, 3	281
$Me_2Sn(CF{=}CF_2)_2$	208	2	282
	25(*cis*); 29(*trans*)	3	

[a] Although only one sign determination has been performed all $^1J_{SnF}$ are probably positive. [b] These designations refer to the positions of ligands in octahedral complexes.

TABLE 10.38 Coupling constants between tin and phosphorus, $J(^{119}Sn, ^{31}P)$

Compound	$^nJ/Hz$	n	Ref.
Me_3SnPH_2	(+) 463	1	271
$Me_3SnPHPh$	(+) 538	1	283
Me_3SnPPh_2	+ 596	1	284
$(Me_3Sn)_2PPh$	+ 724	1	239
$(Me_3Sn)_3P$	+ 832.5	1	239
$Me_3SnPPh_2W(CO)_5$	+ 50	1	239
$(Me_3Sn)_3PCr(CO)_5$	+ 409	1	239
$(Me_3Sn)_3PW(CO)_5$	+ 375	1	239
$SnCl_4(PEt_3)_2$	2383.1	1	285
$Me_3SnW(Cp)(CO)_2PMe_3$	120	2	286
$Me_3SnW(Cp)(CO)_2PPh_3$	103.2	2	286
$Me_3SnPt(C{\equiv}CPh)(PMePh_2)_2$	1650	2	227
$(Me_3Sn)_2SW(CO)_4P(OMe)_3$	26	3	227

TABLE 10.39 Coupling constants between tin nuclei, $J(^{119}Sn, ^{119}Sn)^a$

Compound	$^nJ/Hz^b$	n	Ref.
$Me_3SnSnMe_3$	+4460	1	189
$Me_3SnSnEt_3$	3497	1	187
$Me_3SnSnBu^s_3$	2811	1	187
$Pr^i_2SnSnPr^i_3$	1209	1	187
$Pr^i_2Bu^tSnSnPr^i_2Bu^t$	764	1	187
$Me_3SnSnPh_3$	+4240	1	54
$(Me_3Sn)_2SnMe_3$	+2900	1	245
$(Me_3Sn)_3SnFe(Cp)[P(OPh)_3]_2$	0 ± 20	1	245
$(Me_3Sn)_3SnLi \cdot 3THF$	-5200 to -4445^c	1	245
$(Et_3Sn)_2SnBu^i_2$	450	2	187
$(Bu^n_2ClSn)_2O$	74	2	187
$Et_3Sn(CH_2)_2SnEt_3$	920	3	187

[a] Note that $^nJ(^{119}Sn, ^{119}Sn)$ and $^nK(SnSn)$ are of the same sign. [b] The values of ^1J were calculated by multiplying $J(^{119}Sn^{117}Sn)$ by 1.0465. [c] Depends upon conditions.

TABLE 10.40 Miscellaneous one-bond couplings involving tin or lead

X	Compound (M = Sn or Pb)	$^{1}J(^{119}SnX)$ / Hz	$^{1}J(^{207}PbX)$ / Hz	$^{1}K(SnX)$ / $10^{20}NA^{-2}m^{-3}$	$^{1}K(PbX)$ / $10^{20}NA^{-2}m^{-3}$	$^{1}K(PbX)/^{1}K(SnX)$	Refs
^{1}H	Me_3MH	−1744	+2295	+39.0	+90.8	+2.32	133, 256
^{11}B	$Me_3M\overline{BNMeCH_2CH_2N}Me$	−930	+1330	+64.8	+165	+2.55	224
^{13}C	Me_4M	−340	+250	+30.2	+39.6	+1.31	246, 287
^{15}N[a]	Me_3MNHPh	+2[b]	+261	+0.44[b]	−103	ca. −233	54, 224
^{29}Si[a]	Me_3MSiMe_3	+656	—	+73.7	—	—	54
^{31}P	Me_3MPPh_2	+596	+1335	−33.00	−131.5	+4.0	54, 224
^{77}Se	Me_3MSeMe	+1015	−1170	−119	−244	+2.06	237, 224
^{119}Sn[a]	Me_3MSnMe_3	+4460	−3570	+268	+382	+1.42	54, 224
^{119}Sn[a]	Me_3MSnPh_3	+4240	−2800	+254	+299	+1.18	54, 224
^{125}Te[a]	$(Me_3M)_2Te$	−1385	—	−98.1	—	—	232
^{183}W	$Me_3MW(cp)(CO)_3$	−150	—	+81	—	—	54
^{205}Tl	$[(Me_3M)_4Tl]^{-}Li^{+}$	−11 610	—	+451	—	—	202
^{207}Pb	Me_3MPbMe_3	−3570	+290	+382	+55.3	+0.14	224, 288

[a] γ is negative for this nuclide. [b] Very sensitive to nature of substituents on N.[227]

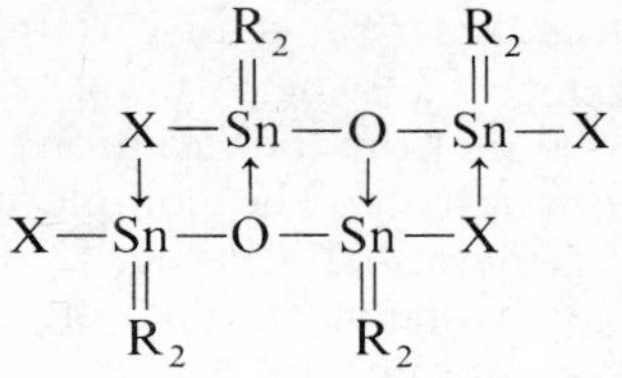

XV

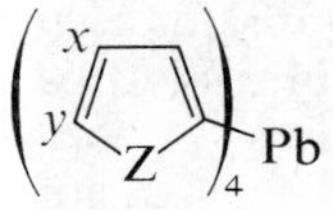

XVI

solvents. In a number of cases however, the stability constant is lower, and in solution there is a rapid reversible equilibrium whose position is sensitive to changes of concentration and composition. The observed tin chemical shift is then a weighted mean and consequently can be used to give the relative proportions of four- and five- or six-coordinate species, provided that $\delta(^{119}Sn)$ is known for the extreme states. Commonly this information can be obtained by extrapolation of data obtained over a range of conditions, or by measurements in either the absence or the presence of a large excess of the donor species. [It should perhaps be noted that the changes in $\delta(^{119}Sn)$ are usually accompanied by variations in $^{n}J(^{119}Sn, H)$ ($n = 2$ or 3), due mainly to redistribution of

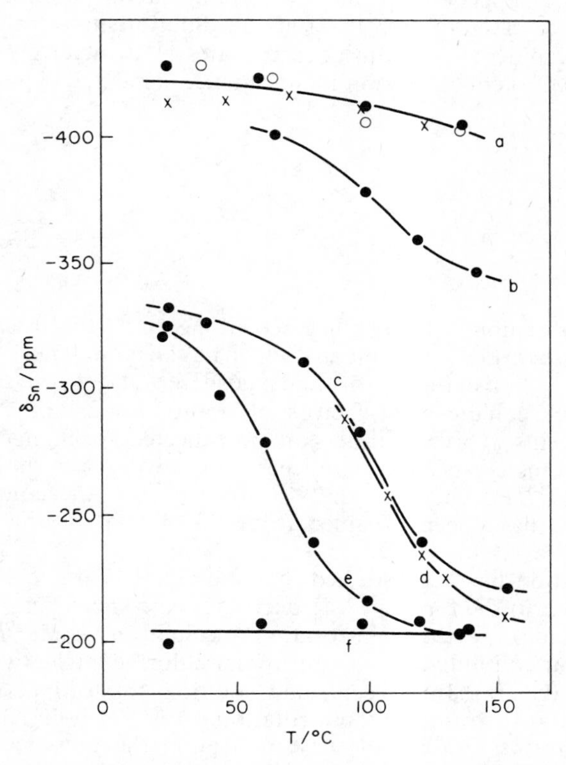

Fig. 10.13 Variation with temperature of $\delta(^{119}Sn)$ in *n*-butyltin trialkoxides, $Bu^nSn(OR)_3$. (a) R = *n*-alkyl, (b) R = Bu^i, (c) R = Pr^i, (d) R = Bu^tCH_2—, (e) R = Bu^s, (f) R = Bu^t. Adapted from ref. 194.

s-character in the tin hybrid orbitals, but the larger range of variation of the shielding makes it more suitable for accurate measurements]. By performing measurements over a range of temperatures it is thus possible to deduce from the variation of $\delta(^{119}Sn)$ thermodynamic parameters for the complexation process. The complexation of Me_3SnBr and Me_3SnCl by such donor molecules as pyridine, acetone, DMSO and trimethyl phosphate has been studied in this way, and a value of ca. 20 kJ/mol has been found for ΔH for an oxygen-tin donor bond in these compounds.[196]

Many species with tin-oxygen bonds are known to associate in the solid state by increasing the tin coordination number,[293] and in the case of tin alkoxides $R^1_{4-n}Sn(OR^2)_n$ this behaviour can persist in solution and has been studied in some detail[199,233] by ^{119}Sn NMR using ^1H-$\{^{119}Sn\}$ observation. The behaviour is typified by the plots[233] in Fig. 10.13 of $\delta(^{119}Sn)$ against temperature for $Bu^nSn(OR)_3$. For R = n-alkyl the tin atom is still essentially 6-coordinate even at 150° whereas for R = Bu^t the tin atom is 4-coordinate even at low temperatures, the bulky Bu^t groups inhibiting the association. Intermediate behaviour is found for species with rather less bulky R: for R = Bu^i the tin coordination number decreases from 6 to 5 over a range of 150° and for Bu^s and Pr^i it decreases from 5 to 4 over the same range. Typical values of ΔH and ΔS obtained in this way are $+66 \pm 4$ kJ mol^{-1} and $+190 \pm 10$ J mol^{-1} K^{-1}, respectively, for the association of $Me_2Sn(OBu^i)_2$ and +115 kJ mol^{-1} and +280 J mol^{-1} K^{-1} respectively[233] for $MeSn(OBu^s)_3$.

In other work the dependence of $\delta(^{119}Sn)$ upon tin coordination number has been used qualitatively to demonstrate the self-coordination in solution of triorganotin carboxylates,[207] cyanides[195] and dithiocarbamates.[234] Conversely, a relatively high-frequency shift allied to concentration independence is indicative of an absence of 5- or 6-coordinate species.

10D LEAD-207

10D.1 General

A surprisingly small amount of work has been done on ^{207}Pb in view of the relatively favourable NMR properties of this nuclide, which has an abundance of 23 %, a receptivity which is ca. 2×10^{-3} that of the proton, and a positive value of γ which leads to potential nuclear Overhauser enhancement factors of up to $\eta = 2.4$ in ^{207}Pb-$\{^1H\}$ double resonance experiments. This situation can be expected to change rapidly as multinuclear FT spectrometers become more generally available although to date about half the published ^{207}Pb chemical shifts[294] and most of the couplings[224] to nuclei other than protons have been obtained from ^1H-$\{^{207}Pb\}$ double resonance experiments.

Lead-207 relaxation has been studied in some detail in liquid $PbCl_4$ and aqueous $Pb(ClO_4)_2$ solution. In the former[295] T_1 decreases with increasing temperature and is apparently dominated by the spin-rotation mechanism, while T_2 is short and is dominated by scalar coupling to the quadrupolar chlorine nuclei (with $|^1J(^{207}Pb^{35}Cl)|$ estimated to be 705 Hz). In concentrated solutions of lead perchlorate $T_1(^{207}Pb)$ receives contributions from the spin-rotation, dipole-dipole, and chemical shift anisotropy mechanisms, the last of these indicating the presence of an appreciable amount of $PbClO_4^+$. In more dilute solutions spin-rotation dominates T_1, which is accurately equal to T_2, but at higher concentrations T_2 is shorter than expected, possibly owing to the presence of quadrupolar ^{17}O at natural abundance in the solvent water.[296,297] Values of T_1 ranging from 0.1 to 2 s have been recorded in organolead compounds, but no mechanisms have been unravelled.[298]

There do not yet appear to be any applications of ^{207}Pb NMR to the solution of chemical problems.

10D.2 ^{207}Pb Chemical Shifts

Selected values, representing about half the available data, are in Table 10.41. Tetramethyl lead seems to be a reasonable choice of reference compound, for which $\Xi(^{207}\text{Pb}) = 20.920597$ MHz. As with ^{119}Sn it appears that solvent and concentration effects are relatively unimportant unless there is specific chemical interaction, although this area has been very little studied.[6] An early theoretical treatment[299] suggested that lead shielding should be dominated by the paramagnetic term, and the broad parallels in behavior with tin shielding serve to confirm this. Figure 10.14 is a plot of $\delta(^{207}\text{Pb})$ against $\delta(^{119}\text{Sn})$ in analogous compounds, which shows a marked degree of correlation[294] (though not as good[300] as in the case of $\delta(^{125}\text{Te})$ plotted against $\delta(^{77}\text{Se})$) and a slope of 3.0. If the bonding is the same in comparable compounds of lead and tin then Eq. 3.36 would predict a slope of $\langle r^{-3}_{6p(\text{Pb})}\rangle / \langle r^{-3}_{5p(\text{Sn})}\rangle$ for four-coordinate species,[209] and this ratio can be estimated[301] as ca. 1.4. There is thus an unexplained factor of

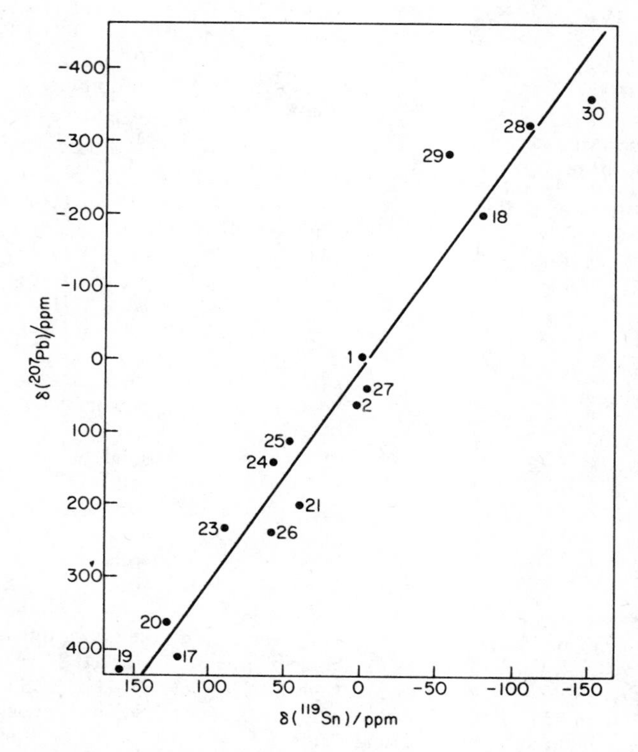

Fig. 10.14 Plot of $\delta(^{207}\text{Pb})$ against $\delta(^{119}\text{Sn})$ in analogous compounds. Points are numbered according to Table 10.41, and the continuous line has a slope of 3.0. Reproduced with permission from ref. 294.

TABLE 10.41 Lead chemical shifts

No.	Compound	$\delta(^{207}Pb)$/ppm[a]	Conditions	Ref.
1	Me_4Pb	0	85% in toluene	294
2	Et_4Pb	$+73.3 \pm 0.1$[b]	neat liquid	294
3	$(Bu^tCH_2)_4Pb$	$+47 \pm 8$	15% in CH_2Cl_2	294
4	Me_3PbPh	-38.5		6
5	Me_2PbPh_2	-80		6
6	$EtPbPh_3$	-114.5	0.5 M in CCl_4	298
7	$Me_3Pb(4\text{-}X\text{-}C_6H_4)$	-53.5 to -41.5[c]		302
8	$Me_3Pb(3\text{-}X\text{-}C_6H_4)$	-52 to -42.5[c]		302
9	$Me_2Pb(4\text{-}F\text{-}C_6H_4)_2$	-63 ± 0.8	neat liquid	302
10	$MePb(4\text{-}F\text{-}C_6H_4)_3$	-102.6 ± 0.8	in CCl_4	302
11	$(4\text{-}F\text{-}C_6H_4)_4Pb$	-153.5 ± 1.5	in CCl_4	302
12	$Me_3PbC{\equiv}CMe$	-139.8 ± 0.8	in C_6H_6	302
13	$Me_2Pb(C{\equiv}CMe)_2$	-304.7 ± 0.8	in C_6H_6	302
14	$(MeC{\equiv}C)_4Pb$	-685.5 ± 0.4	in C_6H_6	302
15	Me_3PbVi	-65.3 ± 0.8	neat liquid	302
16	Vi_4Pb	-251.0 ± 0.8	neat liquid	302
17	$[(CH_2)_4]_2Pb$[d]	$+416.3 \pm 0.2$	neat liquid	294
18	$[(CH_2)_5]_2Pb$[d]	-199 ± 8	50% in CH_2Cl_2	294
19	Me_3PbCl	$+432 \pm 1$[e]	sat. in CH_2Cl_2	294
20	Me_3PbBr	$+367 \pm 4$[f]	sat. in CH_2Cl_2	294
21	Me_3PbI	+203.[illegible] $\pm$ 0.3[f]	10% in CH_2Cl_2	294
22	$Me_3Pb^+(H_2O)_n$	-730	dilute in H_2O	6
23	Me_3PbSMe	$+239 \pm 2$[g]	10% in CH_2Cl_2	294
24	Me_3PbSBu^t	$+161.4 \pm 2$[g]	10% in CH_2Cl_2	294
25	$Me_3PbSeMe$	$+98 \pm 1$	20% in C_6H_6	294
26	Me_3PbNEt_2	$+242 \pm 5$	20% in C_6H_6	294
27	Me_3PbPPh_2	$+40 \pm 4$	20% in C_6H_6	294
28	$Me_3PbSnMe_3$	-324 ± 2	sat. in C_6H_6	294
29	$Me_3PbPbMe_3$	-281 ± 1	sat. in C_6H_6	294
30	$Me_3PbBNMeCH_2CH_2NMe$	-362 ± 2	30% in THF	294
31	$Me_3PbO{\cdot}CO{\cdot}R$	$+260$ to $+410$[h]	in CH_2Cl_2	294
32	$(Bu^tCH_2)_3PbBr$	$+384 \pm 2$	sat. in CH_2Cl_2	294
33	$(Bu^tCH_2)_3PbI$	$+336 \pm 2$	sat. in CH_2Cl_2	294
34	$Pb(O{\cdot}COMe)_2$	-1337.0[i]	1.1 M in H_2O	298
35	$Pb(NO_3)_2$	-2961.2	1.0 M in H_2O	298
36	$Pb(ClO_4)_2$	-2950[j]	in H_2O	303
37	Pb metal	$+11\,150$[j]	solid	303
38	PbO_2	$+4550 \pm 30$[j,k]	powder, 98% pure	304
39	PbO (yellow)	$+3750$[j]	solid	303
	PbO (red)	-5850[j]	solid	303
40	PbS	$+750$[l]	solid	305
41	PbSe	$+1050$[l]	solid	305
42	PbTe	$+50$[l]	solid	305
43	$PbCl_2$	-4750[j]	solid	303
44	$PbSO_4$	-4550[j]	solid	303

[a] To high frequency of Me_4Pb. [b] Variations of several ppm noted with different solvents. [c] Depends on X(CF_3, Cl, F, OMe, Me). [d] Spiroplumbane. [e] Marked solvent variations: +374.8 ($CDCl_3$), +258 (DMSO), +216 (pyridine), +166 (HMPT). [f] Similar solvent variations to those of Me_3PbCl. [g] Concentration dependent. [h] Solvent-dependent and depends upon pKa of parent carboxylic acid. [i] $\delta(^{207}Pb) = -1519$ ppm; 2.1 M in MeCOOH. [j] On the basis that $\delta(^{207}Pb)$ in 0.1 M $Pb(NO_3)_2$ soln. $= -2961$ ppm. [k] Impurities substantially increase $\delta(^{207}Pb)$. [l] On the basis that $\delta(^{207}Pb)$ in solid $Pb(NO_3)_2 = -3750$ ppm.

rather more than 2, which it is difficult to attribute entirely to lower electronic excitation energies in the case of lead compounds. Rather it appears that the electronic distribution about the lead atom is more sensitive than that about tin to changes in its substituents, which is reasonable in view of the greater polarisability of lead. Otherwise, however, in cases where a test is possible using the available data, the same factors[294,298,302] appear to control lead shielding as have been discussed for tin. Thus an increase in the electron-withdrawing ability of the substituents decreases the lead shielding, whilst bulky substituents in most cases increase it. An exception to the latter point is provided by the replacement of tin by lead in passing from $Me_3PbSnMe_3$ [28]* to $Me_3PbPbMe_3$ [29]. The increases $\delta(^{207}Pb)$, as already noted for $\delta(^{119}Sn)$ in the analogous transformation from $Me_3SnSnMe_3$ to $Me_3SnPbMe_3$. Abnormal interbond angles at lead also have similar but considerably larger effects as is shown by the two spiroplumbanes [17] and [18]. In this connection it is perhaps relevant to note the rather large chemical shift of Et_4Pb compared with a much smaller one for Et_4Sn.

The results for solid compounds[303–305] bring out the importance of the effect of coordination number upon lead shielding, and this is confirmed by such solvent studies as have been undertaken. For example, addition of the ligands DMSO or HMPT to solutions of Me_3PbBr or Me_3PbCl in CH_2Cl_2 produces increases in shielding of up to 270 ppm, which can be attributed to the formation of species containing 5-coordinate lead.[294] Interesting behaviour is displayed by Me_3PbI on addition of DMSO, for there is actually a small decrease in shielding, and the chemical shift under these conditions is very similar to those of the bromide and the chloride. One remarkable solvent effect is that $\delta(^{207}Pb)$ decreases by 31 ppm when H_2O is replaced by D_2O in 1 M solutions of lead(II) nitrate.[306]

10D.3 Coupling Constants

Very few coupling constants between lead and nuclei other than protons have been measured: in the case of one-bond couplings this is in part due to the instability or difficulty of preparation of suitable species. In these circumstances the $^1H\text{-}\{^{207}Pb\}$ method is particularly valuable since it facilitates rapid working and the unequivocal identification of species in mixtures by showing how resonances in the proton and lead spectra are related. The data of Table 10.40 were obtained mainly by this method and show that the ratio $^1K(PbX)/^1K(SnX)$ is a very variable quantity, which suggests that the use of the mean excitation energy approximation to describe $^1K(PbX)$ is very dubious. It is reasonable to assume that in analogous molecules α_s^2 should be roughly the same for tin and lead and that $S_{Pb}^2(0)/S_{Sn}^2(0)$ should remain constant at about 1.7, so a major part of the variation is to be attributed to changes in the Pb-X s-electron overlap integral. This is probably the case for $^1J(^{207}Pb, ^{207}Pb)$ in Me_6Pb_2 which has the extraordinarily smaller[183] than $^3J(^{207}PbCCH)$, although this does not necessarily apply when the 700 Hz between $Me_3PbSnMe_3$ and $Me_3PbSnPh_3$. The results so far available for $^1J(^{207}PbX)$ certainly suggest that it is unwise to draw close parallels between tin and lead.

There are rather closer parallels[307] between tin and lead when $^2J(^{207}Pb, H)$ and $^3J(^{207}Pb, H)$ in simple alkyllead compounds are concerned (Table 10.42). The lead couplings are usually some 20–50% larger than the analogous ones involving tin, and are of course opposite in sign because $\gamma(^{207}Pb)$ is positive, the geminal coupling normally being negative and the vicinal positive.[132] Numerically $^2J(^{207}PbCH)$ tends to be smaller[183] than $^3J(^{207}PbCCH)$, although this does not necessarily apply when the

* Numbers in square brackets refer to the lead compounds as listed in Table 10.41.

TABLE 10.42 Coupling constants between ^{207}Pb and ^{13}C or ^{1}H

Compound	$^{1}J_{PbC}$/Hz	$^{2}J_{PbC}$/Hz	$^{3}J_{PbC}$/Hz	$^{2}J_{PbH}$/Hz	$^{3}J_{PbH}$/Hz	Ref.
Me_4Pb	(+)250 ± 1	—	—	−60.5 ± 0.2	—	312
Et_4Pb	—	—	—	−41	+128.6	183
Me_2PbH_2[a]	—	—	—	−76.0 ± 0.5	—	133
Me_3PbCl	—	—	—	(−)70[b]	—	307
Me_3PbBr	(+)246	—	—	(−)68	—	309
Me_2PbCl_2	—	—	—	(−)154.5[c]	—	313
Ph_4Pb	(+)481	68	80	—	—	311
$(4\text{-}MeOC_6H_4)_4Pb$	(+)522	79	89	—	—	311
$Me_3PbPbMe_3$	+28 ± 2	+92 ± 2	—	−42.1 ± 0.1	+22.9 ± 0.1	287, 308
$(Bu^tCH_2)_4Pb$	(+)168	33	48	33.8	—	309
$(Bu^tCH_2)_3PbBr$	(+)133	34	66	28.0	—	309
Vi_4Pb	—	—	—	213.3	125.0 (*cis*), 331.8 (*trans*)	132
structure XVI, Z = O	(+)933	160.4	67.7 (*y*) 59.0 (*x*)	—	—	314
structure XVI, Z = S	(+)656	75.3	39.1 (*y*) 110.3 (*x*)	—	—	314

[a] $^{1}J(^{207}PbH) = +2454.6 \pm 1$ Hz. [b] Measured as saturated solution in $CHCl_3$; value changes to (−)82.5 ± 0.5 in DMSO (ref. 294). [c] Measured in DMSO solution and therefore probably refers to complexed species with co-ordinated solvent.

intervening atoms are other than carbon. This is illustrated by Fig. 10.7 (the proton spectrum of $Me_3PbSnMe_3$) which also shows how suitable many organo-lead compounds are for ^1H-{^{207}Pb} INDOR experiments. Two and three bond couplings between C and Pb have as yet been little examined, but again when the coupling path involves atoms other than sp^3 carbon, interesting variations in sign and magnitude occur. This can happen particularly when small *s*-overlap integrals occur in the bonds involved in the coupling path, as is the case for example in hexaalkyldilead compounds, and has lead to interesting discussions in the literature.[308,309] Couplings to lead over more than three bonds have been very little studied as yet, but the relevant ^{207}Pb satellites are generally rather more easy to detect than the corresponding ones arising from the presence of ^{119}Sn. Some examples are: $|^4J(^{207}Pb, H)| = 5.3 \pm 0.1$ Hz in $(Me_3CCH_2)_4Pb$;[310] $|^4J(^{207}Pb, H_a)| = 15.8 \pm 0.1$ Hz, and $|^4J(^{207}Pb, H_b)| = 18.3 \pm 0.1$ Hz in XVI (Z = O);[254a] $|^4J(^{207}Pb, ^{13}C)| = 20$ Hz in Ph_4Pb;[311] and $|^5J(^{207}Pb, ^{13}C)| = 12$ Hz and $|^6J(^{207}Pb, H)| = 5.4$ Hz in $(4\text{-}MeC_6H_4)_4Pb$.[254b,311]

REFERENCES

1. P. C. Lauterbur, "Determination of Organic Structures by Physical Methods," Vol. 2, pp. 511–515, (F. C. Nachod and W. D. Phillips, eds.) (1962)
2. R. B. Johannesen, F. E. Brinckman and T. D. Coyle, *J. Phys. Chem.* **72**, 660 (1968)
3. B. K. Hunter and L. W. Reeves, *Can. J. Chem.* **46**, 1399 (1968)
4. R. L. Scholl, G. E. Maciel and W. K. Musker, *J. Amer. Chem. Soc.* **94**, 6376 (1972)
5. H.-G. Horn and H. C. Marsmann, *Makromol. Chem.* **162**, 255 (1972)
6. P. R. Wells, "Determination of Organic Structures by Physical Methods," Vol. 4, p. 233, (F. C. Nachod and J. J. Zuckerman, eds.) (1971)
7. H. C. Marsmann, *Chem. Ztg.* **97**, 128 (1973)
8. J. Schraml and J. M. Bellama, "Determination of Organic Structures by Physical Methods", Vol. 6, Ch. 4, (F. C. Nachod, J. J. Zuckerman and E. W. Randall, eds.) (1976)
9. G. C. Levy, J. D. Cargioli, G. E. Maciel, J. J. Natterstad, E. B. Whipple and M. Ruta, *J. Magn. Reson.* **11**, 352 (1973)
10. M. R. Bacon and G. E. Maciel, *J. Amer. Soc.* **95**, 2413 (1973)
11. E. A. Williams, J. D. Cargioli and R. W. LaRochelle, *J. Organometal. Chem.* **108**, 153 (1976)
12. R. K. Harris and R. H. Newman, *J.C.S. Faraday II* **73**, 1204 (1977)
13. R. K. Harris and B. J. Kimber, *J. Magn. Reson.* **17**, 174 (1975)
14. W. McFarlane, *Ann. Rev. NMR Spectry* **1**, 135 (1968)
15. R. H. Newman Ph.D. Thesis, University of East Anglia (1976).
16. G. C. Levy, J. D. Cargioli, P. C. Juliano add T. D. Mitchell, *J. Magn. Reson.* **8**, 399 (1972)
17. R. K. Harris and B. J. Kimber, *Appl. Spect. Rev.* **10**, 117 (1975)
18. G. C. Levy, J. D. Cargioli, P. C. Juliano and T. D. Mitchell, *J. Amer. Chem. Soc.* **95**, 3445 (1973)
19. R. K. Harris and B. J. Kimber, *Org. Magn. Reson.* **7**, 460 (1975)
20. H. C. Marsmann, *Z. Naturforsch.* **29b**, 495 (1974)
21. R. K. Harris and R. H. Newman, *Org. Magn. Reson.* **9**, 426 (1977)
22. M. D. Sefcik, private communication
23. H. C. Marsmann, unpublished results quoted in ref. 8.
24. S. Cradock, E. A. V. Ebsworth, N. S. Hosmane and K. M. McKay, *Angew. Chem.* (*Internat.*) **14**, 167 (1975)
25. H. C. Marsmann and H. G. Horn, *Chem. Ztg.* **96**, 456 (1972)
26. U. Niemann and H. C. Marsmann, *Z. Naturforsch.* **30b**, 202 (1975)
27. E. V. van den Berghe and G. P. van der Kelen, *J. Organometal. Chem.* **59**, 175 (1973)
28. E. V. van den Berghe and G. P. van der Kelen, *J. Organometal. Chem.* **122**, 329 (1976)
29. P. Löwer, M. Vongehr and H. C. Marsmann, *Chem. Ztg.* **99**, 33 (1975)

30. J. Schraml, Nguyen Duc-Chuy, J. Pola, P. Novák, V. Chvalovský, M. Mägi, and E. Lippmaa, unpublished work quoted in ref. 8.
31. G. Engelhardt, M. Mägi and E. Lippmaa, *J. Organometal. Chem.* **54**, 115 (1973)
32. R. K. Harris and B. J. Kimber, *Adv. Mol. Relaxation Processes* **8**, 23 (1976)
33. G. Engelhardt, R. Radeglia, H. Jancke, E. Lippmaa and M. Mägi, *Org. Magn. Reson.* **5**, 561 (1973)
34. G. Engelhardt, H. Jancke, M. Mägi, T. Pehk and E. Lippmaa, *J. Organometal. Chem.* **28**, 293 (1971)
35. J. Schraml, J. Pola, V. Chvalovský, M. Mägi and E. Lippmaa, *J. Organometal. Chem.* **49**, C19 (1973)
36. J. Schraml, V. Chvalovský, H. Jancke and G. Engelhardt, *Org. Magn. Reson.* **9**, 237 (1977)
37. T. Pekhk, E. Lippmaa, E. Lukevits and L. I. Simchenko, *J. Gen. Chem.* **46**, 600 (1976)
38. W. McFarlane and J. M. Seaby, *J.C.S. Perkin II*, 1561 (1972)
39. G. C. Levy and J. D. Cargioli, in "NMR of Nuclei other than Protons", Ch. 17, pp. 251–274, (T. Axenrod and G. A. Webb, eds.). Wiley-Interscience (1974)
40. M. Vongehr and H. C. Marsmann, *Z. Naturforsch.* **31b**, 1423 (1976)
41. H. C. Marsmann, *Chem. Ztg.* **96**, 288 (1972)
42. K. G. Sharp, P. A. Sutor, E. A. Williams, J. D. Cargioli, T. C. Farrar and K. Ishibitsu, *J. Amer. Chem. Soc.* **98**, 1977 (1976)
43. R. K. Harris and B. Lemarié, *J. Magn. Reson.* **23**, 371 (1976)
44. E. A. V. Ebsworth, private communication; J. M. Edward, Ph.D. Thesis, Edinburgh Univ. (1976)
45. H. Muller and H.-J. Kroth, unpublished work
46. Nguyen-Duc-Chuy, V. Chvalovský, J. Schraml, M. Mägi and E. Lippmaa, *Coll. Czech. Chem. Comm.* **40**, 875 (1975)
47. C. R. Ernst, L. Spialter, G. R. Buell and D. L. Wilhite, *J. Amer. Chem. Soc.* **96**, 5375 (1974)
48. R. Radeglia, *Z. Phys. Chem.* (*Leipzig*) **256**, 453 (1975)
49. G. E. Maciel, *Topics in* ^{13}*C NMR Spectry.* **1**, 53 (1974)
50. E. Niecke and W. Flick, *Angew. Chem.* (*Internat.*) **12**, 585 (1973)
51. S. Cradock, E. A. V. Ebsworth, G. D. Meikle and D. W. H. Rankin, *J. C. S. Dalton*, 805 (1975)
52. R. K. Harris and R. J. Morrow, unpublished work
53. H. C. Marsmann and H.-G. Horn, *Z. Naturforsch.* **27b**, 1448 (1972)
54. J. D. Kennedy, W. McFarlane, G. S. Pyne and B. Wrackmeyer, *J.C.S. Dalton*, 386 (1975)
55. a. D. W. W. Anderson, J. E. Bentham and D. W. H. Rankin, *J.C.S. Dalton*, 1215 (1973)
b. S. Cradock, E. A. V. Elsworth, D. W. H. Rankin and W. J. Savage, *J.C.S. Dalton*, 1661 (1976).
56. J. Schraml, J. Pola, H. Jancke, G. Engelhardt, M. Černý and V. Chvalovský, *Coll. Czech. Chem. Comm.* **41**, 360 (1976)
57. M. Mägi, E. Lippmaa, Nguyen Duc-Chuy, J. Pola, P. Novák, V. Chvalovský and J. Schraml, unpublished work quoted by J. Schraml, J. Včelák, G. Engelhardt and V. Chvalovský, *Coll. Czech. Chem. Comm.* **41**, 3758 (1976)
58. J. Schraml, J. Pola, V. Chvalovský, M. Mägi and E. Lippmaa, unpublished work quoted in ref. 56
59. J. Schraml, P. Koehler, K. Licht, and G. Engelhardt, *J. Organometal Chem.* **121**, Cl (1976)
60. D. J. Gale, A. H. Haines and R. K. Harris, *Org. Magn. Reson.* **7**, 635 (1975)
61. A. H. Haines, R. K. Harris and R.-C. Rao, *Org. Magn. Reson.* **9**, 432 (1977)
62. J. Pola, Z. Papoušková and V. Chvalovský, *Coll. Czech. Chem. Comm.* **41**, 239 (1976)
63. J. Schraml, Nguyen-Duc-Chuy, V. Chvalovský, M. Mägi and E. Lippmaa, *Org. Magn. Reson.* **7**, 379 (1975)
64. J. Schraml, V. Chvalovský, M. Mägi, E. Lippmaa, R. Calas, J. Dunogues and P. Bourgeois, *J. Organometal. Chem.* **120**, 41 (1976)
65. G. Engelhardt and J. Schraml, *Org. Magn. Reson.* **9**, 239 (1977)
66. J. Schraml, J. Včelák, V. Chvalovský, G. Engelhardt, L. Vodička and J. Hlavatý, *Coll. Czech. Chem. Comm.* in press (quoted in ref. 65)
67. D. M. Grant and E. G. Paul, *J. Amer. Chem. Soc.* **86**, 2984 (1964)
68. R. K. Harris and B. J. Kimber, *Adv. Mol. Relaxation Processes* **8**, 15 (1976)
69. R. K. Harris, unpublished work

70. T. N. Mitchell, *Org. Magn. Reson.* **7**, 610 (1975)
71. G. Fritz and E. Bosch, *Z. anorg. allgem. Chem.* **404**, 103 (1974)
72. E. Niecke and W. Flick, *Angew. Chem. (Internat.)* **14**, 363 (1975)
73. H. Schmidbauer and M. Heimann, *Angew. Chem. (Internat.)* **15**, 367 (1976)
74. E. Niecke, W. Flick and S. Pohl, *Angew. Chem. (Internat.)* **15**, 309 (1976)
75. H. Schmidbauer, B. Zimmer, F. H. Kohler and W. Buchner, *Z. Naturforsch.* **32b** 481 (1977)
76. U. Klingebiel and A. Meller, *Z. Naturforsch.* **32b**, 537 (1977)
77. O. Schlak, W. Stadelmann, O. Stelzer and R. Schmutzler, *Z. anorg. allgem. Chem.* **419**, 275 (1976)
78. U. Klingebiel and A. Meller, *Angew. Chem. (Internat.)* **15**, 312 (1976)
79. U. Klingebiel and A. Meller, *Angew. Chem. (Internat.)* **15**, 313 (1976)
80. W. Lidy, C. Jackh and W. Sundermeyer, *Z. Naturforsch,* **32b**, 595 (1977)
81. W. Buss, H.-J. Krannich and W. Sundermeyer, *Z. Naturforsch.* **30b**, 842 (1975)
82. D.-L. Wagner, H. Wagner and O. Glemser, *Chem. Ber.* **109**, 1424 (1976)
83. M. Mägi, E. Lippmaa, E. Lukevics and N. P. Erčak, *Org. Magn. Reson.* **9**, 297 (1977)
84. A. Haas and M. Vongehr, *Chem. Ztg.* **99**, 432 (1975)
85. W. Lidy and W. Sundermeyer, *Chem. Ber.* **109**, 2542 (1976)
86. G. Fritz and U. Finke, *Z. anorg. allgem. Chem.* **430**, 121 (1977)
87. S. Cradock, E. A. V. Ebsworth and N. Hosmane, *J.C.S. Dalton*, 1624 (1975)
88. E. A. V. Ebsworth, J. M. Edward and D. W. H. Rankin, *J. C. S. Dalton*, 1673 (1976)
89. K. G. Sharp, P. A. Sutor, T. C. Farrar and K. Ishibitsu, *J. Amer. Chem. Soc.* **97**, 5612 (1975)
90. G. Fritz and M. Hahnke, *Z. anorg. allgem. Chem.* **390**, 157 (1972)
91. G. Fritz, J. W. Chang and N. Braunagel, *Z. anorg. allgem. Chem.* **416**, 211 (1975)
92. G. Fritz and I. Arnason, *Z. anorg. allgem. Chem.* **419**, 213 (1976)
93. G. Fritz and E. Matern, *Z. Anorg. allgem. Chem.* **426**, 28 (1976)
94. D. Seyferth and D. G. Annarelli, *J. Organometal. Chem.* **117**, C51 (1976); *J. Amer. Chem. Soc.* **97**, 2273 (1975)
95. D. Seyferth, R. L. Lambert and D. C. Annarelli, *J. Organometal. Chem.* **122**, 311 (1976)
96. D. Seyferth, D. C. Annarelli and C. Vick, *J. Amer. Chem. Soc.* **98**, 6382 (1976)
97. H. Jancke, G. Engelhardt, M. Mägi and E. Lippmaa, *Z. Chem.* **13**, 435 (1973)
98. A. Blaschette, D. Rinne and H. C. Marsmann, *Z. anorg. allgem. Chem.* **420**, 55 (1976)
99. E. Niecke and W. Bitter, *Chem. Ber.* **109**, 415 (1976)
100. B. Heinz, H. C. Marsmann and U. Niemann, *Z. Naturforsch.* **32b**, 163 (1977)
101. D. W. W. Anderson, E. A. V. Ebsworth and D. W. H. Rankin, *J. C. S. Dalton*, 2370 (1973)
102. H. Bürger, R. Eujen and H. C. Marsmann, *Z. Naturforsch.* **29b**, 149 (1974)
103. H. C. Marsmann and R. Löwer, *Chem. Ztg.* **97**, 660 (1973)
104. D. Kummer and T. Sheshadri, *Z. anorg. allgem. Chem.* **428**, 129 (1977)
105. R. K. Harris, J. Jones and Soon Ng, *J. Magn. Reson.* **30**, 521 (1978)
106. R. W. LaRochelle, J. D. Cargioli and E. A. Williams, *Macromolecules* **9**, 85, (1976)
107. R. K. Harris and M. L. Robins, *Polymer* **19**, 1123 (1978)
108. R. K. Harris and M. D. Wood, unpublished work
109. H. C. Marsmann, unpublished work quoted in ref. 7
110. A. M. Chippendale, unpublished work
111. R. K. Harris and B. J. Kimber, *J. Organometal. Chem.* **70**, 43 (1974)
112. J. Schraml, V. Chvalovský, M. Mägi and E. Lippmaa, unpublished results quoted in ref. 113
113. J. Souček, G. Engelhardt, K. Stránský and J. Schraml, *Coll. Czech. Chem. Comm.* **41**, 234 (1976)
114. D. Hoebbel, G. Garzó, G. Engelhardt, H. Jancke, P. Franke and W. Wieker, *Z. anorg. allgem. Chem.* **424**, 115 (1976)
115. H. Jancke, G. Engelhardt, M. Mägi and E. Lippmaa, *Z. Chem.* **13**, 392 (1973)
116. R. K. Harris, B. J. Kimber, M. D. Wood, and A. Holt, *J. Organometal. Chem.* **116**, 291 (1976)
117. R. K. Harris, J. Jones, E. Lachowski and L. S. Dent Glasser, unpublished work
118. D. Hoebbel, W. Wieker, P. Franke and A. Otto, *Z. anorg. allgem. Chem.* **418**, 35 (1975)
119. R. K. Harris and B. J. Kimber, *J.C.S. Chem. Comm.*, 559 (1974)
120. E. A. Williams and J. D. Cargioli, private communication

121. G. Engelhardt, W. Altenburg, D. Hoebbel and W. Wieker, *Z. anorg. allgem. Chem.* **428**, 43 (1977)
122. G. Engelhardt, H. Jancke, D. Hoebbel and W. Wieker, *Z. Chem.* **14**, 109 (1974)
123. G. Engelhardt, D. Zeigan, H. Jancke, D. Hoebbel and W. Wieker, *Z. anorg. Chem.* **418**, 17 (1975)
124. R. O. Gould, B. M. Lowe and N. A. MacGilp, *J.C.S. Chem. Comm.*, 720 (1974)
125. F. F. Roelandt, D. F. van de Vondel and E. A. van den Berghe, *J. Organometal. Chem.* **94**, 377 (1975)
126. R. Radeglia and G. Engelhardt, *J. Organometal. Chem.* **67**, C45 (1974)
127. R. Wolff and R. Radeglia, *Z. Phys. Chem.* (*Leipzig*) **257**, 64 (1977)
128. R. Wolff and R. Radeglia, *Org. Magn. Reson.* **9**, 181 (1976)
129. R. Wolff and R. Radeglia, *Z. Phys. Chem.* (*Leipzig*) **258**, 145 (1977)
130. G. C. Levy, D. M. White and J. D. Cargioli, *J. Magn. Reson.* **8**, 280 (1972)
131. H. Dreeskamp and K. Hildenbrand, *Liebigs Ann. Chem.* **712** (1975)
132. P. Krebs and H. Dreeskamp, *Spectrochim. Acta* **25A**, 1399 (1969)
133. C. Schumann and H. Dreeskamp, *J. Magn. Reson.* **3**, 204 (1970)
134. G. Pfisterer and H. Dreeskamp, *Ber. Bunsengesell. Phys. Chem.* **73**, 654 (1969)
135. P. A. W. Dean and D. F. Evans, *J. Chem. Soc.* (*A*), 2569 (1970)
136. G. Fritz and H. Schäfer, *Z. anorg. allgem. Chem.* **406** 167 (1970); **409**, 137 (1974)
137. E. Niecke and W. Flick, *Angew. Chem.* (*Internat.*) **13**, 134 (1974)
138. W. Buchner and W. Wolfsberger, *Z. Naturforsch.* **29b**, 328 (1974)
139. R. D. Bertrand, J. W. Rathke and J. G. Verkade, *Phosphorus* **3**, 1 (1973)
140. Sv. Aa. Linde, H. J. Jakobsen and B. J. Kimber, *J. Amer. Chem. Soc.* **97**, 3219 (1975)
141. H. Dreeskamp and K. Hildenbrand, *Liebigs Ann. Chem.* 712 (1975)
142. E. A. V. Ebsworth and J. J. Turner, *J. Chem. Phys.* **36**, 2628 (1962)
143. M. Murray, *J. Magn. Reson.* **9**, 326 (1973)
144. K. Kovačevič and Z. B. Maksič, *J. Mol. Structure* **17**, 203 (1973)
145. K. D. Summerhays and D. A. Deprez, *J. Organometal. Chem.* **118**, 19 (1976)
146. M. D. Beer and R. Grinter, *J. Magn. Reson.* **26**, 421 (1977)
147. M. D. Beer and R. Grinter, *J. Magn. Reson.* **31**, 187 (1978).
148. K. Kovačevič, K. Krmpotič and Z. B. Maksič. *Inorg. Chem.* **16**, 1421 (1977)
149. G. C. Levy, *J. Amer. Chem. Soc.* **94**, 4793 (1972)
150. G. C. Levy, J. D. Cargioli, P. C. Juliano and T. D. Mitchell, *J. Amer. Chem. Soc.* **95**, 3445 (1973)
151. B. J. Kimber and R. K. Harris, *J. Magn. Reson.* **16**, 354 (1974)
152. R. K. Harris, B. Lemarié and Soon Ng, unpublished work
153. G. Engelhardt, *Z. Chem.* **15**, 495 (1975)
154. A. Briguet and A. Erbeia, *J. Phys.* (*Paris*) C5, L-58 (1972)
155. A. Briguet, J.-C. Duplan and J. Delmau, *J. Phys.* (*Paris*) **36**, 897 (1975)
156. P. S. Hubbard, *Phys. Rev.* **131**, 1155 (1963)
157. R. K. Harris and B. J. Kimber, unpublished work
158. R. K. Harris, D. J. Gale and R. C. Rao, unpublished work
159. G. Engelhardt and H. Jancke, *Z. Chem.* **14**, 206 (1974)
160. B. J. Kimber, Ph.D. thesis, University of East Anglia (1974)
161. G. C. Levy and R. A. Komoroski, *J. Amer. Chem. Soc.* **96**, 678 (1974)
162. R. K. Harris and B. J. Kimber, *J. C. S. Chem. Comm.*, 255 (1973)
163. J. W. Fleming and C. N. Banwell, *J. Mol. Spectry.* **31**, 318 (1969)
164. W. Caminati, G. Cazzoli and A. M. Mirri, *Chem. Phys. Lett.* **35**, 475 (1975)
165. A. K. Jameson and J. P. Reger, *J. Phys. Chem.* **75**, 437 (1971)
166. B. D. Mosel, W. Müller-Warmuth and H. Dutz, *Physics and Chem. of Glasses* **15**, 154 (1974)
167. G. R. Holzman, P. C. Lauterbur, J. H. Anderson and W. Koth, *J. Chem. Phys.* **25**, 172 (1956)
168. A. Pines, M. G. Gibby and J. S. Waugh, *J. Chem. Phys.* **59**, 569 (1973)
169. M. Mehring, *NMR: Basic Principles and Progress* **11**, 1 (1976)
170. M. G. Gibby, A. Pines and J. S. Waugh, *J. Amer. Chem. Soc.* **94**, 6231 (1972)
171. J. Schaefer and E. O. Stejskal, *J. Amer. Chem. Soc.* **98**, 1031 (1976)

172. J. D. Kennedy and W. McFarlane, unpublished results from $^{1}H-\{^{73}Ge\}$ double resonance
173. R. G. Kidd and H. G. Spinney, *J. Amer. Chem. Soc.* **95**, 88 (1973)
174. J. Kaufmann, W. Sahm and A. Schwenk, *Z. Naturforsch.* **26a**, 1384 (1971)
175. P. Geerlings and C. van Alsenoy, *J. Organometal. Chem.* **117**, 13 (1976)
176. W. Sahm and A. Schwenk, *Z. Naturforsch.* **29a**, 1763 (1974)
177. A. Schwenk, *Phys. Lett.* **31A**, 513 (1970)
178. J. Banck and A. Schwenk, *Z. Physik* **265**, 165 (1973)
179. C. W. Burges, R. Koschneider, W. Sahm and A. Schwenk, *Z. Naturforsch.* **28a**, 1753 (1973)
180. H. Dreeskamp, *Z. Naturforsch.* **19a**, 139 (1964)
181. D. K. Dalling and H. S. Gutowsky, *Govt. Rep. Announce.(U.S.)* **71**, 65 (1971); [*Chem. Abs.* **75**, 82238 (1971)]
182. P. T. Inglefield and L. W. Reeves, *J. Chem. Phys.* **40**, 2425 (1964)
183. P. T. Narasimhan and M. T. Rogers, *J. Chem. Phys.* **34**, 1049 (1961)
184. J. J. Burke and P. C. Lauterbur, *J. Amer. Chem. Soc.* **83**, 326 (1961)
185. W. McFarlane, *Ann. Rep. NMR Spectry.* **5**, 353 (1972)
186. J. D. Kennedy and W. McFarlane, *J.C.S. Dalton*, 1219 (1976)
187. T. N. Mitchell, *J. Organometal. Chem.* **70**, C1 (1974)
188. T. N. Mitchell, *Org. Magn. Reson.* **8**, 34 (1976)
189. W. McFarlane, *J. Chem. Soc.* (*A*) 1630 (1968)
190. R. R. Sharp, *J. Chem. Phys.* **57**, 5321 (1972)
191. R. R. Sharp, *J. Chem. Phys.* **60**, 1149 (1974)
192. R. R. Sharp and J. W. Tolan, *J. Chem. Phys.* **65**, 522 (1976)
193. P. J. Smith and L. Smith, *Inorg. Chim. Acta Rev.* **7**, 11 (1973)
194. J. D. Kennedy and W. McFarlane, *Rev. Silicon, Germanium, Tin and Lead Compounds* **1**, 235 (1973)
195. A. G. Davies, P. G. Harrison, J. D. Kennedy, R. J. Puddephatt, T. N. Mitchell and W. McFarlane, *J. Chem. Soc.* (*A*), 1136 (1969)
196. V. N. Torocheshnikov, A. P. Tupčiauskas, N. M. Sergeyev and Yu. A. Ustynyuk, *J. Organometal. Chem.* **29**, 245 (1971)
197. J. D. Kennedy and W. McFarlane, *J.C.S. Perkin II*, 146 (1974)
198. P. J. Smith, R. F. M. White and L. Smith, *J. Organometal. Chem.* **40**, 341 (1972)
199. J. D. Kennedy, W. McFarlane, L. Smith, P. J. Smith and R. F. M. White, *J.C.S. Perkin II*, 1785 (1973)
200. A. Tupčiauskas, N. M. Sergeyev and Yu. A. Ustynyuk, *Mol. Phys.* **21**, 179 (1971)
201. G. Gegenbauer, H. Krüger, O. Lutz, A. Nolle, A. Schwenk, and A. Uhl, *Z. Physik* **270**, 255 (1974)
202. J. D. Kennedy and W. McFarlane, unpublished observations
203. W. McFarlane, *J. Magn. Reson.* **10**, 98 (1973)
204. E. D. Becker, R. B. Bradley, and T. Axenrod, *J. Magn. Reson.* **4**, 136 (1971)
205. P. G. Harrison, J. J. Zuckerman and S. E. Ulrich, *J. Amer. Chem. Soc.* **93**, 5398 (1971)
206. J. D. Kennedy, W. McFarlane, G. S. Pyne, P. L. Clarke and J. D. Wardell, *J.C.S. Perkin II*, 1234 (1975)
207. W. McFarlane and R. J. Wood, *J. Organometal. Chem.* **40**, C17 (1972)
208. H. J. Kroth, H. Schumann, H. G. Kuivila, C. D. Schaeffer and J. J. Zuckerman, *J. Amer. Chem. Soc.* **97**, 1754 (1975)
209. C. J. Jameson and H. S. Gutowsky, *J. Chem. Phys.* **40**, 1714 (1964)
210. R. Radeglia and G. Engelhardt, *Z. Chem.* **14**, 319 (1974)
211. J. D. Kennedy, W. McFarlane and G. S. Pyne, *Bull. Soc. Chim. Belg.* **84**, 289 (1975)
212. B. Wrackmeyer, private communication
213. L. Verdonck and G. P. van der Kelen, *J. Organometal. Chem.* **40**, 139 (1972)
214. D. H. Harris, M. F. Lappert, J. S. Poland and W. McFarlane, *J.C.S. Dalton* 311 (1975)
215. W. McFarlane, *Chem. Comm.*, 418 (1970)
216. M. Barnard, P. J. Smith and R. F. M. White, *J. Organometal. Chem.* **77**, 189 (1974)
217. *See for example*: E. A. V. Ebsworth, "Organometallic Compounds of the Group IV Elements", Vol. 1 (A. G. McDiarmid, ed.). Marcel Dekker (1968)

218. E. V. van den Berghe and G. P. van der Kelen, *J. Organometal. Chem.* **26**, 207 (1971)
219. W. McFarlane, J. C. Maire and M. Delmas, *J.C.S. Dalton*, 1862 (1972)
220. J. D. Kennedy, *J. Mol. Struct.* **31**, 207 (1976)
221. E. V. van den Berghe and G. P. van der Kelen, *J. Mol. Struct.* **20**, 147 (1974)
222. J. Mason, *J. Chem. Soc.* (*A*) 1038 (1971)
223. L. Phillips and V. Wray, *J.C.S. Perkin II*, 214 (1972)
224. J. D. Kennedy, W. McFarlane and B. Wrackmeyer, *Inorg. Chem.* **15**, 1299 (1976)
225. A. P. Tupčiauskas, N. M. Sergeyev and Yu. A. Ustynyuk, *Lietvos. Phys. Rink.*, **11**, 93 (1971)
226. A. P. Tupčiauskas, N. M. Sergeyev and Yu. A. Ustynyuk, *Org. Magn. Reson.* **3**, 655 (1971)
227. J. M. Bassett, B. W. Fitzsimmons, P. C. Fowler, D. Harris, J. D. Kennedy, S. Keppie, M. F. Lappert, W. McFarlane, J. Poland, G. S. Pyne and D. S. Rycroft, unpublished observations
228. V. N. Torochesnikov, A. P. Tupčiauskas and Yu. A. Ustynyuk, *J. Organometal. Chem.* **81**, 351 (1974)
229. J. D. Kennedy, *J. Organometal. Chem.* **104**, 311 (1976)
230. W. McFarlane, private communication quoted in R. G. Jones, P. Partington, W. J. Rennie and R. M. G. Roberts, *J. Organometal. Chem.* **35**, 291 (1972)
231. A. G. Davies, L. Smith and P. J. Smith, *J. Organometal. Chem.* **39**, 279 (1972)
232. J. D. Kennedy and W. McFarlane, *J. Organometal. Chem.* **94**, 7 (1975)
233. J. D. Kennedy, *J.C.S. Perkin II*, 242 (1977)
234. A. G. Davies and J. D. Kennedy, *J. Chem. Soc.* (*C*) 759 (1970)
235. A. C. Chapman, A. G. Davies, P. G. Harrison and W. McFarlane, *J. Chem. Soc.* (*C*) 821 (1970)
236. A. G. Davies, L. Smith, P. J. Smith and W. McFarlane, *J. Organometal. Chem.* **29**, 245 (1971)
237. J. D. Kennedy and W. McFarlane, *J.C.S. Dalton*, 2134 (1973)
238. E. V. van den Berghe and G. P. van der Kelen, *J. Organometal. Chem.* **61**, 197 (1973)
239. W. McFarlane and D. S. Rycroft, *J.C.S. Dalton*, 1977 (1974)
240. J. R. Webster and W. L. Jolly, *Inorg. Chem.* **10**, 877 (1971)
241. J. D. Kennedy and W. McFarlane, *Chem. Comm.*, 983 (1974)
242. R. C. Poller, "The Chemistry of Organotin Compounds", Logos Press (1970).
243. J. A. Pople and D. P. Santry, *Mol. Phys.* **8**, 1, (1964)
244. W. McFarlane, *Proc. Roy. Soc.* **A306**, 185 (1968)
245. J. D. Kennedy and W. McFarlane, *J.C.S. Dalton*, 1219 (1976)
246. W. McFarlane, *J. Chem. Soc.* (*A*) 528 (1967)
247. R. Gupta and B. Majee, *J. Organometal. Chem.* **40**, 97 (1972)
248. M. Gilen, M. de Clercq and B. de Poorter, *J. Organometal. Chem.* **34**, 305 (1972)
249. C. E. Holloway, S. A. Kandil and I. M. Walker, *J. Amer. Chem. Soc.* **94**, 4027 (1972)
250. F. J. Weigert and J. D. Roberts, *J. Amer. Chem. Soc.* **91**, 4940 (1969)
251. H. G. Kuivila, J. L. Considine, R. J. Mynott and R. H. Sarma, *J. Organometal. Chem.* **55**, C11 (1973)
252. Y. K. Grishin, N. M. Sergeyev and Yu. A. Ustynyuk, *Org. Magn. Reson.* **4**, 337 (1972)
253. D. Doddrell, I. Burfitt, W. Kitching, M. Bullpit, C. H. Lee, R. J. Mynott, J. L. Considine, H. G. Kuivila and R. H. Sarma, *J. Amer. Chem. Soc.* **96**, 1640 (1974)
254. a. C. Barbieri and F. Taddei, *J.C.S. Perkin II*, 262 (1972); 1323 (1972)
b. W. Kitching, V. G. Kumar-Das and P. R. Wells, *Chem. Comm.*, 356 (1967)
255. T. Birchall and A. Pereira, *Chem. Comm.*, 1150 (1972)
256. N. Flitcroft and H. D. Kaesz, *J. Amer. Chem. Soc.* **85**, 1377 (1963)
257. E. Amberger, H. P. Fritz, C. G. Kreiter and M.-R. Kula, *Chem. Ber.* **96**, 3270 (1963)
258. J. M. Bellama and R. A. Gsell, *Inorg. Nucl. Chem. Letters* **7**, 365 (1971)
259. K. Kawakami, T. Saito and R. Okawara, *J. Organometal. Chem.* **8**, 377 (1967)
260. H. R. H. Patel and W. A. G. Graham, *Inorg. Chem.* **5**, 1401 (1966)
261. D. J. Patmore and W. A. G. Graham, *Inorg. Chem.* **6**, 981 (1967)
262. A. G. Davies and T. N. Mitchell, *J. Organometal. Chem.* **6**, 568 (1966)
263. H. G. Kuivila, J. D. Kennedy, R. Y. Tien, I. J. Tyminski, F. L. Pelczar and O. R. Khan, *J. Org. Chem.* **36**, 2083 (1971)
264. E. W. Randall and J. J. Zuckerman, *J. Amer. Chem. Soc.* **90**, 3167 (1968)
265. A. D. Norman, *J. Organometal. Chem.* **28**, 81 (1971)

266. S. Cawley and S. S. Danyluk, *J. Phys. Chem.* **68**, 1240 (1964)
267. W. L. Wells and T. L. Brown, *J. Organometal. Chem.* **11**, 271 (1968)
268. E. W. Abel and D. B. Brady, *J. Organometal. Chem.* **11**, 145 (1968)
269. W. Kläui and H. Werner, *J. Organometal. Chem.* **54**, 331 (1973)
270. L. Verdonck and G. P. van der Kelen, *Ber. Bunsengesell. Phys. Chem.* **69**, 478 (1965)
271. M.-P. Simonnin, *J. Organometal. Chem.* **5**, 155 (1966)
272. L. Verdonck and G. P. van der Kelen, *Bull. Soc. Chim. Belge* **74**, 361 (1965)
273. E. Amberger and E. Muhlhofer, *J. Organometal. Chem.* **12**, 55 (1968)
274. H. Bürge and V. Groetze, *Angew. Chem.* (*Internat.*) **7**, 212 (1968)
275. A. J. Leusenk, H. A. Budding and J. W. Marsman, *J. Organometal. Chem.* **13**, 155 (1968)
276. J. Lorberth, *J. Organometal. Chem.* **16**, 235 (1969)
277. T. N. Mitchell, *J. Organometal. Chem.* **59**, 189 (1973)
278. C. D. Schaeffer and J. J. Zuckerman, *J. Organometal. Chem.* **47**, C1 (1973); **55**, 97 (1973)
279. R. O. Ragsdale and B. B. Stewart, *Proc. Chem. Soc.* **194** (1964); *Inorg. Chem.* **4**, 740 (1965)
280. P. A. W. Dean and D. F. Evans, *J. Chem. Soc.* (*A*) 1154 (1968)
281. H. C. Clark, N. Cyr and J. H. Tsai, *Canad. J. Chem.* **45**, 1073 (1967)
282. T. D. Coyle, S. L. Stafford and F. G. A. Stone, *Spectrochim. Acta* **17**, 968 (1961)
283. P. G. Harrison, S. E. Ulrich and J. J. Zuckerman, *Inorg. Nucl. Chem. Letts.* **7**, 865 (1971)
284. H. Elser and H. Dreeskamp, *Ber. Bunsengesell. Phys. Chem.* **73**, 619 (1969)
285. J. F. Malone and B. E. Mann, *Inorg. Nucl. Chem. Letts.* **8**, 819 (1972)
286. T. A. George and C. D. Turnipseed, *Inorg. Chem.* **12**, 395 (1973)
287. R. J. H. Clark, A. G. Davis, R. J. Puddephatt and W. McFarlane, *J. Amer. Chem. Soc.* **91**, 1334 (1969)
288. J. D. Kennedy and W. McFarlane, *J. Organometal. Chem.* **80**, C47 (1974)
289. V. S. Petrosyan, *Prog. NMR Spectry.* **11**, 115 (1977)
290. E. V. van den Berghe and G. P. van der Kelen, *J. Organometal. Chem.* **72**, 65 (1974)
291. A. G. Davies, M-W. Tse, M. F. Ladd, J. D. Kennedy, W. McFarlane and G. S. Pyne, *Chem. Comm.*, in press
292. A. Robineau and J-C. Pommier, *J. Organometal. Chem.* **107**, 63 (1976)
293. R. Okawara and M. Ohara, "Organotin Compounds", Vol. 2, p. 253, (A. K. Sawyer, ed.). Marcel Dekker (1971)
294. J. D. Kennedy, W. McFarlane and G. S. Pyne, *J.C.S. Dalton*, 2332 (1977)
295. R. M. Hawk and R. R. Sharp, *J. Chem. Phys.* **60**, 1109 (1974)
296. R. M. Hawk and R. R. Sharp, *J. Magn. Reson.* **10**, 385 (1973)
297. R. M. Hawk and R. R. Sharp, *J. Chem. Phys.* **60**, 1522 (1974)
298. G. E. Maciel and J. L. Dallas, *J. Amer. Chem. Soc.* **95**, 3039 (1973)
299. W. G. Schneider and A. D. Buckingham, *Disc. Faraday Soc.* **34**, 147 (1962)
300. H. C. E. McFarlane and W. McFarlane, *J.C.S. Dalton*, 2416 (1973)
301. R. G. Barnes and W. V. Smith, *Phys. Rev.* **93**, 95 (1954)
302. M. J. Cooper, A. K. Holliday, P. H. Makin, R. J. Puddephatt and P. J. Smith, *J. Organometal. Chem.* **65**, 377 (1974)
303. L. H. Piette and H. E. Weaver, *J. Chem. Phys.* **28**, 735 (1958)
304. J. M. Rocard, M. Bloom, and L. B. Robinson, *Canad. J. Phys.* **37**, 522 (1959)
305. I. Weinberg, *J. Chem. Phys.* **61**, 1112 (1962)
306. O. Lutz and G. Stricker, *Phys. Lett.* **35A**, 397 (1971)
307. H. P. Fritz and K. E. Schwarzhans, *Chem. Ber.* **97**, 1390 (1964)
308. W. McFarlane, *J. Organometal. Chem.* **116**, 315 (1976)
309. G. Singh, *J. Organometal. Chem.* **99**, 251 (1975)
310. G. Singh, *J. Organometal. Chem.* **11**, 133 (1968)
311. D. de Vos, *J. Organometal. Chem.* **104**, 193 (1976)
312. W. McFarlane, *Mol. Phys.* **13**, 587 (1967)
313. G. D. Shier and R. C. Drago, *J. Organometal. Chem.* **6**, 359 (1966)
314. D. Doddrell, K. G. Lewis, C. E. Mulquiney, W. Adcock, W. Kitching and M. Bullpit, *Austral. J. Chem.* **27**, 417 (1974)

11

GROUP V–ARSENIC, ANTIMONY AND BISMUTH

ROBIN K. HARRIS, University of East Anglia, England

Phosphorus-31 is a high-abundance spin-$\frac{1}{2}$ nucleus and has been greatly studied. Nitrogen-14, although quadrupolar ($I = 1$) and with a relatively low magnetic moment, has also received considerable attention. Moreover, FT techniques have caused a boom in the use of ^{15}N ($I = \frac{1}{2}$) by NMR, either using the natural abundance of this rare spin or with enrichment. These three nuclei are briefly discussed in chapter 4. NMR work on the remaining three group V elements has, however, been much more restricted.

11A ARSENIC

In particular, arsenic has been little studied by NMR. This is surprising in view of the 100% natural abundance of ^{75}As, which has a reasonably high magnetic moment. The lack of interest in this nucleus can only be partly traced to its quadrupolar properties, and a moderate increase in the use of ^{75}As NMR may be predicted for the near future.

In fact, the first FT NMR study of ^{75}As has only recently been reported.[1] Table 11.1 lists the chemical shifts observed, using the signal of $[AsF_6]^-$ as a reference (in spite of the splittings due to coupling). In view of the limited variety of compounds studied, the total range observed, 660 ppm, cannot necessarily be considered as typical. The tetraalkylarsonium ions, which comprised the majority of the samples, all resonated between 206 and 258 ppm. Balimann and Pregosin noted[1] the usual high-frequency β- and δ-shifts and low frequency γ-shift due to methyl substitution.

Few coupling constants to ^{75}As are known. That for $[AsF_6]^-$ has been variously reported as being between 870 and 955 Hz. It is doubtful whether the variations are real. Probably the most accurate values are those by Balimann and Pregosin[1] ($|^1J_{AsF}| = 930$ Hz from ^{75}As resonance) and by Arnold and Packer[2] (933 Hz from ^{19}F studies). The value 905 Hz has been reported[3] for the solid state (using the magic-angle spinning method). Lunazzi and Brownstein[4] have shown that for the complex Me_3NAsF_5 the value of $|^1J_{AsF}|$ to the equatorial fluorines (1048 ± 8 Hz) is significantly higher than that to the axial fluorines (840 ± 10 Hz), which is a situation analogous to that for phosphorus (the values of $|^1J_{PF}|$ for Me_3NPF_5 are 848 and 747 Hz respectively). A study of $T_{1\rho}(^1H)$ for liquid AsH_3 has yielded[5] $|^1J_{AsH}| = 92.1 \pm 0.7$ Hz, and the authors comment that for group V hydrides, MH_3, the reduced coupling constants $^1K_{MH}$ do not vary greatly. The value of $|^1J_{AsH}|$ in $[AsH_4]^+$ has been given[1] as 555 Hz.

The relaxation properties of ^{75}As are somewhat better known.[1,2,5–8] Balimann and Pregosin[1] give details of linewidths, some of which are listed in Table 11.1 Variation

TABLE 11.1 Some ^{75}As chemical shifts[a] and linewidths

Compound	δ_{As}/ppm	$W_{1/2}$/Hz[b]
$[AsH_4]^+[Ta_2F_{11}]^-$	−291	115 (at 0°C)
$[AsF_6]^-K^+$	0	94
$[AsMe_4]^+$	206[c]	113[d]
$[AsPh_4]^+$	217	1300
$[AsPr^n_3Me]^+$	225	—
$[AsPr^n_4]^+$	230	230
$[AsPr^n_3Bu^n]^+$	231	250
$[AsBu^n_4]^+$	234	346
$[AsPr^n_3Et]^+$	234	212
$[AsEt_3Me]^+$	242	800
$[AsEt_3Pr^n]^+$	243	203
$[AsEt_3Bu^n]^+$	245	210
$[AsEt_4]^+$	249[e]	194
$[AsPr^i_3Et]^+$	258	750 (at 22°C)
$[AsO_4]^{3-}3Na^+$	369	~1125

[a] Taken from ref. 1. [b] Data are given for 30°C (except where otherwise stated), with the medium conditions as stated in ref. 1. [c] The shift is apparently insensitive to anion. [d] For Cl^- or Br^- anions. [e] Relatively insensitive to solvent.

with molecular size and symmetry, with temperature, and with solvent and concentration are as qualitatively expected for quadrupolar relaxation. The influence of $T_1(^{75}As)$ on bandshapes in ^{19}F or ^1H resonance has been examined by a number of workers.[2,6–8] Arnold and Packer[2,8] studied the ^{19}F bandshape for $[AsF_6]^-$ in detail for a range of solvents, counter-ions and temperatures. They list values of $T_1(^{75}As)$ between 0.057 and 8.35 ms and discuss the origins of the rather efficient quadrupolar relaxation for the symmetrical $[AsF_6]^-$ anion in terms of cation-anion interactions. A detailed NMR relaxation study of the AsH_3 and AsD_3 molecules has been published.[5] The value of $T_1(As)$ for AsH_3 varies from 48 μs at −99°C to ca. 95 μs at +25°C.

The ^{75}As nucleus has been used for NMR applications in solid-state physics (see, for example, ref. 9).

11B ANTIMONY

There have been somewhat more direct NMR studies of this element than of arsenic. Both ^{121}Sb and ^{123}Sb may be observed, and figured among the earliest experiments on NMR.[10–12] Indeed, $[SbF_6]^-$ was one of the first substances to be shown to give a splitting (in the antimony resonance) due to spin-spin coupling,[11,12] though this phenomenon was not at first properly understood. The ^{121}Sb isotope is somewhat more favourable for NMR than ^{123}Sb.

Table 11.2 gives ^{121}Sb chemical shifts, with the resonance of Et_4NSbCl_6 used as a reference. The large shift range (ca. 3500 ppm) is probably less than complete, since only

TABLE 11.2 Some ^{121}Sb chemical shifts[a] and linewidths[a]

Species[b]	$\delta(^{121}Sb)$/ppm	$W_{1/2}$/Hz
$[SbBr_6]^-$	-2436 ± 10	300
$[SbCl_6]^-$	0	300
$[SbF_6]^-$	88 ± 3	—
$Sb(OH)_6^-$	$296^c \pm 15$	1800[c]
$SbCl_5$	380[d]	—
	509 ± 20	8000
$[Me_4Sb]^+$	780[e]	—
$[SbS_4]^{3-}$	1032 ± 10	550

[a] Data from ref. 13 except where otherwise stated.[b] For a discussion of the nature of the species present, and of medium conditions, see ref. 13. [c] Probably involving exchange with other species. [d] Ref. 14. 14. [e] Ref. 15.

a few species have been studied. The increase in shielding in the series $[SbF_6]^- \rightarrow [SbCl_6]^- \rightarrow [SbBr_6]^-$ is typical of many elements. The nature of the species present in solution is in some cases in doubt, and ^{121}Sb NMR offers some hope of investigating equilibria involving Sb, including those in which the solvent participates.[14,16]

Reports of coupling constants involving the antimony nucleus are few, and appear to be limited to the $[SbF_6]^-$ species, for which $|^1J(^{121}Sb, F)|$ is variously quoted as 1843,[17] 1945,[16] 1950[16] and 1934[18] Hz, while $|^1J(^{123}Sb, F)|$ is given as 1055,[16] 1050[16] and 1047[18] Hz. The ratio of the couplings is in accord with the magnetic moments of the two nuclides. The value 1820 Hz has been given[3] for $|^1J(^{121}Sb, F)|$ in solid $KSbF_6$ from measurements with the magic-angle spinning technique. It has been suggested[19] that the sign of the coupling constant $^1J_{SbF}$ in $[SbF_6]^-$ is negative.

Table 11.2 includes some linewidths for ^{121}Sb resonance. Even for symmetrical species these can be appreciable, as for ^{75}As. Kok, Morris and Sharp[15] also report some T_2 data for ^{121}Sb, listing 0.07 to 0.24 ms for Me_4Sb^+, 0.15 ms for $[SbCl_6]^-$ and < 10 μs for $[Ph_4Sb]^+$ and $[SbCl_4]^-$ (resonances could not be observed for the two last-mentioned species—Kidd and Matthews[13] also report failure to see signals due to Sb^{III} species, presumably because of very short T_2 values).

11C BISMUTH

Although the ^{209}Bi resonance of aq. $Bi(NO_3)_3$ was observed[11] at a very early date, bismuth has scarcely received any NMR attention since then. Fukushima[20,21] has reported $|^1J(^{209}Bi, F)| = 2700 \pm 300$ Hz (probably negative) in $KBiF_6$ powder, noting that for $[MF_6]^-$ species with M a group V element $|J_{MF}|$ is proportional to the square of the atomic number of M, from which he deduces that all such reduced couplings, $^1K_{MF}$, should be negative.

REFERENCES

1. G. Balimann and P. S. Pregosin, *J. Mag. Reson.* **26**, 283 (1977)
2. M. St. J. Arnold and K. J. Packer, *Mol. Phys.* **10**, 141 (1966)
3. E. R. Andrew, L. F. Farnell and T. D. Gledhill, *Phys. Rev. Lett.* **19**, 6 (1967)
4. L. Lunazzi and S. Brownstein, *J. Magn. Reson.* **1**, 119 (1969)
5. L. J. Burnett and A. H. Zeltmann, *J. Chem. Phys.* **56**, 4695 (1972)
6. D. W. Larsen, *J. Phys. Chem.* **75**, 3880 (1971)
7. A. G. Massey, E. W. Randall and D. Shaw, *Spectrochim. Acta* **21**, 263 (1965)
8. M. J. Arnold and K. J. Packer, *Mol. Phys.* **14**, 249 (1968)
9. G. J. Adriaenssens, *Ferroelectrics* **12**, 269 (1976)
10. V. W. Cohen, W. D. Knight, T. Wentink and W. S. Koski, *Phys. Rev.* **79**, 191 (1950)
11. W. G. Proctor and F. C. Yu, *Phys. Rev.* **81**, 20 (1951)
12. S. S. Dharmatti and H. E. Weaver, *Phys. Rev.* **87**, 675 (1952)
13. R. G. Kidd and R. W. Matthews, *J. Inorg. Nucl. Chem.* **37**, 661 (1975)
14. P. Stilbs and G. Olofsson, *Acta. Chem. Scand* **28A**, 647 (1974)
15. G. L. Kok, M. D. Morris and R. R. Sharp, *Inorg. Chem.* **12**, 1709 (1973)
16. J. V. Hatton, Y. Saito and W. G. Schneider, *Can. J. Chem.* **43**, 47 (1965)
17. E. L. Muetterties and W. D. Phillips, *J. Amer. Chem. Soc.* **81**, 1084 (1959)
18. R. G. Kidd and R. W. Matthews, *Inorg. Chem.* **11**, 1156 (1972)
19. W. McFarlane and R. J. Wood, *Chem. Comm.* 262 (1969)
20. E. Fukushima and S. H. Mastin, *J. Mag. Reson.* **1**, 648 (1969)
21. E. Fukushima, *J. Chem. Phys.* **55**, 2463 (1971)

12

GROUP VI–OXYGEN, SULPHUR, SELENIUM AND TELLURIUM

CHARLES RODGER,* University of East Anglia, England
NORMAN SHEPPARD,* University of East Anglia, England
CHRISTINA McFARLANE,† City of London Polytechnic, England.
WILLIAM McFARLANE,† City of London Polytechnic, England.

Group VI of the Periodic Table is not one for which nuclear spin properties of the elements are favourable for NMR applications in chemistry, since although both Se and Te possess spin-$\frac{1}{2}$ nuclei in reasonable natural abundance the two lighter elements, O and S, do not. Moreover the dominant isotopes of these elements, ^{16}O and ^{32}S, have zero spin, and the available quadrupolar nuclei, $^{17}O(I = \frac{5}{2})$ and $^{33}S(I = \frac{3}{2})$ have low natural abundances and moderately large nuclear quadrupole moments. Nevertheless, because oxygen is of such widespread importance in chemistry it has been studied extensively by NMR, often using isotopically enriched species. However, little has been achieved for ^{33}S NMR, principally because linewidths are usually rather large. Tellurium has received some attention from NMR spectroscopists and selenium considerably more, both by direct observation and by indirect double resonance methods. Apart from a determination[1] of its nuclear properties polonium has not been studied.

12A OXYGEN-17

12A.1 Introduction

Because oxygen is very important chemically the possibility of the detection of NMR signals from this element is of great interest. However in practice such work is rendered very difficult by the extremely low abundance (0.037 %) of ^{17}O, which is the only nucleus of oxygen with a magnetic moment, and the moderate magnetogyric ratio (Table 1.2). These factors lead to weak NMR signals; the calculated receptivity of ^{17}O nuclei is (Table 1.2) about 10^{-5} of that of ^{1}H nuclei. An additional difficulty, is that the ^{17}O nucleus $(I = \frac{5}{2})$ has a quadrupole moment which (Table 1.2) is of moderate size. Thus, for diamagnetic species the quadrupole relaxation mechanism dominates and leads to

* Dr Charles Rodger and Professor Norman Sheppard wrote Section 12A.
† Drs Christina and William McFarlane wrote Sections 12B, 12C and 12D.

relatively short relaxation times T_1 and T_2; the latter gives rise to large linewidths. For smaller molecules these are typically from several tens to several hundred Hz (cf. typically <1 Hz for ^{1}H). When the liquid or solution is particularly viscous, or when the molecular diameter is large, the molecular tumbling time can become long and linewidths of 1 kHz or over may be obtained.

Despite these difficulties, pioneering investigations[2,3] in the 1950s showed that ^{17}O NMR spectra could be successfully obtained from relatively small molecules. Subsequently Christ and his colleagues,[4,5] and Figgis, Kidd and Nyholm,[6] showed that very useful spectral compilations could be made for series of organic and inorganic molecules respectively, utilising only the natural abundance of ^{17}O. An adequate S/N was obtained by use of broad-line NMR spectrometers using local-field modulation (derivative presentation of the spectra) and "lock in" detectors, and also wide sample tubes.

To date more detailed studies for which better S/N is needed, e.g. for the accurate measurements of linewidths or intensities, have almost invariably been carried out using isotopically enriched samples. As ^{17}O is most readily available in the form of enriched water, it is not surprising that much work of this type has been concerned with aqueous solutions of inorganic species where the solute exchanges oxygen with $H_2{}^{17}O$, or complexes with H_2O as in the formation of ionic solvation shells.

Recently, however, the advent of Fourier transform spectrometers, with their inherently greater sensitivities over traditional CW methods, has provided an alternative approach to obtaining ^{17}O spectra. The spectra are then obtained in the more convenient absorption presentation. However, as has been commented elsewhere,[7] the relative gain in sensitivity compared with CW operation is less for broadline spectra with only a few resonances, than for the many-line and sharp spectra that are typical in ^{13}C NMR studies. This is because the S/N obtainable is proportional to $(\text{time})^{1/2}$. The FT gain is proportional to the square root of the number of spectral elements to be distinguished in the complete spectrum; this is less if the linewidths are large. The broad-line method using modulation and "lock-in" detectors is also a very sensitive one and it remains to be evaluated when the one method is an improvement over the other. Nevertheless, from now on ^{17}O spectra will clearly be more persistently investigated, probably mostly by FT methods, and this is therefore a particularly appropriate time for a review of achievements to date.

This article follows just about 10 years after the previous comprehensive review article by Silver and Luz published[8] in 1966. It attempts a detailed survey of the literature of the intervening period carried out systematically with the help of Chemical Abstracts (particularly using the headings "Nuclear magnetic resonance-oxygen 17" in the General Subject Index, and "Oxygen properties, atomic isotope of mass 17, NMR" of the Chemical Subject Index) up to and including Volume 84. The literature is covered in this systematic way until mid-1976. Some more recent references to mid-1977 from the most commonly used journals have also been included. Some of the more important results obtained before 1966 are also included, particularly in the Tables, when they improve the perspective of the review.

Oxygen-17 NMR spectroscopy has been discussed briefly in two other reviews concerned with "other nuclei";[7,9] Deverill[10] has mentioned results from ^{17}O in a review of the NMR spectra of inorganic electrolytes; and Silver[11] has reviewed the use of ^{17}O NMR in studying reaction mechanisms in organic and inorganic chemistry. A useful review of earlier results on ^{17}O, up to about 1962, is also to be found in the standard textbook by Emsley, Feeney and Sutcliffe.[12]

12A.2 Experimental Methods

Although it is already clear that results as good as those obtained by the broad-line method are obtainable by FT techniques,[13,14,15] more work is needed to evaluate the ultimate sensitivity using the newer method. At present good spectra over the complete range of chemical shifts for organic molecules[5,14] (~700 ppm, or 9000 Hz at a Larmor precession frequency of 13.56 MHz) can be obtained in a reasonable period of time ranging from a few minutes to an hour or so (Fig. 12.1 and 12.2). This is for relatively small molecules studied as neat liquids or concentrated solutions in a sample tube of 12 mm o.d. utilizing a few grams or ml of sample. However difficulties accumulate for larger molecules on account of their weak signals caused by limited molar solubilities or broad lines.[5,14]

For most organic molecules the spectral linewidth is determined by relatively rapid quadrupolar relaxation of the ^{17}O nuclei. The linewidth is then proportional to the correlation time for molecular rotation, and a combination of large molecular size and substantial liquid viscosity can lead to linewidths, at normal temperatures, in the vicinity of 1000 Hz. Because viscosity is a relatively sensitive function of temperature, the possibility of making measurements above room temperature can make important contributions to improving sensitivity in such cases[14] (see Fig. 12.2). An alternative method is to search for a good solvent of low viscosity at ambient temperatures, and more systematic work of this type is needed for different types of molecules.

In most types of spectroscopy it is possible to improve the intensity sensitivity (S/N) at the expense of resolution. In FT NMR spectroscopy such "sensitivity enhancement" is achieved by weighting the beginning of the FID (which determines the main outline of the transformed spectrum) more than the end (which controls the details and hence the

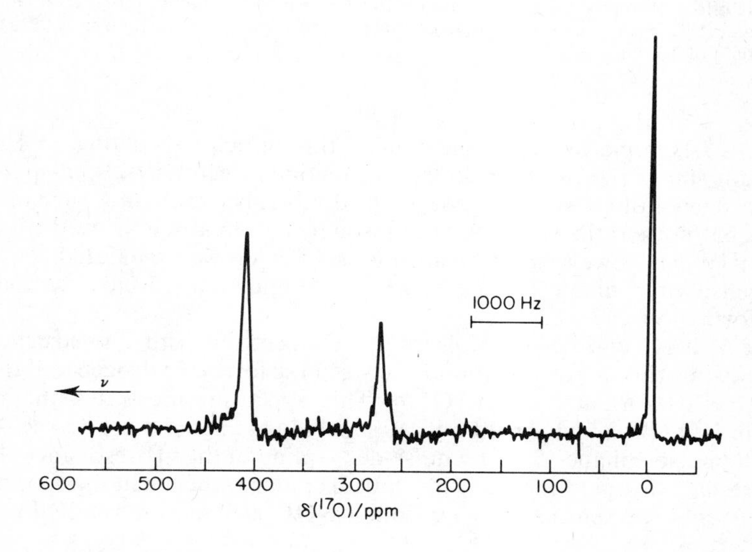

Fig. 12.1 The oxygen-17 FT NMR spectrum of acetic anhydride. The total experimental time was 30 minutes and the operating temperature 55°C. The r.h.s. signal at $\delta = 0$ is from the water reference.

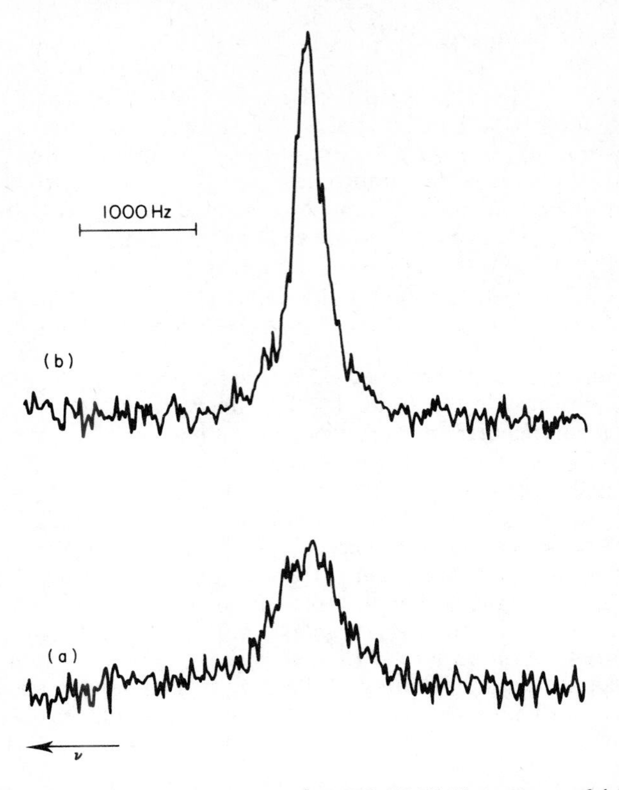

Fig. 12.2 The effect of increased temperature on the ^{17}O FT NMR spectrum of 1,2-ethanediol (a) at 36°C and (b) at 70°C. Both spectra were accumulated with an acquisition time of 0.06 s and a total experimental time of 10 minutes.

resolution). An example of the application of this principle is shown in Fig. 12.3.[14] This spectrum shows the resolved doublet of methanol caused by spin-spin coupling with the hydrogen atom of the OH group. The doublet becomes only partially resolved when the signal/noise ratio for the overall resonance is greatly improved by sensitivity enhancement. When, however, a spectrum consists of a few well-separated peaks without structure, sensitivity enhancement can be substantial and affected without loss of essential information.

Figure 12.4 shows another example of a ^{17}O spectrum with fine structure due to spin-coupling, this time across two bonds between the formyl hydrogen and the "ether" oxygen in methyl formate, H.OC.OMe. This spectrum illustrates the resolution obtainable in ^{17}O NMR by FT methods. As shown in the figure it is possible to collapse this doublet, and so enhance S/N, by noise-decoupling of the ^{1}H resonance.[14,16] However, because the relaxation mechanism is quadrupolar rather than dipolar, there is no useful further enhancement in this case from a NOE such as is normally found in ^{13}C studies.

The repetitive accumulation of FID's in FT spectroscopy requires very stable operation of the spectrometer so field-frequency locking is essential. Using the same coil as for ^{17}O detection, neither ^{2}D or ^{1}H locking is very suitable. This is because ^{2}D and

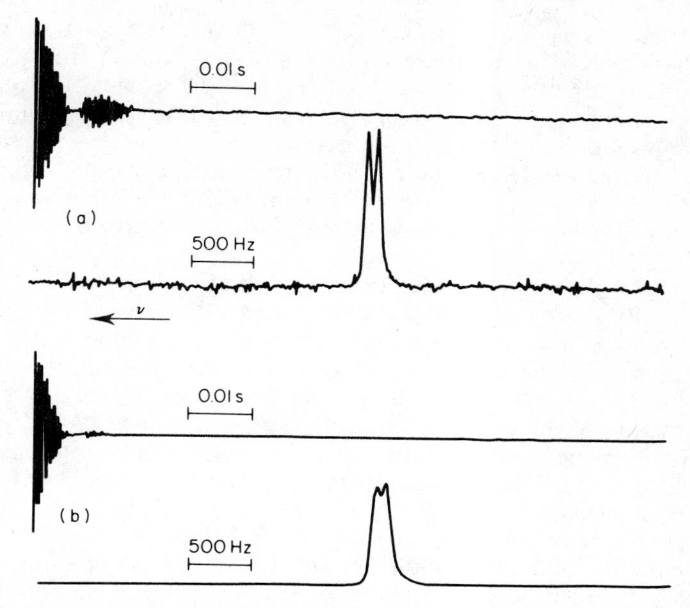

Fig. 12.3 The effect of an exponential weighting factor on the ^{17}O FT NMR spectrum of methanol. (a) no weighting factor applied, (b) sensitivity enhancement of 0.01.

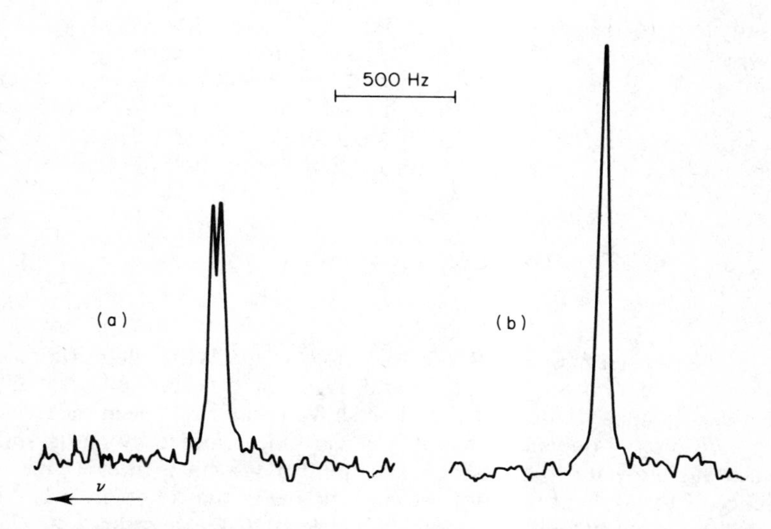

Fig. 12.4 The ^{17}O FT NMR spectrum of methyl formate (ether-type oxygen resonance only) showing (a) the two-bond proton oxygen coupling constant and (b) the effect of proton noise decoupling. This spectrum also illustrates the resolution obtainable in ^{17}O spectra in favourable cases (doublet separation 35 Hz).

^{17}O nuclei resonate at very similar frequencies but ^{17}O and ^{1}H resonate at very different ones. Both situations lead to difficulties in tuning of the probe.[14] However with pulse locking on a shared-time basis the use of an ^{1}H lock is feasible[15] and gives the best method of tuning the field for high resolution. This is clearly less convenient when ^{1}H noise-decoupling is required and, as ^{17}O lines tend not to be very sharp, the resolution-tuning need not be too fine. Under these conditions the best general arrangement seems to be to use an ^{19}F lock with a separate coil and amplifier.[13,14,16] However for variable temperature work the lock sample must be thermally isolated from the sample under investigation.

Oxygen-17 chemical shifts are almost invariably measured relative to that from pure water. The majority of organic compounds have positive δ values on this scale.[5,14] Either sample replacement or simultaneous measurement of the water signal using concentric tubes for sample and reference can be used. Although in principle these procedures require a magnetic susceptibility correction to be made, this is usually only a small fraction of the observed linewidth and is normally considered to be unnecessary.[14] However the position of the water signal can vary with its environment by several ppm and, except for dilute solutions in this solvent, appreciable errors can occur if it is used as an internal reference.

Because ^{17}O resonances are normally broad the FID decays rapidly and, because $T_1 \simeq T_2$ in most cases, it is in principle possible to accumulate an adequate FID in a very short time, e.g. 0.1 s or less.[13,14,15] This means that many pulses can be carried out in a reasonable period of time, so improving signal/noise. However, because the ^{17}O FID signal is weak and high gain has to be used, it may be necessary to use longer-than-normal delay times between the end of the pulse and the commencement of measurement of the FID. If this is not done 'pulse breakthrough' may occur in the form of a residual but relatively strong 'tail' signal from the pulse, superimposed on the first part of the FID.[14,15] Among other things, this initial "spike" transforms into a sinusoidally "rolling" baseline.[14] On the other hand, the delay in measuring the FID leads naturally to reduced signal/noise, and also to phase shifts which vary systematically with the frequency separation of signals.[15] However, in many cases ^{17}O spectra consist of only a few peaks. Canet, Goulon-Ginet and Marchal, who have also discussed obtaining ^{17}O spectra by FT techniques,[15] point out that there is a systematic relationship $\Delta\nu\Delta t = k$ (where k is an integer) between the delay-time, Δt, and the frequency difference, $\Delta\nu$, between two peaks if both are to be recorded in the pure absorption mode, i.e. without phase shifts.

12A.3 ^{17}O Spectra of Aqueous Solutions and Electrolytes

a. General

Spiers *et al.*[17] have used ^{17}O NMR spectroscopy to study the magnitude of the ^{17}O quadrupole coupling constant in H_2O and D_2O in different physical states. The temperature-dependence of the ^{17}O chemical shift in water has been measured by Luz and Yagil,[18] and the pH-dependence of its linewidth determined by Hertz and Klute.[19] The linewidth of pure water reaches a maximum at pH7; this is interpreted in terms of the occurrence of the slowest rate of proton exchange at that temperature.[19] Chemical shifts to low frequency, which are observed at elevated temperatures, are considered to reflect the breaking of hydrogen bonds within the liquid.[18] Two groups of workers[18,20] have studied the chemical shift of ^{17}O in aqueous solutions of a series of diamagnetic electrolytes and non-electrolytes at ambient temperature.

Fister and Hertz[20] also studied the dependence of linewidths (and hence T_2) of water in the same solutions; the relaxation rate was found to increase for structure-forming salts and to decrease for structure-breaking salts, as might have been expected. Hertz and Klute[19] showed that proton exchange is slower for water molecules in NaCl, KI, $NaClO_4$ and $NaNO_3$ solutions than in pure water; for the iodide and perchlorate they observed resolution of the ^{17}O resonance into a triplet, reflecting spin-spin coupling with the two protons. Liquid water does not give such spectral structure at room temperature, but it has been observed for dilute solutions of water in non-aqueous solvents such as acetone.[21]

b. Ionic Solvation—Diamagnetic Cations

The chemical shift measured for water in aqueous solutions of electrolytes represents, of course, an average over different environments, for example of water molecules in the primary solvation spheres of the cations or the anions or of those free from effective interaction with ions. Strong interaction between water molecules and ions in the first hydration sphere is particularly to be expected for small and highly charged cations such as Be^{2+} and Al^{3+}. It was first shown by Jackson, Lemons and Taube[22] that in the latter two cases, and also for Ga^{3+}, separate ^{17}O resonances could be observed from "free" and "bound" water molecules in aqueous solutions containing these ions. The relative integrated areas of resonances from "bound" and "free" water can in principle provide a direct measure of the number of water molecules in the primary solvation shell of the cation, i.e. the hydration number, when the cation concentration is known.

In fact in the above cases it turned out that the chemical shift between "bound" and "free" water molecules was small, leading to overlapping of resonances. However, by the ingenious method of adding small amounts of a paramagnetic salt, such as Co^{II}, which shifted the resonances of the "free" molecules (see later) it was possible to observe separate resonances. Estimates of 4, 6, and 6 for the hydration numbers of Be^{II}, Al^{III} and Ga^{III} were made by area measurements by Jackson, Lemons and Taube.[22] Connick and Fiat[23,24] subsequently confirmed these figures more accurately using water more highly enriched in ^{17}O. Jackson *et al.*[22] pointed to an alternative method of determining hydration numbers from evaluation of the magnitude of the shift of the water resonances caused by the dissolved paramagnetic ion. The paramagnetic shift of the water resonance is measured for a water solution of the cobalt salt as above. Subsequent addition of a known amount of another salt that forms a very stable and hence long-lived hydration shell will remove a proportion of the "free" water molecules previously available. This leads to an increased paramagnetic shift for the remainder. The change can then be used to calculate the number of water molecules made inaccessible in the hydration shell of the added cations. Alei and Jackson[25] applied this method, using Dy^{III} as the paramagnetic ion, to show once again that the hydration numbers of Be^{II} and Al^{III} were 4 and 6 respectively, and this method has been applied in a number of other cases. However the value of ~ 7 obtained for the hydration number of Cr^{III} did not agree with results of ~ 6 derived from other methods, and this led to a critical discussion of the approximations made in the application of the method.

Given separate resonances from "free" and "bound" solvent molecules it is also possible to use linewidth measurements as a function of temperature to estimate the rates of exchange of water molecules between the two environments[23] and the thermodynamic parameters for activation. The values obtained from ^{17}O measurements pertain to the exchange of "whole" water molecules, and may differ from those obtained from analogous measurements of proton resonances where proton-exchange, rather than the making and breaking of metal-oxygen bonds, may be the dominant factor in determining the measured kinetics.

Because it is only possible to lower the temperature of aqueous solutions by a limited amount before they solidify, only very strongly solvated water molecules in practive give separate resonances. Other examples are $[(NH_3)_5CoH_2O]^{3+}$,[22] $[Me_3Pt(H_2O)_3]^{+}$[26] and $[(NH_3)_2Pt(H_2O)_2]^{2+}$, where the NMR measurements have been used to determine the number of water molecules present. Many other unsuccessful attempts to observe separate resonances have doubtless been made.[22,26]

c. *Ionic Solvation—Paramagnetic Cations*

Hyperfine splittings in typical ESR spectra are very large compared with chemical shifts in NMR spectra. Hence a magnetic nucleus interacting with a single unpaired electron would be expected to give a pair of resonances split thousands of Hz on either side of the analogous chemical shift position in a nonparamagnetic situation. However if the relaxation rate, $1/T_e$, of the paramagnetic electron is very short with respect to the hyperfine splitting constant, a_N, then the doublet will collapse, leading to a single peak in the NMR spectrum. Unequal population of the two electronic spin-states means that the collapsed resonance will not be at the "diamagnetic" position but on one side or other of that position depending on the sign of the hyperfine coupling constant. The magnitude of the paramagnetic shift depends on the difference in population of the electronic spin-states, which in turn depends on a_N. Such residual paramagnetic shifts are often very large compared with normal chemical shifts. Therefore, there would seem to be a much greater chance of observing separate resonances from ^{17}O nuclei in "bound" water molecules in the solvation shells of paramagnetic cations than in corresponding diamagnetic cases. This is because the rate of exchange between bound and free sites would have to be much greater in order to give fully averaged spectra. Once such spectra have been found, the magnitudes and temperature dependences of linewidths (or other relaxation time measurements) of the "bound" and "free" resonances, and the chemical shift separations between those resonances, can in principle be used to derive information about a_N, T_e, the rate of the solvent interchange process, and the solvation number. However, because the shifted resonance of the solvated water molecules is also often very broad in comparison with that from the "free" solvent, it is rarely possible in practice to derive the hydration number of the paramagnetic cation from straightforward area measurements; sometimes the resonances are so broad as to be undiscernible. Nevertheless a considerable amount of information, including indirect measurements of the solvation numbers, can be derived from the temperature-dependence of the positions and linewidths of the stronger and sharper "free" resonance observed below the coalescence point, and the single combined resonance observed above coalescence. Such studies have been a major theme of research utilising ^{17}O resonances.

Fiat and his colleagues have made a substantial number of studies of solvated paramagnetic cations by this method, including Fe^{II},[27] Ni^{II},[27–30] Co^{II},[31,32] V^{III},[33] VO^{2+}[34] and Ti^{II}.[35] Although in many cases a separate "bound" resonance was observed, the S/N was such that the solvation number could only be obtained approximately by the indirect method indicated above. Values of 6 for the hydration number were found to be consistent with the experimental results for each of the simple cations listed above; for VO^{2+} 4 was found to be the best value. In the case of V^{III} an atypical negative electron spin density was observed for the bound ^{17}O nuclei.[33] The value for the hydration number of Ni^{II} has been controversial,[27,28,30] and the uncertainties led to a careful scrutiny of the assumptions made in interpreting the experimental data.[29]

Similar studies have been made of the solvation of methanol to Co^{II} [32] and, with the help of proton measurements, to Ti^{III};[35] the indirectly deduced solvation numbers were 6 and 4 respectively. The chemical shifts of solvated water molecules of complexes of formula $[Co(H_2O)_nL_{6-n}]^{2+}$ where L = acetone, MeOH, MeCN, DMSO and DMF have been measured.[36]

d. Kinetics of "Bound" ⇌ "Free" Solvation Processes

A substantial group of publications has been concerned primarily with the temperature-dependence of the bandwidths of ^{17}O resonances, with particular emphasis on the evaluation of the rates of the "bound" to "free" interchange of water molecules, using solutions enriched in $H_2{}^{17}O$. In these cases separate resonances from "bound" molecules were not observed, and the hydration numbers were given assumed values in the analyses of results. The analyses were carried out with the help of theory developed by Swift and Connick,[37] who themselves investigated the Mn^{II}, Fe^{II}, Co^{II}, Ni^{II} and Cu^{II} systems. A substantial series of papers in this area has been published by Hunt and his colleagues, covering $[Ni(H_2O)_6]^{2+}$ [38] and other nickel compounds in which the primary solvation shell of water is shared with various nitrogenous bases such as NCS^-,[39,40] ethylenediamine,[38] ammonia,[41] bipyridyl,[42] multi-ethylenediamines[43] and β,β',β''-triaminotriethylamine. The same laboratory has made studies of rates of water-exchange in $Mn(H_2O)_6^{2+}$ and Mn^{II} with solvation sheaths shared between water and phenanthroline,[45] of complexes of Ni^{II} with water and EDTA,[46] and Cu^{II} with water and β,β',β''-triaminotriethylamine.[44] Studies of hydration kinetics by other authors cover V^{II} and Cr^{III},[47] Cu^{II} with water and ethylenediamine ligands,[48] Ni^{II} with water and 2,2′,2″-terpyridine,[49] Co^{II} with water and chloride ligands,[50] VO^{2+} (where the water exchange was found to be remarkably slow),[51] and dioxide cations UO_2^+, NpO_2^+, NpO_2^{2+} and PuO_2^{2+}.[52] The solvation kinetics of DMF with Co^{II} and Ni^{II} have also been measured.[53]

This type of experimental work, and the chemical significance of the results, have been reviewed relatively recently,[54,55] so that further discussion will not be given here.

12A.4 Other Applications to Inorganic Chemistry

a. Diamagnetic Oxygen-Containing Complex Ions

The most consistently studied ions of this type are those of the family XO_4^{n-}. Figgis, Kidd and Nyholm[6] were the first to study the chemical shifts of such species. An important conclusion of their work was that, for the diamagnetic transition-metal anions of the above formula, there was a very good correlation between the observed chemical shifts and the inverse of the energies of the lowest electronic transitions as measured by visible or UV spectroscopy. This is what is expected if (see below) the local paramagnetic contribution is dominant in determining the ^{17}O chemical shift. Mixing-in of the electronic excited state involves electronic angular momentum, and the effect is likely to be dominated by the lowest-lying excited state.[6] The permanganate ion is, however, an exception in that it falls on the straight-line correlation only when the energy of a higher excited state is used.[6] A similar, but less well-defined, correlation was found by Andersson and Mason[56] for a variety of oxygen- and nitrogen-containing ions and other substances.

The chemical shifts for the XO_4^{n-} ions are collected in Table 12.1, and include some more recent data. Regularities observed are that within an isoelectronic sequence, e.g. VO_4^{3-} to MnO_4^-, or SO_4^{2-} and ClO_4^-, the paramagnetic shift always increases with

TABLE 12.1 ^{17}O NMR chemical shift and bandwidth data for diamagnetic oxyanions

Species	Chemical shift[a] δ/ppm	Bandwidth $W_{1/2}$/Hz
NO_2^{2-}	670(6), 650(56)	
NO_2^{+}	420(56)	
UO_2^{2+}	1115(57)	~12
CO_3^{2-}	192(6)	
NO_3^{-}	420(6), 410(56)	
SO_3^{2-}	~235(58)	
ClO_3^{-}	289(6), 290(59)	
BrO_3^{-}	297(6)	
SO_4^{2-}	167.0(60)	84
ClO_4^{-}	289.5(61), 290(57)	~30[b]
SeO_4^{2-}	204(6)	
VO_4^{3-}	568(60), 571(6)	~20[b]
CrO_4^{2-}	822.1(60), 835(6)	~20[b]
MnO_4^{-}	1230(60,62)	~20[b]
MoO_4^{2-}	532.4(60,63), 560(6)	<5[b]
TcO_4^{-}	749(6)	
WO_4^{2-}	420(6)	
ReO_4^{-}	579(64), 569(6)	

Species	Unit	Chemical shift[a] δ/ppm	Band-width[g] $W_{1/2}$/Hz
$S_2O_3^{2-}$		227.9(60)	89
$Cr_2O_7^{2-}$	$\underline{O}Cr$	1129(65,66), 1090(58)	~115[c]
	$\underline{O}Cr_2$	345(65,66), 338(58)	~200[c]
$V_2O_7^{4-}$		700[d](60)	>50[b]
$V_4O_{12}^{4-}$		925[d](67)	
$V_{10}O_{28}^{6-}$	$\underline{O}V$	~1150[e] (67,68)	bd
	$\underline{O}V_2$[f]	~900[e](67,68)	sh
		~790[e](67,68)	m
		~760[e](67,68)	m
	$\underline{O}V_3$	~390[e](67,68)	m
	$\underline{O}V_4$	~65[e](67,68)	sh
$Mo_6O_{19}^{2-}$	$\underline{O}Mo$	829(68,69)	m
	$\underline{O}Mo_2$	559(68,69)	m(~100)
	$\underline{O}Mo_6$	−27(68,69)	sh
$Nb_6O_{19}^{2-}$	$\underline{O}Nb$	594(68)	
	$\underline{O}Nb_2$	386(68)	
	$\underline{O}Nb_6$	20(68)	
$Ta_6O_{19}^{2-}$	$\underline{O}Ta$	476(68)	
	$\underline{O}Ta_2$	329(68)	
	$\underline{O}Ta_6$	−41(68)	

[a] Ref. given in brackets. [b] These are estimates of linewidths within resolved multiplets. [c] Ref. 65 and 66. [d] Only one resonance observed—assigned to $\underline{O}V$ terminal. [e] pH dependent. [f] Three sites. [g] sh ≡ sharp, m ≡ medium, bd ≡ broad.

the atomic number of the central atom, i.e. with the increase of the negative charge on the ion. Also there are clear indications that in passing from first to second or third transition-element series the corresponding anion is associated with a lower δ value, cf. CrO_4^{2-}, MoO_4^{2-} and WO_4^{2-}, or MnO_4^{-}, TeO_4^{-} and ReO_4^{-}. These same regularities are reflected in the chemical shift data for diamagnetic anions of more complex structure containing several metal atoms (see below).

Figgis, Kidd and Nyholm,[65,66] and also Jackson and Taube,[58] studied the spectrum of $Cr_2O_7^{2-}$ and showed that the terminal oxygens had a chemical shift (δ ~1100) much greater than that of the bridged oxygen (δ ~ 340). More recently this type of regularity has been generalised in a series of papers by Klemperer and his colleagues.[63,67–69] They have shown that ^{17}O can be very valuable in describing the structures of a variety of complex oxygen-containing anions which also contain a number of transition-element atoms. The most complete analysis has been made of the $[V_{10}O_{28}]^{6-}$ species

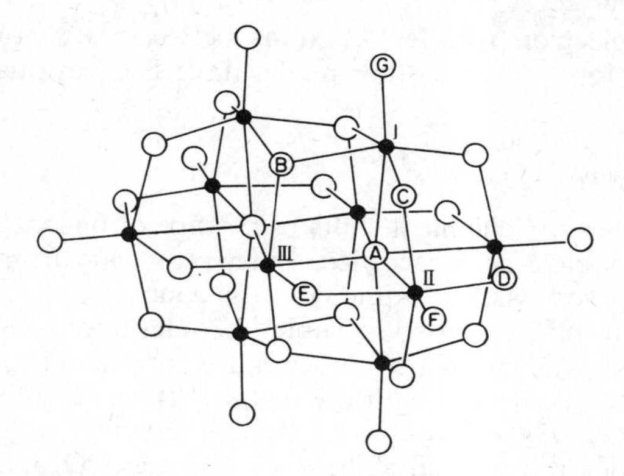

Fig. 12.5 D_{2h} symmetrised structure[68] of $V_{10}O_{28}{}^{6-}$. Small filled circles represent vanadium nuclei and large open circles represent oxygen nuclei. One member of each set of symmetry-equivalent oxygen nuclei is labelled.

whose structure is shown in Fig. 12.5. Exchange of oxygen atoms in the complex ion with those of $H_2{}^{17}O$ in aqueous solution showed that all the different oxygen sites could be separately identified by their NMR chemical shifts.[67,68] If an oxygen atom attached to n vanadium atoms is designated as OV_n then the observed chemical shifts δ relative to H_2O were OV ~1150; OV_2 (3 different sites) ~900, 790 and 760; OV_3 ~390; and OV_4 ~65 ppm. It is seen that there is a systematic reduction in the chemical shift with the increase in the number of vanadium atoms to which the oxygen atom is bound. Other complex anions of formula $[M_6O_{19}]^{n-}$, where M = Mo (n = 2), Nb (n = 8) or Ta (n = 8), show the same qualitative regularities in chemical shifts with $\underline{O}M > \underline{O}M_2 > \underline{O}M_6$.[68] These regularities have been interpreted as a correlation between increasing shift, δ, and increasing π-bonding from metal to oxygen in different sites.[6,66,68]

The regularities persist in the ^{17}O spectra of a variety of more complex anions of lower symmetry[63,69,70,71] in which other metals or non-metals are substituted within an $[M_xO_y]^{n-}$ framework. The spectra of complexes such as $[As_2Mo_6O_{26}]^{6-}$,[63] $[PMo_{12}O_{40}]^{3-}$,[69,71] $[M'W_{10}O_{36}]^{n-}$, where M′ = Ce^{III}, Th^{IV} or U^{IV},[70] and other complex vanadates based on $[XV_{12}O_{40}]^{n-}$ [72] have been measured. The chemical shift data for the parent $[M_xO_y]^{n-}$ species are collected in Table 12.1, together with other data on linewidths.

It is also to be noted that resonances from oxygen nuclei in very symmetrical environments in the complex anions, e.g. $\underline{O}M_4$ or $\underline{O}M_6$, are particularly sharp.[67–69] This is to be expected because the field gradient at these positions will be zero or very small; hence the normally dominant quadrupolar relaxation mechanism will be very weak and the relaxation times correspondingly long. An anomaly concerns a remarkably sharp ^{17}O resonance in MoO_4^{2-} which has been the cause of considerable comment. It is, in fact, one of the sharpest ^{17}O resonances known.[60,73] UO_2^{2+} also has a sharp resonance with some unexplained structure.[57]

The ^{17}O spectra can clearly be used to monitor the rates of slower oxygen-exchange processes between $H_2{}^{17}O$ and the oxyanions,[57] including different sites within the more complex examples. For $[V_{10}O_{28}]^{6-}$ the pH-dependence of the chemical shifts has given information about protonation at different oxygen sites;[67] for $Cr_2O_7^{2-}$ the linewidth dependences on added acid or base have provided information[64,65] about the kinetics of reactions such as $Cr_2O_7^{2-} + H_2O \rightleftharpoons 2HCrO_4^-$; the linewidths of MnO_4^- ions have been used to study electron transfer reactions between that ion and MnO_4^{2-},[58,62] and relaxation times for ^{17}O and other nuclei have been evaluated in a number of cases.[60,62,64]

b. *Uncharged Inorganic Molecules*

Table 12.2 lists the collected chemical shifts of a range of uncharged inorganic species whose ^{17}O resonances have been recorded. Linewidths vary from a few tens to a few hundred Hz except for very viscous liquids such as conc. H_2SO_4.

The largest chemical shifts recorded in Table 12.2 refer to oxyderivatives of transition metals such as RuO_4, OsO_4 and CrO_2Cl_2, and are considered to reflect the low-lying electronic exicted states associated with the metals.[6] It seems to be generally true that bonding of an additional oxygen atom to an atom X to which oxygen is already bonded leads to a lowering of the chemical shift, δ, of ^{17}O e.g. CO and CO_2,[76] —NO and —NO_2,[56] SO_2 and SO_3,[6] ClO_2F and ClO_3F[61] etc. Certain types of functional groups

TABLE 12.2 ^{17}O NMR chemical shift data for non-ionic inorganic compounds

		δ/ppm	Ref.
H	H_2O(liq)	0	74
	H_2O(vap.)	36	75
	D_2O(liq.)	3	5, 14
	H_2O_2	174, 187	5, 6
C	CO	350	76
	$Ni(CO)_4$	362	77
	$Fe(CO)_5$	388	77
	$Mn_2(CO)_{10}$	355	77
	$Fe(CO_2)_2(NO)_2$	418[b]	77
	$Co(CO)_3NO$	377[b]	77
	CO_2	63	76
N[a]	HNO_3	409[b], 414[b]	5, 56
	$O_2N{\cdot}NO_2$	420	56
	$\underline{O_2}N{\cdot}NO$	425	56
	$O_2N{\cdot}N\underline{O}$	855	56
	$Cl{\cdot}NO$	915	56
O[a]	H_2O_2 (see H)		
	F_2O_2 (see F)		
	F_2O_4 (see F)		
F	F_2O_2	647	78
	F_2O_4	971, 1512	78
	CF_3OOOCF_3	321, 479[c]	79

		δ/ppm	Ref.
Si[a]			
P[a]	H_3PO_3	111[b]	5
	H_3PO_4	80[b]	5
	OPF_3	66	80
	$OPCl_3$	216, 215.4	3, 80
	$OPBr_3$	259.2	80
S[a]	SO_2	513	6
	SO_3	188	6
	$SOCl_2$	291, 292	5, 6
	SO_2Cl_2	298, 304	5, 6
	H_2SO_4(conc)	140[d]	5
Cl	$ClOF_2$	311	61
	ClO_2F	384	61
	ClO_3F	301.5	61
Cr	CrO_2Cl_2	1460	6
Ru	RuO_4	1119	6
Os	OsO_4	796	6
Xe	$XeOF_4$	313	81
	$Xe(OH)_6$	278	82

[a] For alkyl derivatives, see Table 12.3. [b] Single resonance reported. [c] Central oxygen. [d] $W_{1/2} = 600$ Hz.

seem to have approximately constant characteristic chemical shifts e.g. M—CO 350–400 in metal carbonyls,[77] N—NO_2 ~ 415, X—NO ~900 ppm etc.[56] However, more data are needed before these correlations may be considered sufficiently well established for structural diagnosis purposes, but they are pointers for the future. The series of compounds OPX_3 is exceptional in that the chemical shift is very small for X=F.

It has been commented that the ^{17}O shifts in metal carbonyls are surprisingly independent of the metal atom,[77] particularly as oxygen directly attached to different transition metals is very sensitive to the latter (Table 12.1). In the case of the acids HNO_3, H_2SO_4 and H_3PO_4 etc.[5] the single observed resonances reflect the mean chemical shifts of H—O—X and X=O (X → O) groups due to rapid proton transfer between the different sites.

Only single resonances are observed for metal nitrosyl carbonyls.[77] Whether these are averaged by oxygen exchange between CO and NO sites, or whether the NO resonances are too broad to be observed is not clear. An attempt has been made to obtain ^{17}O resonances from O_2 complexes of Ir and Rh;[83] it is postulated that the resonances are too broad for detection.

12A.5 Applications to Organic Chemistry

Pioneering broad-line studies of the chemical shifts of over 100 organic molecules were carried out by Christ and his colleagues,[4,5,84] using natural abundance of ^{17}O. A selection of these results, together with recent Fourier-transform measurements by Rodger[14] are listed in Table 12.3. It is seen that the two sets of measurements are in excellent agreement on chemical shifts. Because of the sensitivity of linewidths to sample viscosity there are more variations between the two measurements of $W_{1/2}$, probably in part caused by different mean temperatures within the probes, but once again the general agreement is very satisfactory. Linewidths fall typically between 50 and 200 Hz, but a few results are less than 50 Hz, and those for large molecules which have slow tumbling times, e.g. t-butanol, or form viscous solutions can be much greater, ca. 1 kHz or more.

Christ, Diehl, Schneider and Dahn[5] have discussed the pattern of chemical shifts in some detail and only a summary of some of their conclusions will be given here. First of all, for compounds containing only the elements, H, C, N and O, there is a clear-cut distinction between ^{17}O resonances from ether-like sites, i.e. —O—, and those from C=O groups, the latter occurring at much higher δ values.[5] Primary alcohols and ethers absorb close to the position of the resonance from water, but branching of the attached hydrocarbon groups causes substantial shifts to higher δ values. This means that, for example, useful distinctions can be made between 1°, 2° and 3° alcohols. The formation of saturated rings makes little difference to δ, but adjacent double-bonds, either C=C or C=O, cause substantial shifts of the "ether" resonances to higher δ values.

The keto groups of aldehydes and ketones have δ values close to 600, but these values are substantially reduced by attached oxygen- or nitrogen-containing substituents as in esters and amides. This is presumably because of the participation of alternative resonance forms such as:

$$\mathrm{R{-}\underset{\displaystyle \overset{\|}{O}}{C}{-}NH_2} \quad \text{and} \quad \mathrm{R{-}\underset{\displaystyle \overset{|}{O^-}}{C}{=}\overset{+}{N}H_2}$$

Thus the positions of the carbonyl resonances are largely determined by the mesomeric (+M) effect of substituents.[5] A carboxylic acid gives a single δ value, just about exactly

TABLE 12.3 ^{17}O chemical shifts[a, b] and linewidths[b] for selected organic compounds[c] in the liquid state

Compound	δ/ppm	$W_{1/2}$/Hz	δ/ppm	$W_{1/2}$/Hz
MeOH	37, 37	60, 100		
EtOH	7.4, 6	110, 160		
Pr^iOH	39.4, 38	240, 200		
Bu^tOH	66.2, 70	600, 1000		
Et_2O	12, 15	90, 150		
Pr^i_2O	61, 62	90, 120		
$(MeO)_2CH_2$	8.4, 8	50, 50		
$(MeO)_3CH$	28	160		
MeCHO			596.1, 595	20, 50
Me_2CO			572, 572	45, 45
MeCO · OMe	139, 137	85, 120	362.2, 355	85, 120
MeCO · OVi	204	120	371	120
$(MeCO)_2O$	265.3, 259	130, 170	400.8, 393	130, 170
MeCO · Cl			507	45
$MeCONH_2(H_2O)$			286	200
$MeCONMe_2$			324	50
$EtNO_2$			600	120
Pr^nONO	456	60	803	60
$(Et_2N)_2NO$			683	190
$(EtO)_4Si$	9			
$(MeO)_3P$	52.1, 46	60, 50		
$(MeO)_3PO$	21.9, 18	80	74.6	20
Me_2SO			17, 13	140, 120
$(MeO)_2S$	12	130		
$(MeO)_2SO$	113.7, 115	80, 60	174.0, 176	25, 35
$(MeO)_2SO_2$	101.5, 102	190, 120	145.7, 150	80, 70

[a] All chemical shifts are with respect to liquid water. [b] In most cases two figures are given; the first refers to recent FT measurements[14] and the second to CW broadline measurements.[5] All measurements involve natural abundance of ^{17}O. [c] The left-hand of the table refers to oxygens in "single-bonded" situations, i.e., —O—, and the right-hand half to those in "double-bonded" situations X=O; in the case of acetic acid the rapid proton transfer between O—H and O=C leads to a mean resonance position, with $\delta = 250.7$[14] or 254[5] ppm, and $W_{1/2} = 110$[14] or 140[5] Hz.

half-way between the —O— and C=O resonances of esters, because of averaging caused by rapid proton transfer between:

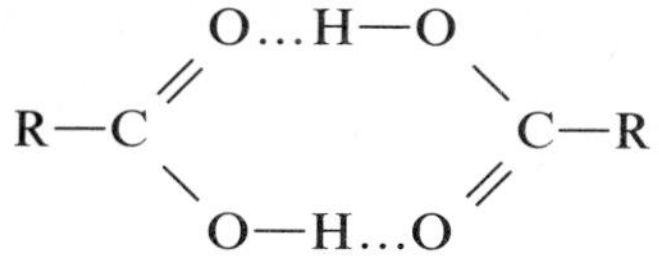

```
       O...H—O                        O—H...O
      //       \                     /       \\
R—C             C—R   and   R—C              C—R
      \        //                   \\       /
       O—H...O                        O...H—O
```

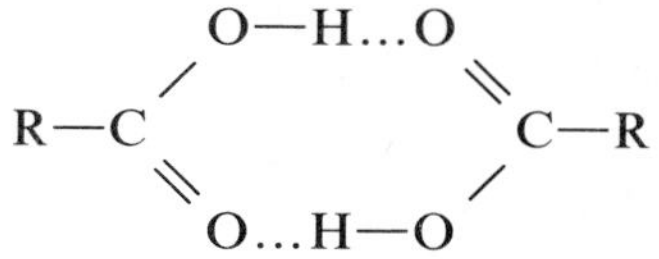

The "double-bond" resonances in phosphorus and sulphur compounds are at much lower values than in carbonyl compounds,[5,14] and the effects of substituents on —O— and =O resonances are different (Table 12.3). This presumably reflects the participation of additional *d* orbitals in the bonding pattern of the third-row elements. The P—O and P=O compounds also gave J_{PO} splitting of the ^{17}O resonances (see later).

There is no doubt, given adequate sensitivity under natural abundance, that ^{17}O resonances could provide much useful information for structure-determination within organic compounds. It will be of interest to see whether the FT technique can turn this into a useful practical area of spectroscopy. However, unfortunately there are strong indications that the resonances from large molecules may be too broad because of slow tumbling-times.[14]

Surprisingly few ^{17}O studies of organic molecules have been made during the last 10 years. Canet *et al.* have recently published FT results for about 10 organic compounds, mainly ethers, lactones and carbonates. Other groups of compounds studied include aliphatic nitro-compounds,[85] nitropyrroles and nitroimidazoles,[86] substituted acetophenones,[87] methylethoxysilanes,[88] larger-ring cyclic ketones where trans-annular interactions of C=O with O and N heteroatoms have been investigated,[89] and acetyl fluoride.[90] Gorodetsky *et al.*[91] have studied equilibria within the enol-forms of asymmetrically disubstituted *β*-diketones, and Irving and Lapidot[92] have used ^{17}O resonances to produce "titration" curves' of amide and carboxylate groups of glycyl-glycines as a function of pH. Finally hydrogen-bonding has been studied for acetic acid at different dilutions in cyclohexane.[93]

12A.6 Applications in Other Areas

a. Physicochemical

A great deal of physicochemical work on rates, equilibria, and relaxation phenomena relating to the hydration etc. of diamagnetic and paramagnetic metal ions in aqueous solutions has been cited above. Also, mention has been made of a review, by Silver, of the use of ^{17}O in studies of reaction mechanisms in organic and inorganic chemistry.[11]

Oxygen-17 NMR spectroscopy has been used to study the mobility of water in clay minerals,[94] and the temperature dependence of the rates of proton exchange in liquid MeOH and EtOH.[95] Niederberger and Tricot[96] have studied water orientation in lyotropic phases by ^{17}O NMR.

b. Biological

A study, listed earlier, of the pH dependence of the ^{17}O signals from amide and carboxylate groups in the dipeptide glycylglycine is of some biological significance.[92] Shporer and colleagues have made a number of studies of ^{17}O relaxation times, using spin-echo techniques, in the striated muscle of the frog,[97] in phospholipid vesicles[98] and in thymus gland lymphocytes[99] of the rat. An ^{17}O study has been made of the reaction of

O_2 with haemoglobin, although the interpretation of the results is not altogether certain.[100] Also, the interaction of Mn^{2+} with ATP, and the rate of exchange of solvated water molecules have been studied.[101] In a review of the use of stable isotopes in NMR studies, some ^{17}O work on staphylococcae has been cited.[102]

12A.7 Internuclear Coupling Involving ^{17}O

Because of the breadth of ^{17}O resonances only relatively large coupling constants, $J(O, X)$, can be measured and hence the majority of those recorded refer to couplings from directly bonded nuclei; however a few ^2J couplings (across 2 bonds) have been recorded. The literature values are collected in Table 12.4, and these have also been converted to the nuclear-independent values, K(O, X), which are of more significance in comparing the electronic structures of bonds when the interacting nuclei are different. (see Sec. 1.6 and 3C.2.)

Some regularities are apparent in the Table. Thus $^1J(O,H)$ in water[14,18,21,103] and the alcohols[14,95] has values close to 80 Hz. The $^1J(P, O)$ couplings involving P=O groups cover a fairly wide range (145 – 210 Hz) and for a series of compounds of type OPX_3 (X = F, O in OMe, N in NMe_2 and C in Me) there is a consistent drop in coupling with decreasing electronegativity of X.[104] This may perhaps be qualitatively attributed to the electronegative substituents removing electrons in outer orbitals from the P atom, leading to a greater effective atomic charge and hence stronger contact interaction between the P nucleus and *s* orbitals. A finite perturbation theoretical treatment within the CNDO/2 formalism has reproduced the observed trends, but not the absolute values.[104]

The "ether" oxygens in P—OMe groups within phosphates give rise to a consistently smaller coupling constant, compared with P=O, of between 90 and 100 Hz; there is an indication of a considerably larger value for phosphites.[14,104]

As was earlier pointed out by Lutz *et al.*,[60] the reduced coupling constants decrease with increasing atomic number of the central metal atoms along the isoelectronic sequence VO_4^{3-}, CrO_4^{2-} and MnO_4^-.

As might have been anticipated, ^2J couplings are normally about an order of magnitude smaller than for the corresponding directly bonded couplings. However, there seems to be a particularly favoured coupling pathway from H to the "ether" O in H.CO.OR systems, probably associated with π-bonding across the planar heavy-atom framework.[14,16]

In some cases coupled resonances cannot be resolved into fine structure but retain an element of width from partially collapsed structure, dependent on the rate of proton transfer. In such cases decoupling of the other nuclei can lead to reduced linewidths and information about the coupling constants.[16]

12A.8 Theory of ^{17}O Chemical Shifts

Coverage of the literature in this section is less comprehensive than elsewhere in this chapter, because the subject of ^{17}O chemical shifts is part of a much larger literature. For nuclei other than ^1H, it is generally considered that nuclear shieldings are determined primarily by "local" diamagnetic and paramagnetic contributions in the Ramsey/Pople formulation. Also, the most widely held view is that the diamagnetic term, although substantial in magnitude, varies much less than the paramagnetic one in different types of chemical environment. Hence, it is considered that the chemical shift pattern is dominated by the paramagnetic term, although this view has been challenged for ^{17}O.[105]

TABLE 12.4 Internuclear coupling constants and reduced coupling constants between ^{17}O and other nuclei, X[a]

(a) Directly-bonded nuclei

X	Compound	$^1J(O,X)/Hz$	$^1K(O,X)/10^{-18}NA^{-2}m^{-2}$	Ref.
^1H	H_2O (gas)	79	4.85	18
	H_2O^b	>74, 82, 83	5.1	21, 103, 14, 15
	$MeOH^b$	76, 85.5	~5.0[e]	95, 14
	$EtOH^b$	81, 83.6	~5.0	95, 14
	Pr^nOH^b	78	4.8	14
^{19}F	F_2O_2	424	27.7	78
^{31}P	OPF_3	188, 184	~28.2	80, 104
	$OPCl_3$	203, 208, 205	~31.1	84, 80, 104
	$OPBr_3$	195, 201	~30.0	80, 104
	$OP(NMe_2)_3$	145	~22.3	80, 104
	$\underline{O}P(OMe)_3$	165, 150	~24.0	84, 14
	$\underline{O}P(OMe)_2H$	160	24.27[e]	14
	$\underline{O}P(OMe)_2Me$	160	24.27	14
	$OPMe_3$	120	18.2	104
	$OP(\underline{O}Me)_3$	90	13.7	84
	$OP(\underline{O}Me)_2H$	98	14.9[e]	14
	$OP(\underline{O}Me)_2Me$	95	14.4	14
	$P(OMe)_3$	154,[c] 140	~22.3[e]	14, 104
^{35}Cl	ClO_4^-	85.5, 81.4, 85.5	~53	60, 61, 59
	ClO_3F	108[a]	67.7	61
^{51}V	VO_4^{3-}	61.6	14.4	60
	$V_2O_7^{4-}$	~50	~11.7	60
^{53}Cr	CrO_4^{2-}	~10	~10.9	60
^{55}Mn	MnO_4^-	28.9 30.2	~7.35	60, 62
^{95}Mo	MoO_4^{2-}	40.5, 40.3	38.2	60, 73
^{129}Xe	$XeOF_4$	692	154	81
^{187}Re	ReO_4^-	~50	~13.5	64
(b) Indirectly-bonded nuclei		$^2J(O,X)/Hz$		
^1H	$\underline{H}\cdot CO\cdot \underline{O}Me$	35, 38	~2.24	14, 15
	$\underline{H}\cdot C\underline{O}\cdot OMe$	~10.5[d]	~0.64	16
	$H\cdot CO\cdot \underline{O}C\underline{H}_3$	~7.5[d]	~0.46	16
	$\underline{H}\cdot CO\cdot \underline{O}Et$	23	1.43	14
^{19}F	OPF_3	31	2.02	104

[a] The nuclei X are arranged in their order in the Periodic Table. [b] In acetone. [c] Quoted in ref. 104.
[d] Obtained indirectly from linewidth analysis of an unresolved multiplet.

Ab initio calculations have been carried out by Ditchfield, Miller and Pople[106] and by Ditchfield.[107] The latter concluded that the use of gauge-dependent atomic orbitals gave a significant improvement in agreement between theory and experiment. Velenik and Lynden-Bell[108] have applied extended Hückel MO methods to a number of cases. More recently Ebraheem and Webb[109] have applied CNDO/S and INDO calculations, using gauge-dependent atomic orbitals in the molecular orbital description, and have calculated the ^{17}O chemical shifts for a large number of molecules. Using either method there is a reasonable general correlation between observed and calculated shifts, but there is a fair scatter of points, and compounds with oxygen bonded to nitrogen give rather poor agreement.

The well-known "average energy" approximation assumes that the dependence of the paramagnetic contribution on the sum of terms $\sum_i \Delta E_i^{-1}$ (ΔE_i is the excitation energy from the ground electronic state to excited states 1, 2, 3 etc. of increasingly high energy) can be approximated to ΔE_1^{-1} if the first excited state has a particularly low value for ΔE. Often this will be a spectroscopically allowed transition, observed as the lowest frequency absorption band in the electronic spectrum of the substance concerned. Carbonyl groups have low-frequency absorptions corresponding to $n \rightarrow \pi^*$ transitions, and at an earlier stage Figgis, Kidd and Nyholm[6] showed that there was a very good correlation between the chemical shifts of C=O groups in different chemical environments and ΔE^{-1} derived from their absorption bands. Rodger has more recently shown that the substantial change in CO chemical shift of acetone (by 55 ppm) as a function of concentration in water also correlates with ΔE^{-1} for the $n \rightarrow \pi^*$ transition.[14]

Andersson and Mason[56] have shown that this relationship is also useful although rather less precise, in correlating chemical shifts of ^{17}O in compounds with NO bonds. The numerical quantity $\langle r^{-3} \rangle$, where r is the mean radius of the atomic p orbitals concerned, plays an important part in evaluating paramagnetic effects. Andersson and Mason[56] showed that, when the same low-energy excited state was associated with an orbital strongly overlapping O and N atoms, there was a fairly good straightline correlation of ^{14}N and ^{17}O shifts. The slope reflected the theoretical values of $\langle r^{-3} \rangle$ for p orbitals on the O and N atoms. Velenik and Lynden-Bell[108] also found a unit-gradient between observed and calculated ^{17}O chemical shifts when values of $\langle r^{-3} \rangle$ for O were evaluated from self-consistent field (SCF) atomic wave functions. Moniz and Paranski[110] have also carried out INDO calculations for ^{17}O in compounds with NO bonds.

Sadlej and Sadlej[111] used CNDO/2 calculations to obtain theoretical values for solute/solvent shifts in electrolyte solutions. The calculated effects were all very small (a few ppm) but gave shielding for $^{17}O \ldots M^+$ (M = metal) and deshielding for $^{17}O \ldots F^-$ interactions. A theoretical study has been made of the hyperfine splittings (and hence the contact chemical shifts) involving ^{17}O in paramagnetic lanthanide complexes of the shift-reagent type.[112] Figgis *et al.*[6] extended their empirical ΔE_1^{-1} correlations to diamagnetic transition metal anions of formula MO_4^{n-} and carried out theoretical calculations to show that the absolute magnitude was reasonable.

12A.9 Conclusion

Where isotopic enrichment is feasible, as in the study of aqueous solutions of inorganic ions, ^{17}O NMR spectroscopy has shown its very substantial value for the study of kinetics and equilibria concerned with oxygen-exchange and hydration processes. Also, for complex oxyanions, it has proved to be very successful for structure determination.

Natural abundance work using broad-line techniques has served to show the very substantial potential of ^{17}O NMR for structure determination of organic and (to a less extent) of inorganic molecules. More recent FT spectroscopic methods have opened up new possibilities, apparently of higher sensitivity, which should aid the transition from potentiality into actuality. There is already little doubt of the success of the NMR method for studying concentrated solutions of small oxygen-containing molecules. The next 5 to 10 years should decide the extent to which work is possible with larger molecules and more dilute solutions, and hence whether ^{17}O NMR will become a widely valued branch of magnetic resonance spectroscopy.

12B SULPHUR-33

The isotope ^{33}S has $I = \frac{3}{2}$, a natural abundance of 0.76 %, and a receptivity to NMR detection which is 17.1×10^{-6} that of the proton; it is therefore not totally unsuited to NMR study. In fact, remarkably little work has been done, the chemically most valuable paper[113] listing only 12 chemical shifts and prognosticating a gloomy future for ^{33}S NMR. The electric quadrupole moment of ^{33}S is similar to that of ^{14}N, and although there is a tendency for sulphur atoms to be at sites of lower electronic symmetry than nitrogen, linewidths should not be too great for satisfactory measurements to be undertaken. In fact, observed linewidths, $W_{\frac{1}{2}}$, have ranged from essentially 0 (in symmetrical species such as SO_4^{2-}) to over 5000 Hz, this latter figure corresponding to 1100 ppm at a measuring frequency of 4.6 MHz ($B_0 = 1.4$ T) or 650 ppm at 7.4 MHz ($B_0 = 2.3$ T). In many compounds, however, linewidths are less than this (see Table 12.5), and it seems reasonable to hope for a precision of ± 10 ppm in measuring the sulphur chemical shift in many cases. This could be improved substantially by performing the measurements with a superconducting magnet.

The data of Table 12.5 were acquired by CW methods,[113] and this severely limited the attainable signal-to-noise ratios and hence the precision of the measurement. Clearly this could be improved by the use of pulsed FT techniques, and some preliminary experiments[115] have indicated that for most compounds without high symmetry at sulphur pulse repetition intervals of ca. 30 μs will be suitable, it thus being possible to accumulate many transients in a short period of time. It has been found that the temperature and concentration dependence of the ^{33}S resonance frequency in aqueous solutions of caesium sulphate are small, and this has therefore been proposed as a suitable reference material.[114] Certainly carbon disulphide is not ideal, although with ^{33}S it does not seem necessary to take into account the possible needs of indirect methods of detection, and so the absence of protons is not relevant.

It is difficult to say much about the rather limited chemical shift data currently available[113] (Table 12.5). Apparently the sensitivity of $\delta(^{33}S)$ to electronic changes is

TABLE 12.5 ^{33}S Chemical Shifts[a]

No.	Species	$\delta(^{33}S)$/ppm	Linewidth/Hz	Conditions	Ref.
1	CS_2	0.0	160	neat liquid	113
2	Et_2S_2	-168 ± 88	5000	neat liquid	113
3	Tetrahydrothiophene	-89 ± 38	2500	neat liquid	113
4	2-Methylthiophene	$+178 \pm 9$	1300	neat liquid	113
5	3-Methylthiophene	$+197 \pm 26$	9000	neat liquid	113
6	2-Bromothiophene	$+134$	1600	neat liquid	113
7	Thiophene	$+220 \pm 6$	650	90 % in CS_2	113
8	DMSO	$+233 \pm 20$	2500	neat liquid	113
9	H_2SO_4	$+225 \pm 32$	2200	neat liquid	113
	H_2SO_4	$+319 \pm 5$	≤ 160	10 N	113
10	Na_2S	-261	1600	Soln. in H_2O	113
11	Sphalerite (ZnS)	-230 ± 6	≤ 70	Solid[b]	113
12	Rb_2SO_4	$+331 \pm 4$	<40	1 M in H_2O	114
13	K_2MoS_4	$+364 \pm 4$	61	1 M in H_2O	60

[a] Measured at a frequency of 4.33 MHz, except for [12] and [13] which were measured at 5.9 MHz. [b] Identical results obtained for powder and single crystal.

less than that of $\delta(^{77}Se)$, as is to be expected in view of the relative sizes of $\langle r_{np}^{-3} \rangle$ for the two atoms, but it is probably large enough for chemically useful work to be done. Thus the S^{2-} ion is highly shielded while the SO_4^{2-} signal is to high frequency, and effects due to changes in substituents in thiophenes are measurable. The only coupling constant measured involving ^{33}S appears to be[116] $|^1J(^{33}S, ^{19}F)| = 251$ Hz in SF_6.

12C SELENIUM-77

12C.1 General

The receptivity (5.26×10^{-4} that of the proton) of ^{77}Se ($I = \frac{1}{2}$, natural abundance 7.6%) implied a necessity for the use of neat liquid samples in the earlier[117–119] CW work on direct selenium NMR, but ^1H-$\{^{77}Se\}$ double resonance experiments subsequently permitted the examination of more dilute solutions.[120] More recently, pulsed Fourier transform methods have been used,[121,122] although unfortunately no details have been given of pulse widths, repetition intervals and other experimental parameters. Two reviews about selenium NMR have appeared, one[123] dealing with ^{77}Se observations, and the other[124] with (^{77}Se, H) coupling constants and proton chemical shifts. In most of its organic compounds ^{77}Se is spin-coupled to protons, and thus considerable improvements in spectral simplicity and signal-to-noise ratio are to be expected under conditions of proton decoupling, the maximum possible ^{77}Se-$\{^1H\}$ nuclear Overhauser enhancement being $\eta = 2.62$. None of the work in which proton decoupling was used[121,122] has actually mentioned this NOE, but a measurement[125] of $T_1(^{77}Se)$ for PhSeH, in which there is a direct selenium-hydrogen bond, indicated that in this case at least, the dipole-dipole contribution to the selenium relaxation is important. In fact the value of $T_1(^{77}Se) = 1.5$ s is the only one yet published for any organoselenium compound, and there is a clear need for further work in this area. The spin coupling of ^{77}Se to protons also facilitates the production of INDOR spectra, and to date this is the method which has enabled data to be acquired for the widest range of compounds;[120] however, for many species with complex proton spectra (e.g. aromatic compounds) the FT approach is clearly superior.

It is perhaps worth noting that ^{77}Se was the first nucleus to which what might be termed the doubly indirect method of detection was applied, ^1H-$\{^{31}P, ^{77}Se\}$ triple resonance experiments on organophosphine selenides being used to transfer information firstly to the phosphorus spectrum, and thence to the proton spectrum.[127]

In view of the importance of selenium in semiconductor technology a fair amount of work has been done on the NMR of ^{77}Se in alloy systems. Most has very little chemical bias and is therefore ignored here.

12C.2 Chemical Shifts

There is some diversity of opinion as to the most suitable reference material for ^{77}Se chemical shifts, H_2SeO_3 aq.,[117] $SeOCl_2$,[118] selenophene,[121] and Me_2Se[120,122,128–130] all having been used. The last of these has been used by more workers than any other, and has the advantage of being suitable for both direct and indirect methods of detection, so it is adopted here. It is also reported[131] to have the surprisingly high LD_{50} for rats of 1.6 g Se/kg body weight, and is somewhat less malodorous than many organoselenium compounds. $\Xi(^{77}Se)$ in Me_2Se is 19 091 523 Hz.[120]

TABLE 12.6 Selenium chemical shifts in inorganic compounds

No.	Compound	$\delta(^{77}Se)$/ppm[a]	Conditions
1	H_2SeO_3	+1282	Sat. in H_2O
2	Na_2SeO_3	+1253	Sat. in H_2O
3	H_2SeO_4	+1001	$+H_2O$
4	K_2SeO_4	+1024	Sat. in H_2O
5	SeO_3	+944	Sat. in H_2O
6	$HSeO_3F$	+1001	"Impure"
7	$HSeO_3Cl$	+1003	Soln. in SO_2
8	$SeOF_2$	+1378	Liquid
9	SeOFCl	+1478.6	Mixture of [8] and [10]
10	$SeOCl_2$	+1479	Liquid
11	$SeOBr_2$	+1559	Liquid
12	SeO_2F_2	+948	Liquid
13	SeF_4	+1092	Liquid
14	$SeF_4 \cdot BF_3$	+1122	Liquid
15	$SeF_4 \cdot SO_3$	+1057	Liquid
16	$SeCl_4$	+1154	Sat. in DMF
17	Se_2Cl_2	+1274	Liquid
18	Se_2Br_2	+1174	Liquid
19	H_2SeCl_6	+1451	In H_2O
20	SeF_6	+610	Liquid
21	H_2Se	−226	Liquid
22	$F_2P(Se)H$	−170	Liquid
23	$(PF_2)_2Se$	+701	Liquid

[a] Chemical shifts in this and the following two tables are given relative to the resonance of Me_2Se. Data for compounds 1–5 are taken from ref. 117 and have been converted on the basis that $\delta(^{77}Se)$ in H_2SeO_3 aq. = 1282 ppm. Data for compounds 6–21 are taken from ref. 118 and have been converted using $\delta(^{77}Se)$ in $SeOCl_2$ = 1479 ppm. The results for compounds 22 and 23 are taken from ref. 129 and 128 respectively.

Representative selenium chemical shifts are given in Tables 12.6, 12.7 and 12.8, the entries having seen selected to illustrate as many general trends as possible, and to embrace a wide range of types of compound.

Where interpretations of the results have been offered it has generally been assumed that the paramagnetic term makes the dominant contribution to the ^{77}Se shielding. Of particular interest in this connection is the variation of $\delta(^{77}\text{Se})$ with temperature in aromatic diselenides noted by Lardon,[119] an increase of ca. 0.4 ppm/K being found. Diselenides are often deeply coloured, implying relatively large values of the $(\Delta E)^{-1}$ factor in the expression for the paramagnetic contribution to the shielding and a corresponding sensitivity to changes in the populations of excited electronic states. It is thus tempting to draw an analogy between the sensitivity to temperature of the selenium shielding in diselenides, and of the metal shielding in many transition metal complexes.

TABLE 12.7 Selenium chemical shifts in organoselenium compounds

No.	Compound	$\delta(^{77}\text{Se})$/ppm[a]	Conditions	Ref.
24	Me_2Se	0	Neat liquid	120
25	Et_2Se	+233	Neat liquid	120
26	Pr^i_2Se	+436	Neat liquid	120
27	Ph_2Se	+402	Neat liquid	119
28	MeSePh[b]	+202	CH_2Cl_2soln.	120
29	$MeSeBu^t$	+294	CH_2Cl_2soln.	120
30	MeSeEt	+108	CH_2Cl_2soln.	120
31	$MeSeCH_2Ph$	+173	$CDCl_3$soln.	122
32	MeSe(naphthyl)	+155	$CDCl_3$soln.	122
33	$(PhCH_2)_2Se$	+333	$CDCl_3$soln.	122
34	$(CF_3)_2Se$	+694	Neat liquid	118
35	MeSeH	−116	Neat liquid	120
36	EtSeH	+42	Neat liquid	120
37	Pr^iSeH	+159	Neat liquid	120
38	Bu^tSeH	+278	CH_2Cl_2soln.	120
39	$PhCH_2SeH$	+107	CH_2Cl_2soln.	120
40	PhSeH	+152	Neat liquid	120
41	$3\text{-}CF_3C_6H_4SeH$	+159	Neat liquid	120
42	$2\text{-}FC_6H_4SeH$[c]	+191	Neat liquid	120
43	$3\text{-}FC_6H_4SeH$[c]	+164	Neat liquid	120
44	$4\text{-}FC_6H_4SeH$[c]	+141	Neat liquid	120
51	$2\text{-}MeC_6H_4SeH$	+112	Neat liquid	120
52	$3\text{-}MeC_6H_4SeH$	+144	Neat liquid	120
53	$4\text{-}MeC_6H_4SeH$	+128	Neat liquid	120
54	$MeSe^-Na^+$	−332	In H_2O	120
55	$EtSe^-Na^+$	−150	In H_2O	120
56	$Pr^iSe^-Na^+$	+9	In H_2O	120
57	$Bu^tSe^-Na^+$	+129	In H_2O	120
58	$Me_3Se^+I^-$	+253	In H_2O	120
59	$Et_3Se^+I^-$	+377	In H_2O	120
60	$Me_2EtSe^+I^-$	+291	In H_2O	120
61	Me_2Se_2	+275	Neat liquid	120
62	Et_2Se_2	+339	Neat liquid	120
63	$Pr^i_2Se_2$	+407	Neat liquid	120
64	$Bu^t_2Se_2$	+493	Neat liquid	120

TABLE 12.7 (*Continued*)

No.	Compound	$\delta(^{77}Se)$/ppm[a]	Conditions	Ref.
65	Ph_2Se_2	+460[d]	Neat liquid	119
66	$(4\text{-}NO_2C_6H_4)_2Se_2$	+465[d]	Neat liquid	119
67	$(4\text{-}BrC_6H_4)_2Se_2$	+468[d]	Neat liquid	119
68	$(4\text{-}ClC_6H_4)_2Se_2$	+471[d]	Neat liquid	119
69	MeSeTeMe	+36	Mixture of [61] + $(MeTe)_2$	132
70	PhSeSeMe	+445, +294	In mixture of [61] + [65]	120
71	$(PhCH_2)_2Se_2$	+402	CH_2Cl_2 soln.	120, 122
72	PhSeSMe	+512	In mixture of [65] + $(CH_3S)_2$	120
73	$(CF_3)_2Se_2$	+528	Neat liquid	118, 120
74	MeSeCN	+125	$CDCl_3$ soln.	122
75	$PhCH_2SeCN$	+291	$CDCl_3$Soln.	122
76	$2\text{-}FC_6H_4CH_2SeCN$[e]	+287	$(CD_3)_2CO$ soln.	121
77	$3\text{-}FC_6H_4CH_2SeCN$[e]	+297	$(CD_3)_2CO$ soln.	121
78	$4\text{-}FC_6H_4CH_2SeCN$[e]	+298	$(CD_3)_2CO$ soln.	121
82	Me_2SeCl_2	+448	CH_2Cl_2 soln.	120
83	Me_2SeBr_2	+389	CH_2Cl_2 soln.	120
84	Et_2SeBr_2	+540	CCl_4 soln.	120
85	$Pr^i_2SeBr_2$	+742	CCl_4 soln.	120
86	$MeSeCl_3$	+890	CH_2Cl_2 soln.	120
87	$EtSeCl_3$	+995	CH_2Cl_2 soln.	120
88	Me_2SeO	+812	H_2O soln.	120
89	Ph_2SeO	+738	–	121
90	$4\text{-}MeOC_6H_4SeC(O)Me$	+649	Neat liquid	120
91	$PhCH_2Se[CH_2]_{1,2,3,4}CO_2H$[f]	+288, +263, +240, +243	$(CD_3)_2CO$ soln.	121
92	$2\text{-}NO_2C_6H_4CH_2SeCH_2CO_2H$	+298	$(CD_3)_2CO$ soln.	121
93	$3\text{-}NO_2C_6H_4CH_2SeCH_2CO_2H$	+299	$(CD_3)_2CO$ soln.	121
94	$4\text{-}NO_2C_6H_4CH_2SeCH_2CO_2H$	+303	$(CD_3)_2CO$ soln.	121
95	$PhCH_2CH_2SeCH_2CO_2H$	+209	$(CD_3)_2CO$ soln.	121
96	PhC(O)SeMe	+445	$CDCl_3$ soln.	122
97	selenophene[g]	+605	$CDCl_3$ soln.	122, 121
98	2-Cl-selenophene[g]	+564	$(CD_3)_2CO$ soln.	121
99	3-Cl-selenophene[g]	+591	$(CD_3)_2CO$ soln.	121
106	2-MeO	+696	$(CD_3)_2CO$ soln.	121
107	3-MeO	+686	$(CD_3)_2CO$ soln.	121
108	structure I	+545	$CDCl_3$ soln.	122
109	structure II	+451	$CDCl_3$ soln.	122
110	structure III	+526	$CDCl_3$ soln.	122
111	structure IV	+584	$CDCl_3$ soln.	122
112	structure V	+157	$CDCl_3$ soln.	122
113	structure VI	+654	$CDCl_3$ soln.	122
114	structure VII	+1013	$CDCl_3$ soln.	122
115	$MeSeO_2H$	+1216	H_2O soln.	120
116	$MeSeO_3^-K^+$	+1045	H_2O soln.	144
117	$(MeO)_2SeO$	+1339	Neat liquid	120
118	$(MeO)_2SeO_2$	+1053	CH_2Cl_2 soln.	120
119	$(Me_3Sn)_2Se$	−547	Neat liquid	133
120	$Me_3SnSeMe$	−277	Neat liquid	134
121	$Me_2Sn(SeMe)_2$	−237	Neat liquid	134
122	$MeSn(SeMe)_3$	−184	Neat liquid	134

(*Continued overleaf*)

TABLE 12.7 (*Continued*)

No.	Compound	$\delta(^{77}Se)$/ppm[a]	Conditions	Ref.
123	$Sn(SeMe)_4$	−127	Neat liquid	134
125	$Me_3SnSePh$	+11.3	Neat liquid	134
126	$(SiH_3)_2Se$	−666	Neat liquid	129
127	$(GeH_3)_2Se$	−612	Neat liquid	129
128	*cis*-$(Me_2Se)_2PtCl_2$	+120	CH_2Cl_2 soln.	120
129	*trans*-$(Me_2Se)_2PtCl_2$	+135	CH_2Cl_2 soln.	120
130	*trans*-$[Pt(SCN)_2(SeMe_2)_2]$	+134	—	128
131	*cis*-$[Pt(NCO)_2(SeMe_2)_2]$	+92	—	128
132	*cis*-$[Pt(NCO)Cl(SeMe_2)_2]$	+118, +92	—	128
133	*trans*-$[Pt(NCO)Cl(SeMe_2)_2]$	+134	—	128

[a] To high frequency of Me_2Se. Data from ref. 118 converted using $\delta(^{77}Se) = 1479$ ppm in $SeOCl_2$; from ref. 119 using $\delta(^{77}Se) = 460$ ppm in Ph_2Se_2; and from ref. 121 using $\delta(^{77}Se) = 605$ ppm in selenophene.[122] [b] Substituted aromatic selenides show a dependence of $\delta(^{77}Se)$ upon the nature of substituent (see the text and Fig. 12.7). [c] Ref. 120 also lists $\delta(^{77}Se)$ for YC_6H_4SeH with Y = 2-Cl [45], 3-Cl [46], 4-Cl [47], 2-MeO [48], 3-MeO [49] and 4-MeO [50]. [d] Temperature dependent, see text. [e] Ref. 121 also lists $\delta(^{77}Se)$ for $YC_6H_4CH_2SeCN$ with Y = 2-Br [79], 3-Br [80] and 4-Br [81]. [f] Many substituted benzyl-selenocarboxylic acids have been studied, see ref. 121. [g] Ref. 121 also lists $\delta(^{77}Se)$ for selenophenes with substituents 2-Br [100], 3-Br [101], 2-CN [102], 3-CN [103], 2-NO_2 [104], 3-NO_2 [105] and many others.

I II III IV

V VI VII

The most general trend which emerges from the data is that an increase in the effective electronegativity of the substituents of selenium leads to an increase in $\delta(^{77}Se)$, this being expected irrespective of whether changes in the paramagnetic or the diamagnetic term dominate. Thus, in substituted aromatic selenols and selenides $\delta(^{77}Se)$ correlates[120] with the Hammett σ-constant of the substituent as shown in Fig. 12.6, the correlation being rather better in the second class of compounds. This may be attributed to variable degrees of hydrogen bonding in the selenols. Similarly, in substituted selenophenes[121] there are excellent correlations with substituent parameters, which in turn are normally related to charge densities, and also with ^{13}C shieldings, although the sensitivity to changes of the ^{77}Se shielding is some six-fold greater. Other classes of compound which display this kind of effect are substituted aromatic diselenides,[119] benzyl selenides, diselenides, and selenocyanates,[121] aryl vinyl selenides,[140] and benzylselenocarboxylic acids of the type $XC_6H_4CH_2Se(CH_2)_nCO_2H$ with X = F, Cl, Br, Me or NO_2.[121] In this last class of compound $\delta(^{77}Se)$ decreases by ca. 20 ppm as n increases from 1 to 2 to 3 and then becomes almost constant.[121]

TABLE 12.8 Selenium chemical shifts and selenium-phosphorus coupling constants in organophosphorus compounds

No.	Compound	$\delta(^{77}Se)$/ppm	$^1J(^{77}Se, ^{31}P)$	Ref.
132	Me_2PSeMe	+58	−205	135
132	$Me_2P(S)SeMe$	+196	−341	135
134	$MeP(SeMe)_2$	+89	−240	136(a)
135	$MeP(S)(SeMe)_2$	+280	−384	136(a)
136	Bu^t_2PSeMe	−35	−231	136(a)
137	$Bu^t_2P(S)SeMe$	+41	−360	136(a)
138	Me_3PSe	−235	−684	127
139	Me_2PhPSe	−272	−710	127
140	$MePh_2PSe$	−277	−725	127
141	Ph_3PSe	—	−736	127
142	$(MeO)_3PSe$	−396	−963	127
143	$(Me_2N)_3PSe$	−366	−805	127
144	$Me(Me_2N)_2PSe$	−327	−767	127
145	$Me_2(Me_2N)PSe$	−279	−720	127
146	Et_3PSe	—	−705	136(b)
147	$(EtO)_3PSe$	—	−935	136(b)
148	$(EtO)_2(EtSe)PS$	—	−477	136(b)
149	$(EtO)_2P(SH)Se$	—	−777	136(b)
150	$(EtO)_2P(SeH)Se$	—	−822	136(b)
153	$(2\text{-}MeC_6H_4)_3PSe$	—	−708	137
154	$(3\text{-}MeC_6H_4)_3PSe$	—	−726	137
155	$(4\text{-}MeC_6H_4)_3PSe$	—	−724	137
156	$(4\text{-}ClC_6H_4)_3PSe$	—	−753	137
157	$EtP(Se)Cl_2$	—	−920	138(a)
158	$EtPhP(Se)Cl$	—	−840	138(a)
159	$Na^+[Se(O)P(OPr^i)_2]^-$	—	−754.6	139
160	$[Se(O)P(OPr^i)_2]_2$	—	−515	139

In contrast to the foregoing behaviour, the selenium shielding has been found to correlate[120] with the Taft σ-constant of the alkyl group in selenols, selenides, diselenides, selenonium cations and selenide anions, but with an increase in the electron withdrawing ability of the alkyl group leading to a *decrease* in $\delta(^{77}Se)$. This is entirely analogous to the behaviour of $\delta(^{31}P)$ in trialkyl phosphines and related molecules,[141] and indeed it is fairly clear that each alkyl group makes an additive contribution to the shielding which depends upon the amount of α-chain branching (see Fig. 12.7). Another analogy with phosphorus is that when an organic selenide forms a selenonium salt, or a metal complex (e.g. [24] → [58] or [128]*) there is an increase in $\delta(^{77}Se)$, just as $\delta(^{31}P)$ increases when R_3P forms R_4P^+ or $R_3P \rightarrow M$, although again the change is much larger in the case of selenium. To some extent (though not entirely) this can be attributed to the larger value of $\langle r_{np}^{-3} \rangle$ in the case of selenium, as indeed can the greater spread of selenium chemical shifts in general.

The lowest selenium shielding appears to arise in species in which the selenium atom retains one lone pair of electrons and has several electronegative atoms (e.g. oxygen or

* Numbers in square brackets refer to the selenium compounds as listed in Tables 12.6–12.8.

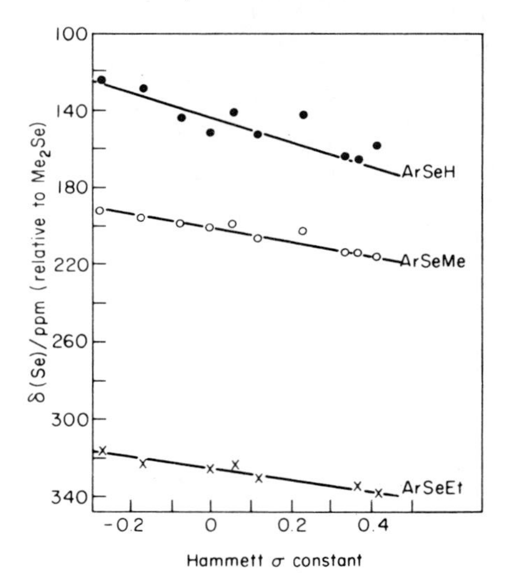

Fig. 12.6 ^{77}Se chemical shifts vs. Hammett σ constants in *meta* and *para* substituted aryl selenols and aryl methyl- and ethyl-selenides. Reproduced with permission from ref. 120.

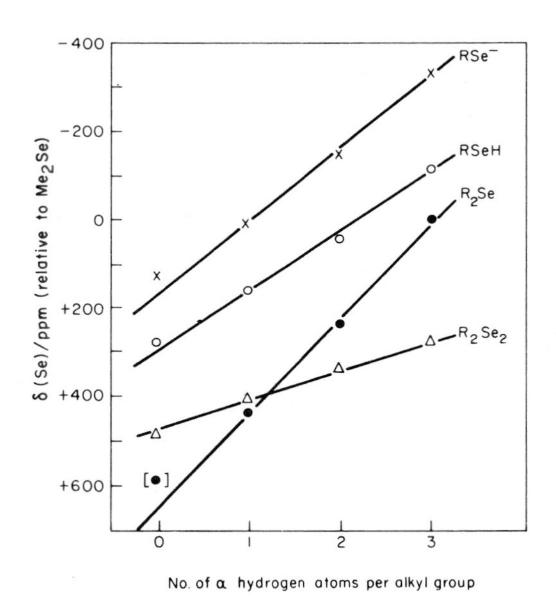

Fig. 12.7 ^{77}Se chemical shifts vs. number of α hydrogen atoms per alkyl group in alkyl selenide-anions, selenols, selenides, and diselenides. The chemical shift of $Bu^t{}_2Se$ ([●]) is an estimate based on the value for $MeBu^tSe$. Reproduced with permission from ref. 120.

halogens) bound to it, for example in compounds [8]–[11] and in [117]. The transformation $(MeO)_2SeO$ [117] → $(MeO)_2SeO_2$ [118], in which the selenium atom loses all its lone pairs, is analogous to $(RO)_3P$ → $(RO)_3PO$, and in each case the shielding of the central atom increases.

Although the effects of altering the atom(s) attached to selenium are complicated, it appears well established[120,127,142] that electropositive elements (e.g. H, Si, Ge, Sn) can produce substantial increases in shielding (it is known[134] that there is a systematic variation of $\delta(^{77}Se)$ with n in $Me_{4-n}Sn(SeMe)_n$, [120]–[123]), and this has led to the suggestion[127] that in organophosphine selenides (VIII) the form (a), which requires negative charge on selenium, predominates, since these species ([138–[145]) have very highly shielded selenium. However, the detailed variation of $\delta(^{77}Se)$ with the nature of R indicated some contribution from form (b).

$$R_3P^+ - Se^- \qquad\qquad R_3P = Se$$
$$(a) \qquad\qquad \text{VIII} \qquad\qquad (b)$$

In summary it can be said that, where it is possible to draw parallels between phosphorus and selenium chemistry, there are analogies in the behaviour of $\delta(^{31}P)$ and $\delta(^{77}Se)$, the sensitivity of the selenium shielding to electronic charges being several times greater. $\delta(^{77}Se)$ also appears to correlate inversely with the (probable) electron density at selenium, but when the atoms directly attached to selenium are charged other factors may supervene. Certainly, it is not yet possible to use changes in $\delta(^{77}Se)$ of only a few tens of ppm to reach structural conclusions except in series of extremely closely related compounds.

12C.3 Coupling Constants

Almost all coupling constants $^nJ(^{77}Se, X)$ have been obtained either by detecting ^{77}Se satellites in the spectrum of X (for X = ^1H, ^{13}C, ^{19}F or ^{31}P) or by ^1H-{X} and ^1H-{^{77}Se} double resonance experiments, the latter approach having the advantage that it often enables the sign of $^nJ(^{77}Se, X)$ to be established, whereas the former does not. The subject of coupling constants between protons and ^{77}Se was reviewed[124] in 1972, and although much more data has accumulated since then, no especially new results have appeared. Table 12.9 summarizes the general patterns of behaviour, and it should be noted that since $\gamma(^{77}Se)$ is positive ^nK and ^nJ are of the same sign for $J(^{77}Se, ^1H)$. As is the case with other elements with electronic lone pairs [e.g. F, P^{III}] the vicinal and the geminal couplings to protons tend to be positive,[146] in contrast to the situation for elements [e.g. Sn, P^V] without lone pairs where the geminal coupling to protons is normally negative. In general the coupling constants show rather small amounts of variation with changes in the substituents on selenium, the most notable effects being the very large values of $^3J(^{77}Se, H)$ = 38 Hz in $EtSeO_3^-K^+$ and 25 Hz in $(MeO)_2SeO_2$, each of these being species in which the selenium atom has no formal electron lone pairs.[148] There is some evidence that selenium-proton coupling constants are sensitive to stereochemical relationships: for example,[149] in (XI) $^3J(^{77}Se, H)$ is 17.5 Hz for Me and 3.8 Hz for CH_2, while in $(PhCH_2O)_2SeO$ (for which prochirality at selenium leads to magnetic inequivalence of the methylene protons) two different values of this coupling have been found,[148] and this also applies to $Pr^i_2MeSe^+I$.[136a] Cyclic species have provided evidence for a "Karplus" type of relationship between $^3J(^{77}Se, H)$ and dihedral angle, typical values being 7.0(H_e) and ca. 1(H_a) Hz in XII(a), 26.4 Hz in XII(b) (with averaging due to ring inversion), and

TABLE 12.9 Coupling between ^{77}Se and ^{1}H or ^{13}C

Compound	$^{n}J(^{77}Se, H)/Hz$ n			$^{n}J(^{77}Se, ^{13}C)/Hz$ n		Ref
	1	2	3	1	2	
H_2Se	(+) 65.4	—	—	—	—	143
MeSeH	+ 44	+10.6	—	−46.0	—	144
Me_3CSeH	+ 43	—	+10.9	−47.6	±10.2	144
PhSeH	+ 56	—	—	−88.3	±9.7	144
$MeSe^-K^+$	—	+6.6	—	−58.2	—	144
Me_2Se	—	−10.5	—	−62	—	144, 145
Et_2Se	—	+10.7	+10.8	−61.2	±8.5	146
$Me_3Se^+I^-$	—	+9.3	—	−50	—	145
$Et_3Se^+I^-$	—	+7.5	+17.4	−56.4	—	144
Me_2SeBr_2	—	+10.0	—	−68	—	144
Et_2SeBr_2	—	+5.3	+17.2	−54.4	—	144
Me_2SeO	—	+11.7	—	−60	—	144
$MeSeO_3^-K^+$	—	−8.8	—	−13	—	144
$EtSeO_3^-K^+$	—	−3.6	+37.9	−18.5	±7.1	144
$(MeO)_2SeO$	—	—	+7.6	—	±0.9	144
$(MeO)_2SeO_2$	—	—	+25.0	—	−12.6	144
MeSeSeMe	—	+11.9	+2.3	−73.7	—	145
$Pt(SCN)_2(SeMe_2)_2$	—	+9.4	—	—	—	128
structure IX	—	—	—	*w* 100 *y* 45.9	*u* 12.3 *x* 13.7 *z* 3.1	147
structure X	—	—	*x* 23.0 *y* 10.5	*x* 25.7 *y* <6	—	147

IX

X

37.8(H_e) and 16.2(H_a) Hz in XIII.[150] In substituted selenophenes the various selenium-proton coupling constants are normally positive (except possibly when two nitro groups are present) and appear to be additive functions of substituent parameters.[151]

Rather little work has been done on coupling between ^{13}C and ^{77}Se, and much of the available data[143–145,147,148,152,153] is included in Table 12.9. The pattern of behaviour seems to be similar to that for coupling between phosphorus and carbon in that loss of *all* the selenium lone pairs leads to a substantial positive increment to $^{1}J(^{77}Se, ^{13}C)$, although the sign remains (just) negative.[154] Longer-range carbon-selenium coupling constants appear to be dependent upon stereochemical relationships,[147] and clearly this type of coupling has great potential in structural work. In particular it has been noted that a magnitude of 45 Hz or more for $J(^{77}Se, ^{13}C)$ is generally indicative of a direct carbon-selenium bond.[144]

Seppelt[154,155] and his group have studied coupling between selenium and fluorine extensively, and a selection of values are in Table 12.10; much of this work has required

TABLE 12.10 Coupling constants between ^{77}Se and ^{19}F

Species	$^1J(^{77}Se, ^{19}F)/Hz^a$	Ref.	Species	$^1J(^{77}Se, ^{19}F)/Hz^a$	Ref.
SeF_6	1432	156	$SeF_5O^-Na^+$	1075, 1195[b]	154
SeF_5OCl	1350, 1453[b]	154	SeF_5Cl	1352, 1258[b]	154
SeF_5OBr	1360, 1413[b]	154	$(SeF_5O)_2Xe$	1338, 1328[b]	154, 155
$(SeF_5O)_2I$	1300, 1320[b]	154	SeF_4[c]	302, 1200	157
SeF_5OXeF	1280, 1300[b]	154, 155	$SeOFCl$	647	158
$SeF_5O^-NO_2^+$	1060, 1203[b]	154	SeO_2F_2	1584	158

[a] All are presumably negative. [b] The first figure refers to axial F, the second to equatorial F. [c] At −140°C when an A_2B_2 spectrum is obtained.

careful analysis of second-order ^{19}F spectra, and in these circumstances care must be taken to distinguish between ^{77}Se satellites and weak lines arising from molecules containing no magnetic selenium. Longer range coupling to ^{19}F has also been observed, e.g. $^2J(^{77}Se, ^{19}F) = \pm 36$ Hz in $(PF_2)_2Se$.[129]

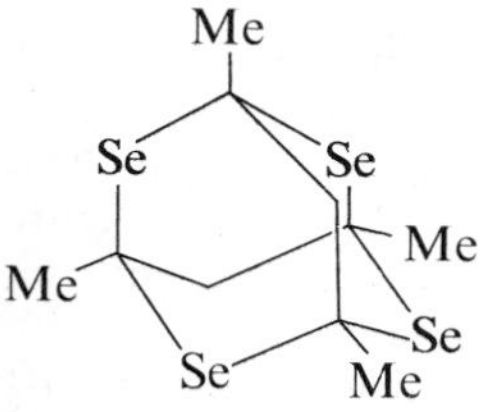

XI

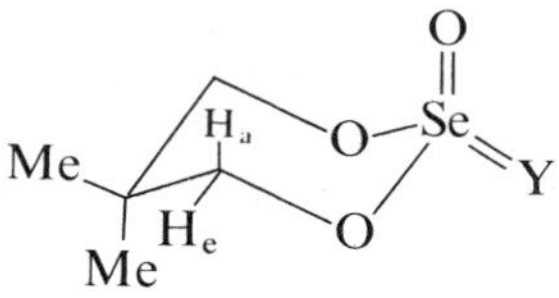

XII (a) Y = lone pair
(b) Y = O

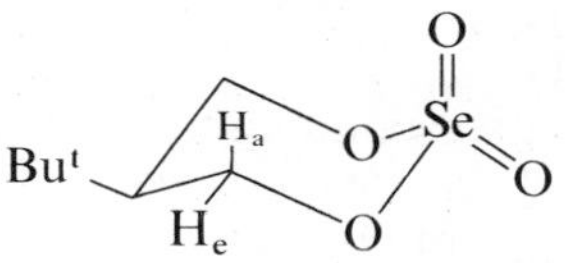

XIII

A surprisingly large amount of work has been done on phosphorus-selenium coupling and a selection of results is in Table 12.8. It appears that the sign of $^1J(^{77}Se, ^{31}P)$ is always negative, as established by double and triple resonance experiments.[127,129,130,135] Values range from −200 to −500 Hz for formally single Se-P bonds from −500 to below −1000 Hz for formally double bonds. This feature has made it possible to assign the structure $(EtO)_2P(SeEt)S$ rather than $(EtO)_2P(SEt)Se$ to [148], and the structure $(EtO)_2P(SH)Se$ rather than $(EtO)_2P(SeH)S$ to [149].[136] As pointed out earlier,[127] the canonical form (VIIIb) probably makes only a small contribution to the resonance hybrid in the organo-phosphine selenides, but this has little effect upon $^1J(^{77}Se, ^{31}P)$, which receives its major contribution from the Fermi contact term which in turn depends only upon the σ-component of the bond. In accord with this the coupling constant is found to correlate quite well with the electronegativity of the other substituents at phosphorus, thus implying a dependence upon $S_P^2(0)$. However, since $^1J(^{77}Se, ^{31}P)$ is negative, it should be noted that the mean excitation energy approximation cannot be valid and that Π_{SeP}, the mutual polarizability term, must be negative. In other species it is not yet possible to systematize the variations of $^1J(^{77}Se, ^{31}P)$. There is evidence that changes in molecular configuration can affect the magnitude of this coupling constant.[159] Values of $^3J(^{77}Se, ^{31}P)$ ranging from −12 to +2 Hz have been measured in the second-order spectra of diphosphorus compounds,[160] and in P_4Se_3 $^2J(^{77}Se, ^{31}P) = (+)\ 57$ Hz.[161]

Other coupling constants involving directly bound selenium include $^{1}J(^{77}Se, ^{119}Sn)$ = +1015 to +1520 Hz in $Me_{4-n}Sn(SeMe)_n$[134] [see Fig. 10.11; note that $\gamma(^{119}Sn)$ is negative]; $^{1}J(^{77}Se, ^{125}Te) = -169$ Hz in MeSeTeMe[132] [note that $\gamma(^{125}Te)$ is negative]; $|^{1}J(^{77}Se, ^{129}Xe)| = 130$ Hz (estimated) in $Xe(OSeF_5)_2$;[157] $^{1}J(^{77}Se, ^{207}Pb) = -1170$ Hz in $Me_3PbSeMe$;[162] $^{1}J(^{77}Se, ^{195}Pt) = +670, +507, +234$ Hz in $[PtX_3SeMe_2]^-$ for X = Cl, Br, I respectively;[163] and $^{1}J(^{77}Se, ^{195}Pt) = +259, +101, -68$ Hz in $[PtX_5SeMe_2]^-$ for X = Cl, Br, I respectively.[163] The final negative value is of especial interest, and is probably related to the presence of a remaining lone pair of electrons in a coordinated dimethyl selenide group.[163]

Structural uses of selenium NMR have been few, apart from those mentioned above, but now that the foundations have been laid there is a good prospect of developments in this area. In organo-phosphorus chemistry in particular the recognition of selenium satellites in a ^{31}P spectrum can have great diagnostic value,[139] and since $|^{1}J(^{77}Se, ^{31}P)|$ is several hundred Hz there is little likelihood of confusion with spinning sidebands or other instrumental artifacts.

12D TELLURIUM-125

With 2.21×10^{-3} of the proton's receptivity to NMR detection the isotope ^{125}Te ($I = \frac{1}{2}$, natural abundance 7.0%) is reasonably promising as a candidate for study by NMR. Against this is the negative magnetogyric ratio, which can produce nuclear Overhauser "enhancements" ranging from $\eta = 0$ to $\eta = -1.58$, the latter figure coreesponding to an inverted signal of just over half normal intensity. In the event, $^{1}H-\{^{125}Te\}$ experiments have dominated the study of organo-tellurium compounds,[164,165] and only one list of results from direct observation on inorganic species is apparently

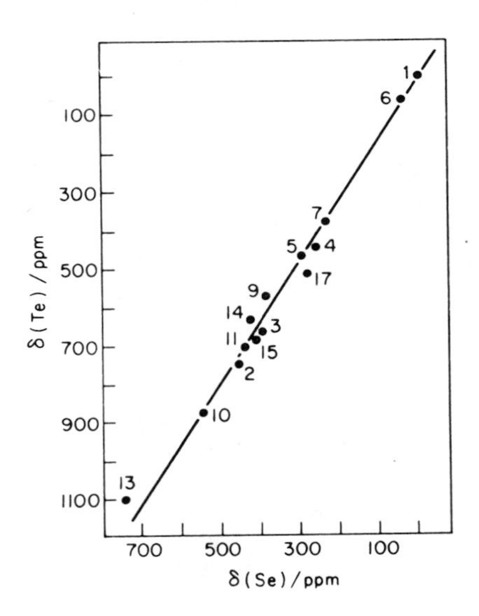

Fig. 12.8 ^{125}Te chemical shifts in organotellurium compounds plotted against ^{77}Se chemical shifts in corresponding species. The value for $(Me_3Sn)_2Te$ lies close to the straight line but is not plotted because it is well off-scale to low frequency. For key to compounds see Table 12.11. Reproduced with permission from ref. 164.

TABLE 12.11 Tellurium chemical shifts

No.[b]	Species	$\delta(^{125}Te)$/ppm[a]	Conditions
1	Me_2Te	0	Neat liquid
2	Me_2TeCl_2	+749	Soln. in CH_2Cl_2
3	Me_2TeBr_2	+669	Soln. in CH_2Cl_2
4	$Me_3Te^+I^-$	+443	Soln. in $(CD_3)_2SO$
5	$Me_2EtTe^+I^-$	+470	Soln. in D_2O
6	Me_2Te_2	+63	Soln. in CH_2Cl_2
7	Et_2Te	+380	Sol. in C_6H_6
8	Et_2Te_2	+188	Soln. in CH_2Cl_2
9	$Et_3Te^+Br^-$	+573	Soln. in D_2O
10	Et_2TeBr_2	+879	Soln. in CH_2Cl_2
11	Pr^i_2Te	+707	Soln. in CH_2Cl_2
12	$Pr^i_2Te_2$	+303	Soln. in CH_2Cl_2
13	$Pr^i_2TeBr_2$	+1105	Soln. in CH_2Cl_2
14	$Pr^i_2MeTe^+I^-$	+630	Soln. in D_2O
15	Ph_2Te	+688	Soln. in CH_2Cl_2
16	$(Me_3Sn)_2Te$	−1214	Soln. in CH_2Cl_2
17	MeSeTeMe	+512	Soln. in C_6H_6
18	$MeSeTePr^i$	+860	Soln. in CH_2Cl_2
19	$(4\text{-}MeC_6H_4)_2Te$	+663	Soln. in CH_2Cl_2
20	Ph_2TeCl_2	+981	Soln. in CH_2Cl_2
21	$PhMeTeI_2$	+698	Soln. in CH_2Cl_2
22	$(4\text{-}MeC_6H_4)MeTeI_2$	+663	Soln. in CH_2Cl_2
23	$(4\text{-}MeOC_6H_4)MeTeI_2$	+664	Soln. in CH_2Cl_2
24	$(4\text{-}EtOC_6H_4)MeTeI_2$	+670	Soln. in CH_2Cl_2
25	$Ph_2MeTe^+I^-$	+595	Soln. in DMSO
26	$(2\text{-}MeC_6H_4)_2MeTe^+I^-$	+580	Soln. in DMSO
27	$PhMe_2Te^+I^-$	+550	Soln. in DMSO
28	$(4\text{-}MeOC_6H_4)Me_2Te^+I^-$	+550	Soln. in DMSO
29	$(4\text{-}EtOC_6H_4)Me_2Te^+I^-$	+542	Soln. in DMSO
30	$Te_2(CH_2COOH)_2$	+538	Soln. in DMSO
31	$(Cl_3TeCH_2CO)_2O$	+864	Soln. in DMSO
32	$CH_2(TeCl_3)_2$	+1198	Soln. in DMSO
33	Ph_2Te_2	+422	0.5M in CH_2Cl_2
34	$TeCl_4$	+1138[c]	1M in Me_2CO
	$TeCl_4$	+1725[c]	1M in THF
35	$TeBr_4$	+1442[c]	1M in DMF
	$TeBr_4$	+1962[c]	1M in THF
36	H_2TeCl_6	+1403	2M in HCl/D_2O
37	H_2TeBr_6	+1356	1M in HBr/D_2O
38	H_2TeI_6	+857	1M in HI/D_2O
39	$[NBu^n_4{}^+]_2TeCl_6^{2-}$	+1329	0.4M in $(CD_3)_2CO$
40	$[NBu^n_4{}^+]_2TeBr_6^{2-}$	+1341	0.3M in $(CD_3)_2CO$
41	H_2TeO_6	+712	1M in D_2O
42	K_2TeO_3	+1732	2M in D_2O

[a] To high frequency of the resonance for Me_2Te. Data for species 1–17, 18, 19–32 and 33–42 are taken from ref. 164, 136, 165 and 166 respectively. [b] Numbering coincides with that used on Fig. 12.8. [c] Values between the two extremes were recorded in other solvents.

TABLE 12.12 Some coupling constants between ^{125}Te and a directly-bonded nucleus X

X	^{1}H	^{13}C	^{13}C	^{19}F	^{19}F
Compound	H_2Te	Me_2Te	Me_2TeCl_2	TeF_6	F_5TeOCl
^{1}J/Hz	(−) 59	+162	(+) 175.3	+3688	(+) 3755[a]
ref.	167	143, 145	136[a]	156, 168	163
X	^{31}P	^{31}P	^{31}P	^{77}Se	^{119}Sn
Compound	XIV	XV	Bu^n_3PTe	MeTeSeMe	$(Me_3Sn)_2Te$
^{1}J/Hz	(+) 1321	(+) 2023	(+) 1720	−169	−1385
ref.	150	160	169	132	133
X	^{129}Xe	^{195}Pt	^{195}Pt		
Compound	$(F_5TeO)_2Xe$	$[PtCl_3TeMe_2]^-$	$[(PtCl_3)_2TeMe_2]^{2-}$		
^{1}J/Hz	±470	−1553	(−) 5923		
ref.	157	163	163		

[a] For the equatorial fluorine.

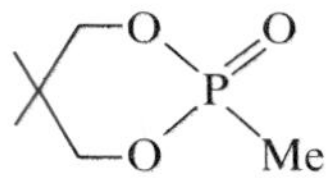

XIV

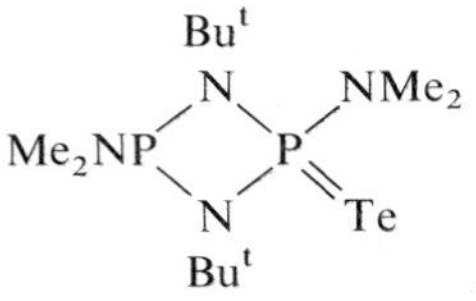

Me_2NP, NMe_2, Bu^t, N, P, Te

XV

available.[166] As with selenium, alloy systems have aroused interest but are ignored here. There is also an isotope ^{123}Te with I = $\frac{1}{2}$ and natural abundance 0.89 %, but there appears to be little reason to study this.

In dimethyl telluride $\Xi(^{125}Te)$ is 31 549 802 Hz[164] and all workers have so far used this as a reference compound. The available data are listed in Table 12.11 and, as can be seen from the plot (Fig. 12.8) of $\delta(^{125}Te)$ against $\delta(^{77}Se)$ in analogous species, the trends in tellurium shielding are so similar to those in selenium shielding that separate discussion is scarcely warranted. It is however significant that λ_{max} and $\delta(^{125}Te)$ are concentration dependent in R_2Te_2. The slope of the plot in Fig. 12.8 is ca. 1.8 and, whilst part of this can be attributed to the ratio $\langle r_{5p}^{-3}(Te)\rangle/\langle r_{4p}^{-3}(Se)\rangle$ = ca. 1.25, there is still an unexplained factor. This is common when chemical shift patterns for lighter and heavier elements in the same group are compared, and can be attributed to greater electronic sensitivity to the effects of substituents on the heavier elements, and to a larger $(\Delta E)^{-1}$ for tellurium compounds.

As a consequence of the negative magnetogyric ratio for ^{125}Te ^nK is generally of opposite sign to ^nJ, but otherwise the rather limited evidence available indicates parallels with the behaviour of ^{77}Se, the couplings involving tellurium being larger by a factor of 2–3 than their selenium counterparts. An exception to this generalization is one-bond coupling constants which happen to be close to zero, for with these a quite small change in electronic structure can produce rather large changes in the coupling constant.

Some typical values of one-bond coupling constants involving tellurium are listed in Table 12.12. The two-bond (^{125}Te, ^{1}H) coupling constant is found to be: −20.7 Hz in Me_2Te;[143,145] −20.9 Hz in $Me_3Te^+ I^-$[145]; −20.9 Hz in Et_2Te;[146] and ca. 90 Hz in tellurophenes.[170] The value of $^3J(^{125}Te, ^1H)$ is −22.7 Hz in Et_2Te[146] and ca. 12 Hz in tellurophenes.[170] The greater magnitudes of $^nJ(^{125}Te, H)$ (n = 2 or 3) compared with $^nJ(^{77}Se, H)$ often facilitate $^1H-\{^{125}Te\}$ experiments; for example, it was possible to use the indirect method to determine $\nu(^{125}Te)$ in Ph_2Te,[165] but not $\nu(^{77}Se)$ in Ph_2Se.[120]

In conclusion it can be said that the prospects for tellurium are at least as good as for selenium NMR, especially as it should be easy to overcome the drawback of a negative magnetogyric ratio by using relaxation reagents or gated decoupling schemes.

REFERENCES

1. C. M. Olsmats, S. Axensten and G. Liljegren *Arkiv. Fysik* **19**, 469 (1961)
2. a. F. Alder and F. C. Yu, *Phys. Rev.* **81**, 1967 (1951)
   b. H. E. Weaver, B. M. Tolbert and R. C. LaForce, *J. Chem. Phys.* **23**, 1956 (1955)
3. S. S. Dharmatti, J. K. Sundara Rao, and R. Vijayoraghan, *Nuovo cim.* **11**, 656 (1959)
4. H. A. Christ, *Helv. Phys. Acta* **33**, 572 (1960)
5. H. A. Christ, P. Diehl, H. R. Schneider and H. Dahn, *Helv. Chim. Acta* **44**, 866 (1961)
6. B. N. Figgis, R. G. Kidd and R. S. Nyholm, *Proc. Roy. Soc.* **A269**, 469 (1962)
7. R. K. Harris, Chem. Soc. Rev. **5**, 1 (1976)
8. B. L. Silver and Z. Luz, *Quart. Rev.* **21**, 458 (1967)
9. R. K. Harris, in "Molecular Spectroscopy", p. 49 (A. R. West, ed.). Institute of Petroleum/Heyden, London (1977)
10. C. Deverill, *Prog. NMR Spectry.* **4**, 235 (1969)
11. B. L. Silver, *Compt. Rend. Colloq. Int. Isot. Oxygene* 88 (1972), published 1975 [Chem. Abs. **84**, 22619 (1976)]
12. J. W. Emsley, J. Feeney and L. H. Sutcliffe, "High Resolution Nuclear Magnetic Resonance Spectroscopy". Pergamon Press, London (1965)
13. R. R. Vold and R. L. Vold, *J. Chem. Phys.* **61**, 4360 (1974)

14. C. Rodger, Ph. D. thesis, University of East Anglia, Norwich, England (1976)
15. C. Canet, C. Goulon-Ginet and J. P. Marchal, *J. Magn. Reson.* **22**, 539 (1976) with an *erratum* in **25**, 397 (1977)
16. W. L. Earl and W. Niederberger, *J. Magn. Reson* **27**, 351 (1977)
17. H. W. Spiers, B. B. Garrett, R. K. Sheline and S. W. Rabideau, *J. Chem. Phys.* **51**, 1201 (1969)
18. Z. Luz and G. Yagil, *J. Phys. Chem.* **70**, 554 (1966)
19. H. G. Hertz and R. Klute, *Z. phys. Chem.* (*Frankfurt*) **69**, 101 (1970)
20. F. Fister and H. G. Hertz, *Ber. Bunsenges. Phys. Chem.* **71**, 1032 (1967)
21. J. Reuben, A. Tzalmona and D. Samuel, *Proc. Chem. Soc.* 353 (1962)
22. J. A. Jackson, J. Lemons and H. Taube, *J. Chem. Phys.* **32**, 553 (1960)
23. R. E. Connick and D. N. Fiat *J. Chem. Phys.* **39**, 1349 (1963)
24. D. Fiat and R. E. Connick, *J. Amer. Chem. Soc.* **88**, 4754, (1966)
25. M. Alei and J. A. Jackson, *J. Chem. Phys.* **41**, 3402 (1964)
26. G. E. Glass, W. B. Schwabacher and R. S. Tobias, *Inorg. Chem.* **7**, 2471 (1968)
27. D. Fiat and A. M. Chmelnick, *J. Amer. Chem. Soc.* **93**, 2875 (1971)
28. R. E. Connick and D. Fiat, *J. Chem. Phys.* **44**, 4103 (1966)
29. J. W. Neely and R. E. Connick, *J. Amer. Chem. Soc.* **94**, 3419 (1972)
30. D. Fiat and A. M. Chmelnick, *Proc. Colloq. Ampere*, **14**, 1157 (1967)
31. A. M. Chmelnick and D. Fiat, *J. Chem. Phys.* **47**, 3986 (1967)
32. D. Fiat, Z. Luz and B. L. Silver, *J. Chem. Phys.* **49**, 1376 (1968)
33. A. M. Chmelnick and D. Fiat, *J. Magn. Reson.* **8**, 325 (1972)
34. J. Reuben and D. Fiat, *Inorg. Chem.* **6**, 579 (1967)
35. A. M. Chmelnick and D. Fiat, *J. Chem. Phys.* **51**, 4238 (1969)
36. A. P. Gulya, V. A. Shcherbakov, *Fiz. Mat. Metody Koord. Khim, Tezisy Dokl. Vses, Soveshch. 5th*, 106 (1974)
37. T. J. Swift and R. E. Connick, *J. Chem. Phys.* **37**, 307 (1962); **41** 2553 (1964)
38. A. G. Desai, H. W. Dodgen and J. P. Hunt, *J. Amer. Chem. Soc.* **91**, 5001 (1969)
39. R. Murray, H. W. Dodgen and J. P. Hunt, *Inorg. Chem.* **3**, 1576 (1964)
40. R. B. Jordan, H. W. Dodgen and J. P. Hunt, *Inorg. Chem.* **5**, 1906 (1966)
41. A. G. Desai, H. W. Dodgen and J. P. Hunt, *J. Amer. Chem. Soc.* **92**, 798 (1970)
42. M. W. Grant, H. W. Dodgen and J. P. Hunt, *J. Amer. Chem. Soc.* **92**, 2321 (1970)
43. D. P. Rablen, H. W. Dodgen and J. P. Hunt, *Inorg. Chem.* **15**, 931 (1976)
44. D. P. Rablen, H. W. Dodgen and J. P. Hunt, *J. Amer. Chem. Soc.* **94**, 1771 (1972)
45. J. P. Hunt, M. W. Grant, and D. W. Dodgen, *Inorg. Chem.* **10**, 71 (1971)
46. M. W. Grant, H. W. Dodgen and J. P. Hunt, *Chem. Comm.* 1446 (1970); and *J. Amer. Chem. Soc.* **93**, 6828 (1971)
47. M. V. Olson, Y. Kanazawa and H. Taube, *J. Chem. Phys.* **51**, 289 (1969)
48. W. B. Lewis, M. Alei, and L. O. Morgan, *J. Chem. Phys.* **45**, 4003 (1966)
49. D. Rablen and G. Gordon, *Inorg. Chem.* **8**, 395 (1969)
50. A. H. Zeltman, N. A. Matwiyoff and L. O. Morgan, *J. Phys. Chem.* **73** 2689 (1969)
51. K. Wuthrick and R. E. Connick, *Inorg. Chem.* **6**, 583 (1967)
52. L. G. Mashirov, D. N. Suglobov, V. A. Shcherbakov, *Radiokhimiya* **17**, 881 (1975) [*Chem. Abs.* **84**, 97280 (1976)]
53. J. S. Babiec, C. H. Langford and T. R. Stengle, *Inorg. Chem.* **5**, 1363 (1966)
54. T. R. Stengle and C. H. Langford, *Coord. Chem. Rev.* **2**, 349 (1967)
55. J. P. Hunt, *Coord. Chem. Rev.* **7**, 1 (1971)
56. L-O. Andersson and J. Mason, *J.C.S. Dalton* 202 (1974)
57. S. M. Rabideau, *J. Phys. Chem.* **71**, 2747 (1967)
58. J. A. Jackson and H. Taube, *J. Phys. Chem.* **69**, 1844 (1965)
59. M. Alei, J. Chem. Phys. **43**, 2904 (1965)
60. O. Lutz, W. Nepple and A. Nolle, *Z. Naturforsch.* **31a**, 978 and 1046 (1976)
61. J. Virlet and G. Tantot, *Chem. Phys. Letters* **44**, 296 (1976)
62. M. Broze and Z. Luz, *J. Phys. Chem.* **73**, 1600 (1969)
63. M. Filowitz and W. G. Klemperer, *Chem. Comm.* **233** (1976)
64. R. A. Dwek, Z. Luz and M. Shporer, *J. Phys. Chem.* **74**, 2232 (1970)

65. B. N. Figgis, R. G. Kidd and R. S. Nyholm, *Canad. J. Chem.* **43**, 145 (1965)
66. R. G. Kidd, Canad. *J. Chem.* **45**, 605 (1967)
67. W. G. Klemperer and W. Shum, *J. Amer. Chem. Soc.* **99**, 3544 (1977)
68. A. D. English, J. P. Jesson, W. G. Klemperer, T. Mamouneas, L. Messerle, W. Shum and A. Tramontano, *J. Amer. Chem. Soc.* **97**, 4785 (1975)
69. M. Filowitz, W. G. Klemperer, L. Messerle and W. Shum, *J. Amer. Chem. Soc.* **98**, 2345 (1976)
70. M. A. Fedotov, A. M. Galubov and L. P. Kazanskii, *Isvest. Khim.* **12**, 2840 (1976)
71. L. P. Kazanskii, *Koord. Khim.* **2**, 11 (1976)
72. L. P. Kazanskii, A. I. Gasanov, V. F. Chuvaev, and V. I. Spitsyn, *Fiz. Mat. Metody Koord. Khim. Tezisy Dokl. Vses. Soveshch. 5th* 98 (1974)
73. R. R. Vold and R. L. Vold, *J. Chem. Phys.* **61**, 4360 (1974)
74. A. E. Florin and M. Alei, *J. Chem. Phys.* **47**, 4268 (1967)
75. O. Lutz and H. Oehler, *Z. Naturforsch*, **32a**, 131 (1977)
76. J. Reuben, quoted in A. Velenick and R. M. Lynden-Bell, *Mol. Phys.* **19**, 371 (1970)
77. R. Bramley, B. N. Figgis and R. S. Nyholm, *Trans. Faraday Soc.* **58**, 1893 (1962)
78. I. J. Solomon, R. K. Raney, A. J. Kacmarek, R. G. Maguire and G. A. Noble, *J. Amer. Chem. Soc.* **89**, 2015 (1967)
79. I. J. Solomon, A. J. Kacmarek, W. K. Sumida and J. K. Raney, *Inorg. Chem.* **11**, 195 (1972)
80. G. Grossmann, M. Gruner and G. Seifert, *Z. Chem.* **16**, 362 (1976)
81. J. Shamir, H. Selig, D. Samuel and J. Reuben, *J. Amer. Chem. Soc.* **87**, 2359 (1965)
82. J. Reuben. D. Samuel, H. Selig and J. Shamir, *Proc. Chem. Soc.* 270 (1963)
83. A. Lapidot and C. S. Irving, *J.C.S. Dalton* 668 (1972)
84. H. A. Christ and P. Diehl, *Helv. Phys. Acta* **36**, 170 (1963)
85. M. Mägi, V. I. Erashko, S. A. Shevelev and A. A. Fainzil'berg, *Eesti Nsv. Tead. Toim. Keem. Geol.* **20**, 297 (1971); M. Mägi *ibid*, **20**, 364 (1971) [*Chem. Abs.* **76**, 79088, 79089 (1972)]
86. E. Lippmaa, M. Mägi, S. S. Novikov, L. I. Khmelmitski, A. S. Prihodko, O. V. Lebedev and L. V. Epishina, *Org. Magn. Reson.* **4**, 153, 197 (1972)
87. D. J. Sardella and J. B. Stothers, *Canad. J. Chem.* **47**, 3089 (1969)
88. R. K. Harris and B. J. Kimber, *Org. Magn. Reson.* **7**, 460 (1975)
89. H. Dahn, H. P. Schlumke and J. Temler, *Helv. Chim. Acta* **55**, 909 (1972)
90. J. Reuben and S. Brownstein, *J. Mol. Spec.* **23**, 96 (1967)
91. M. Gorodetsky, Z. Luz and Y. Mazur, *J. Amer. Chem. Soc.* **89**, 1183 (1967)
92. C. S. Irving and A. Lapidot, *Chem. Comm.* **43** (1976)
93. D. Ziess, U. Jentschura and E. Lippert, *Ber. Bunsengesell. Phys. Chem.* **75**, 901 (1971)
94. V. I. Kulividze and A. V. Krasnushkin, *Dokl. Akad. Sci. URSS* **222**, 388 (1975)
95. H. Versmold and C. Yoon, *Ber. Bunsengesell. Phys. Chem.* **76**, 1164 (1972)
96. W. Niederberger and Y. Tricot, *J. Magn. Reson.* **28**, 313 (1977)
97. M. M. Civan and M. Shporer, *Biochim. Biophys. Acta* **343**, 399 (1974)
98. N. Haran and M. Shporer, *Biochim. Biophys. Acta* **426**, 638 (1976)
99. M. Shporer, M. Haas and M. M. Civan, *Biophysical J.* **16**, 601 (1976)
100. G. Pifat, S. Maricič, M. Petrinovič, V. Kramer, J. Marsel and K. Bunhard, *Croat. Chem. Acta* **41**, 195 (1969)
101. M. S. Zetter, H. W. Dodgen and J. P. Hunt, *Biochemistry* **12**, 778 (1973)
102. Y. Arata, *Nippon Aisotopu Kaigi Hobinshu* **11** 180 (1973) [*Chem. Abs.* **84**, 81771 (1976)]
103. S. W. Rabideau and H. G. Hecht, *J. Chem. Phys.* **47**, 544 (1967)
104. G. A. Gray and T. A. Albright, *J. Amer. Chem. Soc.* **99**, 3243 (1977)
105. R. Grinter and J. Mason, *J. Chem. Soc.* (*A*), 2196 (1970)
106. R. Ditchfield, D. P. Miller and J. A. Pople, *J. Chem. Phys.* **53**, 613 (1970); **54**, 4186 (1971)
107. R. Ditchfield, *Mol. Phys.* **27**, 789 (1974)
108. A. Velenik and R. M. Lynden-Bell, *Mol. Phys.* **19**, 371 (1970)
109. K. A. K. Ebraheem and G. A. Webb, *J. Magn. Reson.* **25**, 399 (1977)
110. W. B. Moniz and C. F. Paranski, Jr., *J. Magn. Reson.* **11**, 62 (1973)
111. J. Sadlej and A. J. Sadlej, *J. Magn. Reson.* **14**, 97 (1974)
112. R. M. Goldring and M. P. Halton, *Austral. J. Chem.* **25**, 2577 (1972)
113. H. L. Retcofsky and R. A. Friedel, *J. Amer. Chem. Soc.* **94**, 6579 (1972)

114. O. Lutz, A. Nolle and A. Schwenk, *Z. Naturforsch.* **28a**, 1370 (1973)
115. H. D. Schultz, C. Karr and G. D. Vickers, *Appl. Spectroscopy* **25**, 363 (1971)
116. P. T. Inglefield and W. T. Reeves, *J. Chem. Phys.* **40**, 2425 (1964)
117. H. E. Wachli, *Phys. Rev.* **90**, 331 (1953)
118. T. Birchall, R. J. Gillespie and S. L. Vekris, *Can. J. Chem.* **43**, 1672 (1965)
119. M. Lardon, *J. Amer. Chem. Soc.* **92**, 1063 (1970); *Ann. N.Y. Acad. Sci.* **192**, 132 (1972)
120. W. McFarlane and R. J. Wood, *J.C.S. Dalton* 1397 (1972)
121. A. Fredga, S. Gronowitz and A.-B. Hörnfeldt, *Chem. Scripta* **8**, 15 (1975); S. Gronowitz, I. Johnson and A.-B. Hörnfeldt, *Chem. Scripta* **3**, 94 (1973); **8**, 8 (1975); S. Gronowitz, A. Konar and A.-B. Hörnfeldt, *Org. Magn. Reson.* **9**, 213 (1977)
122. L. Christiaens, J.-L Piette, L. Laitem, M. Baiwir, J. Denoel and G. Llabres, *Org. Magn. Reson.* **8**, 354 (1976)
123. M. Lardon, "Organic Selenium Compounds: their Chemistry and Biology", p. 933 (D. L. Klayman and W. H. H. Gunther, eds.). Wiley-Interscience, New York (1973)
124. V. Svanholm, "Organic Selenium Compounds: their Chemistry and Biology", p. 903 (D. L. Klayman and W. H. H. Gunther, eds.), Wiley-Interscience, New York (1973)
125. A. Briguet, J.-C. Duplan and J. Delmau, *Mol. Phys.* **29**, 837 (1975)
126. W. McFarlane, *J. Chem. Soc.* (*A*) 670 (1969)
127. W. McFarlane and D. S. Rycroft, *J.C.S. Dalton* 2162 (1973); *Chem. Comm.* 902 (1972).
128. S. J. Anderson, P. L. Goggin and R. J. Goodfellow, *J.C.S. Dalton* 1939 (1976)
129. D. E. J. Arnold, J. S. Dryburgh, E. A. V. Ebsworth and D. W. H. Rankin, *J.C.S. Dalton* 2518 (1972)
130. D. W. Anderson, E. A. B. Ebsworth, G. D. Meikle and D. W. H. Rankin, *Mol. Phys.* **25**, 381 (1973)
131. K. P. McConnell and O. W. Portman, *Proc. Soc. Exptl. Biol. Med.* **79**, 230 (1952)
132. G. Pfisterer and H. Dreeskamp, *Ber. Bunsengesell. Phys. Chem.* **73**, 654 (1969)
133. J. D. Kennedy and W. McFarlane, *J. Organometal. Chem.* **94**, 7 (1975)
134. J. D. Kennedy and W. McFarlane, *J.C.S. Dalton* 2134 (1973)
135. W. McFarlane and J. A. Nash, *Chem. Comm.* 913 (1969)
136. a. H. C. E. McFarlane and W. McFarlane, unpublished results.
b. W. J. Stec, A. Okruszek, B. Uznanski and J. Michalski, *Phosphorus* **2**, 97 (1972)
137. R. P. Pinnell, C. A. Megerle, S. L. Mannatt and P. A. Kroon, *J. Amer. Chem. Soc.* **95**, 977 (1973)
138. a. E. I. Loginova, I. A. Nuretdinov and A. Yu. Petrov, *Teor. Eksp. Khim.* **10**, 75 (1974)
b. I. A. Nuretdinov and E. I. Loginova, *Izv. Akad. Nauk S.S.S.R. Ser. Khim.* **10**, 2360 (1971)
139. C. Glidewell and E. J. Leslie, *J.C.S. Dalton* 527 (1977)
140. D. F. Kushnarev, G. A. Kalabin, T. G. Mannitov, V. A. Mullin, M. F. Larin and V. A. Pestunovich, *J. Org. Chem.* (*U.S.S.R.*) **12**, 1482 (1976)
141. S. O. Grim, W. McFarlane and E. F. Davidoff, *J. Org. Chem.* **32**, 781 (1967)
142. D. F. Kushnarev, G. A. Kalabin and V. A. Pestunovich, *Zh. Obshch. Khim.* **12**, 1482 (1976)
143. H. Dreeskamp and G. Pfisterer, *Mol. Phys.* **14**, 295 (1968)
144. W. McFarlane, D. S. Rycroft and C. J. Turner, *Bull. Soc. Chim. Belg.* **86**, 457 (1977)
145. W. McFarlane, *Mol. Phys.* **12**, 243 (1967)
146. V. Breuninger, H. Dreeskamp and G. Pfisterer, *Ber. Bunsengesell. Phys. Chem.* **70**, 613 (1966)
147. H. J. Reich and J. E. Trend, *Chem. Comm.* 310 (1976)
148. W. McFarlane and D. S. Rycroft, *Chem. Comm.* 10 (1973)
149. K. Olsson and S. O. Almqvist, *Acta Chem. Scand.* **23**, 3271 (1969)
150. D. S. Rycroft, private communication
151. M. P. Simonnin, M. J. Pouet and J. M. Cense, *Org. Magn. Reson.* **8**, 508 (1976)
152. H. J. Reich and J. E. Trend, *Canad. J. Chem.* **53**, 1922 (1975)
153. M. Garreau, G. J. Martin, M. L. Martin, J. Marel and C. Paulmier, *Org. Magn. Reson.* **6**, 648 (1974)
154. K. Seppelt, *Z. Anorg. Allg. Chem.* **399**, 65 (1973)
155. K. Seppelt, *Z. Anorg. Allg. Chem.* **416**, 12 (1975)
156. E. L. Muetterties and W. D. Phillips, *J. Amer. Chem. Soc.* **81**, 1084 (1959)

157. K. Seppelt and H. H. Rupp, *Z. Anorg. Allg. Chem.* **409**, 338 (1974)
158. M. Brownstein and R. J. Gillespie, *J.C.S. Dalton*, 67 (1973)
159. W. J. Stec, *Z. Naturforsch.* **29b**, 109 (1974)
160. R. Keat, D. S. Rycroft and D. G. Thompson, unpublished results
161. R. A. Dwek, R. E. Richards, D. Taylor, G. J. Penney and G. M. Sheldrick, *J. Chem. Soc.* (*A*) 935 (1969)
162. J. D. Kennedy, W. McFarlane and B. Wrackmeyer, *Inorg. Chem.* **15**, 1299 (1976)
163. P. L. Goggin, R. J. Goodfellow and S. R. Haddock, *Chem. Comm.* 1767 (1975)
164. H. C. E. McFarlane and W. McFarlane, *J.C.S. Dalton* 2416 (1973)
165. W. McFarlane, F. J. Berry and B. C. Smith, *J. Organometal. Chem.* **113**, 139 (1976)
166. R. J. Goodfellow, private communication
167. C. Glidewell, D. W. A. Rankin and G. M. Sheldrick, *Trans. Faraday Soc.* **65**, 1409 (1969)
168. G. W. Fraser, R. D. Peacock and W. McFarlane, *Mol. Phys.* **17**, 291 (1969)
169. I. A. Nuretdinov and E. I. Loginova, *Izv. Akad. Nauk S.S.S.R. Ser. Khim.* 2827 (1973).
170. F. Fringuelli, S. Gronowitz, A.-B. Hörnfeldt, I. Johnson and A. Taticchi, *Acta Chem. Scand. B.* **28**, 175 (1974)

13

THE HALOGENS–CHLORINE, BROMINE AND IODINE

BJÖRN LINDMAN, University of Lund, Sweden

STURE FORSÉN, University of Lund, Sweden

Among the halogens, fluorine is unique in possessing (in 100% natural abundance) a spin-$\frac{1}{2}$ nucleus. Fluorine-19 has been exploited a great deal for NMR, since it is strongly magnetic, and its use is discussed in ch. 4. In fact fluorine differs in many ways chemically from the other halogens, and it is of lesser importance in biochemistry. The present chapter is concerned with a survey of the applications of Cl, Br and I NMR in chemistry and biology. Both from a chemical and from a NMR point of view these three elements have very similar properties. All three, especially chlorine, have sizeable natural occurrences in minerals, in sea-waters and in living systems. Biologists and biochemists are well aware of the high intra- and extra-cellular chloride concentrations, and that the interactions of chloride ions with many macromolecules, especially enzymes, have a wide (although badly understood) functional significance. The stable isotopes of chlorine (^{35}Cl and ^{37}Cl) and of bromine (^{79}Br and ^{81}Br) all have $I = \frac{3}{2}$ while the stable iodine isotope, ^{127}I, has $I = \frac{5}{2}$. All these nuclides have sizeable quadrupole moments and therefore quadrupolar effects generally predominate. Indeed, the vast majority of chemical and biological applications make use in some way of the quadrupolar interactions. However, studies of the shielding of the halide ions can also be quite useful for many problems. The sensitivities are reasonably good for the concentration ranges appropriate for many chemical and biological situations, but the possibilities of observation change immensely with the conditions. For example, while a halide ion can be easily studied, a covalently bound halogen generally experiences such large field gradients that it is difficult or even impossible to observe the NMR signal from the pure liquid. A rather comprehensive account[1] of chlorine, bromine, and iodine NMR has recently been published, so only a rather brief survey will be presented in this book. Special emphasis will be placed here on work which appeared after the completion of ref. 1. As regards the general principles, much of what was described for the alkali nuclei in ch. 6 will be directly applicable for the quadrupolar halogens.

13A QUADRUPOLAR EFFECTS

13A.1 Basic Aspects

The appearance of the NMR spectrum, and thus the information provided, is for the halogens strongly dependent on the one hand on the anisotropy of the phase studied and on the timescale of molecular dynamics, and on the other on the origin of the field gradients. With regard to *phase anisotropy and molecular dynamics* it is useful to make a schematic subdivision into the following four categories:

(*a*) solids, with generally very large quadrupole splittings (often only second-order effects are discernible), giving information on properties of chemical bonds, ion-ion distances and molecular motion;

(*b*) liquid crystals with small splittings, giving information on effects of orientation with respect to the liquid crystalline matrix, as well as on liquid crystalline structure, macroscopic orientation effects, ion binding etc.;

(*c*) isotropic liquids or solutions characterized by slow molecular motion giving non-exponential relaxation;

(*d*) isotropic liquids or solutions with rapid molecular motion.

For the biological applications, case (*c*) is generally appropriate, and provided correlation times are not too long the effect of departure from the extreme narrowing situation is mainly to make T_1 and T_2 unequal and dependent on resonance frequency, while non-exponentiality effects are not appreciable. For $\tau_c \gg \omega_0^{-1}$, on the other hand, both longitudinal and transverse relaxation are markedly non-exponential, and spectra consist (for $I = \frac{3}{2}$) of two Lorentzian curves of different widths. Under typical experimental conditions the broader component may be difficult to detect, leading to an apparent loss of 60% of the intensity.

With respect to the *origin of the field gradients* it is suitable to distinguish between the following three cases:

(*a*) covalent compounds like CCl_4, $[HgBr_4]^{2-}$, SnI_4 etc., with the nucleus at a site lacking cubic symmetry. The field gradients are large and of an intramolecular origin and therefore relatively unaffected by environmental effects;

(*b*) the halide ions, where the field gradients are small, of intermolecular origin and highly variable with ion-ion and ion-dipole distances and interactions; and

(*c*) covalent compounds with the nucleus at a site of tetrahedral or octahedral symmetry, as in ClO_4^- and IF_6^+. This situation requires rather detailed considerations since, in addition to intermolecular field gradients, we have to account for intramolecular field gradients becoming operative as a result of distortions from the unperturbed symmetric state.

13A.2 Covalent Compounds

The Cl, Br or I relaxation of diamagnetic covalent compounds is overwhelmingly dominated by quadrupole relaxation effected by the angular variation of the field gradient as a result of molecular reorientation. For ^{35}Cl the relaxation times generally lie in the range 10–100 μs. The most convenient way of obtaining these relaxation times has been to measure the linewidths of continuous-wave spectra but in recent years pulse experiments have also been performed. In a number of cases the scalar relaxation of another nucleus, scalar spin-coupled to the halogen, for example ^1H in $CHCl_3$ or ^{119}Sn in $SnCl_4$, has been used to determine the halogen relaxation time. For Br (and even more so for I) the relaxation times are very short and no direct measurements (pulse methods or linewidths) have yet been accomplished.

Since ^{35}Cl quadrupole coupling constants are relatively easy to determine, ^{35}Cl NMR has been one of the most important sources of information on molecular reorientation in liquids. The object of obtaining these data has in general been to test theoretical models for molecular motion in liquids, and a good example is given by the study of

ClO_3F by Maryott *et al.*[2] These authors found the reorientational correlation times differed markedly from those predicted by the classic Debye equation, while very good agreement with the so-called extended *J*-diffusion model was found. This model is a strong collision model which assumes that both the magnitude and the direction of the angular momentum is randomised by a collision. The angular momentum correlation time needed in the analysis was obtained from ^{19}F relaxation studies.

It is significant that attempts to rationalize relaxation data in general do not take into account intermolecular effects. A recent study[3] dealing with the ^{35}Cl and ^{37}Cl relaxation of, inter alia, CCl_4, $CHCl_3$, $TiCl_4$ and $SiHCl_3$ in liquid mixtures has, however, demonstrated that medium effects on molecular reorientation can be quite pronounced. It seems that Cl NMR provides a convenient method of studying intermolecular friction, and this is also well borne out in the recent study by Sharp and Tolan[4] of ^{35}Cl and ^{127}I relaxation of tin halides.

^{35}Cl NMR has also been successfully used to elucidate reorientational anisotropy in liquids,[5] and another interesting aspect concerns the possibility of monitoring the binding of small molecules to polymers.[6,7] From determinations of the rotational anisotropy of polymer-bound chloroform, Sillescu and co-workers[6,7] deduced information on the mechanism of polymer-chloroform interaction. Chlorine quadrupole splittings have been useful for studying the partial orientation of small chlorinated hydrocarbons in a liquid crystalline matrix of poly-γ-benzyl-L-glutamate.[8,9]

The difficult problem of extending the studies of the quadrupolar halogens to the gas state has recently been overcome by Lee and Cornwell[10] who determined T_2 of ^{35}Cl from linewidths for gaseous HCl in the pressure range 1–35 atm. For the gas phase quadrupole relaxation is effected by molecular collisions changing the rotational state of the molecule, and the correlation time is related to the time between collisions.

13A.3 Halide Ions

(*i*) *Aqueous Solutions of Simple Salts.* It is not surprising that relaxation of the halide ions is by far the most studied aspect of Cl, Br and I NMR; all the range from simple electrolyte solutions to complex biological systems has been investigated. Also, for the halide ions it is only the quadrupole relaxation that we have to consider provided there are no paramagnetic species in the solution. For the difficult problem of establishing a relevant theory of the quadrupole relaxation of monoatomic ions the reader is referred to the treatment of alkali ions; exactly the same considerations apply for the halide ions. For aqueous solutions, the concentration dependence of halide ion quadrupole relaxation is often weak, and the infinite dilution relaxation rates may be obtained with good precision by extrapolation of experimental data. Hertz and co-workers[11,12] obtained the infinite dilution relaxation rates 1050 s^{-1} for $^{81}Br^-$ and 4600 s^{-1} for $^{127}I^-$. Recent determinations of T_1 and T_2 for Cl^- by Reimarsson *et al.*[13] have given an improved infinite dilution $^{35}Cl^-$ relaxation rate of 25 s^{-1}, which is considerably lower than previous linewidth results. A value in close agreement with this has been presented in a recent investigation of T_1 by Holz and Weingärtner.[14]

A comparison of these infinite dilution relaxation rates with those predicted by Hertz' electrostatic theory[11] for the relaxation of weakly solvated ions due to ion-solvent interactions shows a very good agreement. This lends substantial support for Hertz' model. In view of the low accuracy of some of the necessary parameters it is dangerous at present to judge if it is significant that in all three cases calculated relaxation rates are greater (by 20–50 %) than the observed ones.

The concentration dependence of the halide ion quadrupole relaxation varies markedly with the cation present. In principle, data of this type contain unique information on the structure of electrolyte solutions. A good rationalisation of ion-ion contributions to relaxation is provided by a recently-improved theory of Hertz[15] which takes into account quenching effects due to the ion cloud. As regards the ion-ion effects it should be mentioned that two recent reports[13,14] give improved $^{35}Cl^-$ data compared to those of earlier publications. To illustrate the concentration dependence of halide ion relaxation in aqueous solutions we reproduce the results of Holz and Weingärtner for $^{35}Cl^-$ (Fig. 13.1).

For aqueous solutions of most monovalent ions, the concentration dependence of halide ion quadrupole relaxation is weak, but a remarkable exception is given by hydrophobic cation salts, for example alkyl-substituted ammonium halides. In this case, relaxation of both $^{35}Cl^-$, $^{79}Br^-$ and $^{127}I^-$ increases strongly with both increasing number and increasing length of the alkyl chains.[16–19] By detailed characterization of this phenomenon it was established that there is a partial adsorption of halide ions at the surface of the hydrophobic groups. The rapid relaxation of the halide ions interacting with the non-polar groups is mainly a field-gradient effect, the large field gradients being due to the asymmetry caused by the simultaneous contact with water and the low-dielectric groups. These findings are relevant in connection with discussions of ion binding in biological systems. No corresponding effect is observed for the alkali ions.

(*ii*) *Non-aqueous Solutions.* For non-aqueous systems, the halide ion quadrupole relaxation technique provides a very sensitive method for studying both solvation phenomena and ion-pair formation. The theoretical basis for an interpretation in terms of electrostatic effects is given in the work of Hertz[11,20], but it is often much more difficult than for aqueous systems to obtain the experimental data needed. However, two reports[20,21] have given very promising results and provided support for the electrostatic model. Recently a systematic approach to this problem has been given by Weingärtner and Hertz[22] who studied the concentration dependence of ^{35}Cl, ^{81}Br and ^{127}I relaxation in a large number of solvents. From this data Weingärtner and Hertz deduced infinite dilution relaxation rates as well as the ion-ion contribution to relaxation, after appropriate corrections for changes in the solvent correlation time. The infinite dilution relaxation rates given in Table 13.1 show relaxation to depend strongly on solvent properties. A careful analysis of the data showed it to be possible to obtain a good rationalisation on the basis of Hertz' electrostatic theory. Among the results of the analysis may be mentioned the finding of weak halide ion solvation in, *inter alia*, formamide, acetone, acetonitrile and dimethylsulphoxide (see also ref. 21). *N*-methylformamide and *N*,*N*-dimethylformamide gave a somewhat stronger solvation while methanol, ethanol and formic acid are characterised by much more effective relaxation than predicted by the so-called "fully random distribution" (FRD) model applicable for weak solvation (see also ref. 20). The difference between protic and aprotic solvents was presumed to involve hydrogen-bonding in the solvation process in the former case, and a model was developed which considered the effect of a solvation sheath with non-radial dipole orientations.

It is not possible here to go into any detail of the very interesting ion-ion effects on relaxation found by Weingärtner and Hertz[22] but some general remarks should be made. In a number of cases there was a strong asymmetry in the effects for anions and cations, the ion-ion contributions being much stronger for the cations. This was explained in terms of local structural effects and cross-correlation between ions and solvent molecules being much more important for (especially small) cations than for anions. Comparisons with information on ion-pair formation from conductivity studies and the electrostatic treatment[20] of relaxation due to ion-pairing led to the conclusion that solvent-separated

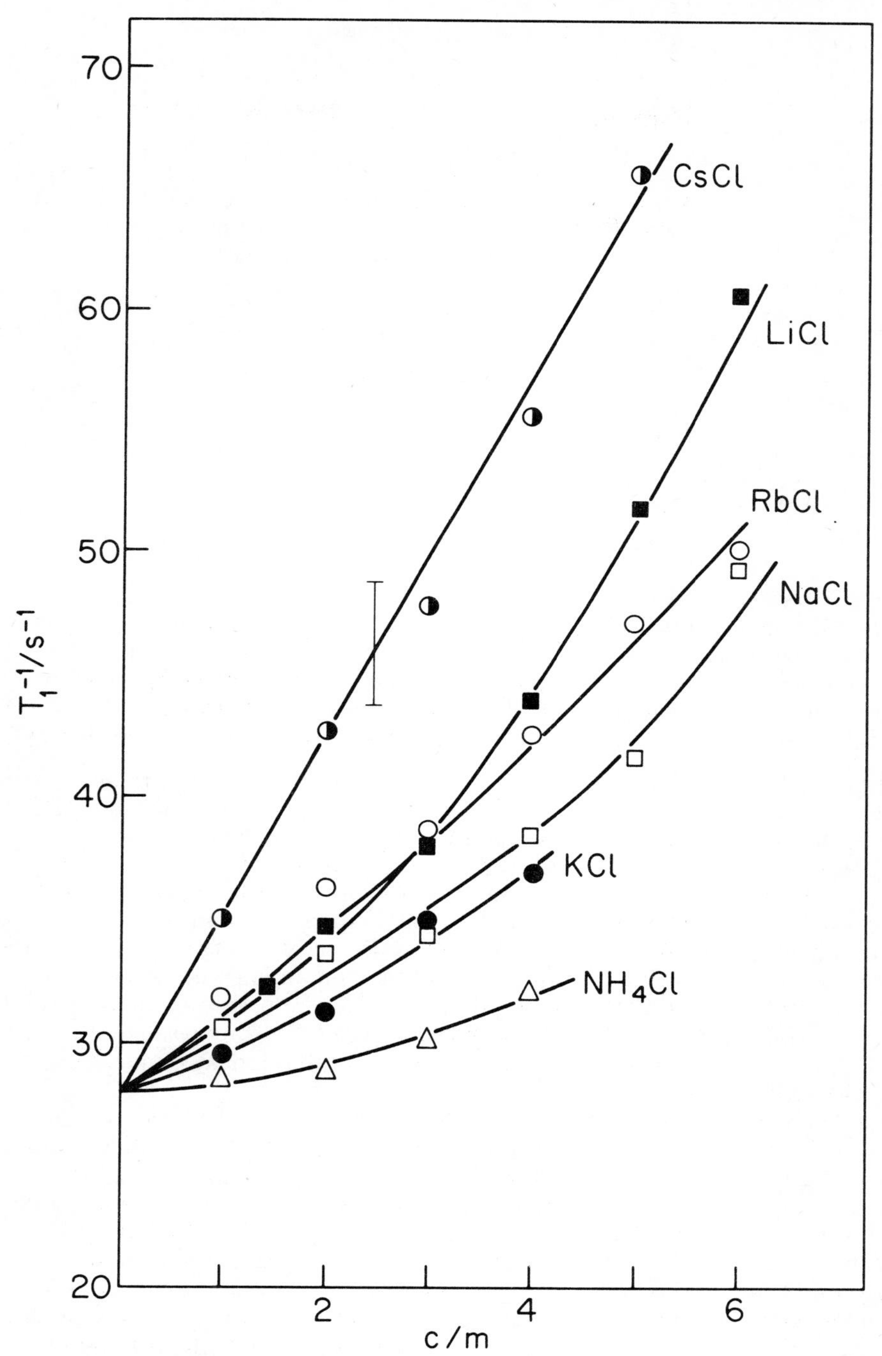

Fig. 13.1 Spin-lattice relaxation of $^{35}Cl^-$ for aqueous solutions of CsCl (◑), LiCl (□), RbCl (○), NaCl (□), KCl (●) and NH_4Cl (△) as a function of electrolyte molality (reproduced with permission from ref. 14).

TABLE 13.1 Infinite dilution $^{35}Cl^-$, $^{81}BR^-$ and $^{127}I^-$ relaxation rates, T_1^{-1}/s^{-1}, for different solvents[22]

Solvent	$^{35}Cl^-$	$^{81}Br^-$	$^{127}I^-$
Water[a]	25	1050	4600
Methanol	400	11 800	46 000
Ethanol	1300	43 000	>100 000
Formamide	250	4200	13 200
NMF	340	11 700	33 000
DMF	800	5400	4700
MeCN	40	700	1200
Formic acid	950	28 000	60 000
Acetone	—	3000	—
DMSO	100	1100	4000

[a] See the text

ion-pairs are formed in these systems, while contact ion-pair formation seems to be relatively insignificant.

(*iii*) *Amphiphilic Systems.* Amphiphilic compounds, possessing at the same time both a strongly polar group and a long hydrophobic chain, have a strong tendency to form colloidal aggregates in the presence of water. Many technically and biologically interesting amphiphiles have positively charged end-groups like $-\overset{+}{N}H_3$ and $-\overset{+}{N}Me_3$, and the interaction of these with halide ions is often of great significance. The great potential of NMR quadrupole relaxation for monitoring these interactions was already established some years ago.[23] The method has been used to study both counter-ion binding to normal micelles formed in an aqueous environment,[24] counter-ion binding to so-called reversed micelles formed at low water contents,[25] and counter-ion binding in liquid crystalline phases.[26] A recent ^{35}Cl and ^{81}Br study[27] has revealed an interesting marked counter-ion specificity in aqueous solutions containing both hexadecyltrimethyl ammonium bromide and the corresponding chloride. It appears that in these solutions there exist both very long rod-like micelles and globular (approximately spherical) ones, and that Br^- ions bind with strong preference to the elongated micelles while Cl^- ions prefer to bind to the small ones.

A halide ion in an anisotropic liquid crystal experiences a rapid local motion but this motion only partially averages the quadrupole interaction. The residual quadrupole interaction gives rise to a quadrupole splitting of the NMR signal, the magnitude of the splitting being determined by macroscopic and microscopic orientation effects, the degree of counter-ion binding to the amphiphilic aggregates and the field gradients the ions experience in the different environments. As can be seen from the discussion of alkali ion splittings in this book the microscopic orientation effects, characterised by an order parameter, may enable quite significant deductions to be made on the geometrical structure of the aggregate surfaces. The quadrupole splitting method for studying ion binding was first presented for $^{35}Cl^-$ and $^{37}Cl^-$ in 1971[28] and has since become frequently employed. A pronounced difference in splitting between $-\overset{+}{N}H_3$ and $-\overset{+}{N}Me_3$

end-groups has been established,[29] and Fujiwara and Reeves[30] have for the $-\overset{+}{N}D_3$ end-group attempted to correlate deuteron quadrupole splittings of the end-groups with $^{35}Cl^-$ splittings of the counter-ion.

Two recent reviews consider in some detail NMR studies of counter-ion binding in amphiphilic systems, one[31] surveys the use of different NMR parameters, and the other[29] gives an account of the quadrupole splitting method.

13A.4 Comment on Exchange Effects

In studies of the interaction between a halide ion and some other species, which may be a metal ion, a micelle or a macromolecule, for example, we must take into account that the halide ion, as a function of time, samples different binding sites characterised by different values of the NMR parameters. In contrast to other nuclei the different sites are, for the quadropolar halide ions, in most cases characterised by different relaxation times rather than different chemical shifts. If chemical exchange between the sites is not rapid compared to the relaxation rates, NMR provides a very useful method to investigate exchange kinetics. Early work of this type can be exemplified by studies of halide exchange with trihalide ions[32,33] (I_3^-, Br_3^-) or transition metal halide complexes[34,35] (for example $[HgI_4]^{2-}$ and $[CdBr]^+$), while presently most interest is focussed on the possibility of monitoring ligand exchange kinetics in biological systems, an aspect exemplified below. In many kinetic studies it is possible to assume a two-site model, one site (in large excess) being the free aqueous ion and the other corresponding to the complexed state. Under slow-exchange conditions only the relatively narrow signal from the free ion is observable, the signal of bound halide being broadened beyond detection. With increasing exchange rate the signal broadens, the relaxation rate ($1/T_1$ or $1/T_2$) being described by the following equation if chemical shift differences are negligible and extreme narrowing conditions apply:

$$\frac{1}{T_{obs}} - \frac{1}{T_f} = \frac{p_b}{\tau_b + T_b}$$

Here p_b is the fraction of halide complexed, τ_b and T_b are the lifetime and relaxation time (T_1 or T_2), respectively, of the halide in the complex, and T_f is the relaxation time of free halide. Expressions valid in more general cases may be found in refs 1 and 36. As T_b and τ_b have opposite temperature dependences, variable temperature studies are useful for indicating the dominance of T_b or τ_b. In kinetic applications a more reliable and convenient test is often to study the isotope effect in relaxation for the $^{35}Cl/^{37}Cl$ and $^{79}Br/^{81}Br$ isotope pairs. Thus the exchange rate can be assumed to have a negligible isotope effect, while the relaxation rates vary in proportion to the quadrupole moment squared. Therefore the isotope ratio of the excess relaxation rates is 1 under slow exchange conditions but given by the quadrupole moments under rapid exchange conditions if quadrupole relaxation dominates. The variation of the isotope effect is illustrated in Fig. 13.2.

For exchange of halide ions in liquid crystalline systems between sites characterised by different quadrupole splittings it is important to realise that both the order parameter and the quadrupole coupling constant may be either positive or negative, but only the absolute value of the splitting is generally observable.[29] Therefore, under rapid exchange conditions, it may occur that the observable splitting is less than that of the individual sites.

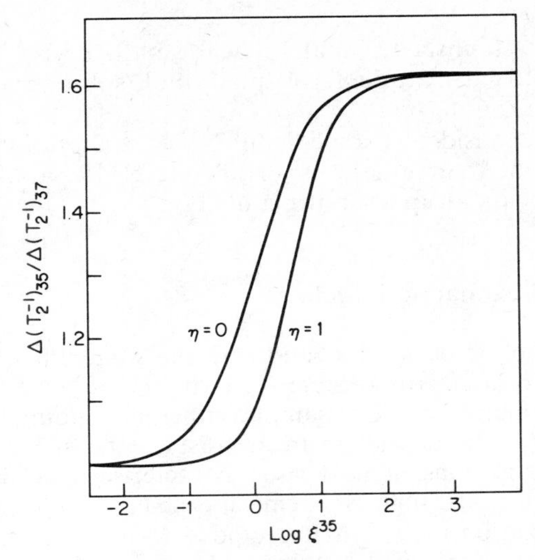

Fig. 13.2 Ratio between the excess ^{35}Cl and ^{37}Cl transverse relaxation rates as a function of the rate of chemical exchange for a two-site system where one site (b) is much less populated than the other. $\xi = T_{2b}/\tau_b$, where τ_b is the lifetime in site b, and $\eta = \Delta\omega \cdot T_{2b}$, where $\Delta\omega$ is the difference in angular resonance frequency between the two sites (reproduced with permission from ref. 1).

13A.5 Halide Ions in Biological Systems

The most prominent trend in Cl, Br and I NMR in recent years has been a rapidly increasing consideration of biological problems; both low molecular weight compounds, like metal-nucleotide complexes and amino acids, and, in particular, macromolecular systems have been investigated. Several factors contribute to the interest in studies of halide ion binding to biological macromolecules. One is that since these systems function in a rather high chloride concentration *in vivo*, chloride-macromolecule interactions certainly influence many aspects of the function, although in a way which remains to be elucidated in most cases. Secondly, it should be observed that proteins, and notably enzymes, frequently display their substrate specificity, coenzyme binding and inhibition pattern in interactions between negative groups on the substrate, coenzyme or inhibitor and groups on the protein. Cases of Cl^- activation of enzymes are known, but Cl NMR studies are mainly of interest because Cl^- generally binds in competition with the functional ligands.

The first condition for the applicability of the NMR method to study halide ion binding to macromolecules (which has recently been reviewed[37,38]) is that there is a reasonably fast chemical exchange between a macromolecular binding site and the free ion in solution. (Because of low sensitivity and extensive line-broadening direct observation of bound halide is not feasible.) The second condition is that the relaxation rate of the bound halide ion should be much larger than that of the free ion. This condition is dictated by the fact that, because of relatively low intrinsic NMR receptivity and a need to keep the macromolecule concentration down (for reasons of limited solubility or the desire to avoid interactions between macromolecules), one has to work with a large excess of free ligand. As the relaxation rate of protein-bound Cl^- is typically 10^3–10^5 times that of free Cl^-, this condition is well fulfilled, and one can work with quite low protein concentrations and still get sizeable effects.

The evaluation of halide ion quadrupole relaxation data of biological systems (reviewed in detail in ref. 1) often requires quite detailed considerations which can only be given here in outline. Three types of primary information are obtainable from the halide relaxation rate, i.e. the fraction of halide ions bound, the rate of halide exchange, and the relaxation rate of bound halide. In general only one of the two latter quantities can be deduced, in combination with a lower limit of the other. We will first consider what use can be made of the information on the concentration of bound halide ions. For a case with several different binding sites on the macromolecule the excess transverse or longitudinal relaxation rate may be written

$$\left(\frac{1}{T_{1,2}}\right)_{\text{ex}} = \frac{1}{T_{\text{obs}}} - \frac{1}{T_f} = C_p \sum \frac{n_i K_{ix}/T_{bi}}{1 + K_{ix} C_x}$$

The different contributions are given by the number of sites (n_i) of each class, the intrinsic relaxation rate ($1/T_{bi}$) and the binding constant (K_{ix}). C_p stands for the protein concentration and C_x for the halide concentration, and it is assumed that $C_p \ll C_x$. From a study of the relaxation as a function of halide concentration, the binding constants may be obtained, although in a number of cases binding is too strong in comparison with presently accessible concentrations. However, halide ion NMR is a good complement to other physical methods, which mostly cannot be applied to study weak or moderately strong binding.

If one adds a second ligand (which may be a coenzyme or inhibitor, for example) which competes with Cl^- for a particular binding site, the release of Cl^- ions from the macromolecule will result in a reduced relaxation rate. The appearance of a titration curve is determined by the binding affinity of the competing ligand. For a ligand which is very strongly bound, the titration curve displays a distinct change in slope, giving the stoicheiometry of binding. For weaker binding the shape of the titration curve gives the binding constant of the competing ligand (L) relative to that of halide, the relaxation rate being given by

$$\left(\frac{1}{T_{1,2}}\right)_{\text{ex}} = C_p \sum \frac{n_i K_{ix}/T_{bi}}{1 + K_{ix} C_x + K_{iL} C_L}$$

A large number of applications of competition experiments may be found in the literature;[1] examples are studies of inhibitor binding to carbonic anhydrase,[39] coenzyme binding to alcohol dehydrogenase,[40] and substrate binding to alkaline phosphatase.[41]

Many enzymes require metal ions for their function and it is often of interest to establish if anion binding occurs at the metal ion or at groups in the amino acid sequence. If the metal-free enzyme can be obtained, it is often illuminating to investigate the effect on halide relaxation of titrating with the metal ion. The very different behaviour in this respect of two zinc enzymes, carbonic anhydrase[39] and alkaline phosphatase,[41] led to the conclusion that Zn is involved in Cl^- binding in the former case but not significantly in the latter. In both cases the anion binding studied is directly connected with the catalytic function.

In their recent review,[37] putting special emphasis on halide exchange studies, Stephens and Bryant discuss the exploitation of a large number of metal ions (inter alia Tl^{3+}, Pb^{2+}, Ag^+ and Cu^{2+}) in halide NMR studies of problems in protein chemistry. The interaction of Cl^- with protein-bound metal ions in chloroperoxidase[42] and bovine superoxide dismutase[43] is discussed (in two recent reports) on the basis of ^{35}Cl linewidth studies. In the case of chloroperoxidase, a Fe^{3+}-containing glycoprotein catalyzing the oxidation of chloride, bromide and iodide ions, it was of interest to elucidate the role of iron in the substrate binding. The ^{35}Cl linewidth was found to parallel closely the enzymic activity and, furthermore, titration with fluoride and azide led only in the first case to strong ^{35}Cl linewidth reduction. On the basis of these findings, and the ability of azide to strongly inhibit the enzyme and to associate with Fe^{3+}, an analysis in terms of different types of chloride ion-enzyme interactions led to the suggestion that direct coordination of Cl^- to iron is not important in the enzymic process.

Garnett *et al.*[70] have recently used ^{79}Br and ^{81}Br T_1 measurements to study the binding of an Hg-labelled substrate, 4-bromo-mercuriocinnamic acid, to α-chymotrypsin, and have deduced detailed kinetic information.

From the magnitude of the decrease in relaxation rate, $1/T_1$ or $1/T_2$, on addition of a chloride-competitive ligand, in combination with data on the number of sites, information on the relaxation rate of a single bound Cl^- ion is obtained. For the rapid-exchange limit one can determine T_{1b} and T_{2b}. Since protein reorientation generally is not rapid compared to the Larmor frequency, it must be taken into account that the NMR signal consists of two Lorentzian curves, and that both longitudinal and transverse relaxation are non-exponential. These effects are described in some detail in the discussion of sodium ion NMR in ref. 71. For Na^+ clear-cut non-exponentiality has been demonstrated both for aqueous solutions of macromolecules and for complex biological systems. For most of the proteins which have been studied by the halide quadrupole relaxation method, the correlation time appears, however, to be small enough so that from an experimental point of view relaxation can be considered to be approximately exponential. Under these conditions, the approximate expressions derived by Bull[36] can be used to determine the quadrupole coupling constant (χ) and the correlation time (τ_c). Details are given in ref. 1. For proteins below about 40 000 in molecular weight no distinction between the two relaxation times is possible at the magnetic field strengths hitherto employed, and the separation of χ and τ_c is not possible without extraneous information (for example, protein reorientation times from other types of experiments or from hydrodynamic theory.) However, for a large number of proteins readily accessible to investigation it is possible to obtain χ and τ_c, and therefore in recent years this possibility of characterising anion binding to proteins has attracted considerable interest. The quadrupole coupling constant is influenced by the proximity of charges and dipoles, and should be especially large if there are covalent contributions to the anion binding, as in the coordination of halide ions to transition-metal ions. An examination of data presently available reveals a large range of variability of the quadrupole coupling constant for $^{35}Cl^-$ from ca. 1 MHz to almost 20 MHz. A noticeable feature is that whereas non-metal proteins—haemoglobin, lactate dehydrogenase, aldolase, serum albumin and others—give very similar χ values (of the order of 1–2 MHz), the variation is over more than one order of magnitude among the metalloproteins. Particularly interesting is the large family of zinc metalloenzymes, where the involvement of Zn in catalytic action and ligand binding is often debated. As the $^{35}Cl^-$ quadrupole coupling constants of *E. coli* alkaline phosphatase and horse liver alcohol dehydrogenase are in the same range as those of the metal-free proteins (very much smaller than the values characteristic of coordination to metal ions) it was deduced that coordination to Zn is not the site of strong Cl^- binding in these two enzymes. For two other Zn-enzymes, carboxypeptidase A and carbonic anhydrase, much larger χ values are obtained, as was also the case for Zn^{2+}- and Hg^{2+}-complexes of a number of proteins. It seems that in general a distinction between metal-coordinative anion binding and anion binding involving amino-acid residues can be made from the χ values, but it is likely that in the future more detailed deductions will become possible.

At least three motions may modulate the field gradients of a macromolecule-bound ion and thus influence the experimentally determined correlation time; these are macromolecule reorientation (τ_{rot}), internal motion (τ_{int}) and rapid halide ion exchange (τ_{ex}). As regards τ_{ex} it can be estimated that an exchange between an ion at a site on a protein and the ion being free in solution probably cannot be rapid enough to account for the results. For globular (approximately spherical) macromolecules, it is easy to give a rough estimate of τ_{rot} using the Debye equation, and in some cases τ_{rot} has been directly measured. For a number of proteins, τ_c of $^{35}Cl^-$ for specific high-affinity binding sites is found to be markedly smaller than τ_{rot}, indicating the importance of rapid internal motion at the anion binding sites in proteins. This applies, for example, to aldolase, pig heart as well as rabbit muscle lactate dehydrogenase, and human serum albumin.[44,45] In a number of cases, direct demonstration of rapid motion at anion binding sites in proteins has been obtained. For example, for aldolase and serum albumin internal motion has been inferred from observations of different correlation times for different ligands competing for the same site in a protein.[45] An interesting example is given by lactate dehydrogenase, an enzyme consisting of four subunits and having a molecular weight of 140 000. Its three-dimensional structure has been determined to high resolution by X-ray techniques. In contrast to another enzyme dependent on the coenzyme NADH, i.e. alcohol dehydrogenase, addition of the coenzyme to lactate dehydrogenase produces an enhanced relaxation. Whereas in the case of alcohol dehydrogenase NADH binding leads to a release of Cl^- ions, NADH binds simultaneously with Cl^- in lactate dehydrogenase. On binding of NADH there is a considerable increase in τ_c of Cl^- both for the rabbit muscle[44] and the pig heart[45] enzyme, while no significant change in χ is observed. As, according to competition experiments, Cl^- binds at the substrate binding site, the ability of bound coenzyme to restrict the motion of the substrate binding group is indicated. These results are in good accordance with the positioning of coenzyme according to the X-ray crystallographic studies.

In the type of studies described, some caution should be made in interpreting a single datum of χ and τ_c in terms of internal motion since such motion may, depending on its rate, affect both quantities as obtained from experimental relaxation rates. Thus in the case of very rapid internal motion, i.e. $\tau_{int} \ll \tau_{rot}, \tau_{ex}$, the main effect of the internal motion is to reduce the quadrupole coupling constant, while the correlation time is essentially given by τ_{rot} (or τ_{ex}). Variable temperature studies generally permit distinction between possibilities. For example, for human serum albumin good evidence for internal motion has been obtained from observations of an increase or constancy in τ_c with increasing temperature. This study, as well as most of the others cited above, is part of an investigation of χ and τ_c values of a large number of proteins.[45] Although the applications as yet are few there is no doubt that the quadrupole relaxation method can give very specific and unique information on internal mobility at functional anion binding sites in proteins. It seems that the presence of rapid motion is sometimes overlooked, and that there is in general a marked flexibility of the amino acid residues of globular proteins.

Investigations of the effect of specific modifications in the amino acid sequence or of varying the pH may be very useful for identifying the anion binding groups in a protein. Studies of haemoglobin have been particularly elucidative in this respect, and of great interest here is also the recent finding of a nonlinear variation of Cl^- relaxation with the degree of oxygenation.[46]

In a quantitative interpretation of halide ion relaxation rates of protein solutions and separation of observed effects into the various possible halide-protein interactions, investigations of simple well-defined model systems like amino acids, small peptides and

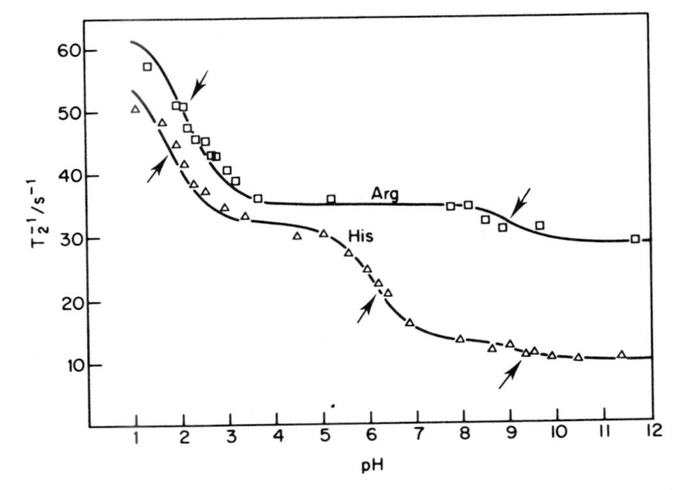

Fig. 13.3 The pH-dependence of the excess $^{35}Cl^-$ transverse relaxation rate of aqueous solutions containing 1.00 M NaCl and 0.50 M arginine (□), or histidine (△); arrows indicate pK_a values of the different titratable groups (reproduced with permission from ref. 47).

polyaminoacids can be expected to be of great value. As can be seen in Fig. 13.3, titrations of the different ionizable groups of histidine and arginine give distinct changes in ^{35}Cl relaxation.[47]

Another type of model system which has been studied is the Cl^- interaction with different polycations and polyampholytes.[48] The ^{37}Cl relaxation was found to respond very sensitively to the ionisation of, *inter alia*, dimethylammonium groups.

Studies aimed at an elucidation of anion binding to proteins vastly dominate the biological and biochemical applications of Cl, Br and I NMR, but several other interesting applications may be foreseen. Plaush and Sharp[49] have recently reported $^{35}Cl^-$ relaxation times of dimethylsulphoxide solutions of a number of nucleosides, and obtained detailed information on the interaction of Cl^- with different functional groups. Lyon *et al.*[50] studied the interaction of Cl^- with bacterial surfaces by ^{35}Cl NMR. For the same system only a small signal intensity reduction was obtained for Cl^- but a large one for Na^+. This may reflect differences in ion binding, but one must also take into account the differences in Larmor frequency between the two nuclei. Complex biological systems, as yet only little considered in this connection, will probably receive much attention in the near future. Here, determinations of T_1 and T_2 as well as investigations of non-exponentiality of relaxation will be needed.

13A.6 The Perchlorate Ion

The perchlorate ion is the chemically most interesting example of a halogen at a site of tetrahedral or octahedral symmetry. Due to symmetry, the field gradient at the Cl nucleus is zero for the free ClO_4^- ion. In solutions, field gradients are created, but as can be seen from the relaxation rate of only 3.7 s^{-1} for infinite dilution in water,[13] these field gradients are smaller than for the Cl^- ion. As has already been mentioned, the field gradients at a ClO_4^- ion can arise in at least two different ways. Firstly, intermolecular field gradients arising from ionic charges or solvent dipoles are present.

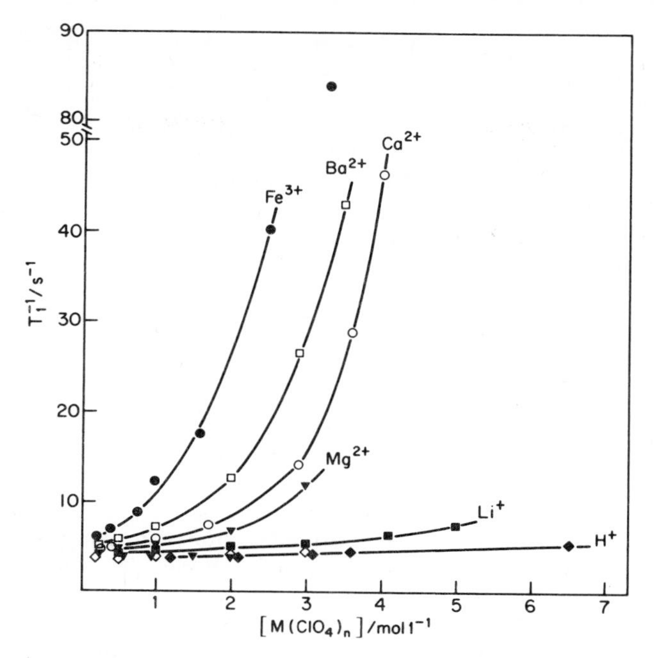

Fig. 13.4 Concentration dependence of the ^{35}Cl spin-lattice relaxation of the ClO_4^- ion in aqueous electrolyte solutions containing various metal cations, M (reproduced with permission from ref. 51).

A considerable difficulty in this case is to account properly for Sternheimer antishielding, as both an external electric field gradient and an external electric field can produce field gradients at the nucleus. Secondly, the large field gradients along the Cl-O bonds become operative if the symmetry is perturbed. Even quite small distortions may have marked effects on relaxation. A discussion of relaxation of aqueous symmetric ions leads to the conclusion that electric-field-induced field gradients are of great significance.[13]

The effect of electrolyte concentration on the ^{35}Cl relaxation of aqueous ClO_4^- is shown in Fig. 13.4.[51] In line with the well-known inertness of ClO_4^- towards interactions with cations, a very weak concentration effect is observed in most cases. However, distinct differences between different cations are clearly observable. For non-aqueous solutions, ^{35}Cl relaxation studies of ClO_4^- should be informative as regards the formation and structure of ion-pairs. A recent ^{35}Cl linewidth study[52] of sodium perchlorate solutions in a number of non-aqueous solvents gave only small effects, suggesting a weak anion-cation interaction; however, instrumental line-broadening was dominating over relaxation. Grandjean and Laszlo,[53] working under better experimental conditions, obtained appreciable effects and also demonstrated the effect of the sugar sorbose on the ion-ion contribution to relaxation.

The use of the relaxation of ClO_4^- to probe into anion binding sites in proteins can for several reasons be believed to be a good complement to, for example, studies of the Cl^- ion. From a chemical point of view, the small tendency of ClO_4^- for coordinating to metal ions should make it a rather specific probe for anion binding to amino acid residues. Furthermore, the different origin of the field gradients, just discussed, may give additional information, and it is probable that this method is of potential value for studying distortion of ligands on binding to a macromolecule. Quadrupole relaxation studies of ClO_4^- binding to proteins have just started but investigations of, *inter alia*, human serum albumin,[54] aldolase, carbonic anhydrase and alkaline phosphatase have clearly complied with expectations. The ^{35}Cl quadrupole coupling constant of protein-bound ClO_4^- was found to be 0.8 MHz for aldolase and 1.5 MHz for human serum albumin, while the correlation times roughly corresponded to those of protein reorientation.[45,54] Titrations of either serum albumin or alkaline phosphatase solutions with Hg^{2+} or Zn^{2+} produced much smaller effects than in the case of Cl^-, verifying the expected small tendency of ClO_4^- to bind to metal ions.

13B SPIN-SPIN COUPLING

The rapid quadrupole relaxation of covalent halogens in general effectively masks fine structure due to scalar spin coupling between the halogen nucleus and other magnetic nuclei. Notable exceptions occur for halogens experiencing very small field gradients due to symmetry effects. Thus for the octahedral IF_6^+ cation the coupling between ^{127}I and the six equivalent ^{19}F nuclei gives a sextet in the ^{19}F NMR spectrum[55] while the ^{19}F NMR spectrum of BrF_6^+ consists of two quartets due to the coupling of ^{19}F to the two Br isotopes of approximately equal abundance.[56] While direct determinations of Cl, Br and I spin couplings for non-symmetric sites are difficult, it is frequently possible to make use of the scalar contribution to the relaxation of a spin $I = \frac{1}{2}$ nucleus coupled to the quadrupolar halogen. Hereby one can compare T_1 and T_2 or can measure $T_{1\rho}$ as a function of the amplitude of the r.f. field. By these methods, spin-spin couplings for a number of compounds have been obtained but little use has been made of them for chemical problems, one reason being the problem of interpretation.[1] Table 13.2 lists some published values of coupling constants to halogen nuclei. In addition, Lee and Cornwell[10] have obtained $^1J(^{35}Cl, ^1H) = 47 \pm 3$ Hz for gaseous HCl from a relaxation study. Oxygen-17 NMR has been used to determine (^{35}Cl, ^{17}O) coupling constants in two cases. Virlet and Tantot[65] obtained $|^1J| = 108$ Hz for ClO_3F and $|^1J| = 85.5$ Hz for ClO_4^- (4M in water), while Lutz *et al.*[66,67] give $|^1J| = 81.4$ Hz for the latter case. A similar value was reported earlier by Alei.[68]

TABLE 13.2 Selected one-bond Cl, Br and I spin-coupling constants

Compound	Spin-spin coupling constant/Hz	Ref.
HCl	$J(^{35}Cl, ^1H) = 41 \pm 2$	57
PCl_3	$J(^{35}Cl, ^{31}P) = 120$	58
$SnCl_4$	$J(^{119}Sn, ^{35}Cl) = 375$	59
$PbCl_4$	$J(^{207}Pb, ^{35}Cl) = 705$	59
ClF_3	$J(^{35}Cl, ^{19}F) = 260$	60
ClF_6^+	$J(^{35}Cl, ^{19}F) = 337$	61
HBr	$J(^{79}Br, ^1H) = 57 \pm 3$	57
$SnBr_4$	$J(^{119}Sn, ^{81}Br) = 920$	62
BrF_6^+	$J(^{79}Br, ^{19}F) = 1575$	63
SnI_4	$J(^{127}I, ^{119}Sn) = 1097$	64
IF_6^+	$J(^{127}I, ^{19}F) = 2730 \pm 15$	55

13C SHIELDING

13C.1 Covalent Compounds

It is immediately realized from the Cl, Br and I relaxation rates of covalent compounds that line-broadening will give a low precision in experimental determinations of Cl chemical shifts, while shielding studies of Br and I are in general impossible. It can be noted that the ^{35}Cl linewidth for a typical small molecule like CCl_4 at room temperature corresponds to almost 2000 ppm at an observing frequency of 10 MHz. This can be compared with the chemical shift range, which extends from the aqueous Cl^- ion, being among the most shielded cases, to ClO_4^- (one of the most deshielded compounds)—a chemical shift difference of ca. 1000 ppm.

Particularly interesting among recent chlorine chemical shift investigations is the study of gaseous HCl presented by Lee and Cornwell,[10] as this provides an absolute reference for chlorine chemical shifts. Experimentally, gaseous HCl was found to be 20 ± 3 ppm less shielded than 6.3 M HCl in water. Using the diamagnetic shielding value obtained from theoretical calculations and the paramagnetic shielding from molecular beam resonance studies, the calculated value of the absolute shielding of gaseous HCl is +953 ppm. The dilute aqueous Cl^- ion was found to be 170 ± 12 ppm less shielded than the gaseous ion. Chlorine-35 chemical shifts of ClO_3F and of aqueous ClO_4^- have recently been determined[65] as $\delta = 978.1$ and 1002.5 ppm respectively (relative to aqueous Cl^-).

It is probably a good approximation to assume that differences in shielding mainly arise from differences in the paramagnetic term, but a detailed interpretation in terms of intramolecular effects only has proved to be difficult.[1] Possibly, intermolecular interactions have a significant effect on the shielding, and in fact medium effects have been observed in certain cases.[3]

13C.2 Halide Ions

Whereas studies of chemical shifts of covalent Cl, Br and I compounds have not been very informative, the situation is considerably better for the halide ions, even if here also the establishment of a good theory causes big problems. Different theoretical models have been proposed, but it seems that the overlap approach advanced by Kondo and Yamashita[69] is generally preferred. Schematically the paramagnetic term, which certainly dominates observed chemical shift effects, can be written:

$$\sigma_p = -8\mu_B^2\mu_0\langle r^{-3}\rangle_p(A_{X-X} + A_{X-M} + A_{X-S})/\pi\Delta E$$

Here ΔE is an average excitation energy and $\langle r^{-3}\rangle_p$ the expectation value of r^{-3} for an outer p electron of the halide ion. The A terms are sums of squares of overlap integrals between the outer p orbitals of the halide ion (X^-) and the orbitals of other halide ions, cations (M^+) and solvent molecules(S). Of great significance for the applications of Cl chemical shifts is the recent establishment of an absolute shielding scale,[10] referred to above; for Br and I such information is still lacking (cf. ref. 1). In the interpretation of variable concentration data of aqueous solutions it is necessary to consider, in addition to direct anion-cation interactions, halide-halide effects as well as a modification of the considerable paramagnetic shielding due to water. In general the sequence of the inter-ionic effects is according to expectation for the overlap model, i.e. increasing overlap from F^- to I^-, and largest overlap with the largest alkali ions. Li^+ behaves somewhat

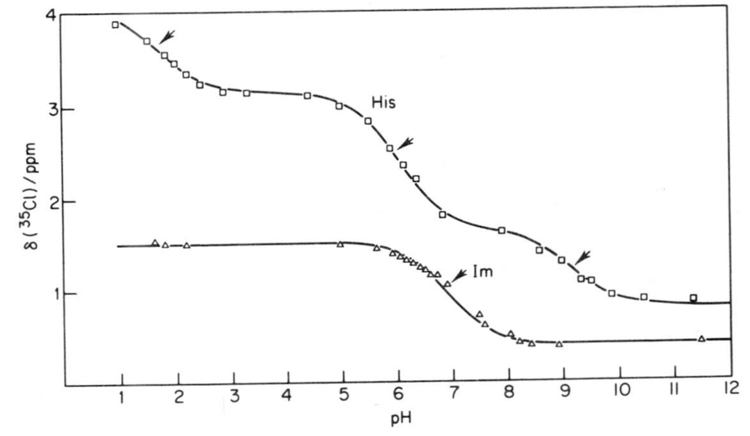

Fig. 13.5 The pH-dependence of the $^{35}Cl^-$ chemical shift of aqueous solutions containing 1.00 M NaCl and 0.50 M histidine (□) or imidazole (△). Arrows indicate pK_a values of the different titratable groups (reproduced with permission from ref. 47).

irregularly in giving a sizeable high frequency shift, which is in contrast to predictions from the overlap model and from results on the alkali halide crystals. It seems that the behaviour of the aqueous Li^+ ion depends on an increased paramagnetic contribution from halide-water overlap, the effect being produced by polarization of the water molecules in the first hydration sheath of the strongly hydrated Li^+ ion.

$^{35}Cl^-$ shielding studies have been employed to investigate chloride binding in amphiphilic systems.[31] Such studies should also be a good complement to halide ion relaxation for investigations of ion binding to synthetic and biological macromolecules, although the line-broadening may in many cases prove to be cumbersome. To have a better basis of understanding of the anion binding to proteins, studies of simple model systems have recently been started. Fig. 13.5 shows the pH dependence of the $^{35}Cl^-$ chemical shift in aqueous solutions containing histidine or imidazole. Titration of the different groups can be seen to be accompanied by quite large chemical shift changes, and the results suggest that, at least for low molecular weight proteins and peptides, it should be possible to study Cl^- ion binding in this way.

REFERENCES

1. B. Lindman and S. Forsén, "Chlorine, Bromine and Iodine NMR. Physico-Chemical and Biological Applications" Springer Verlag, Heidelberg (1976)
2. A. A. Maryott, T. C. Farrar and M. S. Malmberg, *J. Chem. Phys.* **54**, 64 (1971)
3. S. Forsén, H. Gustavsson, B. Lindman and N.-O. Persson, *J. Magn. Reson.* **23**, 515 (1976)
4. R. R. Sharp and J. W. Tolan, *J. Chem. Phys.* **65**, 522 (1976)
5. K. T. Gillen and J. H. Noggle, *J. Chem. Phys.* **53**, 801 (1970)
6. R. G. Brüssau and H. Sillescu, *Ber. Bunsenges. Phys. Chem.* **76**, 31 (1972)
7. R. Eckert, G. Loos and H. Sillescu, in "Molecular Motions in Liquids", p. 385 (J. Lascombe, ed.). D. Reidel, Dordrecht (1974)
8. B. M. Fung, M. J. Gerace and L. S. Gerace, *J. Phys. Chem.* **74**, 83 (1970)
9. D. Gill, M. P. Klein and G. Kotowycz, *J. Amer. Chem. Soc.* **90**, 6870 (1968)

10. C. Y. Lee and C. D. Cornwell, in "Magnetic Resonance and Related Phenomena", p. 261. Groupement Ampere, Heidelberg (1976)
11. H. G. Hertz, *Ber. Bunsenges. Phys. Chem.* **77**, 531 (1973)
12. M. Holz, H. Weingärtner and H. G. Hertz, *J.C.S. Faraday I* **73**, 71 (1977)
13. P. Reimarsson, H. Wennerström, S. Engström and B. Lindman, *J. Phys. Chem.* **81**, 789 (1977)
14. M. Holz and H. Weingärtner, *J. Magn. Reson.* **27**, 153 (1977)
15. H. G. Hertz, *Ber. Bunsenges. Phys. Chem.* **77**, 688 (1973)
16. B. Lindman, S. Forsén and E. Forslind, *J. Phys. Chem.* **72**, 2805 (1968)
17. B. Lindman, H. Wennerström and S. Forsén, *J. Phys. Chem.* **74**, 754 (1970)
18. H. Wennerström, B. Lindman and S. Forsén, *J. Phys. Chem.* **75**, 2936 (1971)
19. H. G. Hertz and M. Holz, *J. Phys. Chem.* **78**, 1002 (1974)
20. C. A. Melendres and H. G. Hertz, *J. Chem. Phys.* **61**, 4156 (1974)
21. B. Lindman, *Acta Chem. Scand.* **A29**, 935 (1975)
22. H. Weingärtner and H. G. Hertz, *Ber. Bunsenges. Phys. Chem.* **81**, 1204 (1977) and H. Weingärtner, Thesis, Karlsruhe (1976)
23. J. C. Eriksson, Å. Johansson and L.-O. Andersson, *Acta Chem. Scand.* **20**, 2301 (1966)
24. G. Lindblom and B. Lindman, *J. Phys. Chem.* **77**, 2531 (1973)
25. G. Lindblom, B. Lindman and L. Mandell, *J. Colloid Interface Sci.* **34**, 262 (1970)
26. G. Lindblom and B. Lindman, *Mol. Cryst. Liquid Cryst.* **14**, 49 (1971)
27. J. Ulmius, B. Lindman, G. Lindblom and T. Drakenberg, *J. Colloid Interface Sci.* **65**, 88 (1978)
28. G. Lindblom, H. Wennerström and B. Lindman, *Chem. Phys. Lett.* **8**, 489 (1971)
29. G. Lindblom, H. Wennerström and B. Lindman, *ACS Symp. Ser.* **34**, 372 (1976)
30. F. Y. Fujiwara, and L. W. Reeves *J. Amer. Chem. Soc.* **98**, 6790 (1976)
31. B. Lindman, G. Lindblom, H. Wennerström and H. Gustavsson, in "Micellization, Solubilization and Microemulsions" (K. L. Mittal, ed.). Plenum Press, New York (1977)
32. O. E. Myers, *J. Chem. Phys.* **28**, 1027 (1958)
33. O. E. Myers, Exchange Reactions Proc. Symp. Upton, N.Y., p. 347. (1965)
34. H. G. Hertz, *Z. Elektrochem.* **65**, 36 (1961)
35. D. E. O'Reilly, G. E. Schacher and K. Schug, *J. Chem. Phys.* **39**, 1756 (1963)
36. T. E. Bull, *J. Magn. Resonance* **8**, 344 (1972)
37. R. S. Stephens and R. G. Bryant, *Mol. Cell. Biochem.* **13**, 101 (1976).
38. S. Forsén and B. Lindman, *Chemistry in Britain* **14**, 29 (1978)
39. R. L. Ward, *Biochemistry* **8**, 1879 (1969)
40. T. E. Bull, R. Einarsson, B. Lindman and M. Zeppezauer, unpublished study cited in ref. 1
41. J.-E. Norne, H. Csopak and B. Lindman, *Arch. Biochem. Biophys.* **162**, 552 (1974)
42. G. E. Krejcarek, R. G. Bryant, R. J. Smith and L. P. Hayer, *Biochemistry* **15**, 2508 (1976)
43. J. A. Fee and R. L. Ward, *Biochem. Biophys. Res. Comm.* **71**, 427 (1976)
44. T. E. Bull, B. Lindman and P. Reimarsson, *Arch. Biochem. Biophys.* **176**, 389 (1976)
45. T. E. Bull, B. Halle, B. Lindman, J.-E. Norne and P. Reimarsson, unpublished work; T. E. Bull, B. Lindman, J.-E. Norne and P. Reimarsson, in "Magnetic Resonance and Related Phenomena", p. 602. Groupement Ampere, Heidelberg (1976), and EUCHEM Conference "NMR on macromolecules", Grasmere (1977)
46. J. E. Norne and S. Forsén, *FEBS Lett.*, in press
47. B. Jönsson and B. Lindman, *FEBS Lett.* **78**, 67 (1977)
48. H. Gustavsson, and B. Lindman "Colloid and Interface Science", Vol. II, p. 339. Academic Press, New York (1976)
49. A. C. Plaush, and R. R. Sharp, *J. Amer. Chem. Soc.* **98**, 7973 (1976)
50. R. C. Lyon, N. S. Magnuson and J. A. Magnuson, in "Extreme Environments, Mechanisms of Microbial Adaption", p. 305. Academic Press, New York (1976)
51. P. Reimarsson and B. Lindman, *Inorg. Nucl. Chem. Lett.* **13**, 449 (1977)
52. M. S. Greenberg and A. I. Popov, *J. Solution Chem.* **5**, 653 (1976)
53. J. Grandjean and P. Laszlo, *Helv. Chim. Acta* **60**, 259 (1977)
54. P. Reimarsson, T. Bull and B. Lindman, *FEBS Lett.* **59**, 158 (1975)
55. M. Brownstein and H. Selig, *Inorg. Chem.* **11**, 656 (1972)
56. R. J. Gillespie and G. J. Schrobilgen, *Inorg. Chem.* **13**, 1230 (1974)

57. R. E. Morgan and J. H. Strange, *Mol. Phys.* **17**, 397 (1969)
58. J. H. Strange and R. E. Morgan, *J. Phys. C, solid state Phys.* **3**, 1999 (1970)
59. R. M. Hawk and R. R. Sharp, *J. Chem. Phys.* **60**, 1009 (1974)
60. M. Alexandre and P. Rigny, *Can. J. Chem.* **52**, 3676 (1974)
61. K. O. Christe, *Inorg. Nucl. Chem. Lett.* **8**, 741 (1972)
62. R. R. Sharp, *J. Chem. Phys.* **60**, 1149 (1974)
63. R. J. Gillespie and G. J. Schrobilgen, *J. C. S. Chem. Commun.* **90** (1974)
64. R. R. Sharp and J. W. Tolan, *J. Chem. Phys.* **65**, 522 (1976)
65. J. Virlet and G. Tantot, *Chem. Phys. Lett.* **44**, 296 (1976)
66. O. Lutz, W. Nepple and A. Nolle, *Z. Naturforsch.* **A31**, 1046 (1976)
67. O. Lutz, W. Nepple and A. Nolle, in "Magnetic Resonance and Related Phenomena", p. 603. Groupment Ampere, Heidelberg (1976)
68. M. Alei Jr., *J. Chem. Phys.* **43**, 2904 (1963)
69. J. Kondo and J. Yamashita, *J. Phys. Chem. Solids* **10**, 245 (1959)
70. M. W. Garnett, T. K. Halstead and D. G. Hoare, *Eur. J. Biochem.* **66**, 85 (1976)
71. H. Gustavsson, B. Lindman and T. Bull, *J. Amer. Chem. Soc.* **100**, 4655 (1978)

14
THE NOBLE GASES

GARY J. SCHROBILGEN, University of Guelph, Canada

Of the noble gases only xenon evokes substantial chemical interest in terms of compound formation. Fortunately xenon is the most favourable noble gas from an NMR point of view, since it has a spin-$\frac{1}{2}$ nuclide with reasonable natural abundance, viz. ^{129}Xe (N = 26.44 %). There is also a quadrupolar xenon nuclide, ^{131}Xe (I = $\frac{3}{2}$, N = 21.18 %), but this has not been used for studies of chemical structure. Helium also has a spin-$\frac{1}{2}$ nuclide (^{3}He), but this has a very low natural abundance (1.3×10^{-4} %) and, in any case, most of the work on this nucleus has involved the solid or super-fluid states and is physics- rather than chemistry-oriented; such work has been adequately reviewed by Landesman[1] and Felter.[2] Argon has no stable nuclide of non-zero spin, a rare situation in the Periodic Table. Neon and krypton possess quadrupolar nuclei (^{21}Ne, I = $\frac{3}{2}$, N = 0.26 % and ^{83}Kr, I = $\frac{9}{2}$, N = 11.55 %) but NMR work has been largely confined to physics. None of the handful of known Kr compounds has been studied by ^{83}Kr NMR, and no (^{83}Kr, ^{19}F) coupling constants have been reported. Consequently a high proportion of the present review concerns ^{129}Xe NMR. This nucleus has a receptivity of 31.8 with respect to ^{13}C, so that modern FT spectrometers allow ^{129}Xe data to be acquired readily.

14A XENON AND KRYPTON CHEMICAL SHIFTS FOR ATOMIC SYSTEMS

The nuclear magnetic resonances of ^{129}Xe and ^{131}Xe in pure xenon gas were observed[3,4] at an early stage in the history of NMR, but only that of ^{129}Xe has been extensively utilised for chemical shift studies. The diamagnetism of the electrons of an isolated xenon atom provides a substantial contribution to the chemical shift from the resonant field of a bare nucleus, amounting to about 5800 ppm.

A density-proportional chemical shift in the ^{129}Xe resonance of xenon gas and liquid has been found by several workers.[5–7] The high-frequency shift due to association with another xenon atom upon increasing the pressure may be conveniently calculated by means of Ramsey's theory of chemical shifts, and is attributed to the negative contribution arising from the paramagnetic term.

Xenon-129 gas has also been used extensively as a probe gas in the study of the intermolecular forces between two molecules approaching one another during collision.[7–15] ^{129}Xe is an ideal probe molecule for many reasons: (1) it is isotropic and has no electrical moments, (2) being monatomic, any changes in electronic distribution are transmitted directly to the ^{129}Xe nucleus, (3) ^{129}Xe is an abundant spin species, (4) it possesses no quadrupole moment, (5) ^{129}Xe chemical shifts are an order of magnitude more sensitive than ^{1}H or ^{19}F chemical shifts and therefore are not dominated by bulk susceptibilities, and (6) xenon has no rotational or vibrational degrees of freedom so that there are no contributions from temperature to the shielding in the isolated molecule.

The results for pure ^{129}Xe suggest that the chemical shift dependence is non-linear beyond a density of approximately 100 amagats.*[7] Least-squares fitting to the virial expansion of chemical shielding in powers of density, ρ_{Xe}, i.e. to:

$$\sigma = \sigma_0 + \sigma_1 \rho_{Xe} + \sigma_2 \rho_{Xe}^2 + \sigma_3 \rho_{Xe}^3 + \cdots \tag{14.1}$$

gave the following values: σ_1(Xe-Xe) $= -0.548 \pm 0.004$ ppm/amagat, σ_2(Xe-Xe) $= (-0.17 \pm 0.02) \times 10^{-3}$ ppm/amagat^2, and σ_3(Xe-Xe) $= (0.16 \pm 0.01) \times 10^{-5}$ ppm/amagat^3. This suggests that at higher densities three-body or higher order interactions become important for xenon. More recently, values of the second virial coefficient have been obtained which are essentially uncontaminated with three-body or higher order interactions by observing samples at low densities.[13] Earlier data[6b] on the density-dependence of the ^{129}Xe chemical shift have been interpreted by Adrian.[10]

For a mixture of xenon and a second gas, A, the chemical shielding of the ^{129}Xe probe nucleus also may be expressed in a virial expansion of chemical shielding in powers of the densities of the ^{129}Xe probe gas (ρ_{Xe}) and gas $A(\rho_A)$ as[7] in Eq. 14.2, where

$$\sigma(\rho_{Xe}, \rho_A, T) = \sigma_0 + \sigma_1(\text{Xe-Xe})\rho_{Xe} + \sigma_1(\text{Xe-}A)\rho_A + \sigma_1(A\text{-}A)\rho_A + \cdots \tag{14.2}$$

σ_0 is an intrinsic property of the xenon probe atom (in general a function of temperature), while σ_1(Xe-Xe), σ_1(Xe-A) and σ_1(A-A) represent the second virial coefficients of chemical shielding (also functions of temperature) and are characteristic of their respective binary interactions. Since the terms σ_1(Xe-Xe) and σ_1(A-A) may be found with good precision, σ_1(Xe-A) is readily determined from NMR measurements. For gases such as Ar, Kr, CO_2, HCl, CH_4, CF_4 and the fluoromethanes, values for the slopes, σ_1(Xe-A), range from -3120 to -12280 ppm mol^{-1} cm^3 (Table 14.1).[7] Unlike ^{1}H and ^{19}F gas-phase chemical shifts, the results cannot be explained in terms of an expression of the type 14.3† coupled with a central-field potential function like Lennard-Jones

$$\sigma_1(\text{Xe-}A) = \tfrac{1}{6}\chi_A - B(E^2 + \overline{F^2}) \tag{14.3}$$

($\frac{1}{6}\chi_A$ is the bulk susceptibility of the medium and is approximately independent of pressure and temperature for diamagnetic materials). Calculations from first principles give variations in the chemical shift of ^{129}Xe gas which are nearly an order of magnitude lower.[7,10] The density dependence due to specific interactions between Xe and A appears to involve to some extent the polarizability of A.

Although ^{129}Xe chemical shifts in the paramagnetic gases O_2 and NO exhibit the usual linear dependence with density of O_2 or NO, the slopes are unusually large (Table 14.1), leading to anomalously high values for σ_1(Xe-A) $- \frac{1}{6}\chi_A$ on a plot of this term vs. polarisability.[8] These results have been attributed to the contact mechanism and possibly an unusually deep Xe-O_2 or Xe-NO potential function.[8] Buckingham and Kollman[16] have interpreted the results in terms of the overlap of the Xe(5s) with the $O_2(\pi_g^*)$ or NO(π^*) orbitals, leading to a net spin density at the nucleus. The temperature

* The amagat is a unit of molar density; 1 amagat is the density of the real gas under consideration at 1 atm (101 325 Pa) and 0°C (273.15 K).

† Magnetic susceptibilities in SI; or $\frac{2}{3}\pi\chi_A$ for the magnetic susceptibility expressed in e.m.u.

TABLE 14.1 Observed σ_1(Xe-A) at 25°C

A	Ar[a]	CO_2[a]	CF_4[a]	CHF_3[a]	CH_2F_2[a]	CH_3F[a]
σ_1(Xe-A)/ppm mol^{-1} cm^3	−3119	−3823	−4327	−4287	−4965	−4314
A	CH_4[a]	Kr[a]	HCl[a]	Xe[a]	NO[b]	O_2[b]
σ_1(Xe-A)/ppm mol^{-1} cm^3	−6201	−6070	−7678	−12283	−17769 ± 256	−21037 ± 216

[a] Ref. 7. [b] Ref. 8.

dependencies of the virial coefficients $\sigma_1(Xe\text{-}O_2)$ and σ_1(Xe-NO) have been determined by measurement of the resonance frequency of ^{129}Xe in gas samples of known densities in Xe and paramagnetic molecules.[12,14] It has been concluded that the temperature dependence of σ_1(Xe-A) can be interpreted in terms of a contact interaction.

More recently, FT NMR spectroscopy has been employed to obtain ^{129}Xe shifts in the presence of the spherical top perturbers CH_4, CF_4, and SiF_4 at low gas densities.[15] A reduced form of the second virial coefficient of chemical shielding has been found, and a method for estimating well depths has thereby been developed. There is also some evidence for the anisotropy of the potential between Xe and the solvent molecules HBr, HCl, CO_2, N_2O, C_2H_2, C_2H_4, C_2H_6 and BF_3, which are rod-shaped and disc-shaped perturbers.[17]

The density-dependence of the chemical shift of ^{83}Kr in the gas phase has been reported and discussed by Brinkmann *et al.*, as summarised in ref. 6b. The same publication also reviews chemical shift measurements for ^{83}Kr, ^{129}Xe and ^{131}Xe in their condensed phases. More recently, Cowgill and Norberg[18] have found the density variation of the ^{83}Kr shift to be continuous upon melting, whereas that of xenon is not. Laurie *et al.*[19] have extended the theory of Adrian[10,20] (formulated for colliding atom-pairs in gaseous xenon) to solid krypton and xenon. The calculated shifts agree reasonably well with experimental values.

14B NUCLEAR SPIN RELAXATION OF ATOMIC NOBLE GASES

Monoatomic gases cannot relax by intramolecular processes, hence intermolecular interactions are responsible for relaxation. Such interactions are provided by interatomic dipolar interactions, strong van der Waals forces, collision with surfaces, or diffusion and adsorption of the gas onto solid material. The spin-lattice relaxation times may be of the order of several days at low pressures.

Relaxation measurements on ^{3}He in the presence of various foreign substances have been made by Gamblin and Carver.[21] In general, T_1 values range from 6.0×10^2 to 2.3×10^4 s at 295°C and 3.1 amagats and indicate that relaxation is moderately independent of wall material. A subsequent, more detailed, study of surface-induced nuclear-spin relaxation in gaseous ^{3}He in glass containers as a function of temperature has been reported by Fitzsimmons, Tankersley and Walters.[22] The results have been interpreted in terms of a phenomenological theory incorporating distinctly different relaxation processes at low and high temperatures, respectively. The low-temperature mechanism involves ^{3}He adsorption on glass surfaces, while the high-temperature mechanism (applicable to Pyrex and quartz surfaces) involves permeation of ^{3}He into the container material. The latter mechanism is eliminated by using relatively impermeable alumino-silicate glass containers, resulting in ^{3}He nuclear-spin relaxation times of 7×10^5 s at low gas densities and moderate temperatures.

Liquid and gaseous xenon are excellent systems for the investigation of the microscopic properties of liquids and gases owing both to the lack of any molecular structure and to the simplicity of the operative spin-relaxation mechanism. Not only are xenon molecules monatomic and spherically symmetric, they are sufficiently massive that intermolecular collisions do not exhibit major quantum effects. Thus, xenon represents the simplest type of classical fluid. The ^{129}Xe spin relaxation is free from complications due to electric quadrupole effects and (because xenon is monatomic) to intramolecular nuclear dipole fields. Only the intermolecular dipole fields, with their time variation determined by diffusion, are expected to provide a relaxation mechanism. In the liquid, Streever and Carr[5] have shown that the temperature dependence of the product of T_1 and ρ_{Xe} can be described by an activation energy of 2.9 ± 0.4 kJ mol.$^{-1}$ In the gas at room temperature between 48 and 73 atm, T_1 was shown to vary as $\rho^{-2.1 \pm 0.6}$. The largest value of T_1 observed was 2600 ± 600 s for a gas sample at 48 atm, and the shortest value was 57 ± 2 s in the liquid at -101°C. A subsequent study by Hunt and Carr[6] showed that the spin-lattice relaxation rate in the gas was proportional to the density, as in Eq. 14.4. For the liquid in equilibrium with its vapour, this same study showed that the relaxation time

$$T_1^{-1}\,\text{s} = (5.0 \pm 0.5) \times 10^{-6} \rho_{Xe}/\text{amagat} \tag{14.4}$$

throughout the temperature range 0° to -72°C was 1000 ± 200 s. The data cannot be accounted for by the direct nuclear dipole-dipole interaction during collision. Such an interaction alone would lead to enormous relaxation times ($\sim 10^6$ s), four orders of magnitude longer than observed. Using the theoretical results obtained by Torrey,[11] it has been shown that the rotation of the diatomic charge distribution, which is present during the period of the collision, is consistent with the ^{129}Xe relaxation and chemical shift data.[6]

Relaxation of gaseous ^{131}Xe has been studied by Staub and co-workers,[4,23] as a function of density, and interpreted by Staub[24] and by Adrian.[20] Brinkmann and Schmid[25] have found that for ^{83}Kr at 156 amagat pressure and room temperature $T_1 = (1.2 \pm 0.5)$ s; the theory of Staub[24] yields $T_1 = 2.5$ s, giving only fair agreement with the experimental value.

Spin-lattice relaxation data have been reported for the quadrupole nuclides of neon,[26] krypton[18] and xenon[27] in their liquid and solid phases as a function of density and temperature. At the respective triple points the values of T_1 for the liquids differ by nearly three orders of magnitude (23.5 s for ^{21}Ne, 0.608 s for ^{83}Kr and 0.041 s for ^{131}Xe). The factors affecting these relaxation times have been discussed in detail.[18]

14C HIGH-RESOLUTION ^{129}Xe NMR OF XENON COMPOUNDS

14C.1 Introduction and Discussion of Structure Determination

Widespread interest in defining structures and properties of xenon fluorides, oxide fluorides, their cations and related covalent species has resulted in several recent studies of their ^{19}F NMR spectra.[28–32] However, ^{129}Xe NMR spectroscopy can, in principle, yield valuable complimentary structural information. Moreover, ^{129}Xe is an ideal species for thorough investigation of a heavy nucleus since it can be obtained in a variety of compounds in a wide range of oxidation states (+2, +4, +6 and +8). Early studies of the ^{129}Xe resonance were confined to ^{19}F-{^{129}Xe} INDOR or "tickling" experiments, and yielded chemical shift values for $XeOF_4$, XeF_4 and XeF_2.[33,34] Recent studies by direct observation FT techniques have given the ^{129}Xe NMR parameters summarised in Table 14.2.[35–38]

In the present discussion little emphasis is given to structural elucidation of xenon species in solution by means of ^{129}Xe NMR, as this aspect is, for the most part, obvious from the tabulated results (Table 14.2). However, a study of the ^{129}Xe resonance is of particular value from the standpoint of gross structural considerations, especially when only one fluorine-on-xenon environment can be observed in the ^{19}F NMR spectrum. Thus, the observed splittings arising from (^{19}F, ^{129}Xe) spin-spin coupling confirm the solution structures of such species as XeF^+, XeO_2F^+, XeO_2F_2, XeF_4, $XeOF_4$ $(FXe)_2SO_3F^+$ and $FXeOS(:O)FOMOF_4$ (M = Mo or W) (Table 14.2).

14C.2 Chemical Shifts—General Theory

The present state of understanding of the chemical shifts of heavy nuclei is unclear. In lieu of a comprehensive theory, the approach to the discussion of ^{129}Xe chemical shifts must remain essentially a semi-empirical one. Where possible, however, attempts have been made to single out those factors which might play a dominant role in determining the observed trends. Of particular interest is the very great sensitivity of the ^{129}Xe chemical shift to (1) oxidation state, (2) ionic character of the Xe-F bond (see below), and (3) solvents (see below).

Useful theoretical approximations have been developed which make it somewhat easier to interpret NMR shifts of heavy nuclei such as xenon in chemical terms. One such approximation represents the shielding of a nucleus like ^{129}Xe by the local terms, σ^d_{Xe} and σ^p_{Xe}, only. These are calculated by Ramsey's theory[39] applied to the electrons on Xe alone.[40] For xenon, which has a large range of chemical shifts, σ^d_{Xe} is often assumed to be not very different from the free-atom value, so that chemical shift relationships are ascribed solely to variations of σ^p_{Xe}. The Jameson-Gutowsky[40] approach has therefore been adopted for ^{129}Xe, and Eq. 3.36 applied with reference to the $5p$ and $4d$ electrons. Jameson and Gutowsky[41] demonstrated that ^{129}Xe chemical shifts could be computed with considerable accuracy in the limited number of cases then known using Eq. 3.36. Their treatment of the problem demonstrated that a localized description of the bonding which employs d-hybridization provides a more satisfactory description than a delocalised model without d-hybridisation. A comparison of their theoretical values for σ^p_{Xe} with the experimental values is given in Table 14.3.

This theoretical treatment shows that ΔE, $\langle r^{-3}\rangle_{5p}$ and $\langle r^{-3}\rangle_{5d}$ can be regarded as essentially constant over the entire range of ^{129}Xe chemical shifts and that P_u and D_u determine variations in the chemical shift. The numerical values of P_u and D_u in turn depend largely upon the coordination number of xenon, the hybridization of the bonding orbitals and the ionicity of its bonds. In the spherically symmetric case of atomic xenon, P_u and D_u both have the minimum value of zero. The maximum for P_u is 2, which occurs when two xenon $5p$-orbitals are filled and one is empty or one filled and two empty. The expression for D_u is more difficult to analyze, but it appears that the maximum value is 12, which occurs when the three t_{2g} $5d$-orbitals are filled and the two e_g orbitals are empty or vice versa.

From Table 14.2 it can be seen that the paramagnetic contribution to the ^{129}Xe chemical shift is extremely sensitive to formal oxidation state of the central xenon

TABLE 14.2 ^{129}Xenon nuclear magnetic resonance parameters[a]

Species	$\delta(^{129}Xe)$/ppm	$J(^{129}Xe, ^{19}F)$/Hz	Solute (molal conc.)	Solvent	T/°C
Xe[c]	−5331	—	Xe	n-C_6F_{14}	25
$Xe(OTeF_5)_2$[b]	−2379	31	$Xe(OTeF_5)_2$	$CFCl_3$	
$F_5SeOXeOTeF_5$[b]	−2289	—	$Xe(OTeF_5)_2$,$Xe(OSeF_5)_2$	$CFCl_3$	
$Xe(OSeF_5)_2$[b]	−2200	37	$Xe(OSeF_5)_2$	$CFCl_3$	
$FXeOTeF_5$[b]	−2067	30	$FXeOTeF_5$	$CFCl_3$	
$FXeN(SO_2F)_2$	−2016	~18, 5572	$FXe(SO_2F)_2N$(~0.5)	BrF_5	−40
$FXeOSeF_5$[b]	−1952	37	$FXeOSeF_5$	$CFCl_3$	
XeF_2	−1905	5630	XeF_2(sat)	SO_2ClF	25
XeF_2	−1750	5616	XeF_2(2.70)	BrF_5	25
XeF_2	−1708	5583	XeF_2(2.70)	BrF_5	−40
$FXeOSO_2F$	−1467	5975	XeF_2(2.81)	HSO_3F	−84
$Xe(OSO_2F)_2$	−1613	—			
$FXeOSO_2F$	−1416	6012	XeF_2(2.81)	HSO_3F	−90
$Xe(OSO_2F)_2$	−1572	—			
XeF_2	−1592	5652	XeF_2(~16)	HF	25
$FXeOSO_2F$	−1666	5830	$FXeSO_3F$(1.20)	BrF_5	−40
$FXeOSO_2F$	−1613	5848	$FXeSO_3F$(1.20)	BrF_5	−77
$FXeOSO_2F$	−1407	6051	$XeF_2 \cdot MoOF_4$(1.32)	HSO_3F	−100
$FXeOS(:O)FOMoOF_4$	−1342	5971			
$FXeOSO_2F$	−1416	6021	$XeF_2 \cdot WOF_4$(1.00)	HSO_3F	−90
$FXeOS(:)FOWOF_4$	−1335	6131			
$FXe{-}F{\cdots}MoOF_4$	−1383	5117, 6139	$XeF_2 \cdot MoOF_4$(0.92)	BrF_5	−80
$FXe{-}F{\cdots}WOF_4$	−1331	5051, 6196	$XeF_2 \cdot WOF_4$(0.86)	BrF_5	−66

$FXe{-}F{\cdots}MoOF_4$	-1441	5076, 6058			
$FXe{-}F{\cdots}MoOF_4(MoOF_4)$	-1338	5036, 6159	$XeF_2(0.98)$, $MoOF_4(2.13)$	SO_2ClF	-118
$FXe{-}F{\cdots}MoOF_4(MoOF_4)_2$	-1321	5026, 6156			
$FXe{-}F{\cdots}WOF_4$	-1315	5000, 6127			
$FXe{-}F{\cdots}WOF_4(WOF_4)$	-1189	4964, 6268			
$FXe{-}F{\cdots}WOF_4(WOF_4)_2$	-1170	4996, 6304	$XeF_2{\cdot}2WOF_4(0.70)$	SO_2ClF	-115
$FXe{-}O{-}WF_5(WOF_4)$	-955	6373			
$FXe{-}O{-}WF_5(WOF_4)_2$	-906	6373			
XeF^+	-911	6703	$XeF^+Sb_2F_{11}^-(0.72)$	HSO_3F	-70
$[(FXeO)_2S(:O)F]^+$	-1258	6428	$(FXe)_2SO_3F^+AsF_6^-(1.00)$	BrF_5	-77
$[FXe{\cdots}F{\cdots}XeF]^+$	-1051	4865, 6740	$(FXe)_2F^+AsF_6^-(1.14)$	BrF_5	-57
XeF^+	-574	7210	$XeF_2(0.71)$	SbF_5	25
$(XeF_6)_4$	-60.8	331.7[d]	$XeF_6(1.44)$	SO_2ClF/CF_2Cl_2[e]	-145
XeF_5^+	-23.9	165, 1377	$F_5XeSO_3F(1.32)$	HSO_3F	-80
$XeOF_4$	0	1123	$XeOF_4(4.48)$	neat	25
$XeOF_4$	-0.1	1127	$XeOF_4(\sim 0.5)$	SO_2ClF/CF_2Cl_2[e]	-145
XeF_5^+	$+12.7$	159, 1400	$XeF_5^+Sb_2F_{11}^-(\sim 4)$	HF	25
XeO_2F_2	$+173.2$	1217	$XeO_2F_2(\sim 0.5)$	HF	25
XeO_3	$+217$	—	XeO_3	H_2O	25
$XeOF_3^+$	$+238$	434, 1018	$XeOF_3^+SbF_6^-(0.21)$	SbF_5	25
XeF_4	$+253$	3823	XeF_4(sat)	BrF_5	25
XeF_3^+	$+595$	2384, 2609	$XeF_4(0.76)$	SbF_5	25
XeO_2F^+	$+600$	95	$XeO_2F^+SbF_6^-$(sat)	SbF_5	25
XeO_6^{4-}	$+2077$	—	$Na_4XeO_6(\sim 0.5)$	H_2O	25

[a] Ref. 38 except where otherwise stated. Spectra were recorded at 22.63 MHz in FT mode, and were referenced with respect to external $XeOF_4$ at 25°C. [b] Ref. 36. No solute concentrations or temperatures reported. [c] Ref. 35. [d] Only 11 of the theoretical 25 lines in this spectrum have been observed. [e] 50 mole % SO_2ClF/50 mole % CF_2Cl_2.

TABLE 14.3 Comparison of calculated and experimental ^{129}Xe chemical shifts

Molecule	^{129}Xe Chemical shift[a]		
	calc.[b]	*expt.*[c]	*expt.*[d]
XeF_2	−3420	−3918	−3930
XeF_6	−5500	−5450	
$XeOF_4$	−5750	−5511	−5511
XeF_4	−5980	−5764	−5785

[a] Given as shielding constants relative to atomic Xe (for which $\sigma^p = 0$), i.e. as $[\sigma(\text{compound}) - \sigma(\text{atomic Xe})]$/ppm. [b] Ref. 41. [c] Ref. 38. [d] Ref. 33 and 34.

atom, extending over a very large range, i.e. −5331 ppm for Xe (in n-C_6F_{14}) to +2077 ppm for XeO_6^{4-} (relative to $XeOF_4$). The large range of chemical shifts in different xenon species is also readily anticipated in view of existing trends among other nuclei. A survey of the existing experimental data reveals that the range of chemical shifts not only increases with atomic number Z for a particular period, but also increases with Z for a particular group.[40,42] This periodic dependence places xenon (Z = 54) in a category among nuclei having one of the greatest ranges of NMR chemical shifts observed to date.

The correlation of $\delta(^{129}Xe)$ to the formal oxidation state of xenon is self-evident, with σ^p and hence $\delta(^{129}Xe)$ increasing, as might be anticipated, with increasing formal oxidation state of the central xenon atom. The only exceptions to this rule are the Xe^{IV} species, XeF_4 and XeF_3^+. The high-frequency positions of XeF_4 and XeF_3^+ relative to Xe^{VI} species are in marked contrast to the corresponding ^{19}F chemical shift trends, which vary monotonically with oxidation state, i.e. $Xe^{VI} < Xe^{IV} < Xe^{II}$.[28] It is, however, of interest to note that the trend observed for the majority of ^{129}Xe chemical shifts, i.e. $Xe^{VIII} < Xe^{IV} < Xe^{VI} < Xe^{II}$ is in agreement with the trend computed by Jameson and Gutowsky[41] for $XeF_4 < XeOF_4 < XeF_6 < XeF_2$ (Table 14.3).

It is also worth noting that in the two series $(XeF_6)_4$, $XeOF_4$, XeO_2F_2, XeO_3 and XeF_5^+, $XeOF_3^+$, XeO_2F^+, a monotonic deshielding of the central xenon atom (additive for the latter series) is observed with increasing oxygen substitution (Table 14.2). This increased net deshielding presumably arises from contributions of the sort $Xe{=}O \leftrightarrow Xe^+{-}O^-$, which serve to increase both P_u and D_u.

14C.3 Spin-Spin Coupling involving Xenon-129

a. Coupling to Fluorine-19

The magnitudes of (^{129}Xe, ^{19}F) coupling constants range from 95 Hz for XeO_2F^+ to 7210 Hz for XeF^+ (Table 14.2). The value of 7210 Hz observed for XeF^+ in SbF_5 solution is among the largest coupling constants ever observed.

Several empirical correlations between ^{19}F chemical shifts and $J(^{129}Xe, ^{19}F)$ exist in the literature. Frame[43] has demonstrated that a plot of (^{129}Xe, ^{19}F) coupling constants versus ^{19}F chemical shifts for xenon fluorides and oxyfluorides yields a relationship

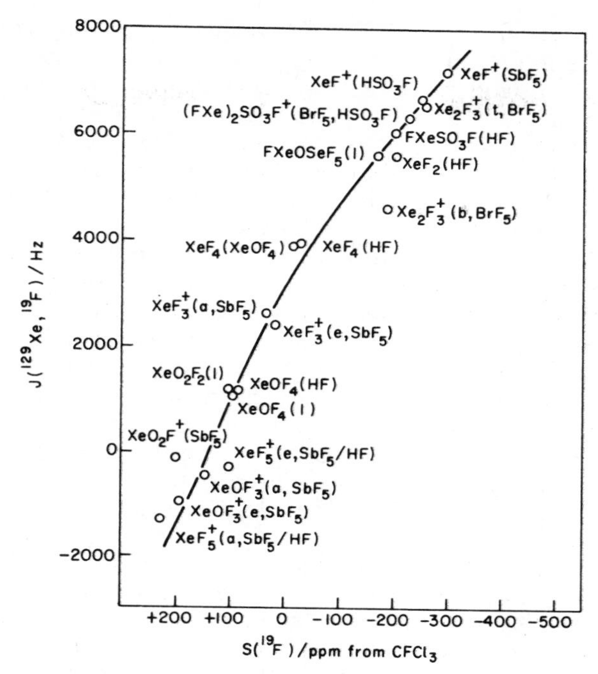

Fig. 14.1 Correlation of ^{19}F chemical shifts and (^{129}Xe, ^{19}F) coupling constants for some xenon compounds. The medium conditions are given in brackets after the molecular formulae. a, axial; e, equatorial; b, bridging; t, terminal; l, neat liquid. (Reproduced with permission from ref. 28.)

in which the modulus of the coupling constant increases monotonically with decreasing chemical shift, δ_{Xe}. Gillespie and Schrobilgen[28] have shown that this correlation can be extended to include all the known xenon compounds and their ions studied by ^{19}F NMR (Fig. 14.1). The correlation provides a rationale for the small (^{129}Xe, ^{19}F) couplings observed for the equatorial fluorines of XeF_5^+ and XeO_2F^+ (Table 14.2). It is assumed that a change in the signs of the coupling constants occurs over the series in the vicinity of the data for the equatorial fluorines of XeF_5^+ and XeO_2F^+; the signs of the axial and equatorial couplings of $XeOF_3^+$, as well as the axial coupling of XeF_5^+, are probably all negative with respect to those of the remaining species. Absolute signs of the various (^{129}Xe, ^{19}F) reduced coupling constants have been derived indirectly by considering the series of isoelectronic Sn^{IV}-I^{VII} hexafluoro-species. A near-linear relationship is obtained when $(K_{XF})^{1/2}$ is plotted *versus* Z and the same sign is assumed throughout the series.[44] The modulus of K_{XF} increases along the series $SnF_6^{2+} < SbF_6^- < TeF_6 < IF_6^+$, and presumably the hypothetical Xe^{VIII} species XeF_6^{2+} would possess an even larger value. Since the signs of both (Sn^{IV}, F) and (Te^{VI}, F) couplings have been shown to be negative, negative signs can also be assigned to K_{XF} for the remaining members of the series including XeF_6^{2+}. The signs of K_{XeF} of high-oxidation state Xe^{VIII} and Xe^{VI} species, with the exception of $XeOF_4$, XeO_2F_2 and possibly XeF_6 and XeO_2F^+, are therefore taken as negative, while the signs for K_{XeF} of the lower oxidation state Xe^{IV} and Xe^{II} species are taken as positive (Fig. 14.1). It is clear that additional data on a Xe^{VIII} compound such as XeO_3F_2 would be useful to confirm the proposed relationship and sign change.

Although the magnitudes of (^{129}Xe, ^{19}F) spin-spin coupling constants might be expected to provide further information on bonding in xenon species, the theories of spin-spin coupling constants, which have been used successfully in the case of lighter elements, are not directly applicable to heavier nuclei. In an early consideration of (^{129}Xe, ^{19}F) spin-spin coupling, Jameson and Gutowsky[41] noted that a localised description seems to be better than the delocalized MO description. If the Fermi contact term is assumed to be dominant, the localized description with *ps*, p^2ds and p^3ds^2 hybrid xenon orbitals in XeF_2, XeF_4 and $XeOF_4$ yields approximate ratios of $\frac{1}{2}:\frac{1}{4}:\frac{1}{6}$ for J_{XeF}, based on relative *s*-characters of the Xe-F bonds. On the other hand, the delocalized MO description of Jortner, Wilson and Rice,[45] using only 5*p* orbitals on xenon, gives coupling constants of zero. The observed (^{129}Xe, ^{19}F) coupling constants for XeF_2, XeF_4 and $XeOF_4$ (Table 14.2) follow the trend predicted for *ps*, p^2ds and p^3d^2s hybridisation, but only qualitatively. Dipolar contributions to J_{XeF} may be in part responsible for the lack of good quantitative agreement.

It is of interest to consider the observed change in $J(^{129}Xe, ^{19}F)$ upon ionization of XeF_2 to give the XeF^+ cation. It may be assumed, in the first instance, that the ionization process corresponds to a change from *sp* to sp^3 hybrid bonding orbitals on xenon. However, the change in $J(^{129}Xe, ^{19}F)$ observed (Table 14.4) is large (ca. 1600 Hz) and opposite to that predicted on the basis of such *s*-character changes in the hybrid bond, assuming a Fermi contact contribution alone. Alternatively, the bonding orbitals on xenon in XeF_2 may perhaps be represented as pure *p*-orbitals, with the lone electron-pairs regarded as residing in sp^2-hybrid orbitals. The increase in *s*-character of the Xe-F bond upon ionization of XeF_2 (assuming sp^3-hybrid orbitals for both the bonding and the three non-bonding electron pairs of XeF^+) is a consistent with the large observed change in $J(^{129}Xe, ^{19}F)$ and with a significant Fermi contact contribution to the coupling constant. The change is also consistent with the increased covalent character (shortening) of the Xe-F bond observed upon ionization (Table 14.4). This trend is clearly shown by

TABLE 14.4 Comparison of Xe-F bond distances and (^{129}Xe, ^{19}F) coupling constants

Species (solvent)	$J(^{129}Xe, ^{19}F)/Hz^a$ bridge	terminal	Xe-F Distance/nm bridge	terminal	Ref.[b]
$XeF_2(BrF_5)$	—	5645	—	0.200^b	46
$XeF_2(HF)$	—	5665			
$FXeOSO_2F(HSO_3F)$	—	6025	—	0.194	47
$FXe{-}F{\cdots}WOF_4(BrF_5)$	5016	6128	0.204	0.189	48
$(FXe)_2SO_3F^+(HSO_3F)$	—	6355	—	0.186	49
$(FXe)_2SO_3F^+(BrF_5)$	—	6470			
$(FXe{\cdots}F{\cdots}XeF)^+(BrF_5)$	4865	6740	0.214	0.190	50
$FXe^+(HSO_3F)$	—	6620	—	*c*	—
$FXe^+(SbF_5)$	—	7210	0.235^d	0.182^d	51

[a] Taken from ^{19}F NMR spectra[28,30,32]. [b] For the Xe-F distance. [c] The XeF^+ cation appears to be strongly coordinated to HSO_3F solvent molecules and/or fluorosulphate-containing anions (which are produced by solvolysis of the anion), resulting in a lowered value for $J(^{129}Xe, ^{19}F)$ and presumably a longer Xe—F bond than that of "free" XeF^+. [d] The bond lengths given are for the $Sb_2F_{11}^-$ compound. Spin-spin coupling is not observed for the fluorine bridge to the anion owing to lability of this bond on the NMR timescale.

plots of ^{19}F shifts, which can also be regarded as a measure of covalent character in the Xe-F bond, versus $J(^{129}Xe, ^{19}F)$ for xenon(II-VI) species (Fig. 14.1) as well as by a direct comparison of the coupling constants with the Xe-F bond lengths for xenon(II) species (Table 14.4). Although the trend of increasing $J(^{129}Xe, ^{19}F)$ with increasing covalent character (decreasing Xe-F bond length) is clear-cut for Xe^{II}, bond length data available for comparison with $J(^{129}Xe, ^{19}F)$ values of higher oxidation state species is sparse. The trend of increasing $J(^{129}Xe, ^{19}F)$ with decreasing bond length seems clear, however, in the two cases for which relative bond length data are available, viz: XeF_3^+ [52] [$J(^{129}Xe, ^{19}F_{ax}) = 2384$ Hz; Xe-F_{ax} distance = 0.191 nm, and $J(^{129}Xe, ^{19}F_{eq}) = 2609$ Hz; Xe-F_{eq} distance = 0.184 nm] and XeF_5^+ [53] [$J(^{129}Xe, ^{19}F_{ax}) = 1400$ Hz; Xe-F_{ax} distance = 0.177 nm, and $J(^{129}Xe, ^{19}F_{eq}) = 159$ Hz; Xe-F_{eq} distance = 0.190 nm].

b. Coupling to Oxygen-17

The ^{17}O NMR spectra have been reported for enriched samples of xenic acid,[54] $Xe(^{17}OH)_6$, and xenon oxidetetrafluoride,[55] $Xe^{17}OF_4$. No ($^{129}Xe, ^{17}O$) spin-spin coupling was observed for $Xe(^{17}OH)_6$. The lack of spin-spin coupling can be attributed to rapid chemical exchange of ^{17}O with water solvent rather than to quadrupole effects from ^{17}O. The only ($^{129}Xe, ^{17}O$) coupling observed to date is that for $Xe^{17}OF_4$, viz: 692 ± 10 Hz. The ratio of the reduced coupling constants, $K_{XeO}/K_{XeF} = 3.2$ to 3.6, for the axial bonds in the isoelectronic C_{4v} species $XeOF_4$ and XeF_5^+ [$J(^{129}Xe, ^{19}F_{ax})$ ranges from 1348 to 1512 Hz, depending upon solvent composition and temperature[28]] is consistent with the relative formal bond-orders of 2 and 1.

14C.4 The Ionic Character of the Xe-F Bond in Xenon(II) Compounds

Xenon(II) species possessing terminal Xe-F groups are of particular interest since both the ^{129}Xe and ^{19}F chemical shifts can be directly correlated to the ionic character of the Xe-F bond. Xenon(II) compounds are ideal for such correlations since the linear geometry F-Xe-F(O)- is conserved throughout the entire series of compounds. In contrast, similar correlations do not exist for series of Xe^{IV} and Xe^{VI} compounds owing, presumably, to large changes in gross geometry in each series.

When considering the ionic character of the *terminal* Xe-F bond in Xe^{II} species, the Xe-F group is regarded as being bridged to either a fluorine, as in XeF_2 and $Xe_2F_3^+$, or to an oxygen, as in $FXeSO_3F$ and $(FXe)_2SO_3F^+$. In the specific case of XeF^+, the cation is not regarded as totally free, but bridged to a solvent fluorine or oxygen atom (even though this bridge is labile on the NMR timescale and possibly extensively dissociated).

Upon ionisation of XeF_2 to yield XeF^+, the ^{129}Xe chemical shift decreases (Table 14.2). The observed change in the ^{129}Xe chemical shift is in accord with the shielding change predicted by the valence-bond structures XeF^+ (ca. bond order 1) and F^- ^+Xe-F, F-Xe^+ F^- (ca. bond order 1/2). The $Xe_2F_3^+$ cation, which can be represented in terms of the contributing valence-bond structures F-Xe-F ^+Xe-F, F-Xe^+ F-Xe-F, F-Xe^+ F^- ^+Xe-F (plus those for XeF_2) would be expected to possess a terminal Xe-F bond order and a corresponding ^{129}Xe chemical shift roughly intermediate between those of XeF_2 and XeF^+. From Table 14.2, it is clear that this is indeed the case.

The near-linear correlation[28,29] between the ^{19}F chemical shift and ($^{129}Xe, ^{19}F$) coupling constant of Xe^{II} species has been used to show that variation in both these NMR parameters is related to the ionic character of the Xe-F bond (see above). For species containing a terminal Xe-F group, the ^{19}F chemical shifts have been shown to *decrease* with *increasing* ionic character of the terminal Xe-F bond. This is consistent with valence bond structures Ia and b, where the bonding electron pair of the terminal Xe-F bond

$$\underset{\text{Ia}}{F{-}Xe^{+}\ {}^{-}X{-}} \qquad\qquad \underset{\text{Ib}}{F^{-}\ {}^{+}Xe{-}X{-}}$$

becomes increasingly more localised with increasing ionic character of the Xe-X (X = F or O) bridge bond.

Plots of the ^{129}Xe chemical shift versus the ^{19}F chemical shift of the terminal fluorine on xenon have been shown to exhibit linear relationships over the entire series of Xe^{II} species studied (Fig. 14.2).[37,38] It is clear from Fig. 14.2 that chemical shifts of Xe-F groups

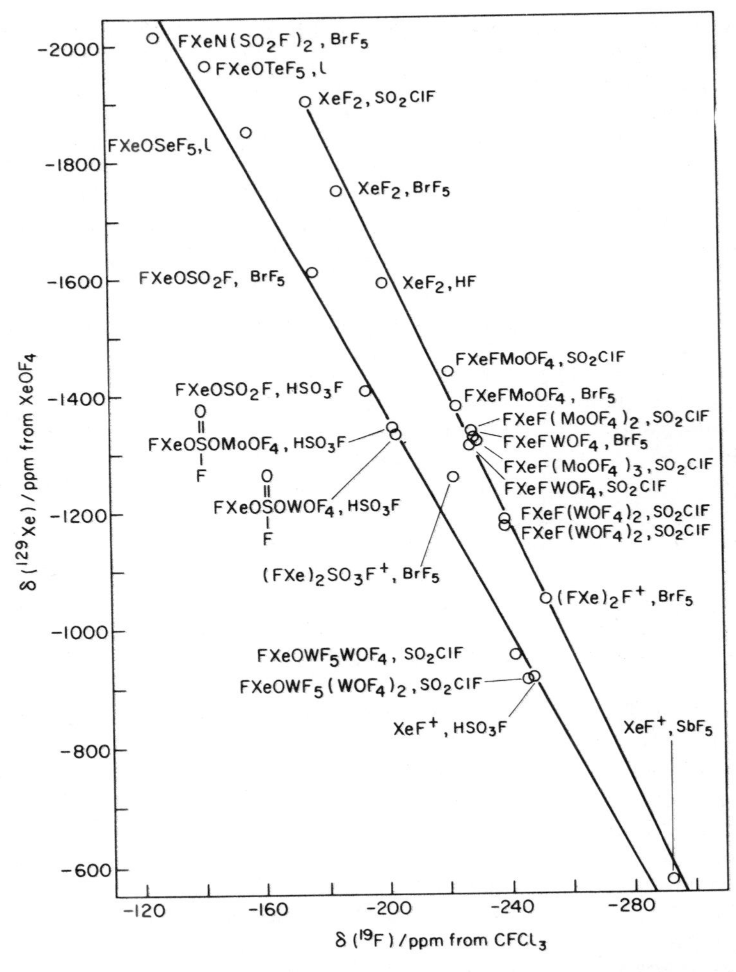

Fig. 14.2 Empirical plot of ^{129}Xe chemical shifts vs. the ^{19}F chemical shifts of the terminal fluorine on xenon for some Xe^{II} species containing F-bridges (lower line) and O-bridges (upper line). The medium conditions are given after the compound formulae. (Reproduced with permission from ref. 38).

form two rather distinct series. In general, the paramagnetic contributions to the ^{129}Xe and ^{19}F chemical shifts of Xe-F groups bonded to oxygen are larger than those of Xe-F groups bonded to fluorine owing to the increased covalent character of the Xe-O bond (increased P_u and D_u terms) as opposed to the Xe—F bridge bond. Within each series the covalency of the terminal Xe—F bond is seen to increase with increasing ^{129}Xe chemical shift and decreasing ^{19}F chemical shift. From Fig. 14.2 it can be seen that both lines converge, as expected, in the vicinity of the XeF^+ cation which, in SbF_5 solution, can be regarded as being only very weakly fluorine-bridged to the Sb_nF_{5n+1} polyanion, thus approximating to a free cation.

Within series of closely related compounds in a common solvent the ^{129}Xe resonance serves as a sensitive probe for accessing the relative fluoride-acceptor strengths of a fluorine-bridge ligand. In the case of the species $F\text{-}Xe\text{-}F\cdots MOF_4(MOF_4)_n$ (n = 0–2, M = Mo or W), which have been recently characterized by ^{19}F NMR spectroscopy,[32,56] the ^{129}Xe resonance frequency in SO_2ClF solvent decreases with increasing chain length, in accord with the anticipated fluoride-ion donor strength of the corresponding $[F\text{-}MOF_4(MOF_4)_n]^-$ anion (Table 14.2). It can also be inferred that the relative acceptor strength rapidly attenuates with increasing chain length. For the oxygen-bridged species, $F\text{-}Xe\text{-}OWF_5(WOF_4)_n$ (n = 0 or 1), the analogous trend in ligand base strength is evident (Table 14.2). The ^{129}Xe chemical shift trends are fully corroborated by the ^{19}F chemical shifts of the terminal fluorines on xenon, which display the opposite chemical shift trends (Fig. 14.1). It is also of interest to note that the ^{129}Xe chemical shift order, $(WOF_4)_{n+1} < (MoOF_4)_{n+1}$ for a given value of n, is in accord with the greater fluoride-ion acceptor strengths anticipated for the tungsten species. Based on the observed order of ^{129}Xe chemical shifts, the relative order of fluoride-acceptor strengths is:

$$(WOF_4)_3 > (WOF_4)_2 > WOF_4 > (MoOF_4)_3 > (MoOF_4)_2 > MoOF_4$$

14C.5 Solvation and Temperature Effects

As far as medium effects are concerned the sensitivity of xenon(II) chemical shifts appears to be far greater than can be explained in terms of non-chemical influences. Thus it has been noted that the ^{19}F chemical shifts of XeF^+ in SbF_5 and in HSO_3F solvents differ considerably.[29] The ^{129}Xe chemical shift of XeF^+ has also been shown to be particularly sensitive to solvent composition,[37,38] as is $J(^{129}Xe, {}^{19}F)$. Data for HSO_3F/SbF_5 solvent mixtures over the whole composition range are presented[38] in Table 14.5. It is clear that the XeF^+ cation interacts rather strongly with the solvent medium. The interaction presumably occurs by formation of a single additional bridge bond with the strongest electron-pair donor available, i.e. a solvent oxygen or fluorine.

Both the ^{19}F and ^{129}Xe chemical shifts have been recorded[38] for XeF_2 in SO_2ClF, BrF_5 and HF solvents at 25°. The decrease in ^{129}Xe chemical shift and accompanying increase in ^{19}F chemical shift (Fig. 14.2) with increasing solvent polarity in the series HF > BrF_5 > SO_2ClF suggests that XeF_2 may be solvated through one of its fluorines (structures IIa-c), giving rise to increased XeF^+ character with increasing solvent polarity.

F—Xe—F···H—F (IIa)   F—Xe—F···BrF_5 (IIb)   F—Xe—F···SO_2ClF (IIc)

TABLE 14.5 Variation of ^{129}Xe chemical shift and (^{129}Xe, ^{19}F) coupling constant for the XeF^+ ion with medium composition and temperature[a]

$[XeF^+]$/M	T/°C	Mole ratio $SbF_5:HSO_3F$	$\delta(^{129}Xe)$/ppm	$J(^{129}Xe, ^{19}F)$/Hz
0.709	25	∞	−574	7210
1.39	25	∞	−547	7214
0.546	25	8.26	−632	7177
0.546	−10	8.26	−604	7159
0.372	25	4.00	−649	7130
0.372	−10	4.00	−640	7145
0.609	25	2.37	−776	6971
0.609	−10	2.37	−765	6975
0.603	−55	1.48	−882	6808
0.695	−55	1.00	−903	6804
0.716[b]	−70	0	−911	6703

[a] Ref. 38. [b] $XeF^+Sb_2F_{11}^-$ dissolved in HSO_3F.

The chemical shift of XeF_2 in BrF_5 is also temperature dependent (Table 14.2), decreasing with decreasing temperature. This observation is consistent with a higher degree of association between XeF_2 and BrF_5 solvent expected at lower temperatures.

Alternatively, the major contribution to the solvation effect may be related to the number of ligands about xenon. It seems evident that the solvation effects for XeF_5^+ (HF and HSO_3F solvents) and $XeOF_4$ (neat and in the mixed solvent SO_2ClF/CF_2Cl_2) are very small when compared to XeF^+ and XeF_2 (Table 14.2). An increase in the number of substituents around xenon produces a "cage" which isolates the central xenon atom from the solvent. If solvation effects stem from weak coordination of the solvent in the xenon valence shell, the solvation effect would be expected to decrease with decreasing numbers of lone pairs (increasing substituents), as noted in the cases of XeF_2, XeF^+ and XeF_5^+, $XeOF_4$.

14C.6 Spin-lattice Relaxation Times

Only a limited amount of data is available for T_1 in chemically bound xenon species (Table 14.6). While a dipole-dipole mechanism is responsible for relaxation in free xenon giving correspondingly long relaxation times ($\sim 10^5$ to 10^6 s), the relaxation times

TABLE 14.6 Spin-lattice relaxation times for some xenon compounds

Molecule	Solvent	T_1/ms	Molecule	Solvent	T_1/ms
XeF_2[a]	CH_3CN	360 ± 30	$XeOF_4$[a]	neat	330 ± 30
XeF_2[b]	$CFCl_3$	360 ± 30	$Xe(OSeF_5)_2$[b]	$CFCl_3$	780 ± 20
XeF_6[a]	BrF_5	285 ± 20	$Xe(OTeF_5)_2$[b]	$CFCl_3$	380 ± 20

[a] Ref. 35. [b] Ref. 36.

in xenon chemically bound to either fluorine or oxygen are considerably shorter (285 to 780 ms).[35,36] Owing to these relatively short relaxation times, it has been possible to obtain FT ^{129}Xe NMR spectra using very high pulse-repetition rates.

REFERENCES

1. A. M. Landesman, *J. Phys.* (*Paris*), *Colloq.* **3**, C3-55 (1970)
2. A. L. Felter, *AIP Conf. Proc.* 29 (1976)
3. W. G. Proctor and F. C. Yu, *Phys. Rev.* **81**, 20 (1951)
4. E. Brun, J. Oeser, H. H. Staub and C. G. Telschow, *Phys. Rev.* **93**, 904 (1954); *Helv. Phys. Acta* **27**, 174 (1954)
5. R. L. Streever and H. Y. Carr, *Phys. Rev.* **121**, 20 (1961)
6. a. E. R. Hunt and H. Y. Carr, *Phys. Rev.* **130**, 2302 (1963)
   b. D. Brinkmann, *Helv. Phys. Acta* **41**, 367 (1968)
7. A. K. Jameson, C. J. Jameson and H. S. Gutowsky, *J. Chem. Phys.* **53**, 2310 (1970).
8. C. J. Jameson and A. K. Jameson, *Mol. Phys.* **20**, 957 (1971)
9. E. Kanegsberg, B. Pass and H. Y. Carr, *Phys. Rev. Lett.* **23**, 572 (1969)
10. F. J. Adrian, *Chem. Phys. Lett.* **7**, 201 (1970); *Phys. Rev. A* **136**, 980 (1964)
11. H. C. Torrey, *Phys. Rev.* **130**, 2306 (1963)
12. C. J. Jameson, A. K. Jameson and S. M. Cohen, *Mol. Phys.* **29**, 1919 (1975)
13. C. J. Jameson, A. K. Jameson and S. M. Cohen, *J. Chem. Phys.* **62**, 4224 (1975)
14. C. J. Jameson, A. K. Jameson and S. M. Cohen, *J. Chem. Phys.* **65**, 3397 (1976)
15. C. J. Jameson, A. K. Jameson and S. M. Cohen, *J. Chem. Phys.* **65**, 3401 (1976)
16. A. D. Buckingham and P. A. Kollman, *Mol. Phys.* **23**, 65 (1972)
17. C. J. Jameson, A. K. Jameson and S. M. Cohen, *J. Chem. Phys.* **66**, 5226 (1977)
18. D. F. Cowgill and R. E. Norberg, *Phys. Rev. B* **6**, 1636 (1972); *B* **8**, 4966 (1973)
19. J. Laurie and G. K. Horton, *Phys. Lett.* **22**, 560 (1966); J. Laurie, J. L. Feldman and G. K. Horton, *Phys. Rev.* **150**, 180 (1966)
20. F. J. Adrian, *Phys. Rev. A* **138**, 403 (1965)
21. R. L. Gamblin and T. R. Carver, *Phys. Rev.* **138**, A946 (1965)
22. W. A. Fitzsimmons, L. L. Tankersley and G. K. Walters, *Phys. Rev.* **176**, 156 (1969)
23. D. Brinkman, E. Brun and H. H. Staub, *Helv. Phys. Acta* **35**, 431 (1962)
24. H. H. Staub, *Helv. Phys. Acta* **29**, 246 (1956)
25. D. Brinkman and H. P. Schmid, Proc. XVII Congress Ampère (1972), p. 333, (V. Hovi, ed.). North-Holland Publishing Co., Amsterdam (1973)
26. R. D. Henry and R. E. Norberg, *Phys. Rev. B* **6**, 1645 (1972)
27. W. W. Warren and R. E. Norberg, *Phys. Rev. B* **6**, 1636 (1972)
28. R. J. Gillespie and G. J. Schrobilgen, *Inorg. Chem.* **13**, 765 (1974)
29. R. J. Gillespie, A. Netzer and G. J. Schrobilgen, *Inorg. Chem.* **13**, 1455 (1974)
30. R. J. Gillespie and G. J. Schrobilgen, *Inorg. Chem.* **13**, 1694 (1974)
31. R. J. Gillespie and G. J. Schrobilgen, *Inorg. Chem.* **13**, 2370 (1974)
32. J. H. Holloway, G. J. Schrobilgen and P. Taylor, *J.C.S. Chem. Comm.* 40 (1975)
33. T. H. Brown, E. B. Whipple and P. H. Verdier, *J. Chem. Phys.* **38**, 3029 (1963)
34. T. H. Brown, E. B. Whipple and P. H. Verdier, "Noble-Gas Compounds", p. 263. (H. H. Hyman, ed.). University of Chicago Press, Chicago, Ill. (1963)
35. K. Seppelt and H. H. Rupp, *Z. Anorg. Chem.* **409**, 331 (1974)
36. K. Seppelt and H. H. Rupp, *Z. Anorg. Chem.* **409**, 338 (1974)
37. G. J. Schrobilgen, J. H. Holloway, P. Granger and C. Brevard, *Compt. Rend.* **282C**, 519 (1976)
38. G. J. Schrobilgen, J. H. Holloway, P. Granger and C. Brevard, *Inorg. Chem.* **17**, 980 (1978)
39. N. F. Ramsey, *Phys. Rev.* **77**, 567 (1950); **78**, 699 (1950); **83**, 540 (1951); **86**, 243 (1952)
40. C. J. Jameson and H. S. Gutowsky, *J. Chem. Phys.* **40**, 1714 (1964)
41. C. J. Jameson and H. S. Gutowsky, *J. Chem. Phys.* **40**, 2285 (1964)
42. J. Mason, Advan. Inorg. *Chem. Radiochem.* **18**, 197 (1976)

43. H. D. Frame, *Chem. Phys. Lett.* **3**, 182 (1969)
44. J. Feeney, R. Hague, L. W. Reeves and C. P. Yue, *Can. J. Chem.* **46**, 1389 (1968)
45. J. Jortner, E. G. Wilson and S. A. Rice, *J. Amer. Chem. Soc.* **85**, 815 (1963)
46. H. A. Levy and P. A. Agron, *J. Amer. Chem. Soc.* **85**, 241 (1963)
47. N. Bartlett, M. Wechsberg, G. R. Jones and R. D. Burbank, *Inorg. Chem.* **11**, 1124 (1972)
48. P. A. Tucker, P. A. Taylor, J. H. Holloway and D. R. Russell, *Acta Cryst.* (B), **31**, 906 (1975)
49. R. J. Gillespie, G. J. Schrobilgen and D. R. Slim, *J.C.S. Dalton* 1003 (1977)
50. N. Bartlett, B. G. DeBoer, F. J. Hollander, F. O. Sladky, D. H. Templeton and A. Zalkin, *Inorg. Chem.* **13**, 780 (1974)
51. J. Burgess, C. J. W. Fraser, V. M. McRae, R. D. Peacock and D. R. Russell, Suppl. to *J. Inorg. Nuclear Chem.*, p. 183, (J. J. Katz and I. Sheft, eds.). Pergamon, Oxford (1976)
52. N. Bartlett, F. Einstein, D. F. Stewart and J. Trotter, *J. Chem. Soc.* (*A*), 1190 (1967)
53. P. Boldrini, R. J. Gillespie, P. R. Ireland and G. J. Schrobilgen, *Inorg. Chem.* **13**, 1690 (1974)
54. J. Reuben, D. Samuel, H. Selig and J. Shamir, *Proc. Chem. Soc.* 270 (1963)
55. J. Shamir, H. Selig, D. Samuel and J. Reuben, *J. Amer. Chem. Soc.* **87**, 2359 (1965)
56. G. J. Schrobilgen and J. H. Holloway, unpublished results

SUBJECT INDEX

References to specific nuclei and compounds are not included due to the book being structured according to the Periodic Table. A reader requiring data on a specific nucleus or a compound containing that nucleus is referred to the section on the particular element, see table of contents, p. xi. References to topics such as chemical shift, coupling constant, and relaxation, which occur frequently throughout the book are only included when of general relevance. Specific examples for a given nucleus may be found *via* the table of contents.